Turbine Aerodynamics, Heat Transfer, Materials, and Mechanics

Turbine Aerodynamics, Heat Transfer, Materials, and Mechanics

EDITED BY

Tom I-P. Shih
Purdue University, West Lafayette, Indiana

Vigor Yang
Georgia Institute of Technology, Atlanta, Georgia

Volume 243
Progress in Astronautics and Aeronautics

Timothy C. Lieuwen, Editor-in-Chief
Georgia Institute of Technology
Atlanta, Georgia

Published by
American Institute of Aeronautics and Astronautics, Inc.
1801 Alexander Bell Drive, Reston, VA 20191-4344

Portions of Chapter 2 are sourced from the U.S. Department of Energy's *The Gas Turbine Handbook*.

Portions of Chapter 3 were previously published as "Sealing in Turbomachinery," *Journal of Propulsion and Power*, Vol. 22, No. 2 (2006) and NASA/TM-2006-214341.

Portions of Chapter 8 were previously published as "Axial Turbine Blade Tips: Function, Design, and Durability," *Journal of Propulsion and Power*, Vol. 22, No. 2 (2006).

In Chapter 15, Figures 2 to 16 are reproduced with permission of ASME from "A Review of Surface Roughness Effects in Gas Turbines," *Journal of Turbomachinery*, Vol. 132 (2010).

American Institute of Aeronautics and Astronautics, Inc., Reston, Virginia

1 2 3 4 5

ISBN 978-1-62410-263-9

FOREWORD

The modern turbine represents the crowning glory of the mechanical arts—the highest levels of practical sophistication in mechanical and aero design, cooling, materials, and fracture mechanics. Its evolution, however, has been long and complex.

In 1939, the U.S. Navy asked the National Academy of Sciences to form the Committee on Gas Turbines to consider the use of gas turbines for ship and aircraft propulsion. The committee had a most distinguished membership, including such technical luminaries as aerodynamicist Theodore von Kármán, Nobel Prize winning physicist Robert Millikan, and automotive inventor and entrepreneur Charles Kettering. The committee's report [1] pointed out the extreme sensitivity of such an engine's thermal efficiency and power output (power to weight) to both turbine inlet temperature and turbine efficiency. In its conclusions, the report was dismissive of gas turbines for aircraft propulsion, saying

> the gas turbine could hardly be considered a feasible application to airplanes mainly because of complying with the stringent weight requirements imposed by aeronautics. . . . The present internal-combustion engine used in airplanes weighs about 1.1 pounds per horsepower, and to approach such a figure with a gas turbine seems beyond the realm of possibility with existing materials.

This report was issued in 1941, a year after the first jet plane had flown in Germany, unbeknownst to the authors. The U.S. report recommended that a test engine be purchased for ship propulsion at the "present limit of 1000°F." The designers of the German aero engines had used air-cooling to move beyond "the realm of possibility with existing materials"; their engines operated at over 1400°F. Thus, we see that at the dawn of the jet age, turbine designers were exploring beyond the realm of the possible through innovative design, setting a pattern that continues to this day.

Thirty years after the first jet flight, the first jumbo jet, the Boeing 747, was in flight test with turbine inlet temperatures above 2000°F. In a university lecture at the time, when this writer was an undergraduate, the chief engineer of the 747 engines held up a turbine blade, calling it the "heart of modern air transportation"—true in 1939, true in 1969, and true today.

In 1969, there were a variety of curiosities in that university laboratory that even an optimistic undergraduate might have regarded as far-fetched in terms of practical application. These included attempts to grow single crystals of nickel in the shape of a turbine blade, layers of ceramic coatings to thermally insulate those blades at over 2000°F (the ceramics kept cracking and spalling off), and the manufacture of tiny metal particles with superior properties. The latter experiment produced what is now known as powder metal, at a rate of one drop per day. At the same time, in the building across the road, a group of engineering researchers worked on advanced measurement techniques for turbine heat transfer and a new numerical simulation-based approach to turbine aerodynamic design, which became computational fluid dynamics (CFD).

In the decades since 1969, gas turbine performance has evolved greatly by perfecting and adopting these and other, then seemingly far-fetched technologies. Most engines now use single crystal blades protected by ceramic thermal barrier and oxidation-resistant coatings to enable increased turbine inlet temperatures and very much improved operating lives. Large aircraft gas turbines employ powder metal turbine disks to reduce weight. The fluid mechanics of turbines have improved as well. Engines operate at higher temperatures made possible by very sophisticated internal and external cooling schemes, enabled in turn by advances in analytical techniques, experimental measurements, and manufacturing technology. Modern analytical techniques facilitate the design of turbine blades with laminar flow along much of the surface, despite the very high levels of turbulence inherent to these machines. The technical curiosities of the 1960s have become the commonplace of today, significantly contributing to the progress that has made the gas turbine the engine of choice for both aircraft and power generation. Large gas turbines are now the most efficient engines on the planet for converting the chemical energy in fuels to useful work.

These advances did not come easily or cheaply. They are the results of decades of intellectual effort from a multitude of talented researchers and engineers fueled by billions of dollars of industrial and government investment. At their current level of sophistication, these are not widely disseminated or easily mastered technologies. The engines they enable and the aircraft and generators they power together represent the wealth and security of nations.

This book is the only modern text to treat all of the important aspects of the engineering of modern turbines. The authors are among the world's experts and leading practitioners. *Turbine Aerodynamics, Heat Transfer, Materials and Mechanics* belongs on the shelves of all serious gas turbine professionals.

Alan Epstein

[1] Committee on Gas Turbines, National Academy of Sciences, "An Investigation of the Possibilities of the Gas Turbine for Marine Propulsion," U.S. Navy Dept., Technical Bulletin No. 2, Jan. 1941.

TABLE OF CONTENTS

Part 2 Aerodynamics and Heat Transfer

PREFACE

The era of the modern gas turbine began with Frank Whittle's 1930 patent of a turbojet engine and Hans von Ohain's patent of the first operational turbojet engine in 1936. The first flight propelled by a turbojet engine took place in 1939, and the first gas turbine for generating electricity was commissioned in the same year. Since that time, tremendous progress has been made in advancing the gas turbine for aircraft and marine propulsion, electric-power generation, and mechanical drives for industrial applications. These advances include substantial improvements in fuel efficiency and performance, reduced pollutant emissions, and greatly enhanced reliability. Today, gas turbines propel almost all commercial and military aircraft with the exception of general aviation vehicles. Gas turbines are also playing an increasingly important role in electric-power generation through aeroderivative engines to meet peak-load demands and through combined cycles in which the hot gas exiting gas turbines serves as a heat source for steam turbines to increase efficiency.

With 70 years of development, the technology associated with gas turbines might be expected to be mature, and even approaching its peak. The reality, however, is quite different. There are still many opportunities to make significant advances in efficiency, performance, service life, environmental friendliness, and affordability. The gas turbine is a high-technology engineered system, and so it can benefit from advances in many fields to enable both breakthroughs and the removal of design constraints. Also, as the gas turbine has advanced, there has been an accompanying increase in understanding that has created new pathways to further advance it. Thus, gas turbines are as exciting today as when they were first introduced in the 1930s, if not more so.

As noted, the gas turbine is a high-technology engineered system. It consists of the following major components: inlet, compressor, fan, combustor, turbine, and nozzle. The inlet seeks to modify the flow that enters the gas turbine so that it will be as uniform as possible when it enters the compressor and the fan. The compressor increases the pressure of the air for the combustor. The fan increases the velocity of the air, without unduly increasing the pressure, for increased thrust. The combustor burns the fuel from the fuel injector and the air from the compressor to raise temperature at nearly constant pressure, with the goals of minimizing pollutant emissions and ensuring the stability and performance of the combustion process. The turbine converts the high-temperature and high-pressure gas from the combustor to mechanical energy to drive the compressor as well as the fan or an external mechanical system. For aircraft, the hot gas exits the turbine through the nozzle for maximum thrust. For power generation, the hot gas from the turbines can be used in a combined cycle to create steam.

Of these components, the turbine is the most expensive. It is also the most prone to failure, and, of course, maintenance costs must be included in the accounting. Its vulnerability stems from the fact that the turbine is subjected to techniques applied to cycle selection and all aspects of the turbine design process that can quickly explore and evaluate the complete design space of interest. Competing effectively in today's marketplace demands computer-based optimization techniques that rely on rapid, parametric-based automated design tools.

I. BACKGROUND

Modern gas turbine engines are available with various configurations to meet customer requirements. Specially designed engines are used to power commercial aircraft and military supersonic fighters and to generate power for ground-based applications such as electrical generators and pumps. The applications are as varied as the types of engines available. The engineering design challenges presented by these types of engines are, however, similar, whether designing a high thrust-to-weight supersonic fighter engine that is capable of vertical flight or a micro-engine used to generate backup power for an apartment complex.

A. GAS TURBINE ARCHITECTURES

Engine architecture is driven by mission and cycle requirements. Commercial engines require very high fuel economy for a variety of missions, ranging from long-haul, intercontinental flights where the engine spends most of its life cycle at high-altitude cruise, to short missions where engine life is dominated by takeoff and climb. Military, high-thrust engines must endure the high-temperature conditions required for strenuous airframe maneuvers. Although it is commonly considered that all gas generator/free turbine arrangements are designed for single-operating-point conditions, this is not the case. Many

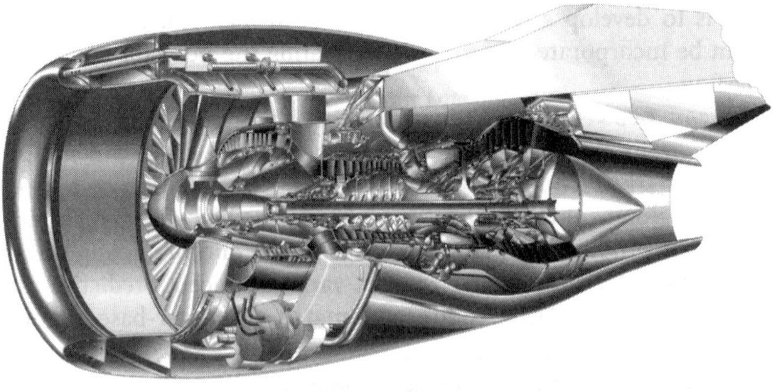

Fig. 1 Pratt & Whitney's PW6000 commercial engine cross section shows high-load turbine design to reduce part count and cost while maintaining high performance standards.

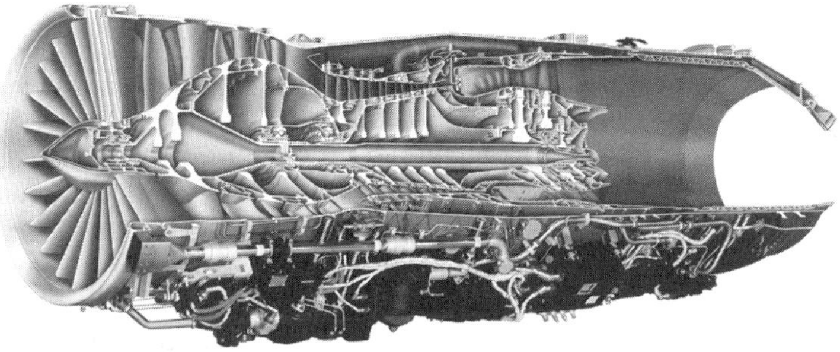

Fig. 2 Pratt & Whitney's F119 military fighter engine cross section shows current, high performance technology for high thrust-to-weight applications.

applications require these engines to have high performance over a wide range of operation (high turn-down from maximum to low power).

Gas turbine engines come in a variety of architectures, from engines that produce thrust from the exhaust flow stream to those that produce shaft power (see Figs. 1–4). All gas turbine engines, however, have a gas-generator section, typically followed by a nozzle or a "free" turbine for shaft power. In the case of high-bypass-ratio commercial aircraft fan engines (Fig. 1), the gas generator usually consists of a number of high- (and sometimes intermediate-) pressure turbine stages, where the fan shaft power is taken from the following low-pressure turbine. Gas turbine engines designed specifically for shaft power output (applications include stationary power, turboprop, helicopter, and ground or marine propulsion) typically have a gas generator consisting of a number of high-,

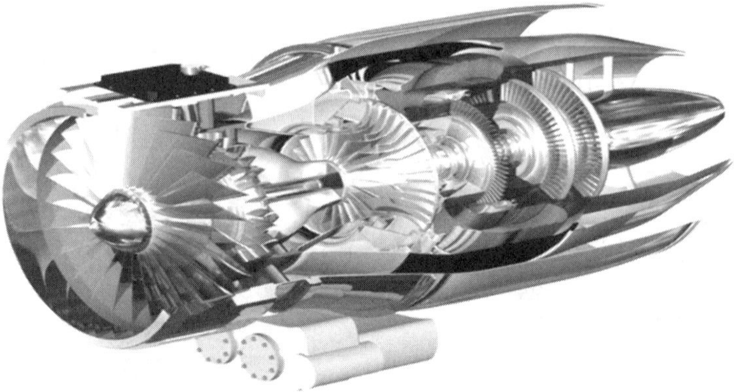

Fig. 3 Pratt & Whitney Canada's JT15 high-bypass engine cross section includes the latest high-performance radial compressor and axial flow turbine technology.

intermediate-, and/or low-spool turbines followed by a free turbine that is defined as a turbine but is not mechanically connected to the gas-generator spools.

The primary components of turbine engines used in commercial and military thrust-specific flight operations are the inlet duct, fan, compressor, combustor, turbine, and nozzle (Fig. 5). The high-pressure turbine of a typical gas turbine engine drives the high-pressure compressor (the combination often referred to as the high spool), an intermediate turbine (if present) drives the intermediate compressor, and a low-pressure turbine drives the low compressor and fan (the low spool). Together with the combustor, these high-pressure, high-temperature, and high-rotational-speed components make up the gas-generator portion of an engine.

Gas turbine engine design has advanced steadily since the early engine development studies pioneered by Frank Whittle and Hans von Ohain as discussed by Hans von Ohain in Mattingly's "Elements of Gas Turbine Propulsion" [1]. Early engines struggled with compression system efficiency, combustor dynamics, and turbine durability problems that limited engine thrust and life. The same challenges are faced today, but at a higher technology level. It is impressive to remember that the early engines were designed without computational fluid dynamics tools or sophisticated airfoil design tools and that fundamental engineering principles were applied to design and build the sophisticated, complex gas turbine engine.

Modern turbine engine design tools have transformed the design process from one in which getting the engine to run was an achievement to one that produces highly efficient, durable engines that can stay "on wing" for years before a major service overhaul. In the aerodynamic field the most significant advances came from the ability to predict the flow around airfoils and estimate losses. Some of the earlier aero analysis tools consisted of transformation methods and potential flow codes for airfoil analysis. Results from these early aero analyses, coupled with boundary-layer methods and empirical data, yielded design systems that produced many of the engines that are still flying. The most powerful tools that have recently advanced turbine aerodynamic design capability are computational fluid dynamics (CFD) tools that model the two- and three-dimensional behavior

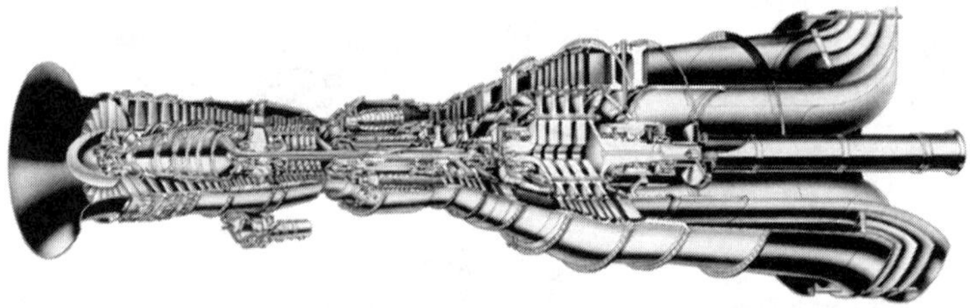

Fig. 4 PW Power Systems (a group company of Mitsubishi Heavy Industries) F18 Stationary Power Generation gas turbine engine provides 25^{+}kW of turnkey power.

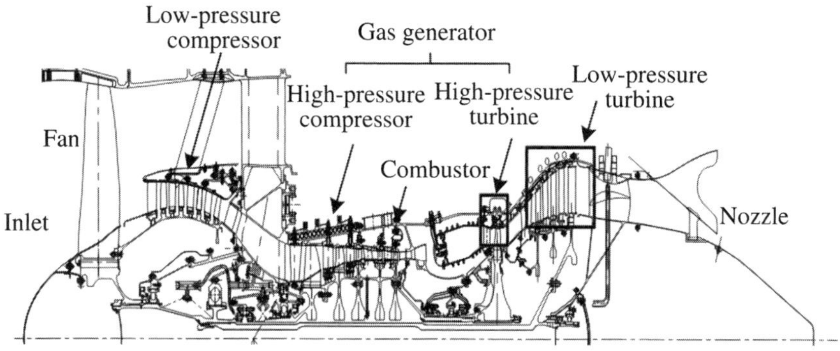

Fig. 5 A typical high-bypass-ratio commercial fan engine consists of a gas generator core and a low-pressure turbine that drives the low compressor and inlet fan (Pratt & Whitney's PW6000).

of the flow through turbine vane and blade rows [2–6]. These tools advanced from two-dimensional Euler to three-dimensional, steady Navier–Stokes (NS) solutions, to time-accurate NS; along with computer memory and speed advancements they have enabled fully coupled, automated turbine designs that far surpass early one-dimensional design methods.

The high-pressure turbine is perhaps the most challenging and interesting engine component, from a design perspective, because it pushes the limits of aerodynamics, cooling and heat transfer, and structures in a corrosive environment that exceeds material melt temperatures. Low-pressure turbines are subjected to cooler gas temperatures and pressures, but require greater attention to unsteady aerodynamics and airfoil boundary-layer development, dominated by lower airfoil Reynolds numbers, in order to maintain engine performance. As it is beyond the scope of this text to consider the design of all classes of turbines, this chapter will focus on the challenging design of cooled high-pressure turbines.

B. TURBINE DESIGN PROBLEM DEFINITION

The primary performance figure of merit for a tactical or fighter aircraft gas turbine engine is engine thrust-to-weight ratio. In the early 1960s, the state-of-the-art gas turbine thrust-to-weight ratio was approximately four to one at a turbine inlet temperature of 1700°F. Today's advanced fighter aircraft engines have thrust-to-weight ratios of eight to one at turbine inlet temperatures greater than 3000°F. Thrust-to-weight ratio projections for future weapon system gas turbines are expected to reach 20 to 1. For transport and commercial engine applications, the primary performance figure of merit is thrust specific fuel consumption (TSFC). The trend toward high-bypass-ratio turbofan engines with improved fuel efficiency and greater range is well established.

In practice, increased thrust-to-weight ratio is achieved by increasing turbine inlet temperature and rotational speeds. Higher temperature tends to reduce the useful life of turbine airfoils, but advances in turbine cooling and materials technology have enabled turbine airfoil life to be extended in spite of the increasingly challenging environment (Fig. 6). Each point on the figure represents a technology plateau corresponding to improved cooling and materials capability. Through each point there exists a line that represents constant technology where airfoil life and turbine temperature can be traded to satisfy a specific design requirement, such as increased life at lower temperature. Gas turbine performance and durability goals for future engines require further development of cooling and materials technologies (Fig. 7) that will enable the airfoils to last the full life of the system while turbine inlet temperatures approach stochiometric limits.

Turbine components, namely, the turbine blades and disks, must be light in weight but still tolerate high temperature and high stress. To illustrate, a turbine design of the PW6000 engine is shown (Fig. 8). This turbine must efficiently convert the energy of the high-temperature flowing gas stream into shaft horsepower to drive the compressor. This requires the most advanced aerodynamic design. Consider also that today's combustor exit temperatures are above the melting point of airfoil alloys and the temperature of the air used for cooling exceeds 1000°F. To provide the efficient cooling required, turbine airfoils have become complex miniature heat exchangers. In addition, consider that the centrifugal pull of a single turbine blade is several tons; this load must be transferred uniformly through the attachment to the supporting disk while the

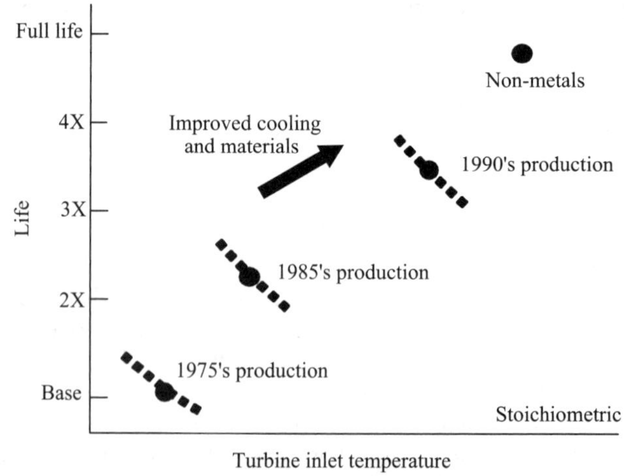

Fig. 6 Improved cooling and materials technology has enabled increases in turbine inlet temperature.

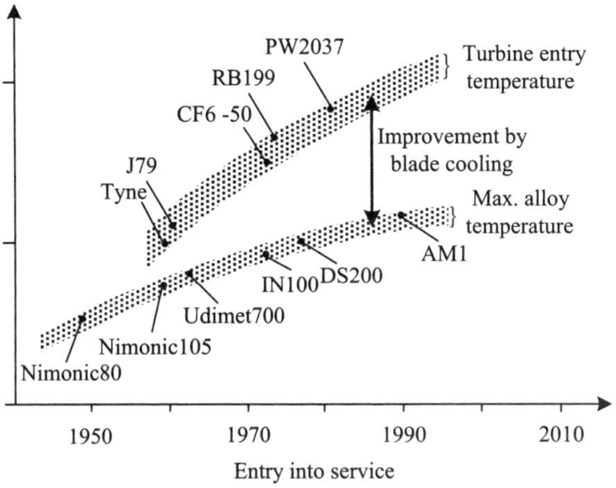

Fig. 7 A combination of materials improvements, along with advancements in cooling, helped achieve performance goals of existing engines.

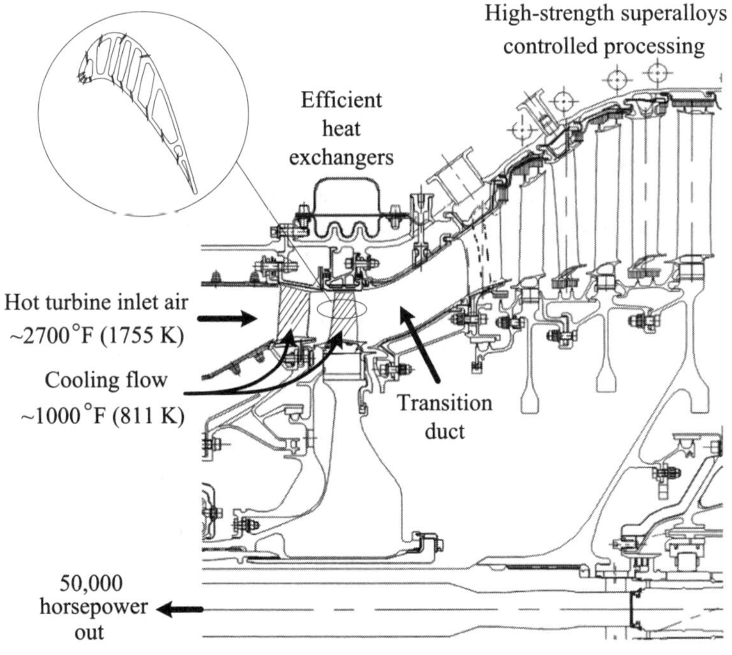

Fig. 8 Turbine design is challenged to meet power requirements by extracting shaft power from a high-temperature gas stream, where many components are cooled with secondary cooling flows.

blade is operating close to melt temperatures. Meticulous care is taken not only to seal all unnecessary leakage, but also to maintain a detailed accounting of all secondary flows and pressures. Without accurate cooling flow management, neither cooling nor performance goals could be achieved. In summary, any improvement in alloy strength, fatigue resistance, or temperature capability can be translated into improved engine performance and reliability.

The design of today's high-performance gas turbine requires the integration of many technical engineering disciplines and relies on detailed knowledge of hot-section material properties and gas properties to ensure the highest efficiency possible given the available materials. The performance of all gas turbines is highly dependent on obtaining the highest level of turbine inlet temperature with the available cooling at the fastest rotating speed possible. The aerodynamic performance of the compression system is also a factor; for example, a highly efficient compressor has lower exit temperatures, which results in cooler cooling air, permitting higher inlet turbine temperatures. Efficient, high-work, high-pressure turbines also enable higher inlet turbine temperatures by way of reduced cooling requirements in latter stages. The design challenge is to produce the hottest, fastest spinning turbine possible with the most efficient components, with minimum coolant flow.

II. TURBINE DESIGN FUNDAMENTALS

Gas turbine engine design is an elegant mix of classical thermodynamics and practical engineering techniques. The classical thermodynamics comes into play in the selection of the optimum cycle, in which thermodynamic relationships and fluid properties determine where on the enthalpy-entropy (or temperature-entropy, T-S) diagram the engine needs to operate for optimum propulsion system efficiency or optimum shaft power efficiency. The gas turbine engine thermodynamic cycle, in principle, is a simple process of compressing air, adding heat at the elevated pressure, and expanding the heated, high-pressure air through a turbine and nozzle system back to the initial pressure, that is, the simple, ideal Brayton cycle (Fig. 9) with isentropic compression and expansion processes. The thermal efficiency of an engine operating with an ideal Brayton cycle increases with pressure ratio. The thermodynamic efficiency of a real engine is, however, governed by the heat added to the cycle, the energy lost to the surroundings by the expulsion of exhaust gases and the inefficiencies in the components (the major components being the inlet, compressor, combustor, turbine, and nozzle). Although the analysis and design of optimum thermodynamic cycles for gas turbines is an exciting subject by itself, the principle focus of this text is the design of efficient turbine components, leading to efficient engines. The reader is encouraged to see associated texts for cycle analysis [1, 7–9].

The intriguing mix of classical thermodynamics and practical engineering results from the thermodynamic process of generating work through the use of

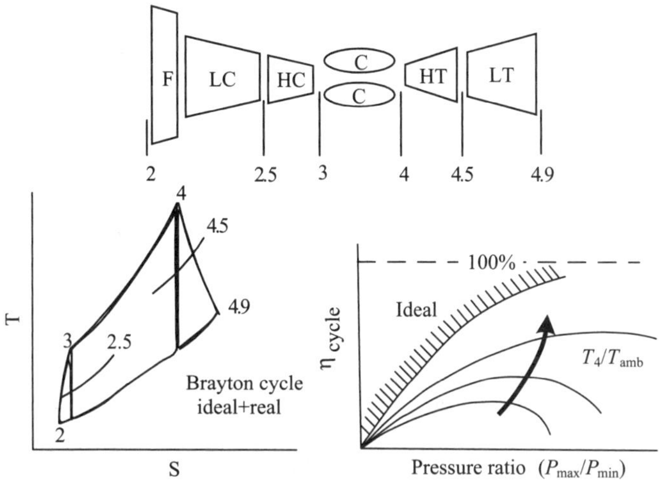

Fig. 9 The gas turbine engine operates on a Brayton cycle, where the performance of an ideal cycle is only dependent on pressure ratio. Component losses make the performance also an effect of turbine inlet temperature, T4.

moving parts, in this case, rotating blade rows. The thermodynamic analyses (shown schematically in Fig. 9, the T-S diagram) conjure the mysterious thermodynamic process of work. The statement that "work happens" during the temperature drop does not, however, help the turbine designer. The practicality of designing a turbine is the need to accelerate and turn flow to create lift forces on moving surfaces, which result in the generation of torque (T) and, because these surfaces are moving, work (W) or in our case power (P), which is the rate of work generation.

Lift forces are generated through an engineering fundamental relationship; a change in momentum is proportional to the forces applied, or in this case, the airfoil lift forces cause a change in momentum. The Euler work equation, where power (P) is proportional to flow times wheel speed (U) times the change in tangential velocity (ΔC_u), is used to define the required momentum change for the prescribed work (Fig. 10). These momentum changes are captured in the customary velocity triangle plots (Fig. 11), which show how the fluid in a turbine stream is turned in each row of airfoils. Variables C and W represent absolute and relative velocities, respectively; α and β can be used to represent absolute and relative gas angles. It is these velocity triangles that link the thermodynamics of a turbine design to the turning an airfoil needs to accomplish to produce shaft power.

Efficient generation of real shaft work determines the extent to which the process is isentropic, that is, minimum entropy is created. A good turbine design will use its pressure ratio efficiently with low-loss airfoil designs that

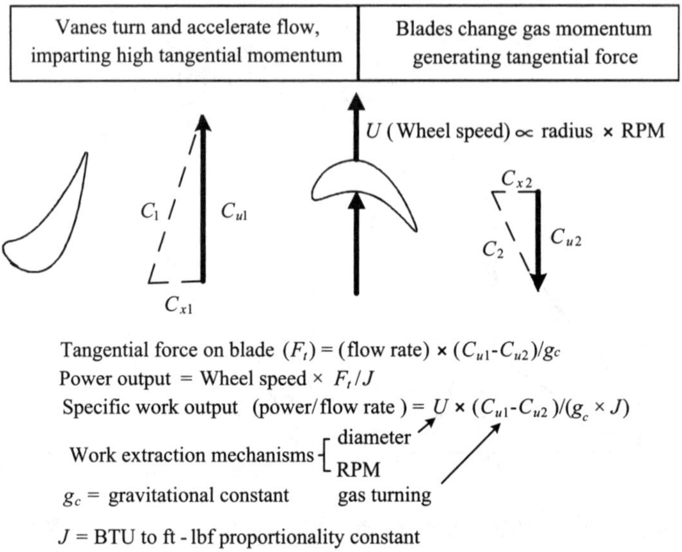

Fig. 10 Power is extracted from the flow by changing the momentum (direction) of the flow to create lift on rotating blades that is translated to torque and power.

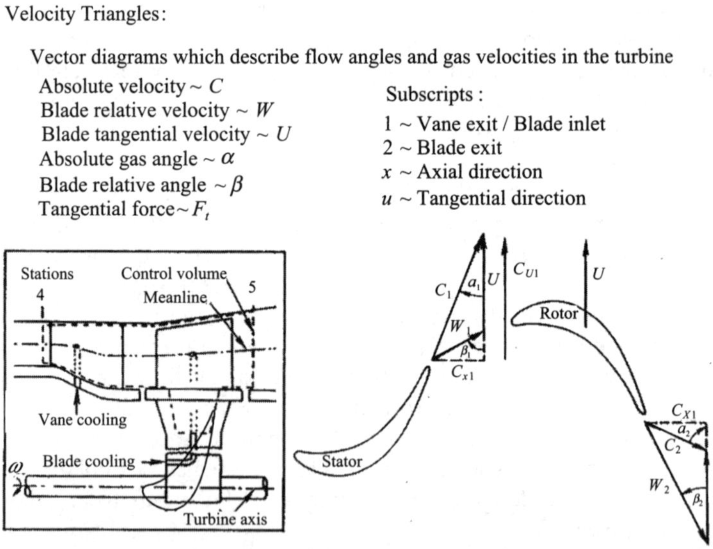

Fig. 11 Typical nomenclature for describing the velocity triangles is needed. This nomenclature can be different depending on the design tool, but the physics is the same.

result in the maximum temperature drop for a given pressure ratio (expansion ratio), resulting in very little entropy generation. Airfoil designs with high losses create more entropy and, therefore, result in less work generated for the same pressure ratio. Therefore, the goal of the turbine designer is to design low-loss airfoils that maximize work output for a given pressure ratio.

The focus of this chapter will be the engineering design process for the aerodynamic design of turbines, principally, cooled, high-pressure turbines. Intermediate, low, and power turbine design is similar in practice, although the focus may be different, requiring less attention to the effects of airfoil cooling and more to airfoil Reynolds number.

A. STANDARD DESIGN PROCESS

The standard turbine aero design process is preceded by thermodynamic cycle studies of the complete engine to identify the optimum engine configuration. These preliminary engine design studies are used to identify component characteristics necessary to match engine components to achieve the desired engine performance and operability. To start the process of designing a gas turbine, the thermodynamic cycle must be evaluated to assess the required specific work, which will, in turn, determine the pressure ratio required for the operation of the gas turbine. The rotational speed can be predetermined by the driven requirements or can be a variable of the structures design, requiring iterations to optimize the design. The higher the fidelity of the system-level optimization ensures better overall system performance.

Selection of the engine thermodynamic cycle fixes the inlet conditions for the turbine and determines the expected power output. The turbine designer has the flexibility to stretch or compress the flow path (geometric annulus dimensions) using results of analytical studies to create the best turbine flowpath for a particular application. In practice, for a given power requirement and flow, the flowpath sets the velocity triangles, and then the velocity triangles determine the row losses and set the efficiency. Then, details of the flowpath design are translated into efficiency, weight, and cost estimates. These factors influence the feasibility of the final engine. When these characteristics are compared to the characteristics of other engine components, adjustments to the thermodynamic cycle might be required to improve the engine.

The design process itself requires increasingly more complete analysis as greater detail is included in optimizing every aspect of the turbine (Fig. 12). Meanline analysis used in preliminary engine design is a solution of the velocity diagrams of the middle stream surface. This analysis determines the flowpath size and the general arrangement of airfoils. Streamline analysis increases the detail by modeling velocity diagrams on extended number of stream surfaces between hub and tip. Adjusting velocity angles in the hub or tip regions can alleviate local problems without affecting other regions. The streamline results influence the flow angles to which airfoils are subsequently designed. Two-dimensional

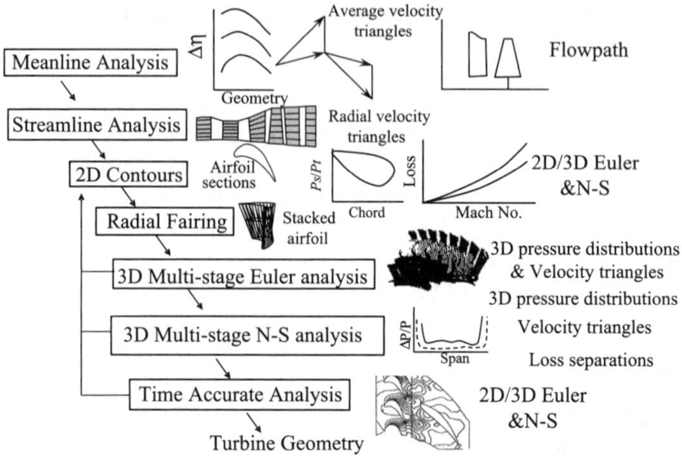

Fig. 12 The turbine design process begins with meanline analysis to optimize flowpath and velocity triangles and ends with time-accurate, three-dimensional Navier–Stokes computational fluid dynamics analysis.

airfoil design and analysis is required to describe the flow through the passage between airfoils and is used to optimize the airfoil shape to achieve the required turning, as dictated by the velocity triangles from the meanline and streamline analyses. Three-dimensional airfoil design analysis is used to identify the best radial stacking of the airfoil sections and to locate regions within an airfoil row where sections should be modified to maximize row and stage performance.

B. MEANLINE DESIGN TOOL

The meanline design tool implements thermodynamic laws, Euler work definitions, and loss modeling systems [10]. Input to this tool is dependent on the desired outcome (design mode or off-design mode). The output of the meanline analysis is always efficiency, but it can also be used to determine the proper velocity triangles to achieve the prescribed work (design mode) or can be used to calculate the work output and flow for a given design (off-design mode).

Meanline analysis is the first step in any turbine design process. The analysis is based on the assumption that flow through the turbine can be represented by the flow at the center (mean) of the flow passage. Aerodynamic losses are modeled in the meanline to optimize the flowpath shape, number of stages, airfoil counts, reactions, and work split between stages (Fig. 13) [10]. Cooling and cavity leakage flows play a significant role in designing cooled turbines; it is highly desirable to include cooling and leakage in the meanline modeling. Thus, the meanline 1) models all of the through flow and therefore determines the airfoil velocity triangles and 2) determines the proper pressure ratio and temperature ratios necessary to achieve the design goals set by the engine thermal cycle analysis.

In much of the solution, one-dimensional gas dynamics is sufficient to resolve the changes in gas properties through the turbine. Properties at span locations other than the meanline can be estimated assuming free vortex radial swirl or other forced vortexing, based on design experience. The interpretation of many performance parameters depends on a well-defined system boundary. Properties and velocities for the mainstream gas and cooling air must be properly interpreted at the system boundary, consistent with the engine cycle analysis, if turbine work is to be accurately determined. Efficiency and other performance terms have different values when different control volumes or system boundaries are used. (The reader is encouraged to review efficiency definitions [11].) Indeed, differences in cooled efficiency definitions can result in a 4% variation in calculated cooled efficiency. Therefore, it is highly desirable to include cooling and leakage flows consistent with the engine cycle model in the efficiency definitions, so as to facilitate comparison of turbine designs with varying cooling levels.

Of critical importance to the meanline design tool is the ability to estimate airfoil row losses with minimal input; principal inputs include thermal properties, Mach number, and flow angles. Therefore, the meanline analysis requires a system for estimating airfoil losses to accurately model gas pressures and overall turbine performance trends (Fig. 14). Profile loss results from friction on the surface of airfoils and is most heavily affected by airfoil turning and loading [12]. Endwall loss occurs due to friction and secondary flows in the endwall region of a passage where aspect ratio and the endwall axial extent are the dominant factors [5–9, 12–25]. Cooling flow entering the mainstream generally causes loss through momentum exchanges and mixing of different temperature streams [24]. Tip leakage flows result from a clearance between the rotating blade tip and the static outer case and result in lost work and nonideal mixing of the leakage

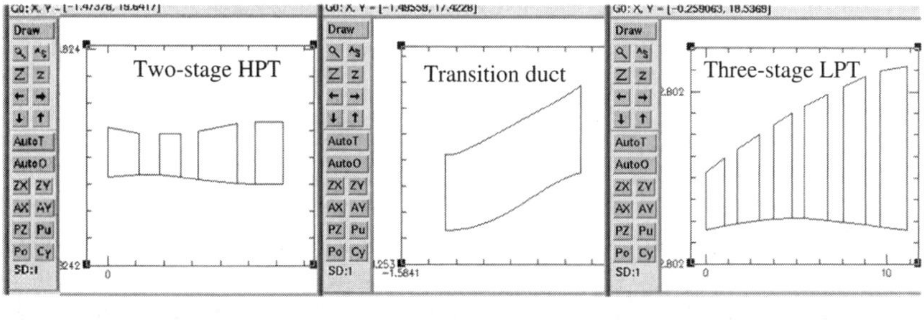

Fig. 13 The meanline is used to optimize the flowpath and stage characteristics based on a few input parameters (Ptin, Ttin, Power, and Flowrate) to maximize turbine performance.

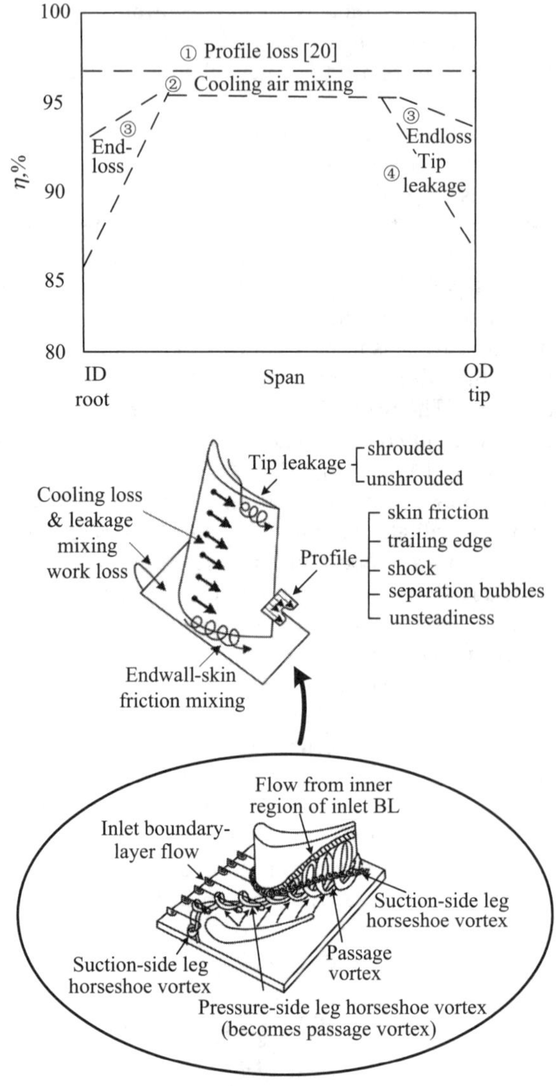

Fig. 14 Typical turbine losses are broken into four categories: profile, endwall, tip leakage, and cooling and leakage. These loss sources affect the flowfield loss differently depending on proximity to the endwalls.

flow with the main stream flow. Tip leakage loss can be estimated by correlating with mixing loss-type models [24]. Shock losses occur when normal and oblique shocks interact with boundary layers or other in-passage compression waves dominated by trailing-edge aerodynamics. These losses are a strong function of airfoil exit Mach numbers and the airfoil design approach.

The relative contribution of losses depends on the turbine configuration (Fig. 15). HPTs typically have large pressure ratios across one or two stages, higher exit Mach numbers, lower aspect ratios and higher gas turnings contributing higher tip leakage losses, endwall secondary flow, and cooling and shock losses relative to LPTs. Performance in LPTs is typically dominated by profile losses because more of the airfoil span is dominated by two-dimensional flow. Endwall losses in LPTs are dominated by lower acceleration in airfoil rows and larger flowpath divergence. To model gas properties and turbine performance trends reasonably well, meanline analyses must accurately estimate all of the losses that will occur in the flowpath in the efficiency calculation (Fig. 16).

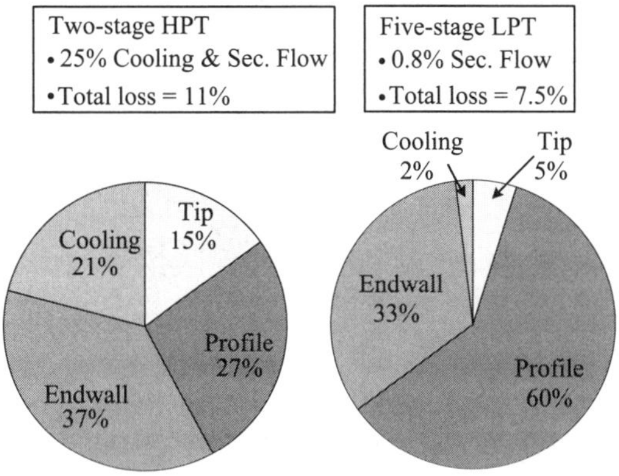

Fig. 15 The breakdown of losses shows that HPTs are dominated by tip clearance, endwall and cooling losses, while the LPTs are dominated by profile.

Several basic turbine parameters become key variables in the process of designing a turbine or comparing a given design with another (Table 1). *Total shaft work,* which is a requirement for the cycle definition, is the energy that must be delivered by the turbine to drive the compressor and/or fan. Work distribution for each stage, or *work split,* also becomes a design variable when the engineer is optimizing a multistage turbine design. For example, reducing work in a first stage of a multistage turbine will reduce airfoil turning, thereby reducing profile loss of that row, but will also increase gas temperatures entering the second stage, thereby increasing the cooling requirements. *Reaction* is a measure of the expansion process shared between the rotor and stators within a stage, that is, trades vane Mach number with blade Mach number. Subsequently, increasing reaction will reduce vane turning but at the same time increase the axial thrust loads on the rotor. *Stage work coefficient* is defined as the reference velocity that the gas would reach if expanded through the full rotor energy drop divided by wheel speed (proportional to square root of average stage work over the wheel speed). Stage work coefficient can be decreased by increasing radius, rpm, or number of stages. *Flow coefficient,*

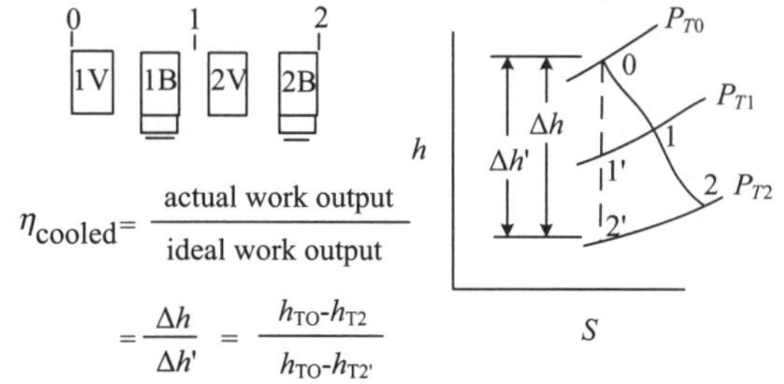

Fig. 16 Calculating cooled turbine efficiency can be a complex process and has generated considerable debate on standard efficiency definition. Shown here is the simplest of efficiency definitions for an uncooled, single stream turbine: actual work (temperature ratio dependent) relative to ideal work (pressure ratio dependent).

TABLE 1 COMMON PARAMETERS OF TURBINES DESCRIBING THE CHARACTERISTICS OF A DESIGN

Parameter	Definition
Work	$U\Delta Cu/gJ$
Reaction	$R = (\Delta Ps)\text{rotor}/(\Delta Ps)\text{stage}$
Velocity Ratio	$VR = U/\sqrt{2gJ\Delta h}$
Load Coefficient	$gJ\Delta h/U^2 = \Delta C_u/U$
Flow Coefficient	C_x/U
Stress Parameter	AN^2
Efficiency	η = Actual Power/Ideal Power

C_x/U, is a similarity parameter used to compare different turbines. Generally, turbines with similar stage work coefficients and flow coefficients have similar velocity triangles. The challenge of the turbine designer is to balance and trade these competing, fundamental turbine parameters to optimize the turbine design, where the ultimate outcome is the gas property states (pressure and temperature), Mach numbers, and flow angles at the inlet and exit of each airfoil row. These engineering parameters are used to design the airfoil sections and ultimately, for a given radial vortexing distribution, the full, three-dimensional airfoil shape that achieves the required airflow turning and shaft work requirements. Therefore, it is the meanline (and the subsequent streamline) analysis that couples and converts the thermodynamic cycle requirements of the gas turbine engine to the flow turning requirements for airfoil design.

The big question is, how does one start when designing a turbine? It is best to start with the experience of others. This collective experience is captured in what is known as the "Smith Chart," the form of which is readily available in the open literature, with a representative version shown in Fig. 17. The Smith Chart [26] is typically a representation of an array of meanline results populated with rig and engine data experience. The Smith Chart plots average stage work coefficient vs flow coefficient, with turbine efficiency plotted as a third parameter. This chart can be used to identify optimal turbine efficiency for given set of aerodynamic and structural constraints. A location on the Smith Chart yields parameter values such as stage work coefficient and flow coefficient that determine the nominal characteristics of the flowpath and velocity triangles.

Selection of a pressure ratio per stage is the first step to sizing a turbine. The approximate pressure ratio per stage, the turbine inlet temperature, and the required specific work (Δh) will set the number of stages. Sizing a turbine is completed by using basic nondimensional turbomachinery parameters of flow coefficient (C_x/U) and stage loading ($\Delta h/U^2$). Selection of an average stage loading ($\Delta h/U^2$) will set the rpm-radius relationship, and based on the Smith Curve, the C_x/U level will set the relationship between flow, radius, and rpm. Finally, incorporating the available flow and rpm, the mean radius and span can be determined.

Meanline analysis of the flowpath for work extraction using loss models for profile, secondary, cooling, and shock losses leads to an estimate of turbine efficiency [5–9, 12–25]. Using the Smith Chart as a roadmap, the turbine designer can see that design considerations such as a decrease in stage work coefficient

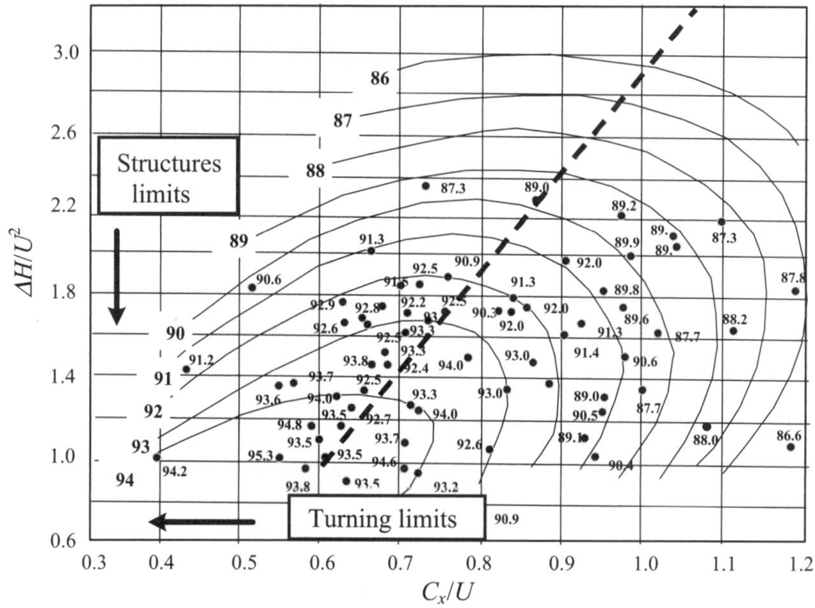

Additional Considerations

Pressure ratio	Stage work split
Reaction	Geometry selection
Clearances	Unsteadiness
Cooling/leakage flows	Flowpath condition

Smith, S.F. "A simple correlation of turbine efficiency", *Journal of Royal Aeronautical Society*, Vol 69, July, pp. 467-470

Fig. 17 The Smith Chart can be used to obtain a starting point for a turbine design. Choosing a point on the Smith Chart with some other constraints, such as rpm and power requirements, can be used to size the flowpath, mean radius, and number of stages. Subsequent meanline modeling and optimization is used to apply personal experience to the design.

reduces gas turning and Mach number requirements through a relative increase in wheel speed with respect to swirl velocity. Major limiting factors for lower stage work coefficient designs are disk rim stresses caused by high wheel speeds and the increase in weight and cost due to increasing the number of stages. Lower C_x/U can be attained through increasing wheel speed by increasing radius or rpm. In addition, lower C_x/U can also be obtained by decreasing axial velocity by increasing annulus area, thereby reducing Mach number. Other benefits of lower C_x/U include increased aspect ratio, which results in reduced endwall loss. Major factors for limiting C_x/U reductions are high gas turning angles, high blade stresses due to the increased annulus or rpm, and the resulting low vane exit angles that cause increased trailing-edge thickness losses. The ultimate turbine design must represent a balance between high efficiency and good engine performance and the difficulty of the airfoil design and risks associated with increased stresses and more complicated cooling configurations.

C. STREAMLINE DESIGN TOOL

Streamline analysis extends the design process by including the effects of radial profiles. Whereas the meanline is considered a one-dimensional design tool, that is, it resolves the turbine design definition in the axial, throughflow direction, the streamline is a two-dimensional design tool resolving design characteristics in the radial and the throughflow directions (Fig. 18). The streamline analysis requires additional input, relative to the meanline analysis, to solve the radial distribution of flow through the turbine, namely, the inlet total pressure and temperature profiles and the desired radial distribution of work (ΔCu or turning).

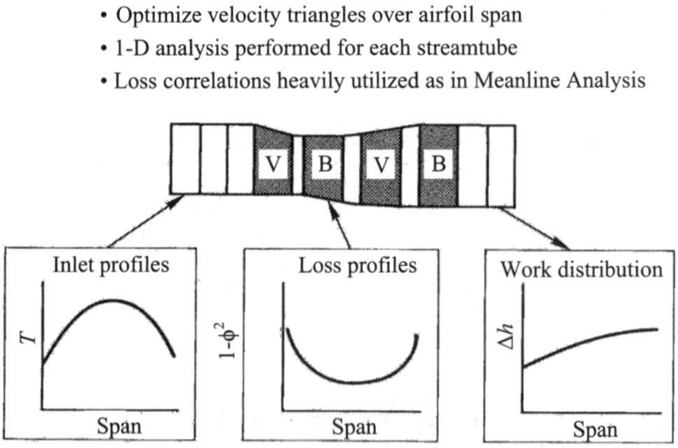

Fig. 18 The streamline design tool is used to optimize the work load (velocity triangle) distribution, taking into account the spanwise distributions of loss associated with turning and Mach number variations.

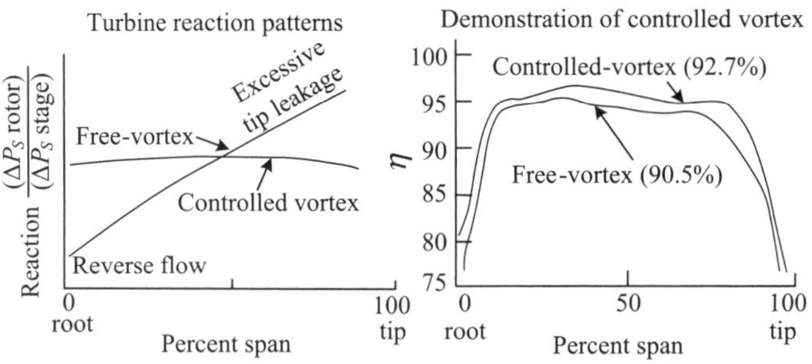

Fig. 19 Forced vortex designs are used to maximize work potential in today's turbine designs.

Using loss modeling, not unlike the meanline analysis, the streamline analysis solves the two-dimensional flow problem in the radial and throughflow directions using the assumption of zero gradients of flow properties in the circumferential direction. The output of the streamline analysis is an estimate of turbine performance with endwall and tip flow considerations not achievable by the meanline analysis, and the radial distribution of flow and velocity triangles for specification of two-dimensional airfoil design.

Inlet profiles of total temperature and pressure are modeled and tracked in the streamline analysis along the meridional surfaces. Airfoil loss profiles are used to realistically represent the effects of endwall loss and tip effects on the local gas triangles. Design variables such as radial work and reaction profiles can be adjusted so that the turbine delivers the shaft power required. First-generation turbines were designed using a free vortex calculation for the radial variation in velocity triangles because of calculation simplicity. Free vortex designs result in rotor reactions that increase rapidly from hub to tip. The resulting low root reactions led to highly loaded root airfoil designs with high losses in these regions. At the tip, high reactions cause excessive fluid leakage because of the high static pressure drop between the front and back of the airfoil. Streamline analysis permits radial control of reaction from hub to tip (Fig. 19). This added dimension in the design process allows the designer to tailor an airfoil row with a controlled-vortex design [27] for lowest loss over the complete span, thereby improving turbine efficiency.

D. TWO-DIMENSIONAL AIRFOIL DESIGN

The two-dimensional airfoil design is the first step toward a full, three-dimensional design. Whereas the meanline and streamline design activities have resolved the overall turbine configuration in the throughflow and radial directions, the two-dimensional airfoil design resolves the airfoil shapes needed in the throughflow and circumferential directions that will create the velocity triangles required from the streamline analysis. A three-dimensional airfoil design is a compendium of two-dimensional airfoil sections defined at selected radii within the flowpath that meets durability and structures requirements (Fig. 20). Consideration of the radial profile information from the streamline analysis is extended to the two-dimensional airfoil design process by supplying the necessary flow properties and swirl boundary conditions for each airfoil section with the ultimate outcome a definition of a three-dimensional airfoil as defined by airfoil section shapes defined along its radial axis.

Airfoil design is generally an iterative process that requires the engineer to shape and reshape each airfoil section until it will deliver the proper gas turning with the lowest aerodynamic loss at a given radius. Pressure distributions are calculated for each airfoil shape, and criteria are invoked to assess the quality of each section design. Results of the pressure distribution calculation are presented in the form illustrated in Fig. 21. Surface static pressure is often normalized by the inlet total pressure relative to the airfoil to facilitate comparison with previous designs. Axial (i.e., throughflow) position in this figure is normalized by the axial projection of the airfoil chord. The area between the suction-side and pressure-side portions of this curve represents the aerodynamic load for the airfoil. This load matches the momentum change experienced by the gas as it goes through the passage as specified by the meanline and streamline analyses.

The biggest challenge for the airfoil designer is to create airfoil shapes that satisfy the turning requirements with minimal loss. Regions of the airfoil that have favorable pressure gradients (decreasing pressure with streamwise distance) will have boundary layers that will remain attached and will be thin, with low loss.

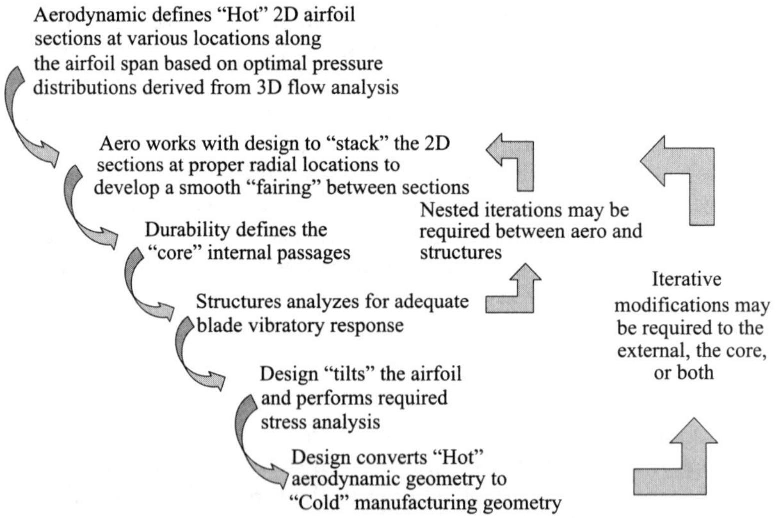

Fig. 20 Airfoil design is a multidisciplinary effort, including input from durability (cooling requirements) and structures (stress and part life).

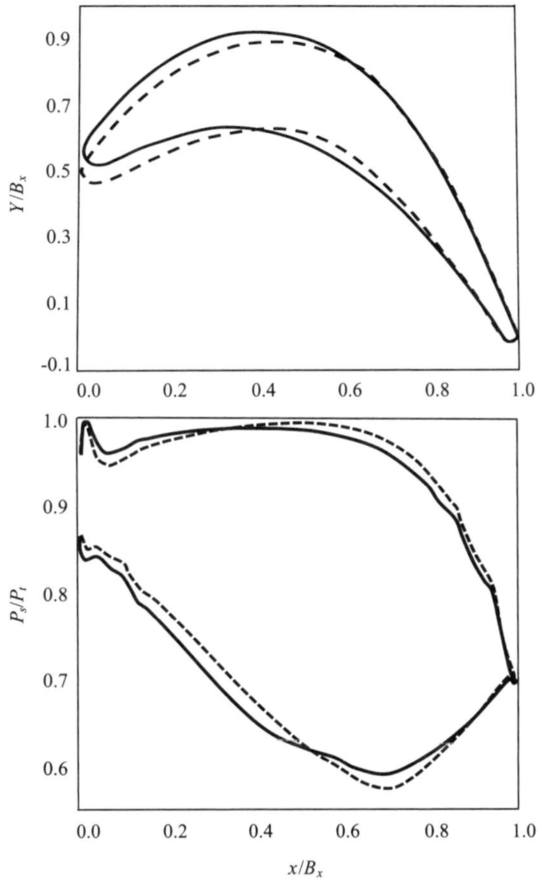

Fig. 21 Two-dimensional airfoil design is conducted through geometry generation and pressure distribution calculation to optimize the airfoil and obtain the required turning based on inlet boundary conditions at minimum loss.

Regions with adverse pressure gradients will have thickening boundary layers that could separate for large adverse pressure gradients. It is the balance of accelerating the flow over the front portions of the airfoil followed by controlled diffusion on the latter portions of the airfoil that will lead to highly loaded airfoil designs that meet high performance design criteria.

The challenge is further increased by the desire to minimize weight and part count to reduce engine costs, thereby leading to highly loaded airfoil designs (Fig. 22). As loading, the area between the two curves, increases, flow on the suction side of the airfoil will eventually become transonic. Transonic flow occurs when the local flow velocity within the airfoil passage approaches or exceeds the speed of sound, that is, where the ratio of static to total pressure falls below about 0.5 for air. Airfoil designs with transonic or supersonic flow can have significant loss increases, and special design methodologies are needed to mitigate these losses by controlling the strength of the shocks emanating from an airfoil and the way the shocks interact with adjacent airfoils or downstream airfoil rows (Fig. 23).

Before the advent of two- and three-dimensional Navier–Stokes computational fluid dynamics analysis tools, two-dimensional airfoil loss was predicted by applying boundary-layer prediction methods to pressure distributions determined from potential flow methods [28]. This approach is still valid, especially

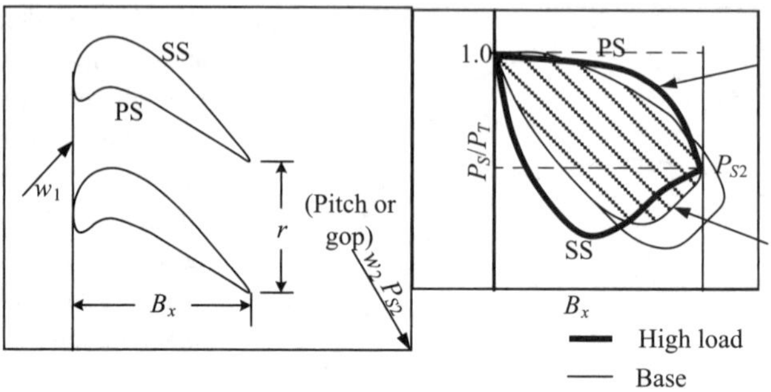

Zweifel Load Coefficient is the turbine parameter used to represent non-dimensional airfoil lift.

$$Zw = f\left(\frac{\tau}{B_x}, \beta_2, \beta_1, Cx_2, Cx_1\right) = \frac{\text{Midspan Airfoil Lift}}{Q_{\text{exit}}}$$

Zweifel Load Coefficient can be calculated on a one-dimensional basis during concept optimization to set foil count.

Fig. 22 High load airfoil designs result from low-velocity-ratio designs and reduced airfoil counts. Highly loaded airfoils have higher suction side Mach numbers and typically have increased losses.

for turbine designs that have high aspect ratios with limited secondary flow losses. Customarily, after an airfoil shape that satisfies good design criteria has been selected, the engineer utilizes the pressure distribution, calculated with potential flow or Euler analyses, to conduct boundary-layer calculations to determine viscous layer parameters and thereby the airfoil profile loss. This step in the design process can be eliminated by utilizing two- and three-dimensional Navier–Stokes analysis tools with turbulence models that account for the effects of boundary-layer transition from the laminar-to-turbulent state. The two-dimensional profile loss and gas turning for each airfoil section in these codes is an output for the two-dimensional airfoil design. These high-fidelity Navier–Stokes tools not only simplify the analysis process relative to the lower fidelity modeling but also enable the automated analysis needed for computer-aided optimization.

E. THREE-DIMENSIONAL AIRFOIL DESIGN

Turbine design ultimately employs three-dimensional design concepts aimed at achieving the spanwise work distribution prescribed by the streamline analysis while maximizing overall turbine efficiency. In addition, physical modeling not available in streamline tools, such as airfoil-to-airfoil interactions, capture additional, physics-based loss mechanisms [24, 29, 30]. The three-dimensional analysis uses the two-dimensional airfoil design section information with the flowpath and inlet flow profile information from the streamline analysis to calculate the complex, three-dimensional flows within each turbine airfoil row. The output is a physical representation of how the turbine will perform in a rotating rig or an engine environment. Turbine performance is estimated, but more importantly the local flow features are determined and can be used to control local diffusion of the flow, thereby minimizing losses and maximizing turbine efficiency. Of primary concern is how the two-dimensional airfoil shapes and the three-dimensional airfoil stacking affects the three-dimensional flow effects, specifically controlling and reducing endwall losses.

Because most of the aerodynamic loss generated in an airfoil passage and heat transfer loading is governed by airfoil and endwall boundary layers, the ability to accurately predict their behavior is a very important part of the turbine design process. Boundary-layer development inside an engine is subjected to many factors that make prediction a difficult task [12, 23]. Among the factors are pressure gradient, temperature gradient, turbulence level, surface roughness, surface curvature, Reynolds number, endwall and tip leakage vortices as well flow unsteadiness. Historically, before rapid CFD tools, the approach taken to resolving the effects of boundary layers has been to experimentally assess the impact of each of these individual factors, either through in-house testing or from data available in the literature, and then to apply these effects through analytical means.

Solution of the Navier–Stokes equations [2–4, 31, 32], resolving both inertia- and viscous-dominated fields, is now industry standard practice. Robust, rapid,

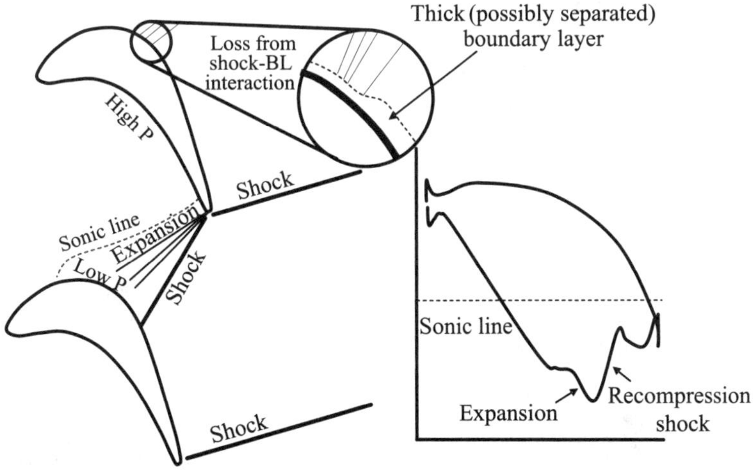

Fig. 23 Transonic and sonic airfoil design must consider shock formation and shock reflection effects on performance. Airfoil design philosophy can be developed to manage shock losses.

pressure-based implicit and explicit time marching procedures to solve the steady-state Navier–Stokes equations have been developed and coupled with rapid grid generators to provide denser meshes near boundaries, where viscous terms are dominant, thereby increasing modeling accuracy. With the advent of rapid three-dimensional flow solvers, design concepts and techniques have been established to provide the aerodynamic designer with greater design flexibility to control the flow development. Using conventional two-dimensional airfoil design and analysis techniques, the designer has only the airfoil thickness and curvature to influence the flow as it passes through the airfoil passage. Divergence or convergence of the endwalls (modeled in two-dimensional as streamtube divergence or convergence) is a global term that improves the two-dimensional analysis, but the three-dimensional effects on the full-span airfoil flow cannot be significantly represented. Using three-dimensional flow solvers, the designer is able to accurately model the additional effects of radial stacking and endwall contour on the flowfield. The added forces acting on the computational domain

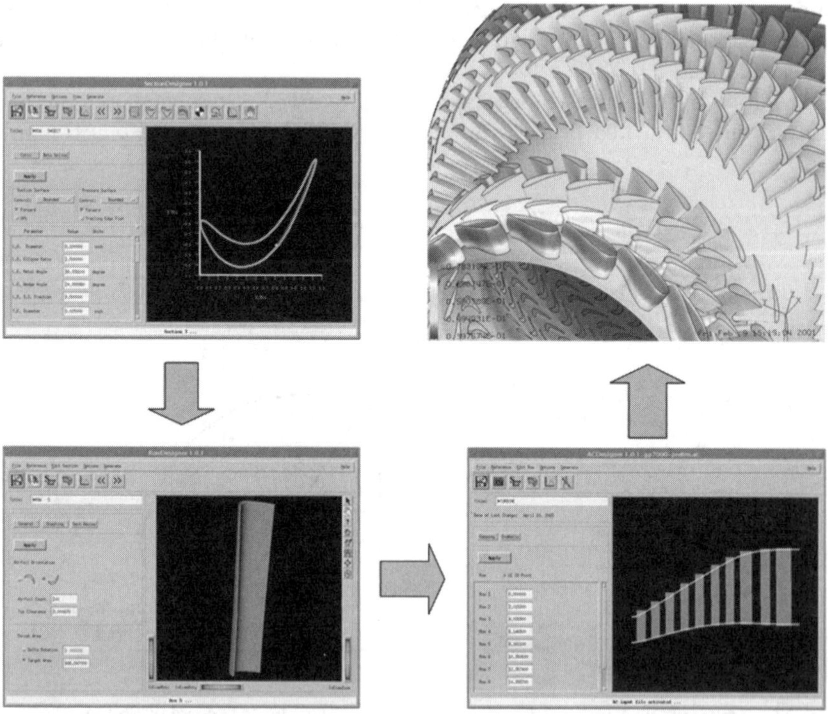

Fig. 24 Automated design tools enable rapid two- to three-dimensional to multistage analyses.

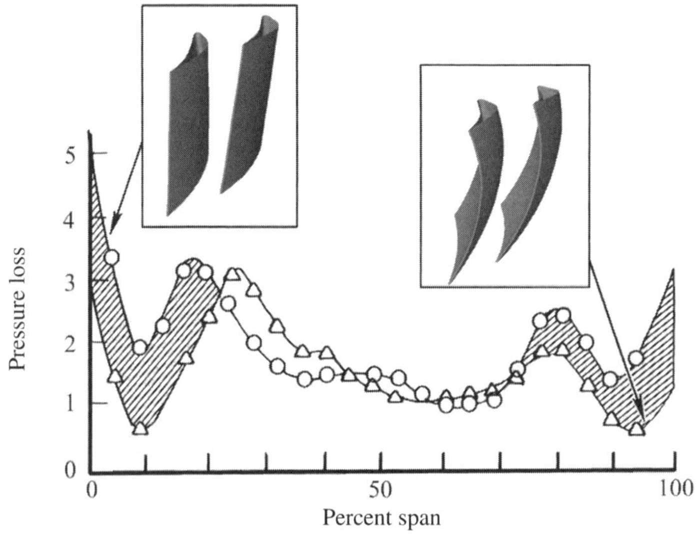

Fig. 25 Three-dimensional design considerations can be used to manage spanwise loss distributions. In this case study, a bowed vane design was used to significantly reduce endwall, interpassage, secondary flow losses.

allow turbine designers to better model the three-dimensional geometry of an engine while giving them additional control of the flow within and downstream of each airfoil row.

Application of three-dimensional modeling analysis to the design process involves rapid calculations, enabled by computing horsepower and automated tools. Viscous flow solvers and the Euler analyses that preceded them have been developed to such an extent so as to permit routine analysis of multirow, multistage turbine models. Rapid, steady-state solutions are obtained by conducting row-by-row analyses coupled through interface planes that transfer average spanwise distributions of air flow properties between adjacent rows of airfoils (Fig. 24). At the interface plane, pressure, temperature, and gas angle are averaged over the pitch to create an axisymmetric flowfield boundary condition into the next row that captures the spanwise distribution and migration of flow. These steady-state analyses capture the complex, salient features of the flowfield within and immediately upstream and downstream of an airfoil row and also transfer the radial realignment of flow characteristics between rows.

Let us consider two examples of the effect of three-dimensional airfoil design on the ultimate performance of a turbine. The first example shows the effects of airfoil stacking on row loss (Fig. 25). This example shows

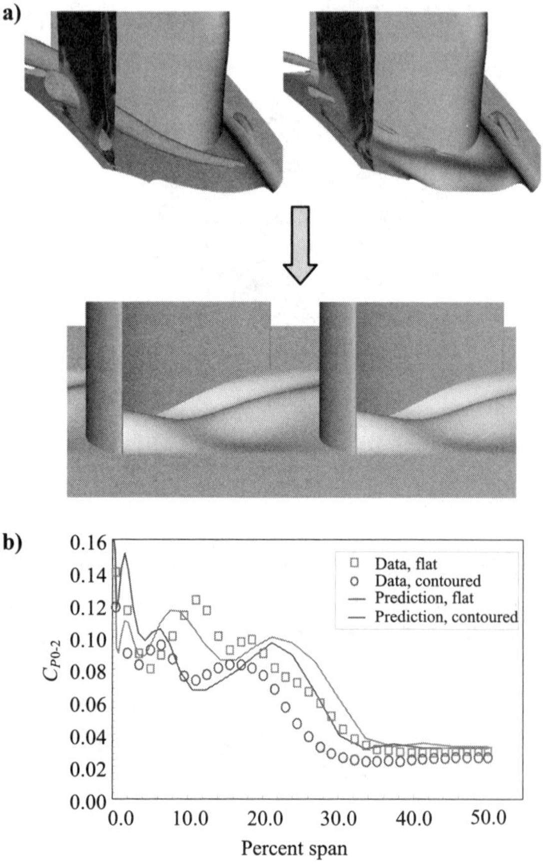

Fig. 26 Automated geometry generation and computer-based optimization techniques can be used to reduce secondary flow passage losses. Concept studies using CFD are used to understand the physics of loss production: a) linear cascades are used to validate the CFD capability b) leading to three-dimensional turbine designs.

back-to-back tests of a vane design with only two-dimensional considerations. Advanced, three-dimensional analysis was used to optimize the vane performance. Significant performance improvement resulted when the contoured three-dimensional vane was tested, which verified the endwall loss-reducing capability of three-dimensional designs.

In the second example, nonaxisymmetric endwall contouring was studied to understand the physical mechanisms of endwall secondary flow loss [33–37] (Fig. 26). Once the loss mechanisms were understood, validation testing in a linear cascade led to a three-dimensional endwall for application in a turbine design for improved efficiency. The improved performance and flow control associated with using these three-dimensional design procedures is beneficial for improving engine performance, including thrust-to-weight ratio and thrust-specific fuel consumption.

F. TIME-ACCURATE COMPUTATIONAL FLUID DYNAMICS

Multirow, time-accurate analyses [5, 6] are used to ascertain row-to-row interaction effects that are not captured by the routine two- and three-dimensional steady-state analyses conducted during the airfoil design process

[38, 39] (Fig. 27). The time-accurate analyses are typically used to provide unsteady pressure field information for resonance stress calculations [40, 41] and also to assess performance effects of row interactions [42]. The primary aerodynamic row interaction effects affecting performance are the effects of airfoil row gapping (i.e., potential flow and wake effects) and the effects of shock interaction and propagation.

Time-accurate analyses are created from the same input files as those used to conduct the three-dimensional, steady-state analyses. It is advisable to use the same flow solver for the time-accurate and steady-state solutions to enable relevant comparisons of performance results. The typical output of time-accurate calculations is time-dependent flowfield information at selected locations/fields within the computational domain. For example, surface pressure field information can be extracted from a converged solution over a computational cycle relevant for the airfoil counts in the solution. In addition, a time-averaged solution can be created for comparison with the steady-state analysis.

Two examples of how time-accurate CFD results can be used are discussed. In the first example, a high-work, high-pressure, single-stage HPT and downstream vane are analyzed (Fig. 28). In this analysis, the primary purpose was to validate the computational capability of the time-accurate CFD to capture unsteady shock reflection effects as measured in a high-speed rig [40]. The time-accurate solver was used to generate unsteady airfoil surface pressures for comparison with data. Fast-Fourier transform pressure field information for the blade passing frequency is shown. The results show excellent agreement with time-resolved pressure measurements; this solver can be used to accurately assess unsteady shock effects for forced response structural analysis.

In the second example, this same CFD modeling tool is applied to a multistage low-pressure turbine configuration [43] (Fig. 29) and results in a better resolution of turbine efficiency. In this case, the time-accurate solver utilizing

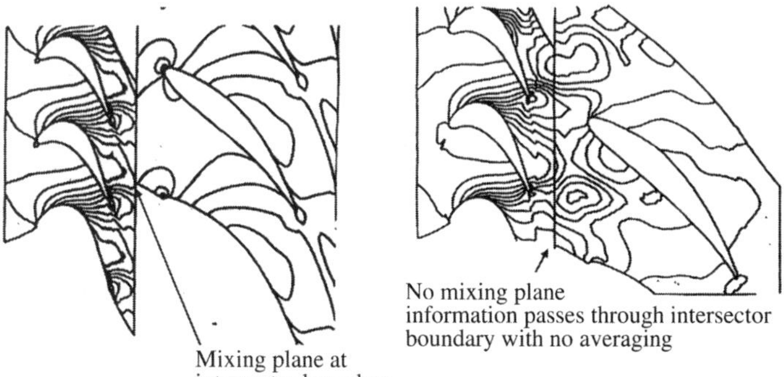

Fig. 27 Time-accurate analyses permit communication of all flow information between airfoil rows to capture impact of unsteady flow interaction.

validated transition modeling capability [44–49] captured an additional loss mechanism (entropy increase) caused by row-to-row wake interaction.

It is anticipated that time-accurate analyses will replace steady-state viscous CFD in the near future and become an essential part of the design process, just as the steady-state viscous tools have replaced potential and Euler analyses.

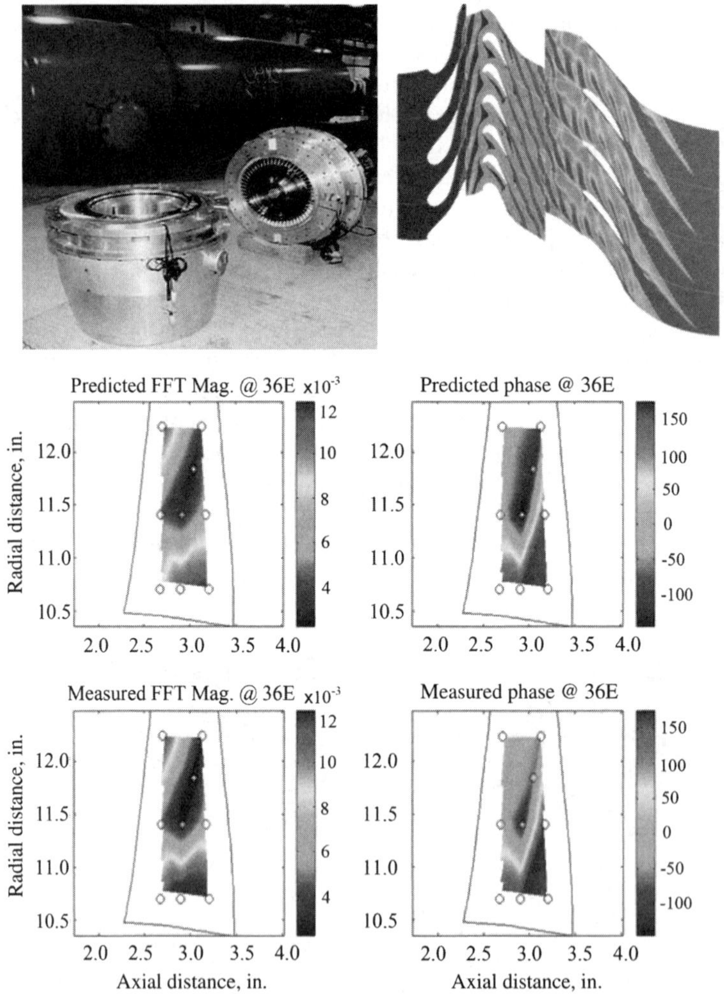

Fig. 28 Three-dimensional unsteady pressure field prediction capability is validated in high-speed rigs with high-response instrumentation.

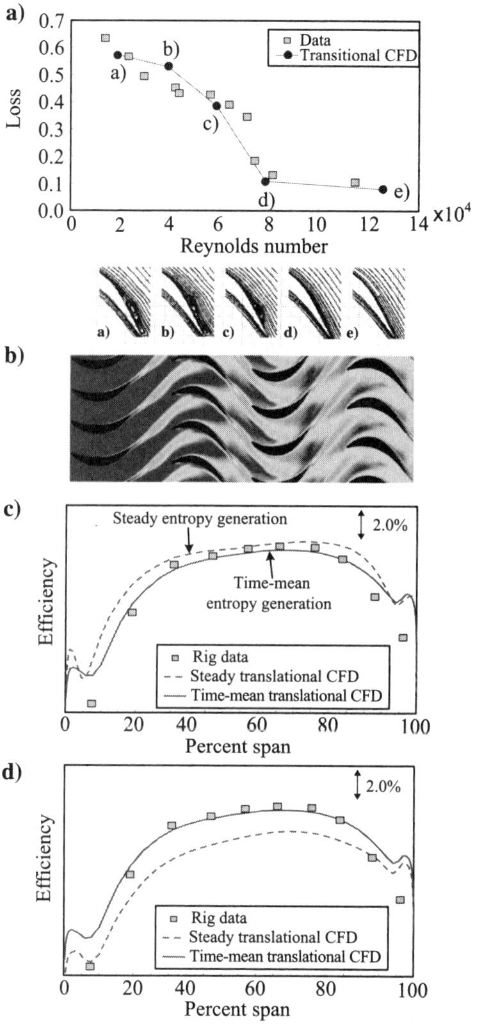

Fig. 29 Implementation of boundary-layer transition models in time-accurate CFD codes provides enhanced insight in the loss generation mechanisms in real turbines.

III. TURBINE DESIGN CONSIDERATIONS

The final turbine design is rarely unaffected by other considerations. Even an uncooled turbine design will usually be influenced by structures concerns, in view of the need to maintain acceptable airfoil and disk attachment stresses. The primary considerations for cooled, high-performance turbine designs are durability and life. Durability design considers the cooling of the turbine vanes and blades while life is addressed in the steady and unsteady load analysis for airfoils in highly corrosive environments.

Both durability and life are strongly governed by the temperature profile exiting the combustor. A successful design of a high-performance cooled turbine balances the need for high average inlet temperatures with allowable combustor exit temperature profile shapes, characterized by profile and pattern factors. The profile and pattern factors affect airfoil and cavity cooling design and ultimately determine turbine life.

A. DURABILITY

Because engine performance is strongly dependent on turbine inlet temperatures, most high-performance turbines operate in an environment with very high gas temperatures. Modern turbines are affected by a combination of effects that reduce gas temperature as the flow exiting the combustor passes through the turbine airfoil rows. Work extraction within blade rows reduces gas temperature entering the following vane rows while cooling air in the first vane mixes with the main stream air, which can reduce the gas temperatures entering the first blade by more than 100 deg. In addition, because of the turbine's high rotational speed, the blade actually experiences relative total temperatures that are less than the absolute total temperature. Therefore, meanline modeling of gas temperatures of the main gas path flow plus cooling is important for an accurate prediction of the durability trends within the engine.

The optimization of a turbine airfoil shape is complicated by the introduction of a cooling system. Design of the external shape focuses on the selection of a contour that will carry the gas loads with the minimum aerodynamic losses. Fine tuning of the contour must be tempered with an understanding of the practical physical constraints associated with cooling and manufacturing requirements. Once the airfoil cooling configuration has been determined, a classical heat exchanger analysis (Fig. 30) is conducted to determine how well the proposed cooling design reduces metal temperatures. For turbine airfoils, this represents a sophisticated interpretation of external boundary layers, internal coolant flow effectiveness, and metal conduction and stress analysis.

As the generation of most of the aerodynamic loss and heat transfer into the airfoil occurs in the viscous boundary layer adjacent to the surface, the ability to predict boundary-layer behavior is a very important part of the turbine design and development process. Boundary layers are subject to many factors that make prediction difficult. Included in the modeling must be effects such as laminar, transitional, and turbulent flows; pressure gradients and airfoil loading distributions; leading-edge separation bubbles; and mechanical effects such as surface roughness. These factors are included in the boundary-layer analysis (two- and three-dimensional CFD modeling) of the airfoil that produces predicted heat-transfer coefficients. Results can be used

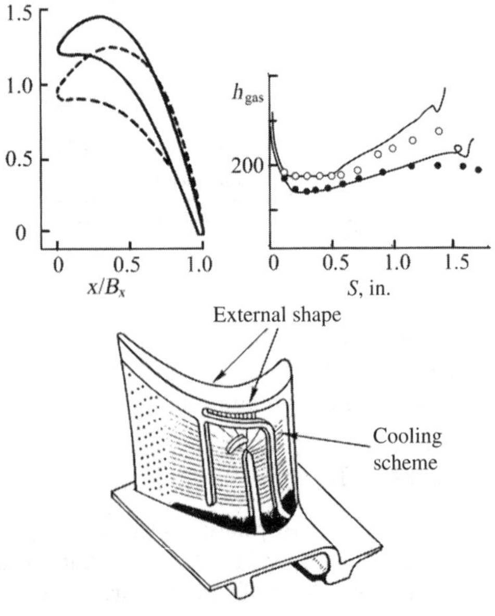

Fig. 30 Airfoil design affects external heat load and, when optimized with internal cooling, produces a low loss, long life airfoil design.

to identify regions of extremely high heat-transfer rates, where mechanisms such as film cooling can be effective. The results can also be used as boundary conditions in subsequent analysis where modeling of the heat load distribution has a strong effect on the calculated and actual metal temperatures.

Analysis of flow inside a cooled airfoil is just as significant as the analysis of external flows and boundary layers. Most high-pressure turbine airfoils in use today are very complicated heat exchangers that use a limited amount of cooling air to prevent melting and to extend airfoil life. Heat transfer and pressure loss models for cooling augmentation features such as impingement jets, trip strips, cross channel cylinders, and turns must be correctly modeled if the designer is to optimize an internal cooling concept. Network models that use one-dimensional correlations along cooling paths and nodes to link paths together are used to analyze the cooling effectiveness of internal geometries. Results include flow distribution within the airfoil, heat pick-up by the coolant, and local internal heat-transfer coefficients. Competitive pressures, rapidly changing technology, and variations in mission from one design to another make the prediction of internal cooling mechanisms a dynamic technical challenge for the designer of modern turbine engines.

Recent progress in the shaping of film holes has resulted in significant improvements in the effectiveness of this airfoil cooling technique. Film cooling protects the airfoil from the hot gas stream by adding cool jets of gas into the airfoil boundary layer by creating a buffer layer of cool air next to the wall. The effectiveness of the film cooling reduces downstream of the hole as the cold air mixes with surrounding hot gas. If the jet exits the holes cut into the surface of the airfoil with a velocity that is too great, it will blow through the boundary layer and be ineffective at cooling the surface. Continued technical development of film hole geometry is expected to lead to significant improvements in film cooling effectiveness.

Conventional high-pressure turbine airfoils are optimized for aerodynamic performance and durability through multidisciplinary design (Fig. 31). Computational-fluid-dynamics prediction methods are now sufficiently advanced to allow consideration of heat transfer in airfoil design. Airfoil shapes are thus modified to enable the necessary cooling schemes and cooling flows to meet airfoil life cycle requirements. Typical airfoil shape adjustments include increasing the cross-sectional area to permit internal cooling mechanisms and changing airfoil shape to minimize heat load on the exterior of the airfoil. Optimizing airfoil design shapes in the very formative stages of a design with consideration of minimizing heat load offers promise as a design technique that will improve turbine airfoil life and reduce engine service intervals.

B. STRUCTURES

Estimating airfoil life requires understanding different failure causes and mechanisms. The metal temperature of the part is an extremely important factor in determining all modes of failure. It is well known that the accurate prediction of average section temperature, local temperature, and the maximum airfoil surface temperature are all important in determining overall airfoil life. A 20-deg change in metal temperature will typically change the life of a part by a factor of two.

In the design process, analytical results generated from evaluation of the airfoil design using a structural stress analysis system are used as the basis for determining airfoil life. Creep is a plastic distortion of the airfoil over a long period of time at the high temperatures that occur in gas turbines. Fatigue is the weakening of a material as it is subjected to repeated cyclic loading, with endurance being measured by the number of cycles encountered before failure. Relatively small changes in the difference between maximum and minimum strain lead to large changes in the endurance of a part. In an engine, thermal effects have a great influence on fatigue endurance. Periodic temperature variations are experienced by the engine as it is throttled through the flight cycle. Typical engines go from idle, to takeoff, to cruise, to descent, to reverse thrust, and back to idle many times a day. Thermal fatigue is the result of these cyclic stresses as the airfoil experiences temperature-driven transients. Temperature

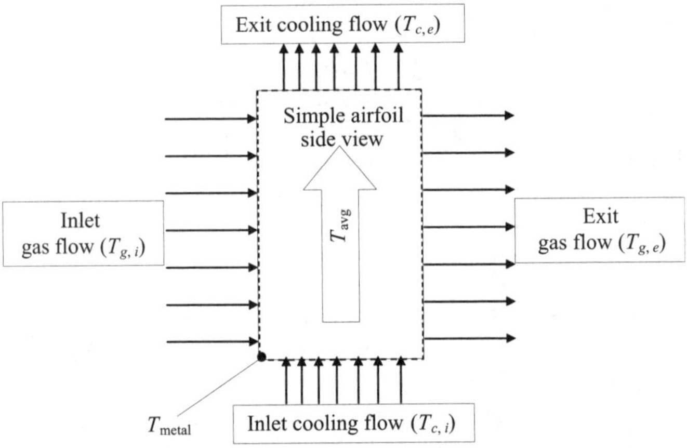

Heat Flux Equations

$$Q=Q_e=Q_i=Q_c=Q_g$$
$$Q_e=\overline{h}_e A_e(\overline{T}_{gas}-\overline{T}_{medal})$$
$$Q_i=\overline{h}_i A_i(\overline{T}_{medal}-\overline{T}_c)$$
$$Q_c=\dot{m}_c c_P(T_{c,e}-T_{c,i})$$
$$Q_c=\dot{m}_g c_P(T_{g,i}-T_{g,e})$$

Assumptions

$$\overline{T}_c \approx \frac{T_{c,i}+T_{c,e}}{2}=T_{c,i}+\frac{Q}{2\dot{m}_c c_P}$$
$$\overline{T}_g \approx \frac{T_{g,i}+T_{g,e}}{2}=T_{g,i}+\frac{Q}{2\dot{m}_g c_P}$$

Fig. 31 A cooled turbine blade is considered a very compact heat exchanger that manages the cooling air to protect the metal surfaces.

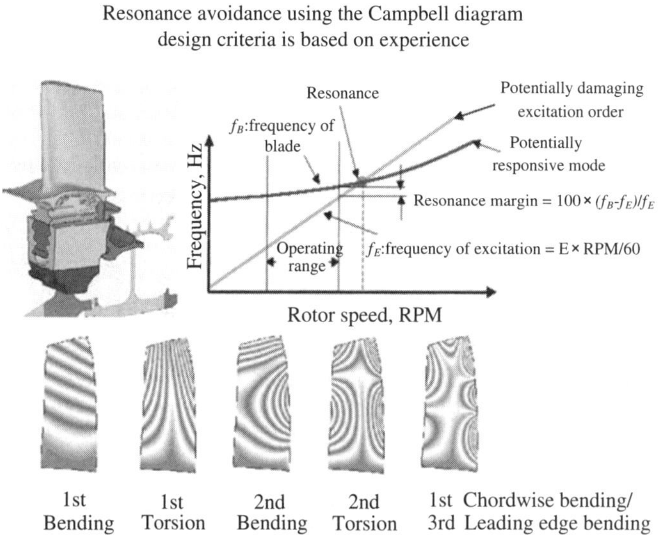

Fig. 32 Structures analysis considering vibratory modes is conducted to maintain acceptable steady stresses and to extend life expectancy.

gradients within the airfoil have the greatest influence on thermal fatigue and are carefully evaluated during the design of the cooling configuration. Indeed, transient operation causes these strains to be even greater than the strain predicted by steady-state stress analysis because cooling flow thermal responses lag mainstream flows when power output is varied.

Conventional high-pressure and low-pressure turbine airfoils are optimized for aerodynamic performance, durability, and life, through multidisciplinary design, considering structures analysis (Fig. 32). Time-accurate computational-fluid-dynamics prediction methods are used to estimate unsteady loads to allow the calculation of forced response resonant stresses, which are used to predict airfoil life as the airfoil is being designed. Airfoil shapes are often modified to change the dynamic characteristics of a vane or blade to extend airfoil life. Typical airfoil shape adjustments include changing the thickness distribution from the leading to the trailing edge and changing the chord length as a function of span to minimize detrimental cyclic stresses. Coupled, automated design systems enable computer-based optimization methods to improve airfoil life in the very earliest stages of a design.

IV. CONCLUSION

It is not unreasonable to presume that the business case for gas turbine engines will require turbines that get hotter, spin faster, and extract more work with fewer stages and airfoil counts, with increased efficiency and longer service intervals. Airport acoustic signature limits are becoming increasingly important and have become a driving requirement for turbine architecture selection. The winners in this field will have superior aero and mechanical design analysis tools and new, lighter-weight materials that require less cooling.

In the near term, it is anticipated that aero design focus will shift from further reductions in subsonic profile losses to reducing interpassage secondary flow losses. Highly three-dimensional airfoil and endwall designs that control these loss mechanisms will dominate. In addition, tip clearance losses will be minimized with new three-dimensional airfoil shapes and tip designs. Mechanical clearances will be reduced with careful attention to the structural designs with new alpha materials and clearance control actuation. Endwall shapes [30, 33–36] combined with new three-dimensional stacking enabled with new materials and three-dimensional structures analysis, plus new clearance designs, will finally drive turbine designs to true two-dimensional flow designs, with no extraneous loss production from airfoil passage secondary flows.

In addition to true two-dimensional flow designs, high-load, high-Mach designs will benefit from unconventional airfoil shaping to control shocks and to minimize losses. These new airfoil shapes will enable high-pressure-ratio single- and two-stage HPTs in the near future and the development of highly loaded, ultra-high-pressure-ratio turbines in the long term.

As a result of the increased overall work requirements, turbine stage work coefficient in recent years continues to increase for commercial and transport engines, as new materials and manufacturing techniques allow higher stress levels in airfoils and disks, balancing the increase in stage work with increases in wheel speed. Commensurate with high-load turbine designs is fewer stages. Reducing the number of stages and airfoils provides significant size, weight, and cost savings. For instance, in the 1980s, a typical turbine consisted of a two-stage, high-pressure turbine and a four-stage low-pressure turbine. The total airfoil count for the turbine was 1193 airfoils. Of this, 218 airfoils were cooled. In the 2000s, a typical turbine consists of a single-stage high-pressure turbine and a three-stage low-pressure turbine. The total airfoil count for the turbine is 648 airfoils. Of this, only 96 airfoils are cooled. This trend towards fewer stages and fewer airfoils per stage will continue as the need for smaller, lighter, and less costly turbines continues to increase.

Finally, and likely most critical, is the continuing development of better materials and coatings, giving the turbine designer an opportunity to increase both annulus area and rpm with higher turbine inlet temperatures. Until the new, high-temperature materials are available, new and exciting cooling technologies will enable advances in the engine cycle. With the new, high-temperature material designs, rotating components will be able to tolerate higher temperatures and higher stresses, thereby driving the turbine design to be high-speed, high-temperature, uncooled, and highly efficient, with minimal secondary flow, shock, and tip leakage losses.

REFERENCES

[1] Mattingly, J. D., *Elements of Gas Turbine Propulsion*, McGraw–Hill, New York, 1996, pp. xxxiii–xxxvii, 240–345.

[2] Ni, R. H., "A Multiple-Grid Scheme for Solving Euler Equations," *AIAA Journal*, Vol. 20, No. 11, 1982, pp. 1565–1571.

[3] Davis, R. L., Ni, R. H., and Bowley, W. W., "Prediction of Compressible, Laminar Viscous Flows Using a Time-Marching Control Volume and Multiple-Grid Technique," *AIAA Journal*, Vol. 22, No. 11, 1984, pp. 1573–1581.

[4] Ni, R. H., and Bogoian, J., "Prediction of 3-D Multi-Stage Turbine Flow Field Using a Multiple-Grid Euler Solver," AIAA Paper 1989-0203, Jan. 1989.

[5] Denton, J. D., "An Improved Time Marching Method for Turbomachinery Flow Calculation," American Society of Mechanical Engineers, Paper 82-GT-239, 1982.

[6] Ni, R. H., and Sharma, O. P., "Using 3D Euler Flow Simulations to Assess Effects of Periodic Unsteady Flow Through Turbines," AIAA Paper 1990-2357, July 1990.

[7] Cohen, H., Rogers, G. F. C., and Saravanamuttoo, H. I. H., *Gas Turbine Theory*, 4th ed., Addison Wesley Longman, Reading, MA, 1996, pp. 1–88.

[8] Shepherd, D. G., *Principals of Turbomachinery*, MacMillan, New York, 1956, pp. 49–99, 151–227.

[9] Horlock, J. H., *Axial Flow Turbines*, Robert E. Krieger, Huntington, NY, 1973.

[10] Kacker, S. C., and Okapuu, U., "A Meanline Prediction Method for Axial Flow Turbine Efficiency," American Society of Mechanical Engineers, Paper 81-GT-58, 1981.

[11] Young, J. B., and Horlock, J. H., "Defining the Efficiency of a Cooled Turbine," *Transactions of the ASME, Journal of Turbomachinery*, Vol. 128, Oct. 2006, pp. 658–667.

[12] Sharma, O. P., Wells, R. A., Schlinker, R. H., and Bailey, D. A., "Boundary Layer Development on Turbine Airfoil Suction Surfaces," American Society of Mechanical Engineers, Paper 81-GT-204, 1981.

[13] Langston, L. S., Nice, M. L., and Hooper, R. M., "Three-Dimensional Flow Within a Turbine Cascade Passage," American Society of Mechanical Engineers, Paper 76-GT-50, 1976.

[14] Langston, L., Nice, M. L., and Hooper, R., "Three Dimensional Flows Within a Turbine Cascade Passage," *Transactions of the ASME Journal of Engineering for Power*, Vol. 99, Jan. 1977, pp. 21–28.

[15] Langston, L. S., "Crossflows in a Turbine Cascade Passage," *Transactions of the ASME Journal of Engineering for Power*, Vol. 102, Oct. 1980, pp. 866–874.

[16] Sharma, O. P., and Graziani, R. A., "Influence of Endwall Flow on Airfoil Suction Surface Mid-Height Boundary Layer Development in a Turbine Cascade," American Society of Mechanical Engineers, Paper 82-GT-127, 1982.

[17] Binder, A., and Romey, R., "Secondary Flow Effects and Mixing of the Wake Behind a Turbine Stator," *Transactions of the ASME Journal of Engineering for Power*, Vol. 105, Jan. 1983, pp. 40–46.

[18] Binder, A., Forster, W., Kruse, H., and Rogge, H., "An Experimental Investigation into the Effect of Wakes on the Unsteady Turbine Flow," American Society of Mechanical Engineers, Paper 84-GT-178, 1984.

[19] Moore, J., and Adhye, R. Y., "Secondary Flows and Losses Downstream of a Turbine Cascade," American Society of Mechanical Engineers, Paper 85-GT-64, 1985.

[20] Moore, J., and Moore, J. G., "Performance Evaluation of Linear Turbine Cascades Using Three-Dimensional Viscous Flow Calculations," American Society of Mechanical Engineers, Paper 85-GT-65, 1985.

[21] Sharma, O. P., and Butler, T. L., "Predictions of Endwall Losses and Secondary Flows in Axial Flow Turbine Cascades," *Transactions of the ASME Journal of Turbomachinery*, Vol. 109, April 1987, pp. 229–236.

[22] Harrison, S., "Secondary Loss Generation in a Linear Cascade of High-Turning Turbine Blades," American Society of Mechanical Engineers, Paper 89-GT-47, 1989.

[23] Sharma, O. P., and Syed, S. A., "Turbulence Modeling in Gas Turbine Design and Analysis," AIAA Paper 91-0514, Jan. 1991.

[24] Denton, J. D., "Loss Mechanisms in Turbomachines," American Society of Mechanical Engineers, Paper 93-GT-435, 1993.

[25] Kang, M. B., and Thole, K. A., "Flowfield Measurements in the Endwall Region of a Stator Vane," American Society of Mechanical Engineers, Paper 99-GT-188, 1999.

[26] Smith, S. F., "A Simple Correlation of Turbine Efficiency," *Journal of Royal Aeronautical Society*, Vol. 69, July 1965, pp. 467–470.

[27] Dorman, T. E., Welna, H., and Lindlauf, R.W., "The Application of Controlled-Vortex Aerodynamics to Advanced Axial Flow Turbines," *Transactions of the ASME Journal of Engineering for Power*, Vol. 98, July 1968, pp. 245–250.

[28] Caspar, J. R., Hobbs, D. E., and Davis, R. L., "The Calculation of Two-Dimensional Compressible Potential Flow in Cascades Using Finite Area Techniques," AIAA Paper 79-0077, Jan. 1979.

[29] Joslyn, D., and Dring, R., "Three-Dimensional Flow in an Axial Turbine: Part 1 – Aerodynamic Mechanisms," *Transactions of the ASME Journal of Turbomachinery*, Vol. 114, Jan. 1992, pp. 61–70.

[30] Joslyn, D., and Dring, R., "Three-Dimensional Flow in an Axial Turbine: Part 2 – Profile Attenuation," *Transactions of the ASME Journal of Turbomachinery*, Vol. 114, Jan. 1992, pp. 71–78.

[31] Adamczyk, J. J., Celestina, M. L., Beach, T. A., and Barnett, M., "Simulation of Three-Dimensional Viscous Flow Within a Multistage Turbine," American Society of Mechanical Engineers, Paper 89-GT-152, 1989.

[32] Denton, J. D., "The Calculation of Three-Dimensional Viscous Flow Through Multistage Turbomachines," *Transactions of the ASME Journal of Turbomachinery*, Vol. 114, Jan. 1992, pp. 18–26.

[33] Harvey, N. W, Rose, M. G., Taylor, M. D., Shahpar, S., Hartland, J., and Gregory-Smith, D. G., "Nonaxisymmetric Turbine Endwall Design: Part 1 – Three-Dimensional Linear Design System," *Transactions of the ASME Journal of Turbomachinery*, Vol. 122, April 2000, pp. 278–285.

[34] Hartland, J. C., Gregory-Smith, D. G., Harvey, N. W., and Rose, M. G., "Nonaxisymmetric Turbine Endwall Design: Part 2 – Experimental Validation," *Transactions of the ASME Journal of Turbomachinery*, Vol. 122, April 2000, pp. 286–293.

[35] Shih, T. I-P., and Lin, Y.-L., "Controlling Secondary-Flow Structure via Leading-Edge Airfoil Fillet and Inlet Swirl to Reduce Aerodynamic Loss and Surface Heat Transfer," American Society of Mechanical Engineers, Paper 2002-GT-603, 2002.

[36] Rose, M. G., "Non-Axisymmetric Endwall Profiling in HP NGV's of an Axial Flow Gas Turbine," American Society of Mechanical Engineers, Paper 94-GT-249, 1994.

[37] Gier, J., Ardey, S., Eymann, S., Reinmoller, U., and Niehuis, R., "Improving 3D Flow Characteristics in a Multistage LP Turbine by Means of Endwall Contouring and Airfoil Design Modification – Part 2: Numerical Simulation and Analysis," American Society of Mechanical Engineers, Paper GT-2002-30353, 2002.

[38] Sharma, O. P., Butler, T. L., Joslyn, H. D., and Dring, R. P., "An Experimental Investigation of the Three-Dimensional Unsteady Flow in an Axial Flow Turbine," AIAA Paper 83-1170, June 1983.

[39] Sharma, O. P., Pickett, G. F., and Ni, R. H., "Assessment of Unsteady Flows in Turbines," American Society of Mechanical Engineers, Paper 90-GT-150, 1990.

[40] Clark, J. P., Stetson, G. M., Magge, S. S., Haldeman, C. W., Jr., and Dunn, M. G., "The Effect of Airfoil Scaling on the Predicted Unsteady Loading on the Blade of a 1 and $\frac{1}{2}$ Stage Transonic Turbine and a Comparison with Experimental Results," American Society of Mechanical Engineers, Paper 2000-GT-0446, 2000.

[41] Clark, J. P., Aggarwala, A. S., Velonis, M. A., Gacek, R. E., Magge, S. S., and Price, F. R., "Using CFD to Reduce Resonant Stresses on a Single-Stage High Pressure Turbine Blade," American Society of Mechanical Engineers, Paper GT-2002-30320, 2002.

[42] Sharma, O. P., Renaud, E., Butler, T. L., Milsaps, K., Jr., Dring, R. P., and Joslyn, H. D., "Rotor-Stator Interaction in Multi-Stage Axial-Flow Turbines," AIAA Paper 88-3013, July 1988.

[43] Praisner, T. J., Clark, J. P., Nash, T. C., Rice, M. J., and Grover, E. A., "Performance Impacts due to Wake Mixing in Axial-Flow Turbomachinery," American Society of Mechanical Engineers, Paper GT2006-90666, 2006.

[44] Roberts, W. B., "Calculation of Laminar Separation Bubbles and Their Effect on Airfoil Performance," *AIAA Journal*, Vol. 18, No. 1, Jan. 1980, pp. 25–31.

[45] Roberts, M. G., and Brown, A., "Boundary-Layer Transition Regions on Turbine Blade Suction Surfaces," American Society of Mechanical Engineers, Paper 84-GT-284, 1984.

[46] Mayle, R. E., "The Role of Laminar-Turbulent Transition in Gas Turbine Engines," ASME 1991 IGTI Scholar Lecture," *Transactions of the ASME Journal of Turbomachinery*, Vol. 113, Oct. 1991, pp. 509–537.

[47] Walker, G. J., "The Role of Laminar Turbulent Transition in Gas Turbine Engines: a Discussion," *Transactions of the ASME Journal of Turbomachinery*, Vol. 115, April 1993, pp. 207–217.

[48] Mayle, R. E., and Schulz, A., "The Path to Predicting Bypass Transition," *Transactions of the ASME Journal of Turbomachinery*, Vol. 119, July 1997, pp. 405–411.

[49] Praisner, T. J., and Clark, J. P., "Predicting Transition in Turbomachinery, Part I – A Review and New Model Development," American Society of Mechanical Engineers, Paper GT-2004-54108, 2004.

[50] Langston, L. S., "Fahrenheit 3600," *Mechanical Engineering – Power & Energy*, Vol. 129, No. 4, April 2007, pp. 34–37.

[51] Cherry, D. G., and Dengler, R. P., "The Aerodynamic Design and Performance of the NASA/GE E3 Low Pressure Turbine," AIAA Paper 84-1162, June 1984.

CHAPTER 2

Turbine Cooling Design

Ronald S. Bunker*
GE Aviation, Cincinnati, Ohio

ACRONYMS

CD	conceptual design
CFD	computational fluid dynamics
DD	detailed design
FEM	finite element model
HGP	hot gas path
HOST	hot-section technology
HPT	high-pressure turbine
NASA	National Aeronautics and Space Administration
NOx	nitrous oxides
OEM	original equipment manufacturer
PD	preliminary design
RPM	revolutions per minute
TBC	thermal barrier coating

The gas-turbine is a specialized engine designed to convert chemical energy into one or more useful forms of energy, such as thrust, shaft work, and process heat. Gas-turbine engines for aviation and marine propulsion, power generation, and combined heat/power applications are most commonly continuously rotating axial turbomachines. Figure 1 shows an overall engine schematic for the CFM56-5B commercial aviation gas-turbine engine. As a thermodynamic Brayton cycle, the efficiency of the gas-turbine engine can be raised substantially by increasing the firing temperature of the turbine. Modern gas-turbine systems are fired at temperatures far in excess of the material melting temperature limits. This is made possible by the aggressive cooling of the hot-gas-path (HGP) components, the use of advanced materials for structural components and protective coatings,

*Consulting Engineer.

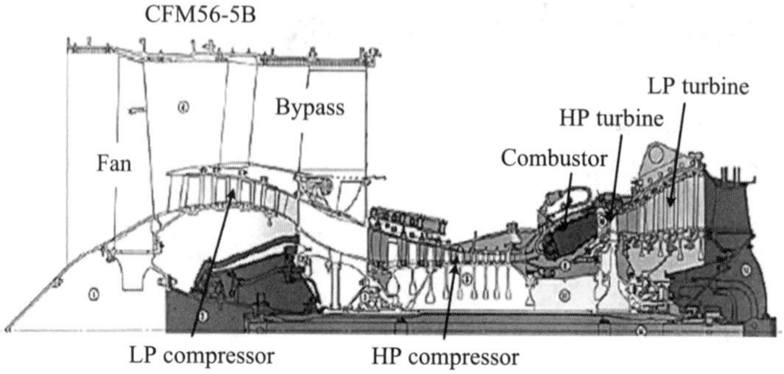

Fig. 1 Aeroengine CFM56-5B.

the application of high-efficiency aerodynamics, the use of prognostic and health monitoring systems, and the continuous development of improved mechanical stress, lifing, and systems interactions and behavioral modeling. These and associated technical areas provide the focus of the present chapter.

The high-pressure turbine (HPT) section of the engine, shown in Fig. 2, encompasses all of these challenges simultaneously. For example, the technology of cooling gas-turbine components via internal convective flows of single-phase gases has developed over the years from simple smooth cooling passages to very complex geometries involving many types of surfaces, architectures, and fluid-surface interactions. The fundamental aim of this technology area is to obtain the highest overall cooling effectiveness with the lowest possible penalty

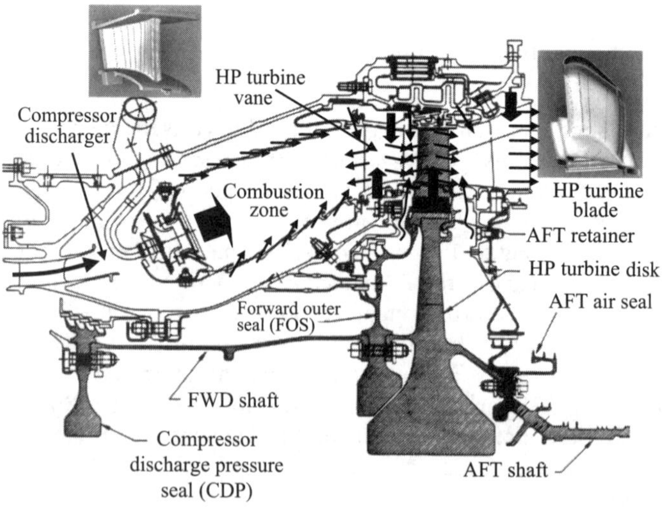

Fig. 2 Aeroengine high-pressure turbine and combustor.

on the thermodynamic cycle performance. The use of 20 to 30% of the available compressed air to cool the HPT presents a severe penalty on the thermodynamic efficiency, unless the firing temperature is sufficiently high for the gains to outweigh the losses. In all properly operating cooled turbine systems, the efficiency gain is significant enough to justify the added complexity and cost of the cooling technologies employed.

In many respects, the evolution of gas-turbine internal cooling technologies began in parallel with heat exchanger and fluid processing techniques, adapting them to the constraints imposed on turbine airfoils: aerodynamics, mechanical strength, vibrational response, and so on. Turbine airfoils are, after all, merely highly specialized and complex heat exchangers that release the cold side fluid in a controlled fashion to maximize work extraction. Actively or passively cooled regions of the hot gas path in both aircraft engine and power-generating gas-turbines include the stationary vanes or nozzles, the rotating blades or buckets of the HPT stages, shrouds bounding the rotating blades, and the combustor liners and flame holding segments. Also included are the secondary flow circuits of the turbine wheelspaces and the outer casings that serve as both cooling and positive purge flows. The ever-present constraints common to all components and systems include but are not limited to, pressure losses, material temperatures, component stresses, geometry and volume, aerodynamics, fouling, and coolant conditions.

Figure 3 presents a generic overview of the HPT design system or method from the "inside-out" perspective, meaning from the viewpoint of the required HPT internal cooling system outward. For the present purposes, the design analysis method is described as a multilevel system. The conceptual design (CD) of the components is largely based on nominal target conditions and is more or less divorced from the surrounding systems constraints and competing requirements.

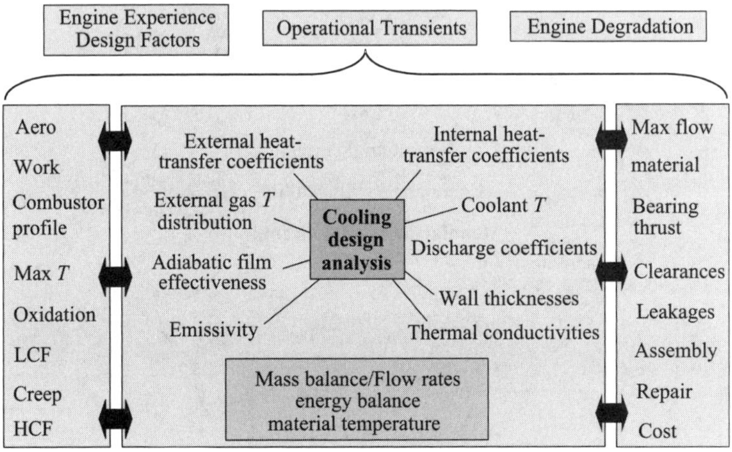

Fig. 3 Cooling design analysis system.

CD analysis can be performed based on one-, two-, or three-dimensional complexities and details and is primarily used to compare design options. Analysis at the CD level must be detailed enough, however, to allow ranking and downselection between options.

Preliminary design (PD) analysis includes much more detailed surrounding effects and constraints from aerodynamics, material properties, mechanical loads, lifing limitations, and clearances, as depicted in the design cycle diagram Fig. 4. PD analyses are often combined thermal-mechanical predictions using very detailed finite element models, at times even submodels of certain component sections. Most PD analyses are performed at one steady-state operating condition, for example, 100% load. The result of PD analysis, after various alterations and iterations, is the basic system design with balanced choices that satisfy the engine design goals.

Detailed design (DD) analysis brings in the operational transient aspects to determine if requirements or constraints are violated under conditions such as normal startup, fast startup, takeoff, thrust-reverse, trips, and hot restarts. DD results can require that additional changes be made with new CD and PD analyses. In all design analysis levels, engine experience design factors and known engine degradation factors must be included. As examples, such factors can include the use of -3σ material properties, knockdown factors on cooling augmentation, and loss of coatings or metal thickness. In addition, there is a pre-PD analysis that sets the overall architecture and preliminary design of the engine. This cycle analysis deals mainly with the mission requirements, such as

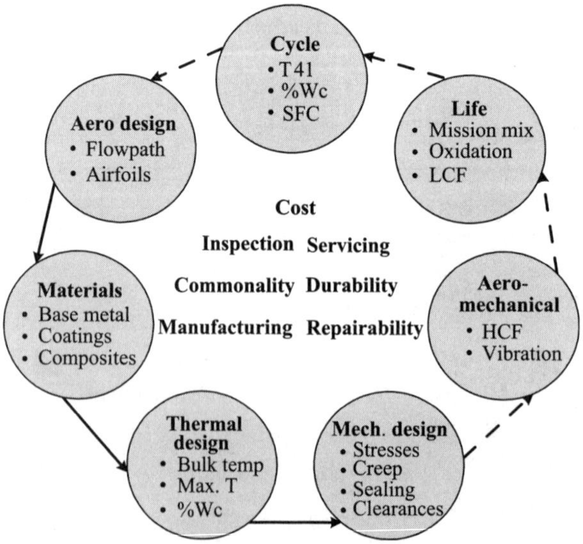

Fig. 4 Turbine engine design cycle.

efficiency, cost-of-electricity, specific fuel consumption, thrust, power sizing, and number of starts. The cycle sets the target goals on the cooling system, including the coolant consumption, turbine airfoil life, and inspection intervals.

Cooling technology, as applied to gas-turbine components such as the high-pressure turbine vanes and blades (also known as nozzles and buckets), is composed of five main elements that must work in harmony: 1) internal convective cooling, 2) external surface film cooling, 3) materials selection, 4) thermal-mechanical design, and 5) selection and/or pretreatment of the coolant fluid. Internal convective cooling is the art of directing coolant via the available pressure gradients into all regions of the component requiring cooling while augmenting the heat-transfer coefficients as necessary to obtain distributed and reasonably uniform thermal conditions. Film cooling is the practice of bleeding internal cooling flows onto the exterior skin of the components to provide a heat-flux-reducing cooling layer. Materials most commonly employed in cooled parts include high-temperature, high-strength nickel, or cobalt-based superalloys coated with yttria-stabilized zirconia oxide ceramics [thermal barrier coating (TBC)]. Today, protective ceramic coatings are actively used to enhance the cooling capability of the internal convection mechanisms. The thermal-mechanical design of the components must marry the first three elements into a package that has acceptable thermal stresses, coating strains, oxidation limits, creep-rupture properties, and aeromechanical response. The last cooling design element concerns the correct selection of the cooling fluid to perform the required function with the least impact on the cycle efficiency. This is usually achieved through the use of compressor air bleed from the most advantageous stage of the compressor, but can also be done using off-board cooling sources such as closed-circuit steam or air, as well as intracycle and intercycle heat exchangers.

I. CONCEPTUAL DESIGN ANALYSIS (COOLING EXAMPLE)

In the very early stages of engine design, the cooling system is completely wrapped up in a single set of performance characteristic curves, usually presented in graphical format, known as cooling technology maps. A generic cooling technology performance chart is shown in Fig. 5 for a turbine airfoil, either a vane or a blade. The technology curves shown on this chart present the gross airfoil cooling effectiveness vs a heat loading parameter, defined as nondimensional quantities:

$$\text{Gross cooling effectiveness} = (T_{\text{gas}} - T_{\text{bulk metal}})/(T_{\text{gas}} - T_{\text{coolant supply}}) \quad (1)$$

$$\text{Heat loading parameter} = (m_{\text{coolant}} * C_{p\ \text{coolant}})/2 * H_{\text{gas}} * A_{\text{gas}} \quad (2)$$

The heat loading parameter is the ratio of the overall hot-gas heat flux (source) delivered to the component vs the overall coolant capability to accept heat flux (sink). Because the gas and coolant temperatures are not in this term, the ratio is not unity, but it does provide a relative scale for placement of past and

Fig. 5 Cooling technology performance chart.

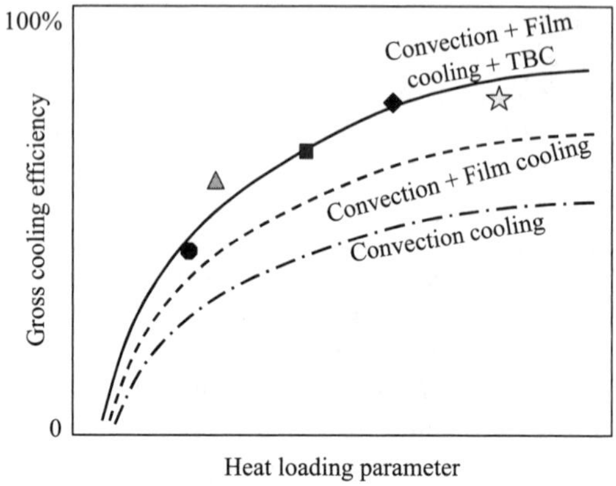

current designs. The symbolic points on the chart represent various engine experience data points for different designs. Several curves will generally be present, showing major levels of cooling technology. Such maps may also present extrapolated design points based on analysis only or target design points for new engines.

In this earliest design phase, cooling analysis is simply a matter of looking up the expected or projected coolant flow rates based on the cycle or mission design goals. Temperatures can be altered by choices of cycles, surface areas by overall power requirements or aerodynamics, coolant-specific heat by selection of cooling fluid, airfoil temperatures by cooling mass flow rate, and so forth. All of these choices lead to differing impacts on overall engine efficiency, emissions, life, and cost. A similar set of performance curves can be used to examine the effect of wheelspace and casing leakage flows from the secondary cooling circuits. Here, variations can be made in the complexity of seals to obtain lower overall leakages flows with potential consequences, such as higher rotor rim material temperatures.

II. PRELIMINARY DESIGN ANALYSIS (COOLING EXAMPLE)

Component design can be performed at any of several depths of analysis, from preliminary estimates, to detailed two-dimensional analyses, to complete three-dimensional computational predictions including the conjugate effects of the convective and radiative environments. Each mode of analysis has its use, as the design progresses from concept to reality. Three-dimensional vane airfoil and endwalls are reduced first to a two-dimensional, constant-thickness cross section of the aerodynamic shape and then again to a one-dimensional basic plate representing flow from the leading-edge stagnation point to the trailing edge. Preliminary design uses primarily bulk quantities and one-dimensional simplified equations to arrive at approximate yet meaningful estimates of temperatures and flow requirements. While the actual airfoil/endwall shape involves many complexities of accelerating and decelerating flows, secondary flows, and discrete film injection holes, a good estimate can be obtained using fundamental flat-plate relations. Two-dimensional design incorporates boundary-layer analyses, network flow and energy balances, and some thermal gradient estimates to refine the results for local temperature and flow predictions suitable for use in finite element stress modeling. Three-dimensional design can use complete computational fluid dynamics and heat-transfer modeling of the internal and external flowfields to obtain the most detailed predictions of local thermal effects and flow losses. Design analyses can, of course, also mix these methods, as when CFD is used to predict the hot-gas-path pressures, velocities, and temperatures for the aerodynamic profile only whereas the internal cooling and film cooling are predicted using semi-empirical correlations.

A. ONE-DIMENSIONAL ANALYSIS

The simplest one-dimensional analysis can be best understood as an iterative sequence of several steps leading to an overall model that is approximately optimized for material thicknesses, cooling configuration, and cooling flow. These steps include 1) estimation of the external heat-transfer coefficient distribution on the airfoil with effects such as surface roughness and freestream turbulence; 2) calculation of the average adiabatic wall temperature caused by film cooling; 3) calculation of the conductive material thermal resistances; 4) estimation of the internal heat-transfer coefficients caused by cooling; 5) calculation of the required aggregate cooling flow rate; and 6) iteration of the solution to achieve target metal temperatures, thermal gradients, or material thicknesses, or to comply with other target constraints. The solution is iterated to account for fluid property changes with temperature, both internal and external to the airfoil, as well as temperature rise in the cooling fluid.

B. TWO-DIMENSIONAL ANALYSIS

The simple one-dimensional model is not of much use in preliminary design unless it is knit into a sectional or complete model representing the cooled airfoil. This means applying the simple analysis to many regions of the airfoil (wall elements) making up a two-dimensional section. This is analogous to a finite element model construction and in many cases can be achieved using a FEM approach. The elements can be disconnected from thermal conduction as a first estimate or simply connected to include axial conduction effects within the airfoil section. Such conduction effects are more important in regions that are not well modeled by a single wall thickness, like the trailing edge. Taking this a step further, many radial sections of the airfoil can be stacked to form a pseudo-three-dimensional model of the nearly complete component (without endwalls, tip, or shank). Again, this can be accomplished with or without complete thermal conduction connections. These are each valid preliminary design modeling approaches with varying levels of accuracy. Note that such approaches do not typically integrate the airfoil and its endwalls, but treat these portions separately by similar analytical means.

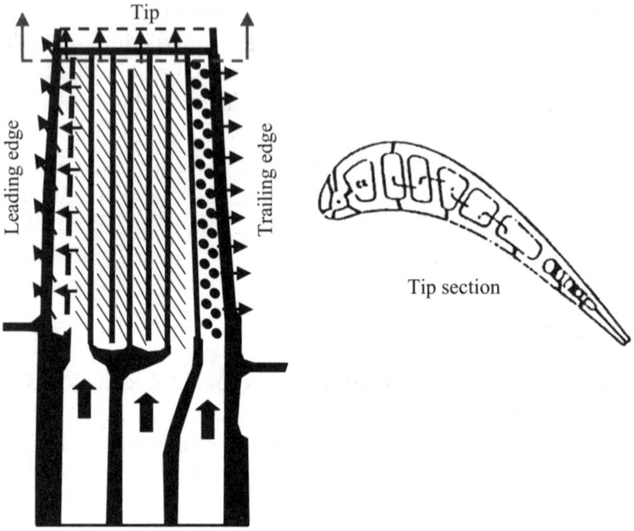

Fig. 6 Cooled HPT blade (bucket).

One might ask why an FEM approach is not always employed for the preliminary design of cooled airfoils and also why the airfoil and endwalls are not always integrated into a single component. The answer is the same for both questions and lies partially in historical design methods and partially in the state-of-the-art of computational analysis. A candidate cooling circuit design, such as that shown in Fig. 6, can be very complex. In this example, the main portion of the blade is cooled using a turbulated five-pass serpentine circuit, the leading edge is cooled using a radial passage impinging through crossover holes into the concave stagnation region, and the trailing edge is cooled with a radial pin-bank array and aft ejection channels. Film cooling is employed heavily in the leading-edge region and tip with additional rows of film holes on both the pressure and suction sides of the blade. The blade has three distinct cooling circuits isolated in the shank cooling supply. This blade design, and or any other, must be analyzed and modified with the following in mind:

1. Typical internal cooling technologies, including turbulators, pin-fins, turns, impingement jets, trailing-edge holes, swirl cooling, vortex cooling, convoluted passages, tip purge holes, and basic number and sizing of passages, must be easily manipulated to investigate design options and their effects on performance. Manipulation includes movement to new locations, change of size, change of number or spacing, and addition to and subtraction from the component. Performance evaluation usually refers to cooling effectiveness and aerodynamic mixing losses at this stage of analysis.

2. Film-cooling holes and rows of holes must be easy to move or alter (including hole shape and angle) in the design process.
3. Rotational cooling circuit differences must be evaluated by altering the general passage layouts.
4. Balancing of flow rates with coolant temperature rises and pressure losses must be easy to perform.
5. Changes in the external heat-transfer coefficient distributions as a result of new estimates of freestream turbulence, surface roughness, film-injection heat-transfer coefficient augmentation, wakes/unsteadiness, hot-streaks/clocking, profile, and pattern factors must be accommodated.
6. Wall thickness and TBC coating thickness can also be changed in design at virtually any location.

These factors and more dictate that complex FEM and CFD analyses of cooled airfoils at the preliminary design phase are simply not practical. In addition to these design manipulation requirements, the majority of current knowledge concerning internal cooling and film cooling is still contained in empirical and semi-empirical correlations. State-of-the-art computational predictions are as yet not sufficiently advanced to provide reliable "data" for the design of cooled airfoils. Preliminary design methods must therefore make use of a multitude of design correlations based on experimental data obtained by the original equipment manufacturers (OEMs) and/or found in open literature.

Putting the foregoing discussion into practice, the two-dimensional or pseudo-three-dimensional cooling analysis of the airfoil portion for a vane or blade is typically performed in the following manner:

1. Given a current prediction of the aerodynamics (static-pressure distributions) and gas temperatures surrounding the airfoil sections, the external heat-transfer coefficient distribution is calculated for each radial section using either boundary-layer analysis or computational heat transfer. The distributions must account for some or all of the influencing factors including the following:
   - Airfoil loading
   - Subsonic boundary-layer laminar and turbulent transitions
   - Bypass transition
   - Transonic shocks
   - Surface roughness distribution
   - Freestream turbulence
   - Freestream approach swirl
   - Rotational effects
   - Boundary-layer disturbances caused by film coolant injection
   - Boundary-layer disturbances caused by coating spallation
   - Periodic unsteadiness and wake passing

- Secondary flow injection in hub and tip regions
- Radiative heat-flux distributions

2. A detailed flow network model of the internal cooling circuits of the airfoil is built using the appropriate coolant supply pressure and temperature with the external airfoil static-pressure distribution as boundary conditions. The network flow model should allow compressible flow effects though some models might be sufficient with incompressible flow only. The flow model is executed with an initial solution guess and iterated to convergence based on the current boundary conditions and internal geometry. This cooling circuit model includes the following:
   - Flow area distributions for each passage
   - Detailed local geometry for each internal feature or repeating feature, such as turbulators and pin-fins
   - Cooling passage aspect-ratio distributions
   - Local impingement cooling geometry definitions
   - Geometry details for all internal cooling holes
   - Film-cooling extraction locations
   - Convective heat-transfer coefficient correlations
   - Coefficient of friction correlations
   - Coefficient of pressure loss correlations for turns, holes, etc.
   - Cooling fluid properties
3. The flow network model can also be arranged to contain a material shell representing the airfoil, such that the model can interact with the external conditions via thermal exchange. This requires definition of the material property tables, as well as wall thickness, internal dividing rib thickness, and protective coating thickness distributions for each section.
4. The flow model is extended to include the flow and discharge of all film-cooling holes. This is done by providing information on film hole or row sizes, shapes, spacing, orientation, and exit locations, as well as correlations for film hole discharge coefficient, adiabatic effectiveness, mixing loss, and internal heat-transfer coefficient.

This total airfoil model can be modified through relatively simple and quick adjustment of the several input distributions and boundary conditions. Execution of the model is straightforward as long as the boundary conditions and geometry parameters are realistic. It must be recognized that such a model contains multiple inlet and exit boundary conditions and parallel flow circuits, of which some flow circuits might be in communication. The complexity of the model must be sufficient to include and resolve all significant pressure losses. The output of the airfoil model can include predictions of all internal heat-transfer coefficients, all flow distributions, individual film hole flow rates and mixing losses, total cooling flow rate, external film temperatures, and, of course, the local material temperatures. This model can be further coupled to a prediction of the external heat-transfer coefficients to update the heat loads for effects of film injection and wall temperature distributions. Once such a model is finalized for a desired design and result, it can then be exercised to further study manufacturing effects on film hole discharge coefficients and turbulated cooling passages, tolerances for material properties, wall thicknesses, hole diameters, and core shifts, and special considerations for harsh operating environments, including surface roughness, TBC spallation, and film hole blockage effects.

C. COOLING DESIGN ANALYSIS CORRELATIONS

A major consideration in the preceding cooling analysis is the availability of good correlations for both internal flows and film cooling under conditions representative of engines. These correlations are numerous, as the variation of internal cooling geometries and film cooling parameters are vast. Because there are so many possible combinations and variations, design analysis is founded on several basic generic correlations from the open literature and augmented by many geometry-specific correlations determined by OEM research. The following is a list of the primary cooling correlation sources from open literature:

1. Impingement jet array heat-transfer coefficients (Nusselt numbers) can be obtained from the correlation of Florschuetz et al. [1] for average jet Reynolds numbers typical in engine design. For square arrays of jets at somewhat lower Reynolds numbers, the graphical data of Kercher and Tabakoff [2] can be used.
2. Impingement cooling that involves the use of individual jets, or slot-type jets, or other nonstandard configurations, can be determined by correlations in the summary paper of Martin [3].
3. Simple, fully developed duct flow turbulent heat transfer can be estimated quite well using the Dittus–Boelter correlation, $Nu = 0.023\ Re^{0.8}\ Pr^{0.33}$, or other variants on this correlation that can be found in any modern textbook. Care should be taken to account for the wall-to-fluid temperature ratio.
4. Most fully developed turbulated duct flow heat-transfer correlations are of the format $Nu = C * Re^n\ Pr^m$. The basic correlations for stationary turbulated ducts with transverse or angled rib rougheners can be found in Han et al. [4]. This research also includes the coefficients of friction.
5. Rotating passage heat-transfer data with and without turbulators is contained in the NASA HOST program data sets [5].
6. Pin bank internal heat transfer and pressure loss correlations are contained in the works by Metzger et al. [6] and Van Fossen [7].
7. Fundamental equations and correlations concerning various cases of idealized slot film cooling, such as might be encountered in various leakage flow paths, are summarized in the review of Goldstein [8].

8. The best source of both adiabatic film effectiveness and heat-transfer coefficient augmentation factors due to film injection for round and shaped holes is contained in the recent series of studies from the Institute for Thermal Turbomachinery at the University of Karlsruhe, Germany [9–12]. Such data are generally put into a simplified form to describe the centerline or laterally averaged adiabatic effectiveness as a function of distance and mass velocity ratio.
9. A broad set of data for discharge coefficients of film-cooling holes is available from the research of Hay and Lampard [13] and also from the University of Karlsruhe group [14, 15].
10. Aerodynamic film injection mixing losses can be estimated using the method of Hartsel [16].

Other excellent sources of summarized data and correlations exist in the open literature, but it is up to the design team to determine what to use and how to use it in analysis. One such source is the Lecture Series accumulated by the von Karman Institute of Fluid Dynamics, Brussels. Specific lecture series that cover turbine cooling include Dailey et al. [17], Harasgama et al. [18], Glezer et al. [19], and Bunker et al. [20].

Although the just-referenced correlations provide a good starting point for the most common methods of cooling, there are dozens of special regions, geometries, and circumstances in cooling design analysis that require case-by-case data. For these cases, the relevant literature is too large to mention here. Most of these cases deal with the so-called "edge" regions of the cooled components, including the endwalls, platforms, airfoil leading and trailing edges, blade tips, interfacial rails, fillets, and any isolated corners. All of these can be treated by the use of similar thermal-flow network models or integrated into the airfoil model as special regions.

Is this level of cooling analysis detail really required? Figure 7 shows the characteristic uncertainties in engine boundary conditions that affect the complete cooling design analysis of a HPT blade. Also shown is the percentage impact of each boundary condition on the final result (these add to 100%). It should be clear that no detail is unimportant here. Also clear is that the accuracy of certain data, such as the adiabatic film-cooling effectiveness distribution, is of very high importance.

D. ADDITIONAL KEY FACTORS

Two additional considerations must be incorporated into the cooling design analyses as indicated in Fig. 3: engine experience design factors, such as film knockdown, coating of film hole interiors, and hole spacing, and engine degradation factors, such as combustor gas profile changes and tip erosion. These factors account for past experience in both test engines and operational engines that

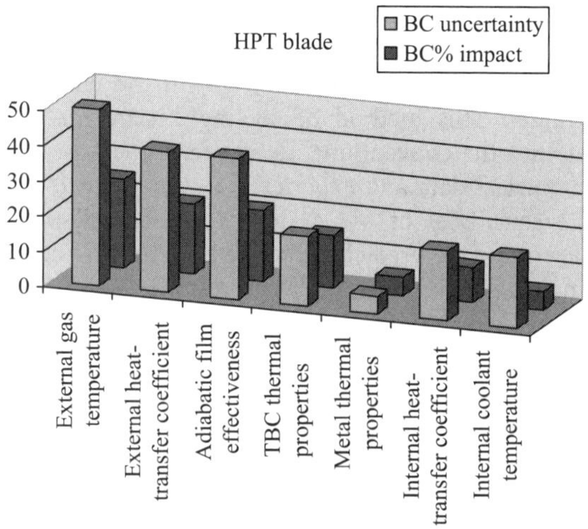

Fig. 7 Impact of boundary condition uncertainties.

cannot be obtained through research and design activities and for the unknown, or poorly understood, conversions from laboratory data and predictions to the reality of complex engine conditions.

In the context of cooling design, experience factors will include film effectiveness realization or knockdown multipliers, film hole diameter reductions caused by protective coating applications, minimum allowable hole spacings to avoid hole-to-hole cracking, reduction of internal heat-transfer coefficients caused by debris collection, typical TBC spallation sizes (if any), surface roughness distribution patterns, and any other generic or design-specific experience. Example engine degradation factors will include alterations to the hot-gas temperature profiles or magnitudes caused by combustor system operation, blade tip erosion, film hole blockages due to deposits, and even material property modification with exposure at elevated temperatures. These additional factors are typically incorporated into the design process by one of two methods. First, the data from engine experience can be "data matched" to the design prediction to arrive at the required adjustment factors. Second, modifications caused by degradation can be carried through the design analysis in a statistical manner to determine magnitudes of change, as well as sensitivity coefficients.

E. THREE-DIMENSIONAL ANALYSIS

The two-dimensional or pseudo-three-dimensional analysis just described is very similar to the simple one-dimensional analysis in format, but includes all of the required detail to perform design manipulations and tradeoff studies and arrive at a "final" cooled component design. Once this iterative process has produced a design that is sufficiently polished, a more precise three-dimensional design analysis can be performed. The three-dimensional analysis primarily adds thermal-mechanical detail through the use of a full, accurate FEM of the component. The FEM is executed using mapped convective boundary conditions of local heat-transfer coefficients and fluid temperatures from the two-dimensional model results. The FEM solution presents the complete temperature distributions of the materials. The three-dimensional analysis can also be performed entirely through the use of computational modeling, with the prediction of external and internal flowfields and heat-transfer coefficients, or coupled with the use of a conjugate CFD heat-transfer model. This method of cooling design analysis must, however, be in agreement with the conventional design result because the latter contains a great deal of empirical data and experience factors. Sufficiency of agreement is defined by the sensitivity of the design to inaccuracies; for example, it is not generally sufficient for the predicted average heat-transfer coefficient in a cooling passage to match the average correlation result.

III. DETAILED DESIGN ANALYSIS (COOLING EXAMPLE)

A. COMPONENT ANALYSES

As indicated in Fig. 3, cooling design analysis must interface with the other component and system goals and requirements. The noncooling subjects noted in Fig. 3 do not comprise an exhaustive list, but do represent the diversity of requirements. These aspects of overall design, manufacturing, and operation apply to all of the cooled hot-gas-path components—vanes, blades, endwalls, platforms, shrouds, supports, and dovetails—so there is no single cooling design analysis method that can be described here. Analyses must pass to and receive results from the other engine design analysis packages in an iterative method until an acceptable total design solution is obtained. This might require many changes to the design with subsequent re-analysis. As one example of the requirements and complexity of this process, the HPT blade tip region design interaction is considered in a later chapter of this volume.

In designing blade tips, both cooled and uncooled, for proper operation within the larger turbine system, one must consider the following major factors (in no particular order):

1. Stage and turbine aerodynamic efficiency
2. Stage thermal efficiency
3. Bulk material temperature limits
4. Maximum local material temperatures
5. Tip sealing methods
6. Casing out-of-roundness
7. Shroud segment variation
8. Approaching and leaving flow disturbances
9. Hot-gas temperature profiles
10. Aeromechanics

11. Mechanical and thermal stresses
12. Operating conditions
13. Operational transients
14. Durability
15. Materials and material loss
16. Cumulative damage
17. Exhaust gas temperature
18. Cost of new parts and cost of repair
19. Blade weight
20. Thrust bearing location and bearing housing distortion
21. Rotor and stator systems thermal matching

Although this summary of system design aspects might appear quite detailed and daunting for such a relatively small region of the turbine, there is one requirement that exceeds all others: the blade tip system design must never cause such severe damage as to liberate blades or pieces of blades in operation. As in the other interacting system relationships within the turbine, prior design and operational experience must guide and temper the design process.

B. COMBUSTOR–TURBINE SYSTEM ANALYSIS

The turbine has a special relationship with the combustion system. Turbine cooling design analysis is directly influenced by the type of combustor system, the combustor exit conditions, and the change in combustor conditions at various cycle points. The combustion system design and operation impact the turbine design in at least six main respects:

1. Hot-gas temperature profiles
2. Hot streak clocking relative to the turbine
3. Turbulence characteristics
4. Airfoil backflow margins
5. NOx emissions
6. Fuel type

Each combustor system design has its own set of characteristic radial and circumferential gas temperature profiles; in any given system the full power radial profile differs from any part-power profile. For example, some systems have annular combustors, some have can-annular combustors, and others have dump combustors. Full annular combustors can be single, dual, or even triple annular systems, depending on the number of fuel nozzle rings present. In such cases, combustor nozzle staging can be used to meet differing power requirements. Another major difference arises between the low NOx systems of power turbine engines and aeroengines, the former employing very little dilution or film flow injection within the combustors and the latter utilizing a great deal of dilution and film injection. Most power generation turbines tend toward very flat radial profiles whereas aero-engines tend to have more peaked radial profiles that can change peaking location with power condition. The radial temperature profile of a power turbine can also change as operation is changed from diffusion mode to premixed mode. The key for turbine cooling design is to know as much as possible about the combustor system exit conditions for all operating conditions and to carry this information through to the design for each cycle point.

Combustion systems have circumferential gas temperature and pressure profiles as well because of the discrete nature of virtually all designs with respect to air/fuel injection and flame holding. While radial profiles are caused by the combined effects of fuel nozzles and combustor dilution/cooling flows, circumferential profiles or pattern factors depend primarily on the number and spacing of the fuel nozzles. Because the turbine inlet vanes are also of a finite number, this leads to the interesting phenomenon of hot streak clocking. The combustor hot streak can be aligned directly on a vane leading edge or midway between two vanes. In fact, the hot streaks can be variable around the entire vane ring, depending on the relative count of fuel nozzles and turbine inlet vanes. Different unsteady gas conditions might be incident upon the rotating blade row. The center hot streak can pass through the passage with little vane interaction while the leading-edge hot streak can be greatly modified by interaction with the vane and its cooling flows. There are, of course, immediate consequences for the vane, but this also translates through to the blade.

As with hot streak effects, combustion system turbulence and swirl flow are additional complicating factors. The turbulence intensity levels, distributions, and length scales will not be the same as those generated by the grids used in simplified studies. The combustor exit flow, in addition to temperature profiles, might also contain significant swirl content. These factors might not be entirely washed out by the inlet vane row. Studies such as those of Van Fossen and Bunker [21, 22] have indicated that combustor exit average turbulence intensity over the entire region is as high as 30%.

These combustor-turbine system interaction issues all point to the requirement that the cooling design analysis must not only be performed for changing conditions because of the combustor, but must in some cases address even vane-to-vane level differences in the cooling analysis.

C. TURBINE SECONDARY COOLING CIRCUIT ANALYSES

While much attention is given to the cooled airfoils of the turbine, the secondary flow cooling circuits deserve equal scrutiny and diligence in the development of a total engine design solution. Figure 2 shows the secondary flow circuits typical of an aeroengine HPT. Secondary circuits of the turbine include the following:

1. Lower wheelspaces or disk cavities inboard of the hot-gas path
2. Supply circuits from the compressor discharge region to the inboard turbine flows
3. Upper wheelspaces, including buffer and trench cavities around the angel wings
4. Supply circuits from the compressor discharge to the outer turbine casing flows
5. Outboard nozzle and shroud cooling air plenums and connections
6. All rotating seals in these areas, for example, labyrinth and brush seals
7. All stationary seals in these areas, for example, labyrinth and cloth seals
8. Component interface leakage pathways and their seals, such as nozzle-to-combustor gaps, nozzle-to-nozzle gaps, shroud-to-shroud gaps, nozzle-to-shroud gaps, and blade-to-blade gaps (spline seals, C-seals, W-seals, leaf seals, etc.)
9. Rotating orifices
10. Stationary orifices
11. Preswirl supply nozzles
12. Inducers and cover plate systems
13. Blade dovetail/shank leakages
14. Bolt leakages
15. Nozzle support leakages
16. Outboard-to-inboard cooling circuits routed through turbine airfoils
17. Nozzle diaphragm chambers
18. Supply flows bled from earlier compressor stages
19. Shroud hanger system flows and leakages

The cooling design analysis for the secondary flow systems is performed in much the same way as that of the turbine airfoils with the main difference being that most of these flow circuits do not directly interact with the hot gas path. Because there is no "external" hot-gas flow involved, the thermal-fluid design analysis of these regions becomes an elaborate flow network model with thermal boundary conditions at the hardware surfaces. Just as in the turbine airfoil analysis, the secondary flow models can be simple or complex depending on the design stage.

Ultimately, some regions will require complex CFD analysis to resolve full details, most obviously at the points at which the secondary flows do interact

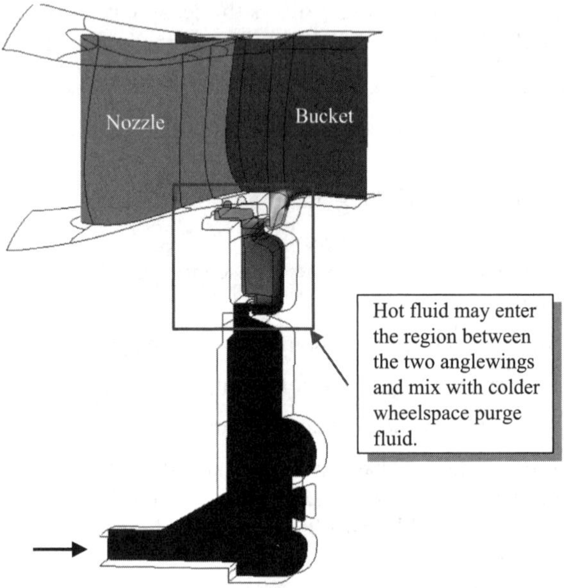

Fig. 8 Purge flow circuit for turbine wheelspace.

with the hot-gas path. One such region is the forward wheelspace sealing cavities between the turbine inlet nozzle and the first stage blade, as depicted in Fig. 8. Cooling air is supplied from the inboard location and routed through the stationary-rotating seal cavities, in this case a buffer cavity and then the trench cavity at the turbine flowpath. Aside from this flow circuit, there are several other leakage pathways influencing the region. In addition, the exit of the flow circuit sees a very three-dimensional flow, which varies in the circumferential direction as a result of nozzle wakes and blade leading-edge effects. Such interaction regions can involve substantial mixing of cold and hot flows. A more detailed understanding of the heat-transfer coefficients and gas temperatures in these regions is required.

Secondary flow design analysis begins with overall, large network models representing the compressor discharge and bleeds to the eventual exit flows into the turbine flowpath, accounting for all key flow areas, lengths, restrictions, and discharge coefficients, using approximate thermal boundary conditions for heat transfer. More detailed models are made to examine separate portions of the flow circuits and add greater fidelity to the boundary conditions. Open literature sources may supply most of the required information concerning flow restrictions, friction coefficients, and discharge coefficients, and some commercial flow network solvers contain correlations for much of this information. Heat-transfer boundary conditions can be estimated by simple forced duct flow and natural convection correlations in most locations other than the radial disk flow, radial cavity flow, and labyrinth seal regions. Good summaries of the flow and heat transfer in radial rotating disk and disk cavity systems for various situations can be found in the works of Owen and Rogers and the subsequent publications of Owen and coworkers [23, 24]. Labyrinth seal flow and heat-transfer data for planar and stepped geometries can be found in the research of the University of Karlsruhe group, such as that of Waschka et al. [25, 26].

The thermal condition of the hardware surrounding the secondary flow circuits must be included in the final design analysis. These boundaries cannot in most cases be treated as adiabatic. For example, the bucket dovetails are connected to the wheel in the disk-posts. While the forward and aft surfaces of the dovetail and disk-post are exposed to the secondary flows of the upper wheel-spaces, the primary cooling flow of the bucket is routed between the bottom of the dovetail and the wheel, and the coolant flows inside the dovetail to the airfoil. This forms an additional network that connects the secondary flow circuit and the coolant circuit of the buckets. This internal cooling of the bucket dovetail and shank will thermally affect the response of the disk-post and wheel. Even the cooling of the bucket airfoils and platforms has an influence on the top portion of the wheel, serving to conduct energy from the hot-gas path down into the wheel. This latter effect is usually analyzed by applying lumped or equivalent thermal masses to the top of the wheels or bucket shanks to act as heat sources. Detailed thermal models of the airfoils, supports, and wheels are rarely, if ever, done in the same model. In a similar manner, the outer shrouds and their hangers must be modeled together to provide the complete prediction of flows and thermal response. Individual wheels, or the entire turbine rotor system, can be modeled. In fact, at some detailed design level, the entire turbine rotor must be thermally analyzed as one in order to correctly predict all clearances. Going one step further, the so-called "unit rotor," which is the combined compressor-turbine or compressor-turbine-generator, rotor must also be analyzed with thermal boundary conditions, albeit with a less detailed application of conditions.

IV. TRANSIENT OPERATIONAL DESIGN ANALYSIS

All of the foregoing cooling design analyses are commonly applied to steady-state operating conditions at some well-defined point in the cycle deck or operational map of the turbine engine. The reality of turbine operation is, however, that both slow and fast transients must be considered in the design process. Slow transients include normal startup and shutdown, climb, and load-following operations whereas fast transients include quick starts for peaking power, takeoff, engine trips, and hot restarts. Within even the normally slow startup procedure for a heavy frame turbine, there are several intermediate operating points and

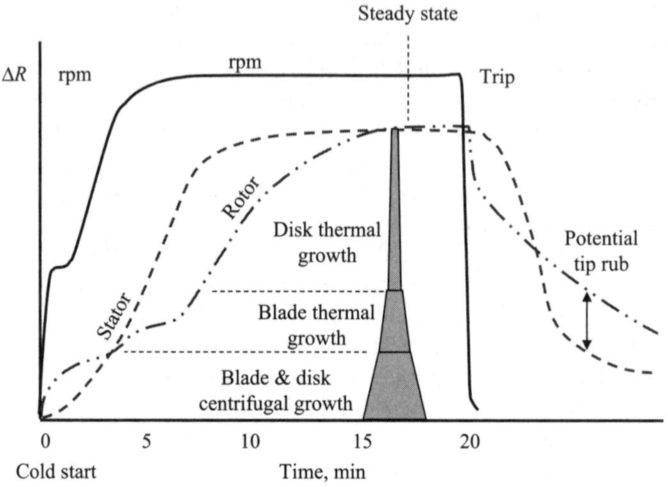

Fig. 9 Transient rotor and stator growth for fast startup.

transients, such as turning gear operation, low rpm holding, and ~80% power point warmup. Other transients can include specific operating domains dictated by the combustion system, water washing, and power augmentation (such as water injection to postcombustion gases).

Figure 9 shows an approximate transient growth behavior for an industrial turbine rotor and stator during a fast start (<30 min). The transient growth of the rotor is a combination of all portions making up the rotor with contributions from centrifugal and thermal effects. The transient growth of the stator and casing outboard of the rotor is thermally dominated, occurring at a different rate than the rotor. The cooling design analysis of all transients is performed using a sufficient number of steady-state analyses and their associated boundary conditions. Each steady-state analysis is performed using the methods discussed in the preceding sections. The boundary conditions of these several operating points, flow rates, pressures, gas temperature profiles, heat-transfer coefficients, and film effectiveness are used to form the anchor points of the transient analysis. Because the number of steady-state analysis points is typically limited, the boundary conditions at several intermediate steps must be obtained by interpolation. As the basic fluid dynamic and thermal domains of the hot-gas and cooling flows also change with operating conditions, these interpolations are performed using explicit or ad hoc rules. The exact nature and definition of these rules are very dependent on the turbine design and operation and, as such, are specific to the OEMs.

Transient analyses of individual components, such as the turbine airfoils, follow the same general guidelines. Usually, the concerns associated with these components are not the same as those of the overall turbine stator and rotor systems. Instead, issues with clearances, leakage gaps, binding, and hot-gas backflow or ingestion are scrutinized. In addition, transient effects on peak material stress and strain are important, as evidenced by the potential for TBC spallation under severe thermal transients. The transient cooling design analysis for hot-gas-path components can therefore focus on certain transients, or portions of transients, known to be of greatest concern.

REFERENCES

[1] Florschuetz, L., Truman, C., and Metzger, D., "Streamwise Flow and Heat Transfer Distributions for Jet Array Impingement with Crossflow," *Journal of Heat Transfer*, Vol. 103, 1981, pp. 337–342.

[2] Kercher, D., and Tabakoff, W., "Heat Transfer by a Square Array of Round Air Jets Impinging Perpendicular to a Flat Surface Including the Effect of Spent Air," *Journal of Engineering for Power*, Vol. 92, 1970, pp. 73–82.

[3] Martin, H., "Heat and Mass Transfer Between Impinging Gas Jets and Solid Surfaces," *Advances in Heat Transfer*, Vol. 13, Academic Press, New York, 1977, pp. 1–60.

[4] Han, J. C., Park, J. S., and Lei, C. K., "Heat Transfer Enhancement in Channels with Turbulence Promoters," *Journal of Engineering for Gas Turbines and Power*, Vol. 107, 1985, pp. 628–635.

[5] Hajek, T. J., Wagner, J. H., Johnson, B. V., Higgins, A. W., and Steuber, G. D., "Effects of Rotation on Coolant Passage Heat Transfer," NASA Report 4396, 1991.

[6] Metzger, D. E., Berry, R. A., and Bronson, J. P., "Developing Heat Transfer in Rectangular Ducts with Staggered Arrays of Short Pin Fins," *Journal of Heat Transfer*, Vol. 104, 1982, pp. 700–706.

[7] VanFossen, G. J., "Heat Transfer Coefficients for Staggered Arrays of Short Pin Fins," *Journal of Engineering for Power*, Vol. 104, 1982, pp. 268–274.

[8] Goldstein, R. J., "Film Cooling," *Advances in Heat Transfer*, Vol. 7, Academic Press, New York, 1971, pp. 321–379.

[9] Gritsch, M., Schulz, A., and Wittig, S., "Adiabatic Wall Effectiveness Measurements of Film-Cooling Holes with Expanded Exits," *Proceedings of the ASME Turbo Expo Conference*, Paper 97-GT-164, 1997.

[10] Gritsch, M., Schulz, A., and Wittig, S., "Heat Transfer Coefficients Measurements of Film-Cooling Holes with Expanded Exits," *Proceedings of the ASME Turbo Expo Conference*, Paper 98-GT-28, 1998.

[11] Saumweber, C., Schulz, A., and Wittig, S., "Free-Stream Turbulence Effects on Film Cooling with Shaped Holes," *Proceedings of the ASME Turbo Expo Conference*, Paper GT-2002-30170, 2002.

[12] Dittmar, J., Schulz, A., and Wittig, S., "Assessment of Various Film Cooling Configurations Including Shaped and Compound Angle Holes Based on Large Scale Experiments," *Proceedings of the ASME Turbo Expo Conference*, Paper GT-2002-30176, 2002.

[13] Hay, N., and Lampard, D., "Discharge Coefficient of Turbine Cooling Holes: A Review," *Journal of Turbomachinery*, Vol. 120, 1998, pp. 314–319.

[14] Gritsch, M., Schulz, A., and Wittig, S., "Discharge Coefficient Measurements of Film-Cooling Holes with Expanded Exits," *Proceedings of the ASME Turbo Expo Conference*, Paper 97-GT-165, 1997.

[15] Gritsch, M., Saumweber, C., Schulz, A., Wittig, S., and Sharp, E., "Effect of Internal Coolant Crossflow Orientation on the Discharge Coefficient of Shaped Film-Cooling Holes," *Journal of Turbomachinery*, Vol. 122, 2000, pp. 146–152.

[16] Hartsel, J. E., "Prediction of Effects of Mass Transfer Cooling on the Blade Row Efficiency of Turbine Airfoils," AIAA Paper 72-11, 1972.

[17] Dailey, G. M., Taslim, M., Rigby, D. L., Sagaut, P., Cakan, M., Han, B., Goldstein, R. J., and Buchlin, J. M., "Aero-Thermal Performance of Internal Cooling Systems in Turbomachines," von Kármán Inst. for Fluid Dynamics Lecture Series VKI-LS 2002-01, 2002.

[18] Harasgama, S. P., Han, J. C., Dutta, S., Iacovides, H., Rau, G., Owen, J. M., and Wilson, M., "Heat Transfer and Cooling in Gas-Turbines," von Karman Inst. for Fluid Dynamics Lecture Series VKI-LS 1996-01, Belgium, 1996.

[19] Glezer, B., Harvey, N., Camci, C., Bunker, R., and Ameri, A., "Turbine Blade Tip Design and Tip Clearance Treatment," von Kármán Inst. for Fluid Dynamics Lecture Series VKI-LS 2004-02, Belgium, 2004.

[20] Bunker, R., Simon, T., Bogard, D., Schulz, A., Baldauf, S., Saumweber, C., Acharya, S., Tyagi, M., and Burdet, A., "Film Cooling Science and Technology for Gas-turbines," von Kármán Inst. for Fluid Dynamics Lecture Series VKI-LS 2007-06, Belgium, 2007.

[21] Van Fossen, G. J., and Bunker, R. S., "Augmentation of Stagnation Region Heat Transfer due to Turbulence from a DLN Combustor," *Journal of Turbomachinery*, Vol. 123, 2000, pp. 140–146.

[22] Van Fossen, G. J., and Bunker, R. S., "Augmentation of Stagnation Region Heat Transfer due to Turbulence from a DAC Combustor Section," *Proceedings of the ASME Turbo Expo Conference*, Paper GT-2002-30183, June 2002.

[23] Owen, J. M., and Rogers, R. H., *Flow and Heat Transfer in Rotating Disc Systems: Volume 1, Rotor-Stator Systems*, Research Studies Press, Somerset, England, 1989.

[24] Owen, J. M., and Rogers, R. H., *Flow and Heat Transfer in Rotating Disc Systems: Vol. 2, Rotating Cavities*, Research Studies Press, Somerset, England, 1989.

[25] Waschka, W., Wittig, S., and Kim, S., "Influence of High Rotational Speeds on the Heat Transfer and Discharge Coefficients in Labyrinth Seals," *Proceedings of the ASME Turbo Expo Conference*, Paper 90-GT-330, June 1990.

[26] Waschka, W., Wittig, S., Kim, S., and Scherer, T., "Heat Transfer and Leakage in High-Speed Rotating Stepped Labyrinth Seals," *Heat Transfer and Cooling in Gas Turbines*, AGARD Conference Proceedings 527, NASA Scientific and Technical Information Div., Washington, D.C., 1993, pp. 26-1–26-10.

CHAPTER 3

Turbomachinery Clearance Control

Raymond E. Chupp*
General Electric Power & Water, Greenville, South Carolina

Robert C. Hendricks†
NASA John H. Glenn Research Center at Lewis Field, Cleveland, Ohio

Scott B. Lattime‡
The Timken Company, Canton, Ohio

Bruce M. Steinetz§
NASA John H. Glenn Research Center at Lewis Field, Cleveland, Ohio

Mahmut F. Aksit¶
Sabanci University, Istanbul, Turkey

ACRONYMS

ACC	active clearance control
APS	air plasma spray
BOM	bill of material
EB-PVD	electron-beam plasma vapor deposition
FEA	finite element analysis
FOD	foreign object damage
FSN	first-stage nozzle
HCF	high-cycle fatigue
HFBS	hybrid floating brush seal
HPC	high-pressure compressor
HPP	high-pressure packing
HPT	high-pressure turbine
IPC	intermediate-pressure compressor
IPT	intermediate-pressure turbine
LCF	low-cycle fatigue

*Principal Engineer (retired), Heat Transfer and Flow Systems Design.
†Senior Technologist, Research and Development Directorate.
‡Specialist, Advanced Modeling.
§Senior Technologist, Materials and Structures Division.
¶Faculty of Engineering and Natural Sciences.

LDV	laser Doppler velocimetry
LPC	low-pressure compressor
LPT	low-pressure turbine
LSAC	low-speed axial compressor
MTBF	mean time between failures
ODS	oxide dispersion strengthened
PSZ	partially stabilized zirconia
T1	first turbine stage
TD	transition duct
TIR	total indicator reading
TO	takeoff
YSZ	yttria-stabilized zirconia

Seals play a major role in load balancing in turbomachinery systems including 1) managing secondary fluid flows enabling the designer to direct cooling and purge flows as needed without incurring excessive leakage or fluid ingestion and therefore minimizing system losses; 2) allowing the designer to effectively address compressor-combustor-turbine pressure load balancing via balance piston seals providing stable operation and preventing bearing overload; 3) promoting proper cooling and purge flows at blade and vane platforms while improving engine efficiency and minimizing chances of compressor unstart; 4) providing for rotordynamic stability; and 5) helping in the bearing compartments to restrict noxious oil vapor from entering the cabin for aircraft applications. Controlling interface clearances is essential for load balancing and enhancing turbomachinery performance. In many instances, sealing interfaces and coatings are sacrificial, like lubricants, giving up their integrity for the benefit of the component. They are subjected to abrasion, erosion, oxidation, incursive rubs, foreign object damage (FOD), and deposits, as well as extremes in thermal, mechanical, aerodynamic and impact loadings. Tribological pairing of materials controls how well and how long these interfaces will be effective in controlling flow.

A variety of seal types and materials are required to satisfy turbomachinery sealing demands. These seals must be properly designed to maintain the interface clearances. In some cases, this will mean machining adjacent surfaces; in many other applications, coatings are employed for optimum performance. Many seals are coating composites fabricated on substructures or substrates, which can be refurbished either in situ or by removal, stripping, recoating, and replacing until substrate life is exceeded.

This chapter presents an overview of turbomachinery sealing. Areas covered include characteristics of gas and steam turbine sealing applications and environments, benefits of sealing, types of standard static and dynamics seals, and advanced seal designs, as well durability and life considerations.

I. SEALING IN GAS AND STEAM TURBINES

Turbomachines range in size from centimeters (the size of a penny) to almost large enough to walk through. The problem is how to control the relatively large changes in geometry between adjacent rotor/stator components from cold-build to operation. The challenge is how to provide geometric control while maintaining efficiency, integrity, and long service life (estimated time to failure or maintenance, and low cost) [1]. Figure 1 shows the relative clearance between the rotor tip and case for an aircraft high pressure turbine (HPT) during takeoff, climb, and cruise conditions [2]. The figure shows blade tip clearances under two scenarios: 1) simple case cooling and 2) active thermal control. With only passive case cooling, a critical clearance event can occur at "cutback" (about 1000 s into climb-out) when takeoff thrust is reduced. Here the disc and blades continue to expand due to thermals; however, with passive thermal clearance the case is at a steady temperature, and a rub event could occur. Using thermal active clearance control (ACC), the running clearance at cruise is drastically reduced while maintaining positive margins during the pinch points as noted. Thermal ACC has produced significant cost savings in fuel reduction and increased service life. Designers are now examining ways to incorporate mechanical active clearance control with clearance sensors to reduce operating clearances even further. Designers must note that changing parameters in critical seals can change the dynamics of the entire engine [3]. These effects are not always positive.

Performance issues are closely tied to engine clearances. Ludwig and Bill [4] determined that improvements in fluid film sealing could lead to an annual energy saving, on a national basis, equivalent to about 37 million barrels

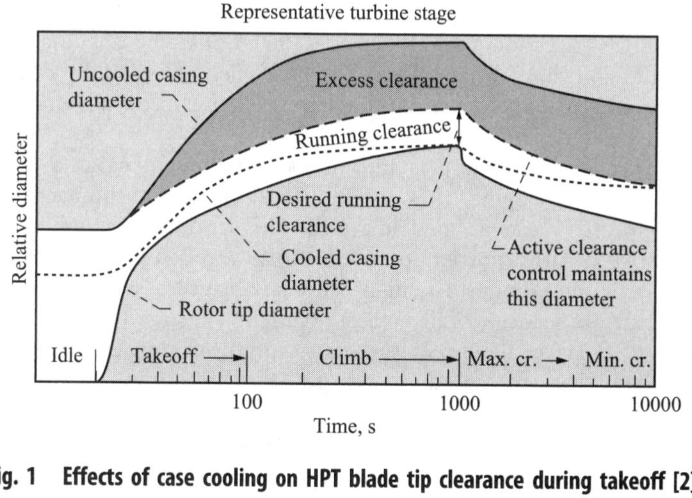

Fig. 1 Effects of case cooling on HPT blade tip clearance during takeoff [2].

(1554 million U.S. gallons) of oil or 0.3% of the total U.S. energy consumption (1977 statistics). In terms of engine bleed, Moore [5] states that a 1-% reduction in engine bleed gives a 0.4-% reduction in specific fuel consumption (SFC), which translates into nearly 33 (1977 statistics) to 55 (2004 statistics) million gallons of U.S. airline fuel savings and nearly 280 million gallons worldwide (2004 statistics), annually. In terms of clearance changes, Lattime and Steinetz [6] state that a 0.0254-mm (0.001-in.) change in HPT tip clearance decreases SFC by 0.1% and EGT (exhaust gas temperature) by 1°C, producing an annual savings of 20 million gallons for U.S. airlines. In terms of advanced sealing, Munson et al. [7] estimate savings of over 500 million gallons of fuel. Chupp et al. [8] estimated that refurbishing compressor seals would yield impressive improvements across the fleet, ranging from 0.2 to 0.6-% reduction in heat rate and 0.3 to 1.0-% increase in power output. For these large, land-based gas turbines, the percentages represent huge fuel savings and monetary returns, with the greatest returns cited for aging power systems.

Key aeroengine sealing and thermal restraint locations cited by Bill [9] are shown in Fig. 2. These include the fan and compressor shroud seals (rub strips), compressor interstage and discharge seals (labyrinth), combustor static seals, balance piston sealing, turbine shroud, and rim-cavity sealing. Industrial engines have similar sealing requirements. Key sealing locations for the compressor and turbine in an industrial engine are cited by Aksit et al. [10] and Camatti et al. [11, 12], and are shown in Figs. 3 and 4, with an overview of sealing in large industrial gas turbines by Hurter et al. [13], Fig. 3c.

Figures 3a and 3b show high-pressure compressor (HPC) and HPT tip seal (abradable) and interstage seal (brush seal) locations, while Fig. 4 shows impeller shroud (labyrinth) and interstage seal (honeycomb) locations for the compressor. Compressor interstage platform seals are of the shrouded type (Figs. 5 and 6). These seals are used to minimize backflow, stage pressure losses, and re-ingested passage flow. Turbine stators, also of the shrouded type, prevent hot-gas ingestion into the cavities that house the rotating disks and control blade and disk coolant flows. Designers need to carefully consider the differences in thermal and structural characteristics, pressure gradient differences, and blade rub interfaces.

In general, the industrial gas turbine can be thought of as a heavy-duty derivative of an aeroengine, although industrial and aeroturbomachines differ in many respects. The most notable are the fan, spools, and combustor. Aeroengines derive a large portion of their thrust through the bypass fan and usually have inline combustors, high- and low-pressure spools, drum rotors, and high exhaust velocities, all subject to flight constraints. Large industrial engines (Fig. 7) have plenum inlets, can-combustors, single spools, through-bolted-stacked disc rotors, and exhaust systems constrained by 640°C (1180°F) combined-cycle (steam-reheat-turbine) requirements. In the two types of engines, core requirements are similar although materials constraints differ.

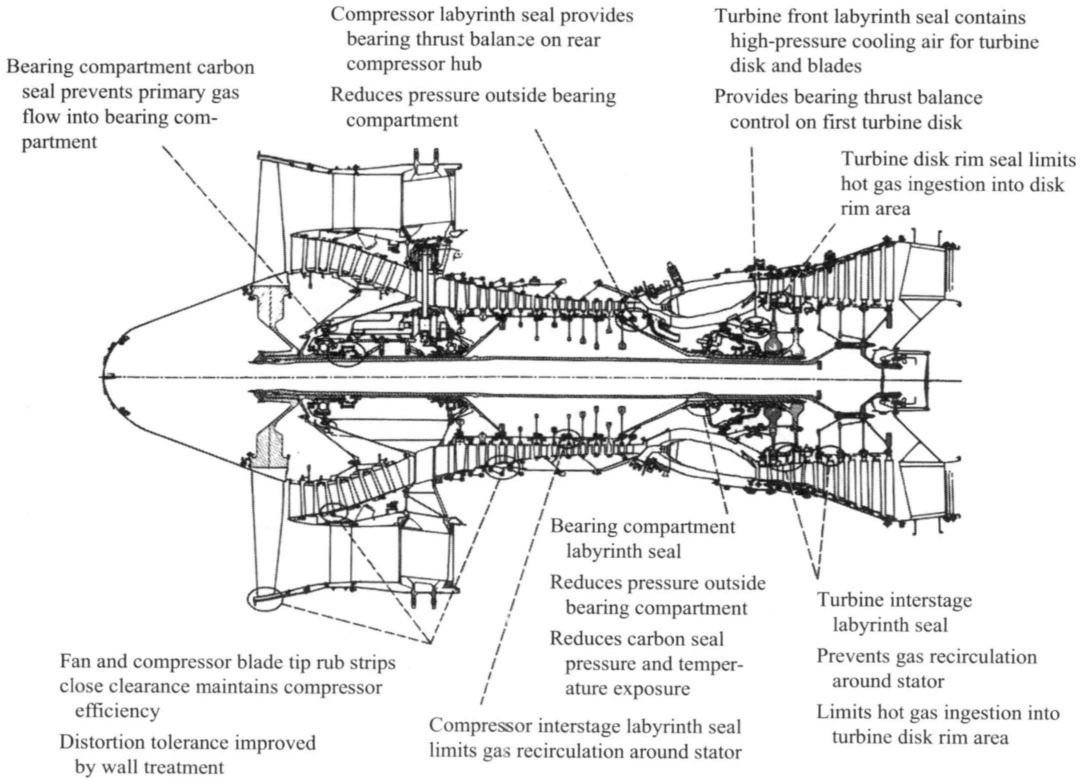

Fig. 2 Key aero-engine sealing and thermal restraint locations [9].

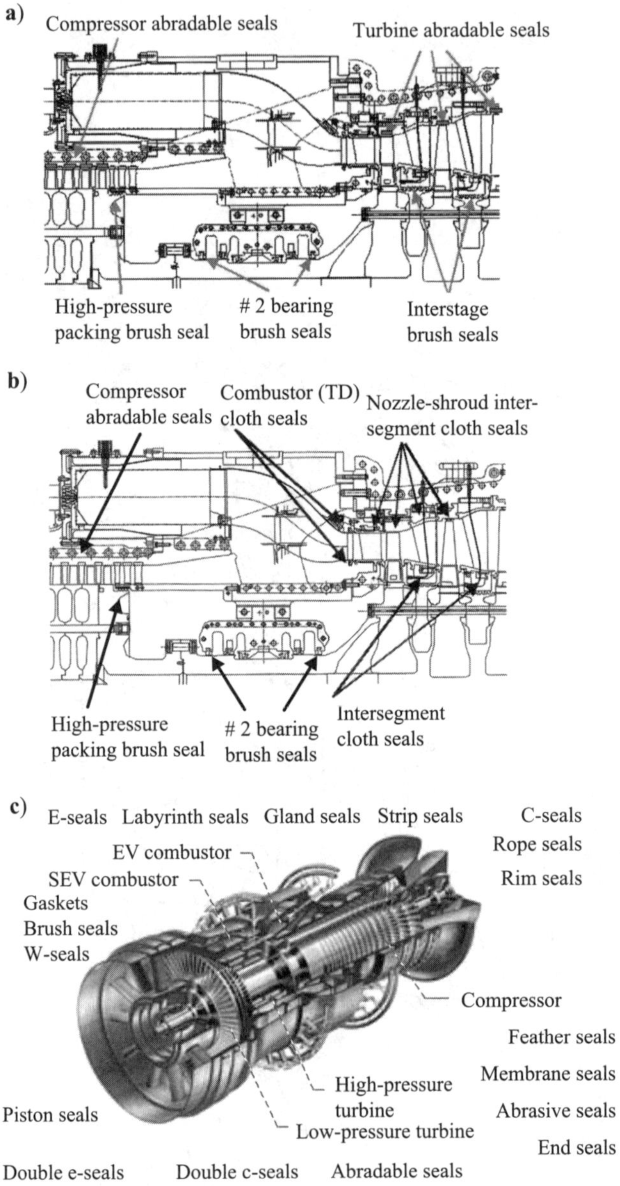

Fig. 3 a) Advanced seal locations in a Frame 7EA gas turbine [10]; b) advanced cloth seal locations in a Frame 7EA gas turbine [10]; c) overview of sealing in large industrial gas turbines [13].

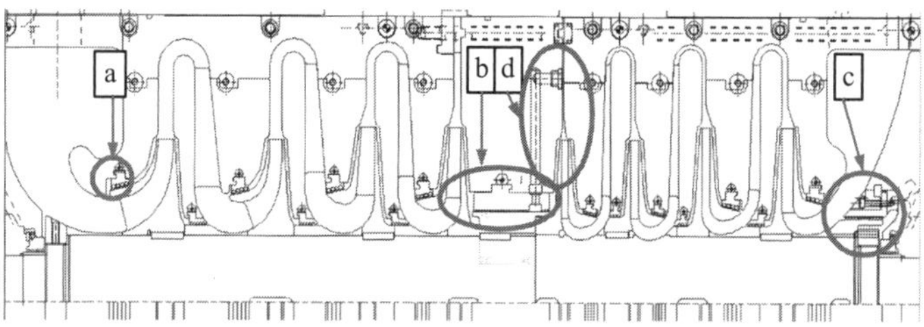

Fig. 4 Compressor cross-sectional drawing showing detail of rotor and seals: a) impeller shroud labyrinth seal, b) honeycomb interstage seal, c) abradable seal, and d) honeycomb interstage seal [12].

Over the years, advances in new base materials, notably nickel-based single crystal alloys, and coatings have allowed increased operating temperatures of turbine engine components. Appropriate to the thermal and pressure profiles, materials used range from steel to superalloys coated with metallics and ceramics. Variations in engine pressure and temperature of the Rolls-Royce Trent gas turbine, which also has an intermediate-pressure turbine (IPT), are illustrated in Fig. 8. (Data available from Sourmail, T., "Coatings for Turbine Blades," University of Cambridge, at http://www.msm.cam.ac.uk/phase-trans/2003/Superalloys/coatings/ [cited 18 May 2005]). The lower temperature blades in the fan and low-pressure compressor (LPC) sections are made of titanium, or composite materials, with corrosion-resistant coatings, which offer high strength and low density. The elevated temperatures of the HPC, HPT, and low-pressure turbine (LPT) require the use of nickel-based superalloys. In the HPT of aeroengines, for example, the first-stage turbine blades can see gas path temperatures around 1400°C (2550°F). To withstand these punishing temperatures for their 20,000-h (and more) service lives, aeroengine designers have turned single crystal blades into highly sophisticated, internally cooled heat exchangers, with

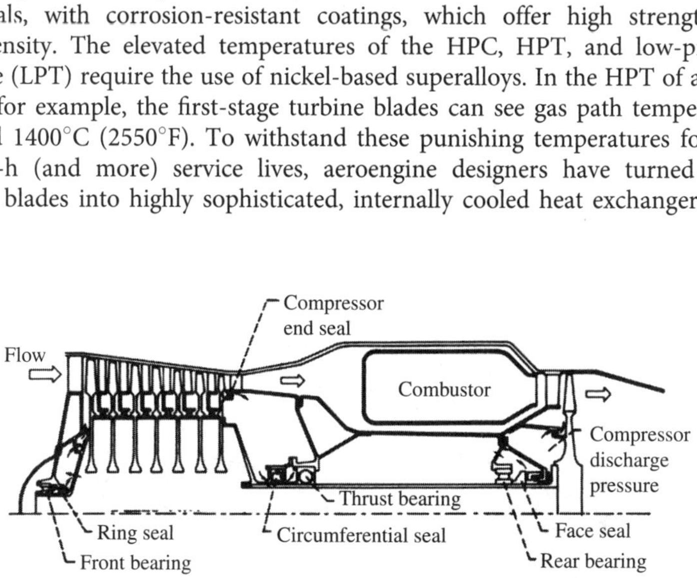

Fig. 5 Engine schematic showing main-shaft seal locations [4].

Fig. 6 Compressor sealing locations: a) blade tip and interstage and b) drum rotor [4].

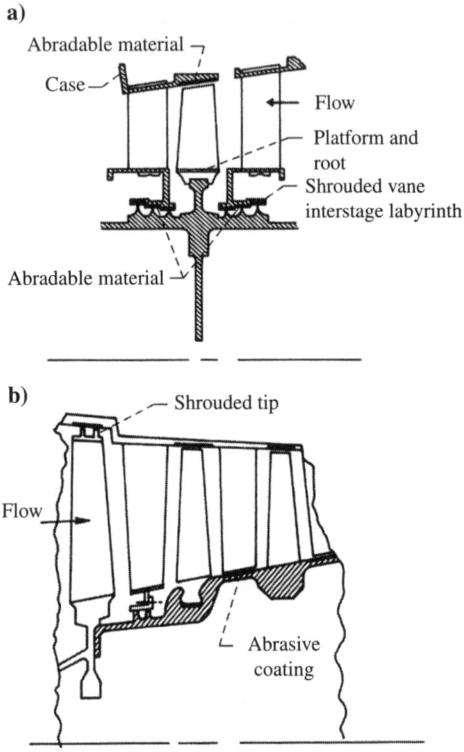

thermal barrier coatings generally using yttria-stabilized zirconia (YSZ) [2]. In this configuration, blade metal temperatures reach 982°C (1800°F) and ceramic surface temperatures reach 1100°C (2010°F).

To improve blade tip sealing effectiveness, squealer tips [approximately 0.8 mm (0.03 in.) high] are integrated into the blade. Depending on engine design, adjacent shroud seals are made of either directionally solidified cast superalloy materials coated with sprayed abradable coatings (YSZ based) [14] or single crystal shroud segments capable of the required operating temperatures.

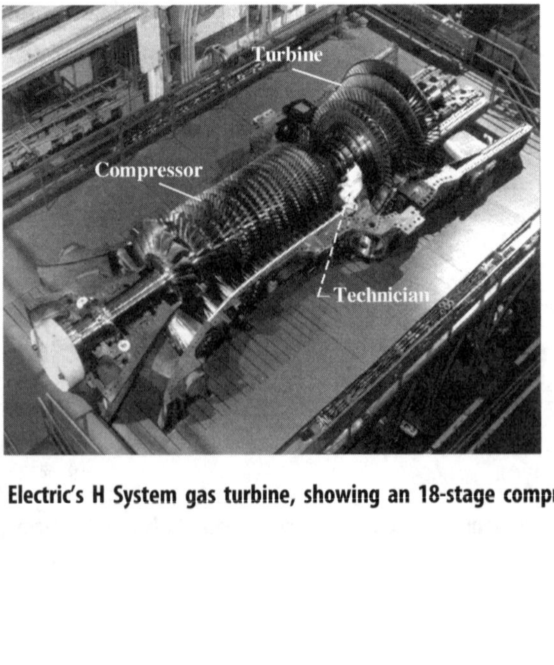

Fig. 7 General Electric's H System gas turbine, showing an 18-stage compressor and 4-stage turbine.

II. STATIC SEALING

Sealing at static or slow-relative-motion interface locations in turbomachinery includes the sealing at interfaces or junctions between the stationary components (such as combustors, nozzles, shrouds, and structure) throughout the internal cooling flow path to minimize or control parasitic leakage flows between turbine components (see Fig. 3b). Typically, adjacent members have to sustain relative vibratory motion with minimal wear or loss of sealing over the design life of the seal. In addition, these types of seals must be compliant to accommodate thermal growth and misalignment. For example, in a cooling application, static (and some structural) seals provide a careful balance of leakage, cooling, and, to some extent, dynamics. If the seals are too tight, improper damping occurs, particularly with low differential pressure loads. Coolant leakage flows then cannot properly purge cavities or offer sufficient cooling to protect themselves against ingested hot gas streaks. Rapid deterioration, oxidizing, and burning can occur. If coolant-seal leakage is excessive, coolant air leaking into the power-stream introduces parasitic air loss and low energy fluid that increases the average passage blockage factor (see also Appendix A). Effective sealing at these static interface locations not only increases turbine efficiency and power output, but also improves the main-gas-path temperature profile. Various types of compliant-interface seals have been developed to address these issues.

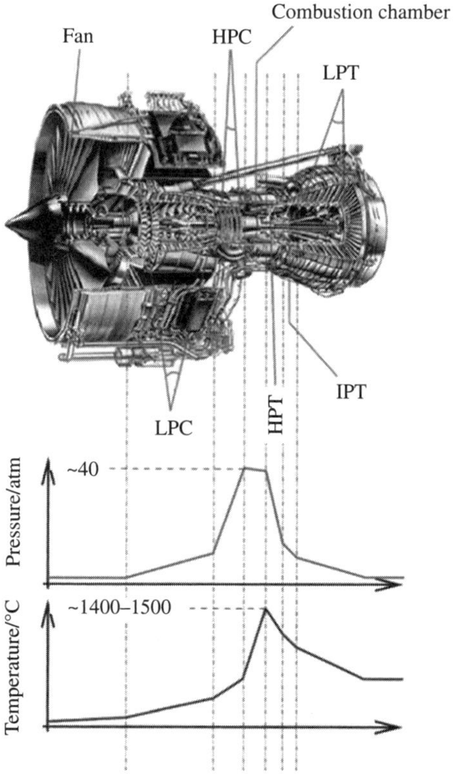

Fig. 8 Temperature and pressure profiles of a Rolls-Royce Trent gas turbine engine (data available from Sourmail, T., "Coatings for Turbine Blades," *University of Cambridge*, at http://www.msm.cam.ac.uk/phase-trans/2003/Superalloys/coatings/ [cited 18 May 2005]).

A. METALLIC SEALS

For smaller gap movements, more conventional seals are used. These seals are metallic for higher temperature and pressure environments

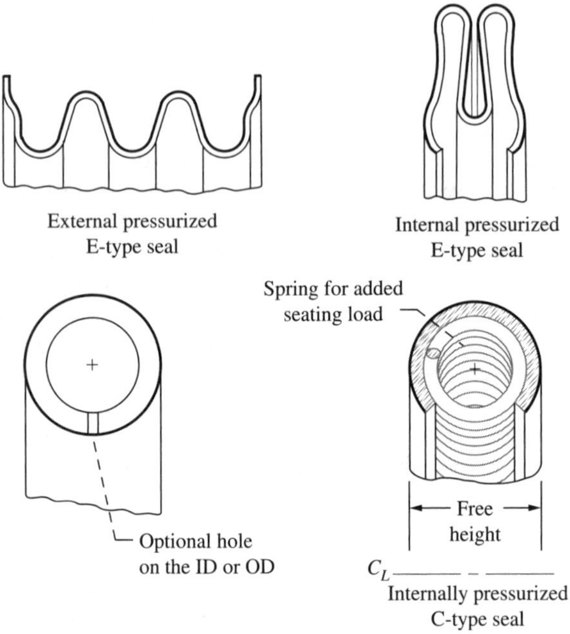

Fig. 9 Some types of metallic seals used in turbomachinery (courtesy of Advanced Products, Parker Hannifin Corp.).

where rubber and polymer seals are not suitable. The wide range of applications in turbomachinery drives the need for multiple configurations, such as the O-, C-, and E-type cross section (see Fig. 9). The type of seal that is best suited for a particular application depends on operating variables such as temperature, pressure, required leakage rate, flange separation, fatigue life, and the load available to seat the seal. Figure 10 shows the locations of some metal seals in an industrial gas turbine. There are many smaller "feather" seals (thin sheet metal) used throughout; all interfaces require sealing of some nature.

In higher-temperature environments, a large amount of thermal growth in surrounding structures is typical. This makes it necessary for the metal seal to maintain contact with the sealing surfaces while the structure moves. A seal's ability to follow the moving structure depends on its spring back and on system pressure to seat the seal. In general, E-type seals (alternatively called W-seals) provide the largest amount of spring back. For this reason, the majority of metal seals found in steam, gas, and jet engines are of the E-type configuration.

The ability of a seal to maintain a low leakage rate is for the most part determined by the force the seal exerts on the mating flange, also called the "seating load." Typically, the leakage rate of a seal will decrease as the seal seating load increases. C-type seals have higher seating loads than E-type seals. To further increase a C-type seal load and spring back, a spring can be inserted around the circumference on the inside of the cross section. The high load of a C seal can be used to enhance sealing performance by the addition of plating such as silver, nickel, gold, and copper. The simple geometry of a C-type seal limits further design possibilities.

The relative complexity and adaptability of the E-type seal cross sections allows for increased design variations with somewhat increased leakage rates as compared to C-seals. The number of convolutions, material thickness, convolution depth, and free height all play a major role in seal performance. Despite the large thermal growth common in turbine engines, a properly designed E-type seal can have millions-of-cycles fatigue life. A majority of the E-type seals used in turbine engines are located between engine case segments, such as the horizontal joint in the combustion section on steam and gas turbines. E-type seals can also be found in the cartridge assemblies of a turbine fuel nozzle. The seal can be cut axially in one or more circumferential locations to accommodate radial growth difference or assembly requirements. Small "caps" can be placed on the seal to span the circumferential gaps to control leakage.

Currently, metallic seals for higher-temperature applications are made from Inconel 718 and Waspaloy nickel-based superalloys, with a temperature limit of

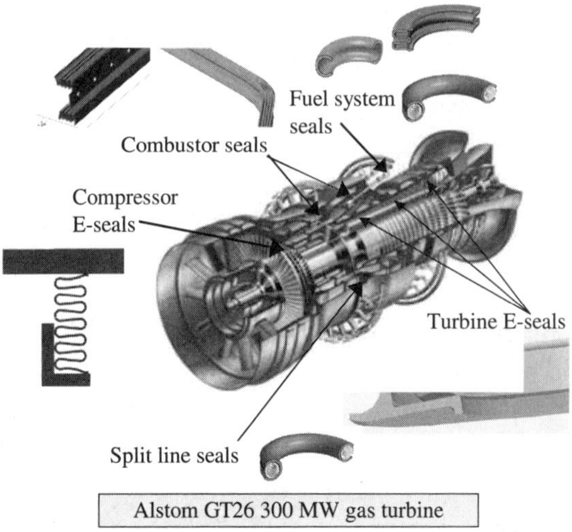

Fig. 10 Typical gas turbine seal locations (courtesy of Advanced Products, Parker Hannifin Corp.).

about 730°C (1350°F) [15]. Above this temperature, such seals under compressive and tensile stresses relax, due to creep, with an attendant loss in sealing performance. Development is in progress to increase the operating temperature range using strengthened [e.g., oxide-dispersion strengthened (ODS)] and refractory alloys. In laboratory tests, new nickel-based superalloy Rene 41 (Allvac, Inc.) seals have exhibited superior performance at 815 to 870°C (1500 to 1600°F) compared to standard Waspaloy seals with the same design. ODS alloys are being tested for temperatures above 870°C (1600°F) [15].

The U-Plex seal (Fig. 11) is another self-energized static seal, similar to a multi-element E seal [16]. The E-seal is a single "folded" element. The U-Plex seal consists of two or more plies of materials nested together that act independently when the seal is compressed, as does a leaf spring, yet function as one under sealing pressure. It will accommodate 2.5 to 5 times more deformation than a single-ply E-seal, is more compliant to surface irregularities, requires one-third the compression force, has enhanced high-cycle-fatigue (HCF) resistance and has comparable leakage rates.

Added compliance (Fig. 12a) is provided for combustor sealing of a large industrial gas turbine, where efficiency and emissions are key drivers for sealing advancements. They offer increases in HCF, an aggressive thermal and pressure environment, and large thermomechanical transients and wear—all of which impact engine operations life (see also Appendix B). Flame stability and combustor emissions become highly dependent on uniformity of coolant and combustion air–fuel mixing. In turn, uniform air supply depends on seals similar to the multiconvolution combustor liner seal, Fig. 12a. Such a seal has been rig tested to withstand wear, oxidation, HCF, fretting, convolution preload loss, and leakage and has been proven in an integrated engine environment [13]. Field data are reported to corroborate the effectiveness of the seals in rig and scaled-up engine testing, but specific data are not provided.

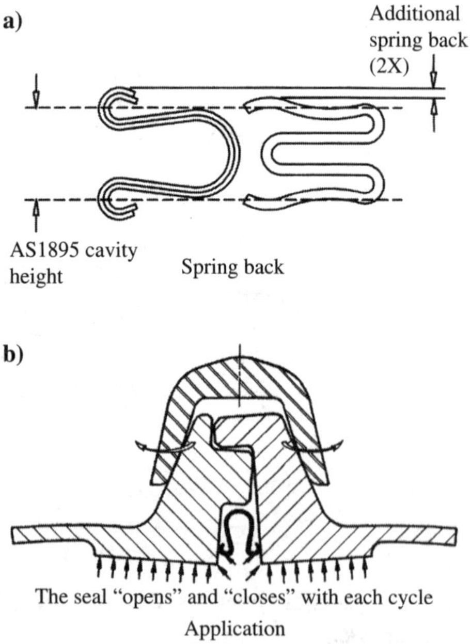

Fig. 11 U-Plex and E-seal geometry: a) springback comparison and b) U-Plex application [16].

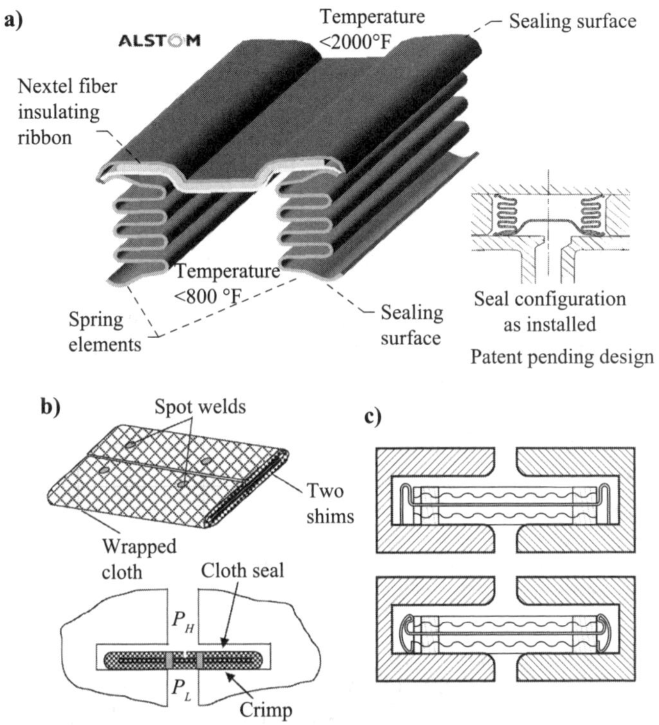

Fig. 12 Structural seal assembly [10, 15]: a) enhanced compliance b) wrapped cloth, and c) alternate crimped cloth.

As noted earlier, it is important to remember that sealing interfaces and coatings are sacrificial. The cost of replacing or refurbishing a seal is minor relative to that of an engine combustor or turbine blade.

B. METALLIC CLOTH SEALS

Metallic and most nonmetallic fibers can be readily fabricated into a variety of configurations that are compliant and responsive to high-speed or lightly loaded systems, as projected by Hendricks et al. [17]. Cloth seals are composite structures that make use of tightly woven, typically metallic, mesh. Most of the issues cited herein are applicable to the design of both metallic and nonmetallic cloth seals—provided proper attention is given to weaves and patterns, such as the multiple strands used in textile fabrics that require modifications for wear volume and contact area calculations. There are differences in applications, some of which are noted in Sec. II.C, "Cloth and Rope Seals."

For large gaps at interfaces with relative motion, rigid metal strips, feather seals, and "dog-bone" shaped strips have been the primary sealing methods. For applications with significant relative motion, these seals can either rock and rotate or jam against the slots in the adjacent components to be sealed. A lack of flexibility can result in poor sealing and excessive wear. Compliance can be attempted by reducing the thickness of the seal strips. The use of thinner foil seals, as in aircraft engine applications, however, results in large stress levels and limited wear life in industrial turbine applications, which involve large interface gaps and demand much longer service life. (Thin foil seals are successfully used in aircraft engines where intersegment gaps and associated stress levels are small and overhaul intervals are much shorter than in industrial applications, where typical expected life may extend to 64,000 hr.) One approach to address seal compliance issues for large relative movements is the development of relatively low-cost, flexible cloth seals. Cloth seals are formed by combining thin sheet metals (shims) with layers of densely woven metal cloth. While shims prevent through leakage and provide structural strength with flexibility, external cloth layers add sacrificial wear volume and seal thickness without adding significant stiffness. As illustrated in Fig. 12b, a typical design requires simply wrapping a layer of metal cloth around thin flexible shims. The assembly is held together by a number of spot welds along the seal centerline [10]. Further leakage reduction can be achieved by a crimped design with exposed and contoured shim ends that enhance enwall sealing (Fig. 12c) [18, 19]. By maintaining contact with the slot surface, crimped shims better reduce the leakage flow. Demonstrated leakage reductions up to 30% have been achieved in combustors and 70% in nozzle segments. The flow savings achieved in nozzle-shroud cloth seal applications translate to large performance gains of up to a 0.50% output increase and 0.25% heat rate reduction in an industrial gas turbine.

Oxidation and wear resistance are the key attributes needed in metallic-cloth fiber material. Likewise, a structural shim must have high-temperature strength and creep and fatigue resistance. A typical metallic-cloth fiber material is cobalt-based alloy Haynes 25, which is used for its superior high-temperature wear resistance. For high-temperature applications, the cobalt-based superalloy Haynes 188 serves well as the shim material. Although high mesh density is preferred for added flow resistance, the small fiber size required for increased density reduces oxidation life. Experience shows that Dutch twill weave with 30×250 fiber density per inch is the best cloth for sealing purposes [20]. To achieve improved wear resistance, the cloth weave should be oriented 45 deg to the dominant relative motion. Diagonal orientation also helps maintaining weave integrity if a local cut is incurred during operation.

Metallic-cloth seal design requires careful engineering to optimize flexibility while maintaining structural strength, resilience, and robustness. A proper design process should include geometric analyses for engagement and jamming, finite element structural and stress analyses, rubbing wear tests, wear analyses, thermal flow analyses, subscale leakage performance tests, and analyses of leakage performance data. A detailed discussion of metallic-cloth seal design and analysis is given in Appendix B.

C. CLOTH AND ROPE SEALS

Rope or gasket seals can be used in various locations in turbomachinery. Table 1 lists the various materials being used. Aircraft engine turbine inlet temperatures and industrial system temperatures continue to climb, however, to meet aggressive cycle thermal efficiency goals. Advanced material systems, including monolithic and composite ceramics, intermetallic alloys (i.e., nickel aluminide), and carbon-carbon composites are being explored to meet aggressive temperature, durability, and weight requirements. To incorporate these materials in the high-temperature locations of the system, designers must overcome materials issues, such as differences in thermal expansion rates and lack of material ductility [21].

Designers are finding that one way to avoid cracking and buckling of high-temperature brittle components rigidly mounted in their support structures is to allow relative motion between the primary and supporting components [22]. Often this joint occurs in a location where differential pressures exist, requiring high-temperature seals. These seals or packings must exhibit the following important properties: operate hot [$\geq$705°C (1300°F)]; exhibit low

TABLE 1 GASKET AND ROPE SEAL MATERIALS

Fiber material	Maximum working temperature*	
	°F	°C
Graphite		
Oxidizing environment	1000	540
Reducing	5400	2980
Fiberglass (glass dependent)	1000	540
Superalloy metals (depending on alloy)	1300–1600	705–870
Oxide ceramics (Thompkins, 1955)[21]	1800*	980
Nextel 312 (62% Al_2O_3, 24% SiO_2, 14% B_2O_3)	2000*	1090
Nextel 440 (70% Al_2O_3, 28% SiO_2, 2% B_2O_3)	2100*	1150
Nextel 550 (73% Al_2O_3, 27% SiO_2)		

*Temperature at which fiber retains 50% (nominal) room-temperature strength.

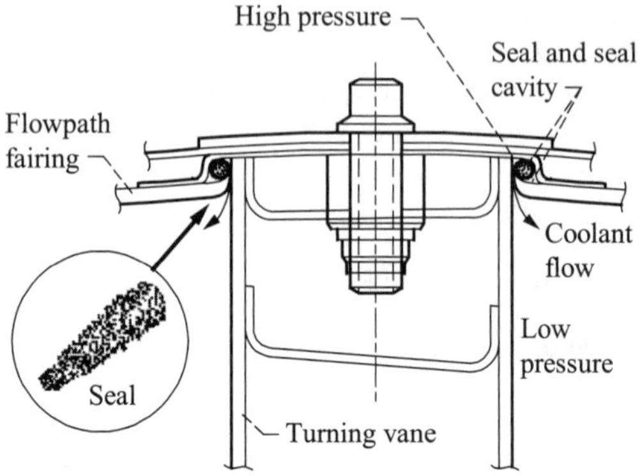

Fig. 13 Cross section of PW F119 engine showing last-stage turning vane with hybrid braided rope seal around perimeter [26].

leakage; resist mechanical scrubbing caused by differential thermal growth and acoustic loads; seal complex geometries; retain resilience after cycling; and support structural loads.

Braided rope seals can be made with a variety of materials and combinations, each having their own strengths and weaknesses. All ceramic designs consist of a ceramic fiber uniaxial core, over braided with ceramic sheath layers [23]. This design offers the potential for very high-temperature 1150°C (2100°F) operation. Researchers have determined, however, that all ceramic seals are susceptible to the vibratory and acoustic loadings present in turbine engines. These seals can also be ejected from the seal gland due to dynamic loading.

To improve upon structural integrity, Steinetz and Adams [22] developed a hybrid braided rope seal design that consists of uniaxial ceramic core fibers over braided with high-temperature superalloy wires. Tests have shown much greater resistance to abrasion and dynamic loadings. Wires made of HS-188 material show promise to 870°C (1600°F) temperatures. This hybrid construction was used to seal the last-stage articulated turning vane of the F119 turbine engine. The seal limits flow of fan cooling air past the turning vane flowpath (or powerstream)/fairing interface and also prevents backflow of potentially damaging high-temperature core air, as shown in Fig. 13 [22].

Researchers at NASA Glenn continue to strive for higher operating temperature hybrid seals. Recent oxidation studies by Opila et al. [24] showed that wires made from alumina forming scale base alloys (e.g., Plansee PM2000) could resist oxidation at temperatures to 1200°C (2200°F) for up to 70 h. Tests showed that alumina-forming alloys with reactive element additions performed best at 1200°C under all test conditions in the presence of oxygen, moisture, and temperature cycling. These wire samples exhibited slow growing and adherent oxide scales.

Dunlap et al. [25] provide experimental data on a 0.62-in.-diam rope seal that consisted of an Inconel X 750 spring tube filled with Saffil insulation (Saffil, Ltd.), and covered by two layers of Nextel 312 (3M Company) fabric wrap for operational temperatures to 815°C (1500°F). Steinetz and Dunlap [26] developed a braided carbon fiber thermal barrier that reduces solid rocket combustion gas leakage [3038°C (5500°F), in a nonoxidizing environment] and permits only relatively cool [<93°C (200°F)] gas to reach the elastomeric O-ring seals. See also Sec. IV.C, "Leaf and Wafer Seals."

In other related developments, Hendricks et al. [17] discussed the modeling and application of several types of brush seals, including hairpin woven or wrapped (hybrid), taconite, self-purging, and buffer.

III. DYNAMIC SEALS

The inherent unsteady nature of turbomachines requires coupled solutions of powerstream and sealing interfaces. For example, rotor-stator-cavity interactions affect seal leakage and passage blockage factor. Nonsteady pressure distribution due to blade-vane interaction perturbs the leakage flows, sucking them into the cavities near blade-vane coincidence and pumping them out near midcircumferential span position. These leakages lead to losses in component efficiency through injection of low momentum fluid into the powerstream, usually with more loss at hub; for smooth interfaces—for example, blade tip sealing—unsteady vortex formations dominate the losses. See also Appendix A for more details.

The flowfield about the tip of the blades is illustrated in Figs. 14 and 15 for the compressor and turbine, respectively [27]. At the leading edge the flow is forced out and around the stagnation region, then joins with the primary leakage zone and extends across the passage toward the low-pressure side, opposing the

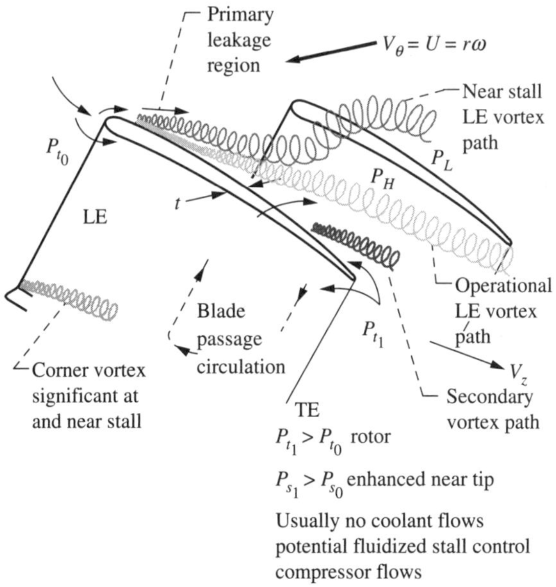

Fig. 14 Tip flow structure for an unshrouded compressor [27].

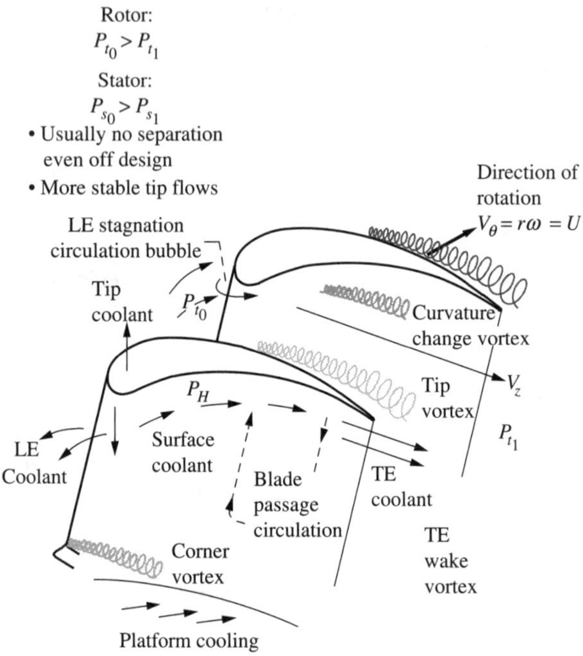

Fig. 15 Tip flow structure for an unshrouded turbine [27].

rotational velocity. These conditions are experimentally verified for tip clearance flows in the transonic compressor rotors and illustrated in Figs. 16 and 17 [28]. Usually, the flow in transonic compressors is subsonic by the time it reaches the third or fourth stage.

Blade tip flows and ensuing vortex patterns lead to flow losses, instabilities, and passage blockage. Without proper sealing, the flowfield can be reversed, resulting in compressor surge, and possible fire at the inlet. Flow losses in static elements, such as vanes in the compressor and turbine, have different sealing requirements, as described next. A few of these dynamic interfaces for aeroengine clearance control are illustrated in Fig. 2. More general flow details are found, for example, in Lakshminarayana [29] and for compressors in Copenhaver et al. [30], Strazisar et al. [31], and Wellborn and Okiishi [32], and for shrouded turbines in Bohn et al. [33].

A. ABRADABLES

For blade and knife tip sealing, an important class of materials known as abradables permit blade or knife rubbing without significant damage or wear to the rotating element while maintaining an effective sealing interface.

Early on, researchers recognized the need for abradable materials for blade tip and vane sealing (see, for example, Ludwig and Bill [4], Bill [9], Bill and Wisander [34], Bill et al. [35], Shiembob [14], Stocker et al. [36], Stocker [37], and Mahler [38]). As an example, schematics of three types of abradable materials with associated incursion types for the outer air-blade tip sealing interface in a compressor are illustrated in Fig. 18. These types of materials usually differ from the platform or inner shroud-drum rotor interface sealing of the compressor, as illustrated in Fig. 19.

As the name suggests, abradable seal materials are worn in by the rotating blade during service. They are applied to the casing of compressors, and gas and steam turbines, to decrease clearances to levels difficult to achieve by mechanical means. Abradable seals are gaining appeal in gas turbines as a relatively simple means to reduce gas-path clearances in both the compressor and turbine. They offer clearance reductions at relatively low cost and minor engineering implications for the service fleet. Abradable seals have been in use in aviation gas turbines since the late 1960s and early 1970s [39]. Although low energy and materials costs and long cycle time have in the past limited applications of abradable seals in land-based gas turbines, current operational requirements demand enhanced heat rate and reduced costs. With increasing fuel prices and advances in materials to allow extended service periods, abradable seals are gaining popularity within the power generation industry [40].

Without abradable seals, the cold clearances between blade or bucket tips and shrouds must be large enough to prevent significant contact during operation. Use of abradable seals allows the cold clearances to be reduced, with the assurance that if contact occurs the sacrificial component will be the abradable material on the stationary surface and

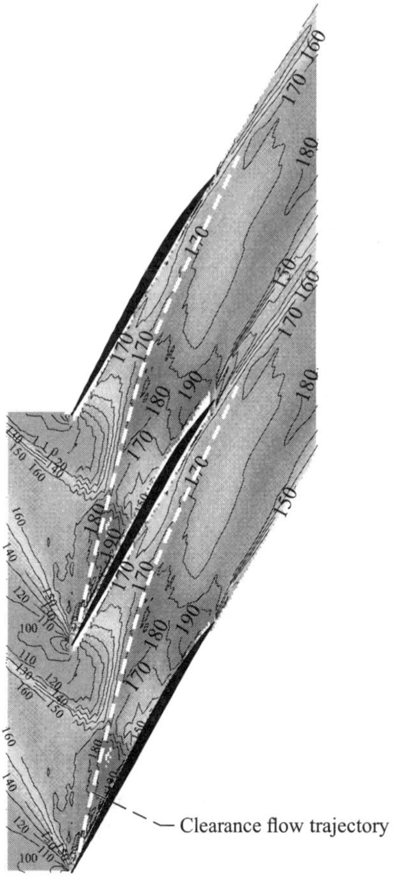

Fig. 16 Contours of axial velocity (m/s) on 92% span stream surface from LDV measurements of a transonic compressor rotor (no frame dependencies) [28].

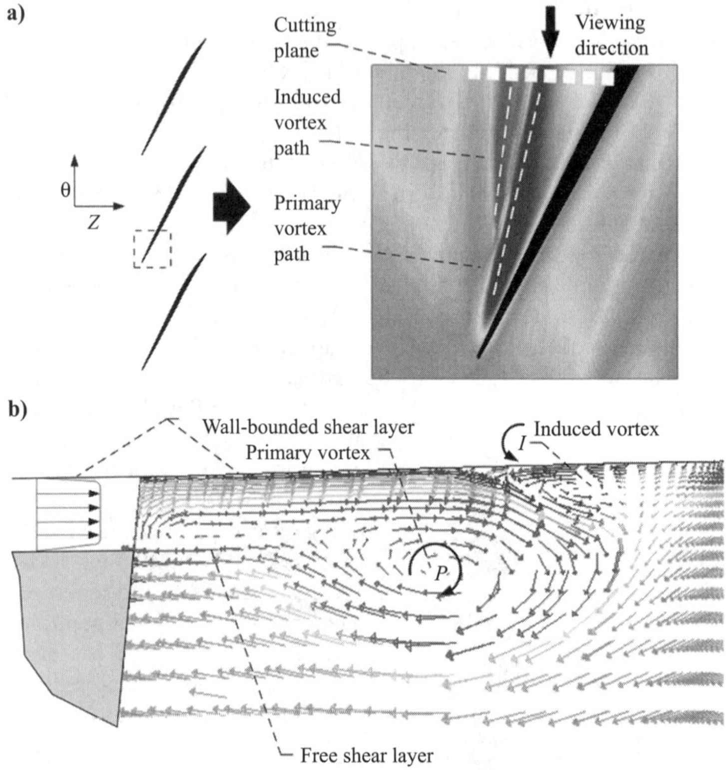

Fig. 17 Visualization of primary and induced clearance vortices in a transonic compressor rotor: a) axial velocity at 6% of clearance gap height from shroud; b) projection of relative velocity vectors on Z-r cutting plane as viewed in positive *u* direction, colored by *u*-component of velocity. Suction surface at right edge of figure. Note that the velocity profile, upper left of figure, is valid only for cascades [28].

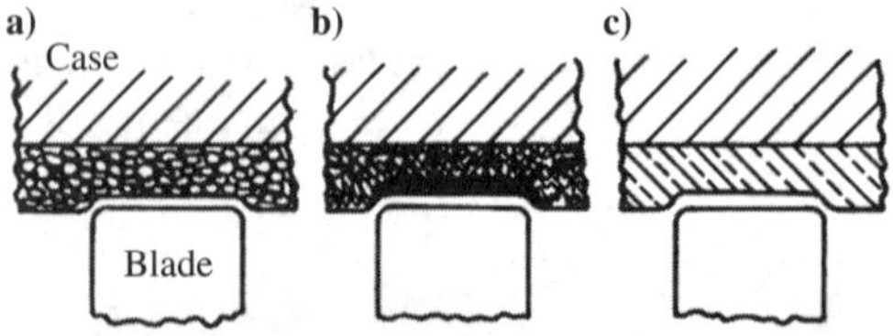

Fig. 18 Illustration of types of materials for interface outer air sealing: a) abradable (sintered or sprayed porous materials), b) compliant (porous material), and c) low shear strength (sprayed aluminum) [4].

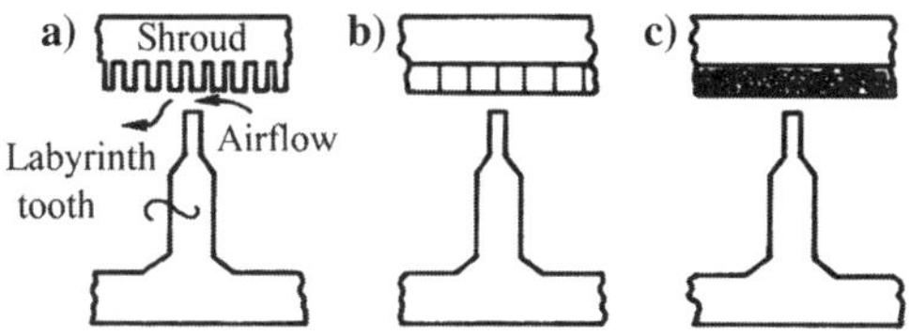

Fig. 19 Inner shrouds for compressor labyrinths: a) striated, b) honeycomb, and c) porous material (abradable or compliant) [4].

not the blade or bucket tips. Abradable seals also allow tighter clearances for common shroud or casing out-of-roundness conditions and rotor misalignment.

1. INTERFACE RUB

For properly designed abradables, if a rub occurs, the blade cuts into the sacrificial seal material with minimal distress to the blade. The abradable seal material mitigates blade wear while providing a durable interface that enhances engine efficiency. Controlled porosity shroud seal materials provide for low-energy material removal without damage to the rotating blade while mitigating leakage and enhancing seal life. Material release, porosity, and structural strength can be controlled in both thermal-sprayed coatings and fiber metals. Filler materials are often used to resist energy input to the shroud seal, mitigate case clearance distortion, and lubricate the wear interface. Worn material must be released to escape sliding contact wear of the blade tip (vs cutting action for an abradable) and plowing of the interface [41]. Asymmetric rubs generate hot spots that can develop into destructive seal drum instabilities. Such modes have destroyed engines and have been known to destroy aircraft with loss of life.

Many attempts have been made to study the wear mechanisms of abradable structures using conventional tribometers [42] or specially designed test rigs [43, 44]. Because of the high relative speeds between the abradable seal and the rotating blade tip surface, >100 m/s (>330 ft/s), however, the mechanisms of wear/cutting in this context differ considerably from low-speed tribology normally associated with machining operations. At high speeds, the removal/cutting of a thermal spray abradable coating is done by release of small particle debris, that is, <0.1 mm (<0.004 in.). In contrast to conventional (low-speed) cutting in machine tools, the particle debris released in abradable materials is ejected at the rear of the moving blade [45]. This, therefore, partly sets the criteria for the design of such materials. It also sets a limiting design criterion for blade-tip thickness. Generally, a cutting element (blade-tip) thickness of less than 1.3 mm (0.05 in.) allows release of the particles from the coating. Thicker tips tend to entrap the loose particles between the blade and the abradable material. As a result, special considerations have to be given to the design of the materials to allow for the cutting mechanisms (for example, altering the base material particle morphology and size).

Certain abradable materials rely more on densification (compaction) of the structure than on particle debris removal [46]. Material compaction limits the functional depth of the abradable material because the compacted material will cause increased wear of the rotating blade tips as the porosity is reduced. These types of seal materials include some of the thermal spray coatings and porous metal fiber structures (fiber metals). Fiber metals can be designed and constructed with varying fiber sizes and densities to alter their tribological behavior [47, 48].

2. INTERFACE MATERIALS

Abradable materials can be classed according to the means by which they are deformed. Some abradables are porous, so that the material wears away when rubbed by the blade tip. Some have a solid lubricant embedded to aid the wear process. Other abradables, such as honeycomb and fiber metal, deform at high speeds, and the cell walls rupture. For honeycomb, rotor wear is most pronounced at the brazed web where cell thickness doubles. Borel et al. [46] mapped incursion velocity as a function of tangential velocity, as shown in Fig. 20. These parameters

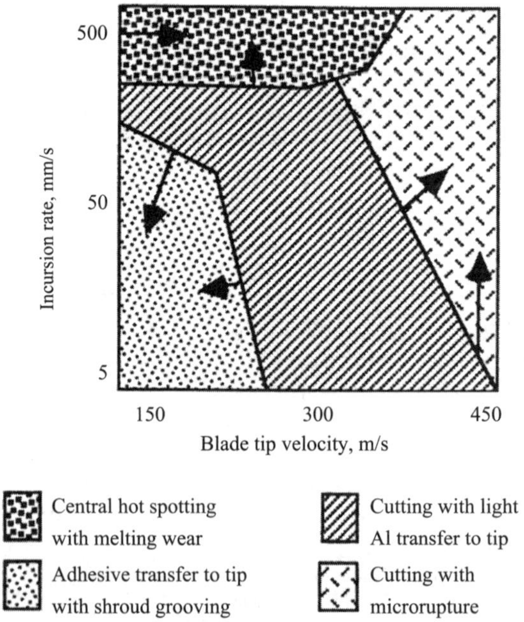

Fig. 20 Aluminum-silicon–polyester coating wear map using a 3-mm (0.12-in.) thick titanium blade at ambient temperature [46].

TABLE 2 ABRADABLE MATERIAL CLASSIFICATION

Temperature	Location/Material	Process
Low amb 400°C (750°F)	Fan or LPC AlSi + filler	Castings for polymer-based materials
Medium amb 760°C (1400°F)	LPC, HPC, LPT/Ni or Co base	Brazing or diffusion bonding for honeycomb and/or fiber metals
High 760°C (1400°F)–1150°C (2100°F)	HPT/YSZ and cBN or SiC	Thermal spray coatings for powdered composite materials

delineate regions of adhesive wear, melting wear, smearing, cutting, and adhesive titanium transfer from blade to interface. Abradable seals are generally classified according to their temperature range [49] but can also be characterized by method of application, as shown in Table 2 [41].

Different classes of thermally sprayed abradable coating materials behave tribologically differently. Traditionally, most of the powder metals available for low-temperature applications, that is, <400°C (<750°F), are aluminum-silicon based. To make them abradable, a second phase is added [49]. This phase is usually a polymeric material or a release agent, and is often called a solid lubricant. The role of the second phase in aluminum-silicon based abradable material is primarily to promote crack initiation within the structure. The size, morphology, quantity, and material of the second phase determine the wear mechanisms and abradability of the seal coating under various tribological conditions. The wear map in Fig. 20 is for an aluminum-silicon-polyester coating. The dominant wear mechanisms are different for various combinations of blade-tip velocity and incursion rate when rubbed by a 3-mm (0.12-in.) thick titanium blade at ambient temperature. The arrows indicate the movement of wear-mechanism boundaries when a polymer that is stiffer than polyester is used as the second phase.

Low-temperature abradables (generally epoxy materials) are used for fan tip sealing. Engine manufacturers' philosophy regarding fan rub strips is engine dependent. The PW4090, for example, uses a filled-honeycomb configuration, shown in Fig. 21a. The uneven rub caused by in-flight maneuvers can become, relatively speaking, quite deep, (tens of mils) which is difficult to see in the photo. The PW4000 and PW2000 have very similar labyrinth-style rub-strips, Fig. 21b. On the other hand, the CFM56 engine uses a smooth surface, which gets repotted during overhaul, and yet is usually not refurbished unless considerable damage has been incurred.

For temperature applications up to 760°C (1400°F), Ni- or Co-based alloy powders are commonly used as the basis of the abradable seal matrix. Other phases are added to the base metal powder to make the material abradable.

Fig. 21 Fan shroud rub strips: a) potted honeycomb PW4090 fan shroud and b) PW2000/4000 fan and rub strip interface (courtesy of Sherry Soditus, United Airlines Maintenance, San Francisco, CA).

a)

b)

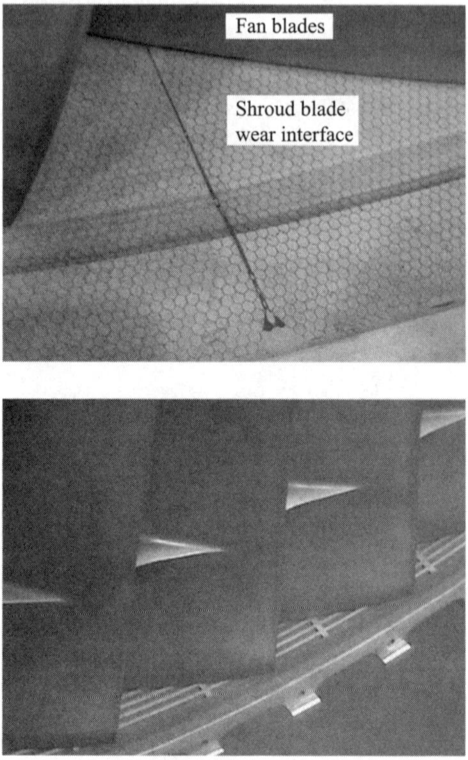

These added phases are polymeric materials that are fugitive elements and generate coating porosity and act as release agents [44, 50].

Figure 22 displays a wear map of a midtemperature coating system abraded at 500°C (930°F) using titanium blades. The map shows the wear mechanism domains vs blade-tip velocity and incursion rate. The arrows indicate the movement of the wear regime boundaries as the polyester level increases. As polyester content and thus porosity increases, cutting becomes the dominant mechanism over the entire range of the speeds and incursion rates. However, increasing porosity has a negative effect on coating cohesive strength and erosion properties.

Fiber metals are a type of midtemperature abradable. Like other abradables, abradability and erosion resistance present conflicting design demands, as illustrated in Fig. 23, and provide the seal designer with some flexibility [47]. Chappel et al. [47, 48] tested different fiber metals against other abradables for high- and low-speed abradability and erosion (see Table 3). The materials were then ranked per their performance (Table 4). Results showed that the high-strength fiber metal had the best performance overall, with the highest abradability and lowest erosion.

For operating temperatures above 760°C (1400°F), common practice is to use porous ceramics as the abradable material. The most widely used material is YSZ, which is usually mixed with a fugitive polymeric phase. There are a number of important considerations regarding porous ceramic abradable materials. To achieve an acceptable abradability, the cutting element/blade tip generally has to be reinforced with hard abrasive grits. Choosing these grits and processes to apply them has been the subject of numerous research activities, and there are

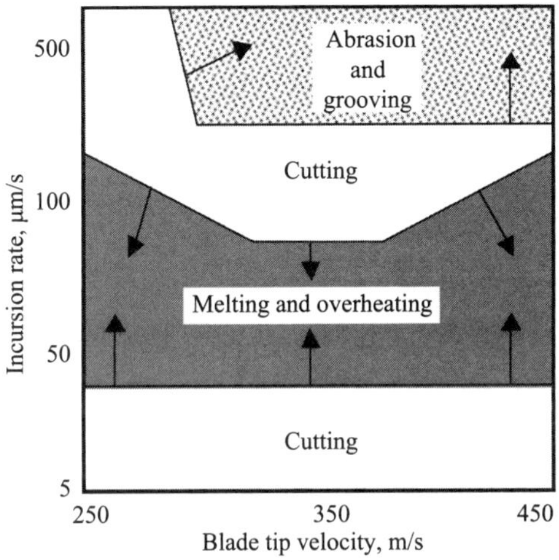

Fig. 22 Midtemperature abradable coating (CoNiCrAlY) wear map at 500°C (930°F) using titanium blades [49].

a number of patents that deal with this aspect of ceramic abradable materials [51–55]. Abrasive grits considered include cubic boron nitride (cBN), silicon carbide, aluminum oxide, and zirconium oxide. Published data suggest that cBN particles of a given size range tend to be the best abrasive medium against YSZ porous ceramics [51, 54]. Cubic boron nitride poses a high hardness (second to diamond) and a high sublimation temperature, >2980°C (>5400°F), which makes it an ideal candidate to abrade ceramic abradable materials. But cBN's relatively low oxidation temperature, ~850°C (~1560°F), allows it to function for only a limited time. This has prompted the use of other abrasives, such SiC [52, 55]. Despite successful functionality of SiC

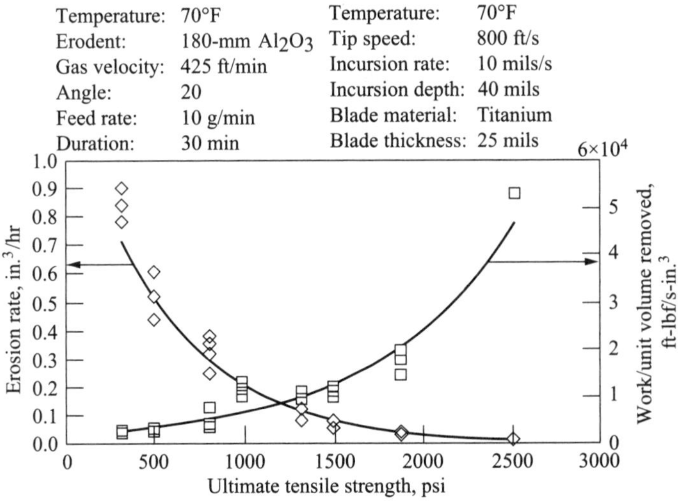

Fig. 23 Erosion and abradability as a function of ultimate tensile strength (courtesy of Technetics Corp.) [47].

TABLE 3 ABRADABLE MATERIALS USED BY CHAPPEL ET AL. [47]

Fibermetal	Density, %	Ultimate tensile strength, psi
1	22	1050
2	23	2150
Honeycomb	Hastelloy-X, 0.05-mm foil, 1.59-mm cell	
Nickel Graphite	Sulzer Metco 307NS (spray)	
CoNiCrAlY/hBN/PE*	Sulzer Metco 2043 (spray)	

*Hexagonal boron nitride (hBN) acts as a release agent; polyester (PE) controls porosity

against YSZ, SiC has been met with limited enthusiasm. SiC requires a diffusion barrier to prevent its reaction with transition metals at elevated temperatures [56]. This adds to the complexity and the cost of the abrasive system.

The ceramic abradable coating microstructure and its porosity are other essential considerations. Clearly, porosity increases the abradability of the coating. However, YSZ is strongly susceptible to high angle erosion because of its brittle nature [57], and adding porosity makes it prone to low angle erosion. Thermally sprayed porous YSZ coatings show different tribological behavior when compared to metallic abradable materials. They tend to show a strong influence of blade-tip velocity on abradability [45] (see Fig. 24). Abradability tends to improve with increasing blade-tip velocity. On the other hand, porous YSZ coatings show less dependency on incursion rate. They tend to have poor abradability at very low incursion rates, <0.005 mm/s (<0.2 mils/s), thus requiring blade-tip treatments.

An example application of a high-temperature abradable has been reported where the bill-of-material (BOM) first-stage turbine gas path shroud seals were

TABLE 4 WEAR RESISTANCE PERFORMANCE RANKINGS OF ABRADABLE MATERIALS [47]

Material	Abradability		Erosion
	High-speed	Low-speed	
1050-psi fiber metal	1	1	3
2150-psi fiber metal	1	1	1
Hastelloy-X honeycomb	2	3	2
Nickel graphite	3	1	2
CoNiCrAlY/hBN/PE	3	3	1

1 = best, and 3 = worst.

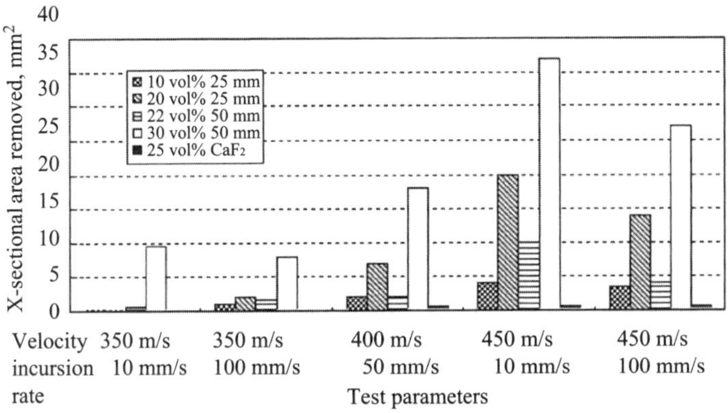

Fig. 24 Abradability of high-temperature materials by SiC tipped blades at 1025°C (1880°F). Right set of data (solid black bar) is for a CaF abradable; other data are for YSZ with polyamide. The *x* axis lists velocity and incursion rates. The legend gives porosity levels and average pore sizes after polyamide is burnt out [56].

coated with a porosity-controlled plasma-sprayed partially stabilized zirconia (PSZ) ceramic, as shown in Fig. 25 [58]. The coating was a 1-mm (0.040-in.) layer of ZrO_2-$8Y_2O_3$ over a 1-mm (0.040-in.) NiCoCrAl-based bond coat onto a Haynes 25 substrate. Characteristic "mudflat" cracking of the ceramic occurred at the blade interface, but backside seal temperature reductions over BOM-seals of 78°C (140°F) were measured, with gas path temperatures estimated over 1205°C (2200°F).

3. DESIGNING ABRADABLE MATERIALS FOR TURBOMACHINERY

Because abradable seals are low-strength structures that wear without damaging the mating blade tips, they are also susceptible to gas and solid particle erosion. Abradable structures intended for use in harsh temperatures occurring in gas turbines can also be prone to oxidation because of the inherent material porosity. These conflicting properties need to be accounted for in designing abradable seals. Abradable seals then have to be considered as a complete tribological system that incorporates

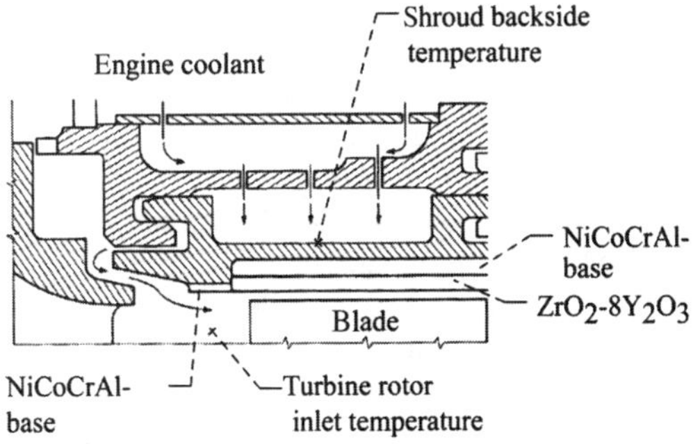

Fig. 25 Schematic of ceramic-coated shroud seal [58].

1) relative motions and depth of cut—blade-tip speed and incursion rate; 2) environment temperature, fluid medium and contaminants; 3) cutting element geometry and material—blade-tip thickness, shrouded or unshrouded blades; 4) counter element—abradable seal material and structure. Manufacturing processes as well as microstructural consistency can have a profound effect on the properties of abradable seals [47, 59].

Other issues to consider in the design of abradables for compressors are the large changes in thermal environment and the fact that titanium fires are not contained. Therefore, rubbing must release particulate matter without producing debris that will impact downstream components or cause a fire. In compressors, an abradable can be combined with intentional grooving to enhance stall margins, but clearance control or fluid injection may be a better method of controlling stall margin.

Considering all the preceding information suggests that the abradable system must be designed to suit the particular application. In other words, despite the availability of many off-the-shelf materials, abradable seals have to be modified or redesigned in most applications to meet the design constraints. More extensive lists of references on abradable seals and their use have been published elsewhere [45, 47, 49].

B. LABYRINTH SEALS

Labyrinth seals and their sealing principles are commonplace in turbomachinery and come in a variety of configurations. The most commonly used configurations are straight, interlocking, slanted, stepped, and combinations (Fig. 26) [60]. By their nature, labyrinth seals, usually mounted on the rotor, are clearance seals that can rub against their shroud interface, such as abradables and honeycomb (Fig. 27) [61]. They permit controlled leakages by dissipation of flow energy through a series of sequential aperture cavities (as sequential sharp edge orifices) with minimum heat rise and torque. The speed and pressure at which they operate is only limited by their structural design.

Principle design parameters include clearance and throttle (tooth or knife) and cavity geometry and tooth number (Fig. 28) [62]. The clearance is set by aerothermomechanical conditions that preclude contact with the shroud allowing for radial and axial excursions. The throttle tip is as thin as structurally feasible to mitigate heat propagation through the throttle-body into the shaft with a sharp leading edge (as an orifice) and is the primary flow restrictor. The angle at which the flow approaches the throttle is usually 90 deg, but slant throttles, into the flow, are more effective seals {Borda inlets, [$C_f = 0.5$, where C_f is the flow coefficient (actual flow/ideal isentropic flow)], are more restrictive than orifice inlets, [$C_f = 0.63$]}. One advantage of 90-deg throttles ($C_f = 0.63$) is the ability to seal flow reversals equally well; slant throttles are less effective handling flow reversals ($C_f = 0.8$ to 0.9). The cavity geometry is nearly 1:1, with axial spacing greater than six times the clearance and often shaped to

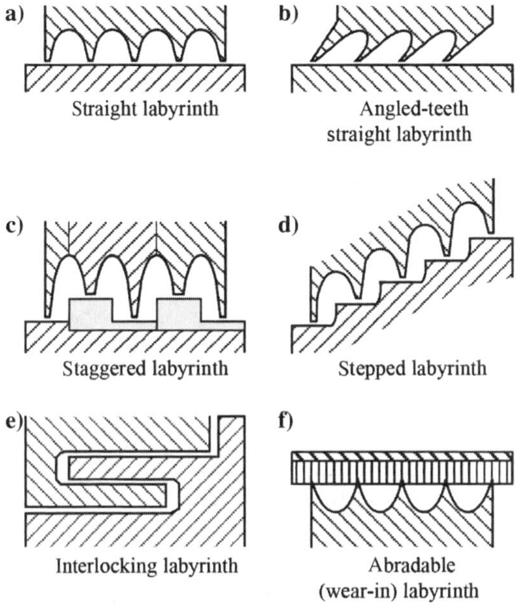

Fig. 26 Generalized labyrinth seal configurations [60].

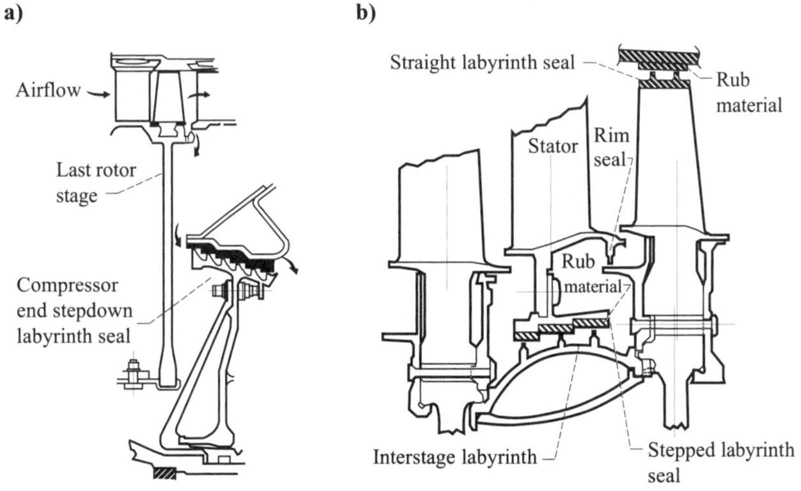

Fig. 27 Drum rotor labyrinth sealing configurations: a) compressor discharge stepdown labyrinth seal [61] and b) turbine interstage stepped labyrinth seal and shrouded-rotor straight labyrinth seal [4].

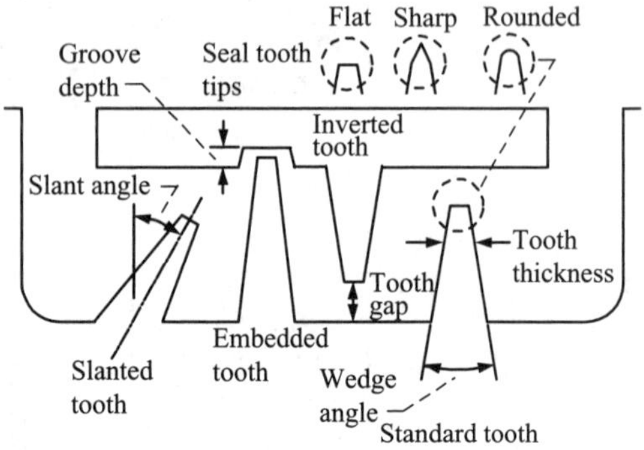

Fig. 28 Generalized schematics of labyrinth seal throttle configurations [63].

enhance flow dissipation through generation of vortices. A relation between the number of knife-cavity modules and leakage for developed cavity flows is given by

$$G_r/G_{r1} = N^{-0.4} \quad (1)$$

where G_r = mass flux and G_{r1} is the mass flux through a single throttle and N the number of throttles or cavity-throttle modules [63]; for gas throttles only, see Egli [64] for an equivalent relation. (Egli's interest was steam, but his work is applicable to gases in general.) Conditions relating the sharpness of the tooth to the ability to restrict flows are given by Mahler [38] (Fig. 29) and have been more recently explored using computational fluid dynamics (CFD) [62, 65].

Labyrinth seals are good at restricting the flow but do not respond well to dynamics and often lead to turbomachine instabilities. These problems have been addressed by several investigators, starting with Thomas [66] and Alford [67]. They recognized that the dynamic forces drove instabilities and heuristically

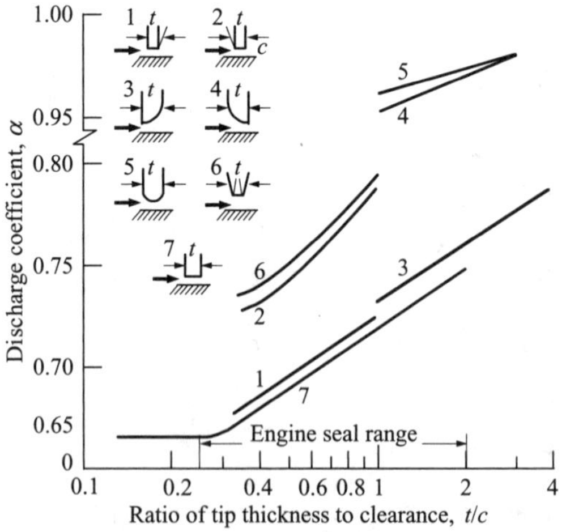

Fig. 29 Discharge coefficient as a function of knife-edge tooth shape [38].

determined stable operating configurations (Fig. 30 [68]). Benckert and Wachter [69], Childs et al. [70], and Muszynska [71] addressed the root causes and introduced the swirl brake at the seal inlet to mitigate the circumferential velocity component within the cavities (Fig. 31). More recently, circumferential flow blocks and flow slots have been introduced to mitigate the circumferential velocity component (Figs. 32 and 33) [12, 72–74].

Labyrinth seals have a lengthy history of proven reliability, with robust operation and well-developed technology, and are well-suited for abradable interfaces. Their tendency to engender instabilities can be controlled by swirl brakes or intracavity slots or blocks and drum dampers. Nearly all turbomachines rely on labyrinth seals or labyrinth sealing principles (Egli [64], Trutnovsky [73], Stocker et al. [36], and Stocker [37]). In general, nearly all sealing applications rely heavily on the essential features of sharp-edge flow restrictors {e.g., 1) the aspirating seal (Sec. IV.F "Aspirating Seals"), which combines a labyrinth tooth and the face-sealing dam [74], Figs. 34a and 34b, and 2)

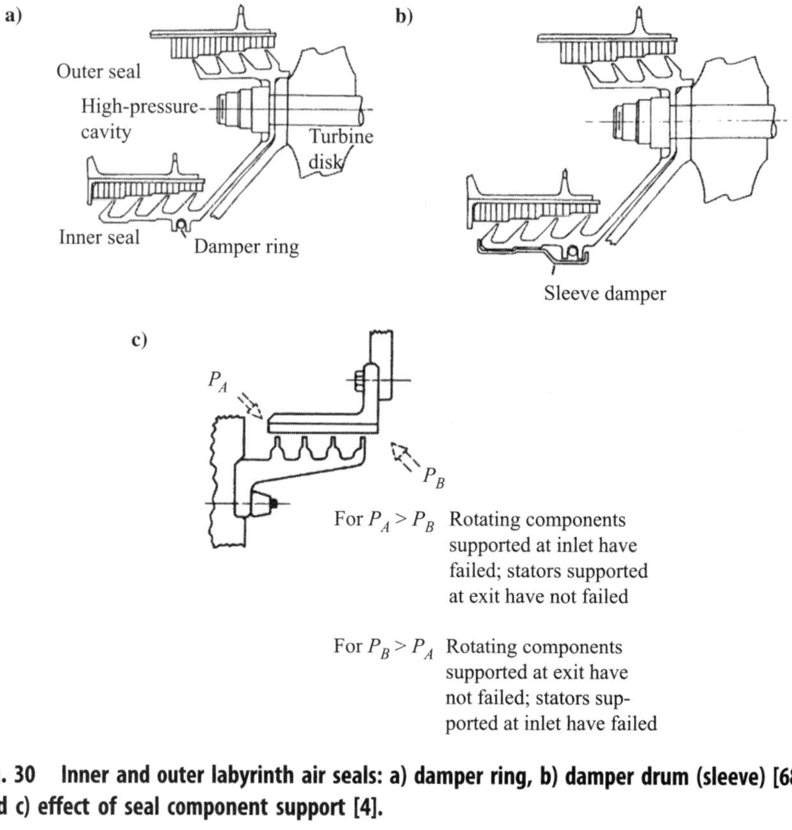

Fig. 30 Inner and outer labyrinth air seals: a) damper ring, b) damper drum (sleeve) [68], and c) effect of seal component support [4].

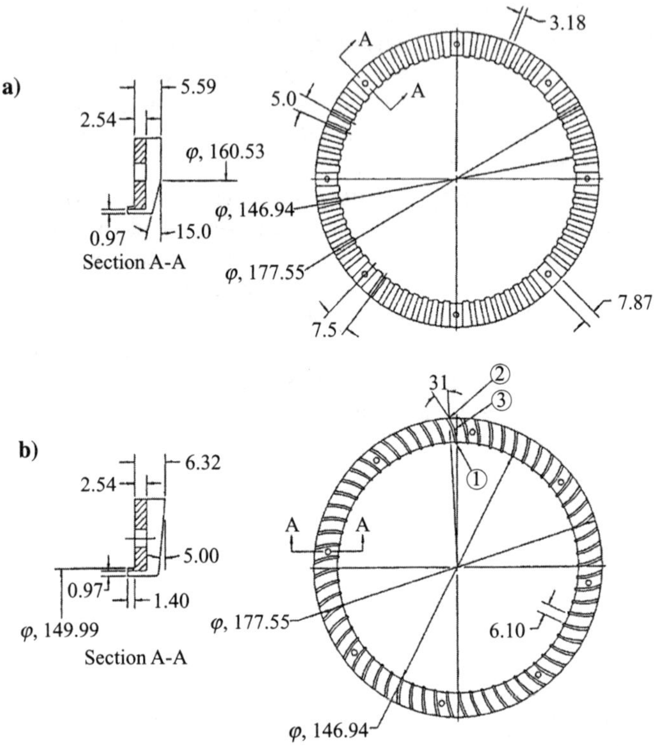

Notes:
1. At point ①, vane face to be tangent to a radial line
2. At point ②, tangent to vane face to be 31 from a radial line as shown
3. Between points ① and ②, vane face should have a constant radius

Fig. 31 Typical swirl brake configurations applied at the inlet to a labyrinth seal: a) radial swirl brake and b) improved swirl brake [70].

dynamic seals with inlet throttles that confine flows over the honeycomb land, Fig. 35) [11, 12]}.

C. BRUSH SEALS

As described by Ferguson [75], the brush seal is the first simple, practical alternative to the finned labyrinth seal that offers extensive performance improvements. Benefits of brush seals over labyrinth seals include 1) reduced leakage compared to labyrinth seals (upwards of 50% possible); 2) accommodation of shaft excursions

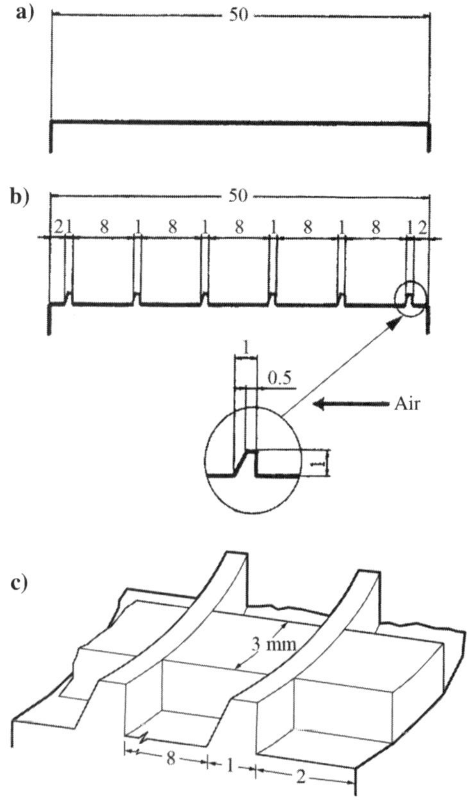

Fig. 32 Labyrinth circumferential flow blocks: a) annular seal, b) labyrinth seal, and c) flow blocks [72].

due to stop/start operations, rotordynamics, and other transient conditions (labyrinth seals often incur permanent clearance increases under such conditions, degrading seal, and machine performance); 3) significantly smaller axial space requirements than labyrinth seal; and 4) more stable leakage characteristics over long operating periods. Brush seals have matured significantly over the past 20 years. Typical operating conditions of state-of-the-art brush seals are shown in Table 5. (Data available at http://www.fluidsciences.perkinelemer.com/turbomachinery and http://www.crossmanufacturing.com.)

Brush seal construction is deceptively simple, requiring the well-ordered layering or tufting of fine-diameter bristles into a dense pack that compensates for circumferential differences between inside and outside diameters (Figs. 36 and 37). This pack is sandwiched and welded (or brazed) between a backing ring (downstream side) and sideplate (upstream side) and then stress relieved to ensure stability and flatness. The weld on the seal outer diameter is machined to form a close-tolerance outer diameter-sealing surface to fit into a suitable housing. The wire bristles

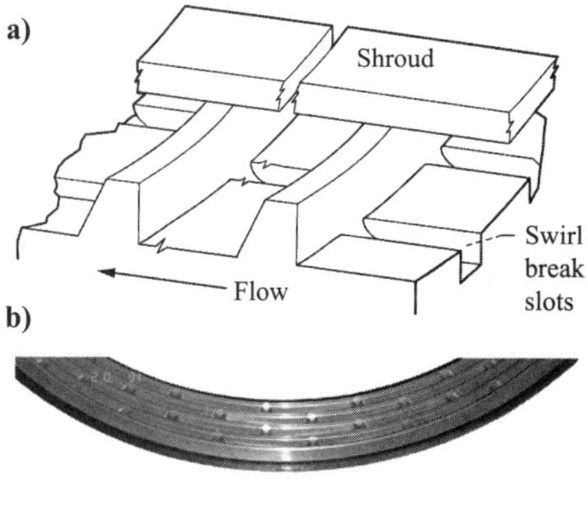

Fig. 33 Web seal with circumferential flow blocking slots: a) conceptual web seal sketch and b) photograph of web seal [12].

Fig. 34 Aspirating seal labyrinth tooth and seal dam sharp-edge flow restrictor: a) at shutdown phase and b) at steady-state operation [74].

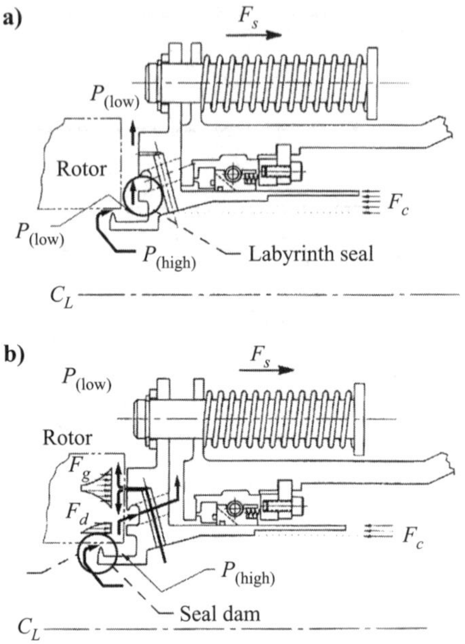

protrude radially inward (shaft-rotor) or outward (drum-rotor) and are machined to fit the mating rotor, with slight interference. Brush seal interferences (preload) must be properly selected to prevent catastrophic overheating of the rotor, seal destruction, and excessive rotor thermal growths.

To accommodate anticipated radial shaft movements, the bristles must bend. To allow the bristles to bend without buckling, the wires are oriented at an angle (typically 45 to 55 deg) to a radial line through the rotor. The bristles are canted in the direction of rotor rotation. The bristle lay angle also facilitates seal installation because of the slight interference between the bristle pack and the rotor. The backing ring provides structural support to the otherwise flexible bristles and assists the seal in limiting leakage. To minimize brush seal hysteresis caused by brush bristle binding on the backplate, features with various forms of relief are added to the backing ring, for example, a recessed pocket and seal dam. The recessed pocket assists with pressure balancing of the seal, and the relatively small contact area at the seal dam minimizes friction, allowing the bristles to follow the speed-dependent shaft growths. The bristle free-radial-length and packing pattern are selected to accommodate radial shaft movements while operating within the wire's elastic range

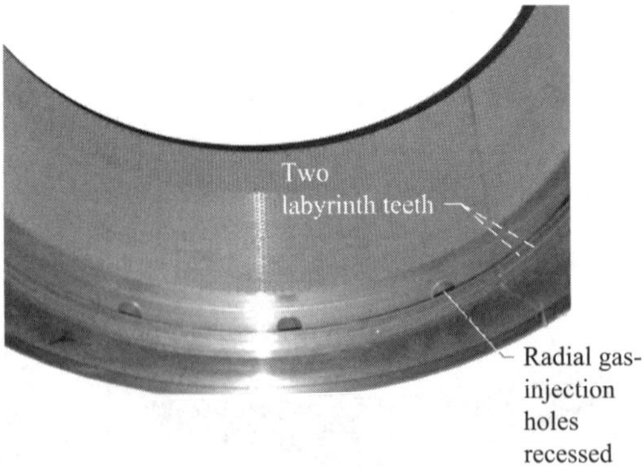

Fig. 35 Primary labyrinth throttle confining flows to the honeycomb journal land [12].

TABLE 5 TYPICAL OPERATING LIMITS FOR STATE-OF-THE-ART BRUSH SEALS

Differential pressure	up to 300 psid per stage	2.1 MPa
Surface speed	up to 1200 ft/s	400 m/s
Operating temperature	up to 1200°F	600°C
Size (diameter range)	up to 120-in.	3.1 m

See http://www.fluidsciences.perkinelemer.com/turbomachinery

at temperature. A number of brush seal manufacturers include some form of flow deflector (e.g., see flexi-front plate in Figs. 36 and 37) on the high-pressure side of the wire bristles. (Data available online at http://www.crossmanufacturing.com.) This element aids in mitigating the radial pressure closing loads (sometimes known as "pressure closing") caused by air forces urging the bristles against the shaft. This element can also aid in reducing installation damage, bristle flutter in highly turbulent flowfields, and FOD.

Brush seals, initially developed for aerogas turbines, have also been used in industrial gas and steam turbines since the 1990s. Design similitude as well as analysis and modeling of brush and woven seals were established earlier in the works of Flower [76] and Hendricks et al. [17]. Here we will address a few sealing types, their locations, and their material constraints. For further details, see Steinetz et al. [77] and Hendricks and et al. [27, 78] and NASA Conference Publications [79, 80]. An extensive summary of brush seal research and development work through 1995 has been published [81, 82] and updated in a more recent summary [40].

1. BRUSH SEAL DESIGN CONSIDERATIONS

To properly design and specify brush seals for a given application, many design factors must be considered and balanced. Comprehensive brush seal design algorithms have been proposed by Chupp et al. [40], Dinc et al. [83], and Hendricks et al. [17]. An iterative process must be followed to satisfy seal basic geometry, stress, thermal (especially during transient rub conditions), leakage, and life constraints to

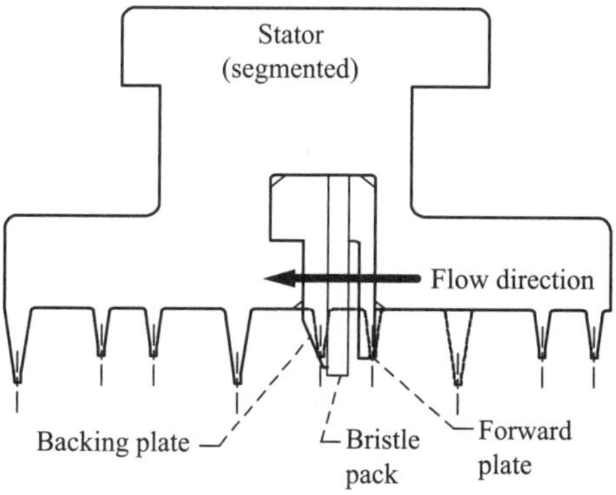

Fig. 36 Typical brush seal configuration and geometric features [40, 83].

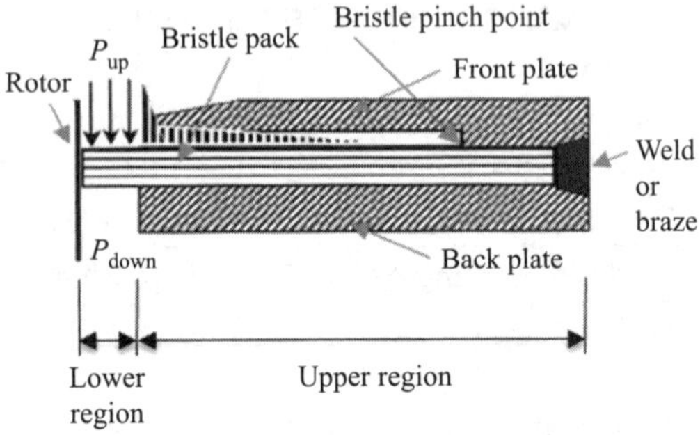

Fig. 37 Brush seal design for steam turbine applications [40, 83].

arrive at an acceptable design within the engine's operations envelope. Many of the characteristics that must be considered and understood for a successful brush seal design are discussed by Dinc et al. [83]. Factors to consider include the following: pressure capability, seal upstream protection, frequency, seal high- and low-cycle fatigue (HCF, LCF) analysis, seal leakage, seal oxidation, seal stiffness, seal creep, seal blowdown (pressure closing effect), seal wear, bristle-tip forces and pressure stiffening effect, solid particle erosion, seal heat generation, reverse rotation, bristle-tip temperature, seal life considerations, rotor dynamics, performance predictions, rotor thermal stability, oil sealing, secondary flow and cavity flow (including swirl flow), and shaft considerations. Design criteria are required for each of the different potential failure modes, including stress, fatigue life, creep life, wear life, and oxidation life. Several important designs parameters are discussed next.

a. Material selection

Materials in rubbing contact in brush seal installations must have sufficient wear resistance to satisfy engine operating envelope and durability requirements. A proper material selection requires knowledge of the rotor and seal materials and their interactions. In addition to good wear characteristics, the seal material must have acceptable creep and oxidation properties.

Metallic bristles. Brush seal wire bristles range in diameter from 0.071 mm (0.0028 in.) (for low pressures) to 0.15 mm (0.006 in.) (for high pressures). The most commonly used material for brush seals is the cobalt-based alloy Haynes 25 because of its good wear and oxidation characteristics. Brush seals are generally run against a smooth, hard-face coating to minimize shaft wear and the chances of wear-induced cracks, which might affect the structural integrity of the rotor. The usual coatings selected for aircraft applications are ceramic, including chromium carbide and aluminum oxide. Selecting the correct mating wire and shaft surface finish for a given application can reduce frictional heating and extend seal life through reduced oxidation and wear. There is no general requirement for coating industrial gas and steam turbine rotor surfaces, as the rotor thicknesses are much greater than in aircraft applications.

Nonmetallic bristles. High-speed turbine designers have long wondered if brush seals could replace labyrinth seals in bearing sump locations. Brush seals would mitigate traditional labyrinth seal clearance opening and corresponding increased leakage. Issues slowing early application of brush seals in these locations included coking (carburization of oil particles at excessively high temperatures), metallic particle damage of precision rolling element bearings, and potential for fires. Development efforts have found success in applying aramid bristles for certain bearing sump locations [82, 84]. While aramid bristles are an improvement over standard bristles in low-temperature applications, they do not have the low leakage, thermal or high pressure drop, afforded by self-acting film riding face seals, for example, Fig. 38. Advantages of the aramid bristles include stable properties up to 150°C (300°F) operating temperatures, negligible shrinkage and moisture absorption, lower wear than Haynes 25 up to 150°C, lower leakage (due to smaller 12-μm-diam fibers), and resistance to coking [82]. Based on laboratory demonstration, the aramid fiber seals were installed in a GE 7EA frame (#1) inlet bearing sealing location. Preliminary field data showed that the nonmetallic brush seal maintained a higher pressure difference between the air and bearing drain cavities and enhanced the effectiveness of the sealing system, allowing fewer oil particles to migrate out of the bearing. These types of nonmetallic brush seals will be discussed further in Sec. IV.E, "Oil Brush Seals" and Appendix C.

b. Seal fence height

A key design issue is the required radial gap (fence height) between the backing ring and the rotor surface. Following detailed secondary flow, heat transfer, and mechanical analyses, fence height is determined by the relative transient growth characteristics of the rotor vs the stator and rotordynamic considerations. This backing ring gap is designed to avoid contact with the rotor surface during any operating condition with an assumed set of dimensional variations. Consequently, the successful design of an effective brush seal hinges on a thorough knowledge of the turbine behavior, operating conditions, and design of surrounding parts.

c. Brush pack considerations

Depending on required sealing pressure differentials and life, wire bristle diameters change [86]. Better load and wear properties are found with larger bristle diameters. Bristle pack thicknesses also vary depending on application: the higher the pressure differential, the greater the pack thickness. Higher-pressure applications require bristle packs with higher axial stiffness to prevent the bristles from blowing under the backing ring. Dinc et al. [83] have developed brush seals that have operated at air pressures up to 2.76 MPa (400 psid) in a single stage. Brush seals have been made in very large diameters. Large brush seals, especially for ground power applications, are often made segmented to

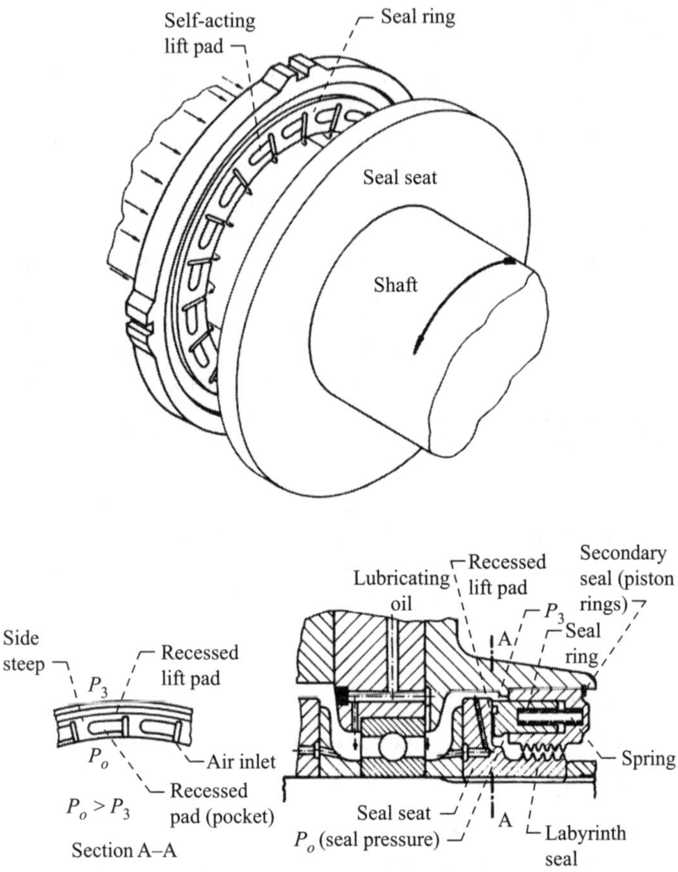

Fig. 38 Self-acting face seal with labyrinth seal presealing [85].

allow easy assembly and disassembly, especially on machines where the shaft stays in place during refurbishment.

d. Seal stress/pressure capability

Pressure capacity is another important brush seal design parameter. The overall pressure drop establishes the seal bristle diameter, bristle density, and the number of brush seals in series. In a bristle pack, all bristles are essentially cantilever beams held at the pinch point by a front plate and supported by the back plate. From a loading point of view, the bristles can be separated into two regions (see Fig. 37): the lower part, fence region, between the rotor surface and the back plate inner diameter (ID), and the upper part, from the back plate ID to the bristle pinch point. The innermost radial portion carries the main pressure load and is the main source of the seal stress [87]. In addition to the mean bending stress, contact stress at the bristle-back plate interface must be considered. Furthermore, bristle stress is a very strong function of the fence height set by the expected relative radial movement of the rotor and seal. Figure 39 shows a diagram illustrating design considerations for seal stress and deflection analysis and includes a list of the controllable and noncontrollable design parameters. As a word of caution, care must be taken in using multiple brush configurations, as pressure drop capability becomes more nonlinear with fluid compressibility and most of the pressure drop or bristle pressure loading is carried by the downstream brush (first and last "throttling" similar to labyrinth seals).

e. Heat generation/bristle tip temperature

As the brush seal bristles rub against the rotor surface, frictional heat is created and must be dissipated through convection and conduction. This is quite similar to the classic Blok problem [88], where extensive heating occurs at the sliding interface. Brush seal frictional heating was addressed by Hendricks et al. [17, 89] and modeled as fin in crossflow with a heat source at the tip by Dogu and Aksit [90]. If the seal is not properly designed, this heating can lead to premature bristle loss, or, worse, the rotor/seal operation could become thermally unstable. The latter condition occurs when the rotor grows radially into the stator, increasing the frictional heating, leading to additional rotor growth, until

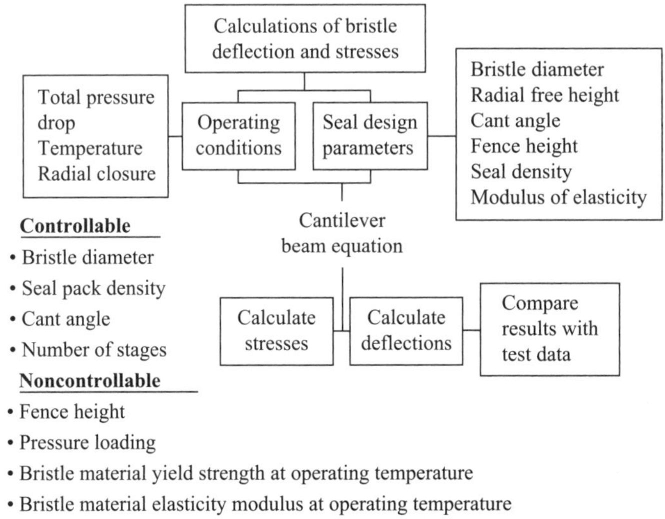

Fig. 39 Bristle stress/deflection analysis [40, 83].

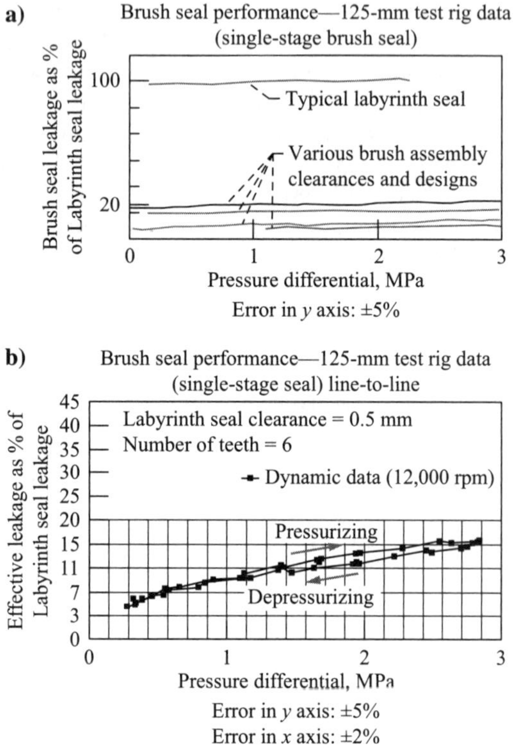

Fig. 40 Brush seal performance as compared to labyrinth seal. Representative brush seal leakage data compared to a typical, 15-tooth, 0.5-mm (20-mil) clearance labyrinth seal. Measured brush seal leakage characteristic with increasing and decreasing pressure drop compared to a typical, 6-tooth, 0.5-mm (20-mil) clearance labyrinth seal [40].

the rotor rubs the seal backing plate, resulting in component failure. In some turbine designs, brush seals are often assembled with a clearance to preclude excessive interference and heating during thermal and speed transients. These mechanical design issues significantly affect the range of feasible applications for brush seals. Many of these issues have been addressed by Dinc et al. [83] and Soditus [91].

f. Seal leakage

Leakage characterization of brush seals typically consists of a series of tests at varying levels of bristle-to-rotor interference or clearance, as shown in Figs. 40 and 41. Static (nonrotating) tests are run to get an approximate level of seal leakage and pressure capability. They are followed by dynamic (rotating) tests to provide a more accurate simulation of seal behavior. Rotating tests also reveal rotor dynamics effects, an important consideration for steam turbine rotors and turbomachines in general, that can be sensitive to radial rubs due to nonuniform heat generation.

Proctor and Delgado studied the effects of speed [up to 365 m/s (1200 ft/s)], temperature [up to 650°C (1200°F)], and pressure [up to 0.52 MPa (75 psid)] on brush seal and finger seal leakage and power loss [92]. They determined that leakage generally decreased with increasing speed. Leakage decreases somewhat with increasing surface speed because circumferential flow is enhanced and the rotor diameter increases; changes in diameter causes both a decrease in the effective seal clearance and an increase in contact stresses (important in wear and surface heating).

g. Other considerations

If not properly considered, brush seals can exhibit three other phenomena deserving some discussion. These include seal "hysteresis," "bristle stiffening," and "pressure closing." As described by Short et al. [86] and Basu et al. [93], after the rotor moves into the bristle pack (due to radial excursions or thermal growths), the displaced bristles do not immediately recover against the frictional forces between them and the backing ring. As a result, a significant leakage increase (more than double) was observed following rotor movement [93]. This leakage hysteresis can exist until after the pressure load is removed (that is, after the engine is shut down). Furthermore, if the bristle pack is not properly designed, the seal can exhibit a considerable stiffening effect with application of pressure. This phenomenon results from interbristle friction loads making it more difficult for the brush bristles to flex during shaft excursions. Air leaking through the seal also exerts a radially inward force on the bristles, resulting in what has been termed "pressure closing" or bristle "blowdown." This extra contact load, especially on the upstream side of the brush, affects the life of the seal (upstream bristles are worn in either a scalloped or coned configuration) and higher interface contact pressure. By measuring baseline seal leakage in a line-to-line (zero clearance) assembly configuration, bristle blowdown for varying loads of assembly clearance can be inferred from leakage data. Blowdown may change with seal design parameters like cant angle, free bristle height, fence height, etc. For the sample case shown in Fig. 41, seal leakage rate does not show much change up to 0.1 mm (4 mils) radial clearance, which indicates some bristle blowdown to close this seal-rotor gap.

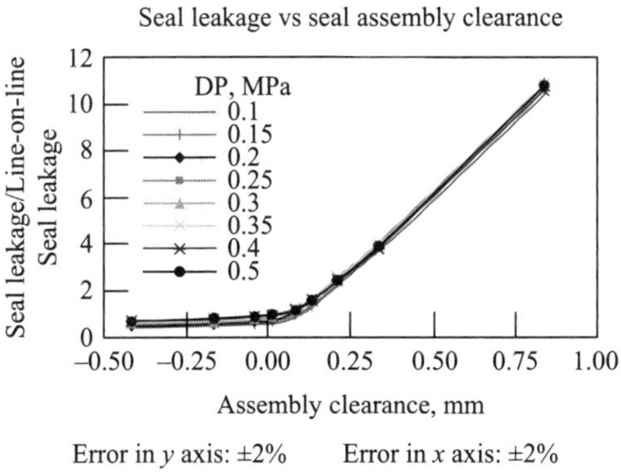

Fig. 41 Measured brush seal leakage for interference and clearance conditions [40].

2. BRUSH SEAL FLOW MODELING

Brush seal flow modeling is complicated by several factors unique to porous structures, in that the leakage depends on the seal porosity, which depends on the pressure drop across the seal. Flow through the seal travels perpendicularly to the brush pack, through the annulus formed between the backing ring bore and the shaft diameter. The flow is directed radially inward towards the shaft as it flows around individual bristles and collides with the bristles downstream in adjacent rows of the pack and finally between the bristle tips and the shaft [17].

A flow model proposed by Holle et al. [94] uses a single parameter, effective brush thickness, to correlate the flows through the seal. Corrections for density ρ and Re must be included:

For laminar flow ($Re_v \leq 100$),

$$G_{\max}^2 = \frac{2g_c}{180(144)}(\Delta P)\rho\, Re_v \frac{HD_{v,m}}{B}\left(\frac{S_{T,m}}{HD_{v,m}}\right)^{0.4}\left(\frac{S_{T,m}}{S'_{L,m}}\right)^{0.6} \tag{2}$$

From the characteristic point 1, the postlaminar friction factor for the interpolation in the transition flow region ($Re_v > 100$ and $Re_b < 5000$)

$$f_{KL_1} = \frac{g_c}{2(144)}\frac{(\Delta P)\rho}{G_{\max_1}^2}\frac{HD'_m}{B} \tag{3}$$

The symbols are defined in Holle et al. [94].

The basics of brush sealing original model was developed by Hendricks et al. [17] for vehicle tribology applications with reference to the models of Flower [76, 95].

Variation in seal porosity with pressure difference is accounted for by normalizing the varying brush thicknesses by a minimum or ideal brush thickness. Maximum seal flow rates are computed by using an iterative procedure that has converged when the difference in successive iterations for the flow rate is less than a preset tolerance.

Flow models proposed by Hendricks et al. [17, 89, 96] are based on a bulk average flow through the porous media. These models account for brush

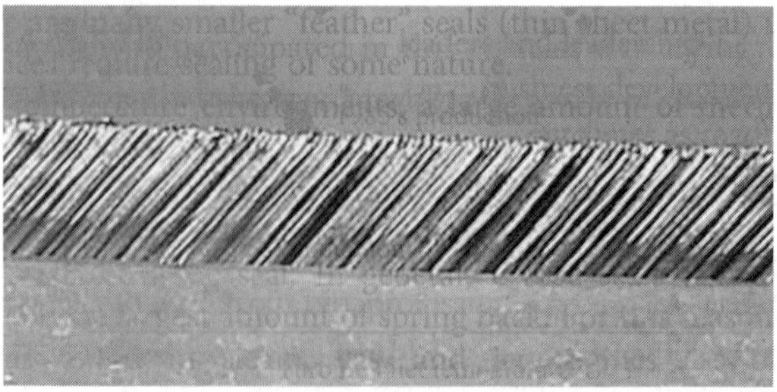

Fig. 42 7EA gas-turbine high-pressure packing brush seal in good condition after 22,000 h of operation [40, 83].

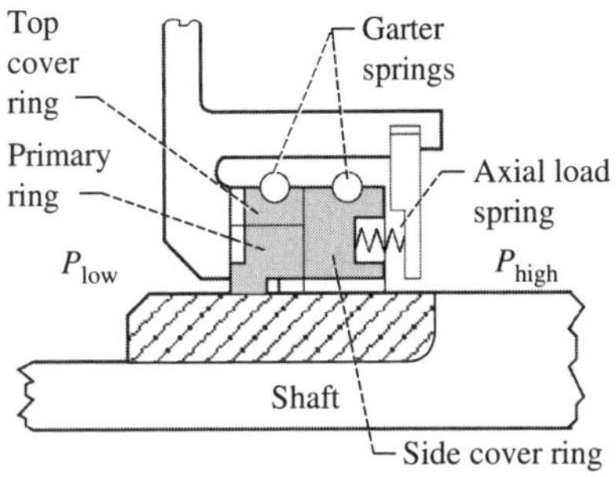

Fig. 43 Shaft riding or circumferential contact seal [107].

porosity, bristle loading and deformation, brush geometry parameters interface heating, and multiple flowpaths. In one model it is shown that flow through a brush configuration can be simulated using an electrical analog with driving potential (pressure drop), current (mass flow), and resistance (flow losses, friction, and momentum) as the key variables. All of the just-mentioned brush flow models require some empirical data to establish correlation constants. Once the constants are established, the models can predict brush seal flow reasonably well.

A number of researchers have applied numerical techniques to model brush seal flows and bristle pressure loadings [97–100]. Though these models are more complex, they permit a more detailed investigation of the subtleties of flow and stresses within the brush pack.

3. APPLICATIONS

Brush seals are currently seeing extensive service in both commercial and military turbine engines. Lower leakage brush seals permit better management of cavity flows and significant reductions in specific fuel consumption when compared to competing labyrinth seals. Allison Engines (now Rolls Royce) has implemented brush seals for the Saab 2000, Cesna Citation-X, and V-22 Osprey. General Electric has implemented a number of brush seals in the balance piston region of the GE90 engine for the Boeing 777 aircraft. Pratt & Whitney has entered revenue service with brush seals in three locations on the PW1468 for Airbus aircraft and on the PW4084 for the Boeing 777 aircraft [91, 101].

Brush seals are being retrofitted into ground-based turbines and are also being combined with labyrinth seals to greatly improve turbine power output and heat rate [40, 83, 102–106]. Dinc et al. [83] report that incorporating brush seals in a GE Frame 7EA turbine in the high-pressure packing location increased power output by 1.0% and decreased heat rate by 0.5%. Figure 42 is a photo of a representative brush seal taken during a routine inspection. The seal is in good condition after nearly three years of operation (~22,000 h). As of this writing, more than 200 brush seals have been installed in GE industrial gas turbines in the compressor discharge high-pressure packing (HPP), middle bearing, and turbine interstage locations. Field data and experience from these installations have validated the brush seal design technology. Using brush seals in the interstage location resulted in similar improvements. Brush seals have proven effective for service lives of up to 40,000 h [83].

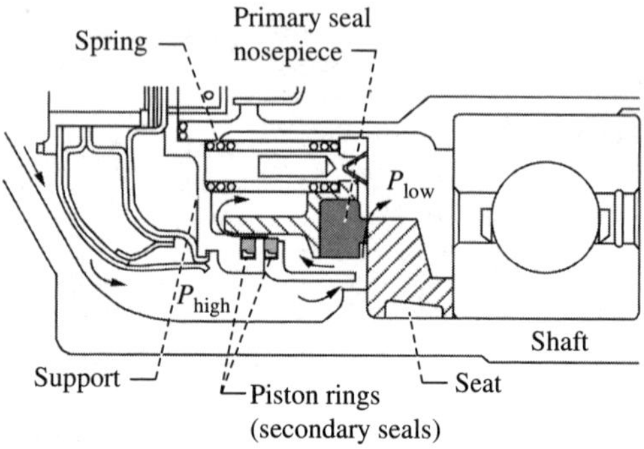

Fig. 44 Positive contact face seal [108].

D. FACE SEALS

Labyrinth seals are less impacted by FOD debris than other types of seals, yet also pass that debris to other components, such as bearing cavities. One of the major functions of face and buffer sealing is to preclude debris from entering bearing or gear-box oil; an equally important function is to prevent oil vapors from leaking into the wheel-space and from entering the aircraft cabin air stream. Debris in the bearing or gear-box oil can radically shorten life, and oil vapor in the wheel space can cause fire or explosions. Oil vapors in the cabin are unacceptable to the air traveler [107].

Face seals are classified as mechanical seals. They are pressure-balanced contact or self-acting seals. The key components are the primary ring (stator) or nosepiece, seat or runner (rotor), spring or bellows preloader assembly, garter or retainer springs, secondary seal, and housing (Figs. 43 and 44) [108, 109]. There is a wealth of information on the experimental data for, design of, and application of mechanical seals in the literature, from technical publications to books (for example, see Ludwig and Bill [4] and Lebeck [110]).

For the face seal, the geometry of the ring or nosepiece is critical. For successful face sealing, the forces due to system pressure, sealing dam pressure, and the spring or bellows must be properly balanced and stable over a range of operating parameters (pressure, temperature, surface speed) (Fig. 45) [111].

Contact seals wear, and are generally limited to surface speeds less than 76 m/s (250 ft/s). To mitigate the wear, prolonging life and decreased leakage, Ludwig [112] and Dini [113] promoted the self-acting Rayleigh step and spiral groove seal (Figs. 38, 46, and 47). A labyrinth seal or a simple projection representing a single throttle is used for presealing to control excessive leakage should the dam of the face seal "pop" open, for example, the labyrinth preseal as is illustrated in Fig. 38 (and in the aspirating seal, see Sec. IV.E). Spiral groove (Fig. 47),

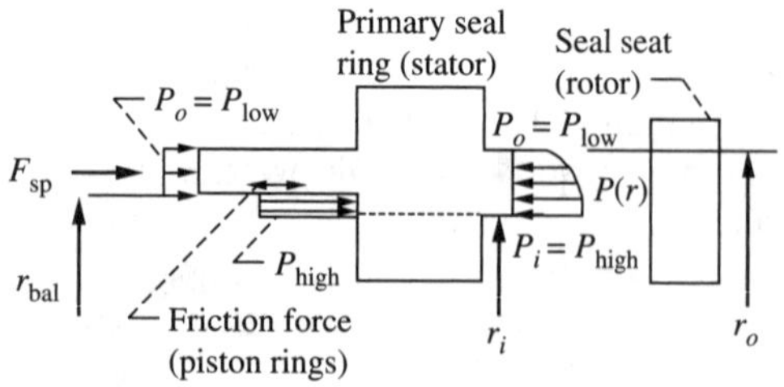

Fig. 45 Pressure balancing forces in face sealing [111].

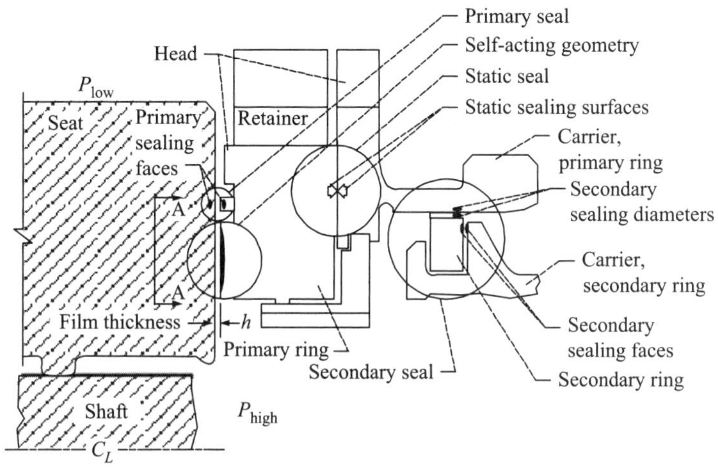

Fig. 46 Component schematic Rayleigh pad self-acting face seal [112].

slot, and T-grooving (bidirectional) are more commonly used than Rayleigh steps to provide more lift at less cost to manufacture.

Self-acting seals permit tighter clearances and better control of the sealing dam geometry as sealing pressure drops are increased, providing lower leakage. Figure 48 provides a comparison of the leakage rates between labyrinth, face-contact, and self-acting seals. Although self-acting face sealing greatly reduces leakage, surface speeds are generally limited to less than 213 m/s (700 ft/s), nearly triple the limits of contact face sealing 61 to 91 m/s (200 to 300 ft/s).

E. OIL SEALS

Gas turbine shaft seals are used to restrict leakage from a region of gas at high pressure to a region of gas at low pressure. A common use of mechanical seals is to restrict gas leakage into bearing sumps. Oil sealing of bearing compartments of turbomachines is difficult. A key is to prevent the oil side of the seal from becoming flooded. Still, oil-fog, and oil-vapor leakage can occur by diffusion of oil due to concentration gradients and oil transport due to vortical flows within the rotating labyrinth-cavities (crude distillation columns). Bearing

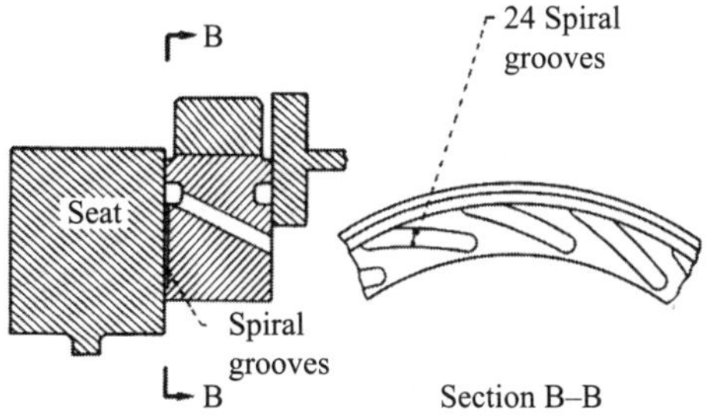

Fig. 47 Spiral groove sealing schematic [112].

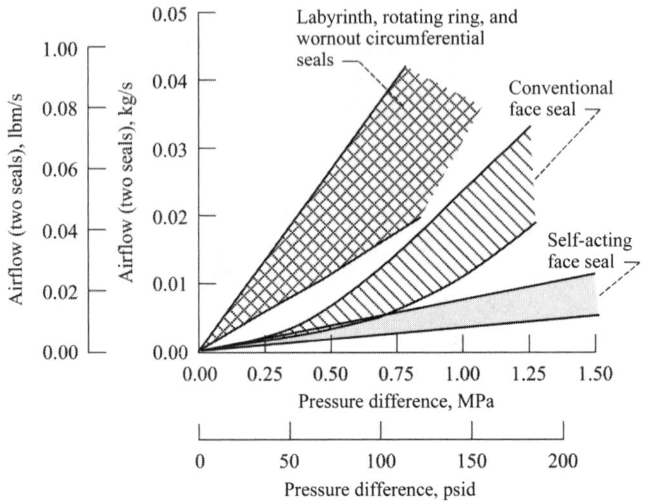

Fig. 48 Comparison of leakage characteristics for labyrinth, conventional (contact) face seal and self-acting face seals [112].

sumps contain an oil-gas mixture at near-ambient pressure, and a minimal amount of gas leakage through the seal helps prevent oil leakage and maintains a minimum sump pressure necessary for proper scavenging. Bearing sumps in the HPT are usually the most difficult to seal because the pressure and temperatures surrounding the sump can be near compressor discharge conditions.

Conventional rubbing-contact seals (shaft-riding and radial face types) are also used to seal bearing sumps. Because of their high wear rates, shaft-riding and circumferential seals (Fig. 43) have been limited to pressures less than 0.69 MPa (100 psi); successful operation has been reported at a sealed pressure of 0.58 MPa (85 psi), a gas temperature of 370°C (700°F), and sliding speed of 73 m/s (240 ft/s) [109].

The ring seal, as described by Whitlock [85] and Brown [108], is essentially an expanding or contracting piston ring. The expanding design is simpler and is illustrated in Fig. 49. Other designs that can be grouped in the ring seal family include the circumferential segmented ring seal and the floating or controlled-clearance ring, as described by Ludwig and Bill [4]. The material requirements for these seals are essentially the same as those for the expanding ring seal. The ring seals are carbon, and they seal radially against the inside diameter of the stationary cylindrical surface, as well as axially against the faces of the adjacent metal seal seats (Fig. 49). The metal seal seats are fixed to, and rotate with, the shaft. The sealing closing force is provided by a combination of spring forces and gas pressures. Ring seals are employed where there is a large relative axial movement due to thermal mismatch between the shaft and the stationary structure. Ring seals are

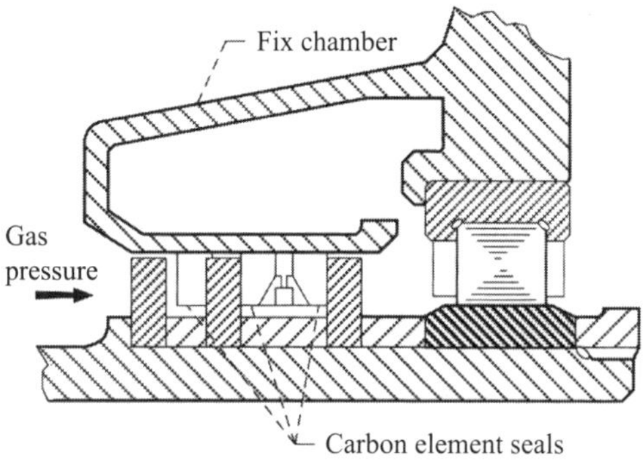

Fig. 49 Expanding ring seal [108].

limited to operation at air pressure drops and sliding speeds considerably lower than those allowed for face seals. They can, however, be used up to gas temperature levels in the same range as for positive-contact face seals, approximately to 480°C (900°F). Generally, a minimum pressure differential of 14 kPa (2 psid) must be maintained to prevent oil leakage from the bearing compartment.

Carbon ring and face sealing of the sumps described by Ludwig and Bill [4], Ludwig [112], Whitlock [85], and Brown [108] are fairly standard. Boyd et al. [114] have investigated a hybrid ceramic shaft seal, which is composed of a segmented carbon ring with lifting features as the outer or housing ring and a silicon-nitride tilt-support arched rub runner mounted on a metal flex beam as the inner ring (Fig. 50). The flex beam adds sufficient damping for stability, and no oil seepage was seen at idle speed down to pressure differentials of 0.7 kPa (0.1 psia), air to oil.

Selecting the correct materials for a given seal application is crucial to ensuring desired performance and durability. Seal components for which material selection is important from a tribological standpoint arc the stationary nosepiece (or primary seal ring) and the mating ring (or seal seat), which is the rotating element. Brown [108] described the properties considered ideal for the primary seal ring: 1) mechanical—high modulus of elasticity, high tensile strength, low coefficient of friction, excellent wear characteristics and hardness, self-lubrication; 2) thermal—low coefficient of expansion, high thermal conductivity, thermal

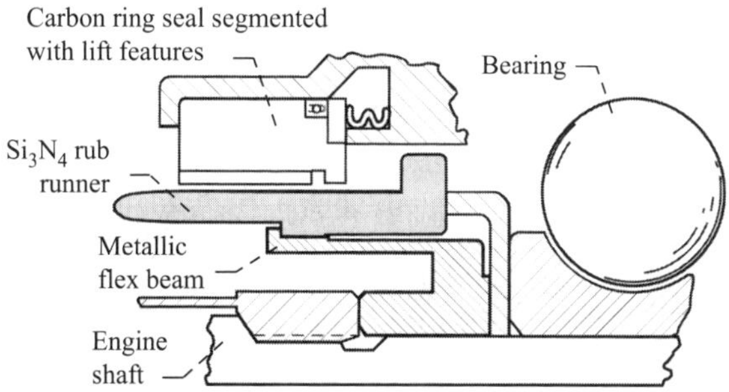

Fig. 50 Hybrid ceramic carbon ring seal [114].

shock resistance, and thermal stability; 3) chemical—corrosion resistance, good wetability; and 4) miscellaneous—dimensional stability, good machinability, and low cost and readily available.

Because of its high ranking in terms of satisfying these properties, carbon graphite is used extensively for one of the mating faces in rubbing contact shaft seals. In spite of its excellent properties, however, the carbon material must be treated in order for it to satisfy the operational requirements of sealing applications in the main rotor-bearing compartment of jet engines.

Seal failures are driven by thermal gradient fatigue or axial and radial thermal expansions during maximum power excursions. Bearing compartment carbon seals will fail from the heat generated in frictional rub. Excessive face wear occurs during transients, and, as mentioned, labyrinth seals can allow oil transport out of the seal and oil contamination by the environment (moisture, sand, etc.) [115].

F. BUFFER SEALING

Public awareness of environmental hazards, well-publicized effect of hazardous leakages (Three Mile Island, Challenger), and a general concern for the environment have precipitated emissions limits that drive the design requirements for sealing applications. Of paramount concern are the types of seals, barrier fluids, and the necessity of thin lubricating films and stable turbomachine operation to minimize leakages and material losses generated by rubbing contacts [107].

A zero-leakage seal is an oxymoron. Industrial practice is to introduce a buffer fluid between ambient seals and those seals confining the operational fluid (Fig. 51), with proper disposal of the buffered fluid mixture [85, 112]. A second example is for shaft sealing, as shown in Fig. 52, where buffer fluids are introduced. In the case of oil sumps, the buffered mixture is vented to the hot-gas exhaust stream and is presumed to be consumed. Within the nuclear industry, this becomes a containment problem, and waste storage becomes an issue. In the case of rocket engines, the use of buffering or inert fluids (such as helium) is

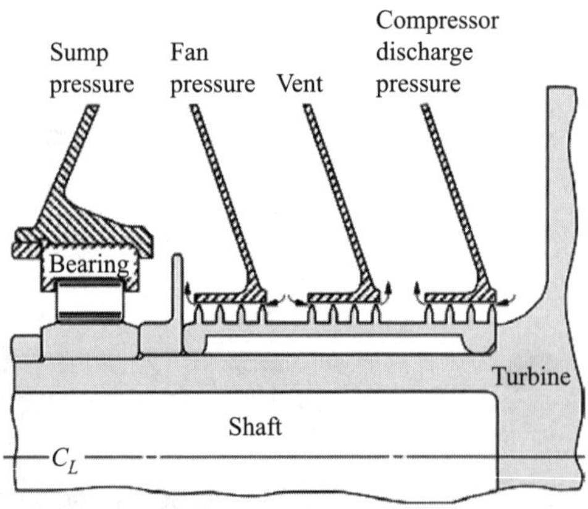

Fig. 51 Schematic of aero-gas-turbine buffer sealing of oil cavity [112].

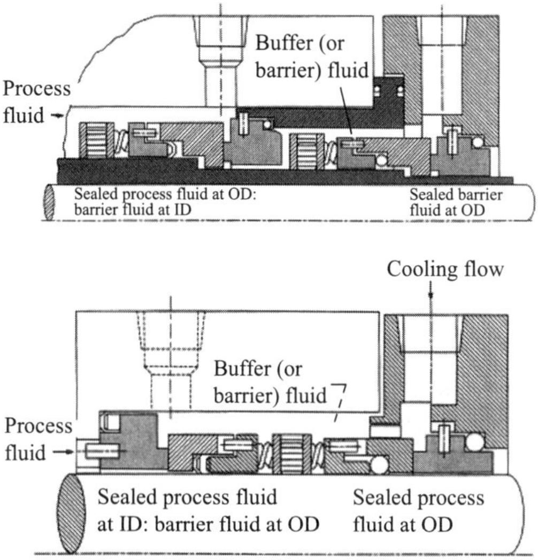

Fig. 52 Schematic of buffer fluid use in system sealing [107].

commonplace to separate fuel and oxidizer-rich environments, as in the space shuttle main engine (SSME) turbomachinery (see Appendix E).

G. RIM SEALING AND DISK CAVITY FLOWS

Turbomachine blade-vane interactions engender unsteady seal and cavity flows in multiple-connected cavities with conjugate heat transfer and rotordynamics. A comprehensive review of seals-secondary flow system developments is presented by Hendricks et al. [116, 117] and NASA Seals Code and Secondary Flow Systems Development publications [80].

Unsteady flows perturb both the power and the secondary flow streams [2]. A T1 turbine disk (first stage of the HPT) can have 76 blades and 46 stators all interacting with unsteady loadings (Fig. 53) [118]. Cavity ingestion of rapidly pulsating hot gases induces cavity heating and increases disk temperature, which in turn limits disk life and can compromise engine safety. Proper sealing confines these gases to the blade platform regions.

Rotordynamic issues further complicate rim seal and interface seal designs. These issues are addressed in the following: Thomas [66], Alford [67, 119, 120], Benckert and Wachter [69], NASA Conference Publications [79], Abbott [68], von Pragenau [121], Vance [122], Childs [123], Muszynska [71], Bently et al. [124], Hendricks et al. [117], and Temis and Temis [125].

Cavity and sealing interface requirements differ between industrial and aero-turbomachines. Major differences include split casings and through-bolted disks

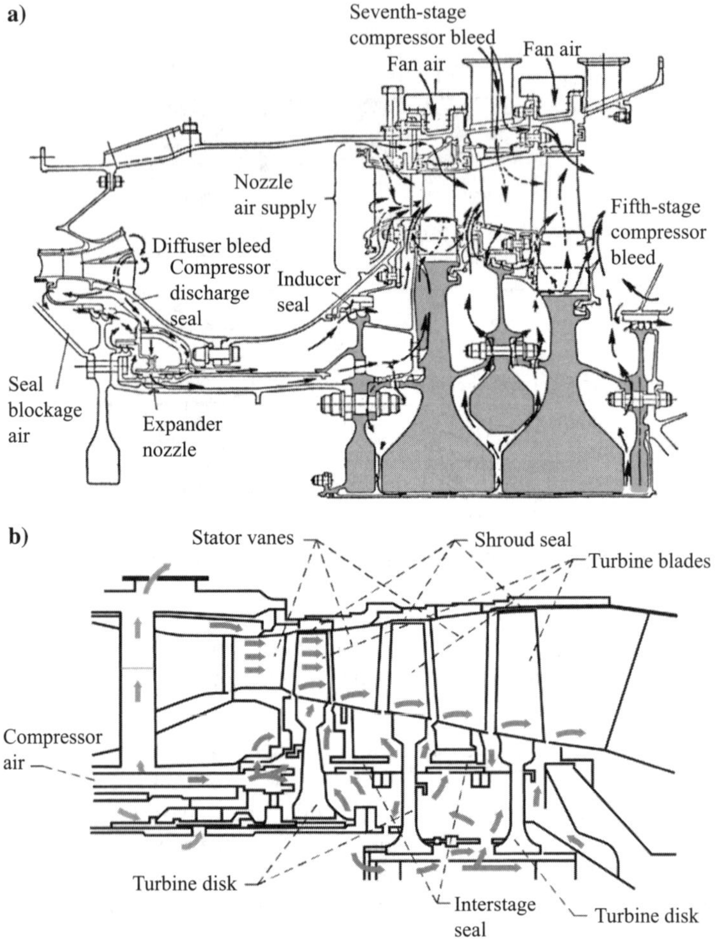

Fig. 53 Typical multistage turbine cavity section: a) energy-efficient engine high-pressure turbine [2] and b) hypothetical turbine secondary-air cooling and sealing (courtesy of AIAA) [118].

and compressors and turbines with common drive shafts for industrial machines vs cylindrical casings and drum rotors on multiple spools for aeromachines. Figure 53 shows a typical aeromultistage turbine cavity section. Several experimental studies have been reported that consider both simplified and complex disk cavity configurations (e.g., Chen [126], Chew [127], Chew et al. [128], Graber et al. [129], and Johnson et al. [130, 131]). Cavity sealing is complex and has a significant effect on component and engine performance and life, but several analytical and numerical tools are available to help guide the designer, experimenter, and field engineer in addressing these challenges including rim and interstage sealing (see Appendix A).

IV. ADVANCED SEAL DESIGNS

A. FINGER SEAL

The finger seal is a relatively new seal technology developed for air-to-air sealing for secondary flow control and gas path sealing in gas turbine engines [132–134]. It can easily be used in any machinery to minimize airflow along a rotating or nonrotating shaft. Measured finger seal air leakage is 1/3 to 1/2 of conventional labyrinth seals. Finger seals are compliant contact seals. The power loss is similar to that of brush seals [135]. It is reported that the projected cost of production for finger seals is estimated to be 40 to 50% that of brush seals.

The finger seal is composed of a stack of several precisely machined sheet stock elements that are riveted together near the seal outer diameter, as shown in Fig. 54. The outer elements of the stack, called the forward and aft coverplates, are annular rings. Behind the forward coverplate is a forward spacer, then a stack of finger elements, the aft spacer and then the aft coverplate. The forward spacer is an annular ring with assembly holes and radial slots around the seal inner diameter that align with feed-through holes for pressure balancing. The finger elements are an annular ring with a series of cuts around the seal inner diameter to create slender curved beams or fingers with an elongated contact pad at the tip. Each finger element has a series of holes near the outer diameter that are spaced such that when adjacent finger elements are alternately indexed to the holes, the spaces between the fingers of one element are covered by the fingers of the adjacent element. Some of the holes create a flowpath for high pressure upstream of the seal to reach the pressure balance cavity formed between the last finger element, the aft spacer and seal dam, and the aft coverplate. The aft spacer consists of two concentric, annular rings. One is like the forward spacer. The second is smaller, with an inner diameter the same as that of the aft coverplate, and forms the seal dam. It is connected to the outer annular ring by a series of radial spokes.

The fingers provide the compliance in this seal and act as cantilever beams, flexing

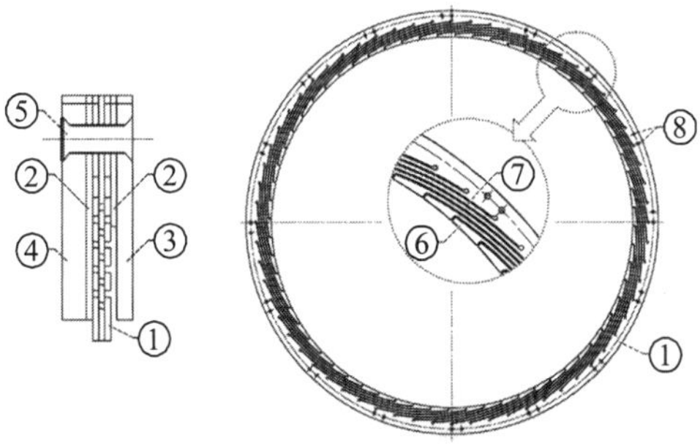

1. Finger element
2. Spacer
3. Forward cover plate
4. Aft cover plate
5. Rivet
6. Finger contact pad
7. Finger
8. Indexing and rivet holes

Fig. 54 Finger seal and detailed components [134].

away from the rotor during centrifugal or thermal growth of the rotor or during rotordynamic deflections. The pressure balance cavity reduces the axial load reacted by the seal dam and hence minimizes the frictional forces that would cause the fingers to stick to the seal dam and cause hysteresis in the finger seal leakage performance. In this seal there are two major leakage paths (similar to brush seals). One is through (along, around, and under) the fingers at the seal/rotor interface. The other is a radial flow across the seal dam. When a pressure differential exists across the seal, the fingers tend to move radially inward toward the rotor. Test results confirm this pressure closing effect. The pressure closing effect is largely due to the pressure gradient under the finger contact pads. The bulk of the radial pressure loads on the curved beam of the finger balance out to a zero net load. Ideally, one would design finger seals to have a line-to-line fit during operation. Most applications, however, involve a range of operating conditions, and seal-to-rotor fits and clearances change due to different coefficients of thermal expansion, centrifugal rotor growth, pressure closing effects, and dynamics of the rotor. Depending on the requirements of the application, it might be desirable to start with an interference-fit at build and allow the seal to wear in, or it might be desirable to have a clearance between the seal and rotor at build and allow the gap to close up. Load balancing is a critical part of seal design.

Finger seals are contacting seals and wear of the finger contact pad is expected. Life is dependent on the materials selected and operating conditions. Arora et al. [133] reported that the seal and rotor were in excellent condition after a 120-h endurance test. Testing of Haynes-25 fingers against a Cr_3C_2 coated rotor resulted in a wear track on the rotor 0.0064 mm (0.00025 in.) deep. The finger seal wore quickly to a near line-to-line fit with the rotor [134].

B. NONCONTACTING FINGER SEAL

Altering the geometry of the basic finger seal concept, Braun et al. [136] and Proctor and Steinetz [137] developed a new seal that combines the features of a self-acting shaft seal lift pad as an extension of the downstream finger with an overlapping row of noncontacting upstream fingers (Figs. 55 and 56). These lift pads are in very close proximity to the

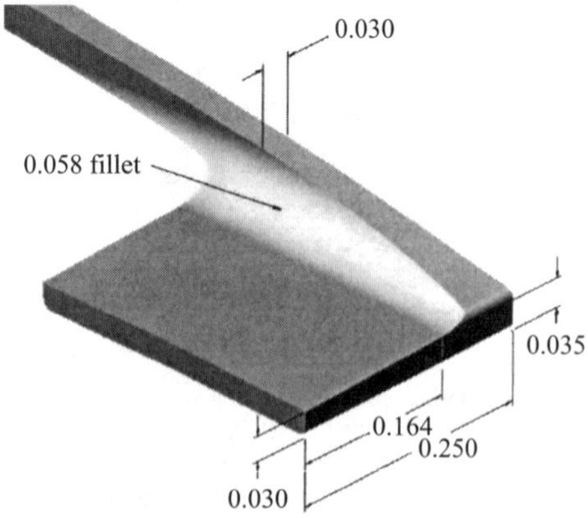

Fig. 55 Illustration of a noncontacting finger seal downstream padded finger [136].

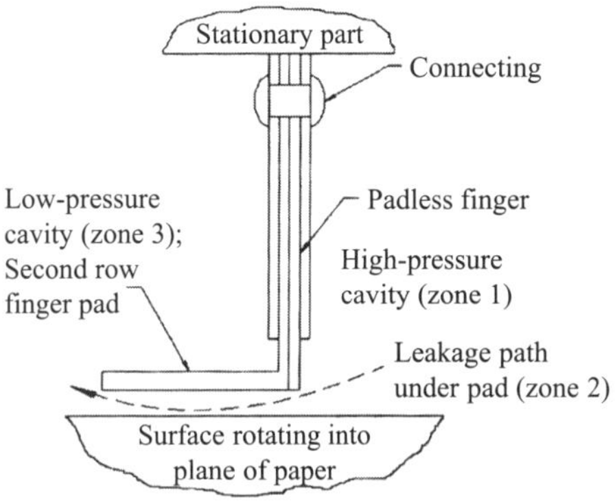

Fig. 56 Cross section of noncontacting finger seal with two rows of padded low-pressure and padless high-pressure fingers [137].

rotor outer diameter, so that hydrodynamic lift can be generated during shaft rotation. The seal geometry is designed so that the hydrostatic pressure between the downstream lift pad and the shaft is slightly greater than that above the pad. Depending on the application and operating conditions, the designer can choose to integrate hydrodynamic lift geometries on either the rotor or pads (taper, pockets, steps, etc., that is, self-acting seals) to further increase liftoff forces. The overlapping fingers reduce the axial and radial flows along the compliant fingers that allow for radial motion of the shaft-seal interface. These seals respond to both radial and axial shaft perturbations and some degree of misalignment with minimal hysteresis.

Finger seal technology is still being developed, but some experimental and analytical work has shown its feasibility [138]. It is expected that noncontacting finger seals will have leakage performance approximately 20% higher than contact finger seals, which is significantly better than conventional labyrinth seals, but have long life because they will not rub against the rotor, except very briefly at start and stop. Both the finger seal and noncontacting finger seal are in the development stage. As of this writing, neither seal has been tested in an engine.

C. LEAF AND WAFER SEALS

Continued efforts in high-temperature rope and wafer sealing are being directed toward improving braid wire, wafer, and preload-spring materials [139]. Silicon nitride (Si_3N_4) is favored for seal wafers (Fig. 57b). Worn-in wafer seals provide 1 to $1\frac{1}{2}$ orders of magnitude less leakage than rope seals, yet consideration must be given to the interface application: leakage, film cooling, or thermal barrier. The wafer seal (Fig. 57) can be preloaded by several methods, including bellows, coil springs, or canted coil springs. Canted coil springs offer near-constant loading characteristics over a range of compression. Coil springs are reported to maintain strength to 1200°C (2190°F); refractory metals as Mo-47.5Re (molybdenum rhenium), Mo-0.5Ti-0.08Zr (TZM), and coatings of iridium and rhodium (oxidation-resistant coatings used in some rocket thruster engine nozzle applications; for example, see Reed and Schneider [141]) are projected to afford superior high-temperature applications [139].

The leaf seal as described by Flower [95] and Nakane et al. [142] (Fig. 57a) is an adaptation of the wafer seal advanced by Steinetz and Sirocky [140],

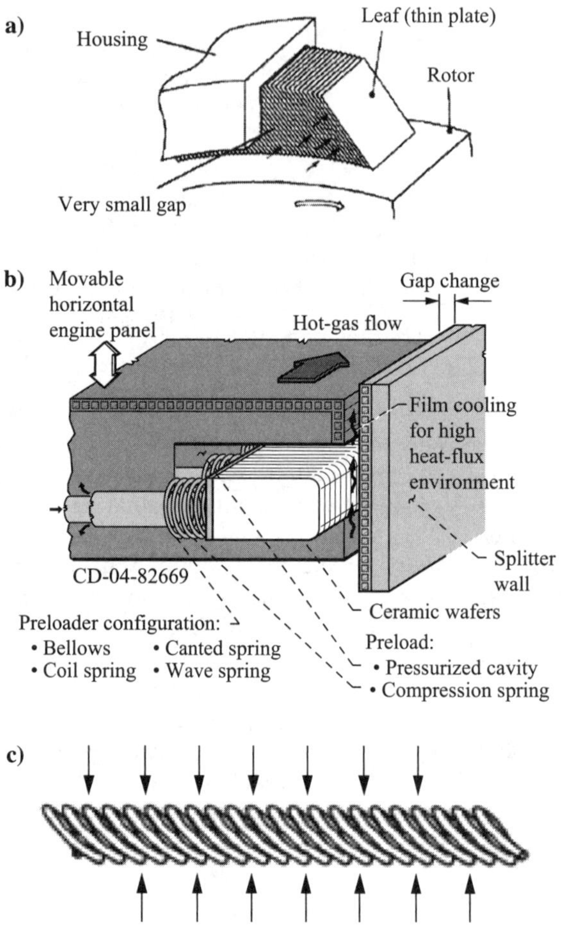

Fig. 57 Basic elements of leaf and wafer seals: a) leaf seal [95], b) wafer seal [139], and c) canted spring preloader.

with principles of operation delineated by Hendricks et al. [27, 89], Steinetz and Hendricks [111] and Nakane et al. [142]. The leaf and wafer seals have similar encapsulation but differ in root attachment and moments of inertia or cross section. The stacked leaves (or wafers) are relatively free to move in the radial direction and are deformable along the length or circumference, providing a compliant restrained two-dimensional motion, as opposed to the brush-seal-bristle, which deforms in three dimensions.

Nakane et al. [142] reported leakage performance of a leaf seal at less than 1/3 that of an equivalent four-stage, 0.5-mm gap labyrinth seal geometry, when run

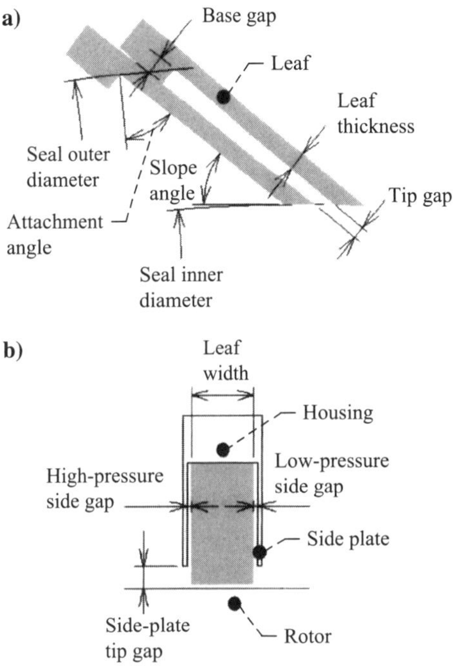

Fig. 58 Leaf seal configuration parameters: a) front view and b) side view [140].

back to back on the same test rotor, with little wear at the smooth-coated rotor interface. Variations in front and back plate gaps are used to control lift of the leaves based on pressure drop and rotor speed. Modeling of a leaf or wafer sealing is similar (Fig. 58) where pressure balances, thickness, length, inclination, housing gaps, and attachment points all require proper treatment for flexure and gap spacing. The latter are difficult to assess, and flow coefficients are most often determined experimentally. With that in mind, computations provided by Nakane et al. [142] are in good agreement with experimental data and CFD results, with little interface wear. Leaf seal, leakage, endurance, and reliability are being evaluated in a M501G in-house industrial gas turbine. Both leaf and wafer seals provide for radial, circumferential, and axial shaft motion. Even though the wafer seal is depicted with a low mobility wall, this restraint is not required.

Evaluation of the rotordynamic stability coefficients for these seals is still required. Stiffness and damping is similar to brush configuration modeling but with an altered cross section. These seals will not respond well to grooved or rough surfaces, which are alleviated when liftoff occurs.

An alternate form of the leaf seal has been advanced by Gardner [143]. The seal is composed of overlapping shaft-riding

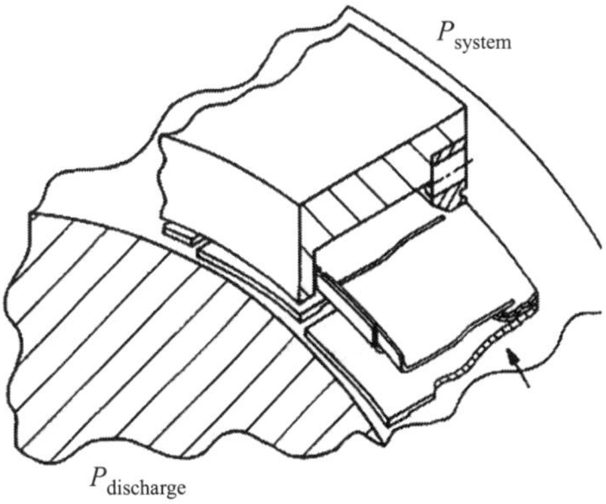

Fig. 59 Pressure-balanced compliant film riding leaf seal [143].

leaves (thin metallic sheets), which extend from the sealing inlet about the shaft, forming shaft-riding fingers (Fig. 59). The cantilevered inner leaves form lifting pads that are overlapped by outer leaves, which appear as cantilevered J-springs. The outer leaf, which sees system pressure, seals the cavity and permits compliant radial excursions. It is similar in configuration to a film-riding compliant foil bearing. This design also allows for hydrostatic operation; when pressure is applied, the interface deforms to maintain an operating film, even without rotation. The overlapping shingled elements float on the fluid film, providing excellent sealing with virtually no wear, but can be somewhat limited by their ability to handle large system pressure and radial excursions without damaging the leaves. Gardner [143] reported hydrostatic and hydrodynamic performance results for a 121-mm (4.75-in.)-diam seal. The pressure-balanced seal permitted hydrostatic liftoff independent of rotor speed with total liftoff and no shaft torque with a 0.48 MPa (70 psid) pressure differential. At a constant speed of 1200 rpm, the seal produced a torque of 0.028 N-m (0.25 in.-lb) with a 0.42-MPa (60-psid) pressure differential and a radial displacement of 0.23 mm (0.009 in.). When compared to a typical industrial and aerospace four-tooth labyrinth seal, with a 0.152-mm (0.006-in.) clearance and 1995 vintage-brush seal configurations, the leaf seal had leakage characteristics 1/4 that of the industrial labyrinth at design conditions and about 1/3 that of the brush (Fig. 60) [143].

D. HYBRID BRUSH SEALS

Justak [144, 145] combined a brush seal with tilt pad-bearing concepts to eliminate bristle wear. He introduced two designs: 1) bristles attached to the pads and 2) pads supported via beam elements and with the bristle tips remaining in contact with the outer surface of the pads (see Figs. 61 and 62, respectively). The brush seal stiffness is based on the design flow conditions and rotational speed. In the spring beam design, the beam

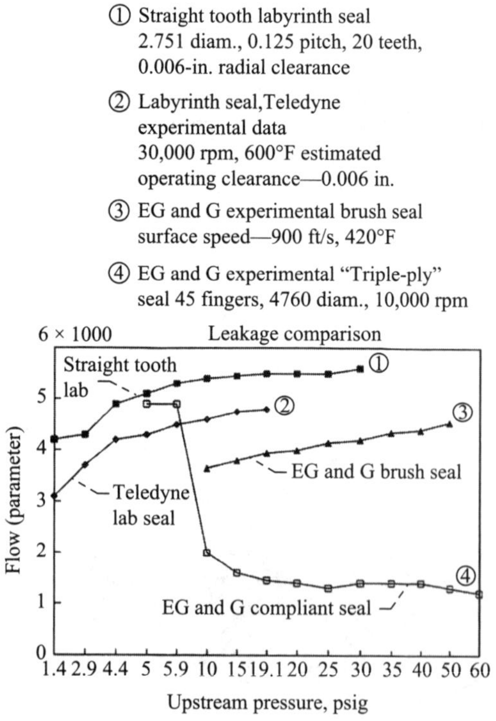

Fig. 60 Leaf-seal leakage comparison with labyrinth and brush seals [143].

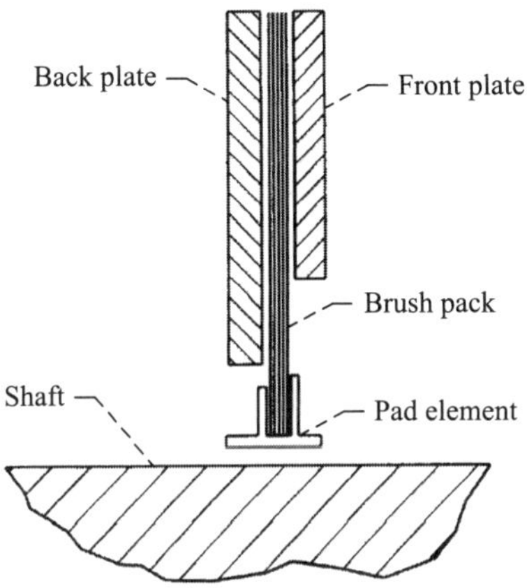

Fig. 61 Illustration of hydrodynamic brush seal (pad elements attached to bristles) [144].

elements are sized to allow radial movement and restrict axial displacements. In either design, seal leakage is controlled by the brush and liftoff by the pads, but leakage is more constant than a conventional brush seal and can accommodate reverse rotation with no bristle wear. Testing with rotor offsets up to 0.51 mm (0.020 in.) showed no wear or temperature rise and 1/3 less torque to maintain speed when compared to a conventional brush seal [138, 145].

Shapiro [146] combined shaft, face, and brush sealing concepts to form a film-riding seal with "L"-shaped annular segments (pads) that are preloaded onto the shaft via the brush seal and held in place by a garter spring (Fig. 63). The brush seal is separated axially from the brim of the cylindrical segments via a spring. This design allows the axial position of the brush seal to alter the preload and hence stiffness of the cylindrical seal segments. Analytical studies show low leakage, compliance, stability, and low wear during seal operation, yet development is required.

The hybrid floating brush seal (HFBS) (Fig. 64) combines a brush seal and a film-riding face seal that allows both axial and radial excursions mitigating interface problems of friction, heat, and wear [148]. The brush seal forms the primary seal and rotates with the shaft while floating against a secondary face seal that acts as a thrust bearing. The HFBS relies on a high interference fit or preload. Major advantages include elimination of wear between rotating and stationary components, improved sealing performance, and the ability to handle large axial and radial shaft excursions while maintaining sealing integrity. For a 71-mm (2.8-in.)-diam HFBS that allowed up to 6.4-mm (0.25-in.) axial

Fig. 62 Hydrodynamic brush seal (spring beam elements) [145].

Fig. 63 Schematic of film riding brush seal: a) assembly, b) joint, and c) installed [146].

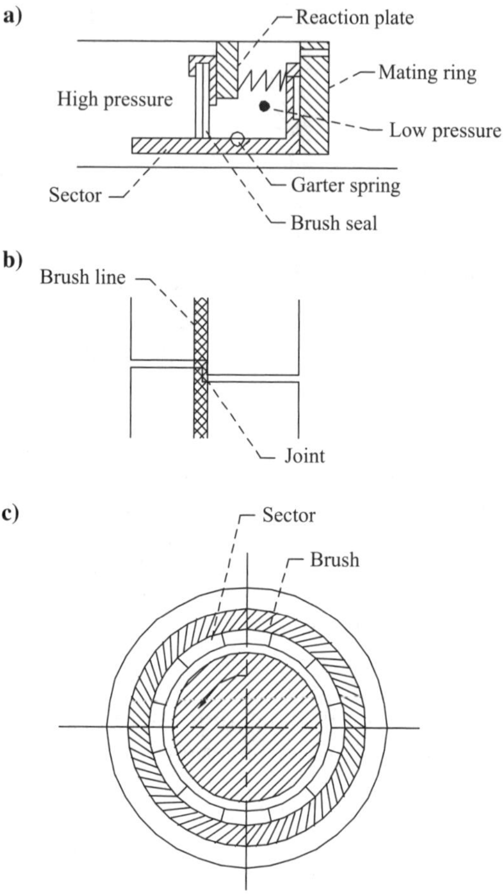

travel, experimental results showed 1/6 the leakage of a standard brush seal at the same operating pressure ratios and rotational speed and an order of magnitude less than numerical predictions of a standard labyrinth seal (see Fig. 65) [147–150].

E. OIL BRUSH SEALS

Brush seals are being considered for oil and oil mist sealing applications in turbomachinery (see Fig. 66) [151]. Brush seals perform very well under rotor transients, owing to the inherent compliance of bristles. Application areas include oil and bearing sumps in gas turbines and aircraft engines and as buffer seals in hydrogen-cooled industrial generator applications; they have also been suggested for vehicle applications [17]. Typically, oil-sump applications require oil mist sealing such as at the front bearing to prevent oil mist ingestion into the compressor [151–153]. Brush seals, however, are primarily contact seals, and in addition to the leakage rate, oil temperature rise and coking become major concerns. The inclined approach at the tip of individual bristles creates small hydrodynamic bearing surfaces at the rotor-bristle interface. The amount of lift determines the critical balance between oil temperature rise and leakage rate. Nonmetallic fibers are generally used in oil brush seals because of concerns about metal particle generation near bearings and thermal requirements. Bristle shrinkage,

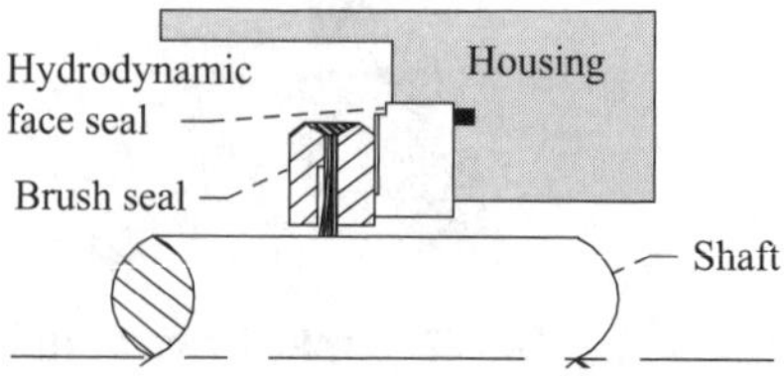

Fig. 64 Hybrid floating brush seal (HFBS) [147].

wear resistance, inertness, and moisture absorption rates are major concerns when selecting bristle fiber material [153]. Although oil applications of brush seals are rather new, their use in gas-turbine front bearing applications have proven successful. Field tests have shown leakage reduction gains and have demonstrated the durability of these seals in field operation [152, 153]. Further discussion on fiber material selection and seal analysis is provided in Appendix C.

F. ASPIRATING SEALS

An aspirating seal is a hydrostatic face seal with a narrow gap to control the leakage flow and a labyrinth tooth to control leakage flow at elevated clearances (Fig. 34a); it forms a high-pressure cavity for the diffused-injected flow in the engine cavity at operating conditions (Fig. 34b) [74].

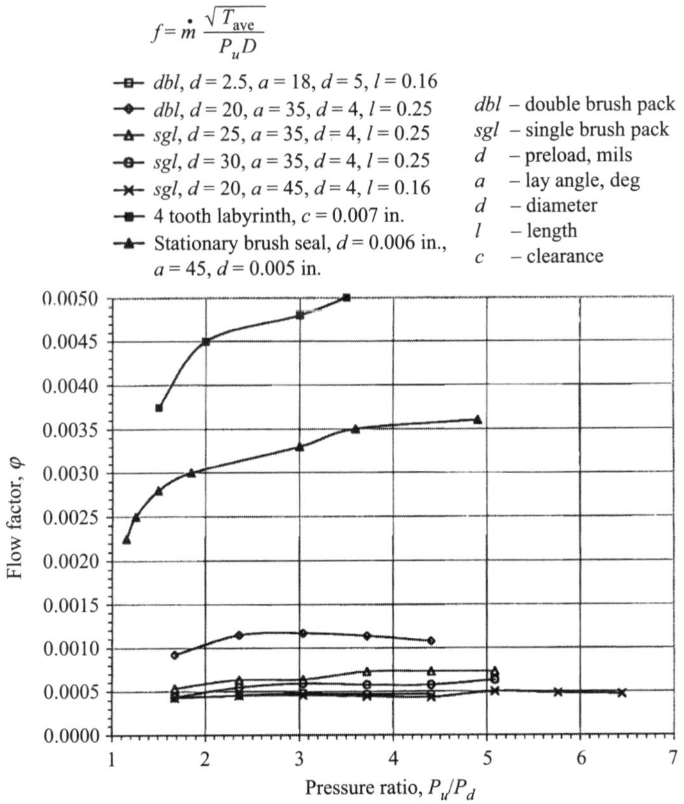

Fig. 65 HFBS performance compared to a stationary brush seal and a labyrinth seal: $\dot{m}$ is the mass flow rate of air (pps), T_{ave} is the average upstream air temperature (°R), P_u is the average upstream air pressure (psia), and D is the shaft outer diameter (in.) [147].

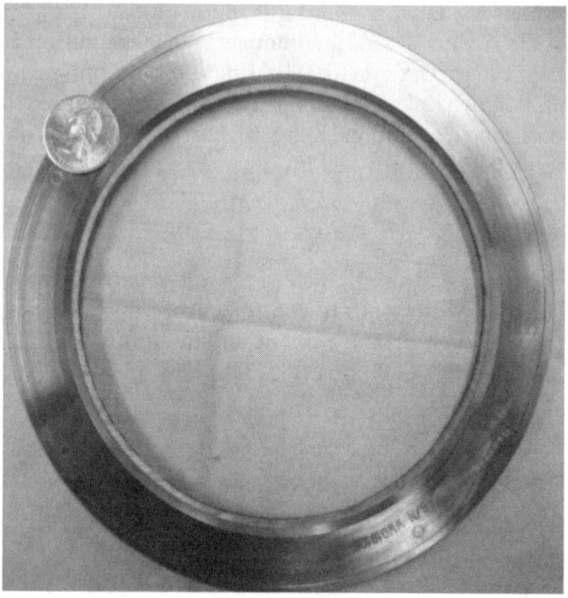

Fig. 66 Sample oil brush seal with nonmetallic bristles [151].

Conventional labyrinth seals are typically designed with a seal/rotor radial clearance that increases proportionally with diameter. Aspirating face seals are noncontacting seals that are designed to establish an equilibrium position within close proximity of the rotor surface regardless of the seal diameter [156–160]. Aspirating seals have a potentially significant performance advantage over conventional labyrinth seals, particularly at large diameters. In addition, these seals are inherently not prone to wear, owing to their noncontacting nature, and so their performance is not expected to degrade over time. Figure 34 shows a cross section of the seal design, which is enhanced by the presence of a flow deflector on the rotor face.

During operation, the aspirating face seal performs as a hydrostatic gas bearing. The gas bearing on the sealing face provides a thin, stiff air film at the interface; as the clearance between the seal face and the rotor decreases, the opening force of the gas bearing increases. The seal face geometry is designed to give an operating clearance of 0.038 to 0.076 mm (0.0015 to 0.003 in.). In operation, the spring forces play a minor role, and the seal is free to follow the rotor on axial excursions. Tolerances between the primary face seal ring and the housing allow the ring to tilt relative to the housing; the seal can thus follow the rotor even if there is an angle between the face and the rotor, such as during a maneuver or because of rotor runout.

When there is an insufficient pressure drop across the seal to maintain an adequate film thickness (such as during startup and shutdown, periods when conventional face seals would touch down), springs retract the seal from the rotor face (Fig. 34a). This ensures that the seal never contacts the rotor, thus providing for long seal life. In the retracted position, the pressure drop across the seal occurs at the aspirator tooth. During startup, as the pressure drop rises, the pressure balance across the primary face seal ring "aspirates" the seal to a closed position (Fig. 34b). The labyrinth tooth provides the required pressure drop to close the seal as well as a fail-safe seal in case of failure of the aspirating face seal.

Tests have been conducted to evaluate prototype performance under a variety of conditions that the seal might be subjected to in an aircraft engine application, including cases of rotor runout and seal/rotor tilt [154–160]. The tests were executed on a full-scale 36-in. (0.91-m)-diam rotary test rig. Analyses were performed using three-dimensional CFD in order to validate test data and to establish the seal design. The full-scale tests demonstrated that with the flow isolation tip, a hydrostatic film forms at the air bearing resulting in a seal/rotor clearance of 0.025 to 0.038 mm (0.001 to 0.0015 in.), with correspondingly low leakage rates. The seal performs effectively with rotor runouts as great as 0.25-mm (0.010-in.) total indicator reading, and the seal was able to accommodate the expected angular misalignment (tilt) of 0.27 deg.

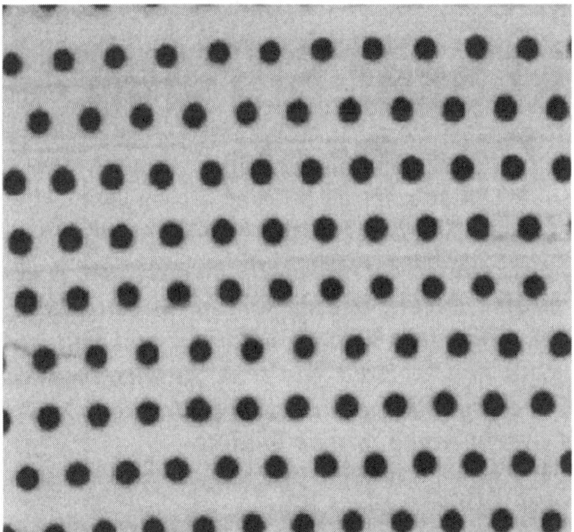

Fig. 67 Microdimpled surface by laser texturing [161].

G. MICRODIMPLE

Laser surface texturing, termed microdimples, (Fig. 67) [161] is a further extension of the damper bearing and seal-bearing work established by von Pragenau [162, 163] and extended by Yu and Childs [164] who found that a hole-area to surface-area ratio of 0.69 is a more effective seal than honeycomb. For the microdimpled seal (Fig. 67) where diameter is 125 $\pm$ 5 μm (4900 $\pm$ 200 μ-in.) and depth 2.5 $\pm$ 0.5 μm (98 $\pm$ 20 μ-in.) with 0.2 μm (8 μ-in.) surface finish, the hole/surface area ratio is 0.3, indicating some potential for improvement. Diamond-like graphite or antifouling coatings were not used, but afford potential improvements.

H. WAVE INTERFACES

Etsion [165] developed a wave pumping face seal, modified by Young and Lebeck [166] to a wave interface; these concepts have been combined and further improved by Flaherty et al. [167] and are considered more debris tolerant (Fig. 68).

I. SEAL-BEARING

Munson et al. [7] describe and provide operations data for room-temperature testing of a seal-bearing concept. The basic concept was advanced by von Pragenau [121, 163]. In the seal configuration of Munson et al. [7], the foil thrust bearing is combined with a mating flat interface to make a device called a foil face seal (Fig. 69). Multiple wave bump foils support the interface foils. With pressure drop and rotation, this interface gives rise to a compliant hydrodynamic film-riding face seal. For a 20,000-lb (89-kN)-thrust class engine, with this technology, an estimated mission fuel burn reduction of 1.85% for a fixed engine and "rubber-airframe" and 3.17% for both engine and airframe being "rubber" was reported. (Here "rubber" refers to allowing for design parameter changes.)

J. COMPLIANT FOIL SEAL

Forming an inexpensive, close tolerance, bell-mouth smooth interface foil, similar to a nozzle inlet, is not an easy task, but it can be accomplished by flow form or shear form spinning. (Data available from Franjo Metal Spinning

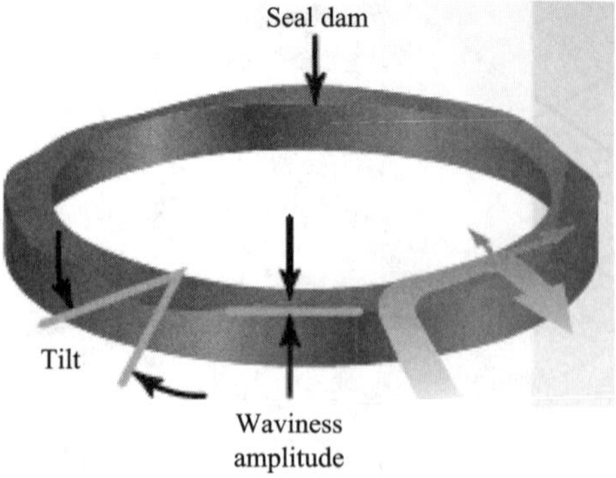

Fig. 68 Wave face seal [167].

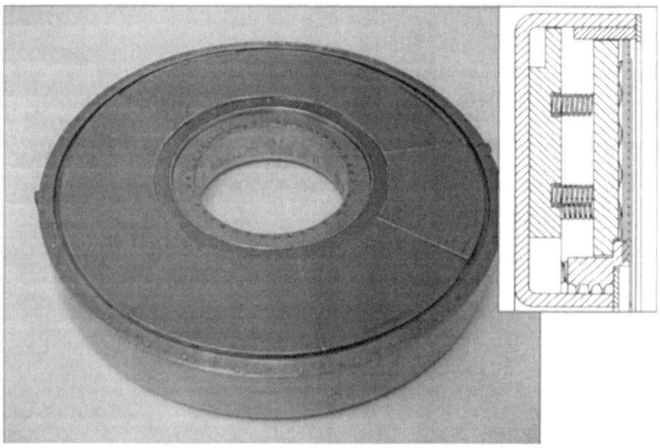

Fig. 69 Proof-of-concept foil face seal (courtesy of Rolls-Royce/Allison) [7].

at http://www.franjometal.com/metal-spinning/flow-forming.html#.) Expanding upon Gardner's leaf seal concept [143], Salehi et al. [168], and Heshmat [169] proposed an extension of their foil bearing work as a seal (Fig. 70); they chose to form the bellmouth-nozzle inlet by cutting radial relief slots to account

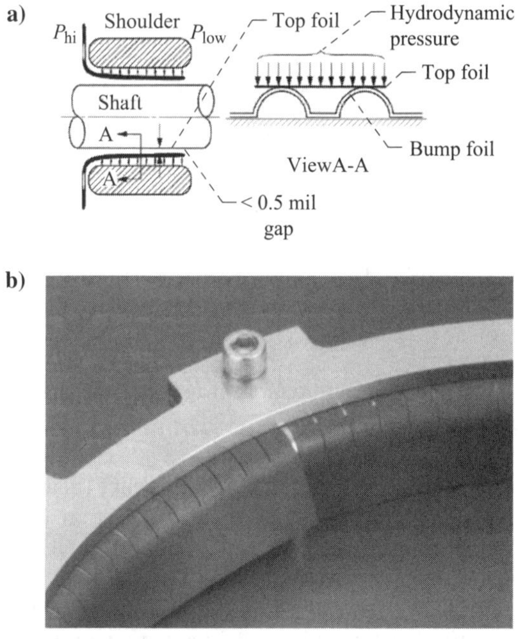

Fig. 70 Foil seal: a) schematic illustrating foil and bump-foil support [166, 167] and b) foil seal "nozzle-inlet or L-shaped" interface at attached and free end.

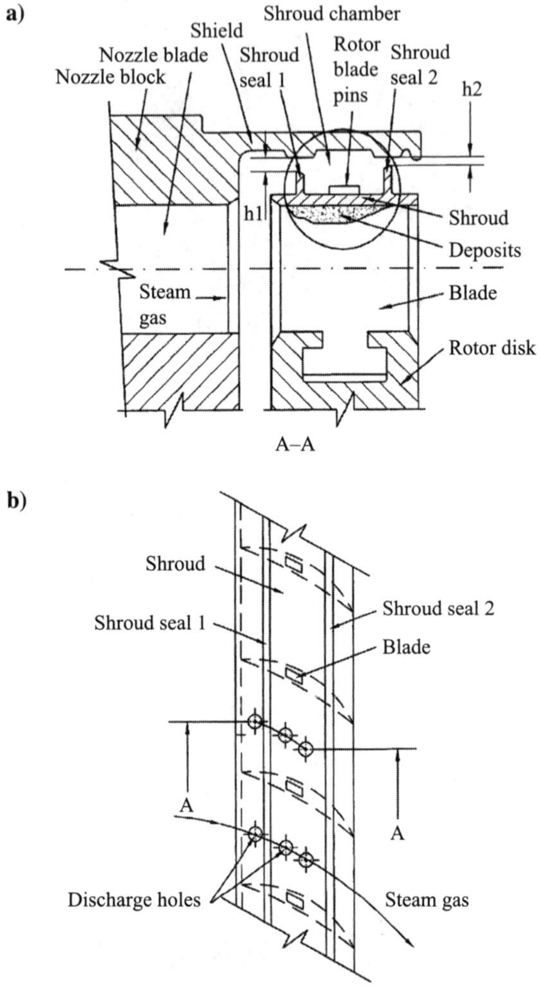

Fig. 71 Turbine shroud ring for deposit control: a) deposits build up in turbine passage and b) shroud discharge hole locations [170].

for the difference between inner and outer circumference (diameters) and bending the tabs to form the bell-mouth or "L-shaped" foil section. The resulting foil is then attached to the housing at one end opposing rotation. The slot relief spacing is dependent on stresses in the bend radius, foil thickness, and seal diameter to prevent significant "pleating" of the seal-interface foil. The resulting near-smooth compliant, noncontacting foil interface rides on a fluid film, typically <0.0127 mm (<500 μ-in.) thickness. The L-shaped section provides blockage for the bump foil opening and must be carefully contoured at the shoulder and

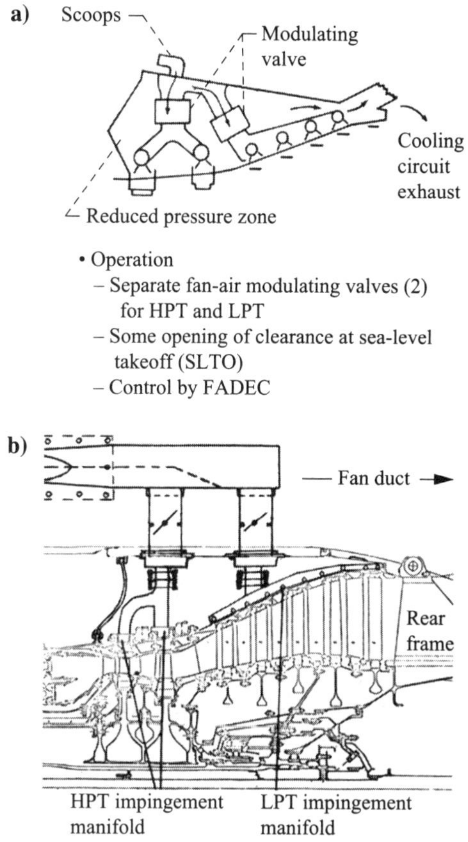

Fig. 72 Thermal active clearance control system: a) scoop design and b) HPT impingement manifold (FADEC: full authority digital engine controller) [2].

spring-pressed against the support structure, forming the necessary secondary seal. Overlapping leaves can help to minimize slot leakages, but it is a difficult area to seal. The interface foil (or foils) are, in turn, supported on a series of bump foils that provide variable stiffness in response to radial shaft excursions. The foils are usually coated with a solid lubricant to minimize startup and shutdown interface contact wear while permitting radial leaf sliding at the inlet. Such seals are low-leakage fluid film devices that are capable of operating at high surface velocities, temperature, and pressure loadings, within limits imposed by the foil materials [for example, 365 m/s (1200 fps) and 600°C (1100°F)].

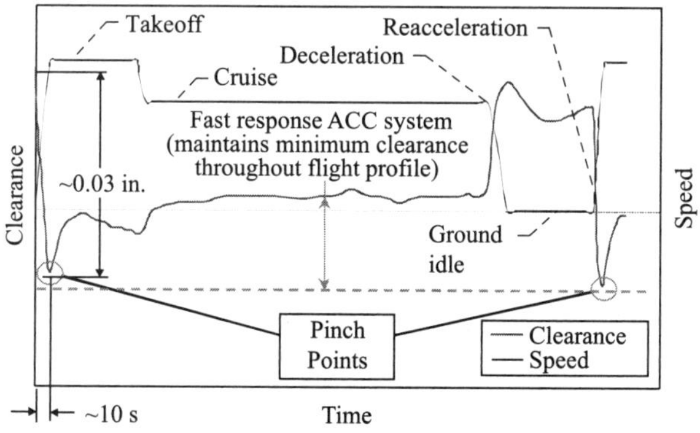

Fig. 73 High-pressure turbine blade tip clearance over given mission profile [6].

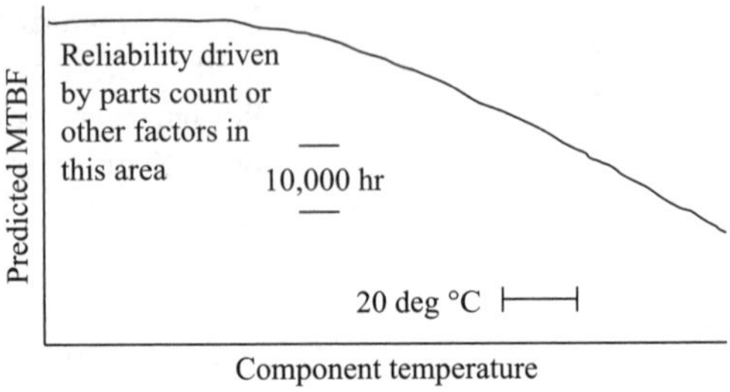

Fig. 74 Effect of component temperature on predicted mean time between failures for typical engine-mounted electronic device (courtesy of The Boeing Company) [172].

K. DEPOSITS CONTROL

For turbines operating in high salt environments, Nalotov [170] introduced strategically placed holes within the turbine shroud ring to provide equalization of circumferential pressures within the labyrinth interface. The concept is shown in Fig. 71. Figure 71a is a sketch representing the cross section of a nozzle and shrouded turbine stage where salt and metal oxides have built up on the shroud. Figure 71b shows the location of the discharge holes. The pressure equalization induces stability inside the shroud chamber, which allows for reduced shroud seal clearance. The flow through the shroud ring produces an obstacle effect to prevent deposits from building up and hence reduces blade passage blockage by accumulated salts. The concept has the net effects of increasing turbine engine efficiency and service life.

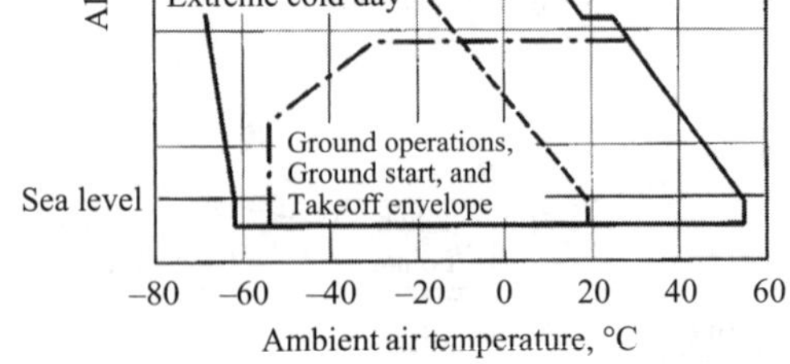

Fig. 75 Engine operating envelope (courtesy of Pratt & Whitney) [173].

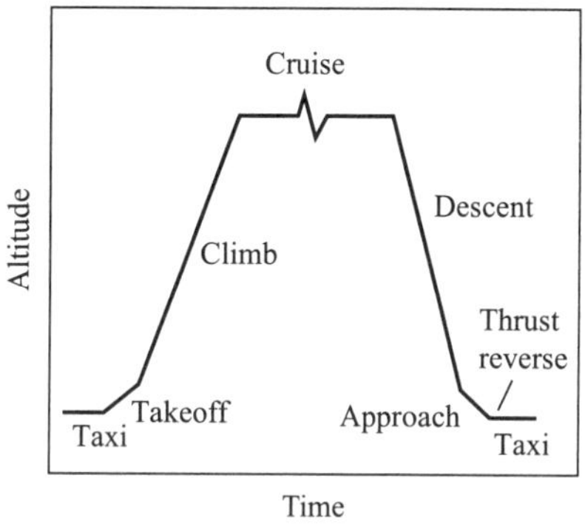

Fig. 76 Flight profile (courtesy of Pratt & Whitney) [173].

L. ACTIVE CLEARANCE CONTROL (ACC)

Most modern turbomachines have variable geometry controls. The compressor, for example, uses inlet guide vanes that are rotated to enhance efficiency at off-design conditions [usually designed to handle takeoff (TO) and set for cruise]. In large aero-engines, case clearance control is used in the turbine. Today's larger commercial engines control HPT blade tip clearances by impinging fan air on the outer case flanges. Systems such as those shown in Figs. 72a and 72b scoop air from the fan bypass duct to cool the outer case flanges, reducing the case diameter and hence shroud clearance. Other engines use a mixture of fan and compressor air to achieve finer HPT tip clearance control. Because these thermal systems are relatively slow, they cannot be used during transient events such as TO and reacceleration. As such, they are generally scheduled for operation during cruise conditions. Lattime and Steinetz [171] are developing fast-response systems that utilize clearance measurement feedback control, enabling true active clearance control at engine startup and throughout the flight envelope (Fig. 73). Active clearance control is not usually used in the compressor, rather the efficiency is enhanced through varying the vane angle and vortex control through fluid injection. Without these systems several points in efficiency are lost, but such close control is not without the potential for blade-shroud and vane-rotor rubbing, where the cited efficiency might be lost.

V. LIFE AND LIMITATIONS

System design conditions for seal-controlled component cooling are driven by compliance to regulatory agencies, reliability and safety standards [172]. The mean time between failures (MTBF) is highly dependent on the thermal loading (Fig. 74) and the aero-engine and flight operations profile (Figs. 75 and 76).

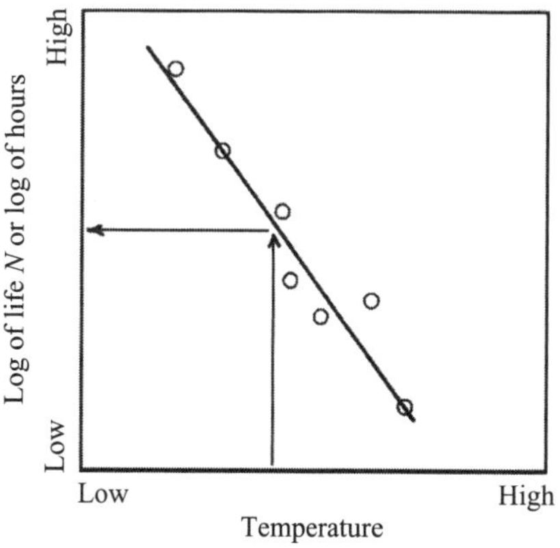

Fig. 77 Material life as a function of temperature relation (courtesy of Pratt & Whitney) [173].

TABLE 6 E3 ENGINE FLIGHT PROPULSION SYSTEM LIFE BASED ON 1985 TECHNOLOGY AND EXPERIENCE [175]

	Service life, h	Total life with repair, h
Combustor	9,000	18,000
HPT rotating structure	18,000	36,000
HPT blading	9,000	18,000
Remainder of engine	—	36,000

For engine component life modeling, the time τ_i and the life L_i under thermomechanical load from the environmental temperature (Fig. 77) and flight envelope profile (Figs. 75 and 76) are used to determine the cumulative loss of component life according to the Palmgren-Miner linear damage rule as used by Stoner [173] and Zaretsky et al. [174].

Zaretsky et al. [174] applied Weibull-based life and reliability analysis to rotating engine structures. The NASA E^3 engine design data served as the basis for the analysis [2, 175]. When limits are placed on stress, temperature, and time for a component's design, the criterion that will define the component's life and thus the engine's life will be either high-cycle or low-cycle fatigue.

Knowing the cumulative statistical distribution (Weibull function) of each engine component is a prerequisite to accurately predicting the life and reliability of an entire engine. The columns in Table 6 show how some of the hot-section component lives correlate to aero-engine maintenance practices without and with refurbishment, respectively. That is, it can be reasonably anticipated that at one of these time intervals, 5% of the engines in service will have been removed for repair or refurbishment for a cause.

There is currently a dearth of data in the open literature on seals and their functional life for basic materials. The classic approach is deterministic and assumes that full and certain knowledge exists for the service conditions and the material strength. This means that specific equations that define sealing conditions are coupled with experience-based safety factors; it is well known, however, that variations with loading can have a significant effect on component reliability. The Weibull-based analysis addresses these issues, but until a sealing database is established, the MTBF will continue to be based on field experience.

VI. CONCLUSION

Sealing in turbomachinery addresses load balancing. Turbine engine cycle efficiency, operational life, and systems stability depend on effective clearance control. Designers have put renewed attention on clearance control, as it is often the most cost-effective method to enhance system performance. Advanced concepts and proper material selection continue to play important roles in maintaining interface clearances to enable the system to meet design goals. No single sealing geometry or material is appropriate for all uses; each interface must be assessed in terms of its operational requirements. Insufficient clearances limit coolant flows, cause interface rubbing, and engender turbomachine instabilities and system failures. Excessive clearances lead to losses in cycle efficiency, flow instabilities, and hot-gas ingestion into disk cavities. Hot-gas ingestion in the turbine cavities reduces critical disk life and in the bearing sump location engenders bearing and materials failures. Reingestion of flow along the compressor drum interface causes unnecessary blockage and can lead to compressor stall.

Materials play a major role in maintaining interface clearances. Abradable sealing materials for the fan shroud are usually polymers. For the LPC compressor, ambient to 400°C (750°F), fiber metals and AlSi + filler can be used, but for the midrange LPC and HPC, ambient to 760°C (1400°F), Ni or Co base can be used (titanium blade fire protection limits). If the blades are Ni-based superalloys, NiCrAl-Bentonite might be a choice. In the HPT, 760°C (1400°F) to 1150°C (2100°F), yttria-stabilized zirconia (YSZ) with controlled porosity and cBN or preferably SiC blade tip abrasive grits can be used, depending on how hot the engine runs; in general, air plasma spray thermal barrier coatings (APS-TBCs) are used in the combustor, and for some engines, first-stage vanes (nozzles) and second-stage blades of the HPT. Electron-beam plasma vapor deposition (EB-PVD) TBCs are used on the HPT T1 or first-stage blades, some second-stage blades, and some first-stage vanes (nozzles). TBCs are not commonly used in the LPT due to lower heat flux and are less effective in decreasing component temperature. APS ceramics are also used on shroud seals (blade outer air seals), where they function as both a thermal barrier for the metallic shroud and abradable seal.

Component life and reliability are closely coupled with the duty cycle, but as energy demands (and emissions regulations) necessitate more time-responsive-controlled engines, the industrial and aero-engine duty cycles become similar.

ACKNOWLEDGMENTS

Sealing in turbomachinery has been the focus of numerous development efforts. Many of the developers have been cited in this chapter. The authors would like to especially acknowledge contributors to this review: Margaret Proctor (finger seals), Norm Turnquist (aspirating seals), Saim Dinc and Mehmet Demiroglu (brush seals), Stephen Stone and Greg Moore (metallic static seals), Farshad Ghasripoor (abradables), and Glenn Holle (an extensive review). We also wish to thank our sponsoring organizations for the time and resources to prepare this sealing review.

REFERENCES

[1] Miller, M., Colehour, J., and Dunkleberg, K., "Engine Case Externals, Challenges and Opportunities," *Proceedings of the 7th International Symposium on Transport Phenomena and Dynamics of Rotating Machinery*, edited by A. Muszynska, J. A. Cox, and D. T. Nosenzo, ISROMAC-7, Bird Rock Publishing House, 1998, pp. 1604–1611.

[2] Halila, E. E., Lenahan, D. T., and Thomas, T. T., "Energy Efficient Engine High Pressure Turbine Test Hardware: Detailed Design Report," NASA CR-167955, June 1982.

[3] Hendricks, R. C., Griffin, T. A., Kline, T. R., Csavina, K. R., Pancheli, A., and Sood, D., "Relative Performance Comparison Between Baseline Labyrinth and Dual Brush Compressor Discharge Seals in a T-700 Engine Test," American Society of Mechanical Engineers, ASME 94-GT-266, June 1994.

[4] Ludwig, L. P., and Bill, R. C., "Gas Path Sealing in Turbine Engines," *ASLE Transactions*, Vol. 23, No. 1, 1980, pp. 1–22.

[5] Moore, A., "Gas Turbine Engine Internal Air Systems. A Review of the Requirements and the Problems," American Society of Mechanical Engineers, ASME 75-WA/GT-1, Nov.-Dec. 1975.

[6] Lattime, S. B., and Steinetz, B. M., "Turbine Engine Clearance Control Systems: Current Practices and Future Directions," *Journal of Propulsion and Power*, Vol. 20, No. 2, 2004, pp. 302–311.

[7] Munson, J., Grant, D., and Agrawal, G., "Foil Face Seal Proof-of-Concept Demonstration Testing," AIAA Paper 2002-3791, July 2002.

[8] Chupp, R. E., Ghasripoor, F., Moore, G. D., Kalv, L. S., and Johnston, J. R., "Applying Abradable Seals to Industrial Gas Turbines," AIAA Paper 2002-3795, July 2002.

[9] Bill, R. C., "Wear of Seal Materials Used in Aircraft Propulsion Systems," *Wear*, Vol. 59, No. 1, 1980, pp. 165–189.

[10] Aksit, M. F., Chupp, R. E., Dinc, O. S., and Demiroghu, M., "Advanced Seals for Industrial Turbine Applications: Design Approach and Static Seal Development," *Journal of Propulsion and Power*, Vol. 18, No. 6, 2002, pp. 1254–1259; also "Advanced Flexible Seals for Gas Turbine Shroud Applications," AIAA Paper 99-2827, June 1999.

[11] Camatti, M., Vannini, G., Baldassarre, L., Fulton, J., and Forte, P., "Full Load Test Experience on the Instability of a High Speed Back to Back Compressor Equipped with a Honeycomb Seal," *Proceedings of the 2nd International Symposium on Stability Control of Rotating Machinery*, edited by A. Gosiewski and A. Muszynska, ISCORMA-2, Max Media, Warsaw, 2003, pp. 617–626.

[12] Camatti, M., Vannini, G., Fulton, J., and Hopenwasser, F., "Instability of a High Pressure Compressor Equipped with Honeycomb Seals," *Proceedings of the 32nd Turbomachinery Symposium*, Texas A&M University, College Station, TX, 2003, pp. 39–48.

[13] Hurter, J., Zierer, T., and Motzkus, T., "Sealing Improvements on the GT24/GT26 Gas Turbine Fleet-Filed Feedback on Combustor Seals," *VDI-Berichte* Nr. 1965, 2006, pp. 221–231.

[14] Shiembob, L. T., "Development of Abradable Gas Path Seals," Pratt & Whitney Aircraft, PWA-TM-5081, East Hartford, CT, 1974; also NASA Contract NAS3-18023, NASA CR-134689, 1974.

[15] More, D. G., and Datta, A., "Ultra High Temperature Resilient Metallic Seal Development for Aero Propulsion and Gas Turbine Applications," NASA/CP-2004-212963, Vol. 1, NASA, Washington, D.C., pp. 359–370.

[16] Layer, J., "Advanced Metallic Seal for High Temperature Applications," NASA CP-10198, NASA, Washington, D.C., 1997, pp. 307–328.

[17] Hendricks, R. C., Braun, M. J., Canacci, V. A., and Mullen, R. L., "Brush Seals in Vehicle Tribology," *Proceedings of the 13th Leeds-Lyon Symposium on Tribology*, E-5712, Leeds, England, 1990, pp. 231–242.

[18] Bagepalli, B. S., Aksit, M. F., and Farrell, T. R., "Gas-Path Leakage Seal for a Turbine," U.S. Patent 5934687, Aug. 10, 1999.

[19] Aksit, M. F., Bagepalli, B. S., Demiroglu, M., Dinc, O. S., Keelock, I., and Farrell, T., "Advanced Flexible Seals for Gas Turbine Shroud Applications," AIAA Paper 99-2827, June 1999.

[20] Dinc, O. S., Bagepalli, B., Aksit, M., Wolfe, C. E., and Turnquist, N., "A New Metal Cloth Stationary Seal for Gas Turbine Applications," AIAA Paper 97-2732, July 1997.

[21] Tompkins, T. L., "Ceramic Oxide Fibers Building Blocks for New Applications," *Ceramic Industry Publications*, Business News Publishing, April 1995.

[22] Steinetz, B. M., and Adams, M. L., "Effects of Compression, Staging, and Braid Angle on Braided Rope Seal Performance," *Journal of Propulsion and Power*, Vol. 14, No. 6, 1998, pp. 934–940; also NASA TM-107504, July 1997.

[23] Steinetz, B. M., "High Temperature Braided Rope Seals for Static Sealing Applications," NASA TM-107233 Rev., 1996; also *Journal of Propulsion and Power*, Vol. 13, No. 5, 1997, pp. 675–682.

[24] Opila, E. J., Lorincz, J. A., and Demange, J. J., "Oxidation of High-Temperature Alloy Wires in Dry Oxygen and Water Vapor," *High Temperature Corrosion and Materials Chemistry V*, Electrochemical Society, Inc., Pennington, NJ, 2005, pp. 67–80.

[25] Dunlap, P. H., Steintez, B. M., Curry, D. M., DeMange, J. J., Rivers, H. K., and Hsu, S. Y., "Investigation of Control Surface Seals for Re-Entry Vehicles," *Journal of Spacecraft and Rockets*, Vol. 40, No. 4, 2003, pp. 570–583.

[26] Steinetz, B. M., and Dunlap, P. H., "Rocket Motor Joint Construction Including Thermal Barrier," U.S. Patent No. 6,446,979 B1 (LEW 16,684-1), Sept. 2002.

[27] Hendricks, R. C., Steinetz, B. M., Zaretsky, E. V., Athavale, M. M., Przekwas, A. J., Tam, L. T., Muszynska, A., and Braun, M. J., "Reviewing Turbomachine Sealing and Secondary Flows Parts A, B, C," *The 2nd International Symposium on Stability Control of Rotating Machinery*, ISCORMA-2003, Max Media, Warsaw, Aug. 2003, pp. 40–91; see also Ref. 113.

[28] Van Zante, D. E., Stazisar, A. J., Wood, J. R., Hathaway, M. D., and Okiishi, T. H., "Recommendations for Achieving Accurate Numerical Simulation of Tip Clearance Flows in Transonic Compressor Rotors," *Journal of Turbomachinery*, Vol. 122, Oct. 2000, pp. 733–742.

[29] Lakshminarayana, B., *Fluid Dynamics and Heat Transfer of Turbomachinery*, Wiley, New York, 1996.

[30] Copenhaver, W. W., Mayhew, E. R., Hah, C., and Wadia, A. R., "The Effect of Tip Clearance on a Swept Transonic Compressor Rotor," *Journal of Turbomachinery*, Vol. 118, No. 2, 1996, pp. 230–239.

[31] Strazisar, A. J., Wood, J. R., Hathaway, A. D., and Suder, K. L., "Laser Anemometer Measurements in a Transonic Axial-Flow Fan Rotor," NASA TP–2879, Nov. 1989.

[32] Wellborn, S. R., and Okiishi, T. H., "The Influence of Shrouded Stator Cavity Flows on Multistage Compressor Performance," *Journal of Turbomachinery*, Vol. 121, No. 3, 1999, pp. 486–498.

[33] Bohn, D., Tummers, C., and Sell, M., "Influence of the Radial Gap on the Flow Field of a 2-Stage Turbine with Shrouded Bladings," ISROMAC-11, Bird Rock Publishing House, 2006.

[34] Bill, R. C., and Wisander, D. W., "Friction and Wear of Several Compressor Gas-Path Seal Materials," NASA TP–1128, Jan. 1978.

[35] Bill, R. C., Wolak, J., and Wisander, D. W., "Effects of Geometric Variables on Rub Characteristics of Ti-6Al-4 V," NASA TP–1835 (AVRADCOM TR 80-C-19), April 1981.

[36] Stocker, H. L., Cox, D. M., and Holle, G. F., "Aerodynamic Performance of Conventional and Advanced Design Labyrinth Seals with Solid-Smooth, Abradable, and Honeycomb Lands," NASA CR–135307 (EDR9339), Nov. 1977.

[37] Stocker, H. L., "Determining and Improving Labyrinth Seal Performance in Current and Advanced High Performance Gas Turbines," AGARD–CP–237 (AGARD–AR–123), Paper 13, Aug. 1978.

[38] Mahler, F. H., "Advanced Seal Technology," Pratt & Whitney Aircraft Report PWA–4372 (Contract No. AD–739922), 1972.

[39] Morrell, P., Betridge, D., Greaves, M., Dorfman, M., Russo, L., Britton, C., and Harrison, K., "A New Aluminum-Silicon/Boron Nitride Powder for Clearance Control Application," ASM Thermal Spray Society, ITSC 98, 1998, pp. 1187–1192.

[40] Chupp, R. E., Ghasripoor, F., Turnquist, N. A., Demiroglu, M., and Aksit, M. F., "Advanced Seals for Industrial Turbine Applications: Dynamic Seal Development," *Journal of Propulsion and Power*, Vol. 18, No. 6, 2002, pp. 1260–1266.

[41] Schmid, R. K., Ghasripoor, F., Dorfman, M., and Wei, X., "An Overview of Compressor Abradables," *Proceedings of the International Thermal Spray Conference*, ITSC 2000, ASM International, Materials Park, OH, 2000, pp. 1087–1093.

[42] Guilemany, J. M., Navarro, J., Lorenzana, C., Vizcanio, S., and Miguel, J. M., "Tribological Behaviour of Abradable Coatings Obtained by Atmospheric Plasma Spraying (APS)," *Proceedings of the International Thermal Spray Conference*, ITSC 2001, ASM International, Materials Park, OH, May 2001, pp. 1115–1118.

[43] Ghasripoor, F., Schmid, R. K., Dorfman, M., and Russo, L., "A Review of Clearance Control Wear Mechanisms for Low Temperature Aluminum Silicon Alloys," *Proceedings of the International Thermal Spray Conference*, ITSC 1998, ASM International, Materials Park, OH, 1998, pp. 139–144.

[44] Nava, Y., Mutasim, Z., and Coe, M., "Abradable Coatings for Low-Temperature Applications," *Proceedings of the International Thermal Spray*

Conference, ITSC 2001, ASM International, Materials Park, OH, 2001, pp. 119–126, 263–268.

[45] Schmid, R., "New High Temperature Abradables for Gas Turbines," Ph.D. Dissertation, Thesis 12223, Dept. of Materials, Swiss Federal Inst. of Technology, Zurich, 1997.

[46] Borel, M. O., Nicoll, A. R., Schlaepfer, H. W., and Schmid, R. K., "Wear Mechanisms Occurring in Abradable Seals of Gas Turbines," *Surface Coating Technology*, Vols. 39–40, Part 1, Elsevier, Atlanta, GA, 1989, pp. 117–126.

[47] Chappel, D., Vo, L., and Howe, H., "Gas Path Blade Tip Seals: Abradable Seal Material Testing at Utility Gas and Steam Turbine Operating Conditions," American Society of Mechanical Engineers, 2001-GT-0583, June 2001.

[48] Chappel, D., Howe, H., and Vo, L., "Abradable Seal Testing: Blade Temperatures During Low Speed Rub Event," AIAA Paper 2001-3479, July 2001.

[49] Ghasripoor, F., Schmid, R., and Dorfman, M., "Abradables Improve Gas Turbine Efficiency," *Journal of the Inst. of Materials*, Vol. 5, No. 6, June 1997, pp. 328–330.

[50] Ghasripoor, F., Schmid, R., Dorfman, M., and Wei, X., *Optimizing the Performance of Plasma Control Coatings up to 850C*, Surface Modification Technologies XII, ASME International, Rosemont, IL, 1998.

[51] Shell, J. D., and Farr, H. J., "Abrasive Ceramic Matrix Turbine Blade Tip and Method for Forming," U.S. Patent No. 5,952,110, Sept. 1999.

[52] Ghasripoor, F., Schmid, R. K., and Dorfman, M., "Silicon Carbide Composition for Turbine Blade Tips," U.S. Patent No. 5,997,248, Dec. 1999.

[53] Benoit, R., Beverly, E. M., Love, C. M., and Mack, G. J., "Abrasive Blade Tip," U.S. Patent No. 5,603,603, Feb. 1997.

[54] Draskovich, B. S., Frani, N. E., Joseph, S. S., and Narasimhan, D., "Abrasive Tip/Abradable Shroud System and Method for Gas Turbine Compressor Clearance Control," U.S. Patent No. 5,704,759, Jan. 1998.

[55] Johnson, G. F., and Schilke, P. W., "Alumina Coated Silicon Carbide Abrasive," U.S. Patent No. 4,249,913, Feb. 1981.

[56] Pan, Y., and Baptista, J., "Chemical Stability of Silicon Carbide in Presence of Transition Metals," *Journal of the American Ceramic Society*, Vol. 79, No. 8, 1996, pp. 2017–2026.

[57] Hutchings, I. M., "Erosion by Solid Particle Impact," *Tribology; Friction and Wear of Engineering Materials*, Edward Arnold, London, 1992, pp. 171–197.

[58] Biesiadny, T. J., McDonald, G. E., Hendricks, R. C., Little, J. K., Robinson, R. A., Klann, G. A., and Lassow, E., "Experimental and Analytical Study of Ceramic-Coated Turbine-Tip Shroud Seals for Small Turbojet Engines," NASA TM X-86881, Jan. 1985.

[59] Wei, X., Mallon, J. R., Correa, L. F., Dorfman, M., and Ghasripoor, F., "Microstructure and Property Control of CoNiCrAlY Based Abradable Coatings for Optimal Performance," *Proceedings of the International Thermal Spray Conference*, ITSC 2000, ASM International, Materials Park, OH, 2000, pp. 407–412.

[60] Burcham, R. E., and Keller, R. B., Jr., "Liquid Rocket Engine Turbopump Rotating-Shaft Seals," NASA SP-8121, Feb. 1979.

[61] Alford, J. S., "Labyrinth Seal Designs Have Benefited from Development and Service Experience," *Society of Automotive Engineers*, SAE Paper 710435, Feb. 1971.

[62] Heidegger, N. J., Hall, E. J., and Delaney, R. A., "Parameterized Study of High-Speed Compressor Seal Cavity Flow," NASA CR–198504, Dec. 1996; also AIAA Paper 96–2807, July 1996.

[63] Hendricks, R. C., and Stetz, T. T., "Flow Rate and Pressure Profiles for One to Four Axially Aligned Orifice Inlets," NASA TP–2460, May 1985.

[64] Egli, A., "Leakage of Steam Through Labyrinth Seals," *Transactions ASME*, Vol. 57, No. 3, 1935, pp. 115–122.

[65] Athavale, M. M., Steinetz, B. M., and Hendricks, R. C., "Gas Turbine Primary-Secondary Flow Path Interaction: Transient, Coupled Simulation and Comparison with Experiments," AIAA Paper 2001–3627, July 2001.

[66] Thomas, R. J., "Unstable Oscillations of Turbine Rotors due to Steam Leakage in the Clearances of the Sealing Glands and the Buckets," *Bulletin Scientifique*, Vol. 71, 1958; also NASA CP–2133, 1958, pp. 1039–1063.

[67] Alford, J. S., "Protection of Labyrinth Seals from Flexural Vibration," American Society of Mechanical Engineers, Paper 63–AHGT–9; also *Journal of Engineering for Power*, Vol. 86, Series A, Apr. 1964, pp. 141–148.

[68] Abbott, D. R., "Advances in Labyrinth Seal Aeroelastic Instability Prediction and Prevention," *Journal of Engineering for Power*, Vol. 103, April 1981, pp. 308, 312.

[69] Benckert, H., and Wachter, J., "Rotordynamic Instability Problems in High-Performance Turbomachinery," NASA CP–2133, 1980, pp. 189–212.

[70] Childs, D. W., Baskharone, E., and Ramsey, C., "Test Results for Rotordynamic Coefficients of the SSME HPOTP Turbine Interstage Seal with Two Swirl Brakes," NASA CP–3122, 1991, pp. 165–178.

[71] Muszynska, A., "The Fluid Force Model in Rotating Machine Clearances Identified by Modal Testing and Model Applications: An Adequate Interpretation of the Fluid-Induced Instabilities, Invited Lecture," *Proceedings of the 1st International Symposium on Stability Control of Rotating Machinery*, edited by D. Bently, A. Muszynksa, and J. A. Cox, ISCORMA–1, Bently Enterprises, Minden, NV, 2001.

[72] Kanki, H., Shibabe, S., and Goshima, N., "Destabilizing Force of Labyrinth Seal Under Partial Admission Condition," *Proceedings of the 2nd International Symposium on Stability Control of Rotating Machinery (ISCORMA–2)*, edited by A. Gosiewski and A. Muszynska, Max Media, Warsaw, Poland, 2003, pp. 278–288.

[73] Trutnovsky, K., "Contactless Seals, Foundations and Applications of Flows Through Slots and Labyrinths," NASA TT F 17, 352, April 1977; also *Beruhrungsfreie Diechtungen, Grundlagen und Anwendungen der Stromung durch Spalte und Labyrinthe*, VDI-Verlag GmbH, Dusseldorf, 1964, pp. 1–300.

[74] Tseng, T., McNickel, A., Steinetz, B., and Turnquest, N., "Aspirating Seal GE90 Test," *2001 NASA Seal/Secondary Air System Workshop*, NASA/CP—2002-211911/Vol. 1, 2002, pp. 79–93.

[75] Ferguson, J. G., "Brushes as High Performance Gas Turbine Seals," American Society of Mechanical Engineers, Paper 88–GT–182, June 1988.

[76] Flower, R., "Brush Seal Development Systems," AIAA Paper 90–21443, July 1990.

[77] Steinetz, B. M., Hendricks, R. C., and Munson, J., "Advanced Seal Technology Role in Meeting Next Generation Turbine Engine Goals," Propulsion and Power

Systems First Meeting on Design Principles and Methods for Aircraft Gas Turbine Engines, sponsored by the NATO Research and Technology Agency, AVT-PPS-Paper No. 11, NASA/TM-1998-206961, E-11109, 1998.

[78] Hendricks, R. C., Chupp, R. E., Lattime, S. B., and Steinetz, B. M., "Turbomachine Interface Sealing," International Conference on Materials Coatings Thin Films, ICMCTF 2005, American Vacuum Society, Paper 608, Feb. 2005.

[79] Childs, D. W., Vance, J. M., and Hendricks, R. C. (eds.), "Rotordynamic Instability Problems in High-Performance Turbomachinery," *NASA Conference Publications*, NASA CP–2133, 1980; NASA CP–2250, 1982; NASA CP–2338, 1984; NASA CP–2409, 1985; NASA CP–2443, 1986; NASA CP–3026, 1988; NASA CP–3122, 1990; NASA CP–3239, 1993; NASA CP–3344, 1997; and "Instability in Rotating Machinery," NASA CP–2409, 1985.

[80] Hendricks, R. C., Liang, A. D., and Steinetz, B. M. (eds.), *Seals Code Development and Seal and Secondary Air Systems Workshops Conference Publications*, NASA CP–10124, 1992; CP–10136, 1993; CP–10181, 1995; CP–10198, 1996; CP–208916, 1998; CP–210472, 2000; CP–211208, 2001; CP–211911, 2002; CP–212458, 2003; and CP–211963, 2004.

[81] Chupp, R. E., and Holle, G. F., "Generalizing Circular Brush Seal Leakage Through a Randomly Distributed Bristle Bed," *ASME Journal of Turbomachinery*, Vol. 118, Jan. 1996, pp. 153–161.

[82] Bhate, N., Thermos, A. C., Aksit, M. F., Demiroglu, M., and Kizil, H., "Non-Metallic Brush Seals for Gas Turbine Bearings," American Society of Mechanical Engineers, Paper GT–2004–54296, June 2004.

[83] Dinc, S., Demiroglu, M., Turnquist, N., Toetze, G., Maupin, J., Hopkins, J., Wolfe, C., and Florin, M., "Fundamental Design Issues of Brush Seals for Industrial Applications," *Journal of Turbomachinery*, Vol. 124, April 2002, pp. 293–300.

[84] Aksit, M. F., Dogu, Y., and Gursoy, M., "Hydrodynamic Lift of Brush Seals in Oil Sealing Applications," AIAA Paper 2004–3721, July 2004.

[85] Whitlock, D. C., "Oil Sealing of Aero Engine Bearing Compartments," AGARD–CP–237, Paper 7, 1978; also AGARD–AR–123, 1978.

[86] Short, J. F., Basu, P., Datta, A., Loewenthal, R. G., and Prior, R. J., "Advanced Brush Seal Development," AIAA Paper 96–2907, July 1996.

[87] Chen, L. H., Wood, P. E., Jones, T. V., and Chew, J. W., "Detailed Experimental Studies of Flow in Large Scale Brush Seal Model and a Comparison with CFD Predictions," *Journal of Engineering for Gas Turbines and Power*, Vol. 122, No. 4, 1999, pp. 672–679; also American Society of Mechanical Engineers, Paper 99–GT–218, June 1999.

[88] Carslaw, H. S., and Jaeger, J. C., *Conduction of Heat in Solids*, 2nd ed., Oxford Univ. Press, Oxford, England, U.K., 1959, pp. 269–270.

[89] Hendricks, R. C., Schlumberger, J., Braun, M. J., Choy, F. S., and Mullen, R. L., "A Bulk Flow Model of a Brush Seal System," American Society of Mechanical Engineers, Paper 91–GT–325, June 1991.

[90] Dogu, Y., and Aksit, M. F., "Brush Seal Temperature Distribution Analysis," American Society of Mechanical Engineers, Paper GT2005–69120, June 2005.

[91] Soditus, S. M., "Commercial Aircraft Maintenance Experience Relating to Current Sealing Technology," AIAA Paper 98–3284, July 1998.

[92] Proctor, M. P., and Delgado, I. R., "Leakage and Power Loss Test Results for Competing Turbine Engine Seals," American Society of Mechanical Engineers, Paper GT-2004-53935, June 2004.

[93] Basu, P., Datta, P., Johnson, A., Loewenthal, R., and Short, J., "Hysteresis and Bristle Stiffening Effects of Conventional Brush Seals," *Journal of Propulsion for Power*, Vol. 110, No. 4, 1994, pp. 569–575.

[94] Holle, G. F., Chupp, R. E., and Dowler, C. A., "Brush Seal Leakage Correlations Based on Effective Thickness," *Fourth International Symposium on Transport Phenomena and Dynamics of Rotating Machinery*, edited by W. J. Yang and J. H. Kim, Vol. A, Begell House, New York, 1992; distributed by CRC Press, Boca Raton, FL, 1993, pp. 296–304.

[95] Flower, R. F. J., "Brush Seal with Asymmetrical Elements," U.S. Patent No. 5135237, Aug. 1992.

[96] Hendricks, R. C., Carlile, J. A., Yoder, D., and Braun, M. J., "Investigation of Flows in Bristle and Fiberglass Brush Seal Configurations," *Fourth International Symposium on Transport Phenomena and Dynamics of Rotating Machinery*, ISROMAC-4, Preprint, Vol. A, Begell House, New York, 1992, pp. 315–325.

[97] Braun, M. J., and Kudriavtsev, V. V., "A Numerical Simulation of a Brush Seal Section and Some Experimental Results," *Transactions of the ASME*, Vol. 117, Jan. 1995, pp. 190–202.

[98] Turner, M. T., Chew, J. W., and Long, C. A., "Experimental Investigation and Mathematical Modeling of Clearance Brush Seals," American Society of Mechanical Engineers, Paper 97-GT-282, June 1997.

[99] Chen, L. H., Wood, P. E., Jones, T. V., and Chew, J. W., "An Iterative CFD and Mechanical Brush Seal Model and Comparisons with Experimental Results," American Society of Mechanical Engineers, Paper 98-GT-372, June 1998; also *Journal of Engineering for Gas Turbines and Power*, Vol. 121, No. 4, 1999, pp. 656–661.

[100] Aksit, M. F, "Analysis of Brush Seal Bristle Stresses with Pressure Friction Coupling," American Society of Mechanical Engineers, Paper GT-2003-38718, June 2003.

[101] Mahler, F., and Boyes, E., "The Application of Brush Seals in Large Commercial Jet Engines," AIAA Paper 95-2617, July 1995.

[102] Chupp, R. E., Johnson, R. P., and Loewenthal, R. G., "Brush Seal Development for Large Industrial Gas Turbines," AIAA Paper 95-3146, July 1995.

[103] Chupp, R. E., Prior, R. J., and Loewenthal, R. G., "Update on Brush Seal Development for Large Industrial Gas Turbines," AIAA Paper 96-3306, July 1996.

[104] Bancalari, E., Diakunchak, I. S., and McQuiggan, G., "A Review of W501G Engine Design, Development and Field Operating Experience," American Society of Mechanical Engineers, Paper GT-2003-38843, June 2003.

[105] Diakunchak, I. S., Gaul, G. R., McQuiggan, G., and Southall, L. R., "Siemens Westinghouse Advanced Turbine Systems Program Final Summary," American Society of Mechanical Engineers, Paper GT-2002-30654, June 2002.

[106] Ingistov, S., "Compressor Discharge Brush Seal for Gas Turbine Model 7EA," *Journal of Turbomachinery*, Vol. 124, No. 2, Apr. 2002, pp. 301–305.

[107] Hendricks, R. C., *Environmental and Customer Driven Seal Requirements*, NASA CP-10136, 1994, pp. 67–78.

[108] Brown, P. F., *Status of Understanding for Seal Materials Tribology in the 80's*, NASA CP-23000, Vol. 2, 1984, pp. 811–829.

[109] Schweiger, F. A., "The Performance of Jet Engine Contact Seals," *Lubrication Engineering*, Vol. 19, No. 6, 1963, pp. 232–238.

[110] Lebeck, A. O., *Principles and Design of Mechanical Face Seals*, Wiley, New York, 1991.

[111] Steinetz, B. M., and Hendricks, R. C., "Aircraft Engine Seals," *Tribology for Aerospace Applications*, edited by Zaretsky, E. V., Society of Tribologists and Lubrication Engineers, STLE SP-37, Park Ridge, IL, 1997, pp. 595–682.

[112] Ludwig, L. P., "Self-Acting Shaft Seals," AGARD–CP–237, Paper 16, 1978; also AGARD–AR–123 and NASA TM–73890 (rev. 1978); http://www.dtic.mil/dtic/tr/fulltext/u2/a060293.pdf

[113] Dini, D., "Self Active Pad Seal Application for High Pressure Engines," AGARD–CP–237, Paper 17, 1978; also AGARD–AR–123, 1978.

[114] Boyd, G. L., Fuller, F., and Moy, J., "Hybrid-Ceramic Circumferential Carbon Ring Seal," *SAE Transactions*, Vol. 111, Pt. 1, 2002, p. 522.

[115] Smith, C. R., "American Airlines Operational and Maintenance Experience with Aerodynamic Seals and Oil Seals in Turbofan Engines," AGARD–CP–237, Paper 5, Aug. 1978; also AGARD–AR–123.

[116] Hendricks, R. C., Steinetz, B. M., Athavale, M. M., Przekwas, A. J., Braun, M. J., Dozozo, M. I., Choy, F. K., Kudriavtsev, V. V., Mullen, R. L., and von Pragenau, G. L., "Interactive Developments of Seals, Bearings, and Secondary Flow Systems with the Power Stream," *International Journal of Rotating Machinery*, Vol. 1, No. 3–4, 1995, pp. 153–185.

[117] "Turbomachine Sealing and Secondary Flows," NASA/TM—2004-211911, Parts 1, 2, and 3; PART 1: "Review of Sealing Performance, Customer, Engine Designer, and Research Issues," Hendricks, R. C., Steinetz, B. M., and Braun, M. J.; PART 2: "Review of Rotordynamics Issues in Inherently Unsteady Flow Systems with Small Clearances," Hendricks, R. C., Tam, L. T., and Muszynska, A.; PART 3: "Review of Power-Stream Support, Unsteady Flow Systems, Seal and Disk Cavity Flows, Engine Externals, and Life and Reliability Issues," Hendricks, R. C., Steinetz, B. M., Zaretsky, E. V., Athavale, M. M., and Przekwas, A. J. (see also [25]).

[118] Allcock, D. C. J., Ivey, P. C., and Turner, J. R., "Abradable Stator Gas Turbine Labyrinth Seals: Part 2 Numerical Modelling of Differing Seal Geometries and the Construction of a Second Generation Design Tool," AIAA Paper 2002–3937, July 2002.

[119] Alford, J. S., "Protecting Turbomachinery from Self-Excited Rotor Whirl," *Journal of Engineering for Power*, Series A, Vol. 87, Oct. 1965, pp. 333–344.

[120] Alford, J. S., "Protecting Turbomachinery from Unstable and Oscillatory Flows," *Journal of Engineering for Power*, Series A, Vol. 89, Oct. 1967, pp. 513, 528.

[121] von Pragenau, G. L., "Damping Seals for Turbomachinery," NASA TP–1987, March 1982.

[122] Vance, J., *Rotordynamics of Turbomachinery*, Wiley, New York, 1988.

[123] Childs, D. W., *Turbomachinery Rotordynamics Phenomena, Modeling, and Analysis*, Wiley, New York, 1993.

[124] Bently, D. E., Hatch, C. T., and Grissom, B. (eds.), *Fundamentals of Rotating Machinery Diagnostics*, Bently Pressurized Bearings Press, Minden, NV, 2002.

[125] Temis, Y. M., and Temis, M. Y., "Influence of Elastohydrodynamic Contact Deformations in Fluid Film Bearing on High-Speed Rotor Dynamics," *Proceedings of the 2nd International Symposium on Stability Control of Rotating Machinery (ISCORMA-2)*, edited by A. Gosiewski and A. Muszynska, Paper 301, Bently Pressurized Bearing Company, Minden, NV, 2003, pp. 150–159.

[126] Chen, J. P., "Unsteady Three-Dimensional Thin-Layer Navier-Stokes Solutions for Turbomachinery in Transonic Flows," Ph.D. Dissertation, Dept. of Aerospace Engineering, Mississippi State Univ., MS, Dec. 1991.

[127] Chew, J. W., "Predictions of Flow in Rotating Disk Systems Using the k-e Turbulence Model," American Society of Mechanical Engineers, Paper 88–GT–229, June 1988.

[128] Chew, J. W., Green, T., and Turner, A. B., "Rim Sealing of Rotor-Stator Wheelspaces in the Presence of External Flow," *Journal of Turbomachinery*, Vol. 114, April 1992, pp. 426–432, 439–445.

[129] Graber, D. J., Daniels, W. A., and Johnson, B. V., "Disk Pumping Test," USAF Wright Aeronautical Lab., AFWAL–TR–87–2050, Dayton, OH, Sept. 1987.

[130] Johnson, B. V., Daniels, W. A., Kaweki, E. J., and Martin, R. J., "Compressor Drum Aerodynamic Experiments with Coolant Injected at Selected Locations," *Journal of Turbomachinery*, Vol. 113, April 1991, pp. 272–280; also Vol. 114, April 1992, pp. 426–432.

[131] Johnson, B. V., Mack, G. J., Paolillo, R. E., and Daniels, W. A., "Turbine Rim Seal Gas Path Flow Ingestion Mechanisms," AIAA Paper 94–2703, June 1994.

[132] Johnson, M. C., and Medlin, E. G., "Laminated Finger Seal with Logarithmic Curvature," U.S. Patent 5,108,116, April 1992.

[133] Arora, G. K., Proctor, M. P., Steinetz, B. M., and Delgado, I. R., "Pressure Balanced, Low Hysteresis, Finger Seal Test Results," NASA/TM—1999-209191, June 1999; also ARL–MR–457, AIAA Paper 99–2686, June 1999.

[134] Proctor, M. P., Kumar, A., and Delgado, I. R., "High-Speed, High-Temperature Finger Seal Test Results," *Journal of Propulsion and Power*, Vol. 20, No. 2, 2004, pp. 312–318.

[135] Proctor, M. P., and Delgado, I. R., "Leakage and Power Loss Test Results for Competing Turbine Engine Seals," NASA/TM-2004-213049, June 2004; also American Society of Mechanical Engineers, Paper GT–2004–53935, 2004.

[136] Braun, M., Pierson, H., Deng, D., Choy, F., Proctor, M., and Steinetz, B., "Structural and Dynamic Considerations Towards the Design of a Padded Finger Seal," AIAA Paper 2003–4698, July 2003.

[137] Proctor, M. P., and Steinetz, B. M., "Non-Contacting Finger Seal," U.S. Patent 6,811,154, Nov. 2004.

[138] Braun, M., Pierson, H., Deng, D., Choy, F., Proctor, M., and Steinetz, B., "Non-Contacting Finger Seal Developments," NASA/CP-2005-213655, Vol. 1, Nov. 2004, pp. 181–208.

[139] Dunlap, P. H., Jr., Steinetz, B. M., and DeMange, J. J., "High Temperature Propulsion System Structural Seals for Future Space Launch Vehicles," NASA/TM-2004-212907, Jan. 2004 [E14304].

[140] Steinetz, B. M., and Sirocky, P. J., "High Temperature Flexible Seal," U.S. Patent No. 4,917,302, April 1990.

[141] Reed, B. D., and Schneider, S. J., "Testing of Wrought Iridium/Chemical Vapor Deposition Rhenium Rocket," NASA TM-107452, 1996.

[142] Nakane, H., Maekawa, A., Akita, E., Akagi, K., Nakano, T., Nishimoto, S., Hashimoto, S., Shinohara, T., and Uehara, H., "The Development of High-Performance Leaf Seals," *Transactions of ASME, Journal of Engineering and Gas Turbines and Power*, Vol. 126, April 2004, pp. 342–350.

[143] Gardner, J., "Pressure Balanced, Radially Compliant Non-Contact Shaft Riding Seal," *NASA Seal/Secondary Flows Workshop*, NASA CP–10198, Vol. 1, Oct. 1997, pp. 329–348.

[144] Justak, J., "Hydrodynamic Brush Seal," U.S. Patent No. 6,428,009 B2, Aug. 2001.

[145] Justak, J., "Non-Contacting Seal Developments," *2004 NASA Seal/Secondary Air System Workshop*, NASA/CP-2005-213655, Sept. 2005, pp. 101–114.

[146] Shapiro, W., "Film Riding Brush Seal," *2002 NASA Seal/Secondary Air System Workshop*, NASA/CP—2003-212458/Vol. 1, 2003, pp. 247–265.

[147] Lattime, S. B., "A Hybrid Floating Brush Seal for Improved Sealing and Wear Performance in Gas Turbine Applications," Ph.D. Dissertation, Dept. of Mechanical Engineering, Univ of Akron, OH, Dec. 2000.

[148] Braun, M. J., and Choy, F. K., "Hybrid Floating Brush Seal," U.S. Patent No. 5,997,004, 7 Dec. 1999.

[149] Kudriavtsev, V. V., Braun, M. J., and Choy, F. K., "Floating Brush Seal: Concept Feasibility Study," NASA SBIR Contractor Report, June 1995.

[150] Lattime, S. B., Braun, M. J., Choy, F. K., Hendricks, R. C., and Steinetz, B. M., "Rotating Brush Seal," *The International Journal of Rotating Machinery*, Vol. 8, No. 2, 2002, pp. 153–160.

[151] Aksit, M. F., Bhate, N., Bouchard, C., Demiroglu, M., and Dogu, Y., "Evaluation of Brush Seal Performance for Oil Sealing Applications," AIAA Paper 2003-4695, July 2003.

[152] Ingistov, S., "Power Augmentation and Retrofits of Heavy Duty Industrial Turbines Model 7EA," *Proceedings of Power-Gen International Conference*, Pennwell Corp. Tulsa, OK, 2001.

[153] Bhate, N., Thermos, A. C., Aksit, M. F., Demiroglu, M., and Kizil, H., "Non-Metallic Brush Seals for Gas Turbine Bearings," American Society of Mechanical Engineers, Paper GT2004-54296, June 2004.

[154] Pope, A. N., "Gas Bearing Sealing Means," U.S. Patent No. 5,284,347, 8 Feb. 1994.

[155] Hwang, M. F., Pope, A. N., and Shucktis, B., "Advanced Seals for Engine Secondary Flowpath," *Journal of Propulsion and Power*, Vol. 12, No. 4, 1996, pp. 794–799.

[156] Wolfe, C. E., Bagepalli, B., Turnquist, N. A., Tseng, T. W., McNickle, A. D, Hwang, M. F., and Steinetz, B. M., "Full Scale Testing and Analytical Validation of an Aspirating Face Seal," AIAA Paper 96–2802, July 1996.

[157] Bagepalli, B., Imam, I., Wolfe, C. E., Tseng, T., Shapiro, W., and Steinetz, B., "Dynamic Analysis of an Aspirating Seal for Aircraft Engine Application," AIAA Paper 96–2803, July 1996.

[158] Turnquist, N. A., Bagepalli, B., Reluzco, G., Wolfe, C. E., Tseng, T. W., McNickle, A. D., Dierkes, J. T., Athavale, M., and Steinetz, B. M., "Aspirating Face Seal Modeling and Full Scale Testing," AIAA Paper 97–2631, July 1997.

[159] Turnquist, N. A., Tseng, T. W., McNickle, A. D., Dierkes, J. T., Athavale, M., and Steinetz, B. M., "Analysis and Full Scale Testing of an Aspirating Face Seal with Improved Flow Isolation," AIAA Paper 98-3285, July 1998.

[160] Turnquist, N. A., Tseng, T. W., McNickle, A. D., and Steinetz, B. M., "Angular Misalignment Analysis and Full Scale Testing of an Aspirating Seal," AIAA Paper 1999-2682, June 1999.

[161] McNickel, A. D., and Etsion, I., "Improved Main Shaft Seal Life in Gas Turbines Using Laser Surface Texturing," NASA/CP—2002-211911/Vol. 1, Oct. 2002, pp. 111–126.

[162] von Pragenau, G. L., "Damping Seals for Turbomachinery," NASA CP-2372, April 1985, pp. 438–451.

[163] von Pragenau, G. L., "From Labyrinth Seals to Damping Seals/Bearings," *Proceedings of the 4th International Symposium on Transport Phenomena and Dynamics of Rotating Machinery (ISROMAC-4)*, CRC Press, Boca Raton, FL, 1993, pp. 234–242.

[164] Yu, Z., and Childs, D., "A Comparison of Experimental Rotordynamic Coefficients and Leakage Characteristics for Hole Pattern Gas Damper Seals and a Honeycomb Seal," NASA CP-3344, May 1997, pp. 77–93.

[165] Etsion, I., "A New Concept of Zero-Leakage Non-Contacting Mechanical Face Seal," *Transactions of ASME Journal of Tribology*, Vol. 106, July 1984, pp. 338–343.

[166] Young, L. A., and Lebeck, A. O., "The Design and Testing of a Wavy-Tilt-Dam Mechanical Face Seal," *Lubrication Engineering*, Vol. 45, No. 5, May 1989, pp. 322–329.

[167] Flaherty, A., Young, L., and Key, B., "Seals Developments at Flowserve Corporation," NASA/CP—2004-212963, Vol. 1, Sept. 2004, pp. 229–238.

[168] Salehi, M., Heshmat, H., Walton, J. F., and Cruszen, S., "The Application of Foil Seals to a Gas Turbine Engine," AIAA Paper 99-2821, June 1999.

[169] Heshmat, H., "Compliant Foil Seal," U.S. Patent 6505837 B1 14 June 2003; 19 Aug. 2003.

[170] Nalotov, O., "Step of Pressure of the Steam and Gas Turbine with Universal Belt," U.S. Patent 6,632,069 B1, 14 Oct. 2003.

[171] Lattime, S. B., and Steinetz, B. M., "Test Rig for Evaluating Active Turbine Blade Tip Clearance Control Concepts," NASA/TM—2003-212533, July 2003; also *Journal of Propulsion and Power*, Vol. 21, No. 3, 2005, pp. 552–563.

[172] Dunkelberg, K., "Commercial Airplane Nacelle Component (Engine Externals) Certification and Typical Temperature Exposure," *Proceedings of the 7th International Symposium on Transport Phenomena and Dynamics of Rotating Machinery (ISROMAC-7)*, edited by A. Muszynska, J. A. Cox, and D. T. Nosenzo, Bird Rock Publ., Minden, NV, Feb. 1998; also NASA/TM—2006-214329/Vol. 1 [NASA Advanced Subsonic Technology (AST) 028, Oct. 1998], pp. 363–380.

[173] Stoner, B. L., "The Importance of Engine Externals' Health," *Proceedings of the 7th International Symposium on Transport Phenomena and Dynamics of Rotating Machinery (ISROMAC-7)*, edited by A. Muszynska, J. A. Cox, and D. T. Nosenzo, Bird Rock Publ., Minden, NV, Feb. 1998, p. 572; also NASA/TM—2006-214329/Vol. 1 [NASA Advanced Subsonic Technology (AST) 028, Oct. 1998], pp. 435–443.

[174] Zaretsky, E. V., Hendricks, R. C., and Soditus, S., "Weibull-Based Design Methodology for Rotating Aircraft Engine Structures," *Proceedings of the 9th International Symposium on Transport Phenomena and Dynamics of Rotating Machinery (ISROMAC-9)*, edited by Y. Tsujimoto, Pacific Center of Thermal-Fluids Engineering, Honolulu, HI, Feb. 2002.

[175] Davis, D. Y., and Stearns, E. M., "Energy Efficient Engine Flight Propulsion System Final Design and Analysis," NASA CR–168219, Aug. 1985.

[176] Athavale, M. M., Przekwas, A. J., Hendricks, R. C., and Steinetz, B. M., "Development of a Coupled, Transient Simulation Methodology for Interaction Between Primary and Secondary Flowpaths in Gas Turbine Engines," AIAA Paper 97–2727, July 1997.

[177] Athavale, M. M., Przekwas, A. J., Hendricks, R. C., and Steinetz, B. M., "Coupled Transient Simulations of the Interaction Between Power and Secondary Flowpaths in Gas Turbines," AIAA Paper 98–3290, July 1998.

[178] Janus, J. M., "Advanced 3–D CFD Algorithm for Turbomachinery," Ph.D. Dissertation, Dept. of Aerospace Engineering, Mississippi State Univ., MS, May 1989.

[179] Janus, J. M., and Horstman, H. Z., "Unsteady Flow-Field Simulation of Ducted Prop-Fan Configurations," AIAA Paper 92–0521, July 1992.

[180] Campbell, D. A., "Gas Turbine Disk Sealing System Design," AGARD–CP–237 (AGARD–AR–123), Paper 18, 1978.

[181] Teramachi, K., Manabe, T., Yanagidani, N., and Fujimura, T., "Effect of Geometry and Fin Overlap on Sealing Performance of Rims Seals," AIAA Paper 2002–3938, July 2002.

[182] Wellborn, S. R., and Okiishi, T. H., "Effects of Shroud Stator Cavity Flows on Multistage Axial Compressor Performance," NASA CR–198536, Oct. 1996.

[183] Hall, E. J., and Delaney, R. A., "Investigation of Advanced Counterrotation Blade Configuration Concepts for High Speed Turboprop Systems," *Task 5—Unsteady Counterrotation Ducted Propfan Analysis Computer Program User's Manual*, NASA CR–187125, Jan. 1993.

[184] Hall, E. J., and Delaney, R. A., "Investigation of Advanced Counterrotation Blade Configuration Concepts for High Speed Turboprop Systems," *Task 5—Unsteady Counterrotation Ducted Propfan Analysis Final Report*, NASA CR–187126, Jan. 1993.

[185] Hall, E. J., and Delaney, R. A., "Investigation of Advanced Counterrotation Blade Configuration Concepts for High Speed Turboprop Systems," *Task 7—ADPAC User's Manual*, NASA CR–195472 (NASA Contract NAS3–25270), 1996.

[186] Adamczyk, J. J., Celestina, M. L., and Greitzer, E. M., "The Role of Tip Clearance in High-Speed Fan Stall," *Journal of Turbomachinery*, Vol. 115, No. 1, 1993, pp. 29–39.

[187] Bohn, D. E., Balkowski, I., Ma, H., Tummers, C., and Sell, M., "Influence of Open and Closed Shrouded Cavities on the Flowfield in a 2-Stage Turbine, with Shrouded Bladings," American Society of Mechanical Engineers, Paper GT 2003–38436, 2003.

[188] Feiereisen, J. M., Paolillo, R. E., and Wagner, J., "UTRC Turbine Rim Seal Ingestion and Platform Cooling Experiments," AIAA Paper 2000–3371, 2000.

[189] Athavale, M. M., Przekwas, A. J., and Hendricks, R. C., "A Numerical Study of the Flow-Field in Enclosed Turbine Disk-Cavities in Gas Turbine Engines," *Proceedings of the 4th International Symposium on Transport Phenomena and Dynamics of Rotating Machinery (ISROMAC–4)*, edited by W.-J. Yang and J. H. Kim, Begell House, Boca Raton, FL, and New York, 1992, pp. 92–101.

[190] Athavale, M. M., Przekwas, A. J., Hendricks, R. C., and Steinetz, B. M., "Numerical Analysis of Intra-Cavity and Power-Stream Flow Interaction in Multiple Gas-Turbine Disk-Cavities," American Society of Mechanical Engineers, Paper 95–GT–325, June 1995.

[191] Virr, G. P., Chew, J. W., and Coupland, J., "Application of Computational Fluid Dynamics to Turbine Disc Cavities," *Journal of Turbomachinery*, Vol. 116, 1994, pp. 701–708.

[192] Ho, Y. H., Athavale, M. M., Forry, J. M., Hendricks, R. C., and Steinetz, B. M., "Numerical Simulation of Secondary Flow in Gas Turbine Disc Cavities, Including Conjugate Heat Transfer," American Society of Mechanical Engineers, Paper 96–GT–67, June 1996.

[193] Athavale, M. M., Ho, Y. H., and Przekwas, A. J., "Analysis of Coupled Seals, Secondary and Powerstream Flow Fields in Aircraft and Aerospace Turbomachines," NASA Contract NAS3–27392, Final Report, Dec. 1999.

[194] Smout, P. D., Chew, J. W., and Childs, P. R. N., "ICAS–GT: A European Collaborative Research Programme on Internal Cooling Air Systems for Gas Turbines," American Society of Mechanical Engineers, Paper GT–2002–30479, June 2002.

[195] Aksit, M. F., Bagepalli, B. S., Burns, J., Stevens, P., and Vehr, J., "Parasitic Corner Leakage Reduction in Gas Turbine Nozzle-Shroud Inter-Segment Locations," AIAA Paper 01–3981, July 2001.

[196] Aksit, M., Bagepalli, B., and Aslam, S., "High Performance Combustor Cloth Seals," AIAA Paper 00–3510, July 2000.

[197] Avallone, A. E., and Baumeister, T., *Marks' Standard Handbook for Mechanical Engineers*, 9th ed., McGraw–Hill, New York, 1987, pp. 8–52.

[198] Archard, J. F., and Hirst, W., "The Wear of Metals Under Unlubricated Conditions," *Proceedings of the Royal Society*, Vol. A236, 1956, pp. 397–410.

[199] Ongun, R., Aksit, M. F., and Goktug, G., "A Simple Model for Wear of Metal Cloth Seals," AIAA Paper 04–3892, July 2004.

[200] Dogu, Y., Aksit, M. F., Bagepalli, B., Burns, J., Sexton, B., and Kellock, I., "Thermal and Flow Analysis of Cloth-Seal in Slot for Gas Turbine Shroud Applications," AIAA Paper 98–3174, July 1998.

[201] Hendricks, R. C., Braun, M. J., and Mullen, R. L., "Brush Seals Configurations for Cryogenic and Hot Gas Applications, Advanced Earth-to-Orbit Propulsion Technology 1990," NASA CP–3092, Vol. II, 1990, pp. 78–90.

[202] Proctor, M. P., Walker, J. F., Perkins, H. D., Hoopes, J. F., and Williamson, G. S., "Brush Seals for Cryogenic Applications Performance, Stage Effects, and Preliminary Wear Results in LN2 and LH2," NASA TP–3536, Oct. 1996.

[203] Aksit, M. F., Dogu, Y., Tichy, J. A., and Gursoy, M., "Hydrodynamic Lift of Brush Seals in Oil Sealing Applications," AIAA Paper 2004–3721, July 2004.

[204] Cetinsoy, E., Aksit, M. F., and Kandemir, I., "A Study of Brush Seal Oil Lift Through Long Bearing Analysis," American Society of Mechanical Engineers, Paper IJTC2006-12370, Oct. 2006.

[205] Bhushan, B., *Modern Tribology Handbook*, CRC Press, New York, 2001.

[206] Harnoy, A., *Bearing Design in Machinery: Engineering Tribology and Lubrication*, Marcel Dekker/Taylor and Francis, New York, 2003, pp. 148.

[207] Duran, E. T., Aksit, M. F., and Dogu, Y., "Effect of Shear Heat on Hydrodynamic Lift of Brush Seals in Oil Sealing," AIAA Paper 2006-4755, July 2006.

[208] Duran, E. T., Aksit, M. F., and Dogu, Y., "Oil Temperature Analysis of Brush Seals," *Proceedings of STLE/ASME International Joint Tribology Conference*, American Society of Mechanical Engineers, Paper IJTC2007-44397, Oct. 2007.

[209] Duran, E. T., and Aksit, M. F., "A Study of Brush Seal Oil Pressure Profile Including Temperature-Viscosity Effects," AIAA Paper 2008-4622, July 2008.

[210] Wheeler, H. T., "Multistage Packing," U.S. Patent: 1996779, 1935.

[211] Mech, C., "Garniture D'etancheite Pour Arbre Rotatif" ["Sealing Gasket for Rotating Shaft"], *Republique Francaise*, Pub. No. 2 650 048, Paris, 1919.

[212] Gardner, J., Basu, P., and Dutta, A., "A New Compliant Seal Concept for Aerospace Applications," *Fourth International Symposium on Transport Phenomena and Dynamics of Rotating Machinery*, preprint, Vol. A., 1992, pp. 593–612.

[213] Gardner, J., "Pressure Balanced, Radially Compliant, Non-Contact, Shaft Riding Seal," *Seals/Secondary Flows Workshop*, 1996, NASA CP-10198, Vol. 1, Cleveland, OH, 1997, pp. 329–348.

[214] Wright, C., "Resilient Strip Seal Arrangement," U.S. Patent 6267381B1, Rolls-Royce, 2001.

[215] Shinohara, T., Akagi, K., Yuri, M., Toyoda, M., Ozawa, Y., Kawaguchi, A., Sakakibara, S., Yoshida, Z., Kunitake, N., Ohta, T., Nakane, H., Ito, E., Kawata, Y., and Takeshita, K., "Shaft Seal and Turbine Using the Same," U.S. Patent 6343792B1, Mitsubishi Heavy Industries, Ltd., 2002.

[216] Hendricks, R. C., Griffin, T. A, Bobula, G. A., and Bill, R. C., "Integrity Testing of Brush Seal in Shroud Ring of T-700 Engine," American Society of Mechanical Engineers, Paper 93-GT-373, May 1993.

[217] Asada, T., Nakano, T., Nishimoto, S., Shinohara, T., Shiral, H., and Uehara, H., "Shaft Seal Mechanism, Shaft Seal Mechanism Assembling Structure and Large Size Fluid Machine," U.S. Patent 7226053 B2, 2007.

[218] Enomoto, K., Konishi, T., Miyawaki, T., Nakano, T., Shinohara, T., Toda, Y., and Yoshida, Z., "Shaft Seal for a Rotating Machine," U.S. Patent 20020117807 A1, 2002,

[219] Akagi, K., Shinohara, T., Uehara, H., and Yuri, M., U.S. Patent 20020105146 A1, 2002.

[220] Deo, H. V., Turnquist, N. A., and Yildirim, B., "Compliant Plate Seal with Self-Correcting Behavior," U. S. Patent No. US2010 0143102 (A1), 2010.

[221] Awtar, S., and Turnquist, N. A., "Compliant Seals for Turbomachinery," U.S. Patent 7419164 B2, 2008.

[222] Awtar, S., and Verma, N. K., "Compliant Plate Seals for Turbomachinery." U.S. Patent 8382119 B2, 2013.

[223] Deo, H. V., "Compliant Plate Seals for Turbomachinery Applications," *Proceedings of the 14th International Mechanical Engineering Congress & Exposition*, American Society of Mechanical Engineers, ASME IMECE 2011-64871, New York, 2011.

[224] Fleming, D. P., "Stiffness of Straight and Tapered Annular Gas Path Seals," *Journal of Tribology*, Vol. 101, No. 3, pp. 349-354; doi 10.1151/1.3453371

[225] Hendricks, R. C., Braun, M. J., Deng, D., and Hendricks, J. A., "Compliant Turbomachine Sealing," NASA TM-2011-214040, May 2011.

[226] Hendricks, R. C., et al., "Brush Seals in Vehicle Tribology," *Vehicle Tribology, Tribology Series 18*, edited by C. Cowson, C. M. Taylor, and M. Godet, Elsevier, Amsterdam, 1990, pp. 231–242.

[227] Flower, R., "Brush Seal with Asymmetrical Elements," U.S. Patent 5,135,237, 1992.

[228] Nakane, H., Maekawa, A., Akita, E., Akagi, K., Nakano, T., Nishimoto, S., Hashimoto, S., Shinohara, T., and Uehara, H., "The Development of High-Performance Leaf Seals," *Journal of Engineering Gas Turbines Power*, Vol. 126, No. 2, 2004, pp. 342–350; doi: 10.1115/1.1615257

[229] Watanabe, E., Tanaka, Y., Nakano, T., Ohyama, H., Tanaka, K., Miyawaki, T., Tsutsumi, M., and Shinohara, T., "Development of High Efficiency Stream Turbine, Mitsubishi Heavy Industries, Ltd.," *Technical Review*, Vol. 40, No. 4, Aug. 2003.

[230] Nakane, H., Maekawa, A., Akita, E., Akagi, K., Nakano, T., Nishimoto, S., Hashimoto, S., Shinohara, T., and Uehara, H., "The Development of High Performance Leaf Seals," *Proceedings of the ASME Turbo Expo 2002*, ASME Paper 2002-GT-30243, June 2002.

[231] Salehi, M., Heshmat, H., Walton, J., and Cruzen, S., "The Application of Foil Seals to a Gas Turbine Engine," AIAA Paper 99-2821, June 1999.

[232] Salehi, M, Heshmat, H., and Walton, J., "High Temperature Performance Evaluation of a Compliant Foil Seal," NASA/CP-2001-211208, Vol. 1, Cleveland, OH, 2001, pp. 171–197.

[233] DeMange, J., Taylor, S., Dunlap, P. H., Jr., Steinetz, B., Delgado, I., Finkbeiner, J., and Mayer, J., "Overview of CEV Thermal Protection System Seal Development," 2008 NASA Seal/Secondary Air System Workshop, 2009, pp. 253–269; also NASA/CP-2009-215677.

[234] Gibson, M., Takeuchi, D., Malak, M., and Hynes, T., "Second Generation Air-to-Air Mechanical Seal Design and Performance," NASA, Nov. 2011; also AIAA Paper 2011-5636, July 2011.

[235] Schweitzer, M., and Ledrappier, F., "Analysis of Metal Seals for Spent Fuel Storage Casks: Statistical Analysis," NASA, Nov. 2011; http://www.tekva.it/pdf/helicoflex.pdf.

[236] Justak, J. F., "Non-Contact Seal for a Gas Turbine Engine," U.S. Patent 8172232 B2, 8 May 2012.

[237] Justak, J. F., "Non-Contact Seal for a Gas Turbine Engine," NASA, Nov. 2011.

[238] Ludwig, L. P., Zuk, J., and Johnson, R. L., "Design Study of Shaft Face Seal with Self-Acting Lift Augmentation. IV-Force Balance," NASA TN D-6568, April 1972.

[239] Fleming, D. P., "Stiffness of Straight and Tapered Annular Gas Path Seals," *ASME Journal of Lubrication Technology*, Vol. 101, No. 3, 1979, pp. 349–355.

[240] Van der Velde Alvarez, D. E., "Test Versus Predictions for Rotordynamic and Leakage Characteristics of a Convergent-Tapered, Honeycomb Stator/ Smooth-Rotor Annular Gas Sea," M.S. Thesis, Dept. of Mechanical Engineering, Texas A&M Univ., College Station, TX, 2006; http://repository.tamu.edu/ bitstream/handle/1969.1/ETD-TAMU-1156/ VAN-DER-VELDE-ALVAREZ-THESIS.pdf?sequence=1.

[241] Hendricks, R. C., "Three-Step Cylindrical Seal for High-Performance Turbomachines," NASA TP 1849, June 1987.

[242] Hendricks, R. C., "Three-Step Labyrinth Seal for High-Performance Turbomachines," NASA TP-1848, 1987.

[243] Hendricks, R. C., "Straight Cylindrical Seal for High-Performance Turbomachines," NASA TP-1850, 1987.

[244] Hendricks, R. C., and Stetz, T. T., "Flow Through Axially Aligned Sequential Apertures of the Orifice and Borda Types," American Society of Mechanical Engineers, Paper 81-HT-79, ASME/AIChE 20th National Heat Transfer Conference, Aug. 1981.

[245] Samuddrata, O., Kumar, S., Wolfe, C., and Chupp, R. E., "Retractable Brush Seals – Design/Optimization/Validation," NASA, Nov. 2011.

[246] Carlile, J. A., Hendricks, R. C., and Yoder, D. A., "Brush Seal Leakage Performance with Gaseous Working Fluids at Static and Low Rotor Speed Conditions," NASA TM-105400, June 1992.

[247] Carlile, J. A., Hendricks, R. C., Hibbs, R. J., McVey, S. E., and Scharrer, J. K., "Preliminary Experimental Results for a Cryogenic Brush Seal Configuration," NASA TM-106236, June 1993.

[248] Ergun, S., "Flow Through Packed Columns," *Chemical Engineering Progress*, Vol. 43, No. 93, 1952, pp. 89–94.

[249] Hendricks, R. C., Flower, R., and Howe, H., "A Brush Seals Program Modeling and Developments," NASA TM 107158, June 1996.

[250] Hendricks, R. C., Wilson, J., Wu, T. Y., Flower, R. L., and Mullen, R. L., "Bidirectional Brush Seals—Post-Test Analysis," NASA TM 107501, Nov. 1997.

[251] Shapiro, W., and Lee, C. C., "Advanced Helium Purge Seals for Liquid Oxygen (LOX) Turbopumps," NASA CR-182105 (MTI-87TR72), 1989.

[252] Turnquist, N., Chupp, R., Baily, R., Burnett, M., Rivas, Bowsher, F., and Crudgington, P. "Brush Seals for Improved Steam Turbine Performance," *NASA Seals Workshop*, NASA/CP—2006-214383/Vol. 1, Nov. 2005, pp. 107–127; http://www.scribd.com/doc/92076582/Brush-Seals.

[253] Hendricks, R. C., Liang, A. D., Childs, D. W., and Proctor, M. P., "Development of Advanced Seals for Space Propulsion Turbomachinery," SAE TP Series 921028, Society of Automotive Engineers, April 1992; also NASA TM–105659, April 1992.

[254] Lemmon, E. W., Huber, M. L., and McLinden, M. O., "REFPROP Reference Fluid Thermodynamic and Transport Properties," NIST Standard Reference Database 23, Ver. 9.0, 2010.

[255] Hendricks, R. C., Kudriavtsev, V. V., Braun, M. J., and Athavale, M. M., "Flows in Pinned Arrays Simulating Brush Seals," NASA TM-107333, Dec. 1966.

APPENDIX A: FURTHER DISCUSSION ON RIM SEALING AND DISK CAVITY FLOWS

Coupling of the powerstream, seals, and cavity flows is a necessary aspect of multistage compressor and turbine design. Athavale et al. [176, 177] have reported detailed descriptions of tools and representative simulations for this application. (See also Janus and coworkers [178, 179]). These works include the interaction of the gas-path flows through the stages (blades and vanes) with those under the platform and within the cavity, as well as the sealing interface (shaft to platform), Fig. A1 [180]. Interstage-labyrinth-rim seal interface design goals are designed to minimize leakage, reduce windage and blockage, and mitigate ingestion and reintroduction of leakage flows.

Teramachi et al. [181] investigated turbine rim interface sealing (Figs. A1 and A2), providing data and some CFD results on four rim seal configurations: (0) T-on rotor, (1) T-on rotor with overlap T-on stator, (2) T-on stator with overlap T-on rotor; and (3) fish mouth on rotor with overlap T-on stator. Dummy stators were introduced, but there were no blades on the rotor. Purge gas flows are directed into the sealing cavities to mitigate/prevent gas-path flows from entering the wheel-space cavities. In this test, to determine sealing effectiveness, a tracer gas (carbon dioxide) was introduced at the gas-path seal interface. Measurements of the concentration ratio of purge gas to ingested gas define sealing effectiveness. (The experimental methods and analysis were similar to the work of Johnson et al. [131] and Graber et al. [129]).

Figure A3 shows the seal effectiveness of these configurations, where flow coefficient $C_w = Q/\nu b$ and $Re_m = Vb/\nu$, where b is the cavity outer radius, V is the mean flow speed, Q is the purge flow rate, and ν is the kinematic viscosity. Configuration (3) is the least affected by changes in overlap and configuration (0) the most; configuration (2) is quite sensitive to overlap. The high effectiveness of configuration (3) is related to the buffer cavity between the two rotor seal teeth with a stator tooth between the rotor teeth. The lowest effectiveness of configuration (1) is due to the large clearance gap although the ingestion is nearly zero. CFD results show the gap recirculation zone where powerstream gas is ingested at on-pitch positions and ejected at

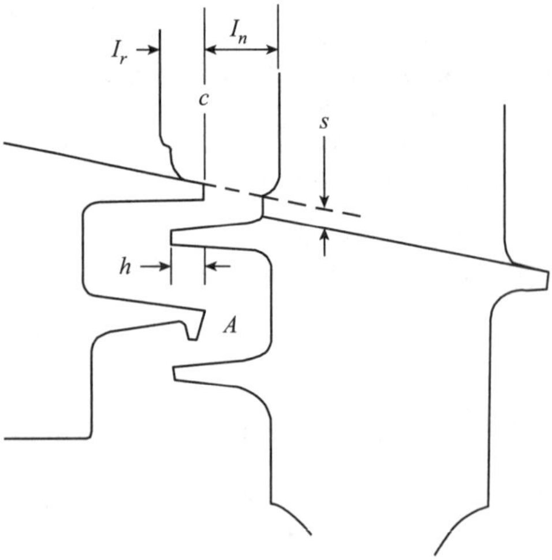

Fig. A1 Generic turbine nozzle rotor gap configuration [180].

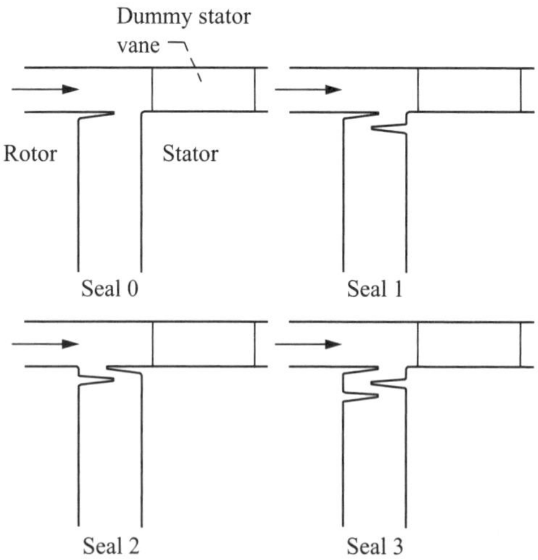

Fig. A2 Experimental rim seal configurations (courtesy of AIAA) [180].

midpitch positions (that is, flow ingestion at the vane leading edge partially returns in the midpitch region mixing with the purge air; see also Wellborn and Okiishi [32, 184] and other investigators [29–31, 32]).

Wellborn and Okiishi [32, 184] investigated the effect of leakage in a four-stage, LPC with blading design based on the NASA E^3 engine. Seal leakages did not affect upstream stages but did progressively degrade performance of the downstream stages. For each 1% change in clearance/span ratio, the pressure rise penalty was nearly 3% with a 1% drop in efficiency. Hall and Delaney [183–185] simulated the low-speed axial compressor (LSAC) experiments with Adamczyk's analysis package

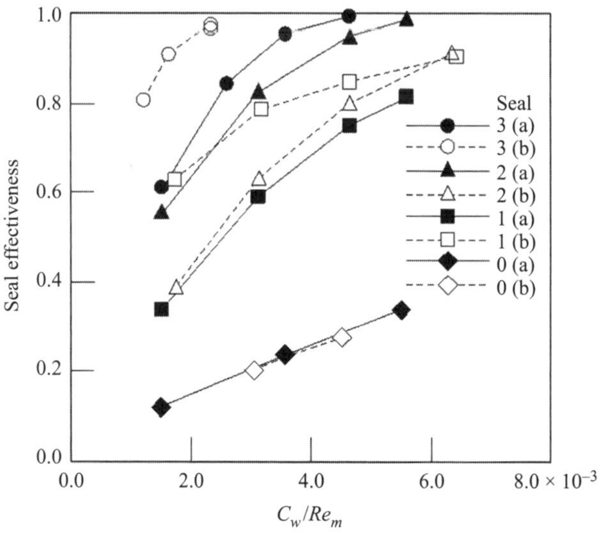

Fig. A3 Comparison of experimental rim seal data at Reynolds numbers a) 2.4 × 10^6 and b) 1.1 × 10^6 (courtesy of AIAA) [181].

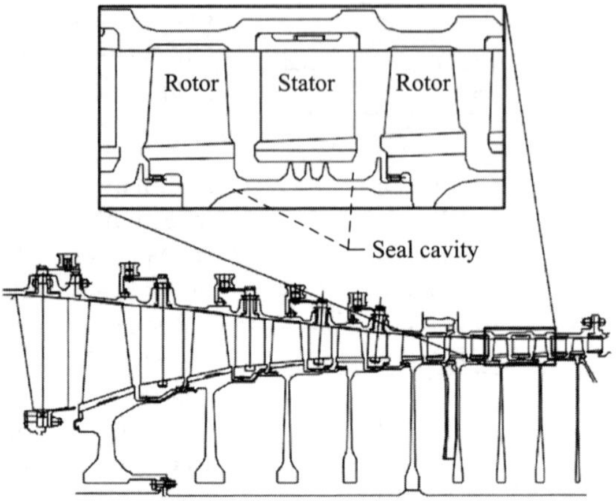

Fig. A4 Schematic of typical high-speed axial compressor with close-up view of seal cavity region under inner-banded stator [62].

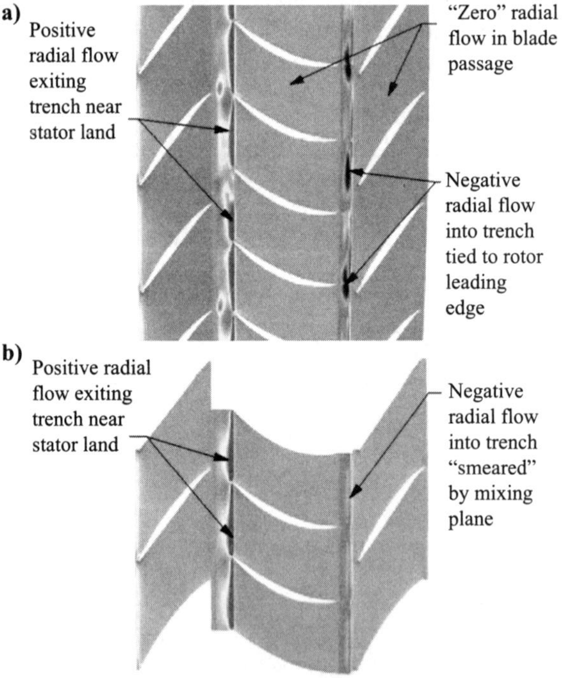

Fig. A5 Contours of radial velocity located one computational cell above hub (rotor) surface for coarse mesh: a) unsteady and b) mixing plane rotor/stator/rotor ADPAC solutions [64].

[186]. They also completed sensitivity studies, but did not address the effects on rotordynamics.

Heidegger et al. [62] presented three-dimensional solutions of the interaction between the powerstream and seal cavity flow in a typical multistage compressor (Fig. A4). Using the Allison/NASA-developed ADPAC code, they performed a parametric study on a three-tooth labyrinth seal/cavity configuration and a sensitivity study to various sealing parameters. Their study shows that the leakage flow out of the seal cavities can affect the powerstream significantly, mainly by altering the inlet flow near the stator blade root area and can potentially affect the performance of the overall compressor (Fig. A5).

Bohn et al. [33] investigated the interaction between the powerstream end-wall cavities and sealing clearances in a two-stage, shrouded-blade turbine. Three cases were considered: A, engineering design, where blade-sealing cavities interactions are neglected; B, end-wall sealing gaps adjusted to near-zero clearance; and C, end-wall radial gaps approximately 0.8 mm (Fig. A6), representing both shroud and hub configurations.

Periodic flow ingestion and circumferential fluid shuttling depends on relative blade-vane circumferential position $\Delta\theta$, as noted in Fig. A5 and implied in Figs. 16 and 17 (for compressors); time-dependent unshrouded turbine

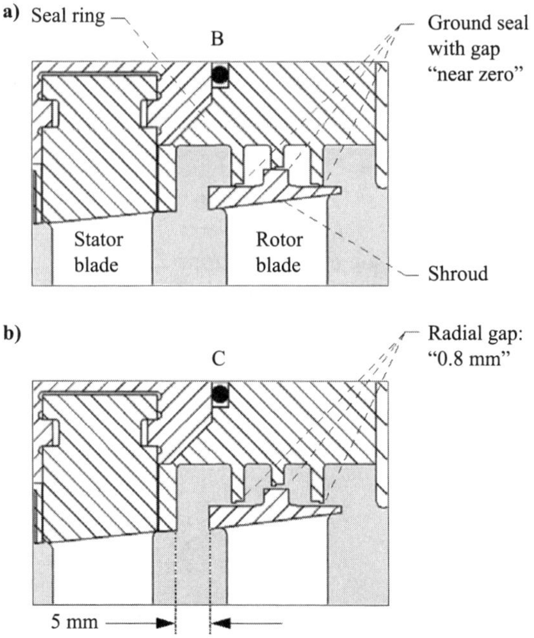

Fig. A6 Shrouded turbine configurations B and C [35]: a) configuration B, with closed radial gap and b) configuration C, with open radial gap.

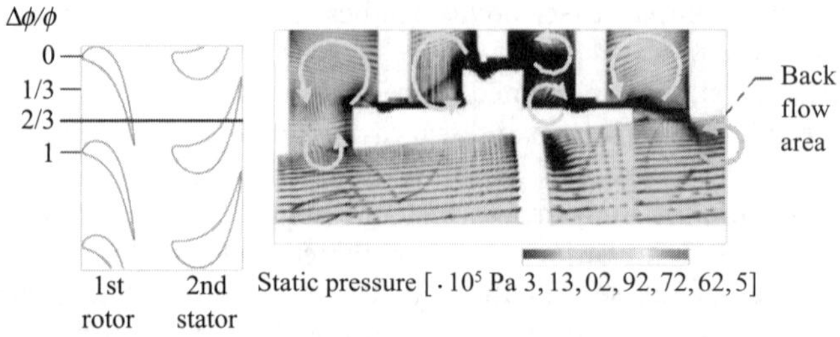

Fig. A7 Pressure distribution and secondary flows at shroud: configuration C [33].

solutions are discussed next. Similar shroud and hub flow patterns are found at the shrouded turbine interfaces, configurations B and C. At the shrouded interface, flows are sucked into the cavities near $\Delta\varphi \approx 0$, shuttled circumferentially, and ejected near $\Delta\varphi \approx$ midspan, Fig. A7. Hub interface flow patterns are illustrated in Fig. A8. In both cases, the injection of low momentum fluid into the powerstream increases passage flow blockage and decreases component polytrophic efficiency; the effect is more detrimental at the hub than at the shroud. These computational results were verified experimentally by Bohn et al. [187], who also investigated unshrouded blading configurations.

Feiereisen et al. [188] completed an experimental study of the primary and secondary flows in a turbine rig. Their study represents an advanced first attempt at understanding this interaction and at generating data for validation. CFD techniques provide detailed flowfield information on complex cavity shapes that cannot be treated with analytical methods (for example, see Athavale et al. [189, 190], Chew and colleagues [127, 128], Virr et al. [191], and Ho et al. [192]).

Ho et al. [192] and Athavale et al. [193] found, in studying the Allison 501D turbine, that ingested fluid could work its way well into the disk space, even though purge fluid flows were substantial. Figures A9–A11 show the calculated

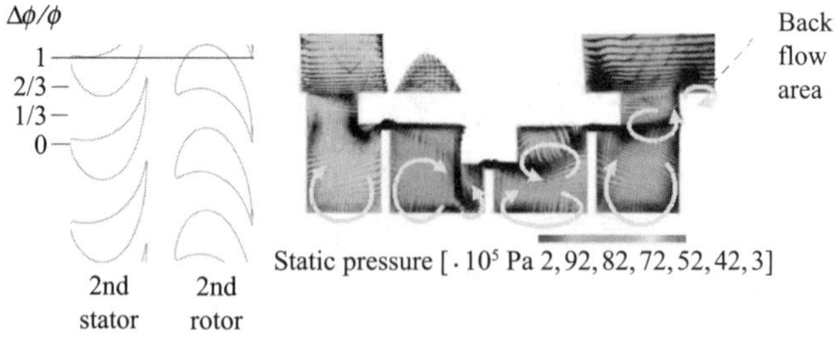

Fig. A8 Pressure distribution and secondary flows at hub: configuration [33].

interaction between powerstream and secondary flows. Without conjugate heat transfer, the calculations would not match the Allison 501D turbine design data. It is most important at powerstream interfaces where the thermal gradient from the platform to the hub is significant. These aerothermomechanical loads can drastically affect disk and engine life.

The coupled codes (SCISEAL and MS–TURBO) [65] have been applied to several experimental test rig data sets showing conditions under which ingested flow can be controlled. The configuration (Figs. A12 and A13) is a 30 deg–pi-sector with four vanes (stators) and five blades (rotors) simulating the stator/rotor set (48/58) with a three-tooth labyrinth seal and overlap rim seals [188]. Athavale et al. [65] found that a recirculation zone in the rim seal was present at the lower purge flow rate but was absent at the higher purge flow rate. The recirculation allows some gas ingestion into the rim seal area. This gas can then travel inside the cavity by both diffusion and convection (Fig. A14). Two important observations can be made: 1) the interface velocities show a tangential component that is lower than the rotor speed, and this slow fluid alters the angle of attack near the roots of the rotor blades and can cause loss of power (turbine) and stall (compressor); 2) the rotor blades have the expected upstream pressure rise, which affects the flow in the rim seal and the cavity (enhances ingestion), although this disturbance is rather small.

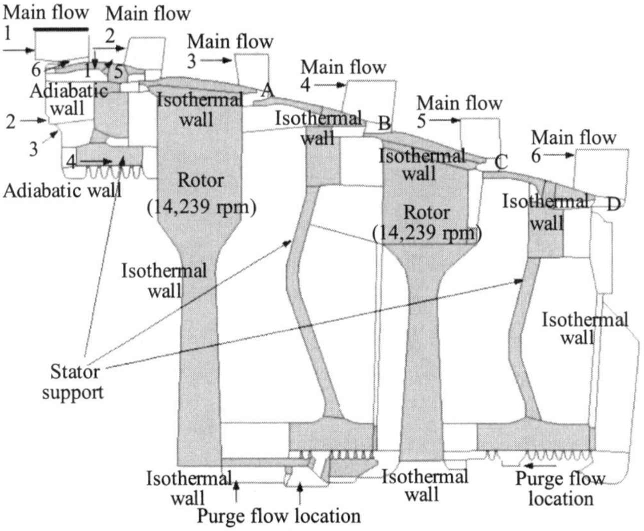

Fig. A9 Flow domain and conjugate heat-transfer calculations of all inner disk cavity pairs. Shaded areas denote conjugate heat transfer. Static pressures are specified at six main flow exits [193].

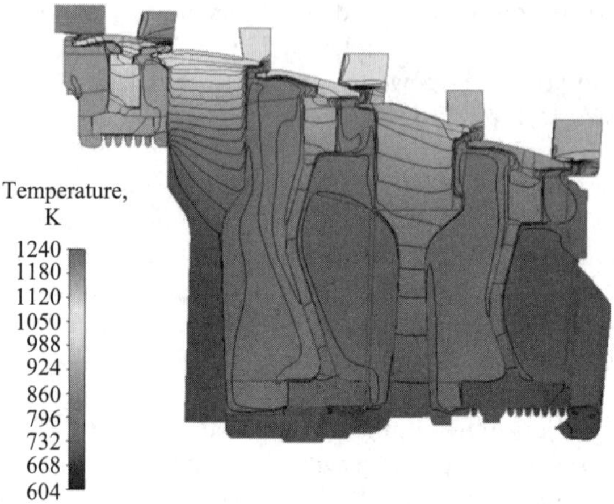

Fig. A10 Temperature field in fluid and solid parts of turbine cavities (absolute frame) [193].

Smout et al. [194] in a CFD analysis of rim sealing cite some collaborative investigations of rotating cavity ventilation, bearing cavity purge and cooling, pressure balance, and sealing rotor/stator gaps. The turbine slinger determines the preswirl of cooling air entering the HPT blades. Controlling preswirl in

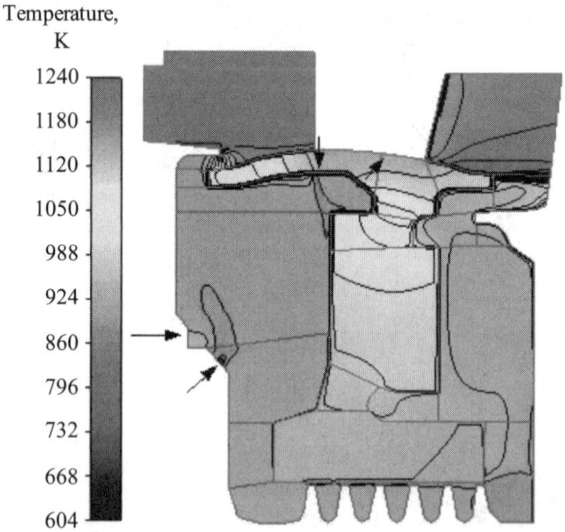

Fig. A11 Details of streamlines and temperatures in stage 1–2 cavities with conjugate heat transfer (absolute frame) [193].

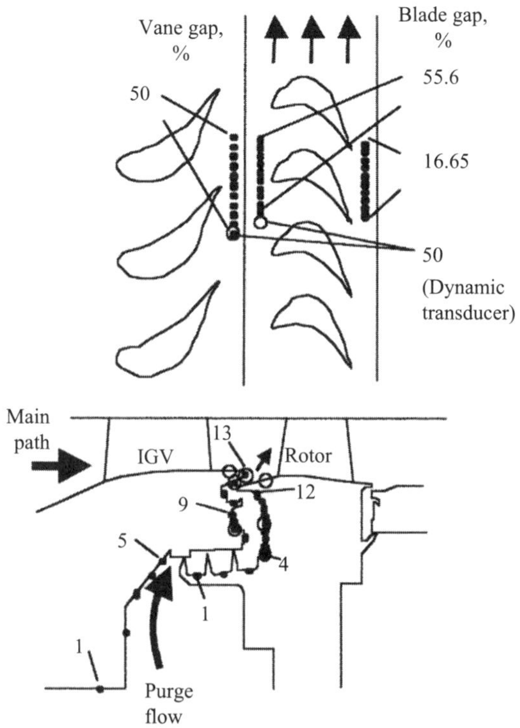

Fig. A12 Locations of pressure taps in United Technologies Research Corp. experimental rig. Dots denote steady-pressure, and circles denote transient pressure measurements [65].

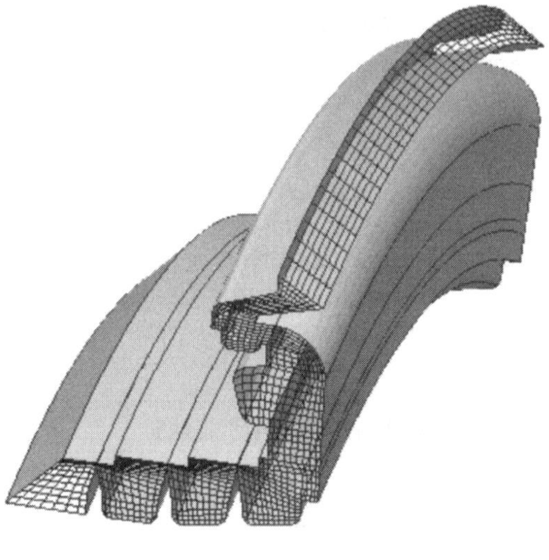

Fig. A13 Computational grid in disk cavity of high-pressure rig [65].

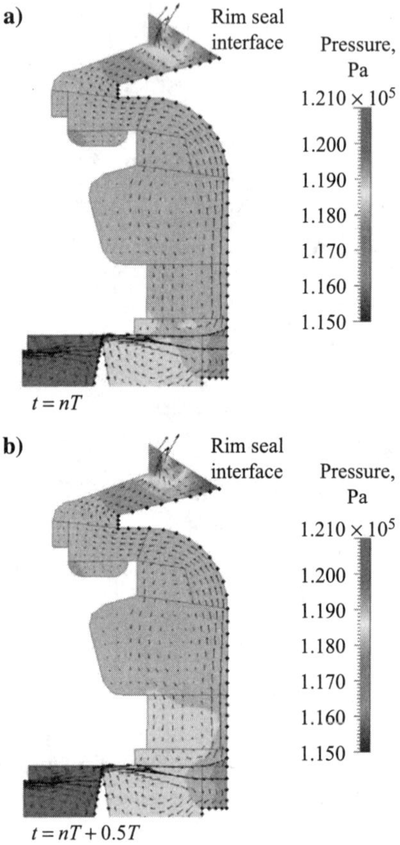

Fig. A14 Time-dependent cavity flows for 0.69% purge flow (absolute frame): a) time-transient pressures and b) velocity vectors in cavity. Here $\eta = Re_{feed} = Re_{turbine}^{0.8} = 0.005$, *t* is time, *n* is cycle number, and *T* is cycle time [65].

power and secondary flow streams are very important for rotordynamics and power-on-demand cycling. For preswirl analysis and control methods, see Thomas [66], Benckert and Wachter [69], NASA Conference Publications [79], von Pragenau [121], Childs [123], Muszynska [71], Bently et al. [124], and Hendricks et al. [117].

APPENDIX B: FURTHER DISCUSSION OF METALLIC CLOTH SEALS

This appendix is devoted to metallic cloth seals, but if proper properties are used, the analysis can be used for other weaves, such as ceramics. Note that if multiple

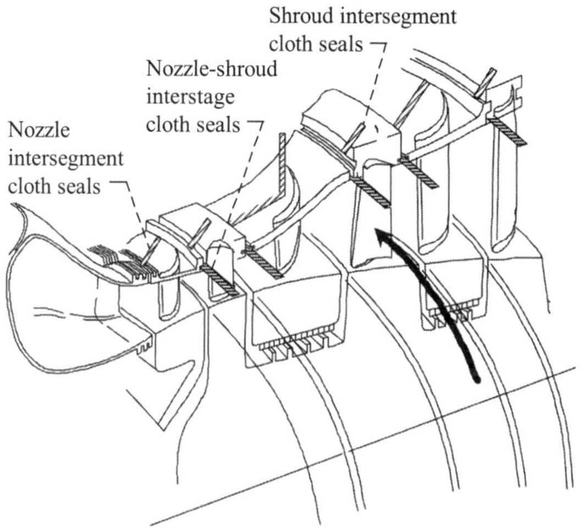

Fig. B1 Advanced flexible seals for gas turbine engines.

strands are used, rather than wires, as in fabrics or textiles, some modifications might be needed in wear volume and contact-area calculations.

Cloth seals are used at the junctions between the components in combustor and turbine sections (Figs. 3b and B1). High temperatures in these sections require intersegment gaps to accommodate large thermal expansions and to aid in cooling [195].

Large varying gaps with relative motion of the mating parts require proper sealing. A seal should be sufficiently thin and compliant to accommodate large misalignments. On the other hand, long inspection intervals and large vibration levels, which are typical of large industrial engines, require thick seal sections to provide sacrificial wear volume. A lack of flexibility can result in poor sealing and excessive wear. Cloth seals, formed by combining thin sheet metals (shims) and layers of densely woven metal cloth, address the compliance and sealing performance issues. Decoupling the structural and wear-related elements of the seal design allows the independent optimization of sheet metal for high-temperature strength while the cloth material and weave can be optimized for maximum wear and oxidation resistance.

A. SEAL CONSTRUCTION

1. INTERSTAGE SEALS

Typically, nozzle and shroud intersegment junctions have slots for seal strips (Fig. B1). As illustrated in Figs. 3, 12, and B2, cloth seal designs for such applications require a simple wrapping of a layer of cloth around thin flexible shims.

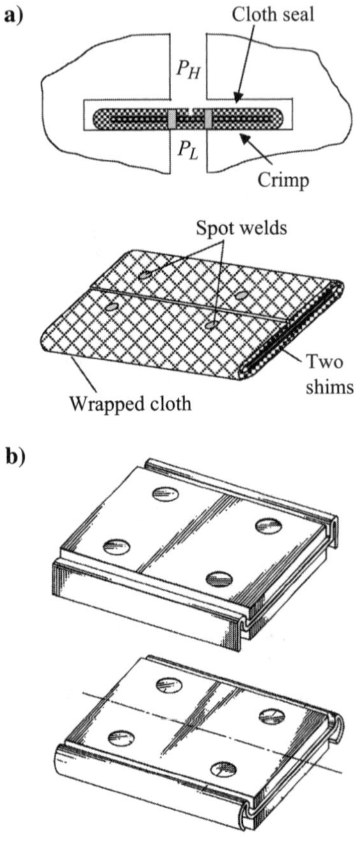

Fig. B2 a) Typical wrapped cloth seal; b) crimped cloth seals.

Further leakage reduction can be achieved by a crimped design, in which shims are crimped over the edge of the cloth layers [10, 18–20, 195, 196]. These composite cloth seal strips replace commonly used nozzle-shroud intersegment metallic splines that are inserted into the deep slots in the mating parts. Properly designed crimps reduce the leakage flow very effectively; care should be taken to ensure that sufficient flow remains, as required for slot and seal cooling.

2. COMBUSTOR SEALS

Combustion dynamics and excessive thermal misalignments make combustor sealing more challenging than the nozzle-shroud intersegment applications. In gas turbines with can-annular combustion systems, the combustor seals are used to seal the gap between the can transition duct (TD) and the first-stage nozzles (FSNs). A typical sealing junction involves two TDs (cans) and multiple FSN segments. Large axial offsets and relative skew misalignments between neighboring cans are quite common. As shown in Fig. B3, these junctions are typically sealed using formed metal strips designed to take relative axial and radial motion by sliding in grooves machined in the TD and FSN [196]. The FSN is, however, made of segments that experience relative misalignments, causing the seal to stick in the FSN slot. Jamming the seal on the FSN side results in wear of the seals by the TD because of the relative dynamic motion. Heavy wear on the seal and in the TD slots is commonplace. Seal failure can cause occasional forced power outages. Combustor cloth seals have addressed this need for flexibility at the TD-FSN junctions [196]. As illustrated in Fig. B4, TD-FSN cloth seals utilize a radial lip formed by a flexible cloth-shim assembly. Cloth seals also incorporate an interference fit providing a uniform seal-slot contact under any condition, thereby providing reduced leakage. When the seal is jammed in the FSN slot, relative vibratory motion is absorbed by flexing of the cloth assembly rather than the wearing on the rigid seal frame.

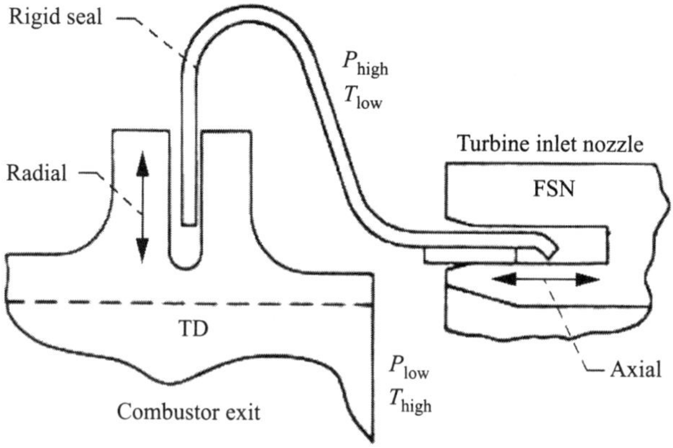

Fig. B3 Typical combustor seal assembly.

B. MATERIALS SELECTION

Oxidation and wear resistance are the key attributes needed in the cloth fiber material for a cloth seal. The structural shim must also have high-temperature strength and oxidation resistance, along with appropriate creep and fatigue properties. A typical cloth fiber material is Haynes 25 (also referred to as L605), which is used for its superior high-temperature wear resistance. For applications beyond 750°C (~1400°F) however, the increased oxidation rate can reduce life drastically. For higher temperatures, Haynes 188 can be considered. Haynes 188 has excellent oxidation resistance, albeit it with higher wear rates, requiring typical engineering tradeoffs (see Secs. III.A.2, "Interface Materials," and III.A.3, "Designing Abradable Materials for Turbomachinery").

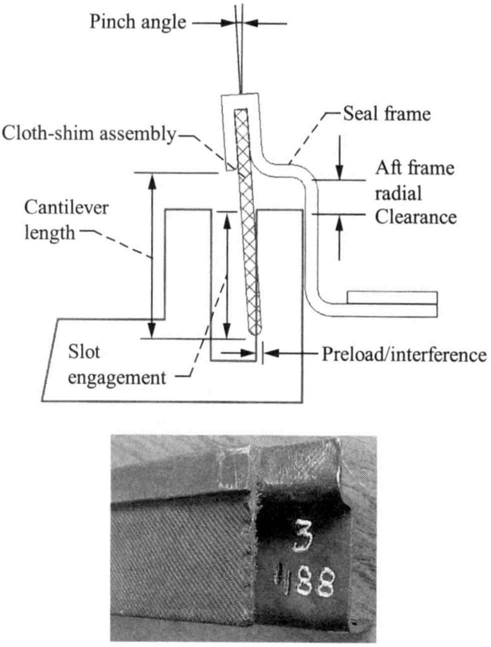

Fig. B4 Combustor cloth seals.

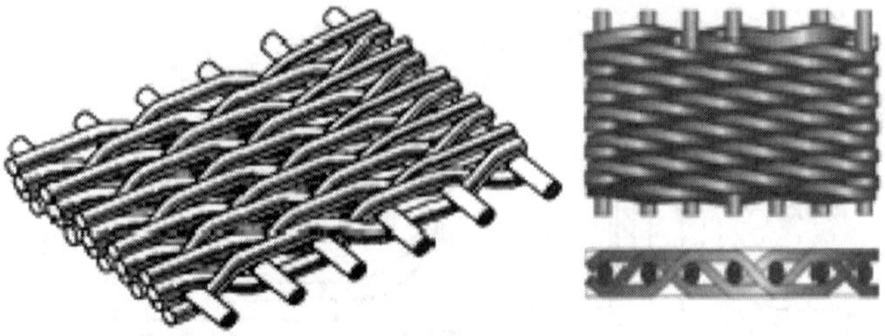

Fig. B5 Dutch twill weave.

Inconel X-750 is the shim material of the choice for applications at below 600°C (~1100°F). It is a precipitation-hardened, high-strength, and fatigue-resistant spring material. Typically, in this temperature range, Inconel X-750 is used for combustor seals. Haynes 188 is used for nozzles and shrouds where seal temperatures run higher. The combustor seals run cooler because, first, coolant air is colder at the combustor section and then gets hotter as it moves through the turbine section, resulting in less efficient cooling, and second, the combustor cloth seals are placed in the cooler transition piece aft frame slot, which is embedded in cold compressor discharge air.

C. WEAVE SELECTION

Proper cloth selection involves cloth weave, mesh density, and orientation. The most common metal cloth constructions can be grouped as plain, twill, Dutch twill, and stranded weaves. Cloth seal weave selection is based on key parameters pertaining to wear and leakage performance, as well as mesh integrity. Dutch twill weave combines unequal wire diameters with staggered and alternating passes (see Fig. B5). It makes the optimal construction for cloth seal designs as it combines the benefits of a high-density mesh with relatively large fiber diameters. Dutch twill cloth also offers an interlocking construction, which results in higher mesh integrity during local cuts. Although there are tighter weave types, like micron weave, these require small wires to achieve high density. High-temperature gas turbine sealing applications require oxidation resistance, making larger wire diameters more beneficial [20].

Cloth selection requires leakage and wear testing of various cloth samples. As illustrated in Fig. B6, there are four relevant orientations for a cloth weave: along the warp wires (1A), along the shute wires (1B), diagonal (2), and normal to the cloth surface (3). The warp wires are the main wires running the length of woven cloth. The shute wires run perpendicular to the warp wires or across the cloth as woven. These are sometimes referred to as “fill” or “weft” wires. Three different leakage tests are performed in the plane of the cloth. Cloth samples are tested parallel and orthogonal to the warp direction of the weave as well as diagonally to the weave direction. An additional leakage test is performed normal to the cloth

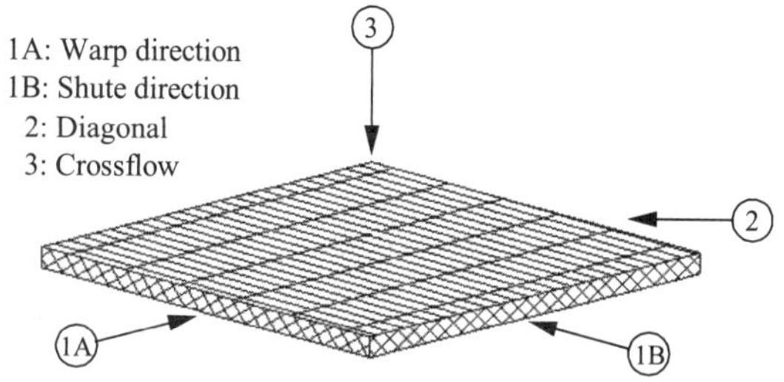

Fig. B6 Cloth seal weave orientation.

surface, measuring the crossflow. Apart from the crossflow, leakage in other orientations depends more on mesh density than the orientation. Experience shows that diagonal orientation yields the best wear performance [20]. Diagonal orientation also helps maintain weave integrity if a local cut is incurred during operation. Finally, using cloth in the diagonal orientation enhances flexibility by allowing the mesh to distort, rather than pull on warp or shute wires during deformations.

D. DESIGN CONSIDERATIONS

Cloth seal design requires careful engineering to optimize flexibility while maintaining structural strength and robustness. The inherent decoupling of structural and wear-related components allows better control over the design. Before the seal design process begins, the operating conditions (or flight design envelope) must be established. Some of the major tasks for a proper cloth seal design include the following:

1. Optimizing seal dimensions to prevent seal jamming or loss of engagement during cold build, startup, steady state, shutdown, and trip-shutdown
2. Ensuring that the differential pressure across the seal will not cause permanent yield deformation or excessive creep over the life span
3. Ensuring that there is adequate pressure load to seat and stabilize the seal
4. Ensuring that leakage airflow will be sufficient to keep seal temperatures at acceptable levels
5. Ensuring that cloth seal wear life will be sufficient under the combined vibration and high differential pressure load
6. Identifying leakage performance improvement and integration into engine flow system.

The analyses and experimentation required to fulfill these tasks include geometric analyses for engagement and jamming, finite element structural and stress analyses, rubbing wear tests, wear analyses, thermal flow analyses, subscale leakage performance tests, and analyses of leakage performance data. Figure B7 presents an outline of the cloth seal design process.

Apart from a good choice of materials and a proper weave, cloth seal design parameters include the number and thickness of the shims for structural sealing loads, the thickness of the cloth layer for sufficient wear performance, and the minimum cloth fiber diameter for adequate wear and oxidation resistance. Typically, shim thickness can be determined through simple analytical calculation. In critical cases, shim stress is calculated with detailed finite element models incorporating all of the pressure, preload, and frictional contact loading. Shim stresses should be checked for worst gap and offset conditions under maximum operating pressure loads. As shown in Fig. B8, for a combustor cloth seal, maximum stress occurs near the pinch point [196]. Shim stresses can be determined under steady-state and transient conditions. Changes in stress as a result of combustion dynamics can also be determined. Analyses can be performed either by beam theory or using a finite element analysis (FEA). In both cases, shims are treated as a sequence of individual laminates made of homogenous material. Ignoring the limited additional support of the woven cloth layer will make stress analyses simpler and more conservative.

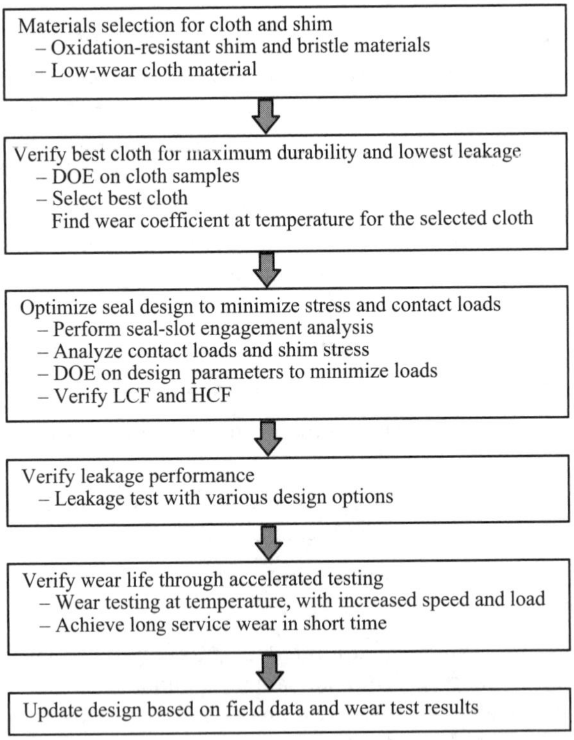

Fig. B7 Cloth seal design methodology.

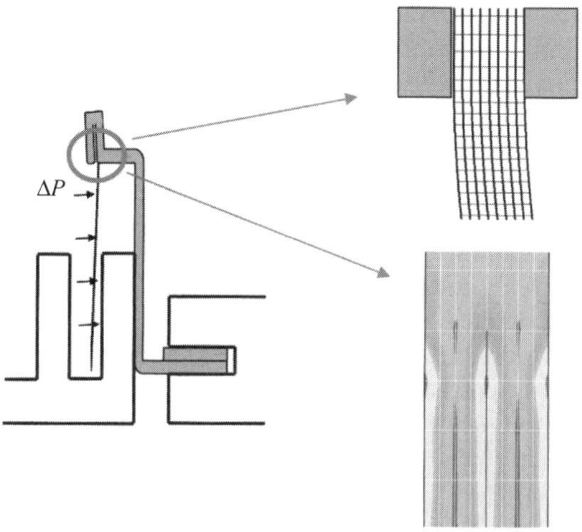

Fig. B8 Finite element analysis of combustor cloth seal shim stresses.

As a cloth assembly is preloaded, vibratory motion of the TD aft frame induces alternating stresses in the shims. The mean and alternating stress components can be determined using FEA. HCF analysis can then be performed using standard formulations [197]:

$$\left(\frac{S_a}{S_e}\right)^m + \left(\frac{kS_m}{S_{\text{ut}}}\right)^p = 1 \tag{B1}$$

where S_a is the alternating stress, S_m is the mean stress, and S_{ut} is the ultimate tensile stress. Material endurance limit S_e must be modified using correction factors for surface, size, load, temperature, reliability, and other like factors. Values for factor k and exponents m and p are determined based on the fatigue-life criteria used in the analysis. If a modified Goodman fatigue-life model is selected as the failure theory, $k = m = p = 1$ (see [197] for other failure criteria), and the preceding relation becomes

$$\frac{S_a}{S_e} + \frac{S_m}{S_{\text{ut}}} = \frac{1}{n} \tag{B2}$$

where n is added in the relation to define a factor of safety against limited cycle life. A typical shim material could be Inconel X750. Mean and alternating stress levels S_m and S_a are determined by the seal design and load levels. Typically, S_m is dictated by the differential pressure load and the deformation due to preload or thermal distortions (if any), and S_a is dictated by the level of pressure oscillations and relative vibration levels of the mating parts. These pressure oscillations can be a result of combustion dynamics as in combustor seals or a result of pressure variations caused by blade passing at shrouds. A typical safety factor n is of the order of 1.5 to 2. If n is less than unity, the seal would have limited life for the given application. Commonly, limited life means below 10^6 cycles. Therefore, n should be kept high as possible by reducing mean and alternating stress levels. Using multiple (thin) shims is a good means to reduce stress levels without adding much stiffness. Rather than deform and comply for uniform slot contact, stiff seals rotate and tiptoe, causing concentrated contact loads, which result in accelerated wear rates.

To ensure durability, a detailed wear analysis is needed. Wear of cloth mesh is much more complicated than contact of two solids. Wear rate of a woven cloth structure depends on weave and orientation and varies through the thickness. Most wear models in the literature are based on the basic relationship developed by Archard and Hirst [198].

$$W = K\left(\frac{FV}{3H}\right)t \tag{B3}$$

where W is wear, K is the wear coefficient, F is the load, V is the rub-interface velocity, H is the material hardness, and t is the time. Dividing both sides by contact area, the wear relation can be represented in terms of PV (contact pressure $\times$ rub-interface velocity), which represents contact severity conditions. Typically, PV charts are available for various materials to determine wear rates. Determining the actual contact area for a cloth mesh, however, involves some detailed geometric analysis.

As illustrated in Fig. B9, when a cloth seal is pressed against the slot surface, the actual contact area of the mesh is much smaller than the nominal slot engagement area [199]. Both flat wire sections and ellipsoidal corner areas should be

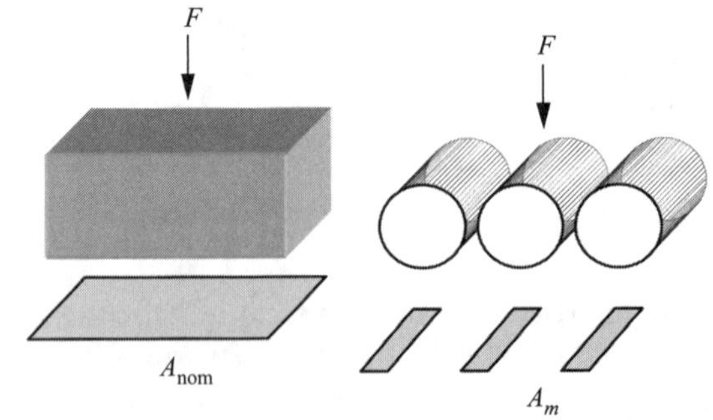

Fig. B9 Nominal and mesh contact areas.

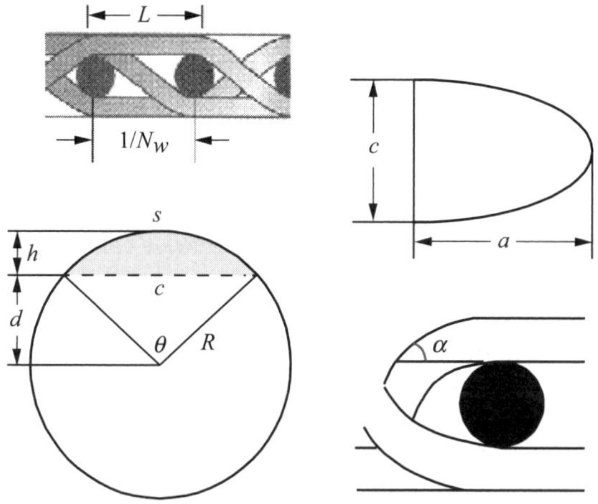

Fig. B10 Cloth wear geometry.

included in the contact-area calculations (see Fig. B10). As wear progresses, the actual mesh contact area changes with wear depth. For a Dutch twill weave actual contact area can be calculated as

$$A_m = \left[L2\sqrt{h(2R-h)} + \pi\left(\frac{h}{\tan\alpha}\right)c\right]\left[n_{\text{shute}}\left(\frac{n_{\text{warp}}}{4}\right)A_{\text{nom}}\right] \tag{B4}$$

where L is the length of the flat wire sections, h is the wear depth, R is the wire radius, η_{warp} is the number of warp wires per length, η_{shute} is the number of shute wires per length, and A_{nom} is the nominal contact area [199]. Variables a, c, and α are defined in Fig. B10. Here α can be calculated from the geometry as

$$a = \frac{h}{\tan\alpha} \tag{B5}$$

For a proper wear analysis, the "volume lost" needs to be determined. For a given wear depth, the volume lost from the mesh can be calculated as [199]

$$V_{\text{mesh}} = (V_{\text{flat wire}} + V_{\text{ellcor}})\left[n_{\text{shute}}\left(\frac{n_{\text{warp}}}{4}\right)A_{\text{nom}}\right] \tag{B6}$$

where

$$V_{\text{flat wire}} = L\left[R^2 \cos^{-1}\left(\frac{R-h}{h}\right) - (R-h)\sqrt{2h(R-h)}\right] \tag{B7}$$

$$V_{\text{ellcor}} = \left(\frac{2}{3}\right)\pi ach \tag{B8}$$

As most of the parameters affecting wear are not constant, detailed transfer functions are needed to evaluate the effects of parameter variations on seal wear life. Using a statistical approach for the design, detailed Monte Carlo simulations can be conducted. Figure B11 shows a sample analysis for estimated seal wear after 12,000 h [196]. This analysis takes into account the variation of key design and operating parameters, such as slot engagement, cantilever length, preload, operating pressure, wear coefficient, and rubbing velocity. The parameters are assumed to vary normally within the allowed tolerance and operating condition limits. A statistical design approach allows for a better understanding of field performance variations. A statistical study also provides valuable sensitivity data.

In most cases, leakage flow also provides cooling air to some critical parts and the seal itself. When achieving a tighter leakage performance, one should also consider the temperature increase in the seal and the surrounding slot surfaces. For critical regions, detailed thermal and flow analyses of the sealing system might be necessary. In such cases, actual leakage flow rates from the seal tests are used for better accuracy. CFD methods are used to analyze whether the temperatures exceed the temperature limitations of the cloth or shims. Figure B12 shows a sample two-dimensional thermal-flow model with symmetrical boundary conditions [200]. The conduction rate through the cloth layer should be reduced because of porosity and leakage convection.

Conduction rate is defined as the change in temperature across a unit thickness of the material in question. In porous materials with increasing porosity, solid

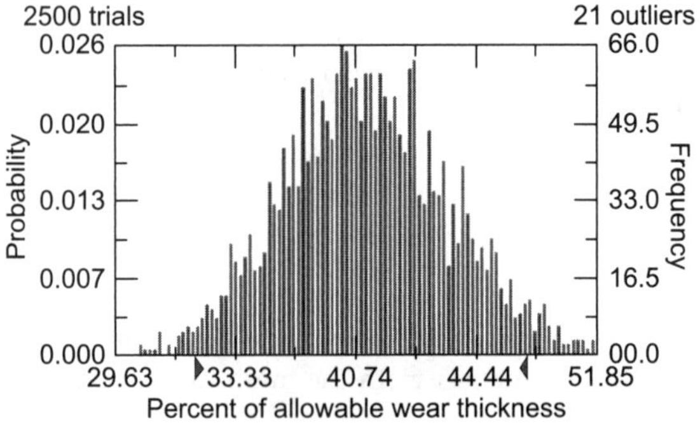

Fig. B11 Sample Monte Carlo analysis for combustor seal wear in TD aft frame slot, with 95% certainty from 33.33 to 48.15% allowable thickness.

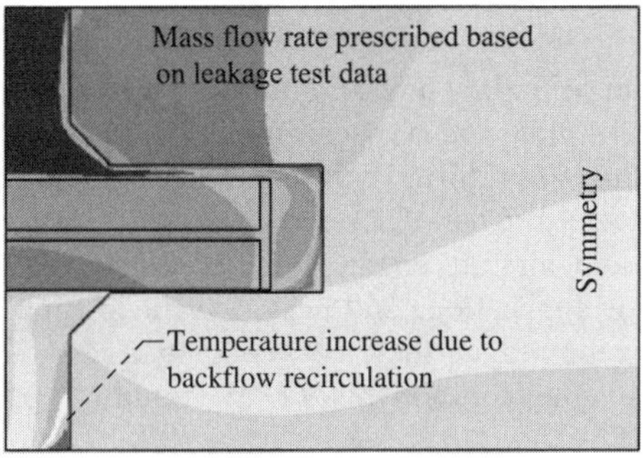

Fig. B12 Two-dimensional thermal flow model of a nozzle-shroud intersegment cloth seal.

material percentage, which defines the available pathways for heat to be conducted, decreases, and thus the heat conduction rate also decreases. Therefore, the conduction coefficient for the cloth fiber material cannot be used for the cloth itself. When conducting a thermal analysis for cloth, one should decrease the conduction coefficient of fiber material with the cloth porosity to define conduction coefficient of cloth layer in the analysis. In the overall heat-transfer process, conduction decreases, yet convection of coolant alters the overall heat into and through the mesh. The mass flow rate should be prescribed across the cloth seal based on the data obtained from leakage performance tests. In general, sealing locations with extreme temperature conditions are selected for thermal flow analyses. Typically, the highest temperature cases occur with the highest inlet flow temperature and the minimum differential pressure to drive the leakage flow. Analyses indicate that the leakage flow through the cloth cools and protects the cloth layer from high slot temperatures. Leakage flow also provides a cool buffer zone against hot streaks. The cloth layer is also effective in protecting the thin structural shim inside.

E. SUMMARY

Cloth seals offer performance improvement through parasitic leakage reduction when applied near the hot-gas path in a gas turbine engine. These locations include nozzle, shroud, and diaphragm intersegment locations, nozzle and shroud interstage locations, and transition piece and first-stage nozzle junctions (see Figs. 3 and B1). The flexibility introduced by cloth seals ensures a uniform slot contact over a range of relative excursions and provides reduced leakage rates. Leakage reductions up to 30% have been achieved in combustors and 70% in nozzle segments. The flow savings have been verified through field tests with General Electric Frame 7E first-stage shroud applications. The flow savings achieved in nozzle-shroud cloth seal applications translate to performance gains of up to 0.50% output increase and 0.25% heat rate reduction in industrial gas turbines. In addition to leakage reduction, introducing flexible combustor cloth seals has demonstrated a potential service life extension of 50% or more. Combustion laboratory tests indicate a 30–35% reduction in leakage. Cloth seals are currently standard for all new Frame 6F and 7F gas turbines. They are also offered as part of an extended life kit for the older E and F class units of Frames 3 to 9 [10].

APPENDIX C: OIL BRUSH SEALS

Brush seals, with proven performance in secondary flow and hot-gas-path sealing applications, are being considered for oil sealing and cryogenic applications in turbomachinery (Fig. 66) [151, 201]. Brush seals perform very well under rotor transients owing to the inherent compliance of bristles. They have been used for oil and bearing sumps in gas turbines and aircraft engines and as buffer seals in hydrogen-cooled industrial generator applications and possible vehicle applications [17]. Tighter clearances are required at these locations to avoid oil contamination of the downstream engine components or to minimize oil consumption levels.

In generator applications, the sealing challenge is accentuated by the presence of explosive cooling gas. Sump applications typically require oil mist sealing. Successful brush seal applications at front-bearing applications prevent oil mist ingestion into the compressor [152, 153]. Oil mist ingestion in gas turbines and aircraft engines causes compressor blade fouling, substantial compressor efficiency loss, and potential contamination of customer air supply. In other applications where a liquid medium needs to be sealed, such as hydrogen generator buffer oil seals or liquid-hydrogen–liquid-oxygen seals in rocket turbopumps [201, 202], the problem gets more complicated as hydrodynamic lift and shear heating will occur.

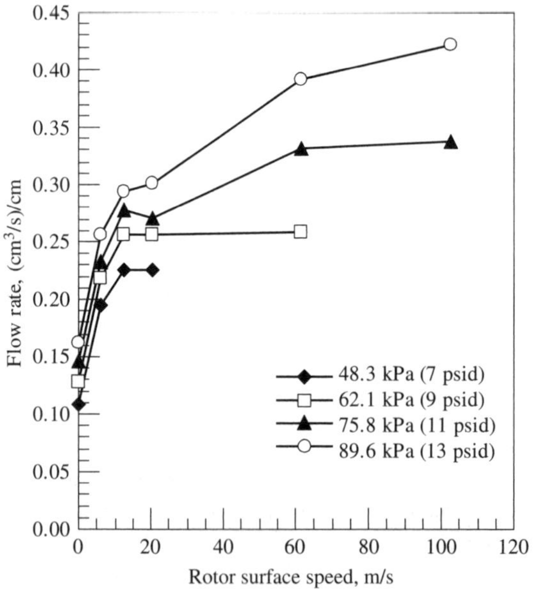

Fig. C1 Hydrodynamic lift of brush seal with speed.

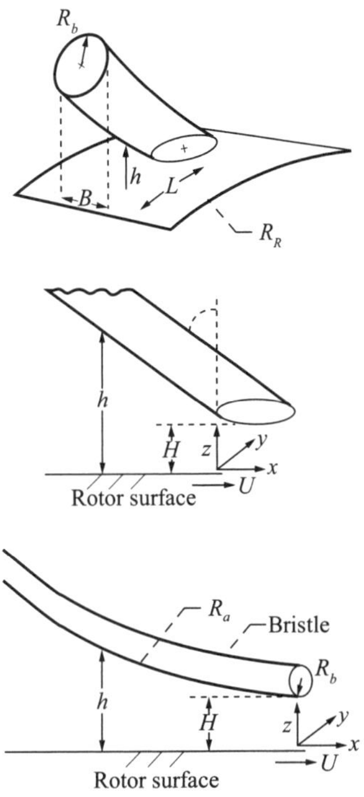

Fig. C2 Considering brush seal bristle tip geometry as convergent wedge/bearing.

Because brush seals are primarily contact seals, oil temperature rise and coking become major concerns, in addition to the leakage rate. Although individual bristles form very small bearing surfaces, hydrodynamic forces generated by viscous sealing medium combined with high surface speeds lift the compliant bristle pack off of the shaft surface depending on bristle stiffness and fluid properties [89, 151]. The amount of lift affects the seal operating clearance, which determines fluid (e.g., oil) temperature rise and leakage rate. Balancing these conflicting performance criteria requires a good understanding of bristle hydrodynamic lift. High-speed oil-brush seal test data presented by Aksit et al. [151] illustrate the presence of the lift (Fig. C1). The data indicate a rapid bristle lift off with rotor speed and then a stabilization of the lift clearance due to a drop in oil viscosity caused by shear heating.

A. BRISTLE LIFTOFF

The inclined approach at the tip of individual bristles creates small hydrodynamic bearing surfaces at brush seal bristle tips, as illustrated in Fig. C2 [203]. The preground or worn-in bristle pack-to-shaft interference of an oil brush seal can be considered as a series of small thrust bearings (one at each bristle tip) having S_T characteristic length, as illustrated in Fig. C3 [204]. The characteristic length S_T and the actual oil lift surface at a single bristle depends on the radial penetration of the oil pumped by the rotating shaft and axial pressure drop. The thin fluid film generated by hydrodynamic lift allows reduction of general Navier–Stokes equations to the well-known Reynolds equations for bearing

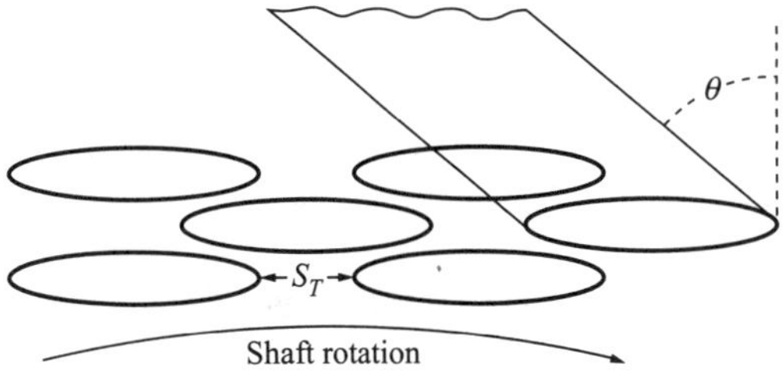

Fig. C3 Bristle spacing S_T characterizes oil lift region at bristle tips.

surfaces. The ratio of the bearing width (bristle diameter in brush seal applications) to bearing length (circumferential length of the wedge) dictates how these tiny microbearings behave.

Depending on seal design and operating conditions, bristles can be packed very tightly, allowing fluid lift pressure to act only at the very tip. In this case, bearing length L is characterized by the tangential bristle spacing S_T shown in Fig. C3, which is an order of magnitude smaller than the bearing width B or bristle diameter. Assuming velocity U as rotational and using fluid bulk properties, the scale differences simplify and allow reduction of Reynolds equations, leading to a simplified solution, which is commonly known in tribology as a "long-bearing," solution [205, 206].

In most applications, bristles deflect under differential pressure load and bloom axially, losing their tight spacing near the rotor. Typical high surface speeds in turbomachinery applications pump sealing fluid strongly into the brush pack. Therefore, actual bearing length of a bristle exposed to the fluid lift pressure is much longer than the bristle spacing S_T. Expressing the clearance h in terms of fixed height H plus flexure and bristle radii R_a, R_b, or

$$h = H + \frac{x^2}{2R_a} + \frac{y^2}{2R_b} \tag{C1}$$

which for the case where the bristle width and characteristic spacing are of the same order becomes

$$\frac{\mathrm{d}h}{\mathrm{d}x} = \frac{x}{R_a} \tag{C2}$$

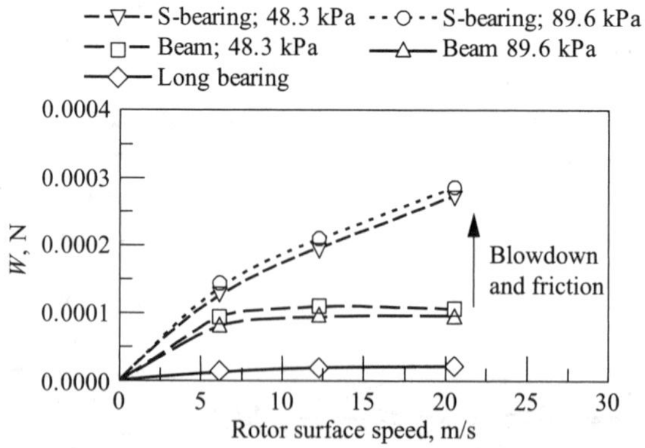

Fig. C4 Comparison of the lift force estimates by long and short bearing theories with beam theory results.

TABLE C1 TYPICAL TURBINE OIL PROPERTIES AT 50°C

Density	$\rho = 884.61\ \mathrm{kg/m^3}$
Specific heat	$c_p = 2030.5\ \mathrm{J/kg\ ^\circ C}$
Dynamic viscosity	$\mu = 0.0195$ Pa-s
Kinematic viscosity	$v = 2.2 \times 10^{-5}\ \mathrm{m^2/s}$
Conductivity	$k = 0.142\ \mathrm{W/m\ ^\circ C}$

Taking advantage of long-bearing length, it is possible to obtain another simplified solution, which is commonly known in tribology as a "short-bearing" solution [205, 206]. The short-bearing solution results in a pressure distribution as [203]

$$P - P_a = \frac{3\mu U R_b}{R_a} \frac{x}{h^2} \tag{C3}$$

where P_a is the ambient sump pressure. Integrating over the bearing area yields the approximate hydrodynamic lift force as [203]

$$W = \frac{6\pi}{\sqrt{2}} \mu U R_b \sqrt{\frac{R_b}{H}} \tag{C4}$$

The long bearing pressure and lift solutions are more complex; they are provided by Cetinsoy et al. [204].

Hydrodynamic lift force is balanced by a reaction force caused by beam/bristle deflection, frictional forces, and so-called "blowdown" forces occurring due to radial pressure gradients within the bristle pack [89, 203]. Figure C4 compares the lift force estimates by short- and long-bearing theory with beam theory results. Analyses are conducted using the typical turbine oil data presented in Table C1 and published experimental oil temperature rise data [151].

Results indicate that long-bearing theory underestimates the hydrodynamic lift. On the other hand, beam theory force results are lower than short-bearing theory estimates. However, when friction and blowdown forces are also considered in addition to beam theory results to represent bristle reaction forces, the short-bearing solution better represents the seal behavior.

In general, the lift force increases with speed, viscosity, and bristle diameter. When the lift-radial clearance increases, the hydrodynamic lift force decreases while the bristle tip force (due to bristle bending, blowdown, and frictional interactions) increases. Multiple bristle interactions and circumferential and axial pressure gradient interactions are not readily modeled or determined without experiments although they contribute to brush leakage, stiffness, and durability as pressure drop increases. The seal operating clearance occurs when forces are balanced.

TABLE C2 TEMPERATURE RISE ALONG *Y* AXIS (FROM UPSTREAM SIDE TO DOWNSTREAM SIDE) FOR DIFFERENT CASES

Rotor surface speed, *u*	Temperature rise of the fluid across the seal	
	$\Delta P = 48.3$ kPa	$\Delta P = 89.6$ kPa
0	0	0
6.2 m/s	5.4°C	5.5°C
12.5 m/s	8.8°C	9.3°C
20.5 m/s	19.9°C	16.6°C

B. SHEAR HEATING

The bristle lift solutions assume constant geometry and viscosity, yet experimental data (Fig. C1) indicate that hydrodynamic lift stabilizes after certain shaft speed because of shear thinning of oil and geometry changes.

Oils are quite sensitive to changes in temperature. For a turbine oil, using the supplier data for coefficient $\beta = 0.0294$ with $\mu_0 = 0.028$ Pa-s at $T_0 = 37.78$°C as the reference point, the viscosity relation becomes

$$\mu = 0.028e^{-0.0294(T-37.78)} \tag{C5}$$

To calculate the average effective fluid temperature at a given rotor speed and lift clearance, a thermal energy equation needs to be solved [207–209]. Based on the experimental leakage data of Aksit et al. [151], flow rate in leakage direction y is taken to be around 0.4 cm^3/s. With this flow rate and other properties of the fluid medium listed in Table C1, the Peclet number, which is the ratio of forced convection to heat conduction, takes a value around 15, indicating that the contribution of heat conduction to energy transfer is small in comparison to convection terms [207]. For the short-bearing solution, convection and viscous dissipation dominate, and the energy equation reduces to

$$\rho c_p v_y \frac{\partial T}{\partial y} = \mu\left[\left(\frac{\partial v_x}{\partial z}\right)^2 + \left(\frac{\partial v_y}{\partial z}\right)^2\right] \tag{C6}$$

After a long process, the temperature distribution is reached as

$$T = T_0 + \frac{1}{2\beta}\ell n\left(\frac{f_1}{f_2} + f_3 \cdot f_4\right) \tag{C7}$$

where

$$f_1 = \exp[2\beta(T_u - T_0)] \tag{C8}$$

$$f_2 = \exp\left[\frac{(2z-H)^2}{\rho c_p(z^2 - zH)}\left(\frac{\Delta P}{w}\right)\beta y\right] \tag{C9}$$

$$f_3 = \frac{4(uw\mu_0)^2}{[H\Delta P(2z-H)]^2} \tag{C10}$$

$$f_4 = \exp\left[\frac{(2z-H)^2}{\rho c_p(z^2 - zH)}\left(\frac{\Delta P}{w}\right)\beta y\right] - 1 \tag{C11}$$

Using the seal data provided by Aksit et al. [151], the calculated temperature rise values, Table C2, compare well with the experimental measurements. Higher sealing pressures derive higher leakage rates and provide more cooling at the same rotor speed. Therefore, fluid temperature rise decreases with increasing pressure load (leakage).

The long-bearing solution is much more complex; it can be found in Cetinsoy et al. [204].

C. FIBER SELECTION

Oil brush seals are located near bearings and sumps. Loose ceramic or metal fibers, and their wear debris, can be hazardous. Therefore, nonmetallic fibers are used in oil seals. Organic fibers, however, are limited in temperature capability and tend to shrink with increase in temperature. Considering the fact that oil or oil mist at bearing cavities can reach temperatures in excess of 150°C (352°F), bristle shrinkage can result in increased leakage. Inertness and moisture absorption rates are the other important considerations. Aramid fibers meet all of these requirements. Their high strength and excellent wear resistance make them a good choice for such applications. Coupon tests indicate they have better wear

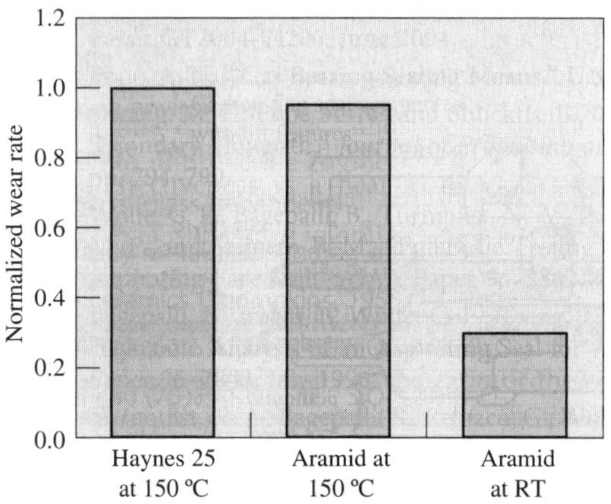

Fig. C5 Wear test results: Aramid fibers against Ni-Cr-Mo-V.

rates than typical Haynes 25 fibers (Fig. C5). In addition, smooth outer surface of the fibers prevents coked oil particles from attaching and sticking fibers together.

Although oil applications of brush seals are rather new, their use in gas-turbine front-bearing applications have proven successful. Field tests have shown leakage reduction gains and have demonstrated durability of these seals in field operation [152, 153]. Additional information is given in Appendix E.

APPENDIX D: LEAF-PLATE SEALS

All seals have interface gaps [δ_i], energy losses, and unbalanced forces. Every gap has an associated area, leakage path, potential gradients (e.g., pressure, temperature, and concentration), interface speed and displacements, dynamic stiffness, and damping. Compliant seals, as the leaf plate, have additional axial and radial stiffness and damping due to the leaf-plate geometry both between leaf-plate elements and response at the rotor interface. Successful long life seal designs optimize these parameters to mitigate system parasite losses and rotor-dynamic instabilities.

Sealing involves controlled leakage. Early forms of leaf-plate sealing can be traced to controlling piston rod leakage using the stuffing boxes, for example, Wheeler [210], Fig. D1. Wheeler combined several sealing concepts found in today's leaf-plate seals: (i) slanted close pack elements between a stationary and surface in motion, with blocking wedges between (ii) end-cap pack elements for adjustment (iii) double ring seals and (iv) interdigitized labyrinth sealing. While Wheeler's application was a reciprocating slider shaft, the sealing of a rotating

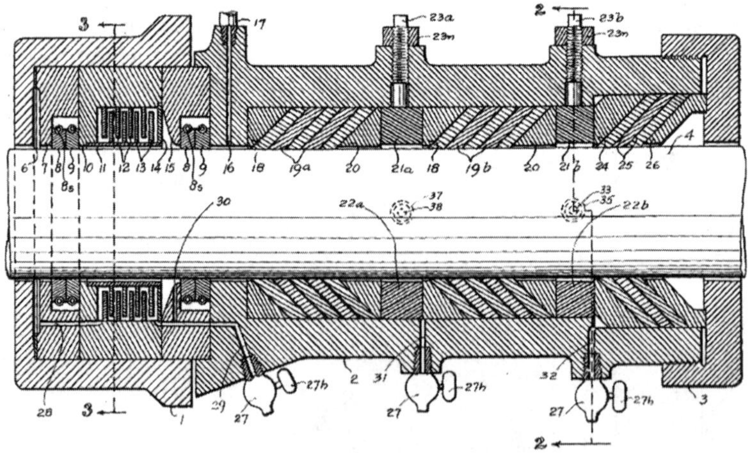

Fig. D1 Leaf-plate sealing [210].

and translating system with eccentric motions has many of the same design requirements.

Today's leaf-plate seal patent literature is rich implying anticipated cost-effective usage in turbomachine applications with anticipated high ROI (investment return). The development of leaf-plate seals resulted from the need for sealing of high-pressure (<14 MPa) industrial power systems with equivalent or lower leakage than for labyrinth seals and brush seals, the brush seals being pressure drop limited to lower pressure systems. Both brush and leaf-plate seals will function as filtration devices and consideration need be given for debris collection and removal, also addressed by Wheeler. The leaf-plate seal differs from the brush, plane leaf, or plate in geometry construction and balance of the pressures (see section on static sealing plates and high-temperature sealing, e.g., Steinetz and Sirocky [140], and Sec. IV.C).

Early high-speed rotating shaft leaf-plate sealing concepts came from many other sources (e.g., Mech [211], Gardner and coworkers [212, 213], Wright [214]), Figs. D2, D3, and D4. Like brush seals, leaf-plate seals have difficulties compensating from difference in shaft (*Rs*) and housing (*Rh*) radii, where given N leaf-plate or wire bristle elements the outer gap spacing Δ is always $\Delta = \pi(Rh\text{-}Rs)/N$ and the total gap is always $N\Delta$. With conventional brush sealing, this is solved by overlapping bristles where the brush thickness at the pinch point ($\sim Rh$) is less than the brush thickness at the shaft (*Rs*). The plurality of leaf-plate patents are directed to overcoming this leakage path. For leaf-plate seals, Δ could be resolved by leaf-plate taper, yet seems to be overlooked in favor of ease of fabrication. Plane and curved leaf-plate seals have been fabricated, Figs. D5 and D6. Designs for these configurations pay close attention to the pressure balance at the front and aft cover clearances, for example, Shinohara et al. [215]. The second difficulty, common to both brush and leaf-plate seals, is interface contact wear against a sacrificial layer on the shaft, which is primarily related to bristle or leaf-plate liftoff and affects both wear and leakage. The liftoff issue in both cases has been largely solved by element lay angle of the bristles or leaf plates with respect to the rotating surface. This requires elements to be ground with tip curvature ($\sim Rs$) that have stiffness sufficient to seal, damp, and follow shaft motions, yet allow leakage flow along the leaf plate (as with brush bristles) to couple with sufficient rotating generated pressures to lift the element from direct contact with the rotor interface surface). A drawback is antirotation, which can open interface clearances or even destroy the seal [216].

The integration of a leaf-plate seal with a labyrinth seal requires significant pressure-balance design requirements, yet provides good sealing and dynamic response. The application would ensure against a complete loss of sealing should the leaf-plate seal fail through rotor excursions, severe wear, loss of leaf-blade elements, or other forms, for example, FOD, antirotation. Typical combinations place the leaf-plate seal upstream or within an N-tooth labyrinth seal (Figs. D7 and D8), a concept also applied with brush seals [3, 216].

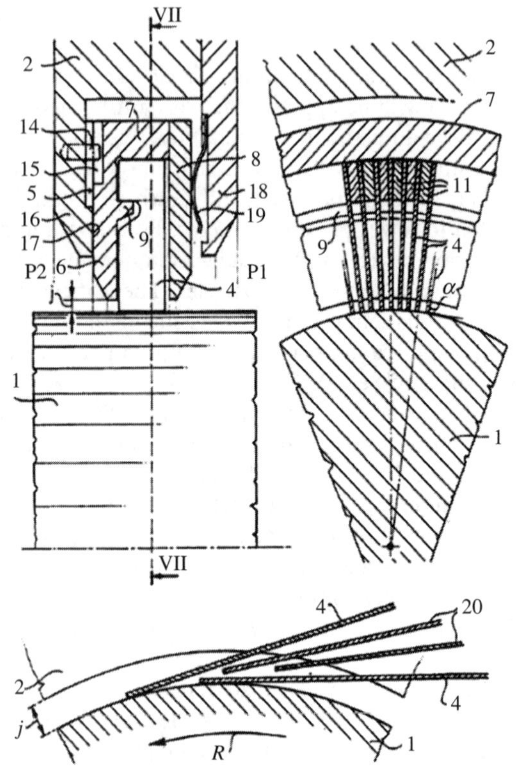

Fig. D2 Sealing gasket for rotating shaft [211].

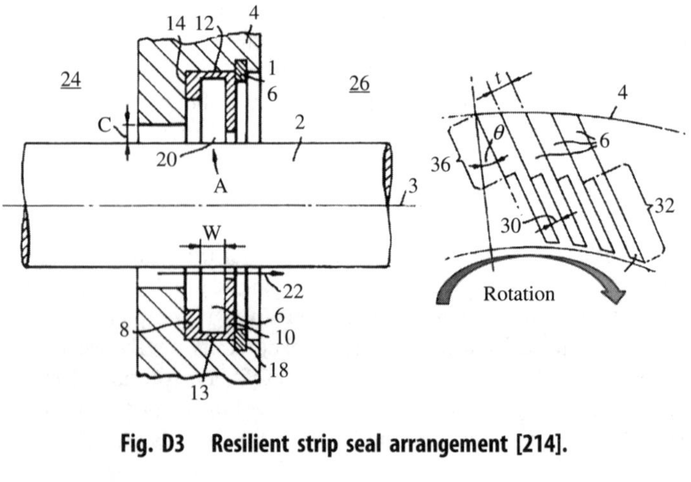

Fig. D3 Resilient strip seal arrangement [214].

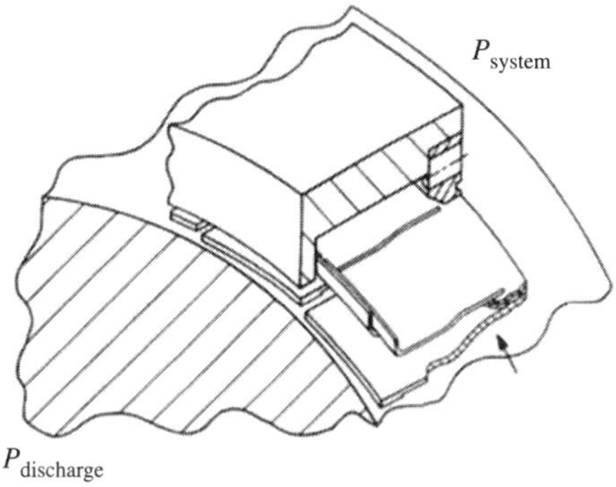

Fig. D4 Pressure balanced, radially compliant, noncontact, shaft riding seal [213].

Integration of the labyrinth, film riding sealing into the leaf-plate seal requires yet another stringent set of pressure balance design requirements such as advanced by Akagi et al. [219], Fig. D9, and Deo et al. and Awtar and colleaguesseal (patents U.S. Patent 2010/0143102 A1 [220]; U.S. Patent 7419164 B2 [221], U.S. Patent 8382119 [222]). This type of seal adapts concepts from the labyrinth, leaf and face seals. As illustrated in Figs. D10 and D11 (also in Figs. D12–D17 discussed later), the seal appears as an E configuration (three-tooth labyrinth) with pressure-balanced labyrinth-tooth encased by film riding U-leaf-plates: in this case, there are three labyrinth-teeth and one U-leaf-plate geometry. To provide sealing and both radial and axial compliance and long life, the leaf-plate length L, thickness t, width w, lay-angle θ, the labyrinth tooth (or teeth) geometry, the axial and radial clearances and tribo-materials must be optimized with respect to the turbomachine operations envelope.

In all seals, clearances are very important to stable operation. In a leaf-plate configuration, these clearances also provide for pressure balance (as in film riding mechanical sealing). Without proper clearances $\delta 0$ – $\delta 8$ (or more if labyrinth-leaf-plate penetrations >3) and shaft-leaf-plate interface, this seal will not function with the leaf plates locking in place resulting in rapid wear, instability, and enhanced leakage.

At the shaft-leaf interface, Fig. D11, close-packed film riding shaft leaves at shaft radius Rs, restricts film interface gap leakage ($\delta 0$) and allows minor interleaf axial leakage; radial leakage into the interface gap is small but cannot be neglected. Both axial and radial leakages are important to control as they minimize interleaf Coulomb friction at surfaces (w, L). With N leaves, the radial spacing at $(Rs + L) \sim 2\pi L \sin \theta/N$ (for angles and interface geometry, see Fig. D6 and following section on design), combined with the labyrinth-tooth, provides leakage flow to pressure balance the film riding leaves. At the high-pressure side, leakage

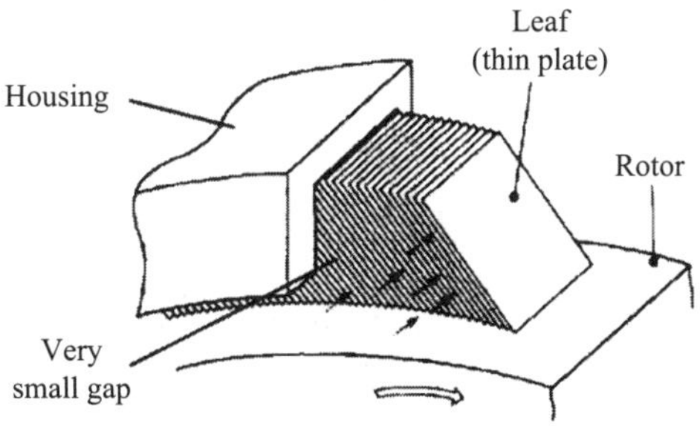

Fig. D5 Leaf-plate turbine seal [215].

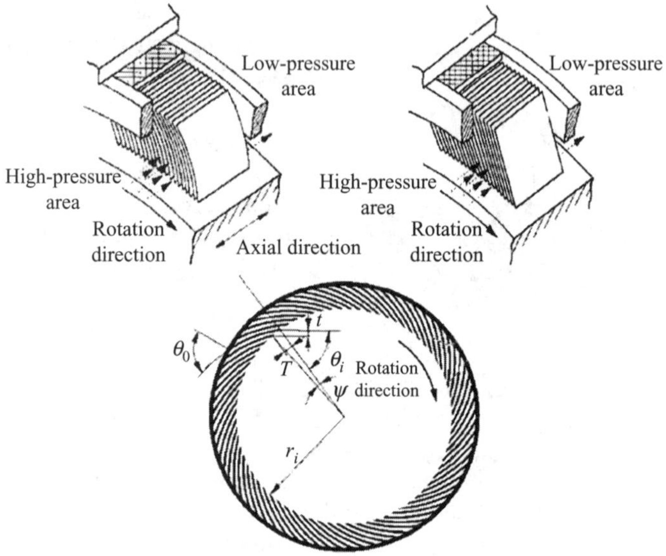

Fig. D6 Leaf-plate sealing, plane, curved leaf-plates accounting for RO, RI differences [213].

flow is throttled by the front tooth (δ1) (and a cover plate to mitigate FOD) and divided between axial and radial flowpaths. The flow proceeds through the interface gap (δ1) and radially outward along the interface between the backside of the front tooth (δ2) continually losing leakage flow into the axial direction through the increasing radial gaps between the leaves (similar to manifold flow) finally reaching stagnation zone at the outer apex of the leaf plate and first labyrinth-tooth, and the inner apex of the labyrinth-tooth where the balance area is greatest. The leakage flow proceeds radially inward along the labyrinth-tooth (δ3) from the higher pressure zone to the lower zone at the labyrinth-tooth tip where it is again throttled (δ4) and the balance area is

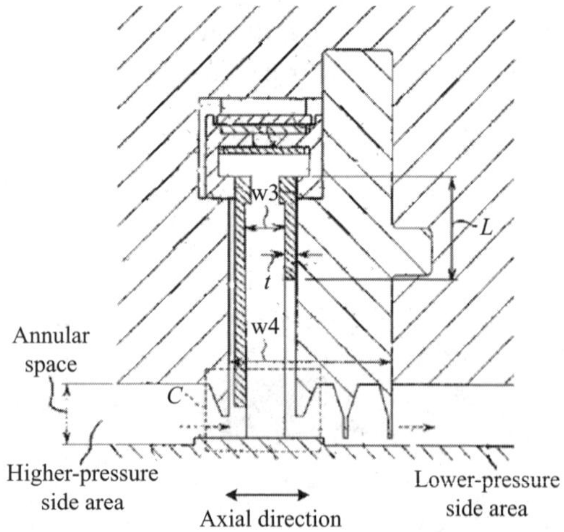

Fig. D7 Integrating shaft labyrinth and leaf-plate sealing [217].

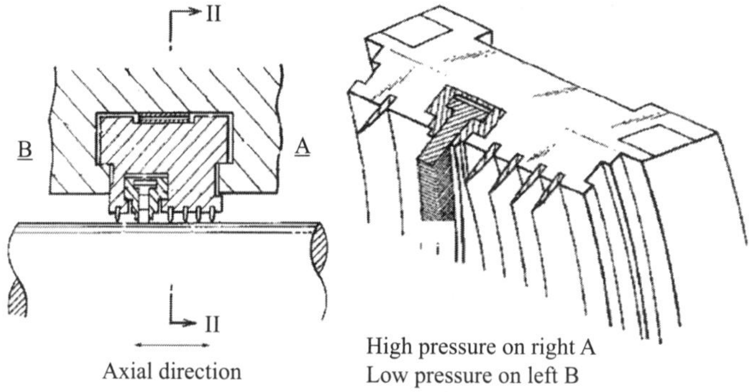

Fig. D8 High-pressure drop integrated labyrinth and leaf-plate sealing [218].

smaller (near minimum). The throttled leakage flow proceeds radially outward along the backside of the labyrinth-tooth ($\delta 5$) toward high-area lower pressure zone continually leaking axially with a small radial leakage. The leakage flow again proceeds inward along the front side of the next labyrinth-tooth ($\delta 6$) to the throttle zone where exit throttled leakage ($\delta 7$) blends radially and axially with the leaf-plate riding gap interface ($\delta 0$) leakage. Figure D12 represents a numerical model of the leaf-plate seal pressure balance.

For stability control, a slight tapering of the interface gap ($\delta 0$) is highly beneficial (cone convergence less than $1.6 \times \delta 0$ at high-pressure side and $\delta 0$ at low-pressure or exit side); see Fleming [224] theory. This gap is often a natural consequence of tip grinding, lay angle, and leaf-plate overlap. Also, extending the attachment concept and length of attachment post (Hendricks et al. [225]) to include the pressure-balanced leaves would enable nondestructive in situ sealing

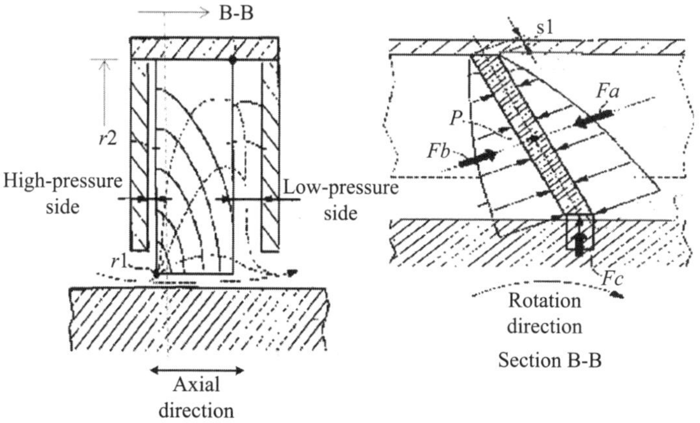

Fig. D9 Leaf-plate flow and pressure balancing [219].

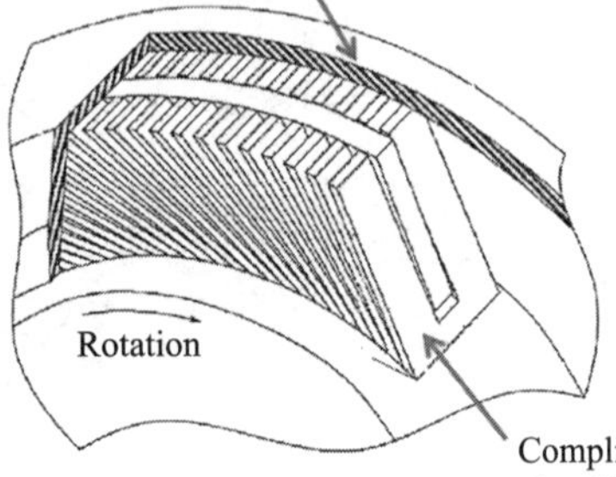

Fig. D10 Leaf-plate sealing with integrated leaf-plate-labyrinth-tooth.

repairs or worn leaf set replacement. However, these added benefits require a rebalance of forces and optimization, including manufacturing and long-term endurance testing.

The combination of film riding (mechanical film riding and leaf seal effect) and the multiple throttled junctures (labyrinth seal effect) and the pressure balance of the leaves (self-acting film riding seal) provide for minimum leakage and wear and elimination of many dynamics problems [e.g., inlet swirl (see swirl brake)] while providing good resolution for axial and radial shaft motion.

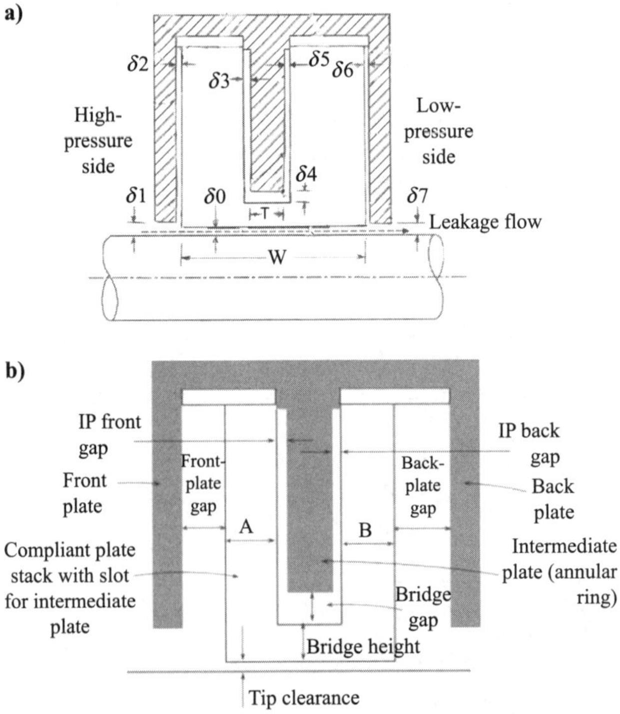

Fig. D11 a) Leaf-plate sealing clearance gaps; b) leaf-plate sealing typical nomenclature.

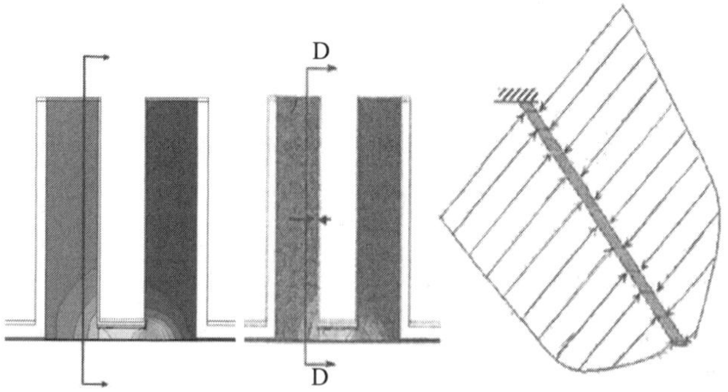

Fig. D12 Example of leaf-plate sealing pressure distribution and the streamlines at a plane upstream of the labyrinth-tooth [220, 223].

Extending the film-riding concept to include multiple labyrinths inserted into leaf-plate compliant elements becomes natural but complicates sealing flows and requires enhanced pressure balancing, Figs. D12–D17. These sealing concepts are similar to filled honeycomb sealing used for fan-tip nacelle interface sealing.

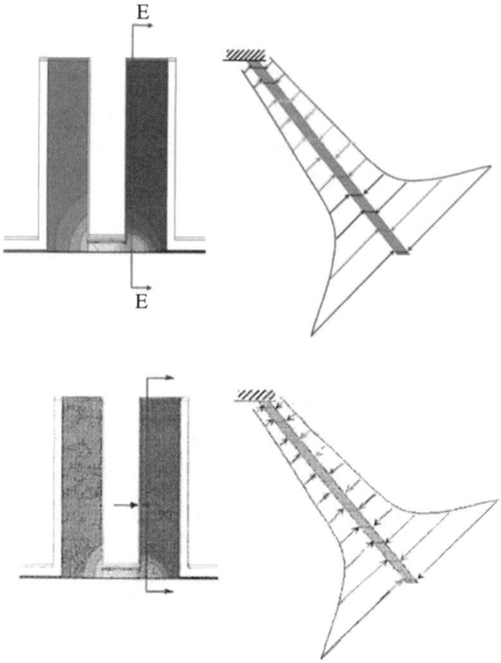

Fig. D13 Example of leaf-plate sealing pressure distribution downstream of labyrinth-tooth [220, 223].

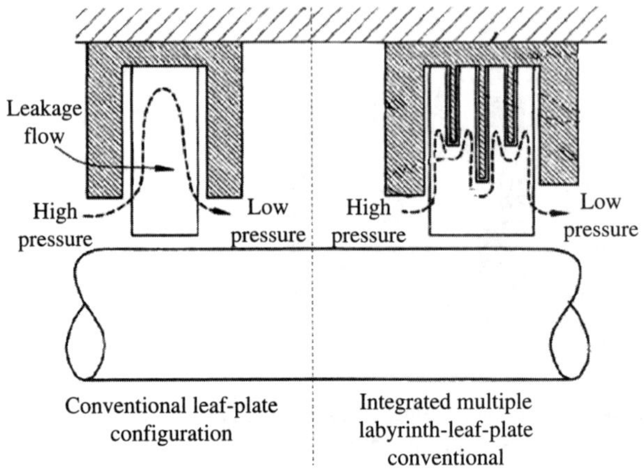

Fig. D14 Conventional leaf-plate sealing and integrated multiple labyrinth-leaf-plate sealing [221, 222].

The design principles of brush, cloth, face, and labyrinth sealing and tribopairing are employed subject to operational environment and turbomachine operations envelope [226, 227]. For example, in selecting leaf materials, 1) for lower temperatures, nonmetallics are considered (e.g., Aramid, plastics); 2) for high-temperature applications, metallic leaves (Hanes 188, Hasteloy X, γ-TiAl), and metallic ceramics are evaluated; and 3) for very high temperatures, ceramics, oxides, and coated metallics are considered depending on the environmental gases (see Abradable sections of this chapter).

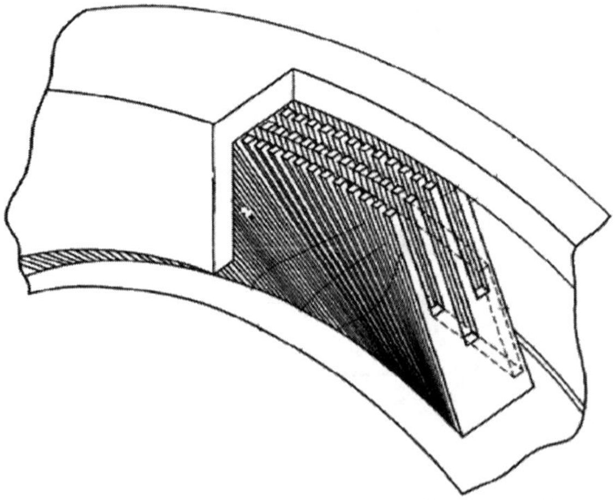

Fig. D15 Integrated leaf-plate sealing with multiple labyrinths [221, 222].

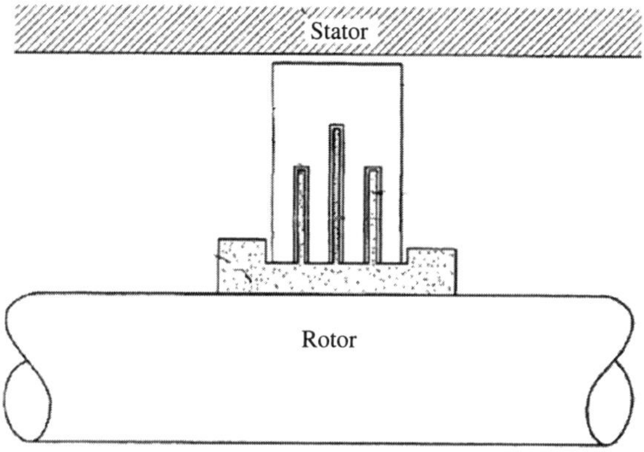

Fig. D16 Leaf-plate sealing with multiple labyrinths and antirotor rub-interface [221, 222].

Leaf-plate seal designs and parameters follow those typical of brush seals with the exceptions of the leaf-plate width and in the case of the E-U leaf-plate, the labyrinth-tooth (or teeth). Some typical E-U and other leaf seal design parameters are as follows:

1. Leaf thicknesses: similar to brush seals: 2–8 mm depends on stiffness damping requirements and rotor interface curvature
2. Leaf width that is dependent on pressure drop, labyrinth-tooth, and seal cavity geometry as well as dynamics and seal diameter; from a few millimeters up to sections, for example, 5- to 50-mm small shaft seals to sections with large engine curvature and diameters
3. Leaf lay angles of 25–55 deg that follow brush seal technology similar to patents of Cross Mfg. leaf-like seal and brush seal designs
4. Leaf length: 1.5 to 2.5 × leaf width, depending more on rotor interface curvature and system dynamics requirements, which is similar to brush seals
5. Labyrinth tooth width: approximately (leaf seal width)/5, with orifice type inlet, labyrinth tooth design, dependent on pressure balance and clearance gap; cannot be tight enough to force all flow along leaf to seal-rotor interface, which would unbalance the seal as well as loose sealing advantages

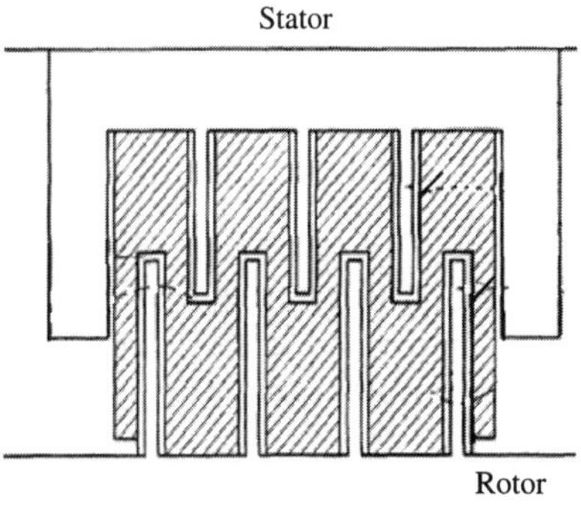

Fig. D17 Interdigitated rotor/stator leaf-plate-labyrinth sealing [221, 222].

According to Fig. D5 and U.S. Patent US6343792 B1 [215], a very small gap T is provided between adjacent leaf plates. Each gap has substantially the same width both at the leaf-plate root and tip. Plating and coatings can be provided near the root as needed. The leaf-plate geometry can be described by the radial coordinate r and the angle θ between each tangent of leaf plate and the line from the center of the center of the shaft to the relevant point of leaf plate. The leaf plates, Fig. D5, are designed based on the following formula, thereby fixing the width of gap between adjacent leaf-plates, at both root and the tip; see Fig. D6.

$$\sin\theta_i = t/T = t/(r_i\psi)$$
$$t/\psi = r_i\sin\theta_i = r\sin\theta$$
$$\sin\theta = (r_i/r)\sin\theta_i$$

where

t = thickness is the sum of thickness ($t_{\text{leaf-plate}} + t_{\text{gap}}$)

ψ = central angle between adjacent leaves

$T = r_i\psi$, width between adjacent leaves along the inner circumference

θ_i = angle of the tip at leaf-plate-shaft interface

θ_0 = angle of the root of leaf plate at housing

r_i = inner seal radius (rotor Rs)

r = radial line between leaf-plate tangent and seal center

Tables D1 and D2 give parameter values and dimensions for a typical design (see Fig. D5).

Table D1 Typical Sealing Parameters [Fig. D5 (Nakane et al. [228] and Watanabe et al. [229])]

Fluid	Air
Rotor speed	5000 rpm
Seal diameter	350 mm
Inlet temperature	Room temperature
Inlet pressure	0.1 $\sim$ 0.4 MPa
Outlet Pressure	0.1 MPa

Leaf-plate sealing has found several applications in the industrial sector. Watanabe et al. [229] describe one application in the development of a new high-efficiency steam turbine where test data show leaf-plate seal leakage of 1/4 to 1/3 that for a 0.5-mm gap, four-stage labyrinth seal with leaf-rotor contact approaching zero at operating speed, Fig. D18.

Table D2 Typical Leaf-Plate Geometries [Fig. D5 (Nakane et al. [230] and Watanabe et al. [229])]

Seal outer diameter	390 mm
Seal inner diameter	350 mm
Slope angle	37.5 deg
Attachment angle	45.4 deg
Seal width	5.0 mm
Leaf thickness	70 μm
Tip gap	8 μm
Base gap	30 μm
Number of leaves	8581 leaves
Side-plate tip gap	1.5 mm
High-pressure side gap	0.05 mm
Low-pressure side gap	0.15 mm
Radial interference	0.6 mm

The clearance gap $\delta 0$ between the rotating shaft and each leaf plate is 10 to 20 μm, compared to a conventional labyrinth seal where the gap is 0.5 to 1 mm. The leakage flow through such small gaps (inter leaf plate and interface) is in most cases laminar with correspondingly high pressure drops and a low level. The compliant leaf-plate tip lifting is due to differential pressures. The lift force is caused by an unbalance between setting force and hydrodynamic film riding interface lifting force caused by rotor rotation.

A leaf-plate seal/bearing configuration suitable for a small aero-engine gas turbine has been successfully tested to 60,000 rpm and 550°C (Salehi et al. [231, 232]).

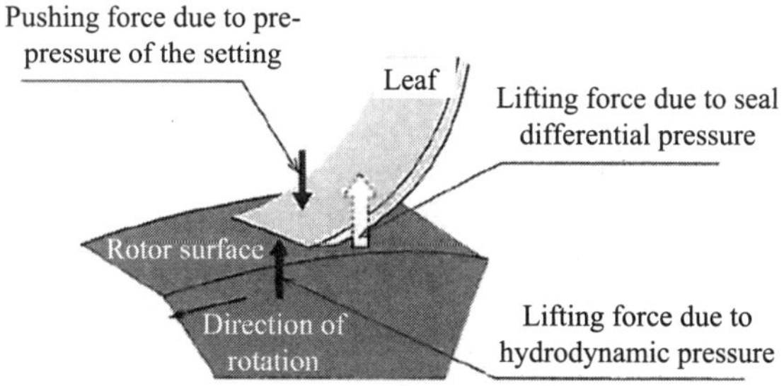

Fig. D18 Leaf-plate sealing steam turbine application; compliant leaf-plate tip lifting due to unbalance between setting force and hydrodynamic lifting force [229].

APPENDIX E. EXTENDED TEMPERATURE AND PRESSURE SEALING DEVELOPMENTS

For rope sealing, Finkbeiner and Mayer [233] successfully fabricated and tested rope-type sealing materials at very high temperature (toward reentry temperatures) to determine static and durability limitations under plasma arc jet conditions.

Gibson et al. [234] successfully designed, fabricated, and tested an air pressure balanced four-element finger seal configurations; the new configurations minimize interelement friction, flow losses, and interface wear.

Spring energized metal-to-metal seals afford low leakage static sealing. Such designs are applicable to fuel storage casks and nonconformal interfaces. The sealing concept itself involves C-shaped elastomer sandwiched between a hard concentric C-shaped, in this case metallic, hard shell and a tight, coiled inner spring. When compressed, the load is resisted by the inner spring, the hard outer shell, and differentials absorbed by the elastomer thereby providing enhanced static sealing. The elastomer provides an enhanced resilience form of spring energized seals. Schweitzer and Ledrappier [235] provided an analysis of spring energized metal seals for spent fuel storage casks with a statistical analysis relating to storage period.

The Advanced Technologies Group (ATG) seal [236, 237] integrates several sealing concepts: 1) film-riding self-acting seal, Ludwig et al. [112, 238]; 2) shaft and face seal, Trutnovsky [73]; 3) conical and planar convergent gap stability, Fleming [239]; 4) forward-slanted labyrinth-teeth, Alvarez [240], as also investigated by Egli [64] and Burcham and Keller [60]; 5) radial and axial compliance, Ferguson [75], Flower [76], Shapiro [146]; 6) floating brush seal radial motion controlled by a cantilevered spring, Justak [144]; and 7) differential pressure dependent on the gap clearance and steps [241–244].

The primary seal floats on the interface flow leakage and closes with increased pressure drop as a result of higher throttle velocities creating lower intertooth cavity pressures. The resulting spring and upper cavity pressure unbalanced forces cause the shoe face to close (or open). The floating element is constrained axially from a fixed ring by cantilever pins. The system provides noncontact sealing (Fig. E1).

The secondary seal is provided by a brush, flexible plate, or sliding interface [236, 237] (also see Figs. 61–63).

Seal-life rotor interface wear in segmented turbine brush seals during startup and shutdown pinch points has been addressed by introducing a pressure-balanced spring-loaded configuration [245].

The T-shaped pressure-balanced sealing segment floats within a C-shaped housing providing both axial and radial compliance. The radial motion principle constraints are the spring, the radial pressure, and the axial pressure drop, which loads or unloads the sliding friction interface between the downstream portion at the T- and C-shaped interfaces (see Fig. E2).

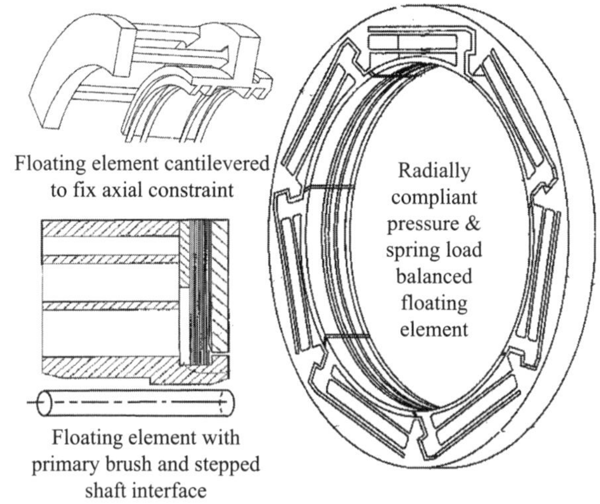

Fig. E1 Noncontact seal for a gas turbine engine [236].

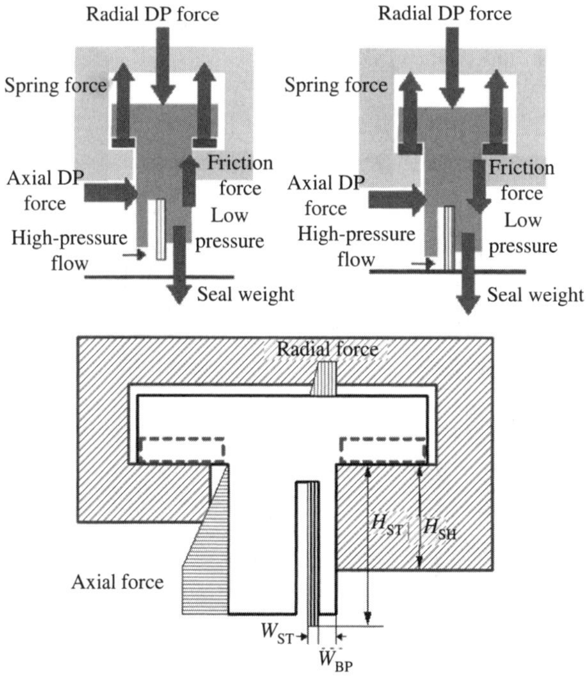

Fig. E2 Pressure-balanced floating brush seal [245].

A. EFFECTS OF LUBRICANTS ON BRISTLE OR LEAF-TYPE SEALING

The effects of lubricants under static and rotating conditions as well as rotor reversal are discussed by Carlile et al. [246, 247]. A potentially simplified method to determine brush seal leakage or pressure drop based on thermodynamic corresponding states model combined with flows through porous media [248] is discussed by Carlile et al. [246, 247] and Hendricks et al. [249].

The development and testing of brush sealing at the outside diameter and sealing at the inside diameter (see Fig. E3) is discussed by Hendricks et al. [250].

B. PURGE SEALING DESIGN DETAILS

Shapiro and Lee [251] provide in detail a high-performance turbomachine buffer seal design analysis and testing.

C. THERMOPHYSICAL PROPERTIES

The NIST code REFPROP [254] provides thermophysical properties for a large variety of single- and multi-component fluids. The code is quite user friendly and highly accurate for engineering requirements. For very accurate scientific use, the reader should contact NIST.

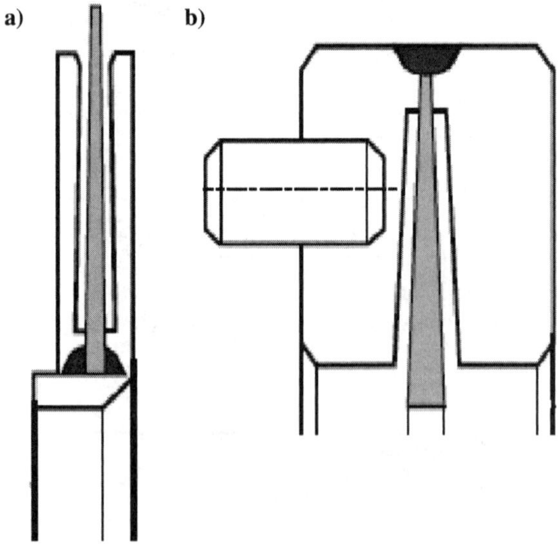

Fig. E3 Brush sealing for a) outside rotor and b) inside rotor [250].

D. POROUS FLOW BRUSH SEAL MODEL

Conservation equations and boundary conditions and fluid state equations governing brush sealing are complex with multiple constraints requiring heuristic assumptions by the modeler. Early innovators and modelers included Ferguson [75], Flower [76], Hendricks [17], Hendricks et al. [253] followed up by many others. Early developments called for cold clearance interference fits with hard coated shaft surfaces; wear, hysteresis, and antirotation were major concerns. As manufacturing technology and modeling gained foothold, it was found that line-line to small clearance operations were most favorable to counter such forces as blowdown (pressure balance forces driving the bristles toward the shaft), bristle liftoff caused by shaft rotation, and blowout (pressure force bending bristles from front to aft) to cite a few. In all cases, the modelers found it necessary to integrate heuristic brush seal data into the model in order to map out and mimic brush seal performance. Nevertheless, useful operational information could be determined. Some models performed quite well knowing two data points close to installed line-to-line operations; such is the case with the Ergun porous medial model [248].

The Ergun porous media flow model [248], adapted to brush sealing [246, 247, 249], provides useful physical insight and a convenient method to predict brush seal flows. The model equation can be written as

$$\psi = \frac{\Delta P \cdot \bar{\rho}}{G_0^2}\left(\frac{1.5d}{\langle t\rangle}\frac{\varepsilon^3}{1-\varepsilon}\right) = 150/(Re/(1-\varepsilon)) + 1.75 \tag{E1}$$

$\mathrm{Re} = 1.5\, G_0 d/\mu \quad G_0 = \rho V_0 = \dot{w}/A \quad A = \pi(d_o^2 - d_i^2)/4 \quad D_p = 1.5d$
C = clearance $\quad d$ = wire diameter $\quad d_o$ = effective brush OD $\quad d_i$ = brush ID
$\langle t\rangle$ = average brush thickness

$$\begin{aligned}\varepsilon &= V_{\text{open}}/V_{\text{total}} = 1 - V_{\text{solid}}/V_t \\ &= 1 - 4N_o d^2/[d_i(1 + d_o/d_i)][\langle t\rangle \cos(\theta + \varphi)] && \text{(E2a)} \\ &= 1 - 4N_o d^2/\{d_i[(1 + d_o/d_i) \\ &\quad + (2C - C^2/d_i)/(d_o - d_i)]\langle t\rangle \cos(\theta + \varphi)\} && \text{(E2b)}\end{aligned}$$

where θ and φ are brush bristle angles relative to the rotor and N_o is the number of bristles. (ε can be determined from standard methods of measure in porous media.)

$$\Delta P = 150\mu_0\left(\frac{\mu}{\mu_0}\right)\left[\frac{\langle t\rangle}{D_p^2}\frac{(1-\varepsilon)^2}{\varepsilon^3}\right]V_0 + 1.75\,\rho_0\left(\frac{\rho}{\rho_0}\right)\left[\frac{\langle t\rangle}{D_p}\frac{(1-\varepsilon)}{\varepsilon^3}\right]V_0^2 \tag{E3}$$

$$\Delta P = a\left(\frac{\mu}{\mu_0}\right)V_0 + b\left(\frac{\rho}{\rho_0}\right)V_0^2 \tag{E4}$$

To determine some effects of $\varepsilon \neq \varepsilon_0$, let $\varepsilon(g) = \varepsilon_0 + g\,\varepsilon_0$

1. $(1-g)\varepsilon_0 \leq \varepsilon \leq (1+g)\varepsilon_0$, where $g \geq 0$, and let

$$a_1 = \frac{(1-\varepsilon)^2}{\varepsilon^3}; \quad b_1 = \frac{(1-\varepsilon)}{\varepsilon^3} \tag{E5}$$

2. For values of $0.14 < \varepsilon_0 < 0.33$ and $0.0014 < (g) < 0.01$, both a_1 and b_1 vary <4%, implying bristle lay geometry effects on a and b (Eq. E4) of less than 4%.

   From Eqs. (E3–E5), with $\varepsilon(g = 0) = \varepsilon_0$, the coefficients a and b of Eq. (E4) become:

$$a = 0.977(150\mu_0)\left[\frac{\langle t\rangle}{D_p^2}\frac{(1-\varepsilon)^2}{\varepsilon^3}\right]; \quad b = 0.975(1.75\rho_0)\left[\frac{\langle t\rangle}{D_p}\frac{(1-\varepsilon)}{\varepsilon^3}\right] \tag{E6}$$

In practical applications, brush seals have small variations in g where $\varepsilon \geq \varepsilon_0$.

For an early vintage Cross brush seal, where the working fluid was air, $a = 0.886$ and $b = 0.0267$ at line-to-line clearance. For other gases, for example, CO_2, the reference properties (μ_0, ρ_0) are evaluated for air and (μ, ρ) evaluated for CO_2, to project a suitable (ΔP, $\acute{V}$) chart for CO_2. Charts for (ΔP, $\acute{w}$) and (ΔP, G_0) follow by substitution from the symbol list into Eq. (E6); see also Eq. (E1). A review of brush sealing in steam-turbine systems provides a recent view (2005) of the technology development and range of applications of more modern day brush seals [252]. Typical metal $d = 0.07$ mm (0.0028 in.).

At standard temperature (298.15 K) and pressure (0.1 Mpa), the values for air are as follows: [$\rho_0 = 1.1685$ (kg/m^3), $\mu_0 = 18.490$ (μPa-s)] and for CO_2 [$\rho_0 = 1.7842$ (kg/m^3), $\mu_0 = 14.932$ (μPa-s)] [254].

Flows in pinned arrays illustrate many of the complexities of brush seal flows [255].

CHAPTER 4

Internal Cooling

Srinath V. Ekkad*

Virginia Polytechnic Institute and State University, Blacksburg, Virginia

NOMENCLATURE

A	cross-sectional area
D_h	hydraulic diameter of impingement channel, 4 A/P
h	heat-transfer coefficient
k	thermal conductivity
L	channel length
Nu	Nusselt number $= hD_h/k$
Pr	Prandtl number
Re	Reynolds number, VD_h/ν
Ro	rotation number, $\Omega D_h/V$
T_b	flow bulk temperature
T_i	initial wall temperature
T_{imp}	impingement jet temperature
T_w	wall temperature
V	mainstream velocity; average velocity in channel
ν	kinematic viscosity of coolant
Ω	rotational speed of channel

Modern gas turbines are pushing the limit on operating temperatures with industrial power generation units at 2000–2500°F, commercial aircraft engines at 2300–2600°F, and military aircraft engines at 3000–3500°F. With increased heat load to the turbine hot-gas-path components becoming a fact of life, the need to design effective cooling schemes to keep the components below their melting point temperatures and obtain reasonable service life is ever more critical. Typically, the hot-gas-path components, such as combustor liners, endwalls, blade and vane surfaces, and blade tips are cooled by extracted coolant air from the compressor bypassing the combustor. This coolant extraction is an added

*Commonwealth Professor, Department of Mechanical Engineering.

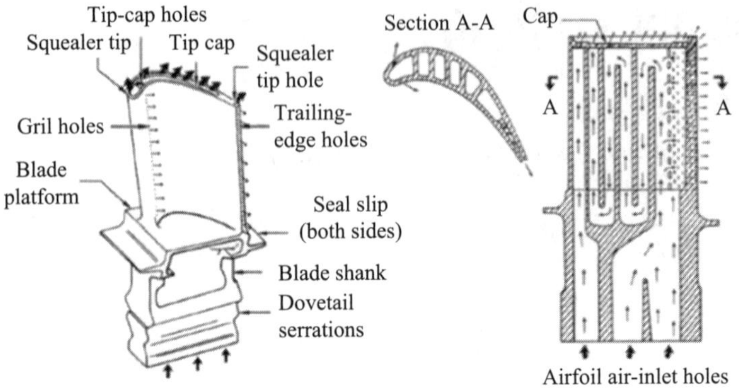

Fig. 1 Typical stage-1 HPT blade for the GE CF6 Engine (Treager et al. [1]).

penalty to the overall engine efficiency. Thus, it is imperative that more cooling efficiency be achieved using less coolant air. In addition, cooling designs need to consider induced thermal gradients and ensure that significant thermal stresses are not generated.

Gas turbine blades are cooled internally and externally. Internal cooling schemes include rib turbulated channels, arrays of impingement holes, pin fin arrays, dimpled surfaces, and more complicated dual or triple combination schemes, depending upon needs. Figure 1 shows a typical first-stage rotor of a GE CF-6 engine. This design is based on 1979 conditions [1]. The blade has multi-pass internal rib turbulated passages of various aspect ratios. The trailing-edge region has pin fin arrays to enhance cooling and structural strength. A more typical design, from Han et al. [2], is shown in Fig. 2. In this design, the blade has a triple-pass rib turbulated channel in the middle of the blade, an impingement cooling channel at the leading edge, and pin fin arrays at the trailing edge. In addition, there is also film cooling for the leading-edge area and the tip. In this chapter, we present updated information on recent developments in internal cooling technology. Several new designs aim to improve cooling efficiency over the current practices in the industry. Several of these designs seem to be likely to be incorporated into current or future designs. This article covers the updated technology beyond what is presented in the book by Han et al. [3].

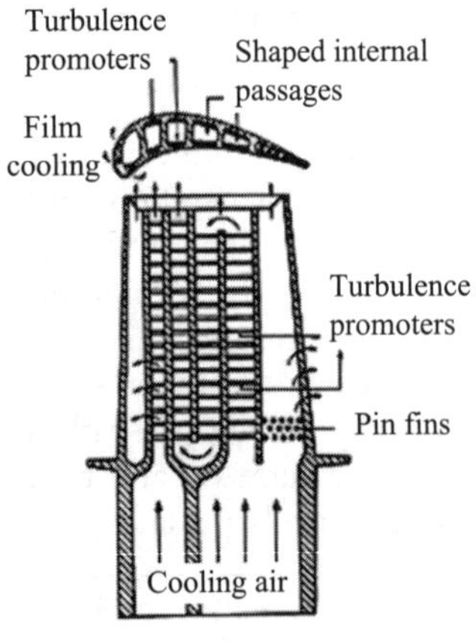

Fig. 2 Typical cooled blade as illustrated by Han et al. [2].

As cooling requirements have become stringent in the face of stronger emissions requirements and higher overall engine efficiency demands, cooling technology has become critical. Reduced coolant usage has a first-order impact on overall engine efficiency. In the past decade, the focus on cooling technologies has been to extract more heat from internal surfaces using highly efficient combination cooling schemes that provide significantly higher heat-transfer enhancement than the single mode methods in the past. Heat-transfer coefficient enhancements up to 5.0 times baseline have been reported.

I. RIB TURBULATED CHANNELS

Periodic rib turbulators inside a channel trip the boundary layer and enhance mixing and thus enhance overall heat-transfer performance. The penalty of increased pressure drop is overcome with the level of enhanced heat-transfer

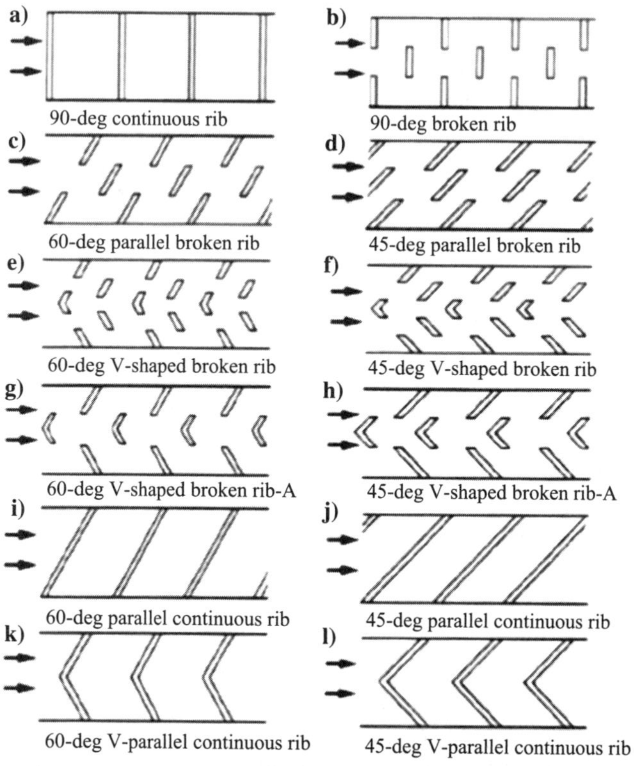

Fig. 3 Various rib configurations tested by Han and Zhang [5].

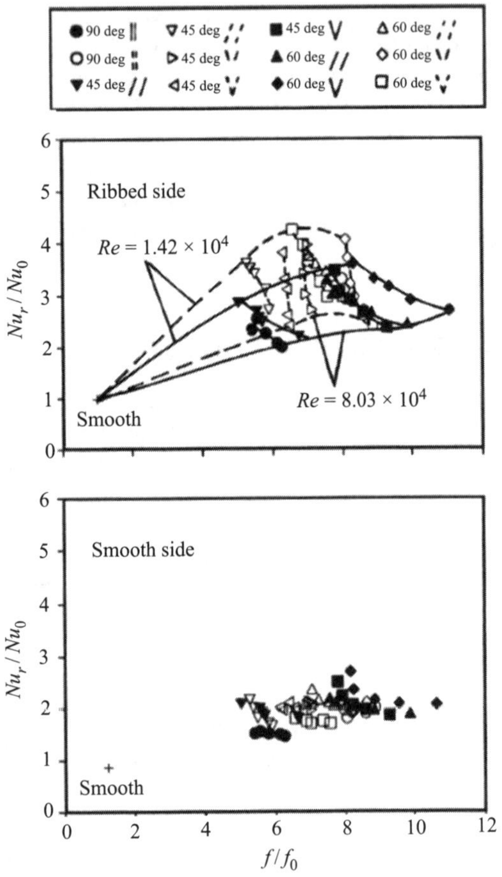

Fig. 4 Overall thermal performance of rib configurations shown in Fig. 3 (Han and Zhang [5]).

as a necessary tradeoff. Han [4] performed the first detailed study on a variety of rib turbulator parameters to identify optimum geometry and develop correlations for predicting heat transfer for design codes. Several parameters, such as channel aspect ratio, rib configuration, and flow Reynolds number were considered. Most of the data presented were for square channels with a 5–10% rib height to channel size ratio and rib-to-rib pitch between 7–10 times the channel hydraulic diameter. Most modern gas turbine blades have irregular cross sections and varying parameters of rib height and pitch. Also, the increased heat-transfer requirement is offset by a smaller increase in pressure availability. The more complex rib turbulator design with relatively small increases in pressure drop and high heat-transfer enhancements have been the focus of research in the recent past.

A. INTERNAL CHANNELS WITH DIFFERENT RIB CONFIGURATIONS

Figure 3 shows a variety of rib configurations tested by Han and Zhang [5]. The 60-deg V-shaped broken rib was found to offer the heat-transfer enhancement, but this configuration can also increase overall pressure drop. It is necessary to find the optimum combination to achieve the highest heat-transfer augmentation with relatively low increases in pressure drop (f/f_0). Figure 4 shows the overall performance comparing all of the geometries in Fig. 3. Han et al. [6] also studied wedge- and delta-shaped ribs, but indicated significantly higher pressure drops for relatively insignificant heat-transfer augmentation. Different channel

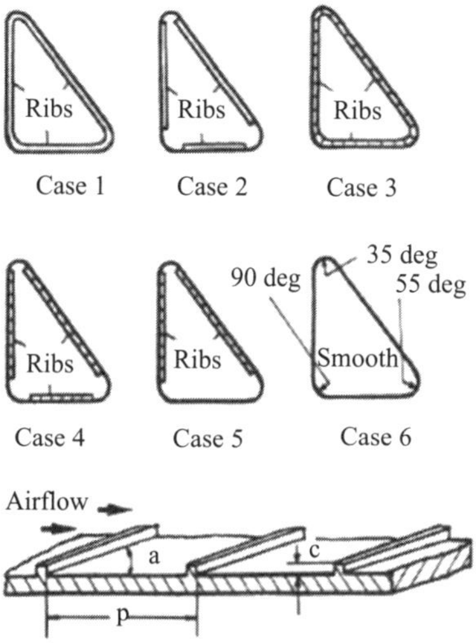

Fig. 5 Triangular channels tested by Zhang et al. [7].

geometries, such as channel aspect ratio, with ratios of 1:1; 2:1; 4:1; and 8:1, have been studied by several investigators. Additional channel geometries, such as trapezoids and triangles, have also been a focus of several studies. Zhang et al. [7] studied triangular channels with full and partial ribbed walls. Figure 5 shows the different triangular channel configurations tested in that study; partial ribs were found to work better on triangle channel walls than full ribs. The triangular channels typically represent the leading-edge channel of the turbine blade. Figure 6 shows the thermal performance comparison for triangular channels as studied by Taslim et al. [8]. More geometry variations and special conditions are summarized in [3].

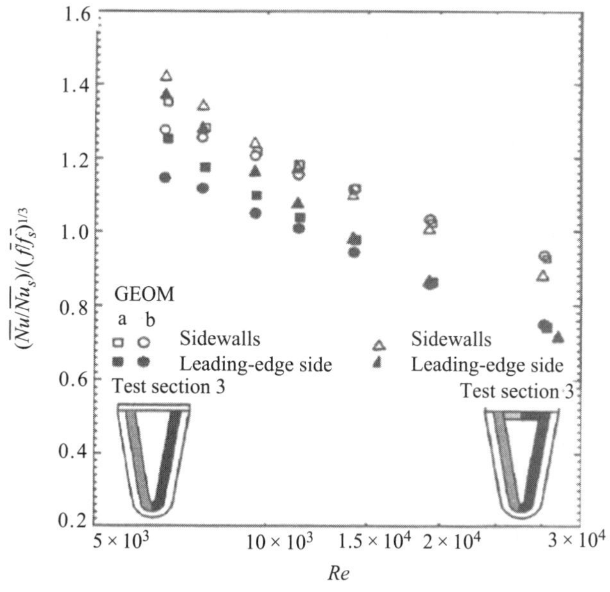

Fig. 6 Thermal performance of the different triangular channel configurations (Taslim et al. [8]).

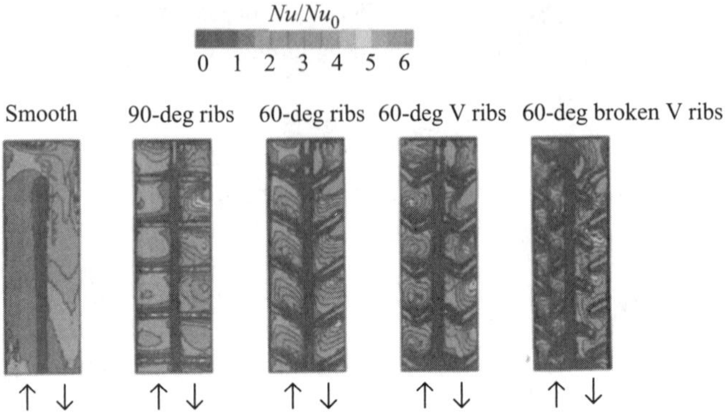

Fig. 7 Detailed heat-transfer distributions in various ribbed channels (Ekkad and Han [9]).

More recently, detailed measurements and accurate numerical prediction of the performance of ribbed channels have been the focus of many studies. Ekkad and Han [9] were the first to provide detailed measurements inside different ribbed channels. Figure 7 shows the measurements that they obtained using a transient liquid crystal technique. The measurements clearly show the local behavior of the secondary flows and their effects on surface heat transfer. Since this study, there have been several recent studies that have focused on detailed heat-transfer measurements in ribbed channels. Hsieh and Chin [10] presented detailed laser Doppler anemometry (LDA) measurements in a rotating two-pass ribbed rectangular channel. Figure 8 shows the streamwise velocity distributions between two consecutive ribs under the effect of rotation ($Ro = \frac{VD}{\Omega}$) where V is the flow velocity through the channel of diameter D and a rotation speed Ω rpm. Such detailed measurements of flow are clearly required for validating CFD predictions. There has been tremendous effort to consider a variety of new rib configurations.

Wright et al. [11] proposed angled, discrete angled, V-shaped, and discrete V-shaped ribs, as well as a new W-shaped and discrete W-shaped ribs and compared them with the standard 45-deg and 60-deg rib configurations. Figure 9 shows the newly proposed rib configurations, and Fig. 10 shows the thermal performance of the rib configurations in the nonrotating channels. It is clear that the performance of the new W ribs is similar to or better than that of the discrete V ribs. The discrete V-shaped ribs showed increased heat-transfer enhancement and decreased frictional losses as compared to the continuous V-shaped ribs. The frictional losses are significantly lower for the continuous V-shaped ribs.

Maurer et al. [12] performed an experimental and numerical investigation to assess the thermal performance of V- and W-shaped ribs in a rectangular channel. They presented detailed measurements using a liquid crystal technique and also performed CFD to capture the secondary flow behavior. They also varied the pitch of the ribs from 5 to 10 rib heights. They showed that the W-shaped ribs with a pitch to rib height (p/e) ratio of 10 clearly outperformed other configurations studied. Figure 11 shows the secondary flow behavior for both V- and W-shaped ribs with a p/e spacing of 5 and 10. The airflow close to the rib

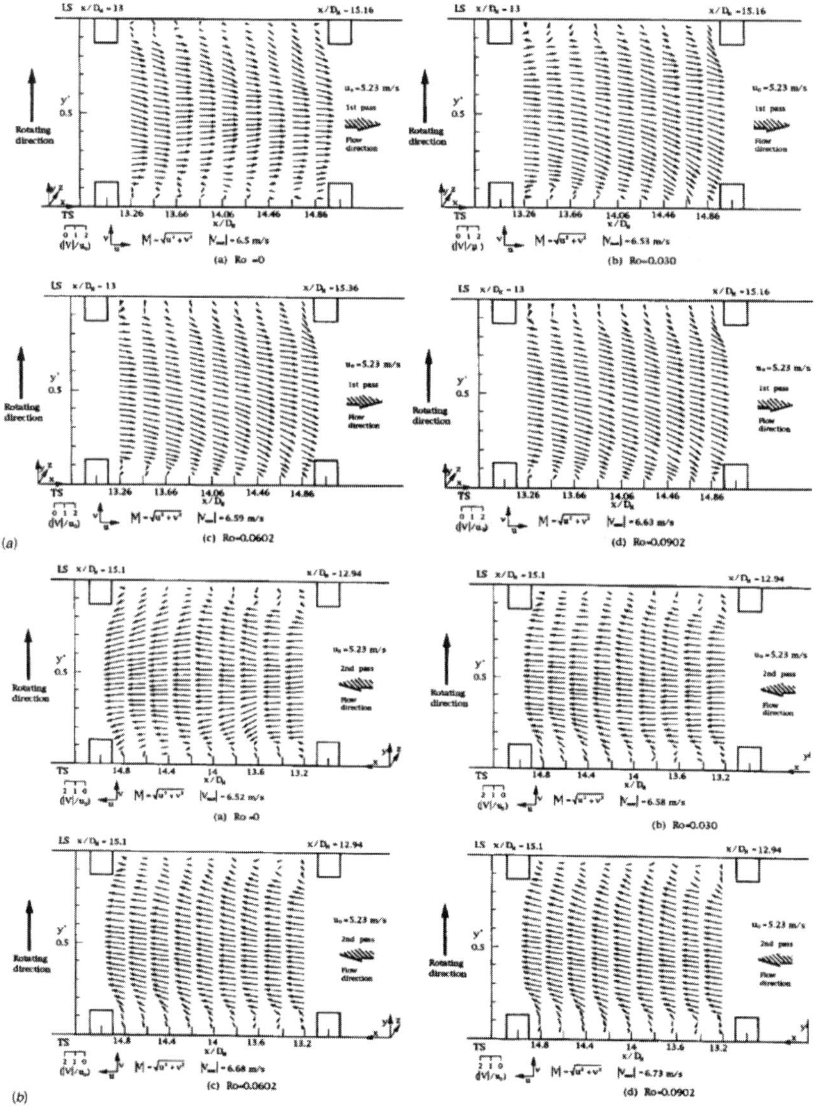

Fig. 8 Streamwise vector plots for rotating ribbed channels [10].

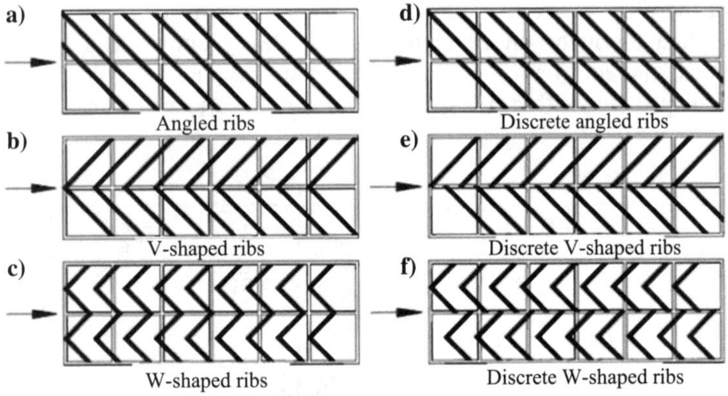

Fig. 9 Newly proposed W-shaped ribs by Wright et al. [11].

is accelerated and detaches once the rib is passed for both V- and W-shaped ribs. For the W-shaped geometry, however, there are a total of four vortices, which start right behind the rib in the midplane between the symmetry plane and channel side wall.

B. INTERNAL CHANNELS WITH BLEED

Most channel passages closer to the blade leading edge have film cooling holes bleeding coolant air for external cooling. Shen et al. [13] presented a heat-transfer study of single-pass channels with coolant holes for film extraction. They showed that film extraction enhances internal cooling heat transfer for small extraction flow rates and reduces heat transfer for large extraction rates. Ekkad et al. [14] also studied film coolant extraction effects for a variety of channels with different rib configurations in two-pass channels. They indicated that the effect of bleed was small on the first pass, except in the region downstream of extraction. In the region downstream of the turn in the second pass, the extraction reduces the heat transfer for the channel. Thurman and Poinsatte [15] investigated the effect of bleed extraction in the first pass of a three-pass test section.

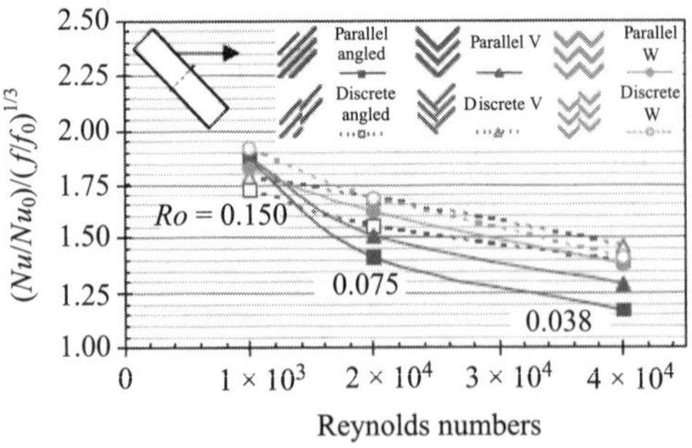

Fig. 10 Thermal performance characteristics of ribs tested by Wright et al. [11].

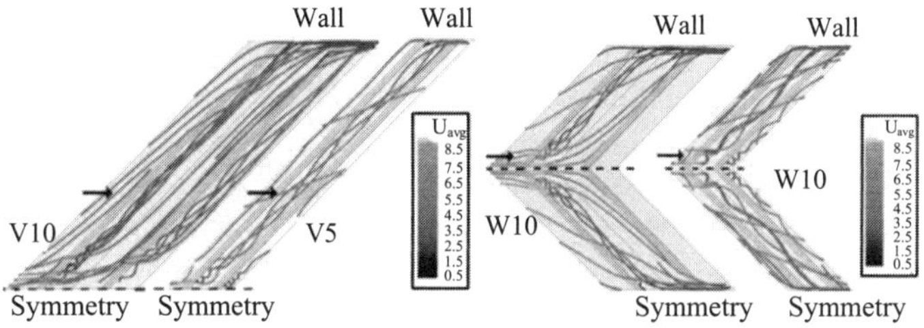

Fig. 11 Secondary flow lines for both new W- and V-rib configurations (Maurer et al. [12]).

They also studied the effect of hole location (midrib and next-to-rib) with respect to the ribs.

Figure 12 presents the effect of bleed hole location on Nusselt-number enhancements. The patterns produced by bleed with ribs between holes are backward-C-shaped compared to the oval shapes for the no-bleed case. These patterns were also observed in Ekkad et al. [14] and Shen et al. [13]. For the uniform bleed case and holes in the middle, there is significantly higher heat transfer near each hole, as well as away from the hole near the ribs and walls. The heat-transfer distributions appear to be periodic, except in the region around hole 1, due to developing flow, and around hole 8, due to the flow turning around the partition. All of the studies also showed that higher bleed rates yield higher heat-transfer values near to and away from the hole. Higher or lower bleed upstream also reduced the heat transfer downstream, away from the hole. For holes near ribs, the patterns produced by bleed are C-shaped, opposite to that produced by the ribs-between-holes configuration. Other than the mirrored patterns, the heat-transfer trends for each case are similar to those for ribs between holes.

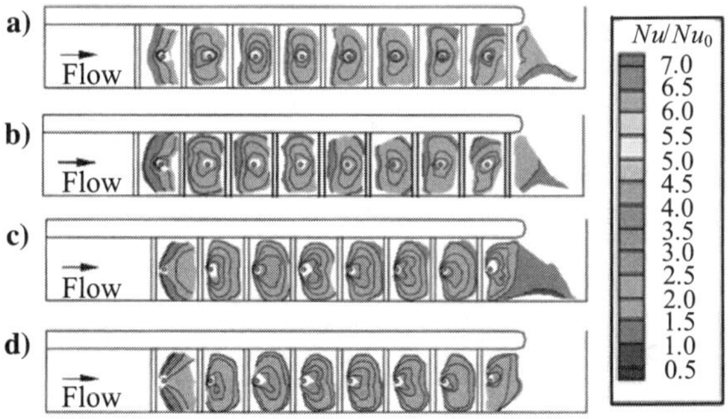

Fig. 12 Effect of bleed hole location relative to rib as shown by Thurman and Poinsatte [15].

Kim et al. [16] studied the effect of rotation on channels with bleed holes. They studied both 90- and 45-deg angled ribbed channels. Figure 13 shows the mean Sherwood number ratios of both the leading and trailing surfaces at tested rotation number and bleed ratios. The 90-deg ribbed channel does not show immediate effect of rotation with a slight reduction in Sherwood numbers for increasing rotation numbers. However, the effect of rotation is significant for the 45-deg ribbed channel. This happens at all rotation numbers. The 45-deg angle creates a secondary flow that directs flow towards the bleed hole, resulting in lower heat-transfer coefficients.

Another interesting aspect of internal channel studies has been the effect of rotation on different rib channel geometries. Han et al. [3] presented the various studies that have focused on the effect of rotation on internal channel heat-transfer characteristics. Several studies have shown the effect of strong Coriolis and rotational buoyancy forces on rotor blade coolant flow and subsequent effect on surface heat-transfer distribution. Dutta and Han [17] correlated the results from several studies to produce a simple correlation for predicting heat transfer in rotational channels. However, rotational results have not been as easy to correlate as for stationary channels. Furthermore, there has been a significant effort recently to predict the flow and heat transfer in rotating ribbed channels with advanced time-resolved CFD models using unsteady RANS and LES schemes.

Wright et al. [18] investigated the effect of rotation on a smooth wedge-shaped channel. This shape is typical of a trailing-edge channel with gill slots for exits. Figure 14 shows the representative geometry tested in their study. The cross

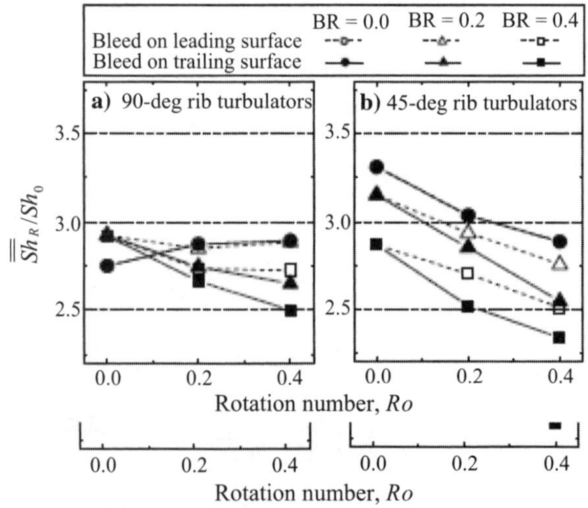

Fig. 13 Effect of bleed on ribbed channels with rotation (Kim et al. [16]).

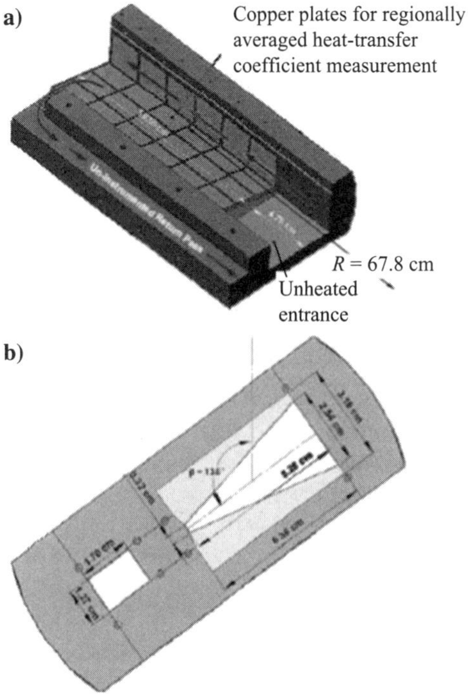

Fig. 14 Wedge geometry as studied by Wright et al. [18].

section of the channel was a wedge (or trapezoid). If the channel were extended to form an isosceles triangle, the apex angle of the triangle was 23.5 deg. To limit the complexities associated with the cooling channel, this study only considers the cooling passage with smooth walls (no turbulence promoters are used to enhance heat transfer).

The spanwise- and streamwise-averaged Nusselt-number ratios are shown in Fig. 15. For all three surfaces, the Nusselt-number ratios increase as the rotation number increases. These results are also present in Fig. 15 with the averaged buoyancy parameters $[Bo = (\Delta\rho/\rho)Ro^2(R_x/D_h)]$. With the very distinct trends present for each surface, correlations were generated for the Nusselt-number ratios as a function of the buoyancy parameter. The data can be represented by Eq. (1), and Table 1 shows the constants for each surface.

$$\frac{Nu}{Nu_0} = A \cdot Bo^m + B \cdot Bo^n \tag{1}$$

Even with the nonsymmetrical wedge-shaped channel, the skewed direction of rotation number, the variable Reynolds number, and the changing rotational speed, the data were represented by a simple correlation expressing the heat-transfer enhancement as a function of the buoyancy parameter.

Typically, heat-transfer coefficient is defined on the basis of temperature differential between the local bulk temperature and the local wall temperature:

$$q'' = h(T_w - T_b) \tag{2}$$

Bulk temperature is calculated based on conducting an energy balance through the channel length.

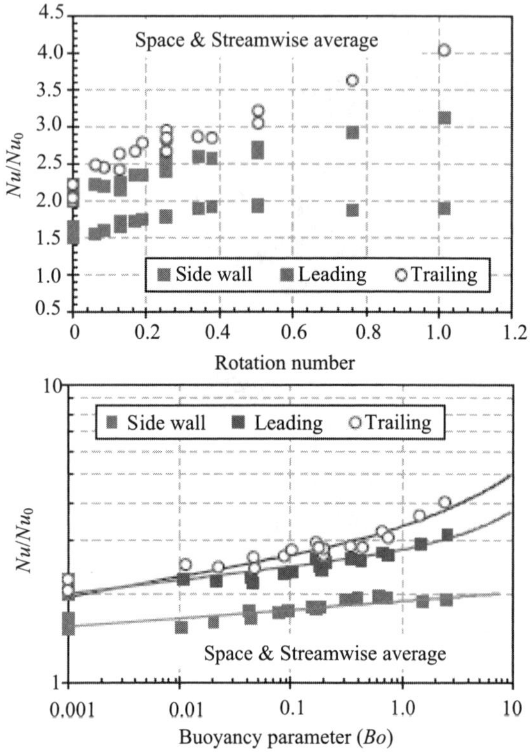

Fig. 15 Overall heat-transfer enhancement with rotation (Wright et al. [18]).

II. IMPINGEMENT COOLING

Jet impingement cooling is considered the most effective cooling technique, and it has been used in a wide range of applications. It is commonly used in regions of high heat load, like the leading edge of airfoils. In gas turbine applications, arrays of jets are used to cool the target surface with the spent flow exiting in one or two directions. Kercher and Tabakoff [19] and Florschuetz et al. [20] provided detailed correlations for predicting heat transfer under an array of jets. These correlations were developed mainly to help designers calculate heat-transfer coefficients as boundary conditions for design codes. However, the negative impact of having arrays of jets with spent flow exiting in one or two directions is the effect of crossflow. Crossflow develops when the spent flow from upstream jets interacts with downstream jets. This has an adverse impact on the impingement and associated heat transfer as the downstream jets are pushed away from the surface and therefore do not impinge completely. This leads to degradation in heat-transfer coefficient for downstream jets. The longer the number of rows in an array, the stronger is the degradation in heat-transfer coefficient. Huang et al. [21] indicated that the

TABLE 1 CORRELATION CONSTANTS FOR AVERAGE HEAT-TRANSFER ENHANCEMENT

	A	*m*	*B*	*n*
Leading surface	2.7	0.043	0.1	0.9
Trailing surface	3.05	0.065	0.3	0.7
Side wall	1.88	0.065	0.0	–

degradation can be reduced by spent flows exiting in both directions. The limiting factor to such a scheme is the number of rows in the spent flow direction.

A. IMPINGEMENT COOLING DESIGN CORRELATIONS

Because of the difficulties involved in crossflow, impingement cooling is primarily used in short arrays for targeted cooling. Several recent studies have focused on impingement cooling designs aimed at reducing the crossflow effects on downstream rows. Impingement holes in real engines are primarily directed to hot spot locations, producing nonsquare arrays. Gao et al. [22] investigated geometries where the spacing between the holes increased in both the streamwise and spanwise direction, simulating the stretching of the hole arrays downstream. Two different arrays were investigated, with the first array having uniform diameter holes through the array placed in a stretched format and the second array having increasing diameter holes. The main goal was to make actual measurements for stretched arrays and compare the results to predictions based on correlations developed by Kercher and Tabakoff [19] and Florschuetz et al. [20]. The linearly stretched arrays were chosen as the simplest example for comparing with existing correlations. The goal was not to compare the two stretched arrays but to compare the predictions from existing correlations to different stretched array configurations.

Figure 16 presents the detailed Nusselt number distributions for the uniform diameter case with an average jet Reynolds number of 6×10^3 and three different jet to target plate distances [22]. The local heat-transfer coefficients are normalized by the hole diameter (1.27 cm). The streamwise flow is from bottom to top. It appears that the interaction between the jets decreases downstream as the jets become sparse from the first row to the eighth row of holes. The jets

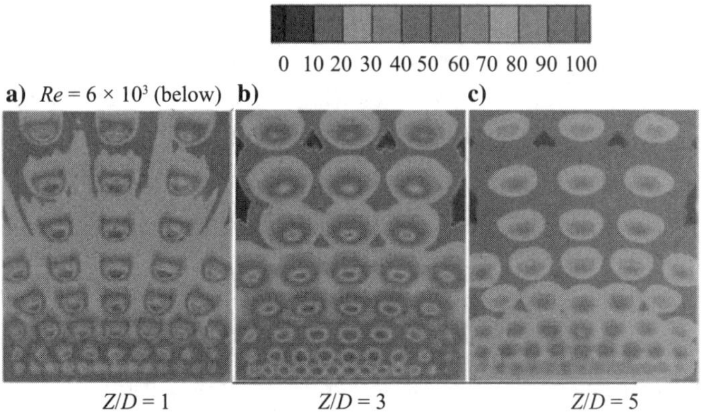

Fig. 16 Detailed Nusselt-number distributions for a stretched array of impinging jets as studied by Gao et al. [22].

also appear to spread outward with increased heat-transfer coefficients in between the holes. The crossflow effect is also evident, with the widened heat-transfer zone behind the impingement locations. The crossflow effect seems strongest for the jet height-to-target wall (Z/D) ratio of 1.0. A Z/D of 3.0 produced the highest heat-transfer coefficients as compared to the other Z/D ratios for this geometry. The detailed heat-transfer coefficient distributions clearly showed that the tendency of increased crossflow is to push the jets away from the wall, resulting in lower heat-transfer coefficients directly underneath the impinging jets, even for the sparse rows downstream. The predictions from the published correlations by Kercher and Tabakoff [19] and Florschuetz et al. [20] locally overpredicted the effect of crossflow on heat transfer and underpredicted the heat-transfer coefficients under strong crossflow effects. Uysal et al. [23] also studied the effect of varying jet size in the spent flow direction. Unlike Gao et al. [22] they studied a single row of jets.

Bailey and Bunker [24] investigated the detailed heat-transfer distributions for a series of square, in-line arrays of impingement jets with no initial crossflow and discharge in one direction. Comparisons with the correlations of [19, 20] showed deviations in the streamwise row heat-transfer behavior, which are sometimes minor, and in several instances major. The detailed surface impingement heat-transfer distributions establish the character of such dense arrays in confined geometries, shedding light on the transitional nature of the flows between impingement-dominated regions and channel-like flows.

B. REDUCED CROSSFLOW IMPINGEMENT

Reduced crossflow affected designs for impingement cooling is currently a major focus area. Several patents exist for low crossflow impingement cooling designs. Correia [25] shows a corrugated wall-type design for turbine shroud cooling through ducts of increasing cross-sectional area to allow spent air removal after impingement. Haumann et al. [26] offer a design similar to a corrugated wall but with trapezoidal channels. Wettstein [27] suggested a baffle cooling arrangement with flared conical ports for turbine blade internal cooling or combustor liner cooling. They showed four different versions of baffle cooled element.

All of the preceding designs were the basic core of the study by Esposito et al. [28]. Two main designs, a corrugated wall and extended ports, were tested. Results show that both geometries reduced the crossflow induced degradation on downstream jets, but the geometries performed better at different Reynolds numbers. The extended port and corrugated wall configurations showed similar benefits at the high Reynolds numbers, but at low Reynolds numbers the extended port design increased the overall level of heat transfer. This is attributed to the developed jet velocity profile at the tube exit. The benefit of the developed jet velocity profile diminished as jet velocities rose, and the air had less time to develop prior to exiting.

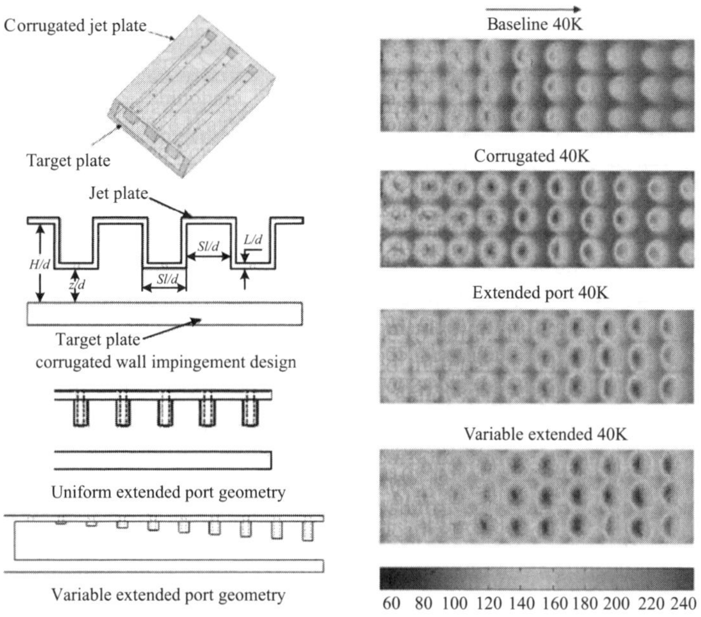

Fig. 17 Low crossflow effect designs as presented by Esposito et al. [28] and detailed heat-transfer distributions underneath the impinging jets.

Figure 17 compares the detailed Nusselt-number distributions for the baseline and low crossflow designs at a jet average Reynolds number of $Re = 4 \times 10^4$. The crossflow effect is still distinct for the corrugated wall design but is relatively lower than for the baseline case. Impingement core heat transfer is strongly evident, with very little mixing between jets indicated by the lower Nusselt numbers between the holes. The extended port design showed lower core Nusselt numbers for the early rows and higher Nusselt numbers for the downstream rows. The distortion of the jets due to the developing spent crossflow is evident. The variable extended port case is similar to the uniform extended ports case, but shows stronger impingement for downstream rows, caused by the longer ports. All three new designs proposed by this study showed the reduction of crossflow effects even at higher Reynolds numbers.

Typically, heat-transfer coefficient is defined on the basis of temperature differential between the fluid temperature upstream of the impingement plate and the local wall temperature:

$$q'' = h(T_w - T_{\text{imp}}) \tag{3}$$

Upstream impingement temperature is measured using thermocouples upstream of the impingement plate.

III. DIMPLES

A. DIMPLE CONFIGURATIONS

Recently, dimples or surface depressions have become a very interesting heat-transfer enhancement technique. Unlike the ribs that protrude into the flow, dimples are shaped depressions on the surface; their purpose is to increase surface area for flow contact. Early studies on effects of dimple cavities on heat transfer and flow structure were primarily by Russian investigators who focused on flow and/or heat transfer either inside or downstream of single or multiple concave depressions on a wall in an internal passage. Gromov et al. [29] described the symmetric and nonsymmetric streamlines and flow patterns produced by such cavities with a variety of sizes. Afanasyev et al. [30] focused on the heat-transfer enhancement mechanism for flows over walls indented with regular arrays of spherical pits. High heat-transfer enhancement levels up to 30–40% without any appreciable pressure losses (compared to a smooth surface) were reported. Belen'kiy et al. [31] showed significant heat-transfer intensification from a tube surface fitted with a staggered array of concave dimples on surfaces of annular internal passages, but indicated high pressure losses. Kesarev and Kozlov [32] presented local distributions of heat-transfer coefficients inside a hemispherical cavity. They described the effects of turbulence intensity of the incident flow on the local heat flux and on the local shear stress on the cavity surface. Terekhov et al. [33] presented experimental measurements of flow structure, pressure fields, and heat transfer in a channel with a single dimple on one surface. Schukin et al. [34] were the first to apply dimples to gas turbine cooling. Average heat-transfer coefficients reported from the measurements on a heated plate downstream of a single hemispherical cavity in a diffuser channel and in a convergent channel provided data on the influences of the mainstream turbulence intensity level and the angles of divergence and convergence on heat-transfer augmentation. Chyu et al. [35] investigated the influences of Reynolds number on local heat-transfer coefficient distributions on surfaces imprinted with staggered arrays of two different shapes of concavities. Their measurements for channel height to dimple diameter ratios of 0.5, 1.5, and 3.0 showed higher heat-transfer coefficients everywhere on the surfaces, as compared to smooth walls. They showed an enhancement of the overall heat-transfer rate of about 2.5 times that of smooth surface values and pressure losses about half the values produced by conventional rib turbulators. Lin et al. [36] presented computational simulations of the flow structures and resulting surface heat-transfer distributions for geometries and flow conditions similar to those of Chyu et al. [35] The flow streamlines and temperature distributions provided insight into flow structural characteristics produced by the dimples. Gortyshov et al. [37] studied spherical dimples placed at different relative positions on the two opposite surfaces of a narrow channel. Moon et al. [38] focused on the effects of channel height to dimple diameter ratios (H/D)

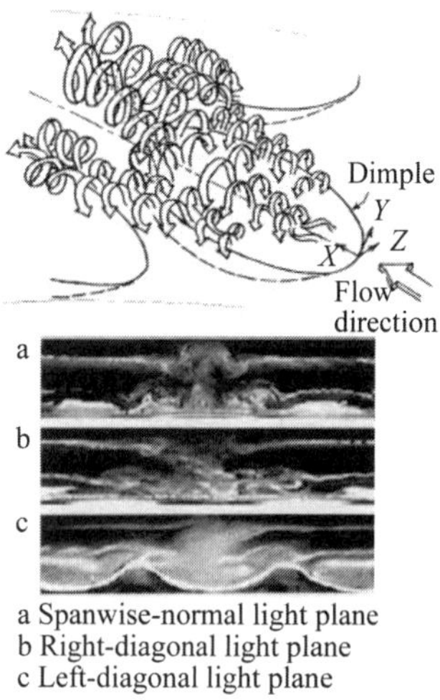

Fig. 18 Flow visualization of fluid structures inside and around a dimple (Mahmood et al. [39]).

from 0.37 to 1.49 on heat-transfer and pressure losses on a surface with a staggered pattern of dimples.

Mahmood et al. [39] investigated influences of dimples on flow and heat transfer and provided spatially resolved distributions of local Nusselt numbers, instantaneous flow structure (from flow visualizations), and distributions of total pressure and streamwise velocity. Figure 18 presents the flow visualization and illustration of flow over dimples. The visualization angles show the complex flow inside and around the dimples. The shedding of fluid from the dimple is indicated as a key feature of the flow structure. Flow visualizations also showed vortical fluid and vortex pairs shed from the dimples. The large upwash region and packets of fluid emanating from the central regions of each dimple, as well as vortex pairs and vortical fluid near dimple diagonals, helped to augment surface heat-transfer levels as they periodically impact the test surface and periodically produce an influx of bulk fluid.

Burgess and Ligrani [40] investigated the effect of dimple depth on both heat-transfer coefficient and friction loss in a channel. They varied the dimple depth to diameter ratio from 0.1 to 0.3. Figure 19 shows increasing globally averaged Nusselt-number ratios as dimple depth to diameter ratio (δ/D) increases from 0.10 to 0.30. Figure 20 also shows the effect of Reynolds number (Re_H varies from 9.54×10^3 to 7.48×10^4) on friction factor ratios, Nusselt-number ratios, and overall thermal performance parameters for dimple depth to diameter ratio (δ/D) of 0.1 and channel height to dimple diameter ratio (H/D) of 1. The dimpled surface friction factor ratios are lower than for several types of turbulated passages, where f/f_o range from 2.5 to 75. They compared their Nusselt-number ratio results with Mahmood et al. [39] ($\delta/D = 0.2$), Burgess et al. [41] ($\delta/D = 0.3$), Chyu et al. [35] ($\delta/D = 0.28$), and Moon et al. [38] ($\delta/D = 0.19$). All of the other studies showed higher values than for $\delta/D = 0.1$. This was surmised to be the effect of the increase in number of dimpled surfaces (i.e., from 1 to 2), or increasing δ/D. Figure 20 also shows globally averaged dimpled channel thermal performance parameters with respect to Reynolds number Re_H comparing results from Burgess et al. [41], Mahmood et al. [39], Chyu et al. [35], and Moon et al. [38]. The data of Burgess and Ligrani [40] compared well with all of the other studies and showed a constant thermal performance value for all Reynolds numbers, which is uncharacteristic compared to typical extended surface results.

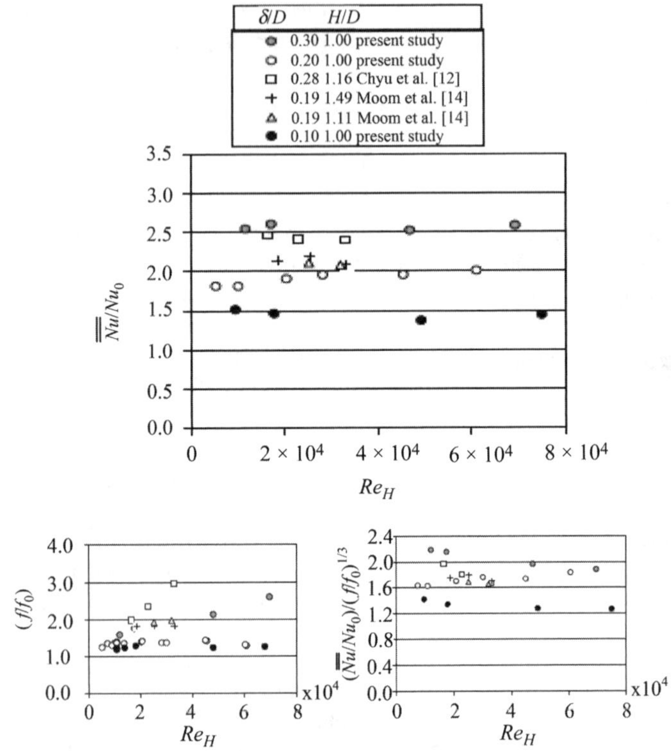

Fig. 19 Overall characteristics of dimple depth on heat-transfer and friction factor characteristics (Burgess and Ligrani [40]).

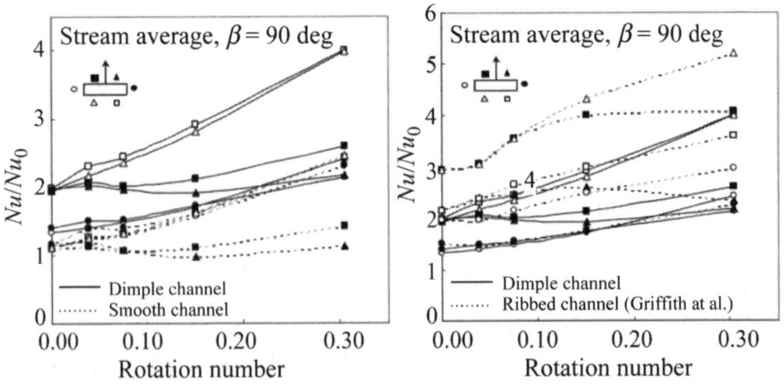

Fig. 20 Effect of rotation on dimpled channels and compared to ribbed channels (Griffith et al. [42]).

B. EFFECTS OF ROTATION ON DIMPLES

Griffith et al. [42] conducted an investigation into determining the effect of rotation on heat transfer in a rectangular channel (aspect ratio = 4:1) with dimples. They studied a range of flow parameters, including Reynolds number ($Re = 5 \times 10^3 - 4 \times 10^4$), rotation number ($Ro = 0.04$–$0.3$), and inlet coolant-to-wall density ratio ($\Delta\rho/\rho = 0.122$). They compared as dimple depth-to-print diameter (δ/Dp) ratio of 0.3 to a smooth channel with and without rotational effects.

Figure 20 shows a comparison between the results of this investigation and the ribbed channel investigated by Griffith et al. [43]. Although the rotational effects are similar on dimpled and ribbed channels, the ribbed channel seems to induce higher heat-transfer enhancement as compared to the dimpled channel. The dimpled channel provides less enhancement on some surfaces, even at the lowest rotation number. The trends are very similar to the ribbed channel with increasing rotational effect. There was less spanwise variation for the orthogonal ($\beta = 90$ deg) dimpled channel under rotation than for the ribbed channel. Spanwise variations were quite significant in the case of the ribbed rotating channel due to the 45-deg rib-angle effect. It is not clear whether the ribbed channel creates greater pressure drop (friction penalty) than the dimpled channel under rotation, as is the case for the stationary channels.

Elyaan and Tafti [44] investigated in detail the flow structure and heat-transfer characteristics in a rotating channel with dimples and protrusions on opposite walls. This is the only detailed study that has been conducted to quantify the behavior of dimpled channels under rotation except for that of Griffith et al. [42] for a dimple–dimple channel. Large eddy simulations were used to perform high-fidelity time-dependent calculations of a rotating channel with dimples on the pressure side and protrusions on the suction side at a nominal channel Reynolds number of 10^3 and rotation numbers of $Ro_b = 0.0$, 0.15, 0.39, and 0.64. Figure 21 shows the velocity streamlines in the domain at a spanwise plane located at $0.2D$ downstream of the dimple (protrusion) at $x/D = -0.2$ for different rotation numbers. The wake region downstream of the protrusion, separation shear layer inside the dimple cavity, and flow impingement at the

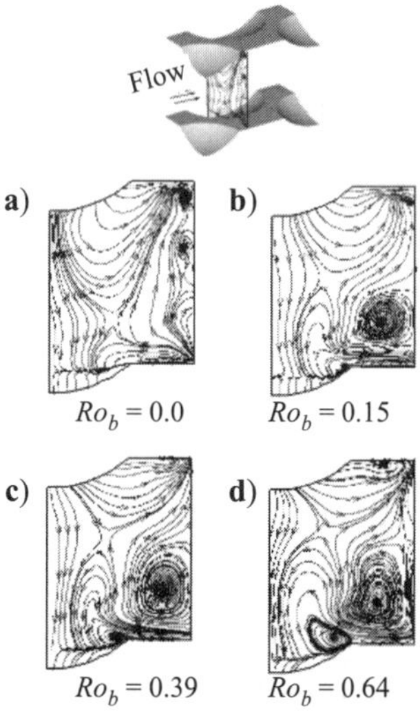

Fig. 21 Secondary flow structure at a spanwise plane located 0.2*D* downstream of first dimple ($x = -0.21$) for: a) $Ro_b = 0.0$, b) $Ro_b = 0.15$, c) $Ro_b = 0.39$, d) $Ro_b = 0.64$ (Elyaan and Tafti [44]).

flat landing downstream of the dimple are seen in the stationary case streamlines. With additional rotation ($Ro_b > 0.0$), the flow structure in the spanwise plane shows two additional large-scale counterrotating structures appearing in the cross section near the dimple (trailing) surface: one above the flat landing and the other above the dimple cavity. Increasing the rotation number shows a gain in definition and strength of these structures. The large structures are caused by rotational Coriolis forces and can be viewed as secondary flows in the cross-section.

Typically, heat-transfer coefficient is defined on the basis of temperature differential between the local bulk temperature and the local wall temperature:

$$q'' = h(T_w - T_b) \tag{4}$$

Bulk temperature is calculated based on conducting an energy balance through the channel length.

IV. PIN FIN COOLING

Pin fins have been used for gas turbine internal cooling systems for decades, especially in the trailing-edge region of airfoils. They also provide mechanical strength by bridging the pressure and suction surfaces in the thin trailing-edge region. Typically, pin fins are attached on both walls and provide added conduction effects to enhance heat transfer and improved cooling effectiveness. A number of publications on this topic are summarized in [3]. Recently, Chyu et al. [45] have investigated different types of pedestals to further enhance heat transfer due to presence of pin fins. Chyu et al. [45] have also investigated pin

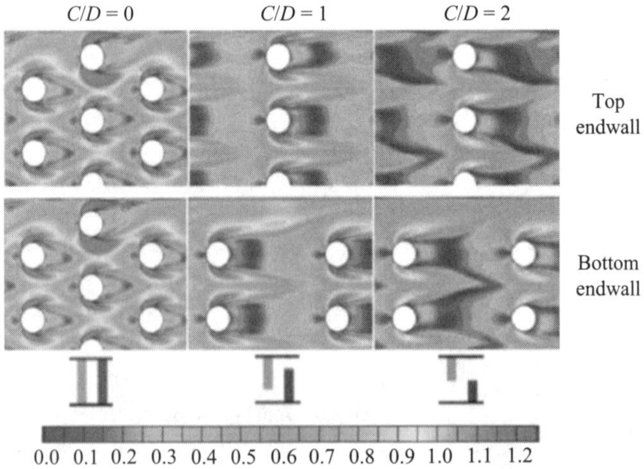

Fig. 22 Shear-stress distributions on top and bottom endwalls of pin finned surfaces [47].

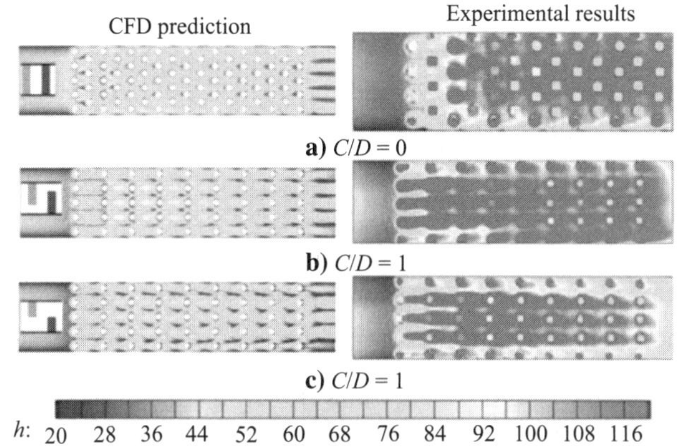

Fig. 23 Heat-transfer coefficient distributions for pin fins with different clearance gaps [47].

fins that are inclined to the flow and studied the heat-transfer enhancement due to the inclination angle and associated flow effects. Chen et al. [46] studied different shaped pedestals for trailing-edge cooling. Siw et al. [47] studied heat-transfer and pressure characteristics of detached pin-fin arrays in a rectangular channel. They studied a 3:1 aspect-ratio channel with three different pin fin height-to-diameter ratios of 4, 3, and 2. Figure 22 presents the wall shear distributions for different pin fins with different clearance combinations. The shear stress on the endwall was high around the full pin due to the horseshoe vortex formation at the leading edge of the pin fin. With the clearance, the shear stress is also higher at the tip-to-endwall clearance due to the presence of separated shear layers and increased turbulence. Figure 23 shows the associated heat-transfer predictions for the same clearance gaps as in Fig. 22. Both CFD and experimental results are presented. CFD results are typically 30–40% lower than the experimental results for the same conditions. A clearance-to-diameter ratio (C/D) of 1.0 provides the highest heat-transfer coefficients; this configuration has the right amount of separated flow mixing and turbulence production. Additional clearance ($C/D = 2$) seems to reduce the heat-transfer enhancement. These types of minor changes to traditional enhancement techniques provide valuable insight into the heat-transfer dynamics and thus drive future research into combination cooling.

V. COMBINATION COOLING

With increasing turbine inlet temperatures, the need for more efficient cooling techniques has become necessary. Traditional cooling techniques have reached the heat-flux removal rate limits. Researchers have been investigating combinations of two highly efficient cooling schemes into one to provide higher cooling rates than those achieved by a single scheme. Some popular schemes tested prior to 2001 are shown in Han et al. [3]. Some of the common combination schemes have been as follows:

1. Impingement with ribbed surfaces – Haiping et al. [48], Gau and Lee [49], Taslim et al. [50]
2. Impingement on pinned and dimpled walls – Chakroun et al. [51], Azad et al. [52]
3. Impingement with fins – Metzger and Fan [53]
4. Swirl and impingement – Glezer et al. [54]

Recently, several studies have used combined cooling schemes to achieve higher heat-transfer coefficients than observed previously.

Ekkad et al. [55], Pamula et al. [56], and Ekkad et al. [57] studied the effect of crossflow-induced swirl and impingement in a two-pass channel with and without rib turbulators. These studies used flow in a typical multipass cooling system from first pass to second pass to be channeled through a series of holes along the divider plate angled in such a way that there would be induced swirl and the spent flow would further induce swirl for downstream jets. Detailed Nusselt-number (*Nu*) distributions were measured on the sidewalls of both passes of the test channel. Results shown are for the 180-deg turn channel and the channels with hole/slot geometries at Reynolds numbers of 2.5×10^4. All local Nusselt numbers on the wall were normalized using the turbulent Nusselt-number correlation for flow inside a pipe. Figure 24 presents the detailed Nusselt-number ratio (Nu/Nu_0) distributions on the sidewall of the tested channels for $Re = 2.5 \times 10^4$ (Pamula et al. [56]). The first pass distributions show little differences for the four cases shown in the figure, except in the case of the slotted divider plate.

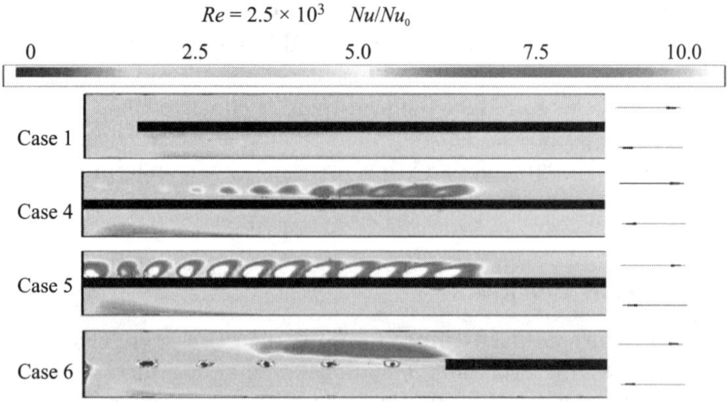

Fig. 24 Detailed Nusselt-number distributions for the channels tested by Pamula et al. [56].

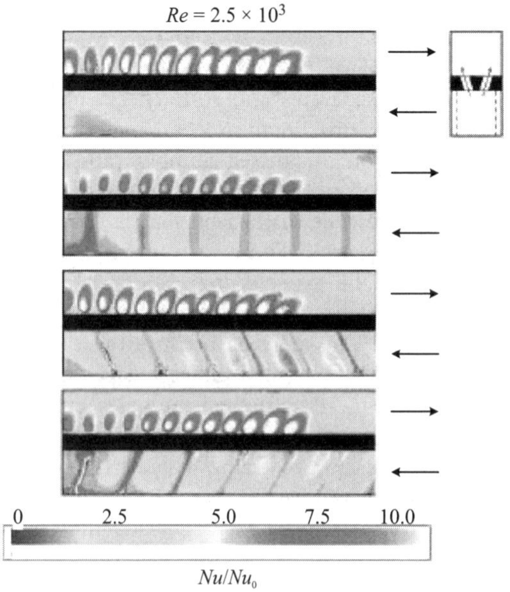

Fig. 25 Detailed Nusselt-number distributions for different rib configurations in first pass as studied by Ekkad et al. [57].

This result indicated that heat-transfer enhancement in the second pass does not affect the first pass heat-transfer distributions. In Fig. 24a, there is strong nonuniformity in Nusselt-number ratio distributions in the second pass immediately after the 180-deg turn. This is because the flow is affected by flow separation, reattachment, and bend effects. Figure 24b shows the case where the flow is injected through a straight hole. Heat-transfer distributions show significant jet impingement effects from the first few holes in the row. In the second pass, the crossflow becomes stronger toward the exit of the channel. This strong crossflow then rides the jets and pushes the impingement location downward for the jets closer to the channel exit. Thus, the roll-cells are displaced by the strong crossflow generated by the upstream jet injection. Nusselt-number ratios are as high as 7–10 inside the impingement core, which is significantly higher than any values obtained with ribbed configurations. Figure 24c shows the case where injection is through inclined holes. The jet impingement is strongly evident, with extremely high Nusselt numbers at impingement locations. Unlike the case with straight holes, the high heat transfer on the sidewall is caused by primary impingement. The spent air after impingement rolls up and swirls into the core of the channel and becomes the crossflow. The crossflow gains strength from the endwall to the channel exit. The stronger crossflow then pushes the impingement location of the jets downward and toward the exit. Figure 24d shows the case where the flow enters the second channel through two-dimensional (2-D) slots. The highest Nusselt numbers are obtained where the slot is narrow. As the slot widens, the local velocity of the flow is reduced, and the resultant Nusselt numbers are lower.

Ekkad et al. [57] added ribs on the first pass to enhance heat transfer on the first pass and keep the holes on the divider walls, as seen in Pamula et al. [56]. Figure 25 presents the detailed Nusselt-number ratio distributions for the angled hole divider wall with different rib configurations at channel

$Re = 2.5 \times 10^4$. Figure 25a presents the no-rib case from Pamula et al. [56]. Figure 25b shows the effect of turbulating the first pass with 90-deg orthogonal ribs. The first pass distributions are typical of rib turbulated channels with periodic high and low regions where flow separates and reattaches. Enhancement levels are around 2.0–3.0 between the ribs. In the second pass, the ribs appear to have caused a significant reduction in Nusselt-number ratios. This may be because of the increased pressure drop in the channel and the flow distributions in the first pass caused by the presence of the ribs. Figure 25c shows the channel with 60-deg forward-facing (FF) ribs. The heat-transfer enhancement in the first pass is higher, as should be expected for 60-deg ribs over 90-deg ribs. However, the second pass heat-transfer ratios are affected only slightly because of the presence of the ribs. This may be because the ribs are angled towards the divider wall and the secondary flows generated by the ribs from the outer wall to the divider wall push the flow toward the holes, creating lesser pressure drop and increasing jet velocities coming out of the holes. Figure 25d shows the channel with 60-deg backward-facing (BF) ribs. The ribs enhance heat transfer in the first pass, as for 60-deg FF ribs. The second pass enhancements are lower near the endwall far from the exit. The secondary flow direction for these ribs is from the divider wall toward the outer wall. This direction is opposite to the main flow direction. The driving pressure difference is for the flow across the holes. The secondary flows impede the main flow through the holes and thus increase pressure drop and reduce jet impingement and Nusselt-number ratios. This effect is more prominent for the jet holes closer to the endwall.

Bailey et al. [58] explored heat-transfer distributions for a combustor liner model utilizing impingement jet cooling, high-Reynolds-number turbulated flow between the liner and flowsleeve, and variable passage geometry. An experimental model provided full-surface heat-transfer distributions under engine-representative conditions. A CFD numerical model precisely representing the experiment provided predictions of full-surface heat transfer, as well as direct comparisons to data with both two-layer and wall-function turbulence modeling methods.

The heat-transfer data for three different liner cooling configurations are shown in Fig. 26. Each case had the same total flow rate. The data were presented as spanwise- or circumferentially averaged heat-transfer coefficients across the middle 50% of the test plate, thereby eliminating any endwall effects. The second configuration was a convection-only case, without impingement, in which 100% of the flow was present at the inlet to the channel (2.67 kg/s). In this condition, the thermal entry region effect was observed to decay to turbulent, fully developed smooth duct flow. For the convection-only case, this showed a combined thermal and hydrodynamic entry length of about 10 passage heights, which is somewhat less than typically observed under more ideal conditions. The downstream turbulated wall is seen to enhance the heat transfer by a factor of about 2. The third configuration uses impingement in the upstream region,

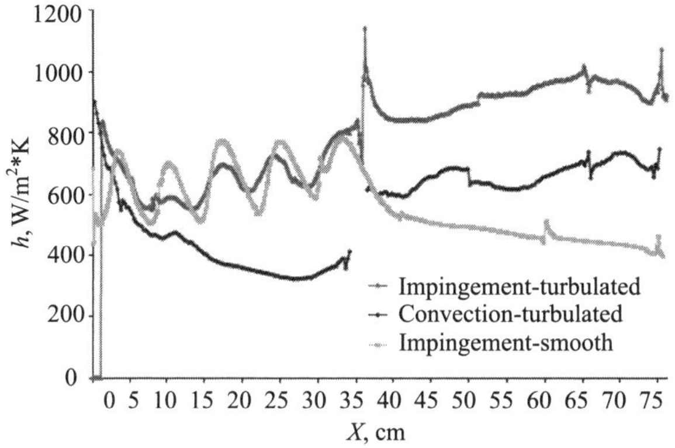

Fig. 26 Comparison of laterally averaged heat-transfer coefficients for three different cooling configurations (Bailey et al. [58]).

but eliminates the turbulators in the downstream region. The most striking feature of these results is the 40 to 50% enhancement in the turbulated region caused by the use of impingement upstream. This enhancement is essentially constant along the channel length and felt to be due to the elevated bulk fluid turbulence levels created by the impingement jets. This clearly shows the advantage of combination cooling schemes.

Impingement flat wall cooling was investigated by Andrews et al. [59] in the presence of an array of interrupted rib obstacles. These obstacles had the form of rectangular pin-fins with a 50% blockage to the crossflow. One side exit of the air was used, and there was no initial crossflow. These results suggest that obstacles of the pin-fin type are only useful with impingement cooling in situations where there is significant crossflow, as in the trailing-edge region of an array of impingement jets with no initial crossflow with minimal extra pressure loss. Son et al. [60] presented shear pattern visualization on the target surface, pressure loss measurements, and heat-transfer coefficient measurements for an impingement cooling system with simply shaped roughness element, namely, cylindrical and diamond pimples. A hexagonal rim was also designed to enhance the complete low heat-transfer coefficient region midway between neighboring jets. The effect of the rim height, cross-sectional shape, and wall angle was studied using a series of heat-transfer and pressure loss experiments. Figure 27 compares the smooth impingement surface and diamond and cylindrical pimpled surfaces for overall averaged Nusselt numbers. The cylindrical pimpled surface produced a 10–15% higher Nusselt number than the baseline no-pimple surface. It appears that the pressure drop requirement also seemed unaffected by the presence of pimples on the surface similar to Andrews et al. [59]. The diamond surfaces produced even higher Nusselt number, at least 20–25% higher than in the case of smooth surface with relatively slight increases in pressure drop.

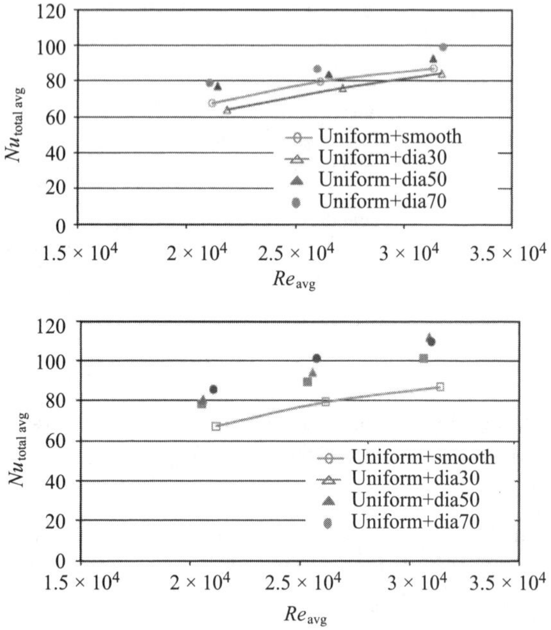

Fig. 27 Overall averaged Nusselt numbers for smooth and cylindrical and diamond dimpled surfaces (Son et al. [60]).

Another technique that has been proposed recently is the latticework geometry. Latticework cooling, also known as vortex cooling or bounded vortical duct cooling, in its application to high-temperature gas turbine components, originated 25 years ago within the former Soviet design bureau engineering system [61]. A Soviet blade design entirely using lattice cooling is shown in Fig. 28. Each channel is inclined to the spanwise direction of the blade and is separated from the next channel by a narrow full passage height rib. Heat-transfer coefficients were measured by Goreloff et al. [62] for lattices with ribs inclined to one another at 30, 60, 90 and 120 deg and found that 60 and 90 deg produced the greatest enhancement in heat-transfer coefficient for a constant pressure ratio and Reynolds number respectively, in the absence of rotation. This method of vane and blade structure and cooling developed into the most recognizable standard within the many Soviet designs. Its application history parallels the Western use of serpentine cooling for turbine blades. Cooled turbine airfoil technology in the former Soviet states (Russia and Ukraine) is still largely based upon vortex cooling.

Bunker [63] described latticework geometry in Fig. 29. The path of one initial subchannel of flow is depicted here. The flow stays within the initial subchannel, with little interaction with the crossing channels, until it encounters the side wall, the interior rib, or the external wall of an airfoil design. Upon hitting the side wall, the flow must turn by the angle 2β as it enters the upper (or lower) crossing channel, that is, it switches from a pressure-side

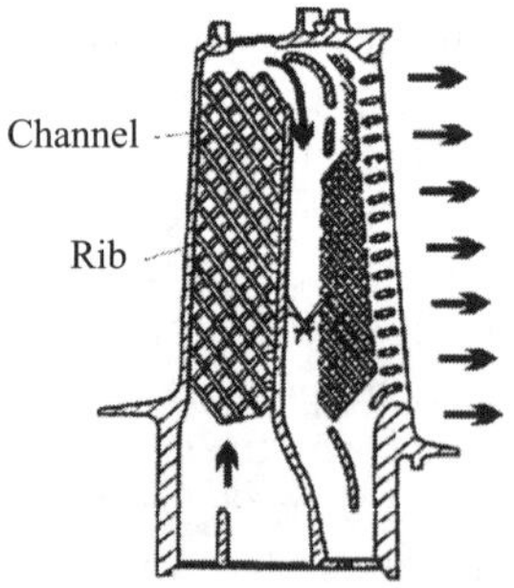

Fig. 28 Soviet blade design with latticework [62].

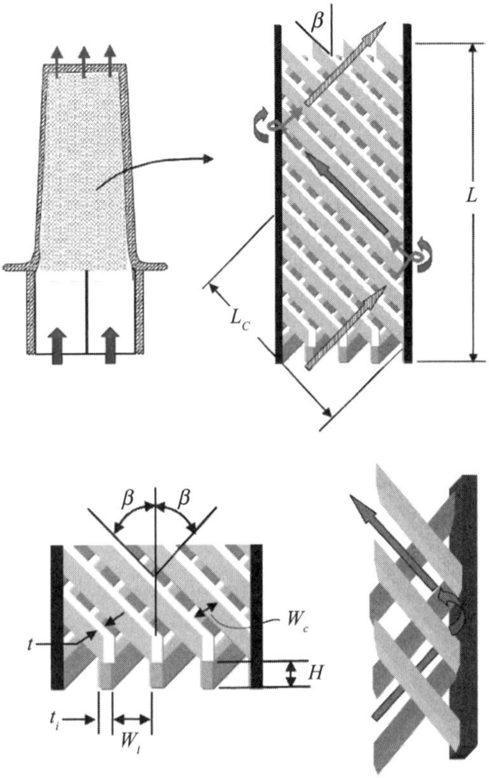

Fig. 29 Latticework geometry studied by Bunker [63].

subchannel to a suction-side subchannel, or vice versa. The flow makes its way in this "switchback" fashion until it leaves the lattice channel by way of film holes or routing to another portion of the airfoil interior, such as another lattice, or trailing-edge holes. The overall motion of the flow in the length of the lattice channel is then that of a "vortex," but one that is forcibly induced by the turning walls involving high pressure losses. The vortex channel shows a uniform heat-transfer coefficient both axially and laterally.

Figure 30 presents the laterally averaged heat-transfer coefficients for each of the vortex channels at each of three Reynolds numbers tested. Aside from a fairly mild entry region effect in each case, the laterally averaged HTCs are very steady with axial location. The dashed lines indicate the Dittus–Boelter correlation values for the same subchannel Reynolds numbers. In most test cases, the vortex channel primary surface heat transfer is about 1.5 times this level. Additionally, the lattice structure adds to heat-transfer surface areas as fins and additional cooling effectiveness is achieved through these fins.

Kontrovitz and Ekkad [64] were the first to study the combined effect of jet impingement and dimples. From their results, it was evident that the resultant combination of jets into dimples created a degradation of heat-transfer coefficient. There was 10–20% degradation in jet impingement heat transfer due to impingement into dimples. Kanokjaruvijit and Martinez-Botas [65] also investigated the heat transfer and pressure information for the jets impinging on the dimpled surface. Various geometric and parametric effects, such as crossflow scheme, dimple geometry, impinging positions, Reynolds number, jet-to-plate spacing H/D_j, dimple depth d/D_d, and ratio of jet diameter to dimple diameter D_j/D_d were considered. Figure 31 compares the streamwise average Nusselt-number distribution of both shapes to that of the flat plate at $Re_{Dj} = 11.5 \times 10^3$,

$H/D_j = 4$. The hemispherical dimples improved the heat transfer by 35%, while the cusped elliptical ones improved 26%. This may be because of the double dimples leading to a double rolling-up recirculation, which disturbed the oncoming jets, and thus reduced their temperature as well as momentum. They concluded that the thermal performance of the dimple impingement was dependent on the heat-transfer results rather than the pressure loss. The pressure results over the dimpled plate were not significantly different from those over the flat plate at the same setup condition. This was thought to be because the vortices produced from the dimples were organized and thus helped enhance the heat transfer with lower pressure loss. The vortices did not significantly increase pressure drop from when the impingement on a flat surface was employed. Overall, this study produced a positive evaluation of impingement on a dimpled plate.

Typically, heat-transfer coefficient is defined on the basis of temperature differential between the local bulk temperature and the local wall temperature:

$$q'' = h(T_w - T_b) \tag{5}$$

Bulk temperature is calculated based on conducting an energy balance through the channel length. In the case of combination with impingement,

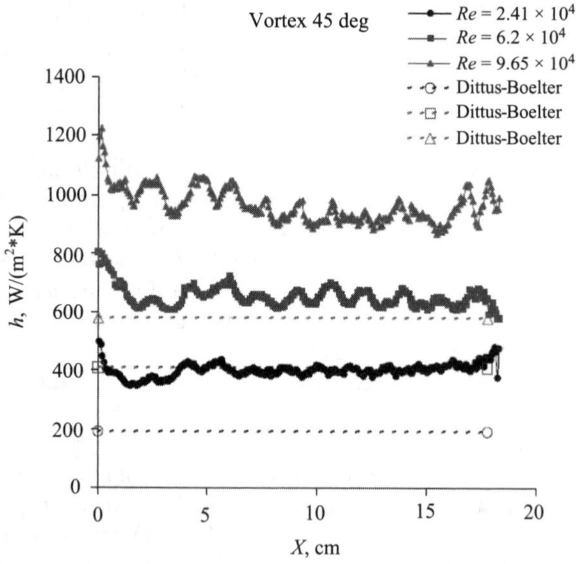

Fig. 30 Heat-transfer coefficient distributions measured by Bunker [63] in latticework channels.

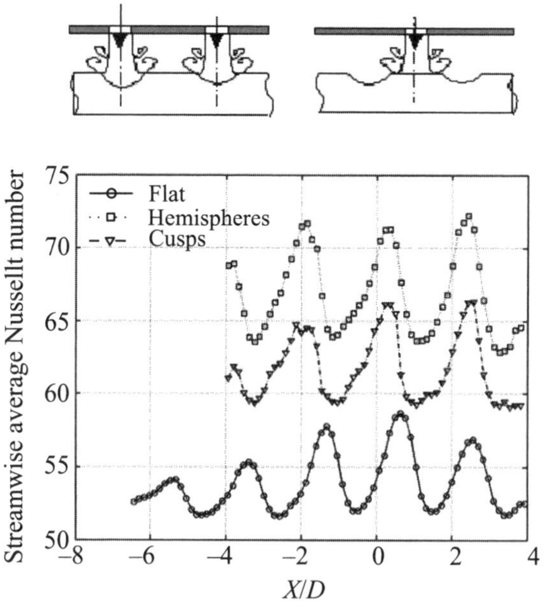

Fig. 31 Heat-transfer characteristics of impinging jets into and around dimples [65].

calculation of bulk temperature becomes complicated and the ratio of crossflow to impingement jet flow is required.

VI. SUMMARY

With increasing turbine inlet temperatures, effective cooling schemes have become critical for maintaining the life and integrity of hot-gas-path components. Several new combination techniques are being considered to achieve higher heat fluxes. We have presented the recent advancements in cooling techniques. Another important aspect is consideration of complex geometries and complex flow situations. Uneven geometries with nonuniform designs create a new challenge when managing surface temperatures to within the limits of operation. All of the new studies are focused on these issues, as shown. Han et al. [3] summarized studies published before 2000; this chapter deals with developments since that time.

REFERENCES

[1] Treager, I. E., "General Electric CF6," *Aircraft Gas Turbine Engine Technology*, 2nd ed., McGraw–Hill, New York, 1979, Chap. 25, pp. 469–525.

[2] Han, J. C., Park, J. S., and Lie, C. K., "Heat Transfer and Pressure Drop in Blade Cooling Channels with Turbulence Promoters," *ASME Journal of Engineering for Gas Turbines & Power*, Vol. 107, No. 4, 1985, pp. 628–635.

[3] Han, J. C., Dutta, S., and Ekkad, S. V., *Gas Turbine Heat Transfer and Cooling Technology*, 2nd ed., CRC Press, New York, 2013.

[4] Han, J. C., "Heat Transfer and Friction in Channels with Two Opposite Rib Roughened Walls," *ASME Journal of Heat Transfer*, Vol. 106, No. 4, 1984, pp. 774–781.

[5] Han, J. C., and Zhang, Y. M., "High Performance Heat Transfer Ducts with Parallel, Broken, and V-Shaped Broken Ribs," *International Journal of Heat and Mass Transfer*, Vol. 35, No. 2, 1992, pp. 513–523.

[6] Han, J. C., Huang, J. J., and Lee, C. P., "Augmented Heat-Transfer in Square Channels with Wedge-Shaped and Delta-Shaped Turbulence Promoters," *Enhanced Heat Transfer*, Vol. 1, No. 1, 1993, pp. 37–52.

[7] Zhang, Y. M., Gu, W. Z., and Han, J. C., "Augmented Heat Transfer in Triangular Ducts with Full and Partial Ribbed Walls," *Journal of Thermophysics and Heat Transfer*, Vol. 8, No. 3, 1994, pp. 574–579.

[8] Taslim, M., Li, T., and Spring, S. D., "Measurements of Heat Transfer Coefficients and Friction Factors in Rib-Roughened Channels Simulating Leading-Edge Cavities of a Modern Turbine Blade," *ASME Journal of Turbomachinery*, Vol. 119, No. 3, 1997, pp. 601–609.

[9] Ekkad, S. V., and Han, J. C., "Detailed Heat Transfer Distributions in Two-Pass Square Channels with Rib Turbulators," *International Journal of Heat and Mass Transfer*, Vol. 40, No. 11, 1997, pp. 2525–2537.

[10] Hsieh, S.-S., and Chin, H.-J., "Turbulent Flow in a Rotating Two Pass Ribbed Rectangular Channel," *ASME Journal of Turbomachinery*, Vol. 125, No. 4, 1999, pp. 609–622.

[11] Wright, L. M., Fu, W. L., and Han, J. C., "Thermal Performance of Angled, V-Shaped, and W-Shaped Rib Turbulators in Rotating Rectangular Cooling Channels (AR 4:1)," American Society of Mechanical Engineers, Paper GT2004-54073, June 2004.

[12] Maurer, M., Von Wolfersdorf, J., and Gritsch, M., "An Experimental and Numerical Study of Heat Transfer and Pressure Losses of V- and W-Shaped Ribs at High Reynolds Numbers," *Proceedings of ASME Turbo Expo 2007*, American Society of Mechanical Engineers, Paper GT2007-27167, May 2007.

[13] Shen, J. R., Wang, Z., Ireland, P. T., Jones, T. V., and Byerley, A. R., "Heat Transfer Enhancement Within a Turbine Blade Cooling Passage Using Ribs and Combination of Ribs with Film Cooling Holes," *ASME Journal of Turbomachinery*, Vol. 118, No. 3, 1996, pp. 428–433.

[14] Ekkad, S. V., Huang, Y., and Han, J. C., "Detailed Heat Transfer Distributions in Two-Pass Smooth and Turbulated Square Channels with Bleed Holes," *International Journal of Heat and Mass Transfer*, Vol. 41, No. 13, 1998, pp. 3781–3791.

[15] Thurman, D., and Poinsatte, P. E., "Experimental Heat Transfer and Bulk Air Temperature Measurements for a Multi-Pass Internal Cooling Model with Ribs and Bleed," American Society of Mechanical Engineers, Paper GT2000-0233, June 2000.

[16] Kim, K. M., Park, S. H., Jeon, Y. H., Lee, D. H., and Cho, H. H., "Heat/Mass Transfer Characteristics in Angled Ribbed Channels with Various Bleed Ratios

and Rotation Numbers," American Society of Mechanical Engineers, Paper GT2007-27166, May 2007.

[17] Dutta, S., and Han, J. C., "Local Heat Transfer in Rotating Smooth and Ribbed Two Pass Channels with Three Channel Orientations," *ASME Journal of Heat Transfer*, Vol. 118, No. 3, 1996, pp. 578–584.

[18] Wright, L. M., Liu, Y.-H., Han, J. C., and Chopra, S., "Heat Transfer in Trailing-Edge Wedge Shaped Cooling Channels Under High Rotation Numbers," American Society of Mechanical Engineers, Paper GT2007-27093, May 2007.

[19] Kercher, D. M., and Tabakoff, W., "Heat Transfer by a Square Array of Round Air Jets Impinging Perpendicular to a Flat Surface Including the Effect of Spent Air," *ASME Journal of Engineering and Power*, Vol. 92, No. 1, 1970, pp. 73–82.

[20] Florschuetz, L. W., Metzger, D. E., and Su, C. C., "Heat Transfer Characteristics for Jet Array Impingement with Initial Crossflow," *ASME Journal of Heat Transfer*, Vol. 106, No. 1, 1984, pp. 34–40.

[21] Huang, Y., Ekkad, S. V., and Han, J. C., "Detailed Heat Transfer Distributions Under an Array of Orthogonal Impinging Jets," *Journal of Thermophysics and Heat Transfer*, Vol. 12, No. 1, 1998, pp. pp. 73–78.

[22] Gao, L., Ekkad, S. V., and Bunker, R. S., "Impingement Heat Transfer, Part I: Linearly Stretched Arrays of Holes," *Journal of Thermophysics and Heat Transfer*, Vol. 19, No. 1, 2005, pp. 57–65.

[23] Uysal, U., Li, P. W., Chyu, M. K., and Cunha, F. J., "Heat Transfer on Internal Surfaces of a Duct Subjected to Impingement of a Jet Array with Varying Jet Hole-Size and Spacing," *ASME Journal of Turbomachinery*, Vol. 128, No. 1, 2006, pp. 158–165.

[24] Bailey, J. C., and Bunker, R. S., "Local Heat Transfer and Flow Distributions for Impinging Jet Arrays of Dense and Sparse Extent," *Proceedings of the ASME Turbo Expo*, American Society of Mechanical Engineers, Paper GT 2002-30473, June 2002, pp. 855–864.

[25] Correia, V. H. S., "Impingement Cooling Apparatus for Turbine Shrouds Having Ducts of Increasing Cross-Sectional Area in the Direction of Post-Impingement Cooling Flow," U.S. Patent # 5480281, 1996.

[26] Haumann, J., Knopfli, A., Sattelmayer, T., and Tresch, R., "Apparatus for Impingement Cooling," U.S. Patent # 5467815, 1995.

[27] Wettstein, H., "Baffle-Cooled Wall Part," U.S. Patent # 5586866, 1996.

[28] Esposito, E., Ekkad, S. V., Kim, Y. W., and Dutta, P., "Impingement Cooling with Extended Jet Ports," American Society of Mechanical Engineers, Paper GT2007-27390, May 2007.

[29] Gromov, P. R., Zobnin, A. B., Rabinovich, M. I., and Sushchik, M. M., "Creation of Solitary Vortices in a Flow Around Shallow Spherical Depressions," *Soviet Tech. Physics Letters*, Vol. 12, No. 11, 1986, pp. 1323–1328.

[30] Afanasyev, V. N., Chudnovsky, Y. P., Leontiev, A. I., and Roganov, P. S., "Turbulent Flow Friction and Heat Transfer Characteristics for Spherical Cavities on a Flat Plate," *Experimental Thermal and Fluid Science*, Vol. 7, No. 1, 1993, pp. 1–8.

[31] Belen'kiy, M. Y., Gotovskiy, M. A., Lekakh, B. M., Fokin, B. S., and Dolgushin, K. S., "Heat Transfer Augmentation Using Surfaces Formed by a System of Spherical Cavities," *Heat Transfer Research*, Vol. 25, No. 2, 1994, pp. 196–203.

[32] Kesarev, V. S., and Kozlov, A. P., "Convective Heat Transfer in Turbulized Flow past a Hemispherical Cavity," *Heat Transfer Research*, Vol. 25, No. 2, 1994, pp. 156–160.

[33] Terekhov, V. I., Kalinina, S. V., and Mshvidobadze, Y. M., "Flow Structure and Heat Transfer on a Surface with a Unit Hole Depression," *Russian Journal of Engineering Thermophysics*, Vol. 5, No. 1, 1995, pp. 11–33.

[34] Schukin, A. V., Koslov, A. P., and Agachev, R. S., "Study and Application of Hemispherical Cavities for Surface Heat Transfer Augmentation," American Society of Mechanical Engineers, Paper 95-GT-59, June 1995.

[35] Chyu, M. K., Yu, Y., Ding, H., Downs, J. P., and Soechting, F. O., "Concavity Enhanced Heat Transfer in an Internal Cooling Passage," American Society of Mechanical Engineers, Paper 97-GT-437, June 1997.

[36] Lin, Y.-L., Shih, T. I.-P., and Chyu, M. K., "Computations of Flow and Heat Transfer in a Channel with Rows of Hemispherical Cavities," American Society of Mechanical Engineers, Paper 99-GT-263, June 1999.

[37] Gortyshov, Y. F., Popov, I. A., Amirkhanov, R. D., and Gulitsky, K. E., "Studies of Hydrodynamics and Heat Exchange in Channels with Various Types of Intensifiers," *Proceedings of 11th International Heat Transfer Congress*, Vol. 6, Taylor & Francis, Philadelphia, 1998, pp. 83–88.

[38] Moon, H.-K., O'Connell, T., and Glezer, B., "Channel Height Effect on Heat Transfer and Friction in a Dimpled Passage," American Society of Mechanical Engineers, Paper 99-GT-163, 1999.

[39] Mahmood, G. I., Hill, M. L., Nelson, D. L., Ligrani, P. M., Moon, H.-K., and Glezer, B., "Local Heat Transfer and Flow Structure on and Above a Dimpled Surface in a Channel," *ASME Journal of Turbomachinery*, Vol. 123, No. 1, 2001, pp. 115–123.

[40] Burgess, N. K., and Ligrani, P. M., "Effects of Dimple Depth on Nusselt Numbers and Friction Factors for Internal Cooling in a Channel," American Society of Mechanical Engineers, Paper GT2004-54232, June 2004.

[41] Burgess, N. K., Oliveira, M. M., and Ligrani, P. M., "Nusselt Number Behavior on Deep Dimpled Surfaces Within a Channel," *ASME Journal of Heat Transfer*, Vol. 125, No. 1, 2003, pp. 11–18.

[42] Griffith, T. S., Al-Hadhrami, L., and Han, J. C., "Heat Transfer in Rotating Rectangular Cooling Channels (AR = 4) with Dimples," *ASME Journal of Turbomachinery*, Vol. 125, No. 3, 2003, pp. 555–563.

[43] Griffith, T. S., Al-Hadhrami, L., and Han, J. C., "Heat Transfer in Rotating Rectangular Cooling Channels with Angled Ribs," AIAA Paper 2001-2820, Jan. 2001.

[44] Elyyan, M. A., and Tafti, D. K., "Effect of Coriolis Forces in a Rotating Channel with Dimples and Protrusions," American Society of Mechanical Engineers, Paper IMECE2008-66677, Nov. 2008.

[45] Chyu, M. K., Oluyede, E. O., and Moon, H.-K., "Heat Transfer on Convective Surfaces with Pin-Fins Mounted in Inclined Angles," American Society of Mechanical Engineers, Paper GT2007-28138, May 2007.

[46] Chen, S. P., Li, P. W., and Chyu, M. K., "Heat Transfer in a Airfoil Trailing Edge Configuration with Shaped Pedestals Mounted Internal Cooling Channel and Pressure Side Cutback," American Society of Mechanical Engineers, Paper GT2006-91019, June 2006.

[47] Siw, S. C., Chyu, M. K., Shih, T. I.-P., and Alvin, M. A., "Effects of Pin Detached Space on Heat Transfer and Pin-Fin Arrays," *ASME Journal of Heat Transfer*, Vol. 134, 081902, 2012, pp. 1–9.

[48] Haiping, C., Daliz, Z., and Taiping, H., "Impingement Heat Transfer from Rib Roughened Surface Within Arrays of Circular Jet: The Effect of Relative Position of the Jet Hole to the Ribs," American Society of Mechanical Engineers, Paper 97-GT-331, June 1997.

[49] Gau, C., and Lee, C. C., "Impingement Cooling Flow Structure and Heat Transfer Along Rib Roughened Walls," *International Journal of Heat and Mass Transfer*, Vol. 35, No. 11, 1992, pp. 3009–3020.

[50] Taslim, M., Li, T., and Spring, S. D., "Measurements of Heat Transfer Coefficients in Rib-Roughened Trailing Edge Cavities with Crossover Jets," American Society of Mechanical Engineers, Paper 98-GT-435, June 1998.

[51] Chakroun, W. M., Al-Fahed, S. F., and Abdel-Rehman, A. A., "Heat Transfer Augmentation for Air Jet Impinging on Rough Surface," *Applied Thermal Engineering*, Vol. 18, No. 12, 1998, pp. 1225–1241.

[52] Azad, G. M., Huang, Y., and Han, J. C., "Jet Impingement Heat Transfer on Dimpled Surfaces Using a Transient Liquid Crystal Technique," *Journal of Thermophysics and Heat Transfer*, Vol. 14, No. 2, 2000, pp. 186–193.

[53] Metzger, D. E., and Fan, C. S., "Heat Transfer in Pin-Fin Arrays with Jet Supply and Large Alternating Wall Roughness Ribs," *Fundamental and Applied Heat Transfer Research for Gas Turbine Engines*, ASME-HTD-Vol. 226, 1992, pp. 23–30.

[54] Glezer, B., Moon, H. K., and O'Connell, T., "A Novel Technique for the Internal Blade Cooling," American Society of Mechanical Engineers, Paper 96-GT-181, June 1996.

[55] Ekkad, S. V., Pamula, G., and Acharya, S., "Influence of Crossflow Induced Swirl and Impingement on Heat Transfer in an Internal Coolant Passage of a Turbine Airfoil," *ASME Journal of Heat Transfer*, Vol. 122, Aug. 2000, pp. 587–597.

[56] Pamula, G., Ekkad, S. V., and Acharya, S., "Influence of Cross-Flow Induced Swirl and Impingement on Heat Transfer in a Two-Pass Channel Connected by Two Rows of Holes," *ASME Journal of Turbomachinery*, Vol. 123, No. 2, 2001, pp. 281–287.

[57] Ekkad, S. V., Kontrovitz, D., Nasir, H., Pamula, G., and Acharya, S., "Heat Transfer in Two-Pass Turbulated Channels Connected by Holes," *Journal of Thermophysics and Heat Transfer*, Vol. 16, July 2002, pp. 404–414.

[58] Bailey, J. C., Intile, J., Fric, T. F., Tolpadi, A., Nirmalan, N. V., and Bunker, R. S., "Experimental and Numerical Study of Heat Transfer in Gas Turbine Combustor Liner," American Society of Mechanical Engineers, Paper GT-2002-30183, June 2002.

[59] Andrews, G. E., Abdul Hussain, R. A. A., and Mkpadi, M. C., "Enhanced Impingement Heat Transfer: The Influence of Impingement X/D for Interrupted Rib Obstacles (Rectangular Pin Fins)," American Society of Mechanical Engineers, Paper GT2004-54184, June 2004.

[60] Son, C., Dailey, G., Ireland, P. T., and Gillespie, D., "An Investigation of the Application of Roughness Elements to Enhance Heat Transfer in an Impingement Cooling System," American Society of Mechanical Engineers, Paper GT2005-68504, June 2005.

[61] Nagoga, G. P., "Effective Methods of Cooling of Blades of High Temperature Gas Turbines," Publishing House of Moscow Aerospace Inst. (Russian language), 1996, p. 100.

[62] Goreloff, V., Goychengerg, M., and Malkoff, V., "The Investigation of Heat Transfer in Cooled Blades of Gas Turbines," AIAA Paper 90-2144, Jan. 1990.

[63] Bunker, R. S., "Latticework (Vortex) Cooling Effectiveness, Part 1: Stationary Channel Experiments," American Society of Mechanical Engineers, Paper GT2004-54157, June 2004.

[64] Ekkad, S. V., and Kontrovitz, D., "Jet Impingement Heat-Transfer on Dimpled Target Surfaces," *International Journal of Heat and Fluid Flow*, Vol. 23, Feb. 2002, pp. 22–28.

[65] Kanokjaruvijit, K., and Martizez-Botas, R. F., "Heat-Transfer and Pressure Investigation of Dimple Impingement," American Society of Mechanical Engineers, Paper GT2005-68823, June 2005.

CHAPTER 5

Film Cooling

David G. Bogard*
University of Texas at Austin, Austin, Texas

Karen A. Thole†
Pennsylvania State University, University Park, Pennsylvania

NOMENCLATURE

c_p	specific heat of gas
DR	density ratio
d	film cooling hole diameter
h	heat-transfer coefficient
I	momentum flux ratio [Eq. (11)]
K	pressure gradient parameter [Eq. (12)]
k	thermal conductivity
M	blowing ratio using local velocity
Ma	Mach number
M^*	blowing ratio using approach velocity
P	hole spacing measured normal to streamwise direction
q''	heat flux
Re	Reynolds number
Re_k	roughness Reynolds number, $u_\tau k/\nu$
S	distance along the vane surface
s	equivalent slot width
T	temperature
Tu	freestream turbulence intensity [Eq. (13)]
U	streamwise velocity
$u_{\rm rms}$	rms of fluctuating U-velocity component
u_τ	friction velocity
V_r	velocity ratio [Eq. (13)]
x	distance downstream of the hole exit
Δq_r	net heat-flux reduction [Eq. (4)]
η	film effectiveness
θ	normalized temperature [Eq. (7)]

*Professor, Department of Mechanical Engineering.
†Professor, Department of Mechanical and Nuclear Engineering.

Λ_f	turbulence integral length scale
υ	kinematic viscosity
ρ	density
ϕ	overall cooling effectiveness [Eq. (6)]

SUBSCRIPTS

aw	adiabatic wall
c	coolant
f	with film cooling
max	maximum
ref	reference
w	wall
∞	freestream conditions

Gas turbine engines are an integral part of our daily lives; we rely on them to propel aircraft, tanks, and large naval ships, and to provide electrical power. Since 1929, when Sir Frank Whittle first applied for a patent on his turbojet engine, complex technologies have been developed to advance turbine engines to meet the needs of our energy-thirsty world. A key development is material and cooling technologies to allow high-temperature gas to enter the first rotor; durability, thermal efficiencies, and power output are a direct function of the inlet temperatures to the turbine rotor.

Since 1960, turbine airfoils have been cooled, so that gas temperatures at the turbine inlet can be allowed to exceed the allowable metal temperatures of the airfoils. Early on, simple convective cooling schemes using high-pressure bleed air from the compressor were used on the internal side of the airfoils. In the 1970s, a new cooling technology was introduced to the engine whereby this bleed air was exhausted from the internal convective passages through small holes drilled into the airfoil surfaces (see Fig. 1). Holes are typically used, rather than porous surfaces or slots, because structural rigidity must be maintained in the face of the large stresses experienced by blades and vanes. This technology is referred to as film cooling, and in today's engines it is applied to all regions of the airfoils, particularly in the first and second stages of the turbine.

Advanced military engines now have turbine inlet temperatures in excess of 1600°C, which can be achieved if 20–30% of the total flow through the engine is used to cool turbine components. Land-based turbines also operate at high turbine inlet temperatures, in excess of 1400°C. Here, too, film cooling is one technology used for turbine airfoils. In the operating range for land-based turbines, improvements in cooling performance that lead to a reduction of airfoil temperatures of just 25°C can increase part life by a factor of two. On the other hand, rather than increasing the part life, engine designers can choose to reduce the required coolant flow. Reducing the required coolant flow results

Fig. 1 Film-cooled turbine vane (from Friedrichs et al. [1]).

in the same airfoil temperature, but improves the turbine efficiency, resulting in lower fuel costs for the same power output.

The ultimate goal in applying film cooling is to reduce airfoil temperatures, but only a small portion of the turbine flow bled from the compressor can be used as coolant without degrading the overall system performance of the engine. Film cooling reduces airfoil temperatures by decreasing the local fluid temperature next to the airfoil surface and making use of convective cooling as the flow passes through the film cooling holes placed in the airfoil surface. This chapter presents a summary of past research in the area of film cooling.

I. FILM-COOLING ANALYSIS METHODS

The total cost to replace a single first vane is on the order of thousands to tens of thousands of dollars, of which a significant fraction is the cost of manufacturing or repairing film-cooling holes in the vane. The actual cost of manufacturing the film-cooling holes depends on the chosen method. Generally, holes are drilled in the surface of an airfoil by either laser drilling or electrodischarge machining (EDM). Machining holes using EDM provides more flexibility in terms of shape and placement, but the cost is significantly higher than laser drilling. Being able to predict local airfoil temperatures is not only important because of the costs incurred to manufacture the cooling holes, but also from the costs incurred for an airfoil failure. Generally, airfoils must be replaced or repaired in an engine because regions such as the leading or trailing edges have burned away while the main body of the airfoil remains intact. Because it is the local variations in metal temperatures that lead to airfoil failure, it is particularly important that models based on the flow physics of the relevant parameters be used to predict local airfoil temperatures.

Overall, the goal in turbine cooling is to reduce airfoil temperatures because it is higher metal temperatures that reduce component life. Turbine airfoil metal temperatures are a result of internal cooling as well as external film cooling. Generally, the approach taken by researchers and turbine designers

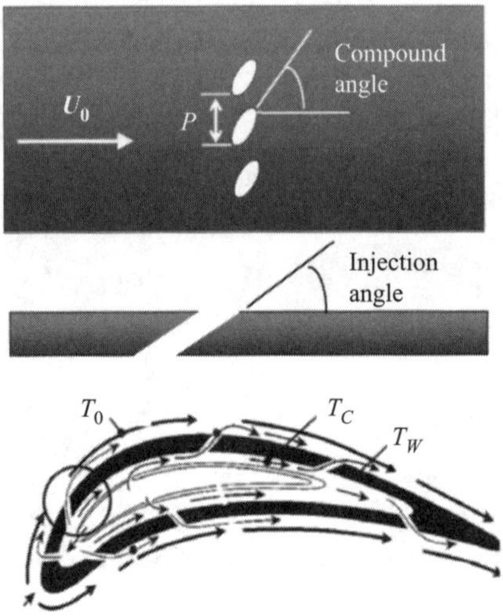

Fig. 2 Schematics of typical film-cooling configuration illustrating the three temperature potentials for a film-cooled surface (lower figure from Gritsch et al. [2]).

is to separately assess internal and external cooling; the present chapter describes the assessment method for film cooling used for external cooling purposes.

Decreasing the local fluid temperature next to the airfoil surface reduces the driving potential for heat transfer to occur. While heat transfer takes place through conduction from the air to the metal, it is modeled with a mechanistic equation using a convective heat-transfer coefficient, as shown in Eq. (1):

$$q'' = h(T_{\text{ref}} - T_w) \tag{1}$$

In this equation, the appropriate reference temperature T_{ref} is not obvious because the film-cooling process involves two temperatures, the coolant temperature T_c and the freestream temperature T_∞, as shown in Fig. 2. As the coolant mixes with the hot freestream fluid, the local fluid temperature varies greatly downstream of the film-cooling injection location. Moreover, the momentum and heat transport in the boundary layer along the airfoil are altered by the coolant injection. If T_∞ is used as the reference temperature, then h will be a function of the flowfield and the temperature of the coolant. To obtain a heat-transfer coefficient for film-cooling flow that is independent of the coolant temperature, T_{ref} should be the driving temperature of the fluid above the surface, regardless of the coolant temperature. Because the adiabatic wall temperature T_{aw} is the fluid temperature immediately above the surface for an adiabatic surface, this is expected to be a good reference temperature for this driving potential. Consequently, the heat-transfer coefficient with film cooling h_f is defined as follows:

$$q''_f = h_f(T_{\text{aw}} - T_w) \tag{2}$$

It is important to remember that the adiabatic wall temperature and the local convective heat transfer vary widely over the airfoil surface, given the discrete nature of the film-cooling holes.

One of the most important driving variables in predicting the airfoil temperatures, as seen in Eq. (2), is the adiabatic wall temperature, which is representative of the fluid temperature just above the surface. Thus, most of the literature characterizing film-cooling performance is reported using a nondimensional film effectiveness (also referred to as adiabatic effectiveness) defined as

$$\eta = \frac{T_\infty - T_{\mathrm{aw}}}{T_\infty - T_{c,\mathrm{exit}}} \tag{3}$$

where $T_{c,\mathrm{exit}}$ is the coolant temperature at the coolant hole exit. It is necessary to evaluate the adiabatic wall temperature as a nondimensional variable so that it can be related to temperatures that would occur at engine conditions. As will become apparent throughout this chapter, film effectiveness values are highly dependent on a number of variables, such as cooling jet-to-mainstream ratios of density, velocity, mass flux, momentum flux, as well as mainstream flowfield effects such as pressure gradients, curvature, freestream turbulence, and others. Moreover, the shape of the cooling hole also affects the overall film cooling performance.

Generally, when a cooling jet is injected into a mainstream flow, the shear layers between the coolant jet and/or the wake behind the jet generate higher levels of turbulence within the boundary layer [3]. These higher levels of turbulence increase the coolant mixing with the hotter surrounding fluid and increase the local convective heat-transfer coefficients. As such, the expected heat-transfer coefficients for a film-cooled boundary layer are generally higher than that for a non-film-cooling boundary layer. Although the increase in h due to film cooling is generally not large, with the exception of film-cooled leading edges, it is an effect that should be considered when predicting airfoil temperatures.

To evaluate the total benefit of film cooling and whether or not to place a film-cooling hole in a particular location on an airfoil, the net benefit needs to be assessed relative to the manufacturing or replacement costs of the part. A simple relationship can be used to indicate the net reduction in the heat flux Δq_r to the airfoil surface through the following:

$$\Delta q_r = 1 - \frac{q_f''}{q_0''} = 1 - \frac{h_f(T_{\mathrm{aw}} - T_w)}{h_0(T_\infty - T_w)} \tag{4}$$

$$\Delta q_r = 1 - \frac{h_f}{h_0}\left(1 - \frac{\eta}{\phi}\right) \tag{5}$$

The net reduction relates the heat transfer that would have occurred on the airfoil with no film cooling q_0'' to that with film cooling q_f'', where h_0 is the heat-transfer coefficient for the flow without coolant injection. Here ϕ is a nondimensional parameter for actual metal temperature of the airfoil, defined as follows:

$$\phi = \frac{T_\infty - T_m}{T_\infty - T_{ci}} \tag{6}$$

where T_m is the “metal temperature” for the actual temperature and T_{c_i} is the coolant temperature in the internal channels prior to the coolant entering the film-cooling holes. The nondimensional parameter ϕ is also referred to as the "overall effectiveness" because the metal temperature for the actual, high-conductivity airfoil is dependent on external and internal cooling. Note that there can be a distinct difference between T_{c_i} used in Eq. (6) and $T_{c,\text{exit}}$ used in Eq. (3) because of the heating of the coolant as it passes through the film-cooling holes.

There are several commonly used methods, including both steady state and transient, to measure h_f and η, the two variables most reported in the literature. Transient methods generally presume a one-dimensional conduction to a semi-infinite solid to determine the heat-transfer coefficients and film effectiveness from the rate of change of the surface temperature. Steady-state methods generally set up different surface boundary conditions to quantify the variables. Film effectiveness can be measured using thermal techniques using a cooled or heated jet, or mass-transfer techniques in which the concentration of a "foreign gas" is measured. For the thermal techniques, the adiabatic wall temperature is measured using a low thermal conductivity material in order to minimize any conductive heat transfer to the surface. For the heat-transfer coefficients with film cooling h_f, the surface boundary condition has a constant heat flux generated by some type of resistive heater. While much of the past literature is limited in terms of spatial resolution, advanced measurement techniques such as infrared cameras and liquid allow characterization of local heat-transfer coefficients and effectiveness values using highly resolved data. For more information on various measurement methods, see Han et al. [4].

II. PHYSICAL DESCRIPTION AND PREDICTION OF FILM COOLING

Ideally, coolant ejected to the surface of airfoils for film cooling would remain attached to the surface of the airfoil and would not disperse to the mainstream. This would provide a gas temperature at the surface of the airfoil equal to the coolant temperature at the exit for the hole, $\eta = 1$, and would minimize heat transfer from the gas to the wall. In reality, the coolant mixes with the mainstream, generally quite rapidly. The dispersion of coolant after it exits a typical film-cooling hole is demonstrated in Fig. 3, which shows measurements of the temperature along the centerline of a coolant jet exiting a hole inclined at 35 deg to the

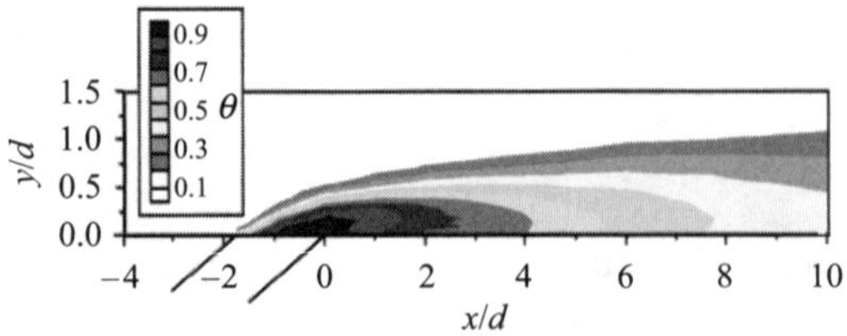

Fig. 3 Thermal profiles of a coolant jet showing the decay of the normalized temperature θ downstream of the hole.

TABLE 1 FACTORS AFFECTING FILM-COOLING PERFORMANCE

Coolant/Mainstream Conditions	Hole Geometry and Configuration	Airfoil Geometry
Mass flux ratio*	Shape of the hole*	Hole location - Leading edge - Main body - Blade tip - Endwall
Momentum flux ratio*	Injection angle and compound angle of the coolant hole*	
Mainstream turbulence*	Spacing between holes, P/d	Surface curvature*
Coolant density ratio	Length of the hole, l/d	Surface roughness
Approach boundary layer	Spacing between rows of holes and number of rows	
Mainstream Mach number		
Unsteady mainstream flow		
Rotation		

* Factors that have a significant effect on predictability of film-cooling performance.

surface. The temperature contours are presented as normalized θ contours, where θ is defined as

$$\theta = \frac{T_\infty - T}{T_\infty - T_c} \tag{7}$$

Note that at the surface the definition of θ is equivalent to η, so that these θ contours also show the η distribution along the surface. Decreasing θ values downstream of the hole represent a measure of the mixing of the coolant jet with the mainstream, that is, a θ value of 0.2 indicates a mixture of 20% original coolant fluid and 80% mainstream fluid. Much of the design of film cooling for turbine airfoils involves the prediction of the η distribution downstream of the coolant holes. This is complicated by the many factors that affect the film-cooling film effectiveness, as listed in Table 1.

The six factors in Table 1 denoted with asterisks have significant effects on film-cooling performance. Each of these factors is not necessarily independent of the other factors, and so every combination of these factors can potentially change film-cooling performance. Consequently, there is an extremely large number of operating conditions that need to be considered, hence the inherent difficulty in predicting film-cooling performance.

Most experimental investigations of film cooling have been done with flat surface test plates, and so this configuration will be used as a baseline. Early studies of film cooling used slots of various configurations to introduce the coolant to the surface (for example, see Goldstein [5]). Although slots are not representative of the discrete holes used in practical film-cooling configurations, slot cooling studies provide a useful reference. A slot introduces coolant in a uniform sheet that has less mixing with the overflowing mainstream than occurs with discrete coolant jets originating from a row of holes. Consequently, the slot provides an ideal performance that can be used as a basis for comparison.

Laterally averaged film effectiveness $\overline{\eta}$ for coolant injection with a slot inclined at 30 deg to the surface were measured by Teekaram et al. [6]. These measurements included a range of coolant mass flux ratios (commonly referred to as blowing ratios) of $M = 0.1$ to 0.7, where M is defined as follows:

$$M = \frac{\rho_c U_c}{\rho_\infty U_\infty} \tag{8}$$

Results from their study, replotted in Fig. 4a, show that for slot injection $\eta = 1.0$ immediately downstream of the slot, but decays farther downstream. The decay rate of η was observed to be inversely proportional to the blowing ratio, and so x/Ms was used as a correlation parameter for film effectiveness from a slot. The collapse of the film effectiveness performance when using the x/Ms parameter is demonstrated in Fig. 4b. These results, showing the scaling of η with x/Ms, give important insights into the film-cooling process. Recognize that total mass flow of coolant per unit span is proportional to Ms, so that the distance for η to decay to a certain level is proportional to the total mass flow of the coolant.

The scaling of η with x/Ms for slot injection is further validated with results presented in Fig. 5, which includes the film effectiveness

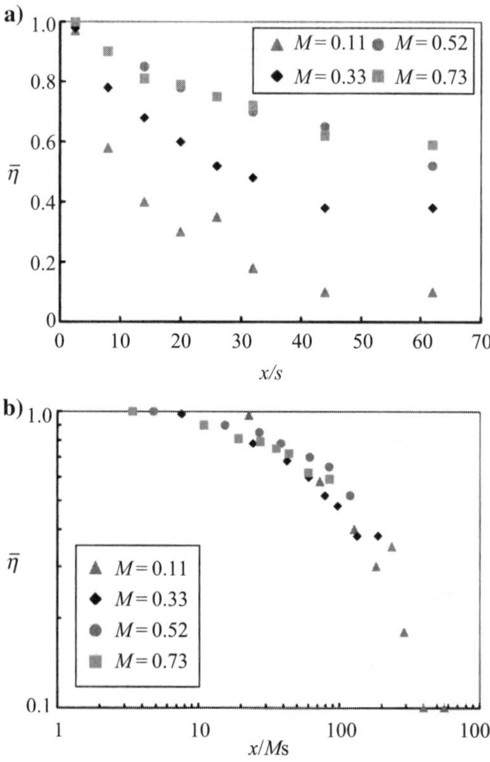

Fig. 4 Comparison of laterally averaged film effectiveness for slot injection using data from Teekram et al. [6]. Presented as a function of a) streamwise distance from injection location and b) the x/Ms scaling.

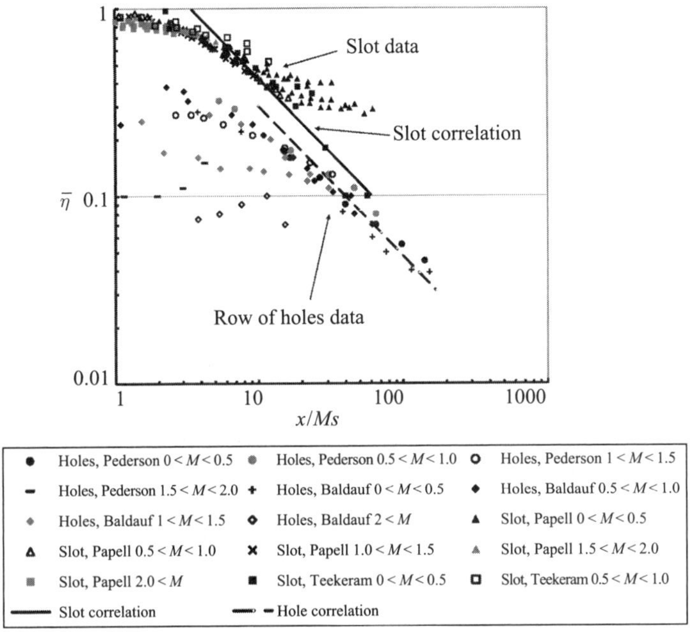

Fig. 5 Comparison of laterally averaged film-effectiveness values for a slot as compared to a row of cooling holes. Data from Teekaram et al. [6], Papell [7], Pedersen et al. [8], and Baldauf et al. [9].

results of Teekaram et al. [6] and Papell [7]. The measurements of Papell presented in this figure are for slot injection with a 45-deg injection angle with blowing ratios that ranged from $M = 0.16$ to 3.7. The downstream decay for these two studies were similar and were consistent with the following correlation for slot injection proposed by Hartnett et al. [10]:

$$\eta = 16.9(x/Ms)^{-0.8} \tag{9}$$

Also shown in Fig. 5 is the laterally averaged film effectiveness $\overline{\eta}$ for film cooling from a single row of holes (Pederson et al. [8] and Baldauf et al. [9]). To present these data for coolant injection from a row of holes, an equivalent slot width s_e is defined as follows:

$$s_e = A_{\text{hole}}/P \tag{10}$$

where P is the pitch between holes. With this definition, the total mass flow of coolant per unit span for the row of holes is equivalent to that for a slot of the same equivalent width. For a single row of holes, $\overline{\eta}$ values are significantly lower than for the slot. There are two primary reasons for this: decreased coverage of the surface by coolant and increased dispersion of the coolant. Immediately downstream of the coolant holes, the surface covered by coolant has a width approximately equal to the hole diameter. If the film effectiveness in this region is maximum, that is, $\eta = 1$, and is zero between coolant holes, the laterally averaged film effectiveness will have a maximum of $\overline{\eta}_{max} = 1/(P/d)$. For $P/d = 3$, this corresponds to $\overline{\eta}_{max} = 0.33$, which is consistent with the observed maximum $\overline{\eta}$ level.

As the coolant moves downstream of the hole, it spreads laterally, covering the surface more completely. The distribution of the coolant is evident in the contours of η presented by Baldauf et al. [11] and shown in Fig. 6. The contours of η in Fig. 6 are for a row of holes with spacing between holes of $P/d = 3$. For $M < 0.85$, some coolant has spread across the full span, but there are still strong lateral variations in η up to $x/d > 25$. The η levels produced by the row of holes for $x/d > 25$ are, however, still significantly lower than that produced by a slot. This can be attributed to the greater dispersion of the discrete coolant

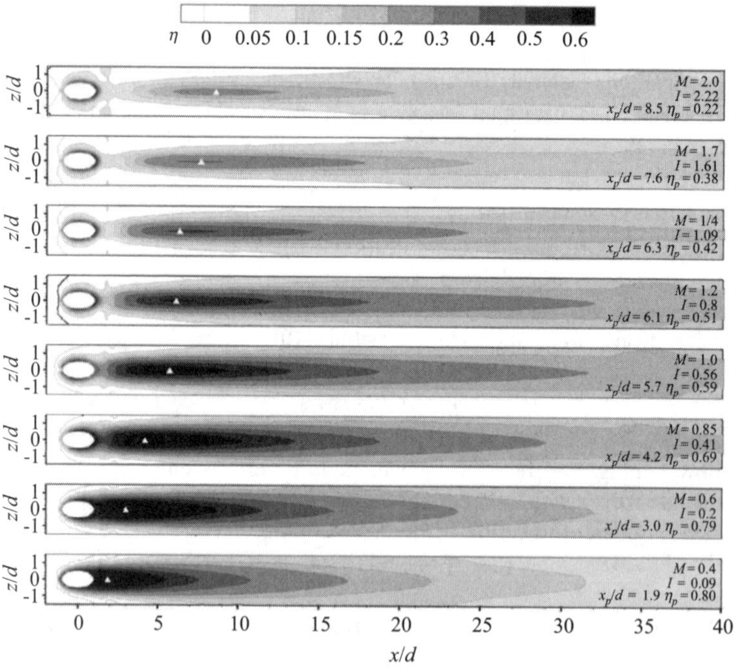

Fig. 6 Spatial distribution of film effectiveness for varying blowing ratios (from Baldauf et al. [11]).

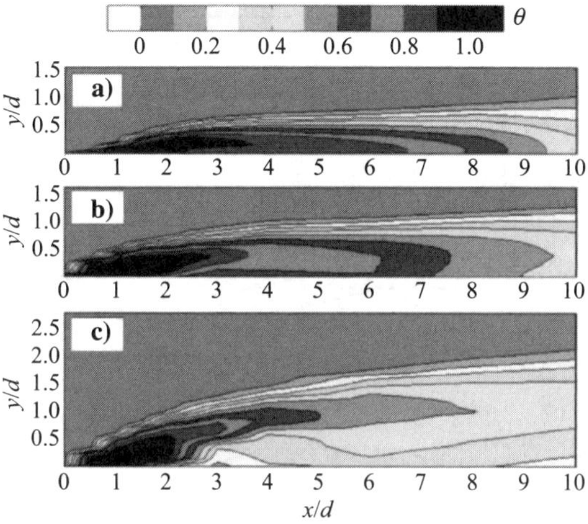

Fig. 7 Thermal field profiles along the centerline of coolant jets illustrating a) a fully attached jet, b) a detached and reattached jet, and c) a fully detached jet (data from Thole et al. [12]).

jets due to the greater contact area with the mainstream and also to vortices generated by the interaction of the coolant jet with the mainstream.

The contours of η in Fig. 6 show that a peak level in η occurs at some distance downstream of the coolant hole ranging from $0.5D$ to $8D$, depending on the blowing ratio. This occurs because discrete coolant jets have a tendency to separate from the surface downstream of the hole exit and reattach farther downstream. This is a very important characteristic of coolant jets and results in a large decrease in film effectiveness at higher blowing ratios.

Coolant jet separation and reattachment was studied by Thole et al. [12] for coolant jets exiting from a row of holes inclined at 35 deg relative to the surface and oriented in the streamwise direction. In this study, thermal profiles of the coolant jets were measured along the centerline of the jets to determine the distribution of coolant above the surface. An important consideration in this study was whether the separation characteristics of the coolant jets scaled with the mass flux ratio M, the velocity ratio V_r, or the momentum flux ratio I. This was accomplished by evaluating the separation characteristics of coolant jets with density ratios varying from $DR = 1.2$ to 2.0. Three distinct regimes were identified: fully attached coolant jets as shown in Fig. 7a, coolant jets that detached then reattached as shown in Fig. 7b, and coolant jets that were fully detached as shown in Fig. 7c. The coolant jet separation characteristics were found to scale with momentum flux ratio I. This is understandable because the dynamics of the force of the mainstream impacting the coolant jet and causing it to turn towards the wall would be expected to be primarily a function of the momentum of the coolant jet relative to the momentum of the mainstream. The coolant jets were found to remain attached to the surface for $I < 0.4$ and were fully detached for $I > 0.8$. For $0.4 < I < 0.8$, the coolant jets initially detached but soon reattached to the surface.

In the following discussion of film-cooling film effectiveness, performance is generally evaluated in terms of laterally averaged film effectiveness $\overline{\eta}$. Because of the high thermal conductivity of the metal airfoils, surface temperature

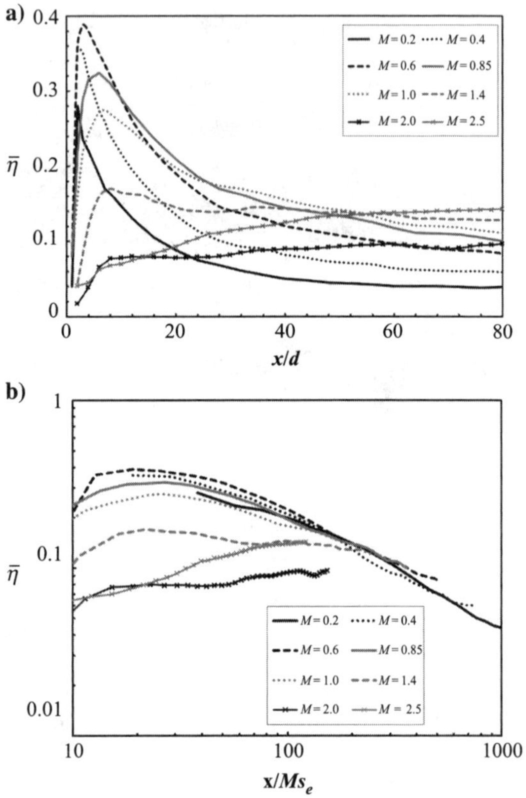

Fig. 8 Distributions of $\overline{\eta}$ for varying blowing ratios presented as a function of a) streamwise distance x/d and b) x/Ms_e parameter (Figs. 2b and 7a in Baldauf et al. [9]).

variations are much less than the lateral variation in adiabatic surface temperatures. Consequently, the ultimate effect of the coolant is well represented by laterally average values [13]. In some cases, however, it is important to examine the spatial distribution of film effectiveness, particularly when deducing physical mechanism, or when evaluating computational predictions.

The effect of coolant jet separation on the film effectiveness performance of a row of coolant holes is evident from the distributions of $\overline{\eta}$ obtained by Baldauf et al. [9] and presented in Fig. 8. Distributions of $\overline{\eta}$ for various M are presented in Fig. 8a as a function of the x/d distance downstream of the hole. For blowing ratios increasing from $M = 0.2$ to 0.6, there is an increase in the overall level of $\overline{\eta}$. As blowing ratio is increased to $M > 0.6$, however, the peak level of $\overline{\eta}$ decreases. This is a consequence of the core of the coolant jet lifting slightly off the surface. Even though the peak values of $\overline{\eta}$ decrease for $M > 0.6$, the levels of $\overline{\eta}$ beyond $x/d = 20$ continue to increase for increasing blowing ratio for $M \leq 1.0$. Over this range of blowing ratios, the tendency of coolant jet separation to reduce $\overline{\eta}$ is offset by the increase in $\overline{\eta}$ caused by increasing coolant mass flow. Eventually the separation effects dominate, and for $M > 1.0$ the level of $\overline{\eta}$ decreases over the full length measured.

Figure 8b shows the $\overline{\eta}$ distributions relative to the x/Ms_e scale. For $M \leq 1.0$ ($I \leq 0.57$), the $\overline{\eta}$ curves collapse to similar curves, indicating that the $\overline{\eta}$ distributions scale with x/Ms_e when the coolant jets do not significantly separate from the surface. As the blowing ratio increases to $M > 1.0$ ($I > 0.57$), there is a continual decrease in $\overline{\eta}$ with increasing M. Hence, $\overline{\eta}$ distributions do not scale with x/Ms_e when the coolant jets begin to detach.

III. SCALING OF FILM-COOLING PERFORMANCE WITH VARYING DENSITY RATIO

In gas turbine operation, absolute coolant temperatures are typically about half that of the mainstream temperature. Consequently, typical coolant-to-freestream density ratios are $DR \approx 2$. However, because this density ratio is difficult to simulate in many experimental facilities, many studies of film cooling have been done with much lower density ratios and even with coolant densities lower than the mainstream densities. Therefore, evaluating the effect of density ratio on film-cooling performance is important.

When using a low-density-ratio coolant in laboratory studies, performance results can be matched to engine conditions at high density ratio by matching either the mass flux ratio M, or the momentum flux ratio I, or the velocity ratio V_r . The I and V_r ratios are defined as follows:

$$I = \frac{\rho_c U_c^2}{\rho_\infty U_\infty^2}, \qquad V_r = \frac{U_c}{U_\infty} \tag{11}$$

The mass flux ratio scales the thermal transport capacity of the coolant because the convective transport is proportional to $c_p \rho U_c$. The momentum flux ratio scales the dynamics of the interaction of the mainstream with the exiting coolant jet as the impact pressure of the mainstream on the coolant jet causes the coolant jet to turn towards the wall. The turning of the coolant jet is a major factor in the cooling performance. If the coolant jet is not turned sufficiently to remain attached to the surface, the bulk of the coolant will be contained in a separated coolant jet and will provide very little cooling of the surface. The velocity ratio scales the shear layer between the coolant jet and the mainstream and hence scales the turbulence production. When testing with a density ratio that does not match engine conditions, only one of these scaling parameters can be matched to the engine condition.

A number of studies have been conducted to evaluate the effects of coolant density ratio on film effectiveness performance, including Pederson et al. [8], Sinha et al. [14], and Baldauf et al. [9]. These studies were conducted using flat surface, zero-pressure-gradient test facilities with a single row of coolant holes inclined 35 deg to the surface and oriented in the mainstream flow direction. These studies showed that, although there are distinct differences between low- and high-density-ratio coolant jet performance, the film effectiveness performance was similar when the appropriate scaling parameter was selected. At very low blowing ratio, $M = 0.2$, Pederson et al. [8] found that $\overline{\eta}$ was essentially the same for coolant density ratios ranging from $DR = 0.8$ to 4. At this low blowing ratio, the coolant jets were well attached for all density ratios, and therefore the film effectiveness performance was dependent on M. At higher blowing ratios, Pederson et al. [8] found better film effectiveness for higher-density-ratio coolant jets operating at the same M. This is because of the lower density coolant jets having higher momentum ratios and hence a tendency to separate from the surface. Baldauf et al. [9] compared film effectiveness for coolant density ratios of $DR = 1.2$ and 1.8 over a range of blowing ratios from $M = 0.2$ to 2.5. The distributions of $\overline{\eta}$ were similar for both density ratios, but the $DR = 1.8$ coolant had a peak film effectiveness of $\overline{\eta} = 0.38$, whereas for $DR = 1.2$ the peak was $\overline{\eta} = 0.32$. This can be attributed to better lateral distribution of the high density coolant, as noted by Sinha et al. [14].

Scaling of film-effectiveness performance on simulated turbine airfoils was investigated by Cutbirth and Bogard [15] for the showerhead and pressure side of a vane and by Ethridge et al. [16] for the suction side of a vane. Showerhead cooling for low and high mainstream turbulence levels, $Tu = 0.5$ and 20% respectively, was studied by Cutbirth and Bogard [15] for coolant density ratios of $DR = 1.2$ and 1.8. They found that film effectiveness for low and high mainstream turbulence levels was most similar when compared at similar M^*, where M^* is the blowing ratio defined using the airfoil approach velocity, rather than the local velocity. However, as shown in Fig. 9, at some blowing ratios the film effectiveness was 10 to 20% lower for the low-density-ratio coolant. On the pressure side of the vane, Cutbirth and Bogard used a row of compound angle holes with injection angles of 30 deg with respect to the surface and 45 deg with respect to the mainstream flow direction. As shown in Fig. 10a, for mainstream turbulence levels of $Tu = 0.5\%$, there was good correspondence in film effectiveness for the two density ratios for position $x/d = -28$, which corresponded to $2d$ downstream of the coolant holes, when compared at similar I. However, at $x/d = -34$ the $DR = 1.2$ coolant had a

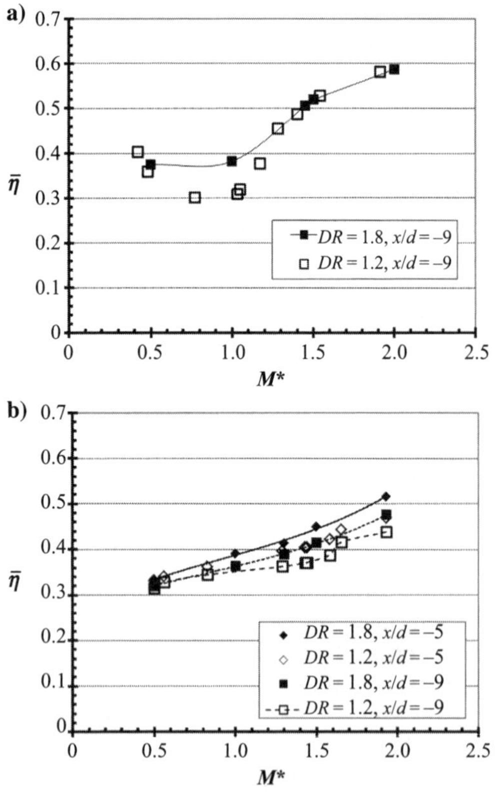

Fig. 9 Comparisons of laterally averaged film effectiveness for different density ratios in showerhead region of a vane for a) $Tu = 0.5\%$ and b) $Tu = 20\%$ (from Cutbirth and Bogard [15]).

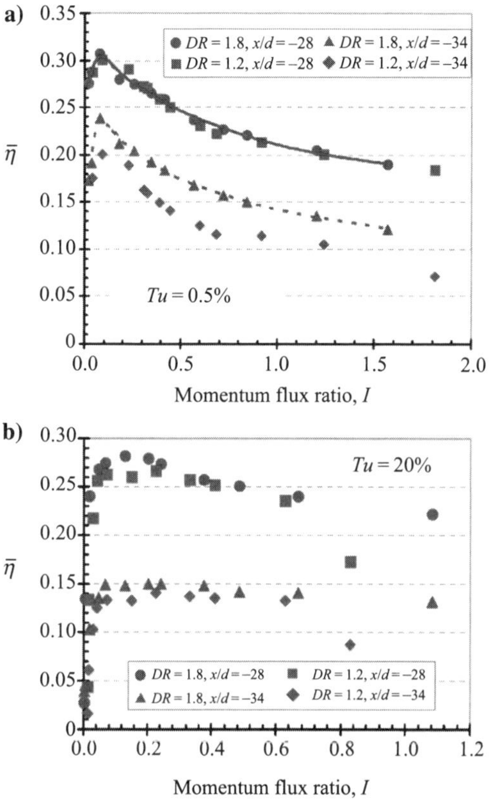

Fig. 10 Comparisons of laterally averaged film effectiveness for different density ratios in the on the pressure side of a vane for a) *Tu* = 0.5% and b) *Tu* = 20% (from Cutbirth and Bogard [15]).

distinctly lower $\overline{\eta}$. For mainstream turbulence levels of $Tu = 20\%$, very similar film effectiveness was obtained from the high- and low-density-ratio coolant when matching I, as shown in Fig. 10b.

Evaluation of the coolant density ratio effect on heat-transfer coefficients for film-cooling injection is particularly important because of the many studies that have used unit density ratio coolant when making measurements of the heat-transfer coefficients. There are only a few studies in which the effects of low- and high-density-ratio coolant on the heat-transfer coefficient have been compared. In a recent study by Baldauf et al. [17], tests were conducted with a flat surface using coolant density ratios of $DR = 1.2$ and 1.8 over a range of blowing ratios from $M = 0.2$ to 2.5. As in most heat-transfer tests, cooling was downstream of the coolant holes. Results for these heat-transfer tests were presented in terms of the augmentation of the heat-transfer coefficients with film cooling h_f relative to the heat-transfer coefficients without film cooling h_0 as h_f/h_0 ratios. For blowing ratios $M < 1.0$, the augmentation of heat-transfer coefficients was less than 10% and was similar for $DR = 1.2$ and 1.8. For $M > 1.4$, the augmentation for the low-density-ratio coolant was slightly larger than for the large-density-ratio coolant.

IV. COOLING HOLE GEOMETRY AND CONFIGURATION

As indicated in Table 1, the shape of the coolant hole and angle of injection has a significant effect on film-cooling performance. Most coolant holes are angled at 25 to 35 deg to the surface, which helps to keep the coolant jets attached to the surface and is a manufacturing limit. In some cases, however, steeper angles of injection are used as a result of manufacturing or geometrical constraints. The surface angle of the holes can be oriented with the mainstream flow direction or inclined at some angle with respect to the mainstream flow direction. Coolant holes that are directed at a nonzero angle from the mainstream flow direction are generally referred to as "compound angle" holes (see Fig. 2 for the definition of the compound angle). Typically, 90-deg compound angle holes, with injection direction perpendicular to the mainstream flow angle, are used on the leading edges of vanes and blades. These leading-edge holes are sometimes referred to as "radial" holes because coolant is injected in the radial direction along a turbine airfoil that resides in a disk. Distinct improvements in film effectiveness are obtained when using "shaped" holes, which have a diffusing expansion at the exit of the hole. The diffusing exit of the coolant holes distributes the coolant over a broader area and reduces the coolant exit velocity, which reduces the tendency of the jet to separate from the surface, even at high blowing ratios. Spacing between holes in a row and interaction between closely spaced rows are configuration variables that affect film-cooling performance. These geometric and configuration variables are described in detail in what follows.

A. COOLING HOLE SPACING

Typical spacing, or pitch P, between coolant holes in the lateral direction for a row of holes is three hole diameters, but spacing can be as much as eight hole diameters. As the hole spacing is decreased, there is greater "coverage" by the coolant, that is, the percentage area of surface covered by coolant increases. Also, for a given mass flow rate from each hole, with decreased hole spacing, the mass flow per unit span increases. When coolant holes are spaced widely apart, each coolant jet acts essentially independently. In this case, the performance by a row of holes can be predicted by a superposition of the performance of a single hole. With closer spacing between holes, the mainstream interaction with the coolant jets is altered because there is more resistance to the mainstream flow between the coolant jets. Consequently, when evaluating the effect of hole spacing on film cooling performance, it is important to recognize whether the spacing is sufficient for the jets to act independently.

Film effectiveness for holes spaced at $P/d = 3$ and 6 was studied by Schmidt et al. [18] using holes angled 35 deg to the surface and oriented with compound angles of 0 and 60 deg to the mainstream. Results from this study showed that the film-effectiveness levels for a hole spacing of $P/d = 3$ was twice that for $P/d = 6$, that is, that film effectiveness for $P/d = 3$ was predictable from a superposition of film effectiveness for $P/d = 6$. Consequently, these results showed that for a spacing between coolant holes as small as $P/d = 3$, the coolant jets were performing as independent jets. These results were consistent with a later study by Baldauf et al. [9], who tested hole spacings of $P/d = 2$, 3, and 5. Baldauf et al. found that for $M > 1.2$ the hole spacing of $P/d = 2$ had increasing film-effectiveness levels with increasing M while spacings of $P/d = 3$ and 5 had decreasing levels of film effectiveness. The decreasing film effectiveness for $P/d = 3$ and 5 is attributable to coolant jet separation. Evidently, for $P/d = 2$ the adjacent jets are close enough that they begin to form a continuous blockage of the mainstream, similar to slot injection. This suppresses the tendency for jet separation. Consequently, for $M > 1.0$ the $P/d = 2$ configuration has a much higher film effectiveness than would be expected based on superposition of the $P/d = 3$ or 5 film effectiveness.

A similar result for higher blowing ratios is evident in the results of Foster and Lampard [19] who used a row of holes angled normal to the surface with spacings of $P/d = 1.5$, 2.5, 3.75, and 5. Although no quantitative comparisons were made for the smallest spacing of $P/d = 1.5$, contour plots of film effectiveness show that the coolant jets had coalesced to form a continuous coolant film. Furthermore, for $P/d \geq 2.5$ and $M = 0.5$, the decrease in film effectiveness with increase in hole spacing appears consistent with superposition (although they do not mention this).

B. DOUBLE ROWS OF HOLES

Sometimes two rows of coolant holes closely spaced in the streamwise direction are used for turbine airfoil cooling. Two rows of holes offer increased coolant coverage without jeopardizing structural strength. Furthermore, for higher blowing ratios two closely spaced rows of holes provide greater film effectiveness than would be expected based on a superposition of the performance of a single row of holes. This was demonstrated by the measurements of Han and Mehendale [20], who compared the performance of a single row of holes and two rows of holes spaced $2.5d$ with spacing between holes of $P/d = 2.5$. Holes in the two row configuration were staggered. All holes had an injection angle of 35 deg relative to the surface and were aligned with mainstream flow direction. To highlight the improved performance for two rows of injection compared to the expected values based on superposition of the single row injection, the results of Han and Mehendale [20] are replotted in Fig. 11. (Han and Mehendale [20] presented local values of η at three spanwise positions; these data were averaged to estimate the $\overline{\eta}$ values used in Fig. 11.) In this figure, the $\overline{\eta}$ values for the single row of holes were multiplied by two to show the expected value based on superposition. Comparisons of $\overline{\eta}$ values for a single row multiplied by a factor of two and $\overline{\eta}$ values for two rows show very similar performance for the lowest blowing ratio of $M = 0.2$. Consequently, the performance for two rows of holes is consistent with the expected performance based on superposition of one row of holes for this case, indicating that the two rows do not interact with each other. For $M \geq 0.5$, the two rows have distinctly higher $\overline{\eta}$ than would be predicted by superposition using results from a single row, with as much as a 60% increase in $\overline{\eta}$ for the $M = 1.0$ case. This improved performance can be attributed to interaction

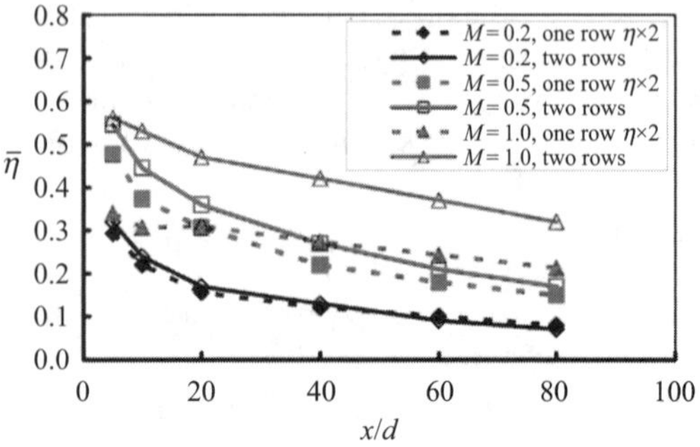

Fig. 11 Evaluation of film effectiveness for two rows injection as compared with superposition of a single row injection (data from Han and Mehendale [20]).

between the two rows of holes, producing a more cohesive coolant film that is less susceptible to dispersion by the mainstream.

C. FULL-COVERAGE CONFIGURATIONS

Full-coverage cooling incorporates multiple rows of coolant holes located over the entire area that is to be cooled. This technique is often used for combustor cooling. Full-coverage configurations have been studied with row spacings from 3d [21] to 14d [22], typically with normal injection, although Sasaki et al. [23] studied 45-deg injection. An investigation of full-coverage cooling with large-density-ratio coolant and high mainstream turbulence levels was conducted by Harrington et al. [24] to determine whether film effectiveness for full-coverage film cooling is predictable using superposition of film effectiveness from a single row of holes. For a range of momentum flux ratios from $I = 0.04$ to 0.59, they found that the film effectiveness reached a maximum level after four to eight rows, depending on blowing ratio, and this asymptotic level was about 15% lower than the superposition prediction. Maximum film effectiveness was $\eta = 0.30$. Using the same coolant hole configuration, Kelly and Bogard [25] found that the heat-transfer coefficient was increased by about 25% by the full-coverage injection for high mainstream turbulence conditions, and this increase was relatively constant for the full length of the full coverage test plate. Maximum net heat-flux reduction of $\Delta q_r = 0.35$ to 0.40 was found for full-coverage film cooling.

D. COOLING HOLE ANGLE EFFECTS: STREAMWISE ORIENTED WITH DIFFERENT SURFACE ANGLES

As already discussed, coolant holes are generally oriented with relatively shallow angles to the surface, but in some cases much steeper injection angles are used. Studies of the effect of the hole injection angle on film effectiveness on flat surfaces generally have found a small reduction in film effectiveness as the injection angle increases. For increasing injection angle there is a greater tendency for coolant jet separation, which causes lower film effectiveness. Kohli and Bogard [26] compared the performance of coolant holes with injection angles of 35 and 55 deg and found a decrease in film effectiveness for the 55-deg holes of 10 and 30% for momentum flux ratios of $I = 0.16$ and 0.63, respectively. (These results are presented in terms of I because this is expected to be the scaling parameter for jet separation.) Hole injection angles of 35 and 90 deg were tested by Foster and Lampard [19], and slightly decreased film effectiveness was found for the 90-deg holes at $M = 0.5$, but improved performance was found for the 90-deg holes for a high blowing ratio of $M = 1.4$. Similar results were found by Baldauf et al. [9] who compared holes with 30-, 60-, and 90-deg injection angles. Their results showed about a 30% decrease in peak $\overline{\eta}$ values for lower blowing ratios for 90-deg injection as compared to 30-deg injection. For higher blowing ratios, $M > 1.2$, there was as much as 60% increase in $\overline{\eta}$, but at these high blowing ratios performance was poor. The increased film effectiveness for 90-deg holes at higher blowing ratios was attributed to more interaction with adjacent jets for 90-deg holes as compared to 30-deg holes.

E. COOLING HOLE ANGLE EFFECTS: COMPOUND ANGLE INJECTION

Coolant holes are often oriented in a direction oblique to the mainstream direction, that is, with a compound injection angle. This orientation presents a greater coverage area downstream of the hole and presents a broader jet profile to the mainstream passing over the coolant jet. Consequently, this orientation is expected to have better film effectiveness, but also causes a greater increase in the heat-transfer coefficient. Because these are counteracting effects, it is important to evaluate the performance of compound injection in terms of the net heat-flux reduction.

The performance of compound injection coolant holes was evaluated by Schmidt et al. [18] and Sen et al. [27] for coolant holes with a 35-deg injection angle with the surface, and oriented 0 and 60 deg with respect to the mainstream flow direction. As shown in Fig. 12a, the 60-deg compound angle hole gave only a small increase in $\overline{\eta}$ for a momentum flux ratio of $I = 0.25$, but essentially doubled $\overline{\eta}$ for a momentum flux ratio of $I = 0.98$. The good performance for the 60-deg compound angle holes is evident in Fig. 12b, showing film effectiveness spatially averaged over the range $3 \leq x/d \leq 15$. The performance of 0-deg compound angle holes (coolant injected parallel with the flow) drops rapidly for $I \geq 0.5$ due to coolant jet separation, but the 60-deg compound angle holes have good performance for momentum flux as high as $I = 4$. The heat-transfer coefficients relative to the no-blowing heat-transfer coefficients $\overline{\overline{h_f/h_0}}$ and the resulting net heat-flux reduction $\overline{\overline{\Delta q_r}}$ were measured for these film-cooling configurations by Sen et al. [27]. As shown in Fig. 13a, the heat-transfer coefficients with 60-deg compound angle holes were about 15% higher than for 0-deg compound angle holes. The larger heat-transfer coefficients for the compound angle injection result from an interaction between the mainstream and the angled coolant jet that causes a strong vertical flow on the opposite side of the coolant jet.

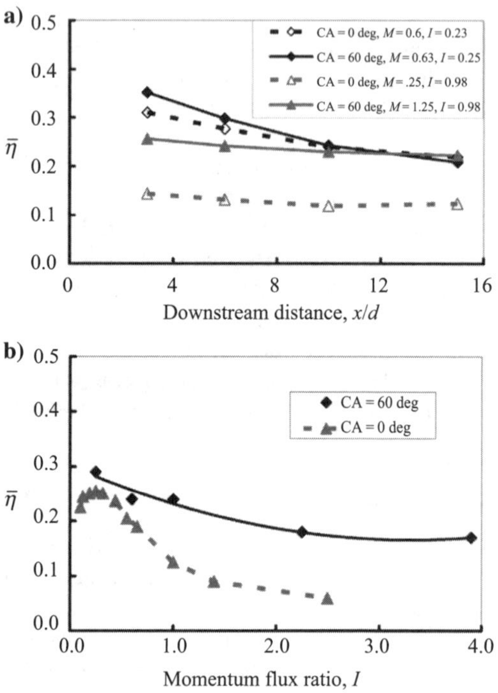

Fig. 12 Film effectiveness for 0- and 60-deg compound angle hole presented as a) streamwise distribution of laterally averaged cooling effectiveness and b) spatially averaged film effectiveness for varying momentum flux ratios (data from Schmidt et al. [18]).

Comparisons of net heat-flux reduction $\overline{\overline{\Delta q_r}}$ for the 0- and 60-deg compound angle holes are presented in Fig. 13b. These results show similar values of $\overline{\overline{\Delta q_r}}$ for the 0- and 60-deg compound angle holes indicating that the improved film effectiveness for the 60-deg compound angle holes is offset by the increased heat-transfer coefficients.

The compound angle results just discussed were for low mainstream turbulence levels. A later study by Schmidt and Bogard [28] examined 0- and 90-deg compound angle holes with mainstream turbulence levels of $Tu = 0.5$ and 17%. Although similar results were obtained for low mainstream turbulence, for $Tu = 17\%$ the film effectiveness for 0- and 90-deg compound angle holes were essentially the same for momentum flux ratios as high as $I = 2$. The 90-deg compound angle holes still caused an increase in heat-transfer coefficients, so the net heat-flux reduction for 90-deg compound angle holes was significantly less than for the 0-deg compound angle holes.

F. SHAPED HOLES WITH STREAMWISE ORIENTATION

Improved film-cooling performance is obtained if the coolant hole is "shaped" towards the exit of the hole with an expansion that diffuses the flow exiting the hole. Examples of hole shapes tested by Saumweber et al. [29] are shown in Fig. 14. Expansion of the hole exit decelerates the coolant jet, resulting in a lower momentum flux and, consequently, less tendency for the coolant jet to separate. Furthermore, the lateral expansion presents a broader jet to the mainstream, so that the mainstream has a greater impact on the jet and more effectively turns the jet towards the wall. The improved performance in terms of film effectiveness for shaped holes is shown by the results of Saumweber et al. [29] and reproduced in Fig. 15. Saumweber et al. [29] used streamwise oriented holes with a 30-deg injection angle and spacing between holes of $P/d = 4$. A coolant density ratio of $DR = 1.7$ was used, with mainstream turbulence levels ranging from $Tu = 3.6$ to 11%. As shown in Fig. 15, the spatially averaged $\overline{\overline{\eta}}$ (averaged from $x/d = 2$ to 22) for blowing ratios from $M = 0.5$ to 2.5 shows much greater film effectiveness for shaped holes as compared to cylindrical holes. With increasing blowing ratio, the shaped hole has increasing film effectiveness while the effectiveness of the cylindrical hole drops sharply. The decreasing film effectiveness for the cylindrical hole is caused by separation of the coolant jet, and so these results indicate that this shaped hole is very effective in reducing the coolant jet separation. Saumweber et al. [29] found that the "laidback fan-shaped" hole (shown in Fig. 14) had essentially the same performance as the "fan-shaped" hole, indicating that the additional streamwise expansion of the hole resulted in no additional benefit.

The effects of the shaped hole coolant injection on increasing heat-transfer coefficients were also measured by Saumweber et al. [29]. In some cases shaped holes were found to have heat-transfer coefficients similar to those of cylindrical holes, but for the highest blowing ratios with $Tu = 11\%$, the shaped holes had 50% greater heat-transfer coefficient than the cylindrical holes. The detrimental effects of this increase in heat-transfer coefficients somewhat offset the improved film-effectiveness performance.

Given the significantly better performance of the shaped holes, it is important to recognize the cost of manufacturing when comparing

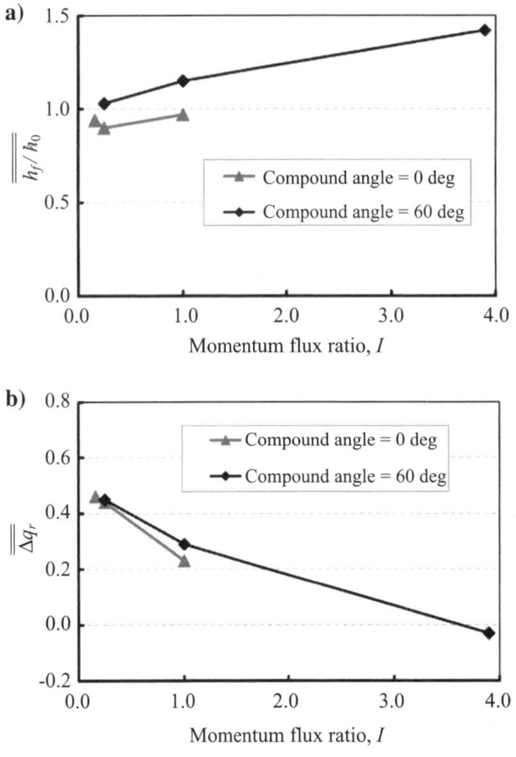

Fig. 13 Comparison of 0- and 60-deg compound angle holes showing a) spatially averaged heat-transfer augmentations and b) spatially averaged net heat-flux reduction as functions of jet momentum flux ratios (data from Sen et al. [27]).

cylindrical and shaped holes. Because of the complexity of forming shaped holes, manufacturing turbine airfoils with shaped holes is considerably more expensive than cylindrical holes. In many cases this additional cost is not warranted, and cylindrical holes are used.

G. SHAPED HOLES WITH COMPOUND ANGLE INJECTION

Studies of film-cooling performance of shaped holes were made by Schmidt et al. [18] and Sen et al. [27] investigating a 60-deg compound angle with 35-deg injection angle relative to the surface. Comparisons were made with cylindrical holes oriented in the streamwise direction and cylindrical holes with 60-deg compound angle injection. The shaped hole had a 15-deg expansion in the direction of the hole orientation. Spatially averaged film effectiveness determined over a range from $x/d = 3$ to 15 showed 30 to 60% higher film effectiveness for shaped holes as compared to the cylindrical 60-deg compound angle holes. At the moderate momentum flux ratio of $I = 1.0$, however, the shaped holes had 25% higher spatially averaged heat-transfer coefficient, so that the net heat-flux reduction for the shaped holes was similar to that for the cylindrical holes. At the high momentum flux ratio of $I = 3.9$, the shaped holes had significantly greater net heat-flux reduction than the cylindrical holes.

Shaped holes with a compound angle of 35 deg were studied by Dittmar et al. [30] using a test facility that simulated the suction side of a turbine vane. The shaped holes had a lateral expansion giving a factor of three increase in exit area. The film effectiveness, heat-transfer coefficients, and net heat-flux reduction performances were compared with shaped holes oriented with 0-deg compound angle and with double

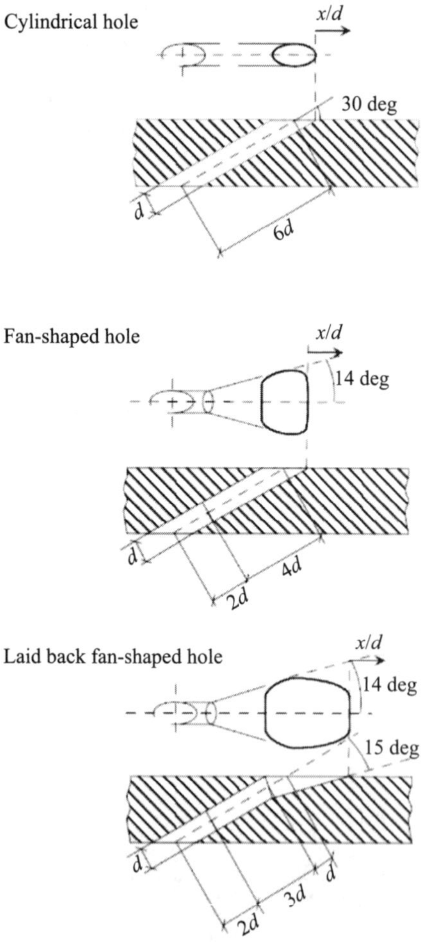

Fig. 14 Schematics of different cooling hole shapes (from Saumweber et al. [29]).

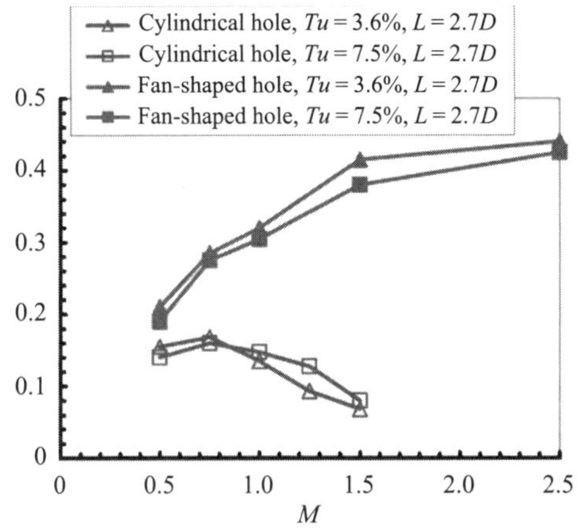

Fig. 15 Comparison of spatially averaged film effectiveness for cylindrical holes and shaped holes (data from Saumweber et al. [29]).

rows of cylindrical holes and double rows of discrete slots. The film-effectiveness performances for 35-deg compound angle holes were similar to the 0-deg compound holes and generally superior to the double rows of cylindrical holes and short slots. The shaped holes were found to cause larger increases in heat-transfer coefficients, resulting in net heat-flux reduction that was generally less than for the round holes. Dittmar et al. [30] speculated that separation within the diffuser part of the shaped hole might have generated increased turbulence, leading to higher heat-transfer coefficients.

V. MAINSTREAM AND SURFACE EFFECTS ON FILM COOLING

There are a number of mainstream and surface variables that influence film-cooling performance on an actual airfoil. These variables include the approach boundary layer, surface curvature, pressure gradients, freestream turbulence, unsteady wakes, rotation, Mach number, and surface roughness. Numerous film-cooling experiments have been done to evaluate these effects on the performance of film cooling; most of them have been flat-plate studies looking at individual effects. These studies, which will be summarized in this section, have provided much insight into the effects of these variables on airfoil film cooling.

A. BOUNDARY-LAYER THICKNESS EFFECTS

As boundary-layer characteristics range from laminar to turbulent and from relatively thin to thick along an airfoil, film-cooling injection is subjected to a range of approach flow conditions. Film-cooling holes for many gas turbine vane and blade applications are on the order of a fraction of a millimeter, while the boundary-layer thickness can range from zero (at the leading edge) to a millimeter. Injecting film-coolant thickens the boundary layer, and downstream coolant injection can be affected.

Kadotani and Goldstein [31] summarized previous work 3 [2–34] on the effects of boundary-layer thickness on film cooling for a fully turbulent boundary layer as follows: there is a large effect on the centerline film-effectiveness levels, with larger boundary-layer thicknesses resulting in lower effectiveness; there is a larger effect on film effectiveness for thinner rather than thicker boundary-layer displacements; and there is a larger effect on film effectiveness when the momentum flux ratio is such that the jet penetration distance is on the order of the boundary-layer thickness. Most of these effects were found on centerline film effectiveness levels in the near-hole region. Although there were apparent effects on the centerline values, the work of Kadotani and Goldstein [31] indicated essentially no effect of boundary-layer thickness on the lateral averaged values of film effectiveness. Using thinner boundary layers that are more consistent with those found on a turbine airfoil, measurements performed by Liess [35] indicated that for displacement-thickness to hole-diameter ratios of less than 0.2, the film-effectiveness levels remained unchanged.

B. WALL CURVATURE EFFECTS

The curvature of a turbine airfoil causes the flow around the airfoil to experience both convex and concave effects. Flow around the leading edge and near-suction surface experiences severe convex curvature whereas flow along the pressure surface generally experiences concave curvature. An inherent characteristic of flow around curved walls is a pressure gradient normal to the streamlines. Thus, when a jet is injected along a curved wall, the normal pressure gradient for a convex wall causes the jet to curve towards the wall. If the jet momentum is less than that of the mainstream, the radius of curvature of the jet will be less than that of the wall radius of curvature, and the coolant jet will be pressed to the surface. However, if the jet momentum is greater than that of the mainstream, the radius of curvature of the jet will be larger than that of the wall curvature, and the jet will move away from the wall. For concave surfaces the normal pressure gradient is opposite to that for convex surfaces, so that the concave surface has the opposite effect on the coolant jets. Furthermore, Taylor–Görtler vortices resulting from flow instabilities developing on concave walls can influence film-cooling performance.

As would be expected from these curvature effects, for low momentum flux ratios, injection along a convex surface improves film effectiveness whereas along a concave surface film effectiveness decreases [36, 37]. The results of Ito et al. [36], shown in Fig. 16, demonstrated the greater film effectiveness for convex walls when $M \leq 1$, but the better film effectiveness for concave walls when $M \geq 1.5$. A direct comparison of convex, flat, and concave walls, presented in Fig. 17 (from Ito et al. [36]) for $M = 0.5$, shows that film effectiveness for convex walls can be as much as 80% larger than over flat walls.

C. PRESSURE GRADIENT EFFECTS

As curvature effects are inherent along turbine airfoils, so are pressure gradient effects, ranging from favorable to adverse. The parameter most commonly used

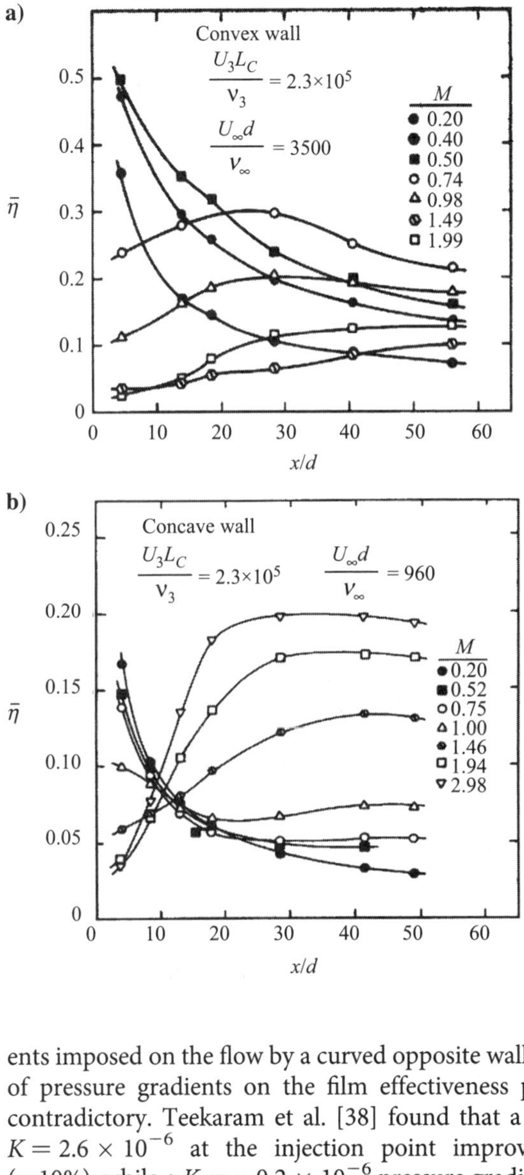

Fig. 16 Comparisons of laterally averaged film effectiveness for a a) convex and b) concave wall (figures from Ito et al. [36]).

to describe the severity of the pressure gradient is the acceleration parameter K, defined as follows:

$$K = (v/U^2)\,\mathrm{d}U/\mathrm{d}s \quad (12)$$

where positive values indicate favorable pressure gradients and negative values indicate adverse pressure gradients. Pressure gradients that have been used to test the effects of film-cooling range from $K = 2.6 \times 10^{-6}$ (Teekaram et al. [38]) to $K = -0.58 \times 10^{-6}$ (Brown and Saluja [39]). For reference, the turbulent boundary layer relaminarizes when there is a favorable pressure gradient greater than $K \approx 3 \times 10^{-6}$.

To isolate curvature effects from pressure gradient effects, tests have been conducted using flat-plate facilities with pressure gradients imposed on the flow by a curved opposite wall (see [38] and [39]). The effects of pressure gradients on the film effectiveness presented in the literature are contradictory. Teekaram et al. [38] found that a favorable pressure gradient of $K = 2.6 \times 10^{-6}$ at the injection point improved film effectiveness slightly ($\sim$10%), while a $K = -0.2 \times 10^{-6}$ pressure gradient decreased film effectiveness slightly. In contrast, Brown and Saluja [39] found that an adverse pressure gradient of $K = -0.58 \times 10^{-6}$ caused over 100% increase in film effectiveness whereas a $K = 1.1 \times 10^{-6}$ pressure gradient decreased film effectiveness by as much as 50%. Liess, using favorable pressure gradients of $K = 1 \times 10^{-6}$ to 2.5×10^{-6}, also found decreased film effectiveness at lower blowing ratios, $M < 0.6$, but

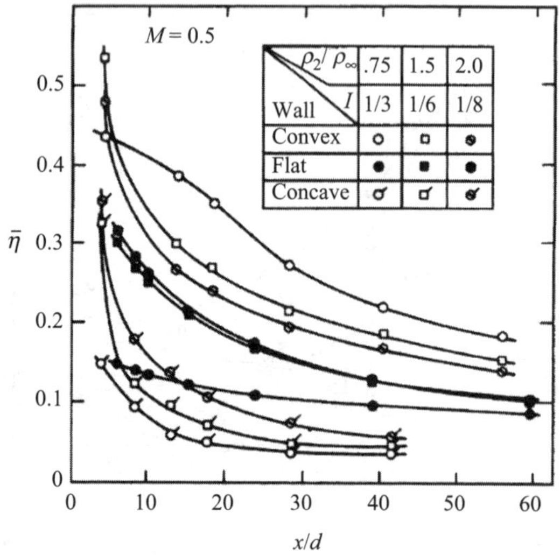

Fig. 17 Comparisons of laterally averaged film effectiveness for a) convex and b) concave wall (figures from Ito et al. [36]).

essentially no effect for higher blowing ratios [35]. (Note that the K values are not given in [35]; these values were calculated using data from the paper.) These differences might be attributable to different approach boundary-layer flows (information about the approach boundary layer is not provided in [38]), or to different variations of pressure gradient downstream of the coolant holes.

The studies just discussed [25, 38, 39] were conducted using low-density-ratio coolant ($0.8 < DR < 1.2$). Effects of a favorable pressure gradient on film effectiveness for a high-density-ratio coolant of $DR = 1.6$ were measured by Schmidt and Bogard [40]. Using a pressure gradient of $K = 1.5 \times 10^{-6}$ at the injection location and blowing ratios ranging from $M = 0.4$ to 1.5, they found only a slight increase in film effectiveness near the hole for $M \leq 1.0$ and no effect for $M > 1.0$.

D. HIGH FREESTREAM TURBULENCE EFFECTS

In gas turbine operations, freestream turbulence originating in the combustor upstream of the turbine can dominate film-cooling performance. Turbulence levels are quantified in terms of velocity fluctuations (root-mean-square levels) divided by the magnitude of the mean velocity.

$$Tu = \frac{u_{\text{rms}}}{U} \tag{13}$$

Simulation of the mainstream turbulence characteristics for turbine sections depends on using appropriate length scales. These are typically quantified using measurements of the integral length scale, a measure of the scale of the largest turbulent eddies.

Most studies of high freestream turbulence reported in the literature have been limited to levels of 8%, which may be representative for industrial turbines. In most laboratory studies, these turbulence levels are generated using a biplanar grid, and the eddies are on the order of the size of the bars within the grid.

Early studies addressing the effects of freestream turbulence on film cooling were reported by Launder and York [41] and Kadotani and Goldstein [31, 42, 43]. They studied turbulence levels ranging from 3 to 8%, with integral length scales of the order $\frac{1}{3}d$ at the cooling hole injection location. Kadotani and Goldstein found that at lower blowing ratios there was a decrease in film effectiveness of as much as 15% whereas at high blowing ratios there was a slight increase in film effectiveness.

More representative turbulence levels for aero engines, and for some land-based turbine designs, are nominally 20%, with turbulence integral length scales on the order of the dilution holes in the upstream combustor. In many combustor designs, the dilution jets are injected at relatively high momentum flux ratios ($\sim$50) through injection holes in the combustor liner that have diameters 10–20 times larger than the airfoil film-cooling holes. This large difference in scales (film-cooling hole to dilution hole diameters) suggests that the appropriate ratios of turbulence length scale to film-cooling hole diameters should be large.

One of the first reported studies with turbulence levels higher than grid-generated turbulence was that of Jumper et al., with a turbulence level of nearly 17% [44]. As might be expected, their flat-plate film-cooling results indicated a more rapid decay in film effectiveness at high turbulence, relative to the low-turbulence case. Interestingly, their results indicated that near the cooling hole the effect of high freestream turbulence was negligible on the film-effectiveness levels.

In the studies just described [41–44], low-density-ratio coolants were used. The significant effects of high mainstream turbulence levels were shown in the film-cooling study by Schmidt and Bogard [45] using a realistic density ratio of $DR = 2$ for the coolant. In this study, film effectiveness and heat-transfer coefficients were measured for freestream turbulence levels of 0.3, 10, and 17%, with turbulence integral length scale of $\Lambda_f = 3d$ for the high turbulence conditions. For $Tu = 0.3\%$, the momentum flux ratio for maximum film effectiveness was $I = 0.2$, similar to many previous studies with low freestream turbulence. At this momentum flux ratio the high freestream turbulence levels of $Tu = 10$ and 17% caused over a factor of two decrease in film effectiveness near the hole and essentially $\eta = 0$ for $x/d > 20$. Furthermore, this study showed that the optimum momentum flux ratio for the high freestream turbulence conditions was $I = 1.1$, almost an order of magnitude larger than for the low freestream turbulence condition. Although the coolant jets would be expected to be detached for this very high momentum flux ratio, the additional dispersion of the jet caused by the high freestream turbulence transports coolant back to the surface.

E. MACH-NUMBER EFFECTS

There has been a relatively limited number of studies of film cooling at supersonic conditions published in the literature. This is particularly the case for exact comparisons between low- and high-speed conditions, with Mach number being the isolated effect that is being addressed. It is difficult to do a one-to-one comparison because most facilities are designed to operate at high- or low-speed conditions, but not both. Moreover, an assessment of Mach-number effects should be done on a flat plate to avoid other effects, such as those of curvature and pressure gradients.

Gritsch et al. [2] studied the effects of external Mach numbers of 0.3, 0.6, and 1.2 on film cooling for single cooling holes that were cylindrical, expanded in the spanwise direction, and fully expanded. Their results indicated that were was little effect of Mach number on measured film-effectiveness levels for the shaped cooling holes. For the cylindrical hole, however, they saw an improved cooling performance at Mach 1.2 for blowing ratios of 0.5 and 1. They attributed this improved performance to the shock structure, which they have hypothesized has a tendency to turn the ejected jet toward the surface.

Similar results were obtained by Liess, who performed film-effectiveness measurements for a row of coolant holes with mainstream Mach numbers of $Ma = 0.3$, 0.6, and 0.9 [35]. These experiments showed that the variation in Mach number had no effect on film effectiveness.

In a test conducted by Juhany et al. [46], the injectant and freestream were at matched pressure, and shock waves in both the freestream and the injected coolant flow were produced to adjust the flow to the same orientation angle. A leading separation shock, an expansion wave, and a recompression shock were observed through schlieren optics. The effect of weak shock waves on the adiabatic wall temperature was found to be insignificant.

F. UNSTEADY FLOW EFFECTS

Main flow unsteadiness is distinctly different from freestream turbulence, in that unsteadiness refers to periodicity in the flow rather than the randomness that is characteristic of turbulence. This unsteadiness generally arises in a turbine environment on rotor blades where the film-cooling jets experience a variation in the mainstream flow as the blades pass through the wakes of the upstream vanes. These variations, in turn, cause a variation in the pressure field external to the film-cooling jet, resulting in a modulation of the film-cooling jets.

Similar to other studies, Bons et al. [47] simulated the modulation due to the vane wake passing through the use of speakers placed in the coolant supply plenum. This produces oscillation in the film-cooling inlet pressures, rather than the freestream pressure. Their results indicated a large reduction in the film-effectiveness values measured at the jet centerline for all of the simulated frequencies. The reduction in centerline effectiveness was more severe at the lower blowing ratios relative to the higher blowing ratios; at a blowing ratio of $M = 1.5$, there was essentially no effect.

Similar results were reported by Seo et al. [48] although they simulated the effect by pulsing the mainstream through the use of a damper placed at the exit of the wind tunnel. Their results also indicated a larger decrease in effectiveness at relatively low blowing ratios ($M = 0.5$) and at short hole length-to-diameter ratios (1.6). In contrast, at a high blowing ratio ($M = 1$), their results actually showed a slight improvement in effectiveness with the short cooling hole length.

G. ROTATING RIG TESTS

Measurements of film-cooling performance on a rotating blade require a sophisticated rotating test rig with a full stage, that is, an inlet nozzle guide vane and a rotor. Because of the complexity and expense of such test rigs, there have been very few studies. Furthermore, to evaluate how rotation affects film-cooling performance, the film-cooling performance for the rotating blade should be compared to that of a stationary cascade blade in which all other conditions have been kept the same. Considering the complex flowfield in the rotating rig test, including vane wakes, a reasonable comparison is very difficult. Film cooling with a rotating rig was experimentally investigated by Dring et al. [49], Abhari and Epstein [50], and Takeishi et al. [51]; in each case comparisons to stationary blade configurations were attempted.

A large-scale, low-speed, rotating rig was used by Dring et al. [49] when testing a rotating blade, but only a single film-cooling hole on the suction side and a single hole on the pressure side of the blade were used. Results were compared to the cascade results from a separate experiment by Ito et al. [36]. No attempt was made to match the airfoil geometry, film-cooling hole configurations, or the flow conditions. Despite the fact that the rotating and stationary blades were not well matched, the authors found similar results for the film effectiveness on the suction side of the blade and small differences on the pressure side.

Abhari and Epstein [50] used a short-duration blowdown tunnel to test a nozzle guide vane and rotor stage. The rotor had two rows of coolant holes on the suction surface and three rows on the pressure surface. Time-resolved measurements of heat flux were made at discrete locations without and with film cooling. Results were compared to the results of Rigby et al. [52], who tested the same rotor airfoil with the same film-cooling configuration in a stationary cascade facility. There were considerable differences in the operating conditions for the rotating tests of Abhari and Epstein [50] and the stationary tests of Rigby et al. [52]. In the rotating tests, coolant was ejected from the upstream nozzle guide vane, and all rows of coolant holes were operational simultaneously, while for the stationary cascade tests the upstream vane row was simulated by rotating bars and only one row of coolant holes was operated in each test. Both rotating and stationary tests showed little change in the heat flux on the pressure side of the blade with coolant injection. On the suction side of the blade, the rotating results showed a greater decrease in heat flux.

Both a stationary cascade and a rotating rig were used by Takeishi et al. [51] to study the effects of rotation on film cooling of a turbine rotor. The rotor had three rows of holes in the leading-edge showerhead, two rows on the pressure side, and a single row on the suction side (but no data were obtained from the pressure-side row of holes for the rotating rig). The cascade model was 6.6 X scale of the rotating model with the same airfoil geometry and film-cooling hole configuration. The cascade facility did not, however, emulate the wakes from upstream vanes that would have occurred for the rotating rig, and matching of the mainstream turbulence levels approaching the rotors was not discussed. Nevertheless, under operation with the showerhead alone and the suction-side row of holes alone, the film effectiveness for the cascade and the rotating blade were very similar.

From these studies, one can conclude that the effect of rotation on film-cooling performance is not well established although the indications are that there is little effect. Perhaps more importantly, examination of the film-cooling performance of the few rotating rig tests in the open literature, and attempts to correlate the results with stationary laboratory tests, shows that the complexity of the flowfield in the actual turbine makes comparisons with laboratory tests very difficult. Similarly, although the various laboratory-generated databases provide valuable insight into film-cooling performance, they generally cannot be expected to provide precise predictions of performance because the databases are not generated in the same highly complex flowfields that occur in the engine.

H. SURFACE ROUGHNESS

Studies of film-cooling performance are generally done with smooth surfaces, which are representative of new turbine airfoils. However, during operation of the turbine engine, airfoil surfaces will be roughened due to deposition, spallation, and erosion [53, 54]. A rougher airfoil surface can lead to early boundary-layer transition, thickening of the boundary layer, and increased turbulent mixing in the boundary layer. These changes due to surface roughness will generally lead to reduced film effectiveness although for high blowing ratios an increase in film effectiveness can result. Increased surface roughness often significantly increases heat-transfer coefficients.

Surface roughness effects on film effectiveness for film cooling using a row of holes on a flat surface were studied by Goldstein et al. [55] and Schmidt et al. [56]. Schmidt et al. also measured changes in heat-transfer coefficient due to coolant injection with a rough surface. For the Schmidt et al. study [56], the roughness consisted of an array of conical elements with a maximum roughness element height of $0.4d$, corresponding to an equivalent sandgrain roughness of $Re_k \approx 100$. Roughness downstream of the coolant holes was found to have a small effect on laterally averaged film effectiveness: less than 10% decrease for low momentum ratios and less than 5% increase for high momentum flux ratios. Although roughness caused a 50% increase in heat-transfer coefficient, coolant injection did not cause a significant change in heat-transfer coefficient, except within $10d$ of the hole, where less than a 10% decrease occurred at low momentum flux ratio, and less than a 10% increase occurred for high momentum flux ratio.

Roughness effects on the suction side of a simulated vane were investigated by Bogard et al. [57] and Rutledge et al. [58]. Roughness upstream and downstream of a row of cylindrical holes were investigated independently, using an array of conical elements 0.25d in height. This roughness configuration was estimated to have an equivalent sandgrain roughness Reynolds number of $Re_k \approx 50$ based on the boundary-layer flow approaching the coolant holes. At low blowing ratios, $M \approx 0.3$, roughness upstream of the coolant holes caused as much as 25% reduction in spatially average film effectiveness. At high blowing ratios, $M > 1.0$, roughness upstream and downstream of the coolant holes caused an increase in film effectiveness. Roughness essentially doubled the heat-transfer coefficients on the suction side of the vane, and film injection did not cause further increase or decrease of the heat-transfer coefficients. Consequently, the net heat-flux reduction with the rough surface was caused by the film effectiveness of the film cooling.

VI. AIRFOILS AND ENDWALLS

Film cooling on vanes and blades generally involves a dense array of coolant holes, referred to as a "showerhead," around the leading edge and more widely spaced rows of coolant holes around the main body of the airfoil. Sometimes coolant is also introduced at the tip of a blade, and arrays of coolant holes are used on the endwalls. Each of these regions has unique film-cooling characteristics.

A. LEADING EDGES

The leading edge of a vane or blade is generally subjected to the largest heat loads as a result of the large heat-transfer coefficients along the stagnation line. Consequently, film cooling of the leading edge is often accomplished using several closely spaced rows of coolant holes. This array of holes around the leading edge is referred to as the "showerhead" and generally consists of six to eight rows of holes for vanes and three to five rows of holes for blades. Holes are typically aligned radially, that is, normal to the mainstream direction, with injection angles relative to the surface ranging from 20 to 45 deg.

Film-effectiveness measurements within and downstream of the showerhead of a simulated vane were made by Polanka et al. [59], Witteveld et al. [60], and Cutbirth and Bogard [61, 62], under conditions of low and high mainstream turbulence levels. The simulated vane tested in these studies, shown schematically in Fig. 18, had six rows of coolant holes, spaced 3.3d apart, in the showerhead region. Holes were oriented radially with an injection angle of 25 deg with respect to the surface, and the pitch between holes was 5.5d. A coolant density ratio of $DR = 1.8$ was used, and blowing ratios up to $M = 2.9$ were tested. Also shown in Fig. 18 is a flow visualization of the coolant distribution around the leading edge of the vane.

Coolant in the showerhead region is projected to a large distance from the surface, extending as much as 5d from the surface even at relatively low blowing ratios. This is because of the lack of a crossflow along the stagnation

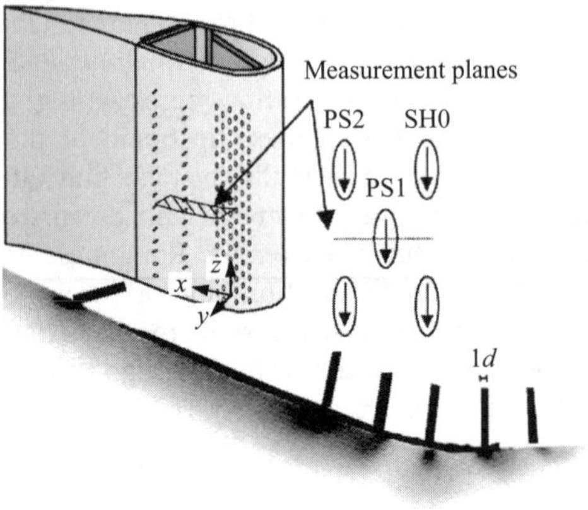

Fig. 18 Flow visualization of coolant flow in the showerhead region for a blowing ratio of $M^*sh = 1.5$ (figure from Cutbirth and Bogard [62]).

line that would tend to turn the coolant jets towards the surface and to the deceleration of the mainstream as it approaches the surface. Flow visualization images (Fig. 19) and thermal field images by Cutbirth and Bogard [61] showed that coolant jets along the stagnation line separate from the surface even at low blowing ratios. Unlike film cooling of flat surfaces, the film effectiveness continued to increase with increasing blowing ratio, as shown in Fig. 20 for a position immediately downstream of the showerhead. The lack of an optimum blowing ratio can be attributed to the separation of the coolant jet even at low blowing ratios.

Showerhead blowing effects on heat-transfer coefficients were investigated by Ames [63] using a simulated vane with five rows of holes in the showerhead spaced 3.8d apart. Coolant holes were oriented radially with an injection angle of 20 deg with respect to the surface, and the pitch between holes was 6.4d. For high mainstream turbulence conditions, the heat-transfer coefficients immediately downstream of the showerhead increase about 20% with showerhead blowing relative to a no-blowing baseline.

A number of different film-cooling configurations for blade leading edges have been tested, including variations in number of rows of holes, inclination angle of the holes, and hole shape. Many simulations of the blade leading edge have used cylindrical or semicylindrical models. A four-row configuration, with rows positioned at ± 15 and ± 40 deg from the stagnation line was tested by Mehendale and Han [64] to determine the film effectiveness, heat-transfer coefficients, and net heat-flux reduction. Coolant holes in their model were aligned radially with a 30-deg injection angle relative to the surface. Spacing between holes in a row was $P/d = 3$. Film-cooling effectiveness, determined using a coolant density ratio of about $DR = 0.92$, was maximum for a blowing ratio of $M = 0.8$, with just a slight decrease for the highest blowing ratio tested, $M = 1.2$. Laterally averaged film effectiveness levels were nominally $\overline{\eta} = 0.4$ downstream of the first row of holes, and $\overline{\eta} = 0.55$ downstream of the second row of holes. Coolant injection caused the heat-transfer coefficients downstream of both rows of holes to more than double.

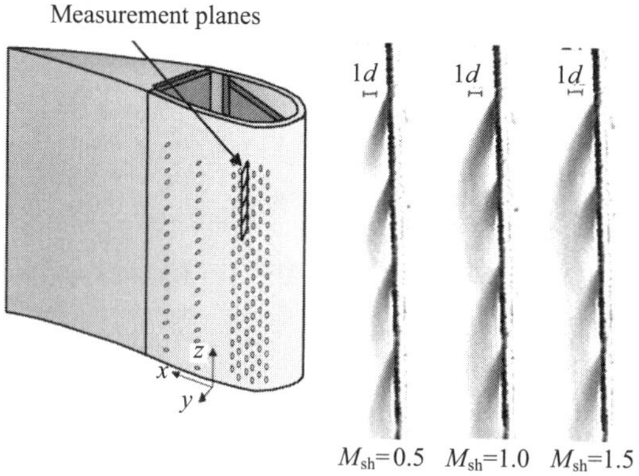

Fig. 19 Flow visualization of coolant flow along the stagnation line showing coolant jet separation for blowing ratios of $M^*sh = 0.5$, 1.0, and 1.5 with $Tu = 0.5\%$ (figure from Cutbirth and Bogard [61]).

Despite the very large increase in heat-transfer coefficient, there was a large net heat-flux reduction because of the high level of film effectiveness.

A five-row configuration, with rows at 0, ± 20, and ± 40 deg, was tested by Reiss and Bölcs [65] to determine film effectiveness, heat-transfer coefficients, and net heat-flux reduction over a range of blowing ratios from $M = 0.6$ to 1.5. The focus of this study was a comparison of the relative performance of cylindrical and shaped holes with two different expansion configurations, "laid back" and "laterally expanded." The holes had a nominal injection angle of 45 deg relative to the surface, and spacing between holes in each row was $P/d = 3.7$. In general, the "laid back" shaped holes, with an average film effectiveness of $\overline{\eta} = 0.4$ to 0.5, had better film effectiveness than the cylindrical holes and the "laterally expanded" shaped holes. All holes induced a large increase in heat-transfer coefficients, over a factor of two in some cases, for all blowing ratios. Maximum net heat-flux reduction was obtained using the "laid back" shaped holes at a blowing ratio of $M = 1.0$.

In contrast to these studies of blade leading-edge cooling in which an optimum blowing ratio of nominally $M = 1.0$ was found, Albert et al. [66] found that film-cooling film effectiveness continued to improve with increasing blowing ratio up to the highest blowing ratio of $M = 4.0$. Albert et al. used a

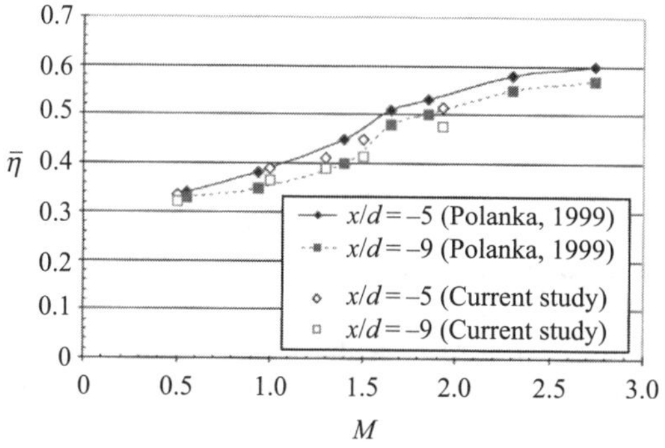

Fig. 20 Laterally averaged film effectiveness within ($x/d = -5$) and immediately downstream ($x/d = -5$) of the showerhead region of a film-cooled vane with $Tu = 20\%$ (figure from Cutbirth and Bogard [61]).

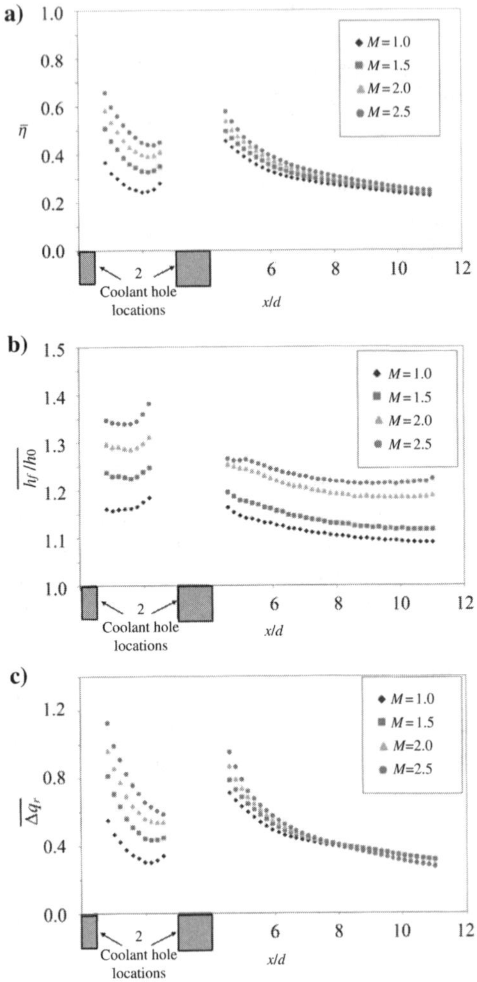

Fig. 21 Film-cooling performance for a simulated blade leading edge with three rows of holes. Mainstream turbulence was $Tu = 10\%$. Stagnation line coolant holes at $x/d = 0$. Performance in terms of a) laterally averaged film effectiveness, b) laterally averaged heat-transfer coefficient augmentation, and c) laterally averaged net heat-transfer reduction (figures from Mouzon et al. [13]).

three-row configuration with "laid back" shaped holes oriented radially, an injection angle of 20 deg, and a spacing between holes of $P/d = 7.6$. Heat-transfer coefficients and net heat-flux reduction for the same configuration were measured by Mouzon et al. [13] for blowing ratios ranging from $M = 1.0$ to 2.5. Results from this study, presented in Fig. 21, show significant increases in film-effectiveness levels and heat-transfer coefficients with increasing blowing ratio. Even though the highest blowing ratio of $M = 2.5$ had the largest heat-transfer coefficients, the maximum net heat-flux reduction (see Fig. 21c) occurred at this blowing ratio.

B. TURBINE BLADE TIPS

Heat-transfer coefficients along the tip of a turbine blade are some of the highest values found on the various surfaces associated with a turbine airfoil. As such, improving the thermal environment along the blade tip is generally accomplished through impingement cooling and film cooling. There are a number of locations and hole shapes through which film cooling can be introduced for the tip; the placement of the holes can be on the blade tip itself or on the pressure surface of the blade. Generally, the cooling holes are placed closer to the pressure surface, either on the tip or on the blade, because the crossflow driving the tip flows from the pressure to the suction surface aids in spreading the coolant across the tip surface.

Putting the cooling holes along the pressure side of the blade insures that coolant passes across the blade tip corner, where high oxidation rates typically occur, as demonstrated by examination of used parts. If the blowing from the holes is too high, however, it can result in either the coolant blowing off the airfoil and along the pressure surface rather than passing through the tip gap, or it can impact the outer shroud rather than attach to the blade tip. Alternatively, placing the holes on the tip fails to cool the corner of the blade tip along the pressure side. It is also possible that with excessive blowing, the coolant will more effectively cool the outer shroud than the blade tip itself. For either hole placement, consideration must also be given to the actual size of the tip gap; results generally indicate that better cooling can be achieved at smaller tip gaps. In a small gap the coolant can fill the entire gap whereas in a large gap there is a larger mass flow of the hot fluid to mix out the coolant in the gap. Moreover, with a large tip gap there is a higher chance that the coolant will impinge on the outer shroud, only to convect along the outer shroud rather than along the blade tip.

In a review paper on tip heat transfer, Bunker [67] states that very little research has been reported in the literature for blade-tip film cooling. In fact, blowing from the tip has been considered by Kim and Metzger [68], Kim et al. [69], Kwak and Han [70, 71], Ahn et al. [72], Christophel et al. [73], Acharya et al. [74], and Hohlfeld et al. [75].

Kim et al. [69] present a summary of the experimental work that Metzger performed on tip blowing, as shown in Fig. 22. In addition to concluding that there is only a weak effect of the relative motion between a simulated blade and shroud on tip heat-transfer coefficient, they stated that there is a strong dependency of film effectiveness on the shape of the hole and injection locations. Four hole configurations are discussed, including the following: discrete slots located along the blade tip, round holes located along the blade tip, angled slots positioned along the pressure side, and round holes located within the cavity of a squealer tip. The studies reported by Kim et al. [69] were performed in a channel that simulated a tip gap, but a blade with its associated flowfield was not simulated. In comparing the discrete slots to the holes, as shown in Fig. 23, their data indicated a substantial increase in film effectiveness using the discrete slots for all blowing ratios tested. Injection from the pressure side holes provided cooling levels of similar magnitude to the holes placed on the tip.

Kwak and Han [70, 71] reported measurements for varying tip gaps with cooling holes placed along the camber line for a flat and a squealer tip geometry. They found a substantial improvement in effectiveness with the addition of a squealer tip. The coolant circulated within the squealer tip cavity, providing a better distribution of the coolant along much of the tip, as compared with no-squealer cases. Only along parts of the suction side was the film effectiveness poor. They found that for the flat tip, good cooling was provided to the trailing

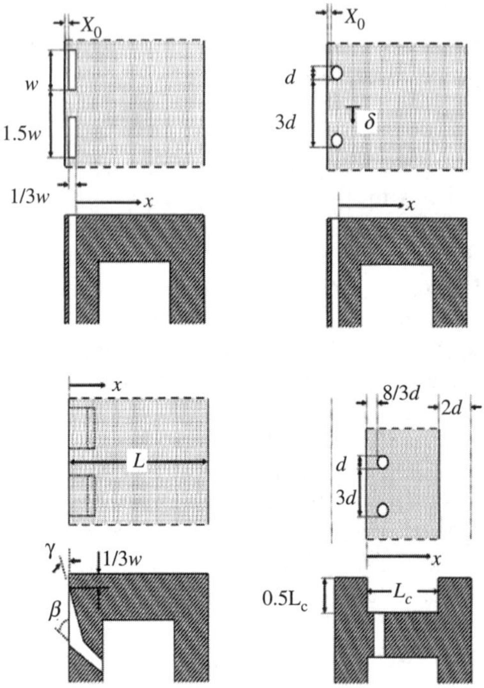

Fig. 22 Summary of tip cooling geometries tested by Kim et al. [69].

edge, resulting from the accumulation of coolant in this area. In a later study from the same group, Ahn et al. [72] found that for the same coolant mass flow injection from either the pressure side or the combined pressure side and tip was highly sensitive to the tip gap, with higher film effectiveness achieved at smaller tip gaps. Their results also indicated a more uniform coolant coverage for the case of a squealer blade tip relative to a flat tip.

Measurements reported by Christophel et al. [73] in which film-cooling holes were placed along the pressure side of the blade indicated that the cooling performance was significantly better for a small tip gap than for a large tip gap, as shown in Fig. 24. Their results did indicate that the cooling pattern was streaky in nature, with very little spreading as the coolant convected across the tip. The results for the small tip gap indicated that the coolant was swept further downstream of the hole prior to entering the tip gap for higher coolant flows, particularly those holes in the leading-edge region. In fact, computational predictions and measured effectiveness levels for the same flow conditions indicated that the jets exited into the pressure-side passage following the pressure side of the blade until the trailing edge of the blade, at which point the coolant entered the tip gap. For high local momentum flux ratios of the jets, the coolant did appear to cool the blade tip. For a large tip gap, their data indicated that the film-effectiveness levels decreased, or remained relatively constant, as the coolant flow was increased. As the coolant flow was increased, the jets impacted and cooled the outer shroud of the large tip gap rather than the blade tip.

Predictions for varying tip gap sizes by Acharya et al. [74] indicated that film-cooling injection alters the nature of the leakage vortex. High film effectiveness and low heat-transfer coefficients were predicted along the coolant trajectory, with the lateral spreading of the coolant jets being quite small for all cases. Acharya et al. studied various leakage reduction strategies for blade tips and found, like Kwak and Han [70, 71], that a single suction-side squealer tip is the best configuration to reduce the heat transfer and leakage flow. Computational results by Hohlfeld et al. [75] indicated that as the blowing ratio is increased for a large tip gap, the tip cooling increased only slightly while the cooling to the shroud increased significantly. With an increased tip gap, the coolant was able to provide better downstream film effectiveness through increased mixing. For the smallest tip gap, the coolant was shown to impinge directly on the surface of the shroud, leading to high film effectiveness at the impingement point. Their predictions indicated that as the gap size increased, the coolant jets were unable to penetrate to the shroud.

C. AIRFOIL ENDWALLS

Endwall regions are another location associated with a turbine airfoil that is relatively difficult to cool because of the complex nature of the flowfield. Secondary flows, in the form of a leading edge and passage vortex,

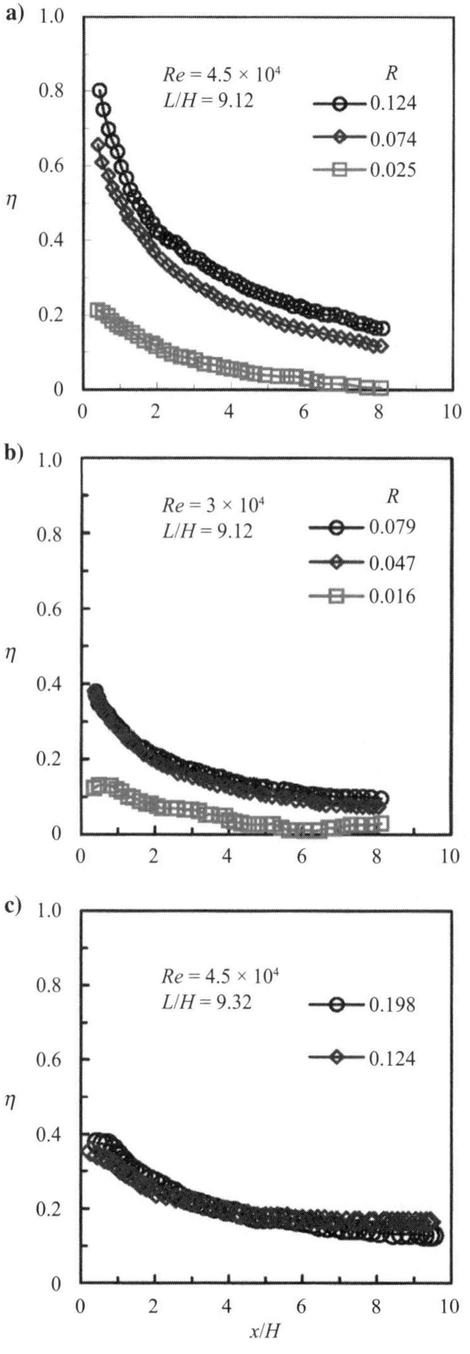

Fig. 23 Comparison of film-effectiveness levels for a simulated tip region using injection from a) a discrete slot, b) a round hole, and c) a pressure-side flared hole (figures reproduced from Kim et al. [69]). Geometries are given in Fig. 22.

Fig. 24 Contours of film effectiveness for film-cooled tips from pressure-side holes for a small (left) and large (right) tip gap with both using 0.68% coolant flow measured relative to the passage flow (from Christophel et al. [73]).

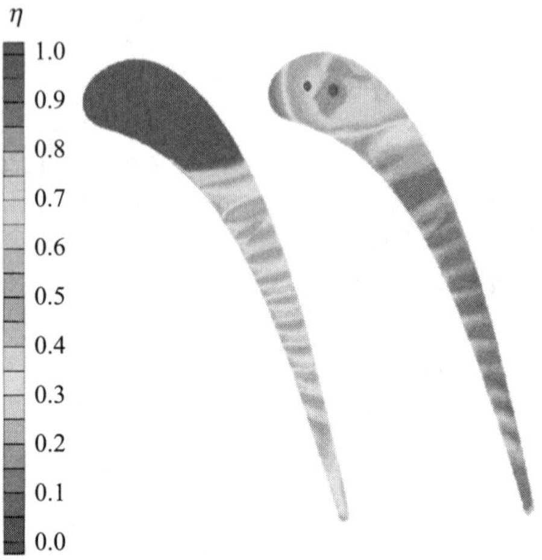

cause much of the coolant injected through the cooling holes to be swept off the endwall surface. There have been a number of studies documenting endwall film cooling and a number of studies documenting cooling from the leakage gap at the turbine-combustor junction. Coolant flow from leakage gaps that are either between the combustor and first vane or between vanes and blades can significantly contribute to cooling the endwalls and can affect the secondary flow pattern.

Detailed endwall film-cooling results have been conducted by Friedrichs et al.; measured results from two geometries from his studies are shown in Fig. 25 [1, 76, 77]. The results of the first study [76], which were all surface measurements or visualization, indicated a strong influence of the secondary flows on the film cooling and an influence of the film cooling on the secondary flows. The data showed that the angle at which the coolant leaves the hole did not dictate the coolant trajectory except near the hole exit. Furthermore, the endwall crossflow was altered so that the crossflow was turned toward the inviscid streamlines, which was caused by the film-cooling injection.

There have also been a few studies that have measured endwall heat transfer as a result of injection from a two-dimensional, flush slot just upstream of the vane. Blair [78] measured film-effectiveness levels and heat-transfer coefficients for a range of blowing ratios through a flush slot placed just upstream of the leading edges of his single passage channel. One of the key findings was that the endwall film-effectiveness distributions showed extreme variations across the vane gap, with much of the coolant being swept across the endwall toward the suction-side corner. Granser and Schulenberg [79] reported similar film-effectiveness results, in that higher values occurred near the suction side of the vane. Colban et al. [80, 81] also showed results for a geometry with a backward-facing step slot. Above the step, there was upstream coolant from simulated film-cooling holes while under the step there was relatively little coolant exiting the slot. Results indicated the presence of a tertiary vortex that developed in the vane passage as a result of a peaked total pressure profile in the near-wall

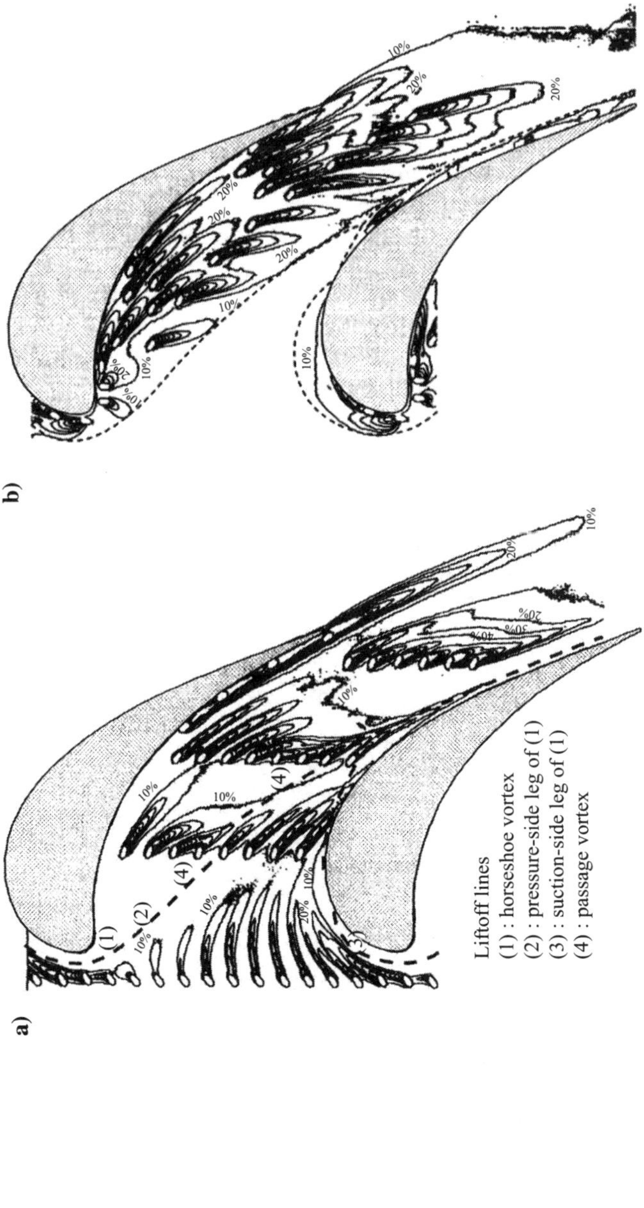

Fig. 25 Film-effectiveness levels for two different film-cooling hole patterns for an endwall as presented by Friedrichs et al. [76, 77].

region. For all of the conditions simulated, the effectiveness contours indicated that the coolant from the slot was swept towards the suction surface.

A series of experiments have been reported by Burd and Simon [82], Burd et al. [83], and Oke et al. [84, 85] for various injection configurations upstream of a nozzle guide vane with a contoured endwall. In these studies, coolant was injected from an interrupted, flush slot that was inclined at 45 deg just upstream of the vane. Like others, they found that most of the slot coolant was directed toward the suction side at low slot flow conditions. As they increased the percentage of slot flow to 3.2% of the exit flow, their measurements indicated that better coverage occurred between the airfoils. Similarly, Zhang and Moon[86] tested a two-row film-cooling configuration upstream of a contoured endwall. Upstream of these two rows of film-cooling holes was placed either a flush wall or a backward-facing step. Direct comparisons between these two configurations showed that measured effectiveness levels were reduced considerably in the case of the backward-facing step configuration. They attributed these reduced effectiveness levels to the increased secondary flows that were present.

The only studies to have combined an upstream slot with film-cooling holes in the passage of the vane were those of Kost and Nicklas [87], Nicklas [88], and Knost and Thole [89, 90]. One of the most interesting results from the Kost and Nicklas and Nicklas studies was that they found that for the slot flow alone, which was 1.3% of the passage mass flow, the horseshoe vortex became more intense. This increase in intensity resulted in the slot coolant being moved off the endwall surface and heat-transfer coefficients increasing to over three times that measured for no-slot-flow injection. They attributed the strengthening of the horseshoe vortex to the fact that for the no-slot injection the boundary layer was already separated, with fluid being turned away from

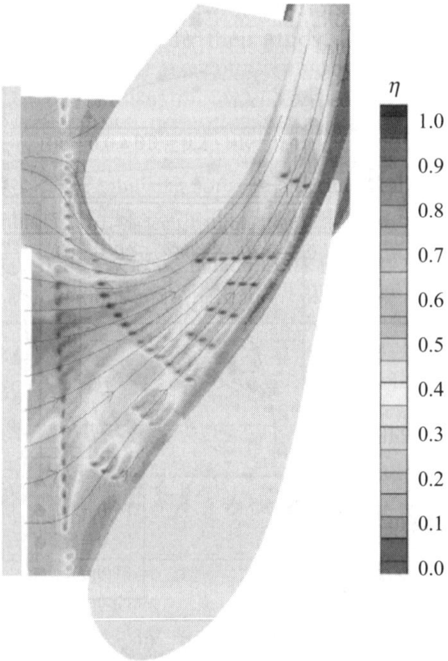

Fig. 26 Measured film-effectiveness levels for a 0.5% slot and 0.5% film-cooling flow where flow percentages are measured relative to the passage flow. Predicted streamlines in the near-wall region are also shown (from Knost and Thole [90]).

the endwall at the injection location. Given that the slot had a normal component of velocity, injection at this location promoted the separation and enhanced the vortex. Their film-effectiveness measurements indicated higher values near the suction side of the vane due to the slot coolant migration. In the studies presented by Knost and Thole [89, 90], the predicted and measured results indicated the presence of a warm ring on the endwall around the vane where no coolant was present despite the combined slot cooling and film cooling, as shown in Fig. 26. As one can see from these results, the film-cooling jet trajectories closely follow the near-wall streamlines in most regions. Their CFD predictions in the near-wall region showed distinct differences that were dependent upon the amount of slot flow exiting the upstream slot. Moreover, their studies indicated a difficulty in cooling the juncture between the pressure side of the vane and the endwall as well as the leading-edge region of the vane.

VII. CFD PREDICTIONS

Whether from stationary cascade or from engine rig film-cooling experiments, considerable expense and time are required to achieve high-quality measurements. Therefore, there has been and continues to be a strong desire to accurately predict the performance of new film-cooling schemes through computational fluid dynamics. There is a wealth of commercially available computational fluid dynamics (CFD) packages available to researchers and engine designers to simulate film cooling in a range of environments. As one might deduce, however, accurate film-cooling predictions are highly dependent upon the calculation of the mixing that occurs with the crossflow. The accuracy of the predictions is therefore highly dependent upon the turbulence model used in the near-wall region. Unlike heat-transfer predictions, in which the temperature gradient at the wall must be accurately predicted, predictions of adiabatic wall temperatures require accurate predictions of the jet trajectory and spreading, given that the temperature gradient at the wall is zero.

Several approaches have been presented in the literature for predicting film-effectiveness levels for film cooling. The most common approach to date is to use the Reynolds-averaged-Navier–Stokes (RANS) equations with some type of turbulence model. Higher-order modeling, such as large eddy simulations (LES) and direct numerical simulations (DNS), is limited by computer power for realistic Reynolds numbers for film-cooling applications. For RANS-type calculations, two-equation eddy-viscosity models, such as k-ε or k-ω. or a second moment closure scheme, such as a Reynolds stress model, are commonly used, each requiring some type of wall treatment. More accurate predictions can be achieved in some instances using a second-moment closure scheme relative to an eddy-viscosity model, but this is at the expense of increased computational time and equation stiffness. For the wall treatment, two approaches have commonly been used: wall function models or two-layer models. Although wall

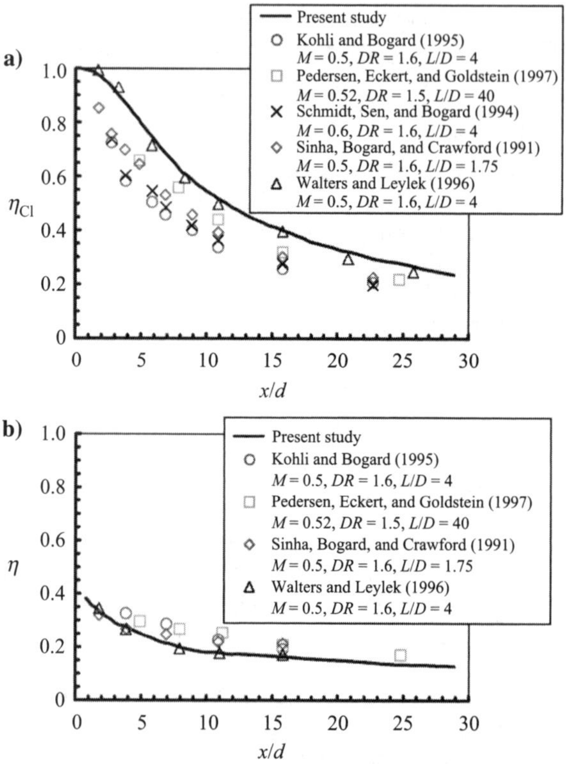

Fig. 27 Measured and predicted film-effectiveness levels for a round film-cooling hole at $M = 0.5$ for a) jet centerline and b) laterally averaged values (from Kohli and Thole [93]).

function models presume that the flow follows the log law near the wall, the two-layer model eliminates the use of wall functions and divides the flow into a viscosity-affected region and a fully turbulent region.

Although the spreading of the coolant is difficult to predict, lateral averages of film effectiveness are predicted relatively well for the case where the film-cooling jet is attached to the downstream surface. This was illustrated by the first paper reporting a full three-dimensional CFD prediction of a film-cooling jet by Leylek and Zerkle [91]. Although their first paper showed an overprediction of the cooling when the jet was attached, their predictions indicated a decay in the film effectiveness similar to that measured. Further refinements in the representation of the cooling hole geometry, grid generation, and discretization illustrated the importance of these factors, as even better predictions were achieved for an attached jet case [92]. Comparisons of predicted and measured centerline and laterally averaged effectiveness levels are given in Fig. 27 for a simple round cooling hole placed in a flat plate for $M = 0.5$. Predictions by two independent research groups [92, 93] using the same CFD code, including a two-equation k-ε turbulence model and wall functions, are shown in Fig. 27. Experiments were conducted by Sinha et al. [14], Pedersen et al. [8], and Schmidt et al. [18]. Although Fig. 27b illustrates a relatively good comparison of the measured and predicted values for the laterally averaged values, Fig. 27a illustrates that there is an overprediction of the effectiveness levels at the jet centerline. Although it is not shown here, the overprediction of the centerline values is compensated by an underprediction of the jet spreading, which results in reasonably predicted laterally averaged values of effectiveness.

The most difficult situation for a CFD model to accurately predict is the case where the cooling jet is separated from the wall. Unsuccessful attempts to predict film-effectiveness levels using a number of turbulence models for a separated jet were shown by Ferguson et al. [94]. The turbulence models evaluated include the standard k-ε with wall functions (KE-WF) and with nonequilibrium wall functions (KE-NE), the renormalization group (RNG) k-ε model with wall functions (RNG-WF) and with nonequilibrium wall functions (RNG-NE), Reynolds stress model with wall functions (RSM-WF) and with nonequilibrium wall functions (RSM-NE), and a k-ε model with a two-layer zonal model. These results showed essentially the same prediction with all turbulence models using wall functions, but a better prediction with the two-layer zonal model. Walters and Leylek [95] also found better predictions with the two-layer zonal model as compared to the wall functions as shown in Fig. 28, showing the centerline film-effectiveness levels for a simple cylindrical hole at a $M = 1$. Both predictions were considerably higher than the experiment.

As with film-cooling experiments, more recent CFD studies have moved towards predicting film effectiveness on actual airfoil geometries. As one would expect, the difficulties are compounded by the fact that airfoil curvature and pressure gradients both have a profound effect on the film effectiveness. Moreover, depending upon where the jets are located, particularly on the suction side of the airfoil, the curvature effects can lead to jet separation even at low blowing ratios. Buck et al. [96], Walters et al. [97], Ferguson et al. [98],

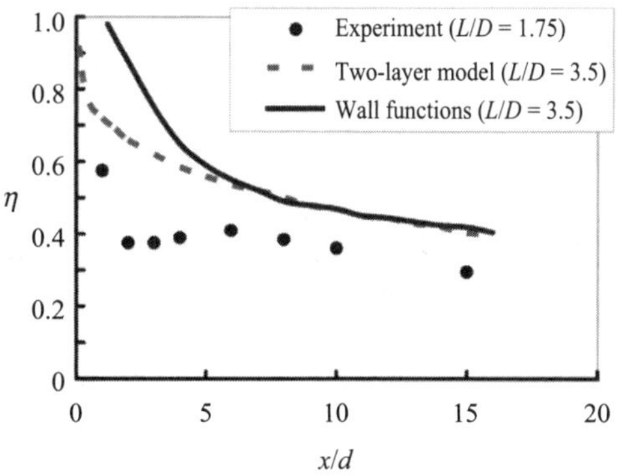

Fig. 28 Comparison of centerline film-effectiveness levels for a number of turbulence models for $M = 1$ with a round film-cooling hole on a flat plate (figure reproduced from Walters and Leylek [95]).

and McGrath et al. [99] had a series of papers that reported on a combined experimental and computational study for a number of different film-cooling hole shapes that were simulated on curved surfaces representing airfoil pressure and suction surfaces. They used a two-layer zonal model in conjunction with the RNG k-ε turbulence model. Given that many of the hole geometries were intended to have an attached jet, the comparisons between the experimental results and computational results generally showed agreement.

VIII. CONCLUSION

As described in the preceding sections, gas turbine airfoil film cooling is influenced by a wide range of variables. The dominant film-cooling configuration used for cooling turbine airfoils and endwalls is rows of discrete coolant holes, and this configuration has been the primary focus of this review. Particular emphasis was placed on identifying which variables have a significant effect and which do not. In each case we have tried to provide an explanation for the effect on the film-cooling performance based on the physical description of the interaction between coolant jets and mainstream.

Film-cooling performance is quantified using the film effectiveness, heat-transfer coefficients, and net heat-flux reduction. A full understanding of the performance requires all three of these parameters. In many cases the film effectiveness dominates, and many studies focus on this measure alone. In some cases, however, the improved film effectiveness is offset by increases in heat-transfer coefficient, which leads to poorer net heat-flux reduction. One example of this is compound angle injection, which provides distinctly improved film effectiveness but ultimately provides a net heat-flux reduction that is equal to or poorer than that for streamwise-oriented holes.

To evaluate the effects of the many variables that affect film-cooling performance, most studies have used relatively simple laboratory test models to isolate the effects of different variables. Although this is appropriate in order to obtain an understanding of the basic physics of the effects of different variables, one should not lose sight of the complicated nature of the actual operating environment for the turbine airfoils. For example, most studies of film-cooling performance have used facilities with relative low mainstream turbulence levels, particularly prior to 1996. As noted in the section on high freestream turbulence effects, the optimum momentum flux ratio for coolant jets is an order of magnitude larger for high freestream turbulence levels as compared to low freestream turbulence levels. Consequently, many of the results found under conditions of low freestream turbulence have to be reevaluated when considering actual turbine operating conditions.

Ultimately, the film-cooling performance is closely linked to whether the coolant jet has separated from the surface. For nominal conditions of a flat surface, low freestream turbulence, and cylindrical holes, the film-cooling performance is reasonably predictable with empirical correlations. Surface curvature, high freestream turbulence, and shaping of the hole exit can, however, greatly change film-cooling performance by significantly affecting the blowing ratio at which the coolant jet separates. Although this review has given indications of how large these effects can be, at this time there are insufficient data to fully characterize the effects of curvature, freestream turbulence level and length scale, and hole shape. CFD predictions, though very useful in providing insight in the spatial details of the film-cooling process, are also limited by the very complex flow conditions that occur for film cooling, particularly when the coolant jets begin to separate. Consequently, the film-cooling performance for actual turbine conditions is often difficult to predict precisely, and this remains a major constraint in the design for the durability of the turbine section of gas turbine engines.

REFERENCES

[1] Friedrichs, S., Hodson, H. P., and Dawes, W. N., "The Design of an Improved Endwall Film-Cooling Configuration," *Journal of Turbomachinery*, Vol. 121, No. 4, 1999, pp. 772–780.

[2] Gritsch, M., Schulz, A., and Wittig, S., "Adiabatic Wall Effectiveness Measurements of Film-Cooling Holes with Expanded Exits," *Journal of Turbomachinery*, Vol. 120, No. 3, 1998, p. 549.

[3] Pietryzk, J. R., Bogard, D. G., and Crawford, M. E., "Effects of Density Ratio on the Hydrodynamics of Film Cooling," *ASME Journal of Turbomachinery*, Vol. 112, No. 3, 1990, pp. 437–450.

[4] Han, J. C., Dutta, S., and Ekkad, S. V., *Gas Turbine Heat Transfer and Cooling Technology*, Taylor and Francis, New York, 2000.

[5] Goldstein, R. J., "Film Cooling," *Advances in Heat Transfer*, edited by T. Irvine and J. P. Hartnett, Academic Press, New York, 1971, pp. 321–379.

[6] Teekaram, A., Forth, C., and Jones, T., "The Use of Foreign Gas to Simulate the Effects of Density Ratios in Film Cooling," *Journal of Turbomachinery*, Vol. 111, No. 1, 1989, pp. 57–62.

[7] Papell, S. S., "Effect on Gaseous Film Cooling of Coolant Injection Through Angled Slots and Normal Holes," NASA TN-D-299, Sept. 1960.

[8] Pedersen, D. R., Eckert, E., and Goldstein, R., "Film Cooling with Large Density Differences Between the Mainstream and the Secondary Fluid Measured by the Heat-Mass Transfer Analogy," *ASME Journal of Heat Transfer*, Vol. 99, No. 4, 1977, pp. 620–627.

[9] Baldauf, S., Scheurlen, M., Schulz, A., and Wittig, S., "Correlation of Film-Cooling Effectiveness from Thermographic Measurements at Enginelike Conditions," *Journal of Turbomachinery*, Vol. 124, No. 4, 2002, pp. 686–698.

[10] Hartnett, J. P., Birkebak, R. C., and Eckert, E. R. G., "Velocity Distributions, Temperature Distributions, Effectiveness, and Heat Transfer for Air Injected Through a Tangential Slot into a Turbulent Boundary Layer," *Journal of Heat Transfer*, Vol. 83, No. 3, 1961, pp. 293–305.

[11] Baldauf, S. M., Schulz, A., and Wittig, S., "High-Resolution Measurements of Local Effectiveness from Discrete Hole Film Cooling," *Journal of Turbomachinery*, Vol. 123, No. 4, 2001, pp. 758–765.

[12] Thole, K. A., Sinha, A., Bogard, D. G., and Crawford, M. E., "Mean Temperature Measurements of Jets with a Crossflow for Gas Turbine Film Cooling Application," *Rotating Machinery Transport Phenomena*, edited by J. H. Kim and W. J. Yang, Hemisphere, New York, 1992, pp. 69–85.

[13] Mouzon, B. D., Terrell, E. J., Albert, J. E., and Bogard, D. G., "Net Heat Flux Reduction and Overall Effectiveness for a Turbine Blade Leading Edge," American Society of Mechanical Engineers, Paper GT2005-69002, June 2005.

[14] Sinha, A., Bogard, D., and Crawford, M., "Film Cooling Effectiveness Downstream of a Single Row of Holes with Variable Density Ratio," *Journal of Turbomachinery*, Vol. 113, No. 3, 1991, pp. 442–449.

[15] Cutbirth, J., and Bogard, D., "Effects of Coolant Density Ratio on Film Cooling," American Society of Mechanical Engineers, Paper GT2003-38582, June 2003.

[16] Ethridge, M., Cutbirth, J., and Bogard, D., "Scaling of Performance for Varying Density Ratio Coolants on an Airfoil with Strong Curvature and Pressure Gradients Effects," *Journal of Turbomachinery*, Vol. 123, No. 2, 2000, pp. 1–7.

[17] Baldauf, S., Scheurlen, M., Schulz, A., and Wittig, S., "Heat Flux Reduction from Film Cooling and Correlation of Heat Transfer Coefficients from Thermographic Measurements at Enginelike Conditions," *Journal of Turbomachinery*, Vol. 124, No. 4, 2002, pp. 699–709.

[18] Schmidt, D., Sen, B., and Bogard, D., "Film Cooling with Compound Angle Holes: Adiabatic Effectiveness," *Journal of Turbomachinery*, Vol. 118, No. 4, 1996, pp. 807–813.

[19] Foster, N. W., and Lampard, D., "The Flow and Film Cooling Effectiveness Following Injection Through a Row of Holes," *Journal of Engineering for Power*, Vol. 102, 1980, pp. 584–588.

[20] Han, J. C., and Mehendale, "Flat-Plate Film Cooling with Steam Injection Through One Row and Two Rows of Inclined Holes," *Journal of Turbomachinery*, Vol. 108, No. 1, 1986, pp. 137–144.

[21] Cho, H. H., and Goldstein, R. J., "Heat (Mass) Transfer and Film Cooling Effectiveness with Injection Through Discrete Holes: Part II-On the Exposed Surface," *Journal of Turbomachinery*, Vol. 117, No. 3, 1995, pp. 451–460.

[22] Mayle, R. E., and Camarata, F. J., "Multihole Cooling Film Effectiveness and Heat Transfer," *Journal of Heat Transfer*, Vol. 97, No. 4, 1975, pp. 534–538.

[23] Sasaki, M., Takahara, K., Kumagai, T., and Hamano, M., "Film Cooling Effectiveness for Injection from Multirow Holes," *Journal of Engineering for Power*, Vol. 101, No. 1, 1979, pp. 101–108.

[24] Harrington, M., McWaters, M., Bogard, D. G., Lemmon, C., and Thole, K., "Full Coverage Film Cooling with Short Normal Injection Holes," *Journal of Turbomachinery*, Vol. 123, No. 4, 2001, pp. 798–805.

[25] Kelly, G. B., and Bogard, D. G., "An Investigation of the Heat Transfer for Full Coverage Film Cooling," American Society of Mechanical Engineers, Paper GT2003-38716, June 2003.

[26] Kohli, A., and Bogard, D., "Adiabatic Effectiveness, Thermal Fields, and Velocity Fields for Film Cooling with Large Angle Injection," *Journal of Turbomachinery*, Vol. 119, No. 2, 1997, pp. 352–358.

[27] Sen, B., Schmidt, D., and Bogard, D., "Film Cooling with Compound Angle Holes: Heat Transfer," *Journal of Turbomachinery*, Vol. 118, No. 4, 1996, pp. 800–806.

[28] Schmidt, D. L., and Bogard, D. G., "Effects of Free-Stream Turbulence and Surface Roughness on Laterally Injected Film Cooling," *ASME Proceedings of the 32nd National Heat Transfer Conference*, HTD-Vol. 350, Vol. 12, edited by K. Vafai and J. L. S. Chen, Baltimore, 1997, pp. 233–244.

[29] Saumweber, C., Schulz, A., and Wittig, S., "Free-Stream Turbulence Effects on Film Cooling with Shaped Holes," *Journal of Turbomachinery*, Vol. 125, No. 1, 2003, pp. 65–73.

[30] Dittmar, J., Schulz, A., and Wittig, S., "Assessment of Various Film Cooling Configurations Including Shaped and Compound Angle Holes Based on Large Scale Experiments," *Journal of Turbomachinery*, Vol. 125, No. 1, 2003, pp. 57–64.

[31] Kadotani, K., and Goldstein, R., "Effect of Mainstream Variables on Jets Issuing from a Row of Inclined Round Holes," *Transactions of the ASME*, Vol. 101, No. 2, 1979, pp. 298–304.

[32] Goldstein, R., Eckert, E., Eriksen, V., and Ramsey, J., "Film Cooling Following Injection Through Inclined Circular Tubes," *Israel Journal of Technology*, Vol. 8, No. 1–2, 1970, pp. 145–154.

[33] Goldstein, R., Eckert, E., and Ramsey, J., "Film Cooling with Injection Through a Circular Hole," NASA CR-54604, May 1968.

[34] Eriksen, V. L., and Goldstein, R., "Heat Transfer and Film Cooling Following Injection Through Inclined Circular Tubes," *ASME Journal of Heat Transfer*, Vol. 96, No. 1, 1974, pp. 239–245.

[35] Liess, C., "Experimental Investigation of Film Cooling with Injection from a Row of Holes for the Application to Gas Turbine Blades," *Journal of Engineering for Power*, Vol. 97, No. 1, 1975, pp. 21–27.

[36] Ito, S., Goldstein, R., and Eckert, E., "Film Cooling of a Gas Turbine Blade," *Journal of Engineering for Power*, Vol. 100, No. 3, 1978, pp. 476–481.

[37] Goldstein, R., Schwarz, S., and Eckert, E., "The Influence of Curvature on Film Cooling Performance," *Journal of Turbomachinery*, Vol. 112, No. 3, 1990, pp. 472–478.

[38] Teekaram, A., Forth, C., and Jones, R., "Film Cooling in the Presence of Mainstream Pressure Gradients," *Journal of Turbomachinery*, Vol. 113, No. 3, 1991, pp. 484–492.

[39] Brown, A., and Saluja, C. L., "Film Cooling from a Single Hole and a Row of Holes of Variable Pitch to Diameter Ratio," *International Journal of Heat and Mass Transfer*, Vol. 22, No. 4, 1979, pp. 525–533.

[40] Schmidt, D. L., and Bogard, D. G., "Pressure Gradient Effects on Film Cooling," American Society of Mechanical Engineers, Paper 95-GT-18, June 1995.

[41] Launder, B., and York, J., "Discrete-Hole Cooling in the Presence of Free Stream Turbulence and Strong Favorable Pressure Gradient," *International Journal of Heat and Mass Transfer*, Vol. 17, No. 11, 1974, pp. 1403–1409.

[42] Kadotani, K., and Goldstein, R., "On the Nature of Jets Entering a Turbulent Flow Part A—Jet-Mainstream Interaction," *Journal of Engineering for Power*, Vol. 101, No. 3, 1979, pp. 466–470.

[43] Kadotani, K., and Goldstein, R., "On the Nature of Jets Entering a Turbulent Flow Part B—Film Cooling Performance," *Journal of Engineering for Power*, Vol. 101, No. 3, 1979, pp. 466–470.

[44] Rivir, R., Jumper, G., and Elrod, W., "Film Cooling Effectiveness in High Turbulence Flow," *Journal of Turbomachinery*, Vol. 113, No. 3, 1991, pp. 479–483.

[45] Schmidt, D. L., and Bogard, D. G., "Effects of Free-Stream Turbulence and Surface Roughness on Film Cooling," American Society of Mechanical Engineers, Paper 96-GT-462, June 1996.

[46] Juhany, K. A., Hunt, M. L., and Sivo, J. M., "Influence of Injectant Mach Number and Temperature on Supersonic Film Cooling," *Journal of Thermophysics and Heat Transfer*, Vol. 8, No. 1, 1994, pp. 59–67.

[47] Bons, J., Rivir, R., MacArthur, C., and Pestian, D., "The Effect of Unsteadiness on Film Cooling Effectiveness," AIAA Paper 95-0306, June 1995.

[48] Seo, H. J., Lee, J. S., and Ligrani, P. M., "The Effect Injection Hole Length on Film Cooling with Bulk Flow Pulsations," *International. Journal of Heat and Mass Transfer*, Vol 41, No. 22, 1998, pp. 3515–3528.

[49] Dring, R., Blair, M., and Joslyn, H., "An Experimental Investigation of Film Cooling on a Turbine Rotor Blade," *Journal of Engineering for Power*, Vol. 102, No. 1, 1980, pp. 81–87.

[50] Abhari, R., and Epstein, A., "An Experimental Study of Film Cooling in a Rotating Transonic Turbine," *Journal of Turbomachinery*, Vol. 116, No. 1, 1994, pp. 63–70.

[51] Takeishi, K., Matsuura, M., Aoki, S., and Sato, T., "An Experimental Study of Heat Transfer and Film Cooling on Low Aspect Ratio Turbine Nozzles," *Journal of Turbomachinery*, Vol. 112, No. 3, 1990, pp. 488–496.

[52] Rigby, M. J., Johnson, A. B., and Oldfield, M. L. G., "Gas Turbine Rotor Blade Film Cooling with and Without Simulated NGV Shock Waves and Wakes," American Society of Mechanical Engineers, Paper 90-GT-78, June 1990.

[53] Bons, J. P., Taylor, R., McClain, S., and Rivir, R. B., "The Many Faces of Turbine Surface Roughness," *ASME Journal of Turbomachinery*, Vol. 123, No. 4, 2001, pp. 739–748.

[54] Bogard, D. G., Schmidt, D. L., and Tabbita, M., "Characterization and Laboratory Simulation of Turbine Airfoil Surface Roughness and Associated Heat Transfer," *Journal of Turbomachinery*, Vol. 120, No. 2, 1998, pp. 337–342.

[55] Goldstein, R. J., Eckert, E. R. G., Chiang, H. D., and Elovic, E., "Effect of Surface Roughness on Film Cooling Performance," *Journal of Engineering for Gas Turbines and Power*, Vol. 107, No. 1, 1985, pp. 111–116.

[56] Schmidt, D. L., Sen, B., and Bogard, D. G., "Effects of Surface Roughness on Film Cooling," American Society of Mechanical Engineers, Paper 96-GT-299, June 1996.

[57] Bogard, D. G., Snook, D., and Kohli, A., "Rough Surface Effects on Film Cooling of the Suction Side Surface of a Turbine Vane," American Society of Mechanical Engineers, Paper EMECE2003-42061, Nov. 2003.

[58] Rutledge, J. L., Robertson, D., and Bogard, D. G., "Degradation of Film Cooling Performance on a Turbine Vane Suction Side due to Surface Roughness," American Society of Mechanical Engineers, Paper GT2005-69045, June 2005.

[59] Polanka, M. D., Witteveld, V. C., and Bogard, D. G., "Film Cooling Effectiveness in the Showerhead Region of a Gas Turbine Vane Part I: Stagnation Region and Near Pressure Side," American Society of Mechanical Engineers, Paper 99-GT-048, June 1999.

[60] Witteveld, V. C., Polanka, M. D., and Bogard, D. G., "Film Cooling Effectiveness in the Showerhead Region of a Gas Turbine Vane Part II: Stagnation Region and Near Suction Side," American Society of Mechanical Engineers, Paper 99-GT-049, June 1999.

[61] Cutbirth, J. M., and Bogard, D. G., "Thermal Field and Flow Visualization Within the Stagnation Region of a Film Cooled Turbine Vane," *ASME Journal of Turbomachinery*, Vol. 124, No. 2, 2002, pp. 200–206.

[62] Cutbirth, J. M., and Bogard, D. G., "Evaluation of Pressure Side Film Cooling with Flow and Thermal Field Measurements, Part I: Showerhead Effects," *ASME Journal of Turbomachinery*, Vol. 124, No. 4, 2002, pp. 670–677.

[63] Ames, F. E., "Aspects of Vane Film Cooling with High Turbulence—Part I: Heat Transfer," *Journal of Turbomachinery*, Vol. 120, No. 4, 1998, pp. 768–776.

[64] Mehendale, A. B., and Han, J. C., "Influence of High Mainstream Turbulence on Leading Edge Film Cooling Heat Transfer," *Journal of Turbomachinery*, Vol. 114, No. 4, 1992, pp. 707–715.

[65] Reiss, H., and Bölcs, A., "Experimental Study of Showerhead Cooling on a Cylinder Comparing Several Configurations Using Cylindrical and Shaped Holes," *Journal of Turbomachinery*, Vol. 122, No. 1, 2000, pp. 161–169.

[66] Albert, J. E., Cunha, F., and Bogard, D. G., "Adiabatic and Overall Effectiveness for a Film Cooled Blade," American Society of Mechanical Engineers, Paper GT2004-53998, June 2004.

[67] Bunker, R. S., "A Review of Turbine Blade Tip Heat Transfer," *Turbine 2000 Symposium on Heat Transfer in Gas Turbine Systems*, Cesme, Turkey, 2000; also *Annals of the New York Academy of Sciences*, Vol. 934, Aug. 2000, pp. 64–79.

[68] Kim, Y. W., and Metzger, D. E., "Heat Transfer and Effectiveness on Film Cooled Turbine Blade Tip Models," *Journal of Turbomachinery*, Vol. 117, No. 1, 1995, pp. 12–21.

[69] Kim, Y. W., Downs, J. P., Soechting, F. O., Abdel-Messeh, W., Steuber, G., and Tanrikut, S., "A Summary of the Cooled Turbine Blade Tip Heat Transfer and Film Effectiveness Investigations Performed by Dr. D. E. Metzger," *Journal of Turbomachinery*, Vol. 117, No. 1, 1995, pp. 1–11.

[70] Kwak, J. S., and Han, J. C., "Heat Transfer Coefficient and Film-Cooling Effectiveness on a Gas Turbine Blade Tip," American Society of Mechanical Engineers, Paper GT2002-30194, June 2002.

[71] Kwak, J. S., and Han, J. C., "Heat Transfer Coefficient and Film-Cooling Effectiveness on the Squealer Tip of a Gas Turbine Blade," American Society of Mechanical Engineers, Paper GT2002-30555, June 2002.

[72] Ahn, J., Mhetras, S., and Han, J. C., "Film-Cooling Effectiveness on a Gas Turbine Blade Tip Using Pressure Sensitive Paint," American Society of Mechanical Engineers, Paper GT2004-53249, June 2004.

[73] Christophel, J. R., Thole, K., and Cunha, F., "Cooling the Tip of a Turbine Blade Using Pressure Side Holes—Part 1: Film Effectiveness Measurements," *Journal of Turbomachinery*, Vol. 127, No. 3, 2005, pp. 270–277.

[74] Acharya, S., Yang, H., Ekkad, S. V., Prakash, C., and Bunker, R., "Numerical Simulation of Film Cooling Holes on the Tip of a Gas Turbine Blade," American Society of Mechanical Engineers, Paper GT-2002-30553, June 2002.

[75] Hohlfeld, E. M., Christophel, J. R., Couch, E. L., and Thole, K. A., "Predictions of Cooling from Dirt Purge Holes Along the Tip of a Turbine Blade," American Society of Mechanical Engineers, Paper GT2003-38251, June 2003.

[76] Friedrichs, S., Hodson, H. P., and Dawes, W. N., "Distribution of Film-Cooling Effectiveness on a Turbine Endwall Measured Using the Ammonia and Diazo Technique," *Journal of Turbomachinery*, Vol. 118, No. 4, 1996, pp. 613–621.

[77] Friedrichs, S., Hodson, H. P., and Dawes, W. N., "Aerodynamic Aspects of Endwall Film-Cooling," *Journal of Turbomachinery*, Vol. 119, No. 4, 1997, pp. 786–793.

[78] Blair, M. F., "An Experimental Study of Heat Transfer and Film Cooling on Large-Scale Turbine Endwalls," *Journal of Heat Transfer*, No. 4, Nov. 1974, pp. 524–529.

[79] Granser, D., and Schulenberg, T., "Prediction and Measurement of Film Cooling Effectiveness for a First-Stage Turbine Vane Shroud," American Society of Mechanical Engineers, Paper 90-GT-95, June 1990.

[80] Colban, W. F., Thole, K. A., and Zess, G., "Combustor-Turbine Interface Studies: Part 1: Endwall Measurements," *Journal of Turbomachinery*, Vol. 125, No. 2, 2003, pp. 193–202.

[81] Colban, W. F., Lethander, A. T., Thole, K. A., and Zess, G., "Combustor-Turbine Interface Studies: Part 2: Flow and Thermal Field Measurements," *Journal of Turbomachinery*, Vol. 125, No. 2, 2003, pp. 203–209.

[82] Burd, S. W., and Simon, T. W., "Effects of Slot Bleed Injection over a Contoured Endwall on Nozzle Guide Vane Cooling Performance: Part I: Flow Field Measurements," American Society of Mechanical Engineers, Paper 2000-GT-199, June 2000.

[83] Burd, S. W., Satterness, C. J., and Simon, T. W., "Effects of Slot Bleed Injection over a Contoured Endwall on Nozzle Guide Vane Cooling Performance: Part II Thermal Measurements," American Society of Mechanical Engineers, Paper 2000-GT-200, June 2000.

[84] Oke, R., Simon, T., Burd, S. W., and Vahlberg, R., "Measurements in a Turbine Cascade Over a Contoured Endwall: Discrete Hole Injection of Bleed Flow," American Society of Mechanical Engineers, Paper 2000-GT-214, June 2000.

[85] Oke, R., Simon, T., Shih, T., Zhu, B., Lin, Y. L., and Chyu, M., "Measurements over a Film-Cooled, Contoured Endwall with Various Coolant Injection Rates," American Society of Mechanical Engineers, Paper 2001-GT-140, June 2001.

[86] Zhang, L., and Moon, H. K., "Turbine Nozzle Endwall Inlet Film Cooling – The Effect of a Back-Facing Step," American Society of Mechanical Engineers, Paper GT-2003-38319, June 2003.

[87] Kost, F., and Nicklas, M., "Film-Cooled Turbine Endwall in a Transonic Flow Field: Part I – Aerodynamic Measurements," *Journal of Turbomachinery*, Vol. 123, No. 4, 2001, pp. 709–719.

[88] Nicklas, M., "Film-Cooled Turbine Endwall in a Transonic Flow Field: Part II – Heat Transfer and Film-Cooling Effectiveness Measurements," *Journal of Turbomachinery*, Vol. 123, No. 4, 2001, pp. 720–729.

[89] Knost, D. K., and Thole, K. A., "Computational Predictions of Endwall Film-Cooling for a First Stage Vane," American Society of Mechanical Engineers, Paper GT-2003-38252, June 2003.

[90] Knost, D. G., and Thole, K. A., "Adiabatic Effectiveness Measurements of Endwall Film -Cooling for a First Stage Vane," American Society of Mechanical Engineers, Paper GT2004-52236; also *Journal of Turbomachinery*, Vol. 127, No. 2, 2005, pp. 297–305.

[91] Leylek, J. H., and Zerkle, R. D., "Discrete-Jet Film Cooling: A Comparison of Computational Results with Experiments," *Journal of Turbomachinery*, Vol. 116, No. 3, 1994, pp. 358–368.

[92] Walters, D. K., and Leylek, J. H., "A Systematic Computational Methodology Applied to a Three-Dimensional Film-Cooling Flowfield," *Journal of Turbomachinery*, Vol. 119, No. 1, 1997, pp. 777–785.

[93] Kohli, A., and Thole, K. A., "A CFD Investigation on the Effect of Entrance Flow Conditions in Discrete Film Cooling Holes," *ASME Proceedings of the 32nd National Heat Transfer Conference*, Vol. 12, edited by K. Vafai and J. L. S. Chen, Baltimore, 1997, pp. 223–232.

[94] Ferguson, J., Walters, D., and Leylek, J., "Performance of Turbulence Models and Near-Wall Treatments in Discrete Jet Film Cooling Simulations," American Society of Mechanical Engineers, Paper 98-GT-438, June 1998.

[95] Walters, D., and Leylek, J., "A Detailed Analysis of Film-Cooling Physics: Part I—Streamwise Injection with Cylindrical Holes," *Journal of Turbomachinery*, Vol. 122, No. 1, 2000, pp. 102–112.

[96] Buck, F. A., Walters, D., Ferguson, J., McGrath, E., and Leylek, J., "Film Cooling on a Modern HP Turbine Blade Part I: Experimental and Computational Methodology and Validation," American Society of Mechanical Engineers, Paper GT2002-30470, June 2002.

[97] Walters, D. K., Leylek, J. H., and Buck, F. A., "Film Cooling on a Modern HP Turbine Blade Part II: Compound Angle Round Holes," American Society of Mechanical Engineers, Paper GT2002-30613, June 2002.

[98] Ferguson, J. D., Leylek, J. H., and Buck, F. A., "Film Cooling on a Modern HP Turbine Blade Part III: Axial Shaped Holes," American Society of Mechanical Engineers, Paper GT2002-30522, June 2002.

[99] McGrath, E. L., Leylek, J. H., and Buck, F. A., "Film Cooling on a Modern HP Turbine Blade Part IV: Compound Angle Shaped Holes," American Society of Mechanical Engineers, Paper GT2002-30521, June 2002.

CHAPTER 6

Endwall Aerodynamics and Heat Transfer

Terrence W. Simon*
University of Minnesota, Minneapolis, Minnesota

Justin D. Piggush†
Applied Systems Engineering, Trane Ingersoll Rand, La Crosse, Wisconsin

This chapter discusses turbine passage three-dimensional flow aerodynamics and endwall heat transfer. It presents recent advances in the description of endwall flows and cooling techniques that have led to improved endwall aerothermal design. Important to this topic are documentation of passage secondary flows and the development of methods for managing them to allow more effective cooling. Over the years, much attention has been given to characterizing and reducing aerodynamic losses associated with passage secondary flows. We have learned that the endwall region flowfield is influenced by 1) stagnation zones established as the endwall boundary-layer flow approaches and meets the airfoil leading edges, 2) curvature of the passages, 3) steps and gaps on the endwall surface ahead of and within the passage, 4) leakage and coolant flows introduced through the endwall surfaces ahead of and within the passage, 5) tip leakage flows between the blades and adjoining shroud segments in the rotor/endwall region, and many more effects.

Recent combustor redesigns have flattened the combustor exit, or turbine inlet, temperature profile, and have raised the turbine inlet temperatures. Temperatures can be extremely high, approaching 1600°C. The net effects are a significant rise in near-endwall passage flow temperatures and greater thermal loading on the endwall and airfoil-to-endwall junction area. This, coupled with a continued need to improve engine durability and availability, has spurred strong interest in a better understanding of the endwall region thermal field and more detailed descriptions of heat-transfer coefficient distributions on the endwall and airfoil near-endwall surfaces. This information can be used to effect new designs with improved thermal control. Thus, the topic of this chapter is particularly important to the continued improvement of efficient and durable modern gas turbine engines.

The turbine section of a gas-turbine engine presents designers with many challenges. For example, the flow through the engine can create large forces on

*Ernst G. Eckert Professor of Mechanical Engineering.
†Mechanical Engineer; justin.piggush@trane.com.

surfaces, particularly those that turn the flow. To maximize engine performance and durability and cut engine losses, care must be taken to guide the fluid through turbine passages in such a way that secondary flow adverse effects are kept to a minimum. The need to cool turbine surfaces and provide sealing flows to leakage paths compounds the problem; cooling and sealing designs must consider the effects on secondary flow, passage aerodynamics, and passage component heat transfer. Consideration of these features in the endwall region constitutes the focus of this chapter.

I. ENDWALL AERODYNAMICS

Early efforts, including those described in 1954 by Herzig et al. [1], identified a dominant passage flow that crosses from the pressure surface to the suction surface in the endwall boundary-layer fluid (see Fig. 1), driven by the pressure difference between the pressure and suction surfaces. The size and strength of this flow, known as the passage secondary flow, are dependent on the amount of turning of the mainstream. Another important study of passage flows considers three-dimensional separation of flow at the junction between a protruding body and a wall. The flow ahead of the junction has a velocity gradient (and therefore a dynamic pressure gradient) normal to the endwall because of the presence of an endwall boundary layer. When the flow stagnates at an airfoil's leading edge, the total pressure gradient becomes an endwall-normal pressure gradient. Boundary-layer fluid on the protruding body, driven by this pressure gradient, is forced toward the endwall, where it migrates upstream slightly as it is rolled into a

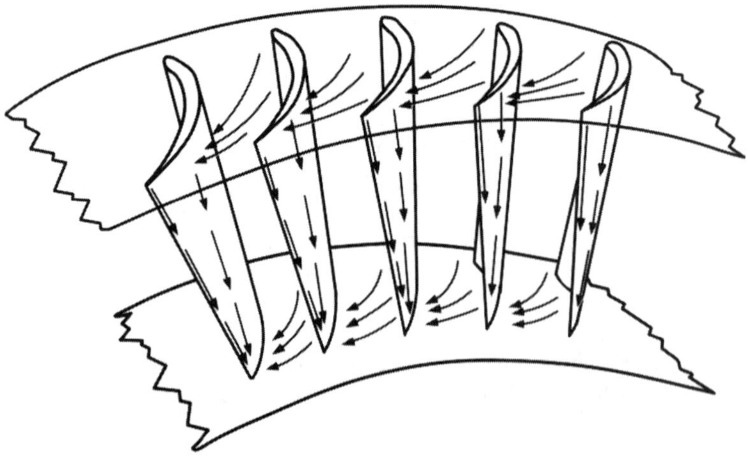

Fig. 1 Endwall secondary flows (Herzig et al. [1]).

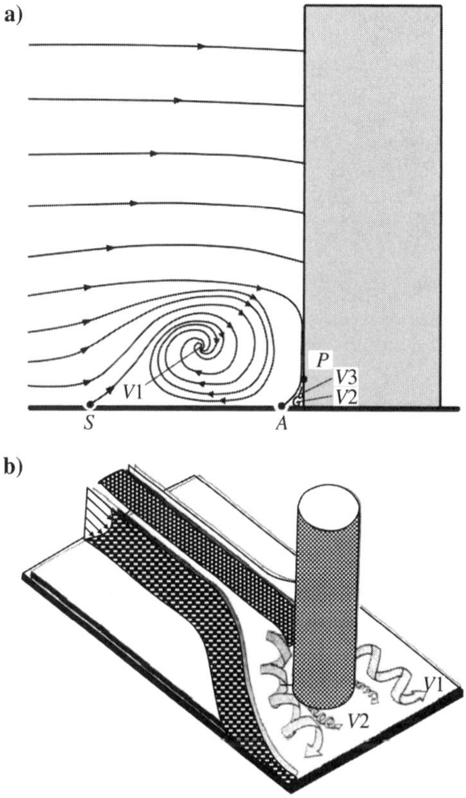

Fig. 2 Secondary flow structure at the juncture between a right circular cylinder and an endwall: a) horseshoe vortex at the plane of symmetry and b) incoming boundary layer and trailing vortices. S and P are separation points, A is an attachment point, and V indicates a particular vortex (from Goldstein and Karni [5]).

vortex upstream of the leading edge (see Fig. 2). This vortex is commonly referred to as the horseshoe vortex. Detailed studies of this flow are available in Pierce and Harsh [2], Eckerle and Langston [3], Pierce and Shin [4], and Goldstein and Karni [5].

The interaction of these two main secondary flows, the passage flow and the horseshoe vortex, along with smaller secondary flows, such as corner vortices that occur in the corner between adjoining walls and the vortices induced by the horseshoe vortex, make the passage secondary flowfield quite complex. Many studies that highlight the main features of the passage flowfield have been presented, including those by Langston et al. [6] and Langston [7](see Fig. 3). They describe the location of a single separation saddle point on the endwall. It is just downstream of the inlet plane and 20 to 50% of the pitch distance toward the suction side (see Fig. 4). Between the saddle point and the junction between the airfoil leading edge and the endwall, another horseshoe vortex similar to those just described is created. One side of the vortex traverses the passage moving toward the suction surface of the neighboring airfoil. Langston's group proposed that the passage flow merges with the pressure side leg of the horseshoe vortex, greatly augmenting this leg's strength as it moves through the passage.

Detailed passage measurements by Langston et al. [6] indicate that the inlet boundary layer separates and a new boundary layer forms within the passage downstream of the separation line. The observed boundary layer at the throat of the passage is consequently very thin. In a low-Reynolds-number study in

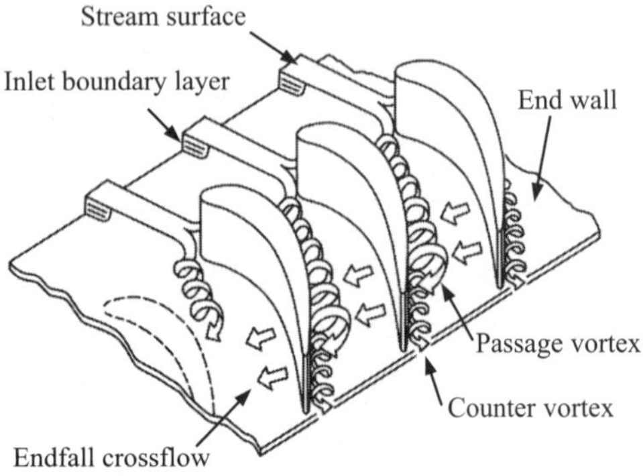

Fig. 3 Passage secondary flows (Langston [7]).

which colored smoke was used to mark the flow, Sieverding and Van den Bosche [8] observed the secondary flowfield previously measured. In addition to causing augmentation and migration of the pressure-side leg of the horseshoe vortex toward the suction side, the passage flow causes the whole of the stream surface to rotate (see Fig. 5). This rotation, now referred to as the passage vortex, tends to entrain flow into the horseshoe vortex structure and increases the portion of the passage affected by secondary flows.

Kawai et al. [9] clarified the passage secondary flow picture using carefully executed endwall oil-film shear-stress direction studies. In addition to identifying all major separation and reattachment lines within the passage, they confirmed many of the observations made by Langston and added new information. The single separation saddle point was confirmed. The pressure side of the horseshoe vortex was observed to roll up into the passage vortex, an Oseen vortex at its core. The suction side of the horseshoe vortex was shown to move up the suction surface and to dissipate as it moves through the passage. Important corner vortices on the pressure side, beginning at the airfoil leading edge, and on the suction side, beginning where the pressure-side horseshoe vortex separation line intersects the suction side of the airfoil, were also noted. This secondary flow picture was confirmed by Chung and Simon [10] and Wang et al. [11]. Additionally, a small but strong vortex was observed in the flow beginning at the intersection of the pressure-side horseshoe vortex with the suction side of the airfoil. It resides above the passage vortex and rotates in an opposite sense. Also, time dependency of the horseshoe vortex was noted. At some instances, the two vortices merge into one as they move through the passage. Otherwise, only a single horseshoe vortex is present. From these works, and others, a fairly clear picture of the secondary flow structure for a straight (noncontoured) endwall has been developed (see Fig. 6). Several review articles dealing with cascade secondary flows are available, including those of Langston [12] and Sieverding [13].

A more recent study by Holley et al. [14] provides experimental and computational data on pressure and shear-stress distributions in a cascade with straight endwalls to demonstrate the computational closure presently achievable. They discuss prospects for accurate loss prediction. Tallman et al. [15] presented

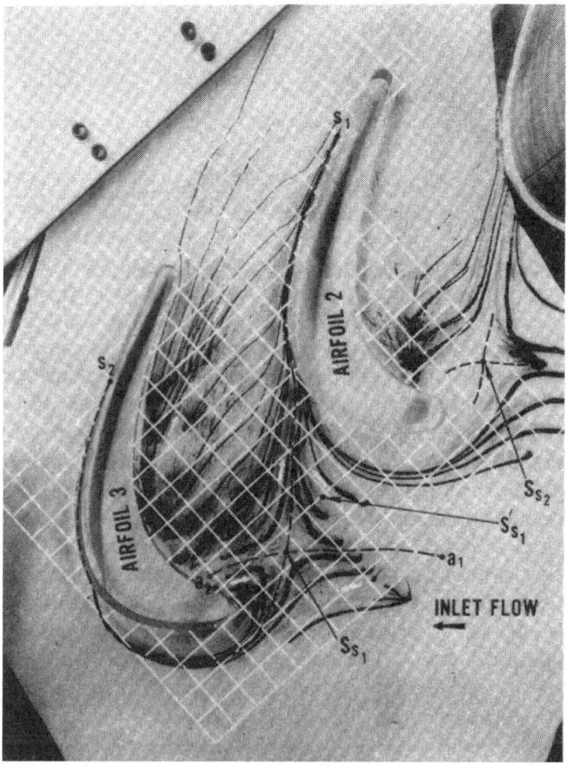

Fig. 4 Endwall secondary flows within a rotor passage. Saddle points are indicated as S_{S2}, separation lines are S_1-S_{S1} and S_2-S_{S1}, and attachment lines are a_1-S_{S1} and a_2- S_{S1} (Langston et al. [6]).

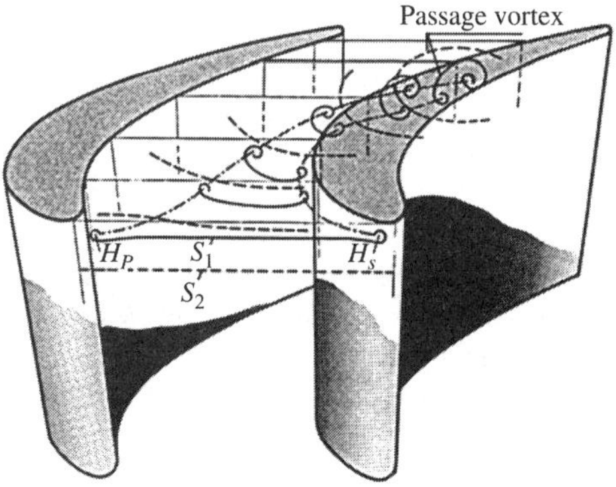

Fig. 5 Secondary flows according to Sieverding and Van den Bosche [8]. The S denotes a stream surface.

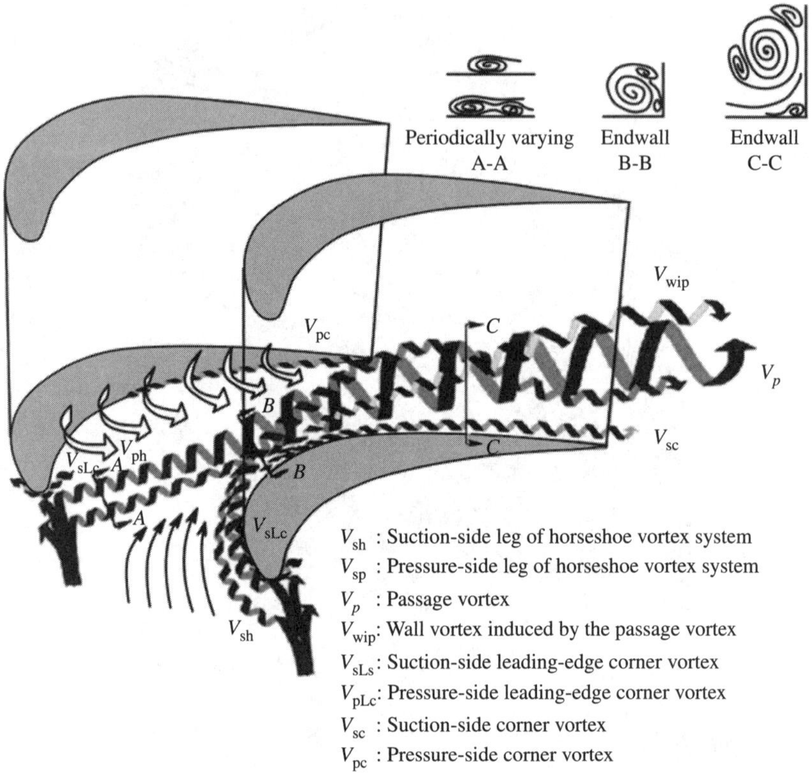

Fig. 6 Turbine passage secondary flows (Wang et al. [11]).

numerical results to show that they can effectively model the endwall region and reproduce experimental results. Pullen et al. [16] showed that endwall secondary flow could be reduced by reconfiguring the nozzle guide vane airfoil so that it is more aft-loaded.

Important to secondary flow pattern in the endwall region are the characteristics of the approach flow. Ames et al. [17] showed measurements of the effects of approach flow turbulence on secondary flow and endwall heat transfer. Kunze et al. [18] and Barringer and coworkers [19, 20] discussed an inlet profile generator for studying the effects of combustor-generated nonuniformities of velocity and temperature. They experimented with such nonuniformities to document their effects on turbine vane heat transfer. Yamada et al. [21] discussed the effects of convected wakes from upstream airfoils on passage secondary flows. They noted that another secondary flow, which is counter-rotating against the passage vortex, is periodically generated by the stator wake passing through the rotor passage.

A. CONTOURED ENDWALL AERODYNAMICS

One method used for minimizing secondary flow losses is to accelerate the endwall boundary-layer fluid as it approaches the airfoil leading-edge plane, or as it moves through the passage, or both. The favorable pressure gradient tends to thin the endwall boundary layer and reduce secondary flow strength. This can be done by contouring one or both endwalls of the passage. Contouring can be either axisymmetric or nonaxisymmetric. Though the cascades discussed next are mostly linear cascades, the term "axisymmetric" is used with reference to the engine geometry.

Several studies have used axisymmetric endwall contouring. One of the initial studies was that of Deich et al. [22], who investigated stages of very low inlet aspect ratios (0.29–0.54) and large-area contractions. The reduction of loss inspired the work of others on cascades of larger aspect ratios. These include the geometry of Morris and Hoare [23], who found that axisymmetric contouring of one endwall could significantly reduce losses in the vane stage, particularly for passages of low aspect ratio. They found that most of the loss reduction was near the noncontoured (straight) wall. They surmised that loss reduction was the result of a general redistribution of the airfoil pressure profile as a result of endwall contouring. Flow curvature on the contoured endwall might have prevented realizing the large loss reduction observed on the noncontoured endwall. They included cases with nonaxisymmetric profiling that were generally unsuccessful. For those nonaxisymmetric cases, losses near the nonprofiled wall were reduced, and losses on the profiled wall were significantly increased. Endwall curvature of the nonaxisymmetric profile contorted the airfoil wake and created a thick region of high loss near the profiled endwall. This loss distribution is quite different from those observed with axisymmetric profiling. Their study showed that care must be taken in endwall profiling, for the consequences of contouring are difficult to predict.

Morris and Hoare's study was with low Mach number; a later study at higher Mach numbers by Kopper et al. [24] confirmed a reduction of secondary losses by contouring. Studies by Boletis [25] and Arts [26] presented measurements and numerical analyses documenting momentum deficits caused mainly by the legs of the horseshoe vortex and the passage flow. Dossena et al. [27] performed a similar study. They noted that the vortex structure on the flat endwall is similar to that for a straight-walled cascade, though the secondary flow strength is reduced near the flat endwall. They stated that "on the profiled endwall, the contraction inhibits the formation of a proper passage vortex and its migration toward midspan; this is the result of intense vortex stretching due to the local acceleration ... " [27]. The study of Burd and Simon [28] characterizes the flowfield of a cascade with endwall contouring. Streamwise and cross-stream velocities, turbulence and other velocity fluctuations, Reynolds shear stresses, total pressure losses, and turbulence kinetic energy losses are all presented and discussed.

The complex profiles of nonaxisymmetric contoured endwall geometries are typically designed with the assistance of computational fluid dynamics (CFD). Representative studies in this category include those of Rose [29], Harvey et al. [30], Hartland et al. [31], Yan et al. [32], Brennan et al. [33], and Rose et al. [34]. The first study, that of Rose [29], was designed to reduce the pitchwise pressure gradients at the exit of the nozzle guide vanes. Uniform static pressure in this region allows better distribution of rim seal coolant at the junction of the nozzle guide vane and blade section. Rose found that some nonaxisymmetric endwall profiles were successful in reducing static-pressure nonuniformity by as much as 70%. A pair of studies by Harvey et al. [30] and Hartland et al. [31] outline a computational design method and experimentally check the results for a nonaxisymmetric profile specifically designed to reduce exit angle nonuniformities. The experimental study confirmed the expected reduction in angle nonuniformities and also produced a 30% reduction in secondary losses at the exit plane, which was not predicted by computation. The studies indicate that properly designed, nonaxisymmetric contoured endwalls lead to decreased secondary losses and reduced variations of flow deviation angles across the passage exit. Several studies have addressed the performance of three-dimensional contoured endwalls: Saha and Acharya [35], Saha et al. [36], and Gustafson et al. [37]. They found that they could reduce pitchwise pressure gradients and shear stresses in the endwall region, weakening the horseshoe vortex. Schobeiri et al. [38] demonstrated experimentally that a three-dimensional, bowed blade design can reduce secondary flow losses.

B. OTHER METHODS OF ENDWALL MODIFICATION

Secondary losses can be reduced also through use of a boundary-layer fence [39–41]. The study of Kawai et al. [39] investigated fences of various heights and positions on the endwall. Their conclusion was that appropriately sized and positioned fences can be used to greatly affect secondary flows within the passage. Flow underturning, secondary kinetic energy, and the thickness of the region of secondary flow vorticity can all be reduced. Studies of Chung et al. [40] and Chung and Simon [41] make use of a fence located on the endwall at midpitch of the passage. The fence was shown to obstruct the migration of the passage flow and turn the pressure side of the horseshoe vortex so that its axis is more in line with the freestream. This removes a mechanism that is responsible for continued, rapid vortex growth.

In unfenced passages, circulation of the pressure leg of the horseshoe vortex is augmented (after it departs from the pressure side of the passage and begins to cross to the suction side) by the skewed endwall boundary-layer flow. The near-endwall portion of the endwall boundary-layer flow moves toward the suction surface whereas the flow in the upper reaches of the endwall boundary layer moves in the direction of the main flow within the passage. The pressure side of the horseshoe vortex embedded in this boundary layer is thus intensified by the high shear component normal to the horseshoe vortex axis acting on the top of the vortex. This is all altered by the fence. When the fence lifts the vortex up into the main flow within the passage and turns the axis of the vortex in the direction of the main flow, the mechanism that augments the vortex strength is removed. As a result, the vortex is weaker when it reaches the downstream end of the passage. It is also displaced off both the endwall and suction surfaces. The magnitude of the passage vortex at the exit plane is greatly reduced by an appropriately positioned fence. The displacement of the vortex off the endwall and the suction surface reduces its augmentation of wall heat-transfer rates and improves the opportunity for film cooling of the surfaces. Other fence designs make use of a small step at approximately the midpassage location (see Kinnear et al. [42]) to produce vortices with rotation counter to the horseshoe vortex. In general, fences show good aerodynamic performance, reduced secondary losses, and decreased strength of the pressure leg of the horseshoe vortex. Unfortunately, the aerodynamic and heat-transfer benefits of having a fence are offset by problems associated with the high local heat-transfer rates on the obstructions (fences) inserted into the engine gas path and excessive surface temperatures. Another method of reducing passage secondary flow, which bears some similarity to the boundary-layer fence studies, makes use of jets located at approximately the midpitch line of the passage. The jets are designed to divert the pressure leg of the horseshoe vortex so that the mainstream flow can carry it downstream. The study of Aunapu et al. [43] indicates some success using this technique. Migration of the pressure side of the horseshoe vortex is retarded although the vortex is observed to not be significantly weakened. The overall effect is an increase in the passage secondary losses because of the added turbulence generated by the interaction of the passage flow and the jets.

C. OFF-DESIGN AND TIME-DEPENDENT AERODYNAMICS

Benner et al. [44] experimentally documented secondary flows under off-design conditions in a turbine cascade. They found that a larger leading-edge diameter gave larger secondary flow losses at off-design conditions, contrary to present design philosophy. They also found that secondary flow losses increase with incidence angle up to 10 deg, but do not increase further for larger incidence angles, to 20 deg. Dossena et al. [45] measured the effects on secondary flowfields and losses of changing the incidence angle, the pitch-chord ratio, and the downstream Mach number in a steam turbine linear cascade. They noted that the incidence angle and the pitch-chord ratio were the most influential. Rehder and Dannhauer [46] introduced leakage upstream of a vane leading edge with two different injection orientations. The first was perpendicular to the mainstream. This method strengthened the horseshoe vortex and amplified the passage vortex, as might be expected. In the second, leakage was injected tangentially to the mainstream. This tended to reduce the horseshoe vortex. When leakage rates were higher, such as 2.0% leakage mass flow rate, the vortex was removed altogether, and heat transfer was reduced.

Schlienger et al. [47] took time-resolved measurements of the flowfield in a rotor and downstream stator passage to document the effects on the stator's secondary flowfield due to convection of the rotor's secondary flowfield through the downstream stator's passage. A time-resolved description of the events was given, and it was noted that the redistribution of high-loss fluid from the wakes and vortices at the rotor hub influences a large portion of the stator's flow area.

D. STUDIES THAT INCLUDE ENDWALL BLOWING (FILM COOLING)

There is a wealth of recent literature on the effects of film cooling and leakage flow on endwall region flow. Effects on the secondary flow of blowing through flat endwalls are first reviewed. One of the pioneering studies investigating the effects of endwall blowing and interaction of secondary flows with film-cooling flows was by Blair [48]. Passage flows and the passage vortex within the cascade were the features of primary importance in this study. The endwall boundary layers were removed just upstream of the vane leading edges, and without a fully developed, turbulent boundary layer entering the passage, it was possible to trip the flow to influence the location of transition. The location of transition on the endwall and the location of film-cooling injection through a slot upstream of the vane leading edge had little effect on the passage vortex. Heat-transfer data indicate that coolant flow is swept by the passage flow across from the pressure side to the suction side. Secondary flows within this cascade were not affected by the introduction of coolant flow.

A study by Granser and Schulenberg [49] indicates that coolant injected from a slot tends to reduce secondary flows by reenergizing the boundary layer. Injection of coolant at a small angle to the endwall surface adds significant streamwise momentum to the boundary layer to retard its growth downstream of the injection slot. Thinner endwall boundary layers produce less intense horseshoe vortices and weaker passage flows. Without a well-developed boundary layer, this effect was not seen in Blair's study.

Other studies focus on the complete secondary flowfield within the passage. Goldman and McLallin [50] found that coolant injection could have a significant effect, decreasing both passage loss and flow angle nonuniformity. A later study by Sieverding and Wilputte [51] discusses data taken with two double rows of discrete-hole injection within the passage and a double row upstream of the leading edge. They conclude that the effects of coolant air injection on secondary flows are more pronounced than the effects documented in the Blair study. Reductions in losses and exit angle nonuniformity along the airfoil axis were confirmed. They suggested that injection of cooling air should be included in an optimal design, noting three important parameters that should be considered: "the coolant-to-mainstream total pressure ratio, the coolant mass flow ratio, and the angles between the coolant flow, main flow, and endwall boundary layer flow."

Work by Bario et al. [52] describes the aerodynamics of jets entering the main flow through the endwall of a turbine cascade. Cooling flow was shown to reduce secondary flow effects. Exit flow angles near the cascade endwalls were reduced with cooling jets. Harasgama and Burton [53] corroborate the findings of Sieverding and Wilputte [51] and Bario et al. [52]. They note (as does Blair [48]) that cooling fluid does little for the pressure-side trailing-edge region because much of it is convected toward the suction side of the passage by the passage flow. Biesinger and Gregory-Smith [54] note the positive effects on loss reduction of a skewed boundary layer at the inlet of an axial-flow compressor, and then proceed to study similar effects within turbine rotor blading. Their cascade was designed so that coolant air is injected with momentum in the direction of the pressure surface to simulate the turbine rotor. The results of the study indicate that low blowing rates tend to thicken the boundary layer, resulting in greater secondary flows and higher losses. At higher rates of injection, the streamwise vorticity of the coolant counteracts that of the secondary flow and secondary kinetic energy is reduced. At very high rates of blowing, the vorticity of the coolant flow persists to the exit, and secondary losses increase. The effects of changing blowing angle were also recorded. The study shows that a lower angle of blowing, 20 deg, is more effective than a higher angle, 34 deg (both measured relative to the surface plane), possibly because cooling fluid does not separate from the wall when the injection angle is small. The study showed also that no net gain on aerodynamic performance is achieved when the energy needed to inject the coolant is included in the thermodynamic availability analysis.

Two studies performed by Friedrichs et al. [55, 56] give a detailed description of the interaction of cooling flows with secondary flows. Surface flow visualization in the first study [55] clearly shows that coolant injection through discrete holes located within and ahead of the passage changes the location of separation lines within the passage. The separation line for the horseshoe vortex appears closer to the leading edge while the separation line of the pressure leg of the horseshoe vortex as it crosses to the suction side appears further downstream. Flow from holes located upstream of the passage and 30% of an axial chord downstream of the leading-edge plane was observed to have the largest effects on secondary flows and was successful in delaying separation, reducing overturning at the passage exit and reducing losses associated with secondary flows. Flow from holes located at 60 and 90% x/Cax downstream of the airfoil leading edges did not reduce secondary flows. Coolant from the holes at 90% x/Cax was observed to thicken the exit boundary layer. The second study [56] varied injection rates from the holes. When injection ratios are high enough that the coolant stagnation pressure is higher than the freestream stagnation pressure, the cooling flow reenergizes the boundary layer, thereby reducing secondary flows and subsequent mixing associated with these flows. The optimum coolant supply pressure gives the coolant a streamwise velocity component similar to that of the freestream.

A study by Liu et al. [57] used discrete holes placed upstream of the leading edge. In contrast to the Biesinger and Gregory-Smith study, they did not attempt to simulate a skewed inlet boundary layer. The results of this experiment generally confirm those of Biesinger and Gregory-Smith [54], with several added conclusions: 1) a decrease of the distance from the injection site to the leading edge or a decrease of the hole inclination angle reduces secondary flows, 2) attachment of the coolant can be improved through the use of a double row of injection holes, and 3) a forced passage vortex with rotation counter to that within the cascade could be created with high blowing ratios.

The geometry used in the study of Roy et al. [58] included three coolant injection ports located ahead of the leading edge of each vane. The coolant thus covered the area directly in front of each vane, though not the area between the vanes. The coolant from these ports suppresses the formation of the junction corner vortices.

Other important studies include those of Kost and Nicklas [59] and Knost and Thole [60]. Both studies investigate the combined effects of injection from slots and holes. The first notes a strengthening of the horseshoe vortex when fluid is injected from an upstream slot. They suggest that this is because of the unique position of their slot relative to the saddle point of the flow on the endwall of the stator passage. In their study, the slot is placed just upstream of the saddle point. Injection is directly into the vortex and therefore increases the quantity of low momentum fluid that can be entrained by the horseshoe vortex. They also theorize that the wall-normal component of the injected flow increases the circulation of the horseshoe vortex. They suggest that moving the slot closer to the leading edge can actually decrease circulation of the vortex. A slight reduction in vortex strength was achieved at low blowing rates by Georgiou et al. [61] using a slot that wraps around the leading edge of a bluff body, but this configuration is impractical in an engine. Kost and Nicklas therefore recommend placing the slot farther upstream, ahead of the saddle point where the streamwise component of the injected flow can reenergize the boundary layer ahead of the vane leading edge. The study also notes that film cooling tends to increase turbulence near the wall, which can enhance heat transfer. Coolant injected from the slot was the major contributor to coolant concentration measured within the passage. Coolant emerging from the holes tended to have a weaker overall effect and only affected the cooling situation near injection holes.

Knost and Thole [60] positioned their slot farther upstream. There was not a large increase in secondary flow with this slot configuration. Film-cooling holes were also included on the endwall within the passage. The interaction of cooling flow from the slot and the cooling flow from the holes was found to affect secondary flow differently from injection from the slot only or from the holes only.

Optimal placement of cooling holes and cooling slots is difficult to determine because performance is sensitive to passage geometry, passage static-pressure distribution, and secondary flow structure, but some important progress on the subject has been made. Clearly, secondary flow strength can be affected by leakage flows. Though an additional loss penalty is incurred because of the added turbulence, the cooling benefits can be significant, improving the overall performance of the engine. Liu et al. [57], Lapworth et al. [62], and Oke et al. [63] experimented with introduction of film coolant ahead of a nozzle. With high momentum film injection, secondary flow strength can be suppressed, providing better thermal protection and reduced aerodynamic losses.

Lampart et al. [64] computed the effects of tip leakage flow at a rotor stage on the endwall boundary-layer secondary flows in a downstream vane row passage. Effects of leakage flow extraction from the passage to the gap and leakage flow injection into the passage from the gap were considered. The tip leakage flow intensifies secondary flows in the endwall region of the downstream stator. A combined experimental and computational study by Paniagua et al. [65] in a high-pressure, transonic turbine documents the effects of leakage through the hub between the stator and rotor stages, and they noted a large blockage of the vane exit flow caused by leakage. They also noted flow unsteadiness, with ingestion and ejection of this leakage flow driven by varying pressure fields corresponding to relative positions of rotor blade to stator vane. The effect of leakage of coolant flow on the rotor is an enhanced migration of secondary flow from the pressure surface to the suction surface and up the suction surface toward the midspan. Gaetani et al. [66] computed steady and unsteady flows in the rotor-stator gap.

II. HEAT TRANSFER

Heat-transfer rates on the endwall are directly related to the structure of endwall secondary flows. Large local variations in heat-transfer coefficients often result from vortices of varying intensities scouring the walls of the passage. Many studies have documented the distribution of heat-transfer coefficients using various methods. Early work includes the studies of Blair [48], who used an array of thermocouples and small heaters (monitoring the power required to maintain each heater at a given temperature); Graziani et al. [67], who used a similar method but with higher resolution; York et al. [68], who monitored heat fluxes through a nickel wall using thermocouples placed on either side of the nickel; and Gaugler and Russell [69].

Blair's study indicates that increased heat transfer can be found near the leading edges of the vanes, the result of the roll-up of the horseshoe vortex. Though the study notes several other trends in endwall heat transfer, the spatial resolution is not sufficient to capture the finer points. Graziani et al. [67] improved spatial resolution to allow a much more complete picture of endwall heat transfer (see Fig. 7). Upstream of the cascade, the boundary layer is essentially two-dimensional, and contours of Stanton number (dimensionless heat transfer coefficient) are parallel to the leading-edge plane. The leading-edge region experiences high heat-transfer rates because of the horseshoe vortex, as noted by Blair. The leading-edge region shows a distinct wedge-shaped area approximately defined by the leading-edge plane, the suction-side leading-edge separation line, and the separation line of the pressure-side leg of the horseshoe vortex. The heat-transfer rates in this area remain approximately equal to those of the incoming boundary layer. Just downstream of the separation line of the pressure-side leg of the horseshoe vortex, a sharp decrease in heat-transfer rates is apparent, and a region of low heat-transfer rates extending all of the way to the trailing edge is formed. Because the inlet boundary layer has been swept up into the horseshoe vortex, a new boundary layer, driven by the cross-passage pressure gradient, must be formed. Heat transfer and secondary flow phenomena in the throat region are complex and apparently depend on inlet boundary-layer thickness. Regardless, a spot of high heat-transfer rates persistently exists in the wake region behind the trailing-edge plane. Stanton numbers remain essentially uniform downstream of the cascade.

Subsequent studies by York et al. [68] and Gaugler and Russell [69] corroborate the work of Blair and Graziani et al. Gaugler and Russell suggest that flow of fluid toward the endwall exists in the vane wake, explaining the region of high heat-transfer rates downstream of the vane trailing edge. Kumar et al. [70] present correlations to determine average heat-transfer rates within a passage. They are segmented into five separate regions. Goldstein and Spores [71] display excellent spatial resolution using mass-transfer sublimation measurements. Their study clearly shows the effects on the endwall of secondary flows (see Fig. 8). Many of the regions of interest identified by Graziani et al. were visualized; interestingly, the wedged-shaped region where heat-transfer rates remain approximately equal to those of the incoming boundary layer was not found to extend as far into the passage as originally suggested by Graziani et al. The effects of corner vortices were seen clearly in the Goldstein and Spores study, and a region of enhanced mass (heat) transfer near the leading-edge plane was identified. The improved spatial resolution indicates that the region of high mass (heat) transfer in the airfoil wake zone is composed of

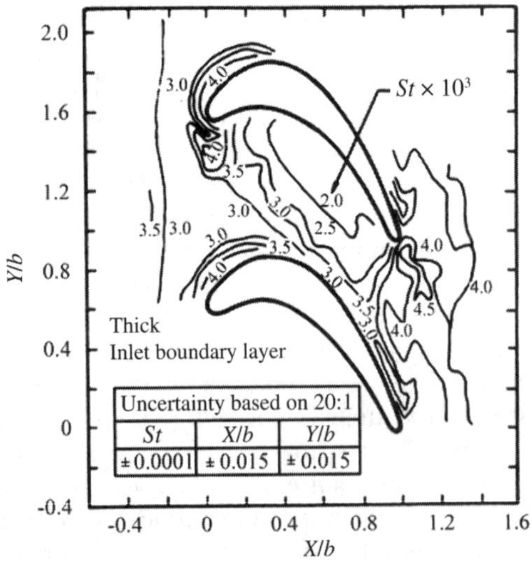

Fig. 7 Endwall Stanton numbers according to Graziani et al. [67].

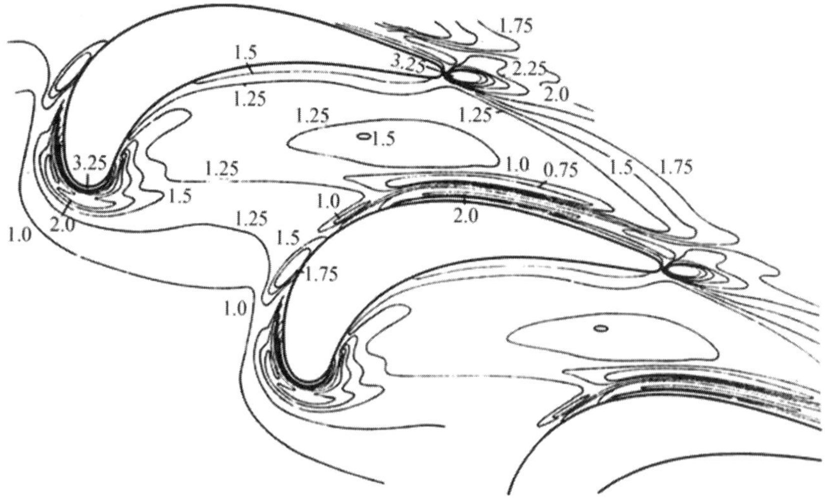

Fig. 8 Mass-transfer increase over flat-plate values for a turbine rotor cascade (Goldstein and Spores [71]). Displayed is *St* scaled on the local *St* value that would exist were there no airfoils.

two peaks. The larger of the two peaks is the result of strong recirculating wakes. The lesser peak, which always resides nearer the suction surface, might be the product of the strong suction-side corner vortex, which continues past the trailing edge and interacts with the wake to produce a region of strong vorticity.

A. EFFECT OF TURBULENCE ON HEAT TRANSFER

It is important to note the effects of turbulence on endwall heat transfer in cascade passages. Engine combustors typically produce high turbulence intensity values and large turbulence length scales. Efforts to understand the effects of turbulence structure on endwall flows continue. In general, higher freestream turbulence can be expected to promote an earlier transition to turbulence in the endwall boundary layer, which will cause an increase in endwall heat transfer. However, large turbulence length scales exhibit lower rates of heat transfer as compared to smaller length scales of turbulence at similar turbulence intensity values [72]. A study by Thole et al. [73] indicates that for high freestream turbulence levels the horseshoe vortex moves closer to the leading edge of the vane. This is because of the turbulence flattening the inlet boundary-layer profile, creating higher velocities near the wall and a shorter region of wall-normal pressure gradient on the leading edge.

Overall, higher turbulence levels act to increase heat transfer throughout the passage, with less of an effect seen near the trailing-edge region. Lee et al. [74] report such results. Their study indicates that separation lines move further downstream when turbulence levels are higher. Turbulence raises heat-transfer levels throughout the passage, but with less of an effect near the leading edge and the trailing edge. Thus, the heat load on the endwall is more uniform. The overall increase in heat transfer was 27%, which compares well with the 25% increase reported by Thole et al. [73]. Ames et al. [75] indicate that high free-stream turbulence causes a weaker augmentation of heat-transfer coefficients at higher Reynolds numbers. Additional studies that characterize the near-endwall flowfield include the work of Kang et al. [76], Kang and Thole [77], and Radomsky and Thole [78]. Sveningsson and Davidson [79] computed secondary flows and heat transfer in a stator vane passage to find that the $v2-f$ turbulence closure model yields better comparisons with experimental results than predictions computed using the realizable $k-\varepsilon$ turbulence closure model.

B. MACH-NUMBER INFLUENCE

Many of the studies reviewed in this paper were collected on low-speed turbine vane or rotor cascades. Engine-representative Mach numbers were not reproduced. Here we discuss cases with Mach-number effects. The study of Bassi and Perdichizzi [80] indicates that Mach number has an appreciable effect on the secondary flow structure in that the passage vortex is shifted towards the endwall. The overall loss, however, is not significantly affected by Mach-number changes. The work of Hermanson and Thole [81] indicates that the subsonic flowfield is similar to the flowfield under transonic flow conditions for the portion of the passage that is upstream of the shock location. A study by Giel et al. [82] investigates the difference between sonic and transonic flow on heat-transfer coefficients on the endwall. They note that increased Mach numbers tend to decrease heat-transfer rates.

C. SURFACE ROUGHNESS EFFECTS

Surface roughness changes endwall heat-transfer rates. Studies by Blair [83] and Guo et al. [84] address the subject. Blair found that a rough wall increases heat-transfer rates over the whole endwall region. Roughness was credited with causing early transition to turbulent flow of the endwall boundary layer. A corresponding increase in heat-transfer rates was observed. Guo et al. observed increased heat-transfer rates caused by roughness, but noted that the heat-transfer coefficient patterns remained unaffected by a change in roughness.

D. HEAT TRANSFER IN BLOWN CASCADES

The introduction of cooling air along vane surfaces has for some time been standard practice in the turbine industry. The use of injected flow through vane endwalls for cooling purposes is relatively new; it became an area of greater interest when recent turbine inlet temperature profiles from the combustor became more flat, and more aggressive endwall cooling became necessary. Several important studies on the subject are reviewed next. One of the first studies to include film cooling on the endwall was that of Blair [48]. Blair's study made use of a large-scale test section with coolant injection through a single slot running the full pitch of the passage. The test section included an endwall boundary-layer bleed slot just ahead of the coolant injection slot, and so the coolant was not injected into a mature boundary layer. Results of the study showed that film-cooling effectiveness is not uniform over the full pitch. Effectiveness near the suction side is high, whereas that near the pressure side is low because the passage flow sweeps the coolant fluid along with the endwall boundary-layer fluid from the pressure side to the suction side. Heat-transfer measurements performed in the cascade show a slight decrease in heat-transfer coefficients with added coolant injection. This was attributed to thickening of the boundary layer by the addition of low-streamwise-momentum coolant flow.

A study by Takeishi et al. [85] made use of a single row of holes located at the inlet plane, a double row of holes approximately midway between the leading edge and the passage throat, and a double row of holes at the passage throat. They noted that passage crossflows move coolant toward the suction side of the passage, reducing film-cooling effectiveness on the pressure side of the endwall. This confirms the observations of Blair [48]. A second zone of low film-cooling effectiveness is found on the suction side near the leading edge. Here, strong roll-up of the endwall boundary layer into a horseshoe vortex removes coolant fluid from the endwall. The authors conclude that passage secondary flows have a strong effect on heat transfer and film cooling within the passage.

Granser and Schulenberg's study [49] identifies the importance of the momentum flux ratio between the injected flow and the main flow in determining both the film-cooling effectiveness and the reduction of potential secondary flow strength. For momentum flux ratios less than unity, coolant does not penetrate to the mainstream, and film-cooling effectiveness increases as the thermal capacity of the injected fluid increases. At higher injection ratios, the film-cooling effectiveness varies with both the momentum flux ratio and the blowing ratio (velocity ratio). The study indicates no gain in additional film-cooling effectiveness at momentum flux ratios higher than 2.5. The studies of Harasgama and Burton [53, 86] and Jabbari et al. [87]confirm the results of the previous studies.

Effectiveness values for film cooling from individual holes were investigated by Freidrichs et al. [56]. The holes were just upstream of the leading edge and at 30, 60, and 90% of the axial chord in the passage. Flow from the coolant hole just upstream of the leading edge tends to produce greater effectiveness values near the suction side of the passage and reduces effectiveness values near the pressure side between the liftoff line of the horseshoe vortex and vane leading edge. Similarly, flow from holes at 30% of an axial chord length downstream of the leading-edge plane produces good film-cooling effectiveness values near the suction surface. Flow from a hole near the pressure surface performs well while flow from a hole located under the liftoff line of the pressure side of the horseshoe vortex performs poorly. The latter creates a noncooled region between rows of holes. Cooling by flow from holes at 60% of an axial chord downstream from the leading-edge plane shows a similar trend although flow from holes near the pressure side was effective, and poor performance was confined to a small region located near the suction side of the vane. Flow from the last row of holes, downstream of where the horseshoe vortex impinges on the suction surface of the vane, provides a uniformly good effectiveness distribution. Flow from a film-cooling hole located near the vane trailing edge produces a large area of elevated effectiveness that could be used to protect the endwall from the high heat transfer rates produced by the trailing-edge vortex.

A particularly well-developed study on a blown cascade is that of Nicklas [88]. The study confirms many of the phenomena previously documented and pays careful attention to the influence of local increases in turbulence on the observed heat-transfer rates.

A study by Roy et al. [58] used injection ports upstream of the leading edge. The coolant from these ports was shown to increase heat-transfer rates in the area downstream of the ports and ahead of the leading edge. This might be because of increased turbulence produced by the injected flow, but the small region of high heat transfer typically observed at the vane leading-edge-to-endwall junction was greatly reduced. Coolant injection was credited with weakening the junction corner vortex. The cooling flow also produces high film-cooling effectiveness values downstream of the slots and upstream of the vane leading-edge plane.

Knost and Thole [60] investigated two hole array geometries. The holes were positioned along the passage isovelocity lines and along lines that would be parallel to the engine axis. Interaction of flow from the cooling holes and cooling slots was noted, and it was concluded that slot cooling alone was not sufficient to protect the whole endwall. Film cooling from holes within the passage was necessary. They measured film-cooling effectiveness values with injection through a slot ahead of the leading-edge plane and through discrete holes between the slot and the leading-edge plane. They found that the coolant from the slot must be considered in an analysis of the coolant coverage. They also found that difficult-to-cool regions near the leading edge and at the pressure surface-endwall junction could be cooled more effectively if the momentum values of the discrete jet flows ahead of the pressure surface were increased to allow the coolant to penetrate the leading-edge vortex, impinge upon the pressure surface, and wash down that surface and onto the endwall. Even higher discrete jet coolant rates led to stronger jet separation and less flow to the pressure surface.

The Knost and Thole [89] study had a slashface gap, or gutter, feature. The slashface gap separates one airfoil's endwall segment from that of a neighboring airfoil. It accommodates differential thermal expansion in this region and, because it is a gap on the endwall surface, it must be sealed with flow from the internal cavity of the engine through the gap and to the passage. Knost and Thole noted that the region around the slashface gap was poorly protected. A study by Cardwell et al. [90] investigates a vane endwall with leakage through the midpassage (slashface) gap and also from the combustor-to-vane interface gap. The importance of the width of the combustor-to-vane interface gap was explored with results indicating that the momentum flux ratio is the important consideration for determining film coolant coverage. The study indicates that some portions of the endwall can be cooled by leakage flows although other areas require dedicated cooling.

A study by Zhang and Jaiswal [91] showed the combined effects, in a turbine vane passage, of upstream slot cooling and downstream film cooling through holes. The upstream slots provided nonuniform cooling along the vane endwall. A region near the pressure side was not covered. Related studies are Zhang et al. [92], Zhang and Pudupatty [93, 94], and Wright et al. [95]. Haselbach and Schiffer [96] discuss a situation in which cooling flow is introduced through a slot ahead of a vane row with various pitch and swirl angles and with various blowing rates. They documented the effects of injection on endwall aerodynamics and heat transfer. Reid et al. [97] showed that turbine efficiency decreases with increasing leakage flow rate, although the penalty can be reduced by swirling the emerging sealant flow. The swirl component can partially replicate the effects of rotation of one airfoil row relative to the next.

Suryanarayanan et al. [98] and Yang et al. [99] studied the effects of rotation on blade endwall film cooling when the coolant is introduced at the stator-to-rotor gap. They found that rotational speed impacts the region of coverage, with lower speeds tending to reorient the film traces toward the suction side of the passage. Film effectiveness tends to decay more rapidly in the streamwise direction with rotation than without rotation. Also, their study confirms observations made in earlier, stationary test sections; coolant coverage improves with increasing mass flow ratio, and the passage vortex sweeps coolant off the endwall, making it impossible to film cool the endwall with coolant introduced solely at the stator-to-rotor gap. Measurements by Pau et al. [100] show, in a rotating rig, how the stator-to-rotor leakage flow interacts with the mainstream flow. They were able to document the occurrence of pressure asymmetries near the wheel-space cavity that cause both ingression and ejection at various portions of the rotor's revolution. Ong et al. [101] measured endwall effectiveness distributions and noted that the stator-to-rotor leakage flow is largely consumed by the hub vortex thus restricting its cooling effect over the upstream portion of the passage.

Colban et al. [102] and Barigozzi et al. [103, 104] evaluated fan-shaped holes on the endwall. Sundaram and Thole [105] showed how the effectiveness of endwall film cooling is changed as holes are partially blocked with surface deposition.

E. HEAT TRANSFER IN CONTOURED, BLOWN CASCADES

Clearly, flow phenomena influencing heat transfer within vane cascades are driven by a complex interdependence between endwall and vane geometries, main flow features, cooling flow characteristics and locations, and angles of coolant injection. As suggested, it is difficult to isolate the effects of any one of these or to predict their cumulative effects. Next, cases with endwall contouring and blowing through the endwall are discussed. Though general design rules are difficult to draw from the following specific cases, they do illustrate combined effects of contouring and endwall blowing on aerodynamics and heat transfer.

One of the original studies that simulated many of these complex interactions was that of Burd et al. [106]. The study was performed in a two-passage cascade with bleed flow entering through a single, nearly continuous slot located ahead of the vane leading-edge plane. The endwall was contoured within the passage beginning at $x/Cax = 0.5$ and ending at the trailing-edge plane. Data include leakage cooling effectiveness values at three planes within the passage for several different blowing flow rates. The study shows that coolant entering the passage at low flow rates (under 2.0% of the passage mass flow rate) does not fully cover the endwall, leaving the downstream portion of the pressure surface near the endwall with no cooling protection. Coolant accumulates in the corner between the suction surface and the endwall, having been carried by the passage flow. Higher blowing flow rates, 3.2–4.5%, provide the coolant flow with enough momentum to overcome the cross-stream secondary flow and remain near the pressure side of the passage. This provides better thermal protection as a result of both better coverage and a larger mass flow rate of coolant. At flow rates higher than 3.2%, the coolant is observed to collect near the pressure-side wall, covering as much as 25% of the pressure surface span. The suction surface is not similarly well covered. Effectiveness values in the suction-surface-endwall corner are reduced from those seen with lower blowing rates. The study offers two possible explanations for this behavior. The first explanation notes that high blowing is seen to shift the location of the passage vortex, moving it down the suction side of the vane toward the endwall with increased blowing (data shown in Burd and Simon [107]). The effective mixing of the passage vortex is responsible for the low effectiveness values in the corner. According to the second theory, streamwise acceleration imposed by the contoured endwall thins the endwall boundary layer, weakens the passage flow, and reduces its effectiveness in carrying coolant toward the suction surface.

A similar study by Oke et al. [108] produced corroborating results in a slightly different cascade configuration. For the Oke et al. study, endwall contouring began ahead of the airfoil leading-edge plane and continued through the passage to the airfoil trailing-edge plane. They observed coolant accumulation similar to that described by Burd and offered a variation on Burd's coolant flow model. It was speculated that the component of coolant flow momentum normal to the endwall might be sufficient to carry this flow over the top of the horseshoe vortex. This flow then impinges upon the vane pressure surface and is carried with the passage secondary flow down the pressure surface toward the endwall. They offered a second hypothesis that (as noted in Granser and Schulenberg [49]) the emerging coolant energizes the boundary layer ahead of the leading-edge plane, thereby reducing the strength of the vortex that forms at the leading edge of the vane endwall junction. This weakens secondary flows ahead of and within the passage and allows less mixing of the coolant with the main flow near the pressure wall. The net effect is more efficient cooling of the airfoil pressure surface near the endwall when blowing rates are sufficiently high.

Oke and Simon [109] noted that acceleration caused by endwall contouring can help keep the coolant near the endwall whereas the higher momentum associated with single-slot injection tended to increase the uniformity of cooling flow injection and of endwall cooling effectiveness near the leading-edge plane and throughout the passage. The single-slot case and the double-slot comparison case had the same total mass flow rate. Increasing the injection mass flow rate further, with either single- or double-slot injection, tends to increase the overall effectiveness, but with little improvement in uniformity of effectiveness.

This latter result is somewhat consistent with the results of Liu et al. [57] taken in a noncontoured passage. They found that injection through a double row of holes (and therefore double the flow rate) was more effective than with single-row injection. Oke et al. determined that moving the slot nearer to the vane leading-edge plane produced higher effectiveness values, but reduced the uniformity of pressure-side coverage. This claim was based on data taken at a single measurement plane near the leading edge and that a more complete study might be needed to support it. Numerical studies by Lin et al. [110] and Shih et al. [111] investigated the contoured endwall geometry of Oke et al. [109] and another geometry where contouring was complete ahead of the leading-edge plane These studies include blowing on the contoured and flat endwalls of the cascade through slots ahead of the vane leading edges. With the contouring of the Oke et al. studies, film-cooling coverage was found over much of the first half of the passage. The study with contouring completed ahead of the leading-edge plane had high adiabatic effectiveness values just downstream of the slot, but poor coverage within the passage. Adiabatic effectiveness value distributions on the flat endwall were similar to those on the contoured endwall for both configurations. A later study performed by Oke and Simon [109] investigated the effects of different geometries of the film-cooling injection slots made by partially blocking the full-length slots. Partial blocking of the slots increased the coolant momentum, which was valuable for the lower blowing flow rate cases. Modification of the slot geometry was shown to allow placement of coolant where it was most needed and to control secondary flow within the passage. One problem noted in this study is that a partial slot leads to partial blockage of the mainstream approach flow by the emerging coolant. This, in turn, causes a streamwise-oriented vortex originating at the edge of that blockage, which, unfortunately, is effective in mixing the coolant with the mainstream and negating the benefits derived by the controlled placement of the coolant.

The study by Pasinato et al. [112] also considered contoured endwall geometry. In this study, the injection ports were located only in the pitchwise vicinity of the vane leading edge (similar to the geometry of Roy et al. [58]). The ports were embedded in the endwall, creating backward-facing steps. Endwall axial contouring was credited with causing a roll-up of the horseshoe vortex ahead of the leading edge, limiting the degree to which the vortex mixed hot passage fluid with the endwall boundary-layer fluid. The study showed that the lowest heat-transfer rates were upstream of the passage throat. An increase in heat transfer was observed as the flow was accelerated through the throat and the boundary layers were thinned. Pasinato et al. [113] made comparisons between measurements and computed heat transfer, pressure loss, and film-cooling effectiveness values. They also extended the investigation of Lin et al. [110].

III. STUDIES IN MORE COMPLEX GEOMETRIES

Heat transfer and secondary flows within cascades of several specific geometries were documented thoroughly in the studies just discussed. Because both secondary flows and heat transfer in engines are so highly geometry dependent, general design guidelines are difficult to construct. We often use CFD to bridge from one geometry to the next. To verify the CFD codes, experimental studies of high accuracy and resolution made within simulations of representative engine geometries that include important aspects of leakage and film-cooling flow injection are needed. Particularly important would be accurate heat-transfer measurements on contoured endwalls with blowing.

Another area deserving attention is misalignment of components along the gas path. One location requiring attention is the gap at the combustor interface with the stator section endwall. This gap is designed to accommodate manufacturing variations and differential thermal expansion. Though many studies investigating the effects of backward- or forward-facing steps on heat transfer are available, relatively few exist for cascade geometries with the steps just ahead of the airfoil row, particularly when leakage fluid is introduced through the gap. One such study, but without downstream airfoils, is by Chyu et al. [114]. The test section includes a gap with blowing and misalignment. As one might expect, the study indicates a large difference between the heat-transfer rates downstream of a forward-facing step and heat transfer downstream of a backward-facing step. The forward-facing step produces a slight reduction of heat-transfer rates ahead of the gap leading edge and then an increase downstream of the gap. The backward-facing gap produces a large area of decreased heat-transfer rates downstream of the step in the recirculation zone. Blowing through the gap that has a forestep or backstep does not change these trends although heat-transfer rates downstream of the gap tend to increase.

Yu and Chyu [115] studied the influence of slot leakage downstream of injection cooling holes. Moderate film cooling upstream of the slot provides better protection than no cooling; increased leakage flow leads to decreased cooling protection. Zhang and Moon [116] investigated the effects of a backward-facing step with blowing in a cascade. The step is shown to create an unstable boundary layer that reduces the effectiveness of film cooling. They later showed that the effect of the backstep could be reduced, and acceptable film-cooling effectiveness could be attained, by proper choice of the injection velocity [117]. Colban et al. [118, 119] also investigated the effects of steps and leakage flow on endwall aerodynamics and cooling. Their work shows sensitivities to various parameters and indicates the importance of fully characterizing the inlet flow. Hada and Thole [120] demonstrated how leakage flow through the combustor-turbine gap and a midpassage (slashface) slot, as well as misalignment of the midpassage gap, affect endwall film coverage. Kost and Mullaert [121] showed that placement of the combustor-turbine gap too near the airfoil leading-edge plane can strengthen secondary flows in the passage. In a series of papers, Wu and coworkers [122 – 124] investigated the effects of a step on the endwall, similar to the misalignment that might be present in an industrial gas turbine. Two of the studies include a forward-facing step [123, 124]. With an entrance step height of 4% of chord length, separation and reattachment of the boundary layer over the step causes an area of higher heat transfer just downstream of the leading edge. The step is also credited with weakening the horseshoe vortex. Wu et al. [124] investigated endwall film-cooling effectiveness with a similar forward-facing step ahead of the leading-edge plane and ahead of the injection holes. They found that the step causes a very significant reduction in film-cooling effectiveness, particularly in the forward part of the passage. The latest study [124] investigates a backward-facing step. A region of high heat transfer corresponding to the reattachment of the flow was noted downstream of the step. Reid et al. [125] showed how the inter-platform (slashface) gap might be redesigned for improved engine performance. A numerical study by Rubensdorffer and Fransson [126] showed how various changes in the geometry of the gap region affect secondary flow and endwall heat-transfer patterns.

The slashface gap (gutter) on the endwall between individual vanes deserves particular attention. Piggush and Simon [127] provide measurements of aerodynamic losses where steps, gaps, and leakage flows are added at the transition section and at the slashface. They apply n-factorial experimental design to learn that leakage through the slashface gap is more important than the other effects documented; the effects of having a step at the transition section are also significant. The study of Yamao et al. [128] simulates the vane slashface gap in a flat wall cascade. Their study indicates little effect on the passage loss for the slashface flow and little effect on the passage adiabatic effectiveness values between a case with blowing and one with no blowing. The discrepancy in measured effect of slashface blowing between the Yamao et al. study and the Piggush and Simon study might be caused by the difference in the no-blowing base cases. The Yamao et al. base comparison case is an open-slot, no-blowing case whereas the Piggush and Simon base case is a smooth-slot (covered) no-blowing case. With their open gap, flow can enter the gap upstream, where the static pressures are high, and travel through the gap to emerge in the downstream portion of the passage where the static pressures are low. Also, the Yamao et al. study had blowing through the transition section gap as well as through the slashface gap whereas the Piggush and Simon study isolated the slashface gap effect. Finally, the

Piggush and Simon study involves a contoured endwall so that the comparison case is one of particularly low losses.

Adiabatic film-cooling effectiveness values were measured by Ranson et al. [129] in a rotor cascade constructed with straight endwalls having a slot ahead of the airfoils, three slot segments on the endwall, and a slot downstream of the airfoils (see Fig. 9). Some benefit in adiabatic effectiveness in the upstream portions of the endwall attributable to flow through the upstream slot was recorded. Increased flow did not lead to increased effectiveness. The endwall slot showed little benefit in cooling effectiveness and no improvement with increased blowing. Computed results showed higher effectiveness values on the endwall than measured. The study of Aunapu et al. [43] (discussed earlier), in which there was blowing within the passage, has some features akin to slashface gap step disturbances although it was performed in a straight-walled cascade and heat-transfer measurements were not made.

Some studies have focused on the airfoil-to-endwall transition geometry. Modification of the wing-fuselage junction has been employed in the aircraft industry to reduce the overall drag experienced by aircraft [130–132]. Such modifications typically include a large fillet or similar geometry to fair out the right angle of the junction. As discussed frequently, the junction of a bluff body with a wall creates a horseshoe vortex (Fig. 2). A separation point ahead of the leading edge of the body is created on the endwall. Modifications of the wing-fuselage junction, such as fillets, are effective because the fluid approaching the leading edge of the bluff body accelerates locally as it moves up the fillet. The effects of the total pressure deficit created by the endwall are reduced by this imposed acceleration.

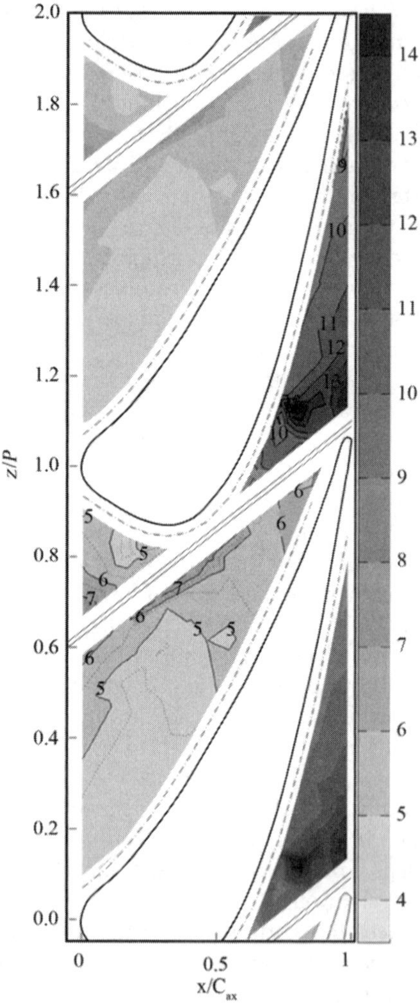

Fig. 9 Adiabatic film-cooling effectiveness values measured by Ranson et al. [130].

The use of a vane-endwall modification in gas turbines (similar to the wing-fuselage modification) has recently received increasing attention. In the turbine, the flow is more complicated than flows found on the aircraft body. Although most experiments have shown a reduction of loss when fillets are introduced, the exact mechanism responsible for the reduction is poorly understood. The work of Sauer et al. [133] in low-pressure turbine applications has shown loss reductions of up to 50%. Similar modifications in compressor cascades have shown similar loss reductions [134]. Sauer makes use of a leading-edge bulb, where the leading-edge radius is increased locally near the endwall. This strengthens the horseshoe vortex and in particular the suction side of the horseshoe vortex. Sauer et al. theorize that the enlarged suction-side vortex displaces the pressure side of the horseshoe vortex off the suction surface of the vane. The sense of rotation of the suction-side vortex is opposite to the pressure side, and therefore the vortices will not combine. This prevents interaction of the passage flow with the pressure-side vortex. Though the horseshoe vortex is strengthened, preventing interaction between the passage flow and the pressure side of the horseshoe vortex creates a loss reduction of a much greater magnitude, and the net effect is a loss reduction. Work by Becz et al. [135, 136] investigating a leading-edge bulb of a geometry similar to that of Sauer et al. [134] did not show the same magnitude of loss reduction as reported by Sauer. The Becz et al. studies indicate that more research is needed to identify the optimum geometry for loss reduction and to identify the actual mechanisms of loss reduction.

Zess and Thole [137] investigated the effects of a leading-edge fillet. Shih and Lin [138] computationally investigated the performance of particular designs of leading-edge fillets and documented the effects of changing the degree of inlet swirl. They noted that introduction of a fillet or inlet swirl can reduce both the aerodynamic losses and endwall heat transfer. They also noted that when there is swirl, leading-edge fillets become less effective. This shows the importance of optimizing the fillet design for implementation of swirl. They suggested that a description of the mechanisms responsible for aerodynamic losses and surface heat-transfer changes might require more than simply considering changes in intensity of secondary flows. The results indicate that the characteristic horseshoe vortex was not formed at the leading edge. This is because of local acceleration. Also, unsteadiness in that region, usually associated with the horseshoe vortex, was not observed. Turbulence kinetic energy in the passage was reduced by 80%. They theorized that these changes might be responsible for loss reductions observed with the use of fillets. The fillet had a sharp leading edge. Because the magnitude of the horseshoe vortex scales on the leading-edge radius, the vortex was weak. This sharpening would, however, increase off-design incidence losses. A study on endwall fillet heat transfer by Lethander et al. [139] reports a slight improvement on passage cooling with a fillet due to reduction in passage surface area when the fillet, endwall, and vane surfaces are considered and reduction in secondary flow strength and the leading edge, which prevents hot fluid from being driven to the endwall at the vane leading edge. Han and Goldstein [140] presented measurements that show the effects of modifying airfoil leading-edge geometry (filleting) on the endwall heat (mass) transfer coefficient distribution.

IV. CONCLUSIONS

Studies on endwall flow and heat transfer have begun to resolve the complex flow patterns in the endwall regions of turbine passages. With this new understanding, researchers are finding methods for improving aerodynamic performance of turbines and methods to manage the high thermal loads in this region of modern engines. The results are increased aerodynamic efficiencies, improved cooling performance with reduced cooling flows and greater engine durability. Success has been demonstrated with such methods as endwall contouring, leading-edge modification, strategic application of film cooling, and optimum use of leakage flows for sealing and film cooling. As more is learned, real effects such as leakage flow interactions, endwall steps and gaps, and flow features like embedded vorticity and turbulence from upstream regions are being integrated into the analysis and design.

Further data are needed on Mach-number effects, the effects of rotation, and the benefits of airfoil-to-endwall junction modification and three-dimensional airfoil and passage geometries. Though the experimental work has been and will continue to be valuable in documenting flow and heat transfer and in evaluating new methods for improving the endwall regions, CFD has had a major impact on this topic. The flowfields are complex, but can be computed rather accurately. Thus, CFD has given great insight into this complex problem and has given a means for testing proposed aerothermal improvement schemes. This is a complex flow with separation, stagnation, transition, embedded vorticity, and strong pressure gradients. Thus, improvements in CFD modeling are still needed. As with many engineering flows, more is to be learned about turbulence modeling and transition to turbulence.

REFERENCES

[1] Herzig, H. Z., Hansen, A. G., and Costello, G. R., "A Visualization Study of Secondary Flows in Cascades," NACA Report 1163, Feb. 1953.

[2] Pierce, F. S., and Harsh, M. D., "Three-Dimensional Turbulent Boundary Layer Separation at the Junction of a Streamlined Cylinder with a Flat Plate," *Flow Visualization III, Proceedings of the Third International Symposium on Flow Visualization*, edited by W. J. Yang, Hemisphere, Washington, DC, 1985, pp. 331–335.

[3] Eckerle, W. A., and Langston, L. S., "Horseshoe Vortex Formation Around a Cylinder," *Journal of Turbomachinery*, Vol. 109, April 1987, pp. 278–285.

[4] Pierce, F. J., and Shin, J., "The Development of a Turbulent Junction Vortex System," *Journal of Fluids Engineering*, Vol. 114, No. 4, 1992, pp. 559–565.

[5] Goldstein, R. J., and Karni, J., "The Effect of a Wall Boundary Layeron Local Mass Transfer from a Cylinder in Crossflow," *Journal of Heat Transfer*, Vol. 106, May 1984, pp. 260–267.

[6] Langston, L. S., Nice, M. L., and Hooper, R. M., "Three-Dimensional Flow Within a Turbine Cascade Passage," *Journal of Engineering for Power*, Vol. 99, Jan. 1977, pp. 21–28.

[7] Langston, L. S., "Crossflow in a Turbine Cascade Passage," *Journal of Engineering for Power*, Vol. 102, No. 4, 1980, pp. 866–874.

[8] Sieverding, C. H., and Van den Bosche, P., "The Use of Coloured Smoke to Visualize Secondary Flows in a Turbine Cascade," *Journal of Fluid Mechanics*, Vol. 134, 1983, pp. 85–89.

[9] Kawai, T., Shinoki, S., and Adachi, T., "Visualization Study of Three-Dimensional Flows in a Turbine Cascade Endwall Region," *JSME International Journal, Series 2: Fluids Engineering, Heat Transfer, Power, Combustion, Thermophysical Properties*, Vol. 33, No. 2, 1990, pp. 256–264.

[10] Chung, J. T., and Simon, T. W., "Three-Dimensional Flow near the Blade/Endwall Junction of a Gas Turbine: Visualization in a Large-Scale Cascade Simulator," American Society of Mechanical Engineers, Paper 90-WA/HT-4, Nov. 1990.

[11] Wang, H. P., Olson, S. J., and Goldstein, R. J., "Flow Visualization in a Linear Turbine Cascade of High Performance Turbine Blades," *Journal of Turbomachinery*, Vol. 119, No. 1, 1997, pp. 1–8.

[12] Langston, L. S., "Secondary Flows in Axial Turbines—A Review," *Turbine 2000 Workshop, Proceedings of the International Symposium on Heat Transfer in Gas Turbine Systems*, Cesme, Turkey, Annals of the New York Academy of Sciences, edited by R. J. Goldstein, Vol. 934, NY Academy of Sciences, New York, pp. 11–26.

[13] Sieverding, C. H., "Recent Progress in the Understanding of Basic Aspect of Secondary Flows in Turbine Blade Passages," *Journal of Engineering for Gas Turbines and Power*, Vol. 107, April 1985, pp. 248–257.

[14] Holley, B. M., Becz, S., and Langston, L. S., "Measurement and Calculation of Turbine Cascade Endwall Pressure and Shear Stress," *Journal of Turbomachinery*, Vol. 128, 2006, pp. 232–240.

[15] Tallman, J. A., Haldeman, C. W., Dunn, M. G., Tolpadi, A. K., and Bergholz, R. F., "Heat Transfer Measurements and Predictions for Modern, High-Pressure, Transonic Turbine, Including Endwalls," American Society of Mechanical Engineers, Paper GT2006-90927, May 2006.

[16] Pullen, G., Denton, J., and Curtis, E., "Improving the Performance of a Turbine with Low Aspect Ratio Stators by Aft-Loading," American Society of Mechanical Engineers, Paper GT2005-68548, June 2005.

[17] Ames, F. E., Johnson, J. D., and Fiala, N. J., "The Influence of Aero-Derivative Combustor Turbulence and Reynolds Number on Vane Aerodynamics Losses, Secondary Flows, and Wake Growth," American Society of Mechanical Engineers, Paper GT2006-90168, May 2006.

[18] Kunze, V. R., Wolff, M., Barringer, M. D., Thole, K. A., and Polanka, M. D., "Numerical Insight into Flow and Thermal Patterns within an Inlet Profile Generator Comparing to Experimental Results," American Society of Mechanical Engineers, Paper GT2006-90276, May 2006.

[19] Barringer, M. D., and Thole, K. W., "Experimental Evaluation of an Inlet Profile Generator for High Pressure Turbine Tests," American Society of Mechanical Engineers, Paper GT2006-90401, May 2006.

[20] Barringer, M. D., Thole, K. A., and Polanka, M. D., "Effects of Combustor Exit Profiles on High Pressure Turbine Vane Aerodynamics and Heat Transfer," American Society of Mechanical Engineers, Paper GT2006-90277, May 2006.

[21] Yamada, K., Hiroma, K., Hirano, Y., Funazaki, K., Tsutsumi, M., and Matsuo, A., "Effect of Wake Passing on Unsteady Aerodynamic Performance in a Turbine Stage," American Society of Mechanical Engineers, Paper GT2006-90783, May 2006.

[22] Deich, M. E., Zaryankin, A. D., Filippov, G. A., and Zatsepin, N. F., "Method of Increasing the Efficiency of Turbine Stages with Short Blades," Translation No. 2816, Associated Electrical Industries Ltd., Manchester, England, U.K., Feb. 1960, pp. 18–24.

[23] Morris, A. W. H., and Hoare, R. G., "Secondary Loss Measurements in a Cascade of Turbine Blades with Meridional Wall Profiling," American Society of Mechanical Engineers, Paper 75-WA/GT-13, July 1975.

[24] Kopper, F. C., Milano, R., and Vanco, M., "Experimental Investigation of Endwall Profiling in a Turbine Vane Cascade," *AIAA Journal*, Vol. 19, No. 8, 1981, pp. 1033–1040.

[25] Boletis, E., "Effects of Tip Endwall Contouring on the Three- Dimensional Flow Field in an Annular Turbine Nozzle Guide Vane: Part 1—Experimental Investigation," *Journal of Engineering for Gas Turbines and Power*, Vol. 107, Oct. 1985, pp. 983–990.

[26] Arts, T., "Effects of Tip Endwall Contouring on the Three-Dimensional Flow Field in an Annular Turbine Nozzle Guide Vane: Part 2—Numerical Investigation," American Society of Mechanical Engineers, Paper 85-GT-108, March 1985.

[27] Dossena, V., Perdichizzi, A., and Savini, M., "The Influence of Endwall Contouring on the Performance of a Turbine Nozzle Guide Vane," *Journal of Turbomachinery*, Vol. 121, April 1999, pp. 200–208.

[28] Burd, S. W., and Simon, T. W., "Flow Measurements in a Nozzle Guide Vane Passage with a Low Aspect Ratio and Endwall Contouring," American Society of Mechanical Engineers, Paper 2000-GT-0213, May 2000.

[29] Rose, M. G., "Non-Axisymmetric Endwall Profiling in the HP NGV's of an Axial Flow Gas Turbine," American Society of Mechanical Engineers, Paper 94-GT-249, June 1994.

[30] Harvey, N. W., Rose, M. G., Taylor, M. D., Shahpar, S., Hartland, J., and Gregory-Smith, D. G., "Non-Axisymmetric Turbine End Wall Design: Part Three Dimensional Linear Design System," American Society of Mechanical Engineers, Paper 99-GT-337, June 1999.

[31] Hartland, J., Gregory-Smith, D. G., Harvey, N. W., and Rose, M. G., "Non-Axisymmetric Turbine End Wall Design: Part II Experimental Validation," American Society of Mechanical Engineers, Paper 99-GT-338, June 1999.

[32] Yan, J., Gregory-Smith, D. G., and Walker, P. J., "Secondary Flow Reduction in a Nozzle Guide Vane Cascade by Non-Axisymmetric End-Wall Profiling," American Society of Mechanical Engineers, Paper 99-GT-339, June 1999.

[33] Brennan, G., Harvey, N. W., Rose, M. G., Fomison, N., and Taylor, M. D., "Improving the Efficiency of the Trent 500HP Turbine Using Non-Axisymmetric End Walls: Part 1 Turbine Design," American Society of Mechanical Engineers, Paper 2001-GT-0444, June 2001.

[34] Rose, M. G., Harvey, N. W., Seaman, P., Newman, D. A., and McManus, D., "Improving the Efficiency of the Trent 500HP Turbine Using Non- Axisymmetric Endwalls: Part 1 Experimental Validation," American Society of Mechanical Engineers, Paper 2001-GT-0505, June 2001.

[35] Saha, A. K., and Acharya, S., "Computations of Turbulent Flow and Heat Transfer Through a Three-Dimensional Non-Axisymmetric Blade Passage," American Society of Mechanical Engineers, Paper GT2006-90390, May 2006.

[36] Saha, A. K., Mahmood, G. I., and Acharya, S., "The Role of Leading-Edge Contouring on End-Wall Flow and Heat Transfer: Computations and Experiments," American Society of Mechanical Engineers, Paper GT2006-91318, May 2006.

[37] Gustafson, R., Mahmood, G., and Acharya, S., "Aerodynamic Measurements in a Linear Turbine Blade Passage with Three-Dimensional Endwall Contouring," American Society of Mechanical Engineers, Paper GT2007-28073, May 2007.

[38] Schobeiri, M. T., Suryanarayanan, A., Jermann, C., and Neuenschwander, T., "A Comparative Aerodynamic and Performance Study of a Three-Stage High Pressure Turbine with 3-D Bowed Blades and Cylindrical Blades," American Society of Mechanical Engineers, Paper GT2004-53650, June 2004.

[39] Kawai, T., Shinoki, S., and Adachi, T., "Secondary Flow Control and Loss Reduction in a Turbine Cascade Using Endwall Fences," *JSME International Journal Series II—Fluids Engineering, Heat Transfer, Power, Combustion, Thermophysical Properties*, Vol. 32, No. 3, 1989, pp. 375–387.

[40] Chung, J. T., Simon, T. W., and Buddhavarapu, J., "Three-Dimensional Flow near the Blade/Endwall Junction of a Gas Turbine: Application of a Boundary Layer Fence," American Society of Mechanical Engineers, Paper 91-GT-45, June 1991.

[41] Chung, J. T., and Simon, T. W., "Effectiveness of the Gas Turbine Endwall Fences in Secondary Flow Control at Elevated Freestream Turbulence Levels," American Society of Mechanical Engineers, Paper 93-GT-51, May 1993.

[42] Kinnear, I. S., May, A. L., and Wall, R. A., "Secondary Flow Control in Axial Fluid Flow Machine," U.K. Patent Application GB 2 042 675 A, 1980.

[43] Aunapu, N. V., Volino, R. J., Flack, K. A., and Stoddard, R. M., "Secondary Flow Measurements in a Turbine Passage with Endwall Flow Modification," *Journal of Turbomachinery*, Vol. 122, No. 4, 2000, pp. 651–658.

[44] Benner, M. W., Sjolander, S. A., and Moustapha, S. H., "Measurements of Secondary Flows Downstream of a Turbine Cascade at off-Design Incidence," American Society of Mechanical Engineers, Paper GT2004-53786, June 2004.

[45] Dossena, V., D'Ippolito, G., and Pesatori, E., "Stagger Angle, and Pitch-Chord Ratio Effects on Secondary Flows Downstream of a Turbine Cascade at Several off-Design Conditions," American Society of Mechanical Engineers, Paper GT2004-54083, June 2004.

[46] Rehder, H.-J., and Dannhauer, A., "Experimental Investigation of Turbine Leakage Flows on the Three-Dimensional Flow Field and Endwall Heat Transfer," *Journal of Turbomachinery*, Vol. 129, July 2007, pp. 608–618.

[47] Schlienger, J., Kalfas, A. I., and Abhari, R. S., "Vortex-Wake-Blade Interaction in a Shrouded Axial Turbine," American Society of Mechanical Engineers, Paper GT2004-53915, June 2004.

[48] Blair, M. F., "An Experimental Study of Heat Transfer and Film Cooling on Large-Scale Turbine Endwalls," *Journal of Heat Transfer*, Vol. 96, Nov. 1974, pp. 524–529.

[49] Granser, D., and Schulenberg, T., "Prediction and Measurement of Film Cooling Effectiveness for a First-Stage Turbine Vane Shroud," American Society of Mechanical Engineers, Paper 90-GT-95, June 1990.

[50] Goldman, L. J., and McLallin, K. L., "Effects of Endwall Cooling on Secondary Flows in Turbine Stator Vanes," AGARD, CPP-214, 1977.

[51] Sieverding, C. H., and Wilputte, P., "Influence of Mach Number and Endwall Cooling on Secondary Flows in a Straight Nozzle Cascade," American Society of Mechanical Engineers, Paper 80-GT-52, May 1980.

[52] Bario, F., LeBoeuf, F., Onvani, A., and Seddini, A., "Aerodynamics of Cooling Jets Introduced in the Secondary Flow of a Low Speed Turbine Cascade," American Society of Mechanical Engineers, Paper 89-GT-192, June 1989.

[53] Harasgama, S. P., and Burton, C. D., "Film Cooling Research on the Endwall of a Turbine Nozzle Guide Vane in a Short Duration Annular Cascade: Part 1—Experimental Technique and Results," *Journal of Turbomachinery*, Vol. 114, Oct. 1992, pp. 734–740.

[54] Biesinger, T. E., and Gregory-Smith, D. G., "Reduction in Secondary Flows and Losses in a Turbine Cascade by Upstream Boundary Layer Blowing," American Society of Mechanical Engineers, Paper 93-GT-114, May 1993.

[55] Friedrichs, S., Hodson, H. P., and Dawes, W. N., "Distribution of Film Cooling Effectiveness on a Turbine Endwall Measured Using the Ammonia and Diazo Technique," American Society of Mechanical Engineers, Paper 95-GT-1, June 1995.

[56] Friedrichs, S., Hodson, H. P., and Dawes, W. N., "Aerodynamic Aspects of Endwall Film-Cooling," *Journal of Turbomachinery*, Vol. 119, Oct. 1997, pp. 786–793.

[57] Liu, S., Liu, G., Xu, D., Lapworth, B. L., and Forest, A. E., "Aerodynamic Investigation of Endwall Film-Cooling in an Axial Turbine Cascade Part I: Experimental Investigation," XVI ISABE International Symposium on Air Breathing Engines, Paper 99-7080, 1999.

[58] Roy, R. P., Squires, K. D., Gerendas, M., Song, S., Howe, W. J., and Ansari, A., "Flow and Heat Transfer at the Hub Endwall of Inlet Vane Passages—Experiments and Simulations," American Society of Mechanical Engineers, Paper 2000-GT-198, May 2000.

[59] Kost, F., and Nicklas, M., "Film-Cooled Turbine Endwall in a Transonic Flow Field: Part I—Aerodynamic Measurements," *Journal of Turbomachinery*, Vol. 123, Oct. 2001, pp. 709–719.

[60] Knost, D. G., and Thole, K. A., "Computational Predictions of Endwall Film-Cooling for a First Stage Vane," American Society of Mechanical Engineers, Paper GT2003-38252, June 2003.

[61] Georgiou, D. P., Papavasilopoulos, V. A., and Alevisos, M., "Experimental Contribution on the Significance and the Control by Transverse Injection of the

Horseshoe Vortex," American Society of Mechanical Engineers, Paper 96-GT-255, June 1996.

[62] Lapworth, B. L., Forest, A. E., and Liu, S., "Aerodynamic Investigation of Endwall Film Cooling in an Axial Turbine Cascade Part II: Numerical Investigation," International Symposium on Air Breathing Engines, Paper 99-7193, 1999.

[63] Oke, R. A., Burd, S. W., Simon, T. W., and Vahlberg, R., "Measurements in a Turbine Cascade over a Contoured Endwall: Discrete Hole Injection of Bleed Flow," American Society of Mechanical Engineers, Paper 2000-GT-214, May 2000.

[64] Lampart, P., Yershov, S., Rusanov, A., and Szymaniak, M., "Tip Leakage/Main Flow Interactions in Multi-Stage HP Turbines with Short-Height Blading," American Society of Mechanical Engineers, Paper GT2004-64047, June 2004.

[65] Paniagua, G., Denos, R., and Almeida, S., "Effect of the Hub Endwall Cavity Flow on the Flow-Field of a Transonic High-Pressure Turbine," American Society of Mechanical Engineers, Paper GT2004-53458, June 2004.

[66] Gaetani, P., Persico, G., Dossena, V., and Osnaghi, C., "Investigation of the Flow Field in a HP Turbine Stage for Two Stator-Rotor Axial Gaps: Part I - 3D Time-Averaged Flow Field," American Society of Mechanical Engineers, Paper GT2006-90553, 2006; also "Part II - Unsteady Flow Field," American Society of Mechanical Engineers, Paper GT2006-90556, May 2006.

[67] Graziani, R. A., Blair, M. F., Taylor, J. R., and Mayle, R. E., "An Experimental Study of Endwall and Airfoil Surface Heat Transfer in a Large-Scale Turbine Blade Cascade," *Journal of Engineering for Gas Turbines and Power*, Vol. 102, April 1980, pp. 257–267.

[68] York, R. E., Hylton, L. D., and Mihelc, M. S., "An Experimental Investigation of Endwall Heat Transfer and Aerodynamics in a Linear Vane Cascade," *Journal of Engineering for Gas Turbine and Power*, Vol. 106, Jan. 1984, pp. 159–167.

[69] Gaugler, R. E., and Russell, L. M., "Comparison of Visualized Turbine Endwall Secondary Flows and Measured Heat Transfer Patterns," *Journal of Engineering for Gas Turbine and Power*, Vol. 106, Jan. 1984, pp. 168–172.

[70] Kumar, G. N., Jenkins, R. M., and Sahu, U., "Regionally Averaged Endwall Heat Transfer Correlations for a Linear Vane Cascade," American Society of Mechanical Engineers, Paper 85-GT-19, March 1985.

[71] Goldstein, R. J., and Spores, R. A., "Turbulent Transport on the Endwall in the Region Between Adjacent Turbine Blades," *Journal of Heat Transfer*, Vol. 110, Nov. 1988, pp. 862–869.

[72] Wang, H. P., Goldstein, R. J., and Olson, S. J., "Effect of High Free-Stream Turbulence with Large Length Scale on Blade Heat/Mass Transfer," *Journal of Turbomachinery*, Vol. 121, April 1999, pp. 217–224.

[73] Thole, K. A., Radomsky, R. W., Kang, M. B., and Kohli, A., "Elevated Freestream Turbulence Effects on Heat Transfer for a Gas Turbine Vane," *International Journal of Heat and Fluid Flow*, Vol. 23, No. 2, 2002, pp. 137–147.

[74] Lee, S. W., Jun, S. B., Park, B., and Lee, J. S., "Effects of High Free-Stream Turbulence on the near-Wall Flow and Heat/Mass Transfer on the Endwall of a Linear Turbine Rotor Cascade," American Society of Mechanical Engineers, Paper GT-2002-30187, June 2002.

[75] Ames, F. E., Barbot, P. A., and Wang, C., "Effects of Aeroderivative Combustor Turbulence on Endwall Heat Transfer Distributions Acquired in a Linear Vane Cascade," *Journal of Turbomachinery*, Vol. 125, April 2003, pp. 210–220.

[76] Kang, M. B., Kohli, A., and Thole, K. A., "Heat Transfer and Flowfield Measurements in the Leading Edge Region of a Stator Vane Endwall," *Journal of Turbomachinery*, Vol. 121, July 1999, pp. 558–568.

[77] Kang, M. B., and Thole, K. A., "Flowfield Measurements in the Endwall Region of a Stator Vane," American Society of Mechanical Engineers, Paper 99-Gt-188, June 1999.

[78] Radomsky, R. W., and Thole, K. A., "Detailed Boundary Layer Measurements on a Turbine Stator Vane at Elevated Freestream Turbulence Levels," *Journal of Turbomachinery*, Vol. 124, Jan. 2002, pp. 107–118.

[79] Sveningsson, A., and Davidson, L., "Computations of Flow Field and Heat Transfer in a Stator Vane Passage Using the $v2-f$ Turbulence Model," American Society of Mechanical Engineers, Paper GT2004-53586, June 2004.

[80] Bassi, F., and Perdichizzi, A., "Secondary Flow Downstream of a Transonic Cascade," *Proceedings of the 1987 Tokyo Gas Turbine Conference, International Gas Turbine Society of Japan*, Oct. 1987, pp. 11-123–11-130.

[81] Hermanson, K. S., and Thole, K. A., "Effect of Mach Number on Secondary Flow Characteristics," *International Journal of Turbo and Jet Engines*, Vol. 17, 2000, pp. 179–196.

[82] Giel, P. W., Thurman, D. R., Van Fossen, G. J., Hippensteele, S. A., and Boyle, R. J., "Endwall Heat Transfer Measurements in a Transonic Turbine Cascade," American Society of Mechanical Engineers, Paper 96-GT-180, June 1996.

[83] Blair, M. F., "An Experimental Study of Heat Transfer in a Large-Scale Turbine Rotor Passage," *Journal of Turbomachinery*, Vol. 116, Jan. 1994, pp. 1–13.

[84] Guo, S. M., Jones, T. V., Lock, G. D., and Dancer, S. N., "Computational Prediction of Heat Transfer to Gas Turbine Nozzle Guide Vanes with Roughened Surfaces," American Society of Mechanical Engineers, Paper 96-GT-388, June 1996.

[85] Takeishi, K., Matsuura, M., Aoki, S., and Sato, T., "An Experimental Study of Heat Transfer and Film Cooling on Low Aspect Ratio Turbine Nozzles," *Journal of Turbomachinery*, Vol. 112, July 1990, pp. 488–496.

[86] Harasgama, S. P., and Burton, C. D., "Film Cooling Research on the Endwall of a Turbine Nozzle Guide Vane in a Short Duration Annular Cascade: Part 2—Analysis and Correlation of Results," *Journal of Turbomachinery*, Vol. 114, Oct. 1992, pp. 741–746.

[87] Jabbari, M. Y., Marston, K. C., Eckert, E. R. G., and Goldstein, R. J., "Film Cooling of the Gas Turbine Endwall by Discrete-Hole Injection," *Journal of Turbomachinery*, Vol. 118, April 1996, pp. 278–284.

[88] Nicklas, M., "Film-Cooled Turbine Endwall in a Transonic Flow Field: Part II—Heat Transfer and Film-Cooling Effectiveness," *Journal of Turbomachinery*, Vol. 123, Oct. 2001, pp. 720–729.

[89] Knost, D. G., and Thole, K. A., "Adiabatic Effectiveness Measurements of Endwall Film-Cooling for a First Stage Vane," American Society of Mechanical Engineers, Paper GT 2004-53326, June 2004.

[90] Cardwell, N. D., Sundaram, N., and Thole, K. A., "The Effects of Varying the Combustor-Turbine Gap," *Journal of Turbomachinery*, Vol. 129, Oct. 2007, pp. 756–764.

[91] Zhang, L. J., and Jaiswal, R. S., "Turbine Nozzle Endwall Film Cooling Study Using Pressure Sensitive Paint," *Journal of Turbomachinery*, Vol. 123, Oct. 2001, pp. 730–738.

[92] Zhang, L. J., Baltz, M., Pudupatty, R., and Fox, M., "Turbine Nozzle Film Cooling Study Using the Pressure Sensitive Paint (PSP) Technique," American Society of Mechanical Engineers, Paper 99-GT-196, June 1999.

[93] Zhang, L. J., and Pudupatty, R., "Turbine Nozzle Leading Edge Film Cooling Study in a High Speed Wind Tunnel," *ASME 33rd National Heat Transfer Conference*, NHTC99-234, Aug. 1999, pp. 1–9.

[94] Zhang, L. J., and Pudupatty, R., "The Effect of Injection Angle and Hole Exit Shape on Turbine Nozzle Pressure Side Film Cooling," American Society of Mechanical Engineers, Paper 2000-GT-247, May 2000.

[95] Wright, L. M., Gao, Z., Yang, H., and Han, J-C., "Film Cooling Effectiveness Distribution on a Gas Turbine Blade Platform with Inclined Slot Leakage and Discrete Film Hole Flows," American Society of Mechanical Engineers, Paper GT2006-90375, May 2006.

[96] Haselbach, F., and Schiffer, H. P., "Aerothermal Investigations on Turbine Endwalls and Blades (AITEB)," American Society of Mechanical Engineers, Paper GT2004-53078, June 2004.

[97] Reid, K., Denton, J., Pullan, G., Curtis, E., and Longley, J., "The Interaction of Turbine Inter-Platform Leakage Flow with the Mainstream Flow," American Society of Mechanical Engineers, Paper GT2005-68151, June 2005.

[98] Suryanarayanan, A., Mhetras, S. P., Schobeiri, M. T., and Han, J. C., "Film-Cooling Effectiveness on a Rotating Blade Platform," *Journal of Turbomachinery*, Vol. 131, Jan. 2009, pp. 011014-1:12.

[99] Yang, H., Chen, H-C., and Han, J-C., "Numerical Study of a Rotating Blade Platform with Film Cooling from Cavity Purge Flow in a 1-1/2 Turbine Stage," American Society of Mechanical Engineers, Paper GT2006-90322, May 2006.

[100] Pau, M., Paniagua, G., Delhaye, D., and de la Loma, A., "Aerothermal Impact of Stator-Rim Purge Flow and Rotor-Platform Film Cooling on a Transonic Turbine Stage," American Society of Mechanical Engineers, Paper GT2008-51295, 2008.

[101] Ong, J. H. P., Miller, R. J., and Uchida, S., "The Effect of Coolant Injection on the Endwall Flow of a High Pressure Turbine," American Society of Mechanical Engineers, Paper GT2006-91060, May 2006.

[102] Colban, W., Thole, K. A., and Haendler, M., "A Comparison of Cylindrical and Fan-Shaped Film-Cooling Holes on a Vane Endwall at Low and High Freestream Turbulence Levels," American Society of Mechanical Engineers, Paper GT2006-90021, May 2006.

[103] Barigozzi, G., Benzoni, G., Franchini, G., and Perdichizzi, A., "Fan-Shaped Hole Effects on the Aero-Thermal Performance of a Film Cooled Endwall," American Society of Mechanical Engineers, Paper GT2005-68544, June 2005.

[104] Barigozzi, G., Fanchini, G., and Perdichizzi, A., "Endwall Film Cooling Through Fan-Shaped Holes with Different Area Ratios," American Society of Mechanical Engineers, Paper GT2006-90684, May 2006.

[105] Sundaram, N., and Thole, K. A., "Effects of Surface Deposition, Hole Blockage, and TBC Spallation on Vane Endwall Film-Cooling," American Society of Mechanical Engineers, Paper GT2006-90379, May 2006.

[106] Burd, S. W., Satterness, C. J., and Simon, T. W., "Effects of Slot Bleed Injection over a Contoured Endwall on Nozzle Guide Vane Cooling Performance: Part II—Thermal Measurements," American Society of Mechanical Engineers, Paper 2000-GT-200, May 2000.

[107] Burd, S. W., and Simon, T. W., "Effects of Slot Bleed Injection over a Contoured Endwall on Nozzle Guide Vane Cooling Performance: Part I— Flow Field Measurements," American Society of Mechanical Engineers, Paper 2000-GT-199, May 2000.

[108] Oke, R. A., Simon, T. W., Shih, T., Zhu, B., Lin, L., and Chyu, M., "Measurement over a Film-Cooled, Contoured Endwall with Various Coolant Injection Rates," American Society of Mechanical Engineers, Paper 2001-GT-0140, June 2001.

[109] Oke, R. A., and Simon, T. W., "Film Cooling Experiments with Flow Introduced Upstream of a First Stage Nozzle Guide Vane Through Slots of Various Geometries," American Society of Mechanical Engineers, Paper 2002-GT-30169, June 2002.

[110] Lin, Y.-L., Shih, T. I.-P., Chyu, M. K., and Bunker, R. S., "Effects of Leakage on Fluid Flow in a Contoured Turbine Nozzle Guide Vane," American Society of Mechanical Engineers, Paper 2000-GT-0555, May 2000.

[111] Shih, T. I.-P., Lin, Y.-L., and Simon, T. W., "Control of Secondary Flows in a Turbine Nozzle Guide Vane by Endwall Contouring," American Society of Mechanical Engineers, Paper 2000-GT-0556, May 2000.

[112] Pasinato, H. D., Liu, Z., Roy, R. P., Howe, W. J., and Squires, K. D., "Prediction and Measurement of the Flow and Heat Transfer Along the Endwall and Within an Inlet Vane Passage," American Society of Mechanical Engineers, Paper GT-2002-30189, June 2002.

[113] Pasinato, H. D., Squires, K. D., and Roy, R. P., "Measurements and Modeling of the Flow and Heat Transfer in a Contoured Vane-Endwall Passage," *International Journal of Heat and Mass Transfer*, Vol. 47, No. 26, 2004, pp. 5685–5702.

[114] Chyu, M. K., Hsing, Y. C., and Bunker, R. S., "Measurements of Heat Transfer Characteristics of Gap Leakage Around a Misaligned Component Interface," American Society of Mechanical Engineers, Paper 98-GT-132, June 1998.

[115] Yu, Y., and Chyu, M. K., "Influence of Gap Leakage Downstream of the Injection Holes on Film Cooling Performance," *Journal of Turbomachinery*, Vol. 120, July 1998, pp. 541–548.

[116] Zhang, L., and Moon, H. K., "Turbine Nozzle Endwall Inlet Film Cooling—The Effect of a Back-Facing Step," American Society of Mechanical Engineers, Paper GT-2003-38319, June 2003.

[117] Zhang, L., and Moon, H.-K., "Turbine Nozzle Endwall Inlet Film Cooling: The Effect of a Back-Facing Step and Velocity Ratio," American Society of Mechanical Engineers, Paper IMECE2004-59117, Nov. 2004.

[118] Colban, W. F., Thole, K. A., and Zess, G., "Combustor Turbine Interface Studies—Part 1: Endwall Effectiveness Measurements," *Journal of Turbomachinery*, Vol. 125, April 2003, pp. 193–202.

[119] Colban, W. F., Lethander, A. T., Thole, K. A., and Zess, G., "Combustor Turbine Interface Studies—Part 2: Flow and Thermal Field Measurements," *Journal of Turbomachinery*, Vol. 125, April 2003, pp. 203–209.

[120] Hada, S., and Thole, K. A., "Computational Study of a Midpassage Gap and Upstream Slot on Vane Endwall Film-Cooling," American Society of Mechanical Engineers, Paper GT2006-91067, May 2006.

[121] Kost, F., and Mullaert, A., "Migration of Film-Coolant from Slot and Hole Ejection at a Turbine Vane Endwall," American Society of Mechanical Engineers, Paper GT2006-90355, May 2006.

[122] Wu, P.-S., Lin, T.-Y., and Lai, Y.-M., "Heat Transfer Coefficient Distribution in the Endwall Region of a Vane with a Finite-Step Entrance Condition," *Proceedings of the 3rd Pacific Symposium on Flow Visualization and Image Processing*, edited by T. Kobayashi, Paper No. F3073, Pacific Center of Thermal-Fluid Engineering, Maui, Hawaii, March 2001.

[123] Wu, P.-S., and Lin, T.-Y., "Effects of a Forward-Facing Entrance Step on Film Cooling Effectiveness in the Endwall Region of a Vane," *Recent Progress in Transport Phenomena*, Dept. of Mechanical Engineering, Univ. of Indonesia, Bali, Indonesia, July 2003, pp. 573–578.

[124] Wu, P.-S., Chiu, Y.-C., Chen, L.-Y., and Chang, S.-F., "Effects of a Backward-Facing Entrance Step on Heat Transfer and Film Cooling of the Endwall Surface of a Vane Passage," *6thWorld Conference on Experimental Heat Transfer, Fluid Mechanics, and Thermodynamics*, edited by N. Kasagi, S. Marugama, H. Yoshida, and T. Inove, Heat Transfer Society of Japan and Inst. of Fluid Science, 2005.

[125] Reid, K., Denton, J., Pullan, G., Curtis, E., and Longley, J., "Reducing the Performance Penalty due to Turbine Inter-Platform Gaps," American Society of Mechanical Engineers, Paper GT2006-90839, May 2006.

[126] Rubensdorffer, F. G., and Fransson, T. H., "Numerical Investigations of Geometric Design Parameters Defining Nozzle Guide Vane Endwall Heat Transfer," American Society of Mechanical Engineers, Paper GT2006-90177, May 2006.

[127] Piggush, J. D., and Simon, T. W., "Flow Measurements in a First Stage Nozzle Cascade Having Endwall Contouring, Leakage and Assembly Features," American Society of Mechanical Engineers, Paper GT-2005-68340, June 2005.

[128] Yamao, H., Aoki, S., Takeishi, K., and Takeda, K., "An Experimental Study for Endwall Cooling Design of Turbine Vanes," *IGTC-67 Proceedings of the International Gas Turbine Congress, Gas Turbine Society of Japan*, Vol. III, 1987, pp. III-171–III-177.

[129] Ranson, W. W., Thole, L. A., and Cunha, F. J., "Adiabatic Effectiveness Measurements and Predictions of Leakage Flows Along a Blade Endwall," American Society of Mechanical Engineers, Paper IMECE2004- 62021, Nov. 2004.

[130] Kubendran, L. R., and Harvey, W. D., "Juncture Flow Control Using Leading-Edge Fillets," AIAA Paper 85-4097, Oct. 1985.

[131] Sung, C., and Lin, C., "Numerical Investigation on the Effect of Fairing on the Vortex Flows Around Airfoil/Flat-Plate Junctures," AIAA Paper 88-0615, Jan. 1988.

[132] Devenport, W. J., Simpson, R. L., Dewitz, M. B., and Agarwal, N. K., "Effects of a Strake on the Flow past a Wing-Body Junction," AIAA Paper 91-0252, Jan. 1991.

[133] Sauer, H., Müller, R., and Vogeler, K., "Reduction of Secondary Losses in Turbine Cascades by Leading Edge Modifications at the Endwall," American Society of Mechanical Engineers, Paper 2000-GT-473, May 2000.

[134] Sauer, H., Müller, R., Vogeler, K., and Hoeger, M., "Influencing the Secondary Losses in Compressor Cascades by a Leading Edge Bulb Modification at the Endwall," American Society of Mechanical Engineers, Paper GT-2002-30442, June 2002.

[135] Becz, S., Majewski, M. S., and Langston, L. S., "Leading Edge Modification Effects on Turbine Cascade Endwall Loss," American Society of Mechanical Engineers, Paper GT-2003-38898, June 2003.

[136] Becz, S., Majewski, M. S., and Langston, L. S., "An Experimental Investigation of Contoured Leading Edges for Secondary Flow Loss Reduction," American Society of Mechanical Engineers, Paper GT-2004-53964, June 2004.

[137] Zess, G. A., and Thole, K. A., "Computational Design and Experimental Evaluation of Using a Leading Edge Fillet on a Gas Turbine Vane," *Journal of Turbomachinery*, Vol. 124, No. 2, 2002, pp. 167–175.

[138] Shih, T. I.-P., and Lin, Y.-L., "Controlling Secondary-Flow Structure by Leading-Edge Airfoil Fillet and Inlet Swirl to Reduce Aerodynamic Loss and Surface Heat Transfer," *ASME Journal of Turbomachinery*, Vol. 125, Jan. 2003, pp. 48–56.

[139] Lethander, A. T., Thole, K. A., Zess, G., and Wagner, J., "Optimizing the Vane-Endwall Junction to Reduce Adiabatic Wall Temperatures in a Turbine Vane Passage," American Society of Mechanical Engineers, Paper GT 2003-38939, June 2003.

[140] Han, S., and Goldstein, R. J., "Influence of Blade Leading Edge Geometry on Turbine Endwall Heat (Mass) Transfer," American Society of Mechanical Engineers, Paper GT2005-68590, June 2005.

CHAPTER 7

Trailing-Edge Cooling

Frank J. Cunha*
Pratt & Whitney Aircraft, East Hartford, Connecticut

Minking K. Chyu†
University of Pittsburgh, Pittsburgh, Pennsylvania

NOMENCLATURE

A	area
C_p	fluid specific heat at constant pressure
c	specific heat
D	diameter
d	pin diameter
E	cooling parameter
f	friction factor
g_c	gravitational constant
H	trailing-edge height at the base
h	heat-transfer coefficient
I_0	Bessel function of second kind, order zero
J	mechanical to thermal energy conversion factor
k	thermal conductivity
L	length
M	slope parameter
m	flow rate
N	number of cooling openings
n	number of pins
Nu	Nusselt number
P	perimeter
Pr	Prandtl number
Re	Reynolds number
r	radius
S	sink parameter
s	slot height

*Principal Engineer; fjcunha@att.net.
†Leighton and Mary Orr Chair Professor of Engineering, Mechanical Engineering and Materials Science Department.

T	temperature
t	time or thickness
Q	total heat transfer
q	heat flux
U	velocity
u	trailing-edge slope parameter
V	velocity
W	width
x	axial distance
y	$L - x$
z	distance normal to the surface
Δ	finite change
δ	small change
η	film effectiveness
θ	dimensionless temperature
ρ	density
ϕ	cooling effectiveness
ω	speed

SUBSCRIPTS

aw	adiabatic wall
B	bottom
b	base
c	coolant
d	pin diameter
f	coolant flow
g	gas
h	holes
L	longitudinal
l	longitudinal
m	main flow
o	baseline
P	transverse
p	along the longitudinal pitch
R	reference
S	surface
T	top

In modern high-pressure turbine airfoils, the aft section of the airfoil, denoted as the trailing-edge section, requires a high degree of attention in the research and design processes. This is particularly true because of the fundamental role that airfoil trailing edges have in the overall design of turbines in modern aircraft engines and electrical generating powerplants.

Airfoil trailing edges provide a degree of blockage in the gas flowpath, contributing to the overall degradation of turbine performance and efficiency. Along the turbine airfoil external walls, there are pressure profiles on the pressure (concave) and suction (convex) sides, which lead to pressure differences between the pressure and suction sides at the airfoil trailing edge. These pressure profiles result in expansion waves in the blade passage, trailing-edge normal shocks, and wake shedding, as illustrated in Fig. 1 [1, 2]. All of these effects contribute to increasing turbine losses. Thus, the elusive theoretical goal of the aerodynamic Kutta condition [3], requiring the gas flow to leave the airfoil smoothly at the trailing edge, is impossible to achieve with today's technology and cooling requirements. The thickness of airfoil trailing edges is directly related to the turbine aerodynamic performance, as it affects the gas-path aerodynamic losses.

Airfoil cooling in this region is also extremely difficult to achieve as it must be implemented effectively in a relatively small area of the airfoil. Without cooling, however, the trailing-edge metal temperatures would be higher than the melting temperatures of most superalloys.

Innovative ideas are required to obtain the most practical and effective means to cool these airfoil regions. Low or inadequate trailing-edge cooling can lead to high thermal strains as airfoil trailing edges respond thermally faster than other parts of the airfoil. This is because of the heat capacity of relatively small trailing-edge mass when compared to the rest of the airfoil. Resulting thermal mismatch within the airfoil walls can lead to excessive thermal-mechanical cyclic loading, which can be exacerbated with long dwell times at high temperatures during climbing, cruising, and continuous power generation for aircraft and land-based turbines.

Current cooling technology for gas-turbine trailing edges relies primarily on the coolant air induced from the airfoil main-body upstream. As the coolant reaches the trailing edge, it may have limited heat removal capability. As a result, passive heat-transfer enhancement features must be implemented in this

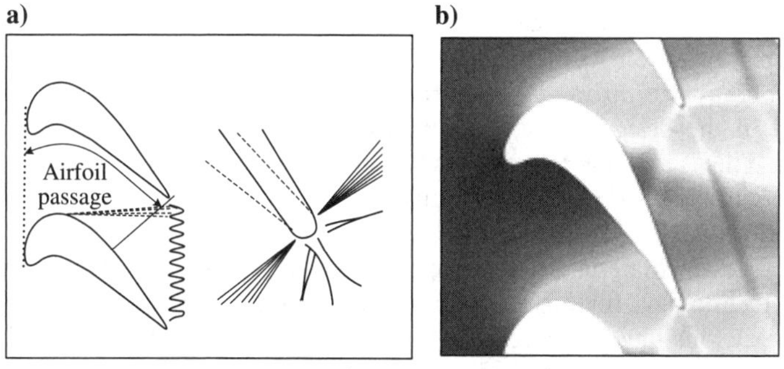

Fig. 1 Typical shock structures in high-pressure turbine airfoils (from a) Kuhne [1] and b) Eisemann [2]).

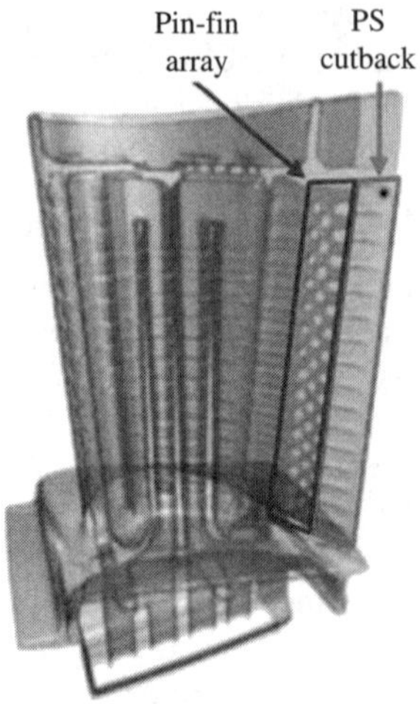

Fig. 2 Cutaway view of turbine blade (from Martini et al. [14]).

region. One type of cooling channel near the trailing edge is an open chamber with cylindrical pedestals connecting opposed suction and pressure-side walls. These pedestals are generally arranged in a staggered array to promote convective heat transfer in the chamber, but they may be subject to concentrated thermally induced stress. Because the pressure-side external thermal load is higher than that of the suction side, there is more thermal expansion on the pressure side. This is particularly the case near the root region immediately adjacent to the platform, where the temperature differential between the two airfoil side surfaces is more substantial. To alleviate such stress, while still maintaining sufficient level of heat-transfer enhancement, some of the cylindrical pedestals are replaced by shaped, filleted, and often simply larger pedestals.

Figure 2 shows a cutaway view of a blade illustrating the trailing-edge configuration. As depicted in this figure, the part to the right of the exit of the internal cooling chamber is the pressure-side cutback area, which is equally spaced by the oblong-shaped features, also known as trailing-edge exit teardrops.

These oblong-shaped teardrops also provide integrity for the blade. From a phenomenological point of view, creep growth can occur close to the apex of the trailing edge. Nearby thermal strains give rise to stress intensities and subsequent trailing-edge thermal-mechanical cracking nucleation. As the centrifugal force remains constant, the colder portions of the airfoil begin sharing the load as the hottest portions of the blade start the process of stress-relaxation. This interaction is commonly referred to as the airfoil load shakedown [4]. Load sharing could eventually lead to a state of potential overstress in the cold regions of the airfoil, with substantial plastic deformation, and the interaction can eventually culminate in catastrophic structural failure. Attention is thus required in the design process to avoid exceeding the endurance limits of the material. This is a major concern for turbine designers and researchers as airfoils are generally designed using deterministic means for crack initiation before crack propagation [5].

Advances in material science have continued, along with advances in blade cooling technology, for airfoil trailing-edge designs. Materials of construction for these airfoils have, therefore, become an important part of material science development and testing. As the rotor inlet temperatures have increased in the last 20 years, there have been considerable advances in alloy composition and casting technology. Early equiaxed nickel and cobalt alloys have been displaced by single crystal nickel superalloys for the most advanced military and commercial gas turbine engines. Even in land-based power generation turbines, directionally solidified and single crystal alloy castings are now found as first-stage turbine components. Similarly, metallic and thermal barrier coatings are well-developed and are considered prime-reliant features in all advanced turbine designs. Even with these advances, the oxidation, creep, and fatigue resistance of trailing edges are still being challenged by the adverse and compounding effects of vibrational stresses. These dynamic effects occur in extreme hot rotating environments with harmonically varying pressure fields.

To bring trailing-edge designs into the available design space, several fabrication methods are currently used. These include 1) internal cooling schemes to bring the metal temperature within design limits; 2) internal aluminide coatings to prevent internal oxidation; 3) external metallic bond coatings to prevent external oxidation and aluminum content depletion; 4) thermal barrier coatings to reduce external thermal loading; 5) structural design features and devices integrated in the cooling schemes to reduce thermal, bending, panel, and vibratory stresses; and 6) optimization of trailing-edge thickness and airfoil wedge angles to minimize film-cooling degradation and aerodynamic losses. These techniques are used to balance the design and satisfy stringent heat transfer, structures, and performance requirements.

I. PREVIOUS STUDIES

Unlike the case for airfoil main body cooling, the research concerning trailing-edge cooling has to date been rather limited. This is particularly the case for geometry involving pressure-side cutback. Taslim et al. [6] investigated the film-cooling effectiveness downstream of trailing-edge slots disrupted by lands in the transverse direction. The influence of density ratio, lip thickness, slot width, and ejection angles was examined. In line with some earlier findings [6–10], they suggested that lip-to-slot ratio is a key parameter for film cooling on cutback surfaces. The other important finding observed in their study is that the adiabatic effectiveness for a given blowing ratio is virtually insensitive to density ratio and slot width, but it is sensitive to the injection angle, with an optimum angle of about 8.5 deg. More recently, Uzol and coworkers [11, 12] studied the discharge behavior of different trailing-edge slots at various lengths of the cutback surface in a scaled-up subsonic cascade. Their results, from PIV and total pressure measurements in the wake region, suggested that trailing edges with pressure-side cutback induce smaller aerodynamic loss than those without pressure-side cutback. Holloway et al. [13, 14] numerically investigated flow and heat transfer in a trailing-edge slot with disrupted lands under realistic engine conditions. Their simulation revealed periodic vortex shedding when the lip-to-slot ratio $t/s = 0.9$. This vortex shedding might be involved in the relatively fast decay of film-cooling effectiveness for slots with larger t/S. Chyu et al. [15] reported a study concerning internal heat transfer in a triple-cavity structure trailing edge. They reported that the heat-transfer enhancement level in such a configuration is nearly 50% lower than that for the pedestals or pin fins when $Re > 2 \times 10^4$. Martini and Schulz [16] examined a trailing-edge slot equipped with an internal rib array featuring a small lateral pitch of 2. Their experimental and numerical efforts showed regrouping of coolants jets resulted in a significant impact on trailing-edge pressure-side cutback film-cooling effectiveness. Kim et al. [17] studied the effect of mainstream acceleration on the film-cooling effectiveness and heat transfer in a trailing edge with pressure-side cutback. Their results indicated that both local heat-transfer coefficients and film-cooling effectiveness exhibited a strong dependency on blowing ratio and mainstream acceleration. Recently, Martini et al. [18] measured the cooling effectiveness and heat transfer on the trailing-edge cutback of gas-turbine airfoils with different internal cooling designs. The designs included double in-line rib array, equilaterally staggered cylindrical pin arrays, and equilaterally staggered cylindrical pin arrays with lip overhang. Their results indicated that the extension of the core region where η is close to unity is influenced strongly by the internal cooling design whereas the decay of η downstream of the core region is similar for all trailing-edge cooling slots. The internal cooling design mainly affects the region $x/s < 5$ while the influence of the blowing ratio extends further downstream. Cunha et al. [19] presented closed-form, analytical solutions for temperature profile for four of the most common trailing-edge configurations, which will be detailed here for reference. They also discussed the overall and detail design parameters for the preferred trailing-edge cooling configurations.

One of the most unique features inherited in the present problem associated with the cooling mechanism depicted in Fig. 2 is a combination of both internal and external cooling. The internal cooling takes place in the cooling chamber that houses the pedestal array; the external cooling prevails in the cutback area. For research investigations, the heat transfer in the cutback area subjected to slot film protection is characterized as essentially a "three-temperature" convection problem as the thermal transport in the region is determined by 1) wall temperature, 2) mainstream temperature, and 3) slot-film discharge temperature. In addition to heat-transfer coefficient, a new reference temperature, such as adiabatic temperature or dimensionless film effectiveness, needs to be determined. A detailed method for resolving three-temperature problems can be found in [20–24].

In the present treatment, attention will be focused on two fundamental building blocks for assessing cooling in gas-turbine trailing edges. The first is an analytical development; the second is an experimental setup and procedures currently used for assessing the heat transfer and film effectiveness associated with gas-turbine trailing-edge regions. In the analytical development, four of the most representative configurations will be introduced and their relative merits considered, in terms of durability. The interplay between geometrical features and airfoil metal temperatures will be described and explained using closed-form, analytical models. Trailing-edge cooling schemes will be presented in the context of metal temperature or resulting cooling effectiveness as it affects all other design parameters, including creep, thermal-mechanical fatigue, oxidation, spallation, and performance. The four configurations studied are as follows: 1) solid wedge shape without discharge, 2) wedge with slot discharge, 3) wedge with discrete-hole discharge, and 4) wedge with pressure-side cutback slot discharge.

In the experimental domain, a "hybrid" measurement technique is presented. The hybrid denomination arises from the contribution of two different test procedures: 1) the one-dimensional transient liquid-crystal technique is used for surface uncovered by pedestals, and 2) one lumped heat capacity liquid-crystal technique is used for the heat transfer over pedestal surfaces.

II. ANALYTICAL MODELS

The underlying consideration for the models developed in this section is the one-dimensional nature of the heat transfer verified at the airfoil trailing-edge regions of operational blades and vanes. Thus, the heat-flux vectors are considerably normal to the trailing-edge surface with very little lateral heat flux. These considerations have been confirmed in hardware design. The setup of thermal boundary conditions for the model can be a very laborious process. On the external side, computational fluid dynamics (CFD) can be used to set up external pressure and Mach-number distributions with a set of turbulence intensity levels, as shown in Fig. 1. Boundary programs are used to establish the external heat-transfer coefficients based on expected wall roughness and airfoil curvature. The evaluation of these coefficients is not the purpose of this chapter, and as a result, representative values are provided as boundary conditions to the analytical models. The same procedure is followed for the gas temperatures, which are usually determined from previous film cooling tests. Different coating systems can be modeled as modified equivalent heat-transfer coefficients. Inside the airfoil, flow network tools can be used to determine internal pressure drops, internal fluid temperatures, and Mach-number distributions. Heat-transfer correlations based on previous testing can be used to provide the means to evaluate internal heat-transfer coefficients. With known external and internal conditions, the following analytical models can be used to study different airfoil trailing-edge configurations. In these configurations, the effects of airfoil curvature are considered second-order effects, as these sections are relatively short, on the order of less than 0.35 cm (0.138 in.) in a true airfoil scale. The value of the analytical models presented here is to provide the basis for determining analytically the geometrical effects relevant to each trailing-edge configuration. This analytical approach also offers a platform for benchmarking subsequent experimental testing.

A. CASE 1: SOLID TRAILING EDGE WITHOUT INTERNAL COOLING

By way of introduction, the simple wedge model shown in Fig. 3 is considered. In this figure, the airfoil trailing edge is represented by a wedge-fin type of configuration without internal convective cooling. Different cooling schemes will be considered in subsequent models.

If an infinitesimal control volume of Fig. 3 is considered, the following mathematical expression states that the energy stored in the control volume is equal to the heat transferred to the system by external convection and conducted away from the system by thermal conduction:

$$\underbrace{-\frac{\partial}{\partial x}\left(-kA\frac{\partial T}{\partial x}\right)\delta x}_{\text{Conduction}} + \underbrace{h_g(T_g - T)P\delta x}_{\substack{\text{External}\\ \text{Convection}}} = \underbrace{\rho A c\frac{\partial T}{\partial t}\delta x}_{\substack{\text{Energy}\\ \text{Storage}}} \tag{1}$$

where

$$A = (HW)\left(\frac{L - x}{L}\right)$$

$$P = 2\left[W + H\left(\frac{L - x}{L}\right)\right]u$$

$$u = 2\left\{1 + \left[\frac{H/2}{L}\right]^2\right\}^{1/2}$$

In expression (1), the conduction cross-sectional area A and the corresponding perimeter P of the control volume are functions of the axial distance x. Other geometrical attributes are depicted in Fig. 3. Variables T and T_g denote metal temperature and external gas temperature, respectively. The thermophysical variables k, ρ, and c denote the metal thermal conductivity, density, and specific heat capacity properties, respectively. On the right-hand side, time is denoted by variable t. The right-hand side term of Eq. (1) will vanish if one considers the steady-state condition only for the control volume. Further, simplification for

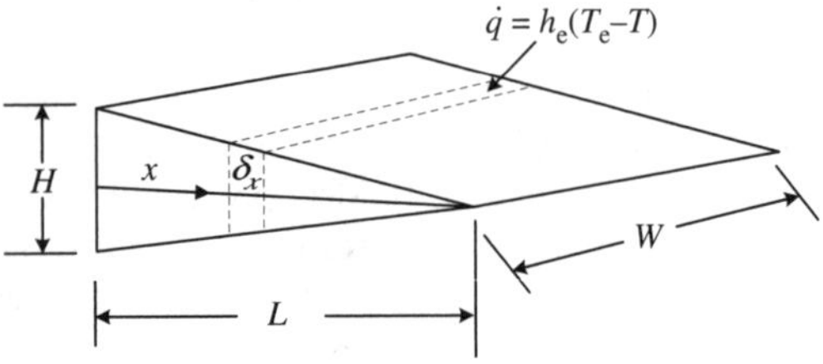

Fig. 3 Trailing-edge analytical model without internal convective cooling.

large W leads to the following equation:

$$\frac{\mathrm{d}}{\mathrm{d}x}\left[(L-x)\frac{\mathrm{d}T}{\mathrm{d}x}\right]+\left[\left(\frac{h_g}{k}\right)\left(\frac{L}{H}\right)(2u)\right](T_g-T)=0 \tag{2}$$

Equation (2) is further simplified by a change of variables, using $\theta = T - T_g$ and $y = L - x$, to obtain the fundamental modified Bessel equation of order zero, Eq. (3):

$$y^2\frac{\mathrm{d}^2\theta}{\mathrm{d}y^2}+y\frac{\mathrm{d}\theta}{\mathrm{d}y}-E^2y\theta=0 \tag{3}$$

$$E^2=\left[\left(\frac{h_g}{k}\right)\left(\frac{L}{H}\right)(2u)\right]$$

To solve this second-order differential equation, two boundary conditions are required. The first boundary condition is located at the base of the wedge model of Fig. 3 at $x = 0$ and requires that $T = T_b$ or $\theta = \theta_b$, where the subscript b stands for a base quantity. The other boundary condition comes from the requirement that the metal temperature at the apex of the wedge model, at $x = L$, be finite. The solution of Eq. (3) then becomes

$$\frac{\theta(x)}{\theta_b}=\frac{I_0\left[2E(L-x)^{1/2}\right]}{I_0(2EL^{1/2})} \tag{4a}$$

$$T(x)=T_g-(T_g-T_b)\frac{I_0\left[2E(L-x)^{1/2}\right]}{I_0(2EL^{1/2})} \tag{4b}$$

Solutions (4a) and (4b) are equivalent. Either solution is fundamental in that it assumes a form that will appear again as the homogeneous solution for problems involving different cooling schemes. These fundamental solutions will always follow the form of solution (4), with different interpretations for the argument E, of the Bessel function of second kind and order zero I_0. Variations of the model of Fig. 3 with different internal cooling schemes are now considered.

B. CASE 2: TRAILING EDGE WITH CENTERLINE SLOT DISCHARGE

The next model, depicted in Fig. 4, is analyzed in similar fashion. The governing equation is derived as per Eq. (3) with the addition of the nonhomogeneous term S, which accounts for the slot cooling effect. As noted before, the homogeneous solution is

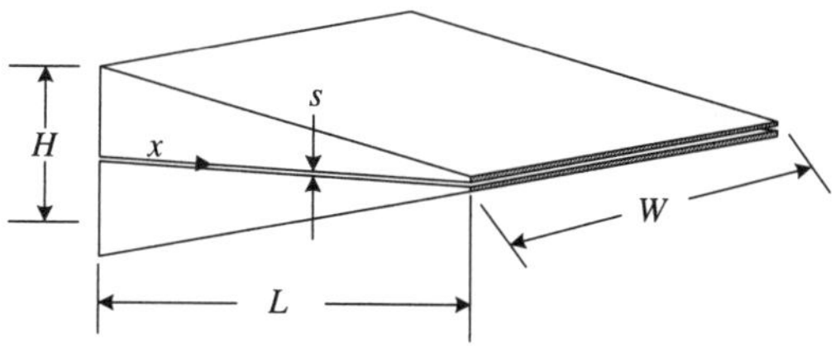

Fig. 4 Trailing-edge model with centerline slot.

provided by Eq. (4). In this case, argument E is modified to account for the internal cooling effect of the centerline slot.

The internal heat-transfer coefficient, denoted by h_c, is assumed to be constant. In subsequent sections, expressions to relate this internal heat-transfer coefficient to geometrical attributes of the cooling passage will be discussed. It is sufficient to say here that the coefficient is considered to be known and that the sink effect is the particular solution of Eq. (5), defined by the source term S, as follows:

$$y\frac{\mathrm{d}^2\theta}{\mathrm{d}y^2}+\frac{\mathrm{d}\theta}{\mathrm{d}y}-E^2\theta=S \tag{5}$$

where

$$y=L\left(1-\frac{s}{H}\right)-x$$

$$E^2=\left(\frac{h_g}{k}\right)\left(\frac{L}{H}\right)(2u)+2\left(\frac{h_c}{k}\right)\left(\frac{L}{H}\right)$$

$$S=2\left(\frac{h_c}{k}\right)\left(\frac{L}{H}\right)\left[T_g-T_c(x)\right]$$

The complete solution of Eq. (5) will be presented in a subsequent section with more details about the coolant temperature distribution.

C. CASE 3: TRAILING EDGE WITH ARBITRARY OPENINGS IN A CENTERLINE DISCHARGE

In the cooling configuration shown in Fig. 5, discrete cooling holes with arbitrary shapes are considered. The formulation follows that considered for the model of Fig. 3. The resulting governing equation is similar to Eq. (5) with parameters modified to account for the new cooling arrangement.

In this treatment, the geometrical parameter A_H denotes the cross-sectional area of an arbitrarily shaped cooling hole with P_H as the perimeter of the cooling hole. The term N_H denotes the number of cooling

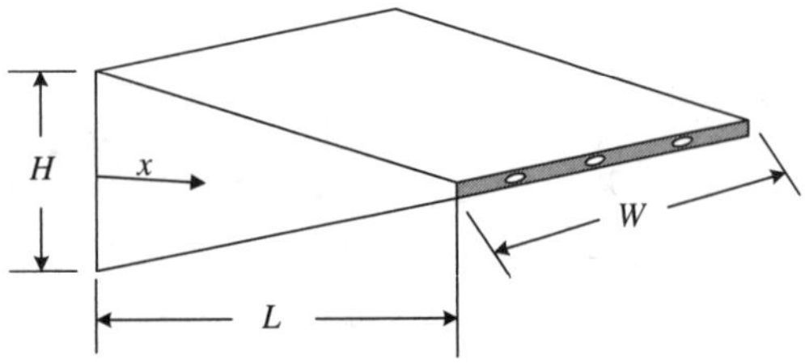

Fig. 5 Trailing-edge model with cooling holes.

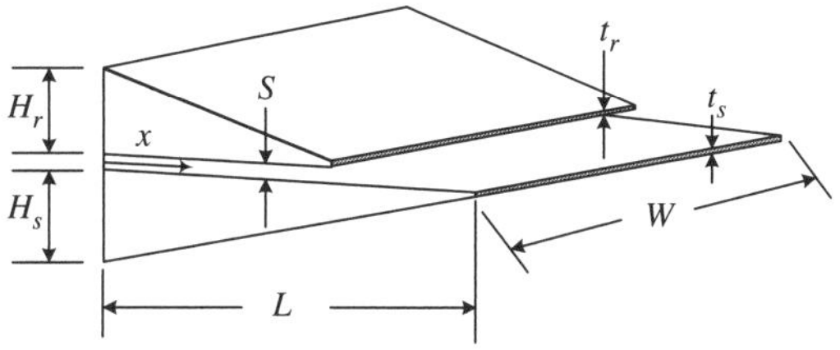

Fig. 6 Cutback trailing-edge model with slots.

holes in the radial span. The new governing parameters for this configuration become

$$y = L\left(1 - \frac{A_H N_H}{WH}\right) - x$$

$$E^2 = \left(\frac{h_g}{k}\right)\left(\frac{L}{H}\right)(2u) + \left(\frac{P_H N_H}{W}\right)\left(\frac{h_c}{k}\right)\left(\frac{L}{H}\right)$$

and

$$S(x) = \left(\frac{P_H N_H}{W}\right)\left(\frac{h_c}{k}\right)\left(\frac{L}{H}\right)\left[T_g - T_c(x)\right]$$

The complete solution of Eq. (5) will be presented in a subsequent section when more details about the coolant temperature distribution are provided.

D. CASE 4: TRAILING-EDGE CUTBACK DESIGN WITH PRESSURE-SIDE DISCHARGE

This cooling configuration deviates from centerline discharge as the pressure side of the trailing edge has a cutback leading to pressure-side ejection of the coolant into the main gas stream. The model shown in Fig. 6 has the trailing edge of the airfoil extended all of the way on the suction (bottom) side with a cutback located on the pressure (top) side. The pressure-side cutback is probably one of the most utilized cooling configurations in today's airfoil design because it permits a significantly thinner trailing edge, approximately 0.76 mm (0.030 in.), as opposed to 1.27 mm (0.050 in.). Therefore, the aerodynamic losses associated with the cutback design attain the lowest values among all of the cases analyzed here.

The governing equation is similar to that given by Eq. (5). For simplicity, the defining governing parameters for this configuration are divided into two parts: one for a side defined up to $x < L_T$ and the other for the suction side defined by $L_T \le x \le L_B$.

For $x < L_T$:

$$y = \frac{H}{M} - x$$

$$M = \left(\frac{H_T - t_T}{L_T}\right) + \left(\frac{H_B - t_B}{L_B}\right)$$

$$E^2 = \left[\frac{\bar{h}_g(u_T + u_B) - 2h_c}{kM}\right]$$

$$S(x) = \left(\frac{2h_c}{kM}\right)[T_g - T_c(x)]$$

with

$$\bar{h}_g = (h_{g,T} + h_{g,B})/2$$

$$u_T = \left[1 + \left(\frac{H_T - t_T}{L_T}\right)^2\right]^{1/2}$$

$$u_B = \left[1 + \left(\frac{H_B - t_B}{L_T}\right)^2\right]$$

and for $L_T \le x \le L_B$:

$$y = \frac{H_B}{M} - x$$

$$M = \left(\frac{H_B - t_B}{L_B}\right)$$

$$E^2 = \left[\frac{h_{g,B}(u_B) - h_c}{kM}\right]$$

$$S(x) = \left(\frac{h_c}{kM}\right)[T_g - T_c(x)]$$

For the particular solution of Eq. (5), it is necessary that the term S be obtained. In this case, the external conditions T_g and h_g are assumed to be known constants, for simplicity and convenience, while the coolant temperature T_c must be determined.

If one considers a fluid control volume at the same axial distance as that for the control volume shown in Fig. 3 and assumes that the heat flux q'' is a known constant, again for simplicity and convenience, then an expression for the coolant heat pickup can be written as follows:

$$\rho\, VA\, c_P \frac{\mathrm{d}T_C}{\mathrm{d}x}\delta x = q''(x)\delta A_{C,S} \tag{6}$$

where

$$q''(x) = q_T''(x) + q_B''(x) \qquad x < L_T$$

$$q''(x) = q_B''(x) \qquad L_T \le x \le L_B$$

The heat fluxes q_T'' and q_B'' denote the heat flux from the top and bottom parts of the cooling passage. The quantity $\delta A_{C,S}$ denotes the control volume internal surface area of the cooling passage. The governing Eq. (6) has a single inlet boundary condition at $x = 0$ as $T_c = T_{c,i}$ or $\theta = \theta_{c,i}$, and the solution becomes

$$\theta_c = \theta_{c,i} + \Delta\theta_c(x) \tag{7}$$

where

$$\Delta\theta_c = \frac{q''\Delta A_{C,S}}{\rho V A c_P}$$

Even though the calculation of the trailing-edge metal temperature is an important step in the durability design process, it is equally important to determine the cooling flow rate through the trailing edge. After all, the cooling flow rate is directly related to the internal heat-transfer capability of the cooling design. To estimate the trailing-edge cooling flow, one needs to consider the inlet coolant flow rate, at the root of the supply cavity, denoted here as m_{Root}, as well as the tip-cooling flow rate at the tip, denoted as m_{Tip}, and flow variation through the trailing-edge feed cavity. The linear function of cooling flow rate distribution from the root to the tip as flow exits the trailing edge leads to the following flow relationship:

$$m(r) = m_{\mathrm{Root}}(1 - ar) \tag{8a}$$

where

$$a = \frac{1}{L_H}\left(1 - \frac{m_{\mathrm{Tip}}}{m_{\mathrm{Root}}}\right) \tag{8b}$$

with L_H as the blade span radial height. Note that if there is no communication between the tip holes and the trailing-edge supply cavity, the term a in Eq. (8) is equal to the reciprocal of the blade span radial height, and the trailing-edge flow distribution varies linearly from a finite mass flow rate at the root section to a no-flow condition at the tip.

The temperature of the coolant inside the feed cavity is obtained by considering an infinitesimally small control volume dr in the supply cavity. This consideration leads to the following energy balance:

$$m c_P \frac{\mathrm{d}T_c}{\mathrm{d}r} = \frac{\mathrm{d}Q}{\mathrm{d}r} + \frac{m r \omega^2}{J g_c} \tag{9}$$

where Q denotes the total heat transfer onto the corresponding trailing-edge section of radial span L_H.

The trailing-edge coolant flow rate of Eq. (8) is then introduced in Eq. (9). The resulting expression is integrated from the root to any other radial section of the airfoil. This yields an expression for the coolant temperature increase in terms of heat transfer and rotational pumping, as a function of coolant flow rate. The result is

$$T_c(r) = T_{c,\text{Root}} + \left(\frac{Q_{\text{Total}}}{aL_H m_{\text{Root}} c_p}\right) \ell n \left(\frac{1 - ar_{\text{Root}}}{1 - ar}\right) + \frac{(r^2 - r_{\text{Root}}{}^2)\omega^2}{2\, Jg_c c_p} \tag{10}$$

This expression allows for calculation of the temperature increase of the coolant in the supply cavity before it enters the trailing-edge passages, and it should be used as $T_{c,i}$ for Eq. (7), presented earlier.

Finally, when the particular solution Eq. (7) is substituted in Eq. (5) for evaluating source term S, the following complete solution is obtained:

$$\theta(x) = \frac{I_0[2E(L-x)^{1/2}]}{I_0[2EL^{1/2}]}\left[\theta_b - \left(\frac{E_c}{E}\right)^2 \theta_{c,i}\right] + \left(\frac{E_c}{E}\right)^2 [\theta_{c,i} + \Delta\theta_c(x)] \tag{11a}$$

or

$$T(x) = T_g - \left\{ \begin{array}{l} \dfrac{I_0[2E(L-x)^{1/2}]}{I_0(2EL^{1/2})} \left\{ [(T_g - T_b] - \left(\dfrac{E_c}{E}\right)^2 (T_g - T_{c,i}) \right\} \\ + \left(\dfrac{E_c}{E}\right)^2 [(T_g - T_{c,i}) + \Delta T_c(x)] \end{array} \right\} \tag{11b}$$

Solutions (11a) and (8b) are equivalent and applicable to all of the configurations studied. The parameters E^2 and E_c^2 for each configuration are summarized in Table 1.

TABLE 1 PARAMETERS FOR DIFFERENT MODELS

Configuration	E^2	E_c^2
Solid wedge model (Fig. 3)	$\left[\left(\frac{h_g}{k}\right)\left(\frac{L}{H}\right)(2u)\right]$	0
Wedge with slot cooling (Fig. 4)	$\left[\left(\frac{h_g}{k}\right)\left(\frac{L}{H}\right)(2u)\right] + 2\left(\frac{h_c}{k}\right)\left(\frac{L}{H}\right)$	$2\left(\frac{h_c}{k}\right)\left(\frac{L}{H}\right)$
Wedge with cooling holes (Fig. 5)	$\left(\frac{h_g}{k}\right)\left(\frac{L}{H}\right)(2u) + \left(\frac{A_{S,H}N_H}{W}\right)\left(\frac{h_g}{k}\right)\left(\frac{L}{H}\right)$	$\left(\frac{A_{S,H}N_H}{W}\right)\left(\frac{h_g}{k}\right)\left(\frac{L}{H}\right)$
Cutback wedge with slot cooling (Fig. 6)	$\left[\frac{\bar{h}_g(u_T + u_B) - 2h_c}{kM}\right]$ $x < L_T$ $\left[\frac{h_{g,B}(u_B) - h_c}{kM}\right]$ $L_T \le x \le L_B$	$\left(\frac{2h_c}{kM}\right)$ $x < L_T$ $\left(\frac{h_c}{kM}\right)$ $L_T \le x \le L_B$

TABLE 2 BOUNDARY CONDITIONS FOR TRAILING-EDGE EXAMPLES

Configuration	Solid	Centerline Slot	Centerline Holes	Cutback PS Ejection
Figure	2	3	4	5
T_g	2300	2300	2300	2300
$\Delta T_{C,T}$, ΔT_C	50	50	50	50
$\Delta T_{C,B}$	—	—	—	150
T_{cjn}	1600	1600	1600	1600
T_{base}	1900	1900	1900	1900
h_g or $h_{g,T}$	2000	2000	2000	2000
$h_{g,B}$	—	—	—	1000
h_C	2000	2000	2000	—
$h_{C,T}$	—	—	—	2000
$h_{C,B}$	—	—	—	2500
k	12	12	12	12
L_T or L	0.12	0.12	0.12	0.06
L_B	—	—	—	0.12
H_T or H	0.06	0.06	0.06	0.03
H_B	—	—	—	0.03
t_T	—	—	—	0.01
t_B	—	—	—	0.029
P_H	—	0.02	—	—
d_h	—	—	0.015	—
W	—	—	2.5	—
S	—	0.015	—	0.015

Notes:

h_C, h_g: $\dfrac{\mathrm{BTU}}{\mathrm{hr-ft^2-F}}$, k: $\dfrac{\mathrm{BTU}}{\mathrm{hr-ft^2-F}}$;

$\dfrac{\mathrm{BTU}}{\mathrm{hr-ft^2-F}} = 5.68\dfrac{W}{\mathrm{m^2k}}$, $\dfrac{\mathrm{BTU}}{\mathrm{hr-ft-F}} = 1.73\dfrac{W}{\mathrm{m\,k}}$

All length dimensions are in inches, and temperatures are in degrees Fahrenheit. 1 in. = 0.0254 m; $F = 32 + \left(\frac{9}{5}\right)C$

E. ANALYTICAL SOLUTIONS AND THEIR IMPLICATIONS FOR TRAILING-EDGE DESIGN

To illustrate the realistic characteristics of each design, Table 2 gives representative magnitudes of key variables associated with the parameters listed in Table 1.

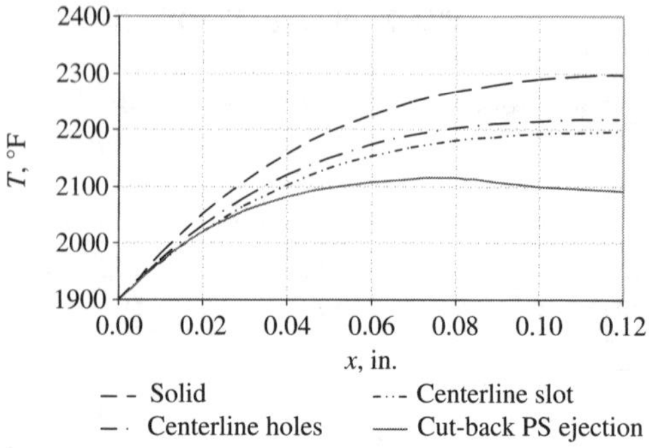

Fig. 7 Metal temperature results for different configuration.

With these data, the temperature distributions based on Eq. (11b) for all four configurations are given in Fig. 7. One notable finding is that the solid trailing edge yields the highest metal temperatures, approaching 1260°C (2300°F). As per solution (4), using the ratio of Bessel functions requires that the metal temperature profile increase monotonically.

The same trend is observed for the centerline discharge cases of the slot and discrete cooling openings. Based on the input data from Table 2, the maximum temperature occurs at the end of the trailing edge with 1202°C (2196°F) and 1214°C (2219°F) for the slot and discrete cooling openings, respectively. This decrease in metal temperature is a sole function of the sink effect provided by the cooling configuration as the base temperature and all other conditions are kept constant for all cases considered. For the centerline discharge configurations, the cooling design affects the metal temperatures monotonically. That is, if the metal temperature is considered elevated for a particular design application, then the only variable to be iterated upon is the coolant passage heat-transfer coefficient to reduce the metal temperature at the end of the trailing edge. This implies that either the size of the coolant passage should be increased or the features inside the passage must be modified to enhance the magnitudes of internal heat-transfer coefficient. Alternatively, coolant flow rates can be increased to enhance cooling. This approach is viable only if sufficient supply pressure is available.

For the configuration with a pressure-side ejection cutback trailing edge, the limitations associated with the thicker trailing edge in the centerline designs are removed. As shown in Fig. 7, the metal temperature profile does not increase monotonically for pressure-side ejection designs. This allows for several alternative design options to optimize the cooling configuration: 1) increase in the size of the cooling passage, 2) selection of internal cooling features inside the cooling passage to enhance internal heat transfer, 3) reduction of trailing-edge thickness, 4) reduction of pressure-side lip thickness, 5) increase in pressure-side land roughness to augment coolant heat transfer, and 6) increase in slot-film coverage and thus effectiveness. From this set of design features, only two features, namely 1 and 2, can be used realistically for centerline discharge designs whereas all features, 1 through 6, can be used for pressure-side ejection cutback trailing-edge designs. There are also benefits in improved aerodynamic performance with thinner trailing edge. In addition, the thermal-mechanical fatigue and creep life capability are also improved as metal temperature distributions are improved for the entire trailing-edge region of the airfoil.

III. EXPERIMENTAL SETUP AND PROCEDURES

The experimental procedure described here uses a "hybrid" measurement technique based on transient liquid crystal imaging. The hybrid denomination is based on two different test procedures: 1) the one-dimensional transient model for surface uncovered by pedestals and 2) one-lumped heat capacity model for the heat transfer over pedestal surfaces. Both of these models will be discussed in detail in this section. The hybrid technique employs the transient conduction model in a semi-infinite solid and the lumped capacitance method for resolving the heat-transfer coefficient for the endwall surface uncovered by the pedestals and for the pedestal surface, respectively. A brief description of the hybrid method and data reduction is also presented in this section.

The test section is a scaled-up model, made of Plexiglas® of the actual trailing edge. Figure 8 is a schematic view of the configuration. The internal cooling chamber is a convergent, wedge-shaped duct, with a four-row staggered pin fin arrays. Near the exit of the duct, oblong-shaped features are used for connecting the pressure side and the suction side. All pedestals or pins are positioned orthogonally relative to the mainstream. The suction side wall is inclined 10.5 deg relative to the horizontal line, as shown in Fig. 8. The pin diameter is $d = 6.0$ mm. The pin spacing along the longitude and transverse direction is the same: $s_l/d = s_p/d = 2.0$. The pins and the oblong-shaped features are made of aluminum, which has high thermal conductivity, for the lumped capacity model. On the other hand, the surfaces modeling the pressure and suction sidewalls are made of Plexiglas®, whose low thermal conductivity is required for the use of the transient semi-infinite solid conduction model. The pressure-side cutback area partitioned by the oblong-shaped features is also shown in Fig. 8. The land section subjected to film-cooling measurement is highlighted in blue color for reference.

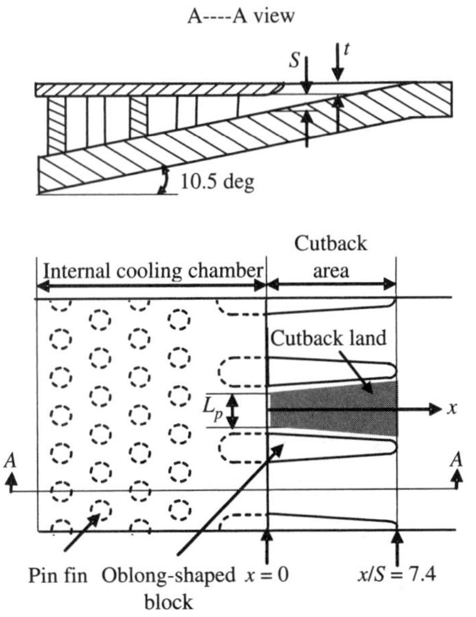

Fig. 8 Schematics of the trailing-edge configuration.

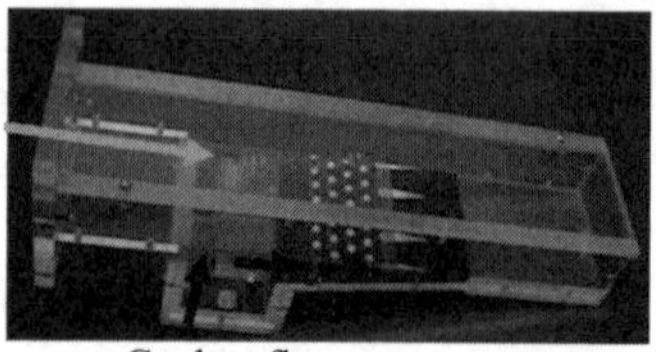

Fig. 9 Photo of trailing-edge test rig.

Figure 9 shows a photo of the actual test rig. The rig, made of Plexiglas®, is of rectangular cross section, 38.1 mm (1.5 in.) high and 76 mm (3 in.) wide. To model the film cooling, mainstream is provided outside the pressure side. The coolant enters through a settling chamber underneath the test rig, then turns rightward to the pedestal cooling channel, and injects on the cutback land. For the internal heat-transfer study, the main flow is not introduced.

The test surface is coated with a thin layer of thermochromic liquid crystal (TLC) about 0.1 to 0.3 mm (0.004 to 0.012 in.) thick. An airbrush is used to spray the coating. The temperature calibrated for the particular TLC used in the present study is 36°C (96.8°F) for the maximum green color intensity. The maximum green intensity displayed by TLC is used as the surface temperature tracer during a transient measurement.

Figure 10 shows a schematic of the test setup. A laboratory compressor supplies both the main stream and film-cooling flows. Flow rates of the compressed air are measured by standard American Society of Mechanical Engineers (ASME) orifices. After metered, each stream is routed through a tubular in-line heater controlled by an autotransformer, and its temperature can be accurately

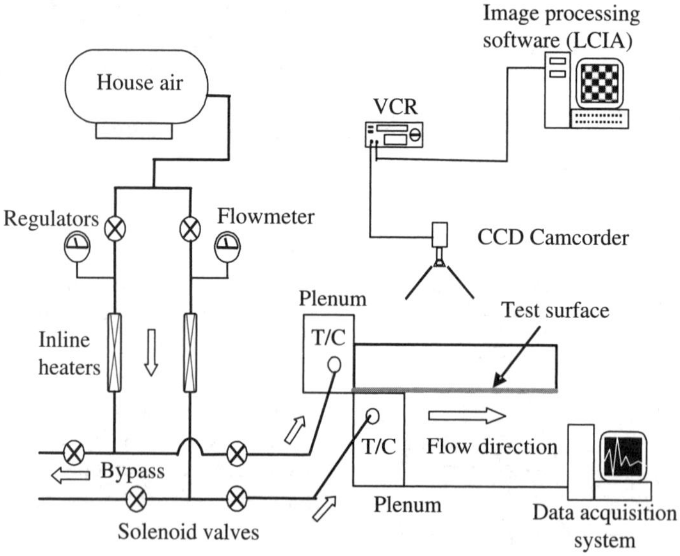

Fig. 10 Test apparatus.

set to the desired level. Downstream of the heater, two flows are initially diverted away from the test section by a three-way ball valve before the test is started.

Before the test is started, the flows are diverted to bypass so that the test section remains at the laboratory ambient temperature. When the flow rate and the temperature have reached steady state, the valves are suddenly switched to introduce the flows to the test section. Initiation of the test triggers an automated data acquisition system for recording thermocouple readings at the flow inlet, as well as the exit. Simultaneously, a charge-coupled device (CCD) camera starts to record the video images of the TLC coated on the test surface. The video provides the follow-on data-reduction procedure with the lap time of TLC changing from colorless to green at every pixel of the viewing domain. The postrun data-reduction procedure uses a custom-developed software package in a personal computer to obtain the local heat-transfer coefficient [25]. As mentioned before, only the film-cooling flow line is used whereas main flow line is closed for the internal cooling chamber heat-transfer tests; otherwise, the experimental procedure is basically the same as that just described.

A. HYBRID METHOD AND DATA REDUCTION

For the internal cooling chamber, the local heat-transfer coefficient over the uncovered endwall surface is obtained by using one-dimensional transient conduction over a semi-infinite solid while the regional-averaged heat-transfer coefficient over the pin surface is resolved using the lumped capacity model. The following are brief introductions of both approaches.

1. ONE-DIMENSIONAL TRANSIENT MODEL

The one-dimensional transient model treats the substrate beneath the surface as a semi-infinite solid domain, as illustrated in Fig. 11. When the surface is suddenly exposed to a forced flow stream, with a steady convective heat-transfer coefficient h, prescribed on the surface, the temperature field in the solid domain can be modeled by the following one-dimensional, transient heat-conduction equation:

$$k\frac{\partial^2 T}{\partial z^2} = \rho C\frac{\partial T}{\partial t} \tag{12}$$

with boundary and initial conditions

$$-k\left.\frac{\partial T}{\partial z}\right|_{z=0} = h(T_w - T_r) \tag{13}$$

$$T|_{z=\infty} = T_i \tag{14}$$

$$T|_{t=0} = T_i \tag{15}$$

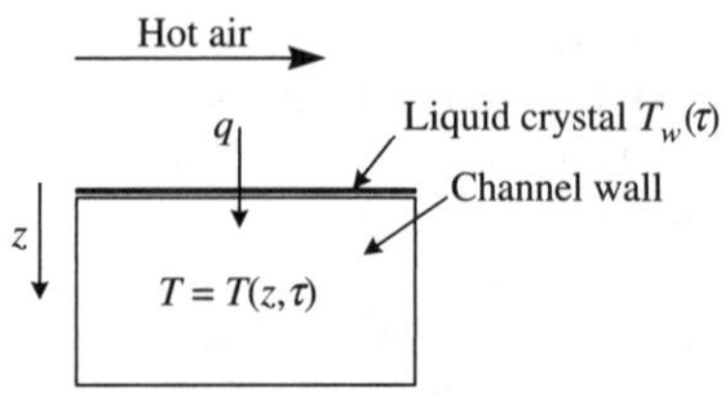

Fig. 11 Schematic of one-dimensional transient heat-transfer model.

where T_i is the initial temperature of the test section, T_w is the local surface temperature, and T_r is the flow reference temperature. The difference between the temperatures T_w and T_r represents the driving potential for the convective heat transfer in the system. Equations (12–15) lead to a solution of T_w expressed as

$$\frac{T_w - T_i}{T_r - T_i} = 1 - \exp\left[\frac{h^2 \alpha t}{k^2}\right] erfc\left[\frac{h\sqrt{\alpha t}}{k}\right] \tag{16}$$

The reference temperature T_r is equal to the temperature of the hot air or bulk flow. As the time-varying TLC images can provide a relation between temperature T_w and time t over the entire viewing domain, the distribution of local heat-transfer coefficient h can be resolved from the preceding equation.

In an actual experiment, a perfect step change of the applied flow temperature is usually not possible, and the reference temperature is, in fact, a function of time. This can be accounted for by modifying the solutions via superposition and Duhamel's theorem. The solution becomes

$$T - T_i = \sum_{i=1}^{N} U(t - \tau_i)\Delta T_r \tag{17}$$

where

$$U(t - \tau_i) = 1 - \exp\left[\frac{h^2}{k^2}\alpha(t - \tau_i)\right] erfc\left[\frac{h}{k}\sqrt{\alpha(t - \tau_i)}\right] \tag{18}$$

The local heat-transfer coefficient on the uncovered endwall is obtained by using the preceding two equations (16) and (17).

2. LUMPED-HEAT-CAPACITY MODEL

Consider a test element with a mass m and initial temperature T_i that is suddenly exposed to a flow stream that has a steady temperature T_r and imposes a convective heat-transfer coefficient h on the element's surface. If the element is of low Biot number, its temperature as function of time $T(t)$ can be modeled by an initial value problem:

$$hA(T - T_r) = -mC\frac{dT}{dt} \tag{19}$$

and

$$T|_{t=0} = T_i \tag{20}$$

where A is the effective heat-transfer area and C is the heat capacitance of the element. The solution to the preceding equation is

$$\frac{T - T_i}{T_r - T_i} = 1 - \exp\left(-\frac{hAt}{mC_p}\right) \tag{21}$$

For imperfect step changes in flow temperature, superposition and Duhamel's theorem is applied as in Eqs. (17) and (18).

In present study, the regional-averaged heat-transfer coefficient over the pedestal surface is resolved using this method.

For the test method and data reduction for the film-cooling study, two tests, referred to as "hot test and cold test," are needed, and two coupled equations similar to Eq. (16) are solved to obtain film-cooling effectiveness and heat-transfer coefficient h. Detailed information can be found in [23].

B. GOVERNING PARAMETER DEFINITIONS

For the internal cooling chamber investigation, the channel is convergent, and as a result, the flow area changes along the flow direction. Therefore the minimum area A min is set as the first row of pins and used for Reynolds-number definition

$$Re = \frac{\rho Q d}{A \min \mu} \tag{22}$$

where Q is the volume flow rate and d is the diameter of the circular pin. For Nusselt number, pin fin diameter d is used as characteristic length, yielding the definition

$$Nu = \frac{hd}{k} \tag{23}$$

For the film-cooling performance on cutback land study, the definition of blowing ratio is given as shown here:

$$M = \frac{(\rho U)_f}{(\rho U)_m} \tag{24}$$

where subscripts m and f represent main flow and film cooling, respectively. The density ratio in the present study is equal to one. Hence, the existing blowing ratio is essentially the flow velocity ratio between the film-cooling flow and the main flow. The Reynolds numbers for main flow and film-cooling flow are defined as

$$Re_m = \frac{\rho\, UD_h}{\mu} \tag{25}$$

and

$$Re_f = \frac{\rho\, MUd_h}{\mu} \tag{26}$$

where D_h is the hydraulic diameter of the rectangular duct and d_h is the hydraulic diameter of the film-cooling injection slot. For the present configuration, its value is calculated from the cross-section area at the exit of the internal cooling chamber. The variable U denotes the mean velocity of main flow.

IV. RESULTS AND DISCUSSION

A. HEAT TRANSFER IN INTERNAL COOLING CHAMBER

The heat transfer in such a cooling configuration combines the effect of flow acceleration due to gradually decreasing cross-section area and pin fin effects with varying height. Thus, the heat-transfer values could be very different from the results obtained from either a general pin fin or smooth wedge shape duct heat-transfer study. Five cases with different Reynolds numbers ranging from 3.5×10^3 to 9.5×10^3 have been tested. The results are compared directly to pin fin effects studied earlier by Chyu [25]. The value of the Nusselt number Nu is estimated to have an uncertainty of less than $\pm 7\%$, based on an uncertainty analysis technique developed by Kline and Mclintock [26].

The heat-transfer coefficients are presented as the row average for pin, endwall, and both combined based on the designated heat-transfer regions as illustrated in Fig. 12. In particular, the endwall area of each row is the area covered by that row, subtracting the area occupied by pins or shaped blocks in the row.

The overall averaged Nusselt number for each row or entire wetted surface

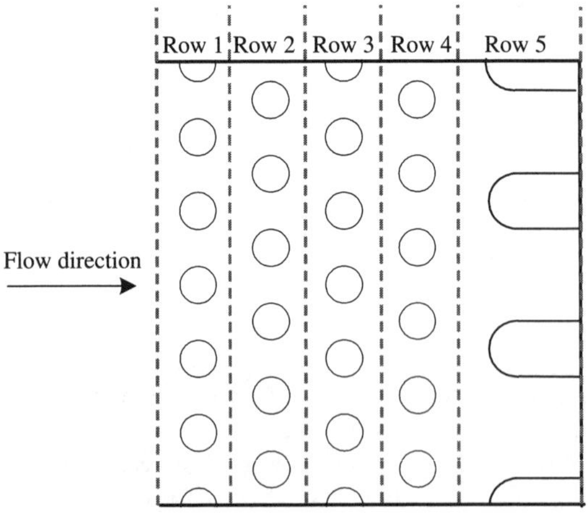

Fig. 12 Covered area for each row.

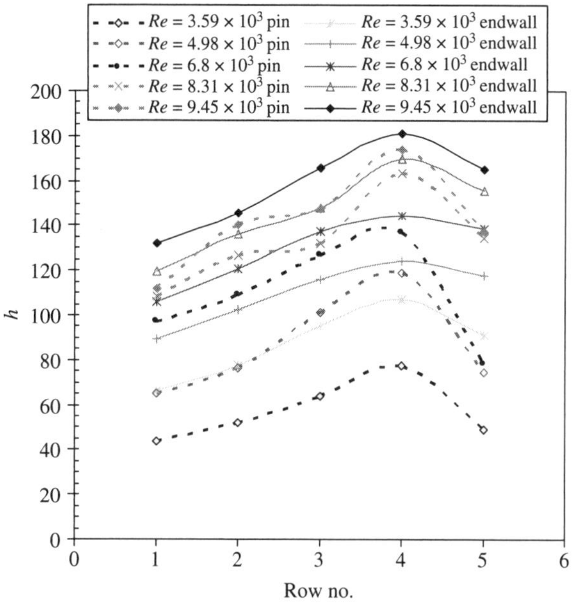

Fig. 13 Row averaged heat-transfer coefficient.

domain is obtained using the formula in the form of

$$Nu = \frac{Nu_{\text{endwall}} \times A_{\text{endwall}} + Nu_{\text{pin}} \times A_{\text{pin}}}{A_{\text{endwall}} + A_{\text{pin}}} \tag{27}$$

where A_{pin} represents the effective heat-transfer area of the pin-fin surface. Because both ends of a fin element are attached to the endwall, the effective heat-transfer area is only the circumferential area of the fin element:

$$A_{\text{pin}} = n \cdot \pi \cdot d_{\text{pin}} \cdot H \tag{28}$$

To demonstrate the heat-transfer enhancement effect, the Nusselt number is normalized by the corresponding Nusselt number of a smooth duct Nu_o at the corresponding Reynolds number Re_o.

1. ENDWALL AND PIN-AVERAGED HEAT-TRANSFER COEFFICIENT

Figure 13 presents the endwall and pin-averaged heat-transfer coefficient of each row for five different Reynolds numbers. One noticeable feature revealed in the figure is the increase of the averaged heat-transfer coefficient, which reaches its maximum at the fourth row. This is consistent with the result of previous general pin-fin heat-transfer studies [24, 25]. Because the fifth row of pin has a different configuration and geometrical arrangement, the heat-transfer coefficient shows a decrement ranging from 9 to 43% compared to that of the fourth row. The mechanisms for this significant drop in the heat-transfer coefficient may come from two aspects. On one hand, the increase of the heat-transfer coefficient is minor, or nonexistent, after the fourth row of pins. On the other hand, the flow velocity tends to decelerate after the fourth row of pins because of the enlarged free space. This can significantly reduce the turbulence and thus affect the heat transfer in the fifth row.

Another feature revealed from Fig. 13 is the higher average heat-transfer coefficient of the endwall area as compared to that of the pins. The contracting feature on the channel in the present configuration allows the pin-generated turbulence to have a stronger effect to the endwall heat transfer.

2. ROW OVERALL AVERAGED NUSSELT NUMBER

Figure 14 shows the overall row-averaged Nusselt number normalized by Nu_o for each row at the five tested Reynolds numbers. All rows have some degree of heat-transfer enhancement with the ratio of Nu/Nu_o ranging from 1.9 to 3.9. The magnitude of heat-transfer enhancement gradually increases from the first row to the fourth row, which agrees well with the trends of the variation of the row-averaged pin and endwall heat-transfer coefficient. The fifth row has the same level of heat-transfer enhancement as that of row 3 though there is a decrease relative to the fourth row.

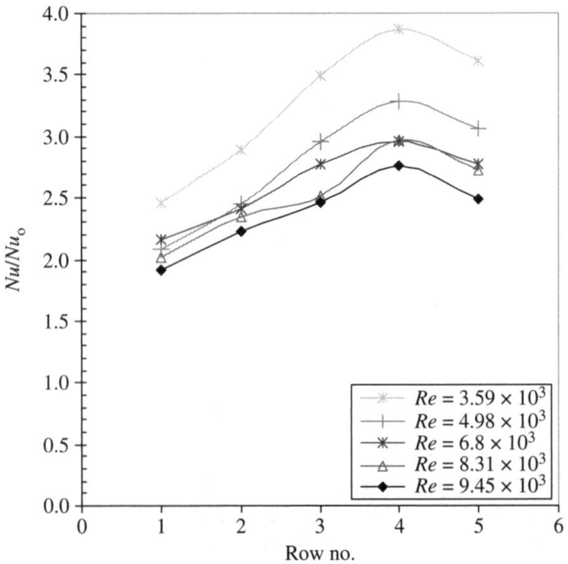

Fig. 14 Nu/Nu_o **value by row.**

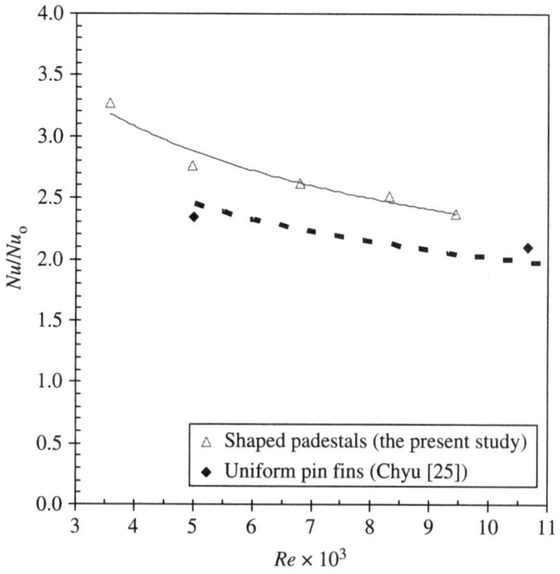

Fig. 15 Comparison of heat-transfer enhancement for shaped pedestals vs pin fins.

3. ENTIRE DOMAIN-AVERAGED NUSSELT NUMBER

The entire domain-averaged Nusselt-number values for all of the tested Reynolds numbers are also normalized by Nu_o. Nu/Nu_o vs Re is plotted in Fig. 15 to show the effect of Reynolds number on the level of heat-transfer enhancement. The magnitude of Nu/Nu_o decreases exponentially with the Reynolds number; this is similar to typical cases involving turbulent flow impingement and separation.

Also plotted are the results of uniform pin fins, from Chyu et al. [25], for comparison. The value of the enhancement for the present configuration ranges from 2.4 to 3.3 for all tested Reynolds numbers. The enhancement is comparable to that of the uniform pin-fin structure and at least 10% higher than those of uniform pin-fin structures at the tested Reynolds numbers.

B. FILM-COOLING PERFORMANCE ON CUTBACK LAND

The film-cooling performance over the cutback land is expected to be different from that in the case of clear slot, as the internal cooling chamber structure significantly affects the coolant flow. By introducing elements inside, the present configuration generates strong turbulence in the coolant flow before it exits the slot. In addition, the oblong-shaped features, laterally interrupting the cutback region, make the main flow three dimensional. Such a complicated interaction between the turbulent coolant flow and the three-dimensional main flow could be very detrimental to the film cooling on the cutback land. To examine the film-cooling performance on the cutback land of this configuration, four blowing ratios ($M = 0.7$, 1.0, 1.32, and 1.96), with fixed main flow Reynolds number $Re_m = 3.74 \times 10^4$ and corresponding coolant flow Reynolds number $Re_f = 3.03 \times 10^3$, 4.25×10^3, 5.76×10^3, and 8.34×10^3 were tested. The geometrical parameters are illustrated in Fig. 8.

1. LOCAL FILM-COOLING EFFECTIVENESS

Figure 16 presents the local film-cooling effectiveness over the cutback land. As expected, the film effectiveness increases with blowing ratios.

2. SPANWISE AVERAGED FILM-COOLING EFFECTIVENESS AND HEAT-TRANSFER COEFFICIENT

Figures 17 and 18 show the spanwise averaged film-cooling effectiveness and heat-transfer coefficient, respectively. For all tested blowing ratios, the film-cooling effectiveness decreases monotonically as x/s increases. Value of the film effectiveness is above 0.4 for $x/s < 7.4$, which is the length of the cutback land.

The ratio of lip thickness t to slot height S for the present configuration is $t/s = 0.432$. For comparison purposes, the results of close t/S ratios in Goldstein's study [27] for clear, two-dimensional slot injection film cooling are plotted together. As shown in Fig.16, the film-cooling effectiveness for the current configuration at $x/s = 11.5$, which is already out of the actual trailing-edge end (where $x/s = 7.4$), is much lower than that of the clear slot case. The turbulence of the coolant flow caused by the pedestals inside the cooling chamber probably contributes to the decreasing values [13, 15].

As shown in Fig. 18, the value of h is typically relatively low immediately downstream of the slot injection, rises to a local maximum at x/S around 7.5, and then decreases rapidly downstream.

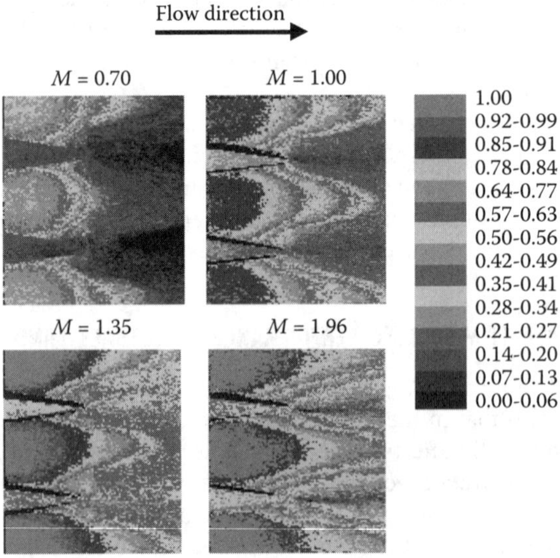

Fig. 16 Local film-cooling effectiveness distribution.

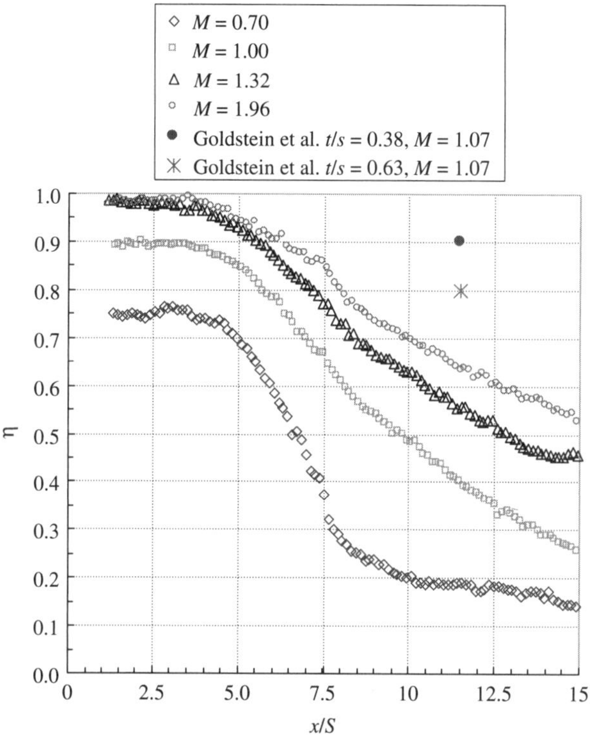

Fig. 17 Spanwise-averaged film-cooling effectiveness.

Starting from x/S about 9, it has a relatively stable value. Geometrically, the lip affects the location where a shear layer is formed by two separated emerging streams. As in flow over a backward-facing step, the turbulence and mixing level in the region directly beneath the shear layer, or immediately behind the step, are relatively low. This, in turn, results in a low heat-transfer coefficient. As the width of shear layer expands and eventually reaches the surface downstream, the magnitude of heat transfer is expected to reach a maximum as a result of the reattachment effect. Further downstream, the magnitude of heat-transfer coefficient decreases as the boundary layer grows thicker.

3. HEAT TRANSFER RELATIVE TO UNCOOLED SITUATION

The wall heat-transfer rate per unit area, or heat flux, with film cooling can be determined by

$$q = h(T_{\mathrm{aw}} - T_w) \tag{29}$$

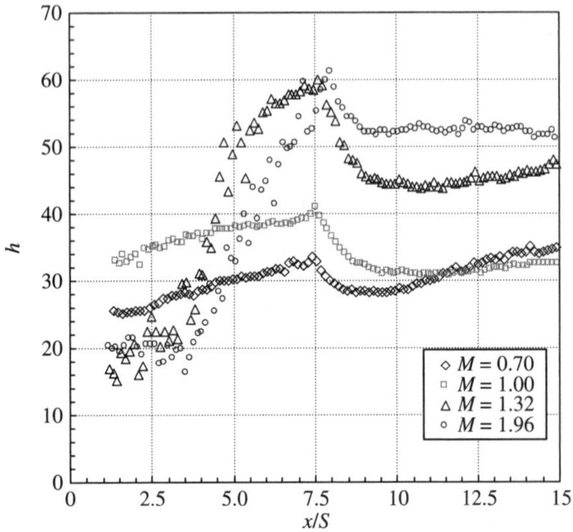

Fig. 18 Spanwise-averaged heat-transfer coefficient.

where T_{aw}, the adiabatic wall temperature, is related to the film effectiveness η via the definition of $\eta = \dfrac{T_{aw} - T_m}{T_f - T_m}$. According to Metzger et al. [28], Sen et al. [29], and Ekkad et al. [21, 22], the ratio of heat flux on a film-protected surface to the corresponding baseline value without film cooling q_o can be expressed as

$$q/q_o = (h/h_o)(1 - \eta/\varphi) \tag{30}$$

where φ is the overall cooling effectiveness given by $\varphi = \dfrac{T_w - T_m}{T_f - T_m}$. For any effective film protection, the value of q/q_o should be less than one. The situation when $q/q_o = 0$ represents a limiting case in which the local surface is fully protected by the cooling film. While the typical value of φ ranges from 0.5 to 0.7 in the main-body section [23], the value of $\varphi = 0.8$ is used in the calculation for the present evaluation.

The q/q_o values are plotted in Fig. 19. For all tested blowing ratios, the magnitude of q/q_o is less than 15% for $x/s < 7.5$. Particularly for $M > 1.32$, the protection is very effective for $x/s < 7.5$, as $q/q_o = 0$. A noticeable feature revealed in the plots is that for $M = 0.7$, q/q_o has a relative high value for $x/s < 2.6$. This phenomenon might result from nonuniform coolant distribution, as the blowing ratio is relatively small.

V. COOLING CONSIDERATIONS IN TRAILING-EDGE DESIGN

A. COMBINED CENTERLINE DISCHARGE AND PRESSURE-SIDE CUTBACK

High-pressure turbine airfoils are typically cooled internally with compressor air that enters a blade inlet plenum just before flowing into the blade internal cavities. Some designs may use cooling media other than air. In closed-loop cooling combined-cycle powerplants, for instance, the cooling medium is steam

[30, 31]. Regardless of the cooling medium, the overall design procedures that govern the airfoil trailing-edge cooling remain the same.

To illustrate various facets of trailing-edge design, this part of discussion begins by considering a typical high-pressure turbine blade configuration from Lee [32], shown in Fig. 20. Air enters an inlet plenum located underneath the blade. Cooling air passes onto the blade as rotating forces pump the flow towards blade internal cavities and eventually to the trailing edge. In the embodiment of Fig. 20, the trailing-edge configuration is made up of two cooling arrangements: 1) the pressure-side cutback with slots, located in the upper portion of the blade; and 2) the centerline discharge with round openings located in the lower portion of the blade. The combination of these two trailing-edge cooling schemes in one component is the essence of this design. In line with the material presented in the preceding sections, the combination of slots and round openings at the trailing edge provides a performance improvement because of the thin trailing edge in the upper portion of the airfoil.

The use of centerline discharge with cooling holes in the lower portion of the blade is practical as the gas-path temperatures can be considerably reduced at the lower radial portion of the airfoil. This effect is aided by hot gas-path migration to the blade tip and stage profile attenuation. Structural considerations might require a thicker trailing edge in the lower portions of the blade. In general, high centrifugal stresses exist in the lower regions, and it might be necessary to add material to the blade at the trailing edge to, among other considerations, decelerate the thermal response of the airfoil trailing edge relative to the rest of the airfoil and endwalls. In this way, both creep and thermal-mechanical material fatigue resistance can be increased. Thus, it should be recognized that a combination of trailing cooling schemes might provide preferred arrangements for specific designs, as suggested here for the model shown in Fig. 20. Other trailing-edge design configurations are listed in [30–39].

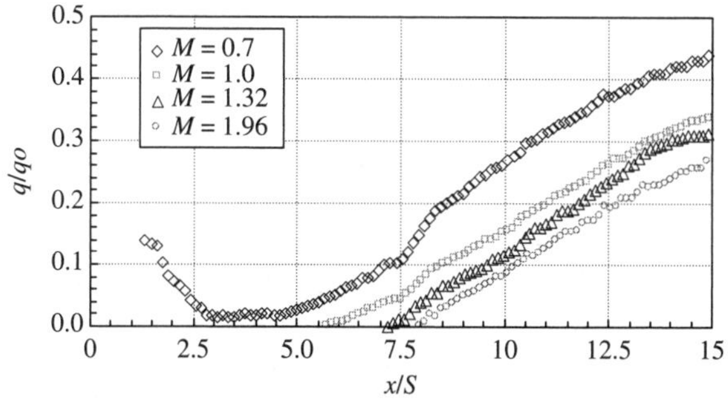

Fig. 19 Spanwise-averaged q/q_o.

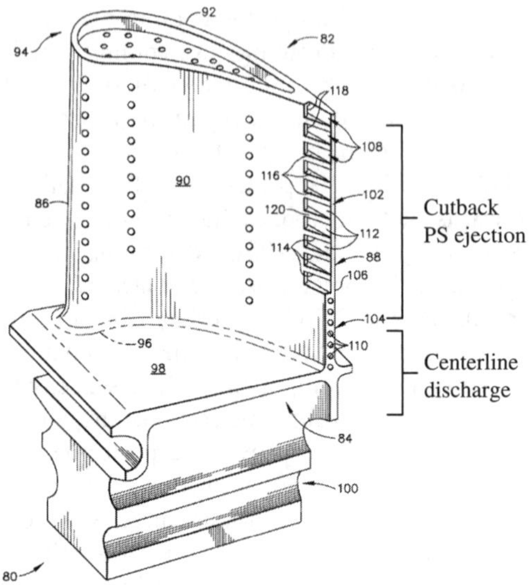

Fig. 20 Typical high-pressure turbine blade showing trailing-edge cooling openings and pressure-side ejection slots (from Lee [32]).

B. INTERNAL COOLING NEAR TRAILING EDGE

Figure 21, from Hill et al. [40], highlights other relevant details associated with trailing-edge design. This figure illustrates a cross-sectional area of a typical high-pressure turbine blade and shows a supply cavity that provides cooling air through a series of impingement crossover holes. The cooling circuit ends at the trailing-edge pressure-side slots in a cutback arrangement. Note that the tip pressure-side film holes are in communication with the trailing-edge supply cavity. Tip cooling is done through film-cooling holes located close to the blade tip. Even though the function of these tip holes is to cool the blade tip with air film, preventing tip oxidation, it reduces the flow to the trailing edge. Clearly, such flow interaction is accounted for in the trailing-edge cooling requirements as expressed by Eq. (11).

Inside the trailing-edge internal cavities, there are a large number of cooling features that can be explored in the design space to enhance internal cooling. For instance, internal trailing-edge cooling configurations can consist of more than one set of impingement rows with internal cooling features, such as pedestals and/or trip strips or turbulators.

Impingement cooling configurations have been used extensively in gas turbine blades. In general, cooling air is allowed to pass through crossover openings (circular holes, racetrack holes, elliptical holes, or slots), leading to jet impingement on the downstream airfoil ribs and surrounding walls. In these designs, coolant flow acceleration is high through crossover impingement openings. As a result, coolant flow Mach-number profiles, which are related to the coolant local static pressure, almost always assume nonuniform, or step-wise, characteristics as the coolant flow crosses these openings. Step-wise or nonuniform coolant profiles are undesirable as they lead to spatial nonuniformity in internal heat-transfer coefficients at the walls of the blade. Such nonuniformity is evident in a recent study by Chyu et al. [15], as shown in Fig. 22. The experimental study on the triple impingement configuration revealed that segment subjected to

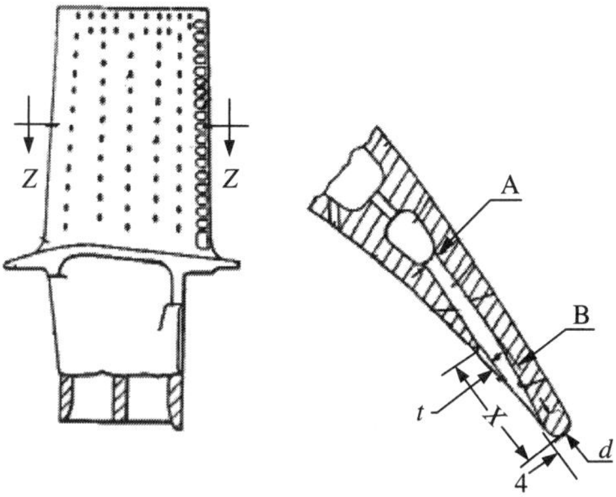

Fig. 21 Typical high-pressure turbine blade showing cross-sectional area and trailing-edge detail (from Hill et al. [40])

direct impingement of the accelerated flow exiting from the first set of crossover holes experiences the highest overall heat transfer, nearly three to four times the Nu/Nu_o of those surfaces without direct impingement. This implies that there are regions in the blade walls that attain relatively lower metal temperatures because of high internal heat-transfer coefficients. Other areas can attain relatively higher metal temperatures because of lower internal heat-transfer coefficients. Thus, step-wise profiles of coolant jets could lead to step-wise metal temperature differences, which, in turn, can lead to high thermal strains. In parallel to

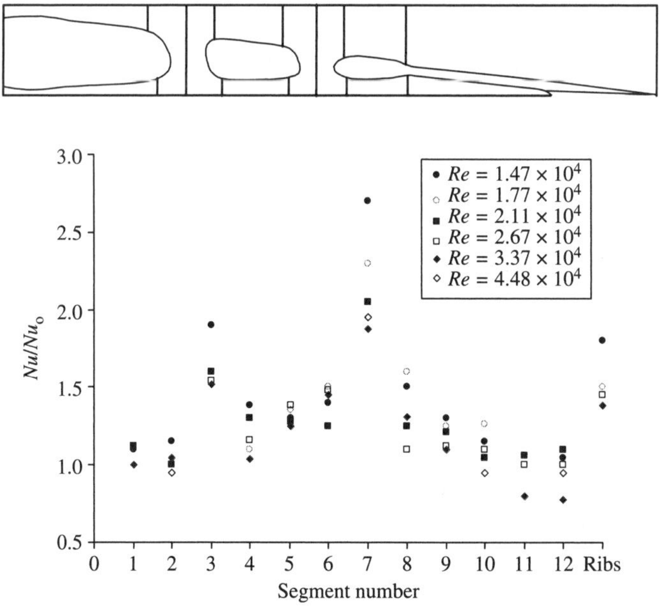

Fig. 22 Localized heat-transfer enhancement on different segments in trailing-edge cavities (from Chyu et al. [15]).

these temperature and heat-transfer differences, relatively fast blade trailing-edge thermal response during takeoff or power loading can exacerbate further undesirable thermal-mechanical effects of metal fatigue at the airfoil trailing edge.

It is, therefore, necessary to design a trailing-edge cooling configuration that improves the internal profiles for Mach number, static pressure, and internal heat-transfer coefficient distributions. In this regard, internal features, such as pedestals, provide distinct advantages. Optimal designs can include internal pedestals with many different cross-sectional areas, such as circular, oval, racetrack, square, rectangular, diamond, clover-leaf cross sections, to mention only a few. In this context, Zukauskas [41] suggested that the Reynolds number, based on pin or pedestal diameter, can be used to characterize the flow passing through the trailing-edge internal chambers in terms of corresponding Nusselt numbers for assessing the internal heat-transfer capability. However, a certain correction measure is needed here, as Zukausas's correlations are primarily for two-dimensional tube bundles and do not account for pin-endwall interaction. The corresponding internal heat-transfer enhancement or heat multiplier, HM, is then normalized by the smooth channel Dittus–Boelter correlation, to assess the effectiveness of the cooling design. Similarly, in this context, the pressure drops through a bank of pedestals, lead to the notion of the friction multiplier (FM), normalized with the Blasius resistance formula, from the experimental results of Ishida and Hamabe [42].

The relationships provided by [41, 42] for the heat and friction multipliers, HM and FM, fully characterize the trailing-edge cooling design with pin-fins, in terms of cooling effectiveness. The spacing and location of these internal cooling features are then optimized to ensure the elimination of step-wise distributions of metal temperatures in the airfoil trailing-edge walls. These features are presented here as representations of a typical design procedure; however, there are many other features or combinations of features that could be used effectively as well. In all design cases, experimental correlations are needed to assess a multitude of cooling feature characteristics. Some of these correlations are compiled by Han et al. [43].

The overall procedure for designing airfoil trailing edges with optimum heat-transfer characteristics relies on the knowledge of the corresponding heat and friction multipliers. In many instances, plots of FM/HM vs *Re* are used to compare the performance of different configurations. One such plot is shown in Fig. 23, based on mass-transfer measurements conducted by Chyu [25]. Using a well-established analogy between heat transfer and mass transfer [44], the results of these experiments for different cooling features can be used to assess the heat-transfer characteristics of the overall cooling design. This information, also known as performance index, is useful in the design process to relate heat-transfer characteristics of different cooling features with available pressure gradients and flow requirements.

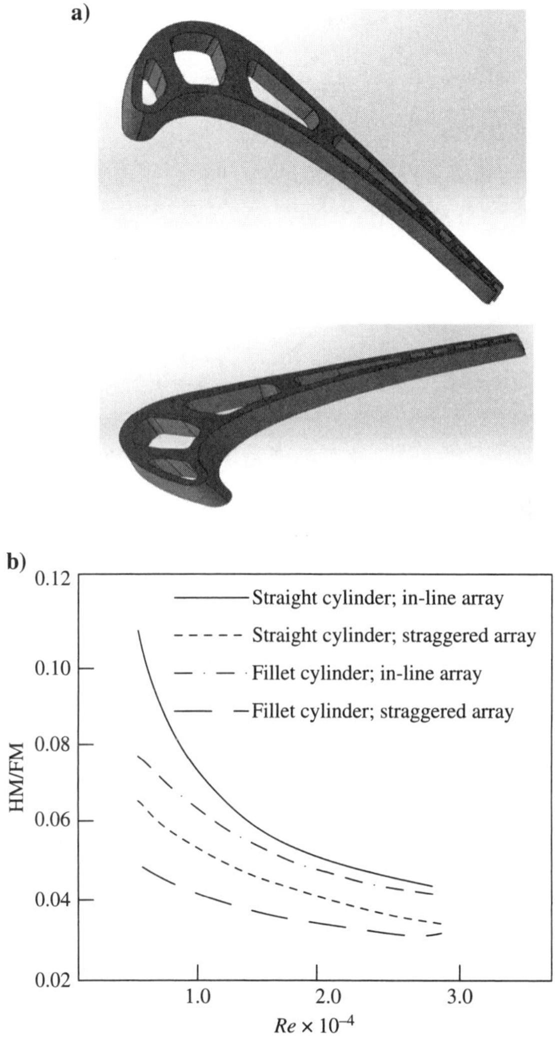

Fig. 23 Pedestal trailing-edge configuration and test results: a) transparent view of blade showing pedestal bank at trailing-edge b) effect of Reynolds number on performance index (HM/FM) in straight pins with fillets (from Chyu [25]).

C. EXTERNAL COOLING NEAR TRAILING EDGE

In conjunction with the internal heat transfer, the heat-transfer characteristics on the external airfoil surface near the trailing edge must also be evaluated in detail. As mentioned earlier, the performance of the airfoil is significantly affected by the thickness of the trailing edge. It is, therefore, necessary to have an airfoil trailing edge with minimum thickness and with a cooling configuration that allows for lower coolant flow rate while not exceeding the metal temperatures required for the metal durability in terms of oxidation, creep, and fatigue. In addition, features should be cast rather than machined, to the extent possible, to avoid machining cost and process time.

In examining possibilities for film cooling for the design in Fig. 21, one recognizes that the airfoil has trailing-edge cavities and slots in communication with the supply cavity. The slot outlet is disposed at a cutback downstream edge of the pressure-side wall with thickness *t*. A number of channels with width *s* as the

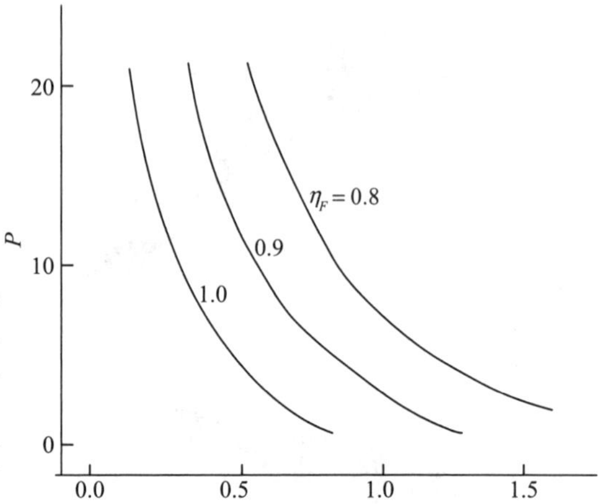

Fig. 24 Dimensionless flow parameter *P* as function of ratio of cutback lip thickness-to-slot for different film-cooling effectiveness (from Hill et al. [40]).

slot outlet allows for discharge cooling air over the extended back-surface suction-side wall or the so-called land. If a film-cooling parameter P is defined as a dimensionless air flow parameter directly proportional to the cutback distance x and inversely proportional to the slot s and the cooling flow rate ratio, one obtains the relation $P = x/(Ms)$ with M being the blowing parameter defined as the ratio of coolant-to-gas mass flux ratio $M = (\rho_c v_c)/(\rho_g v_g)$. This film-cooling parameter P is critical in film-cooling effectiveness calculations, as it has been used to correlate the film-cooling effectiveness η_F at the trailing edge, as shown in Fig. 24.

Higher values of the film-cooling parameter P imply greater cutback distances and less airflow for an equivalent film-cooling effectiveness η_F. It has been recognized that high film effectiveness can be maintained over significantly longer cutback distances with less air if the ratio t/s is low [40]. According to Fig. 24, the value of η_F can remain constant as x increases if the value of the ratio t/s is decreased. This implies that if all other parameters affecting P are held constant, the cutback distance x could be increased significantly without loss of film-cooling effectiveness over the length of the cutback portion if the ratio t/s is decreased. Alternatively, or in combination, the coolant flow rate could be reduced, and the cutback distance increased. In this context, Goldstein [27] also presented experimental results for film effectiveness of various t/s ratios and the blowing parameter M of unity. These results are shown in Fig. 25.

To maximize the benefits of this configuration, two geometrical attributes are particularly desirable: 1) reduction of pressure-side lip thickness t and 2) enlargement of slot s, provided that both are consistent with structural strength requirements. The combination of geometrical attributes 1 and 2 can then lead to minimum cooling flow rates for the pressure-side cutback trailing-edge configurations. The limit on the cooling flow rate would then come from the metal temperature ahead of the cutback distance. Because the metal temperature increases monotonically with distance from the supply cavity to the cutback distance from the trailing edge, the effect of trailing-edge film cooling takes place after this point.

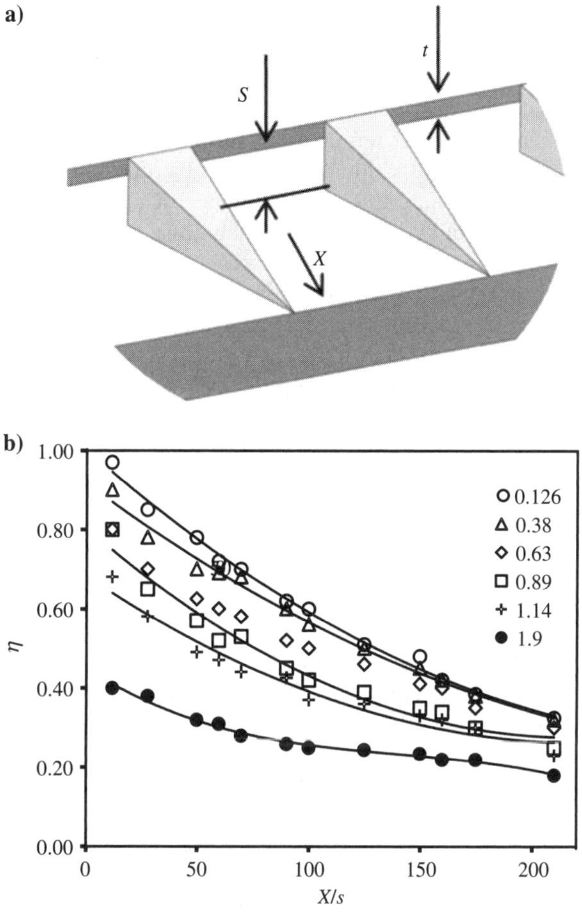

Fig. 25 Cutback trailing-edge configuration and test results: a) schematic showing locations of relevant geometrical parameters b) effect of x/s and t/s parameters on trailing-edge film effectiveness (from Goldstein [27]).

The metal temperature profile with pressure-side cutback might not increase monotonically toward the trailing edge as in centerline discharge configurations. The surface of the pressure-side land could also be made aerodynamically rough to enhance the coolant heat-transfer coefficient on that surface. Thus, the preferred selection of distance x, lip thickness t, slot s, and blowing parameter M leads to a pressure-side cutback trailing-edge configuration, which is attractive in terms of both airfoil durability and overall aerodynamic performance. This chapter has summarized the characteristics affecting the design of trailing-edge configurations for high-pressure turbine airfoils.

VI. CONCLUSIONS

Two sets of conclusions are derived from this treatment, analytical and experimental. The analytical results include closed-form, analytical models of airfoil temperature for the four most representative trailing-edge configurations. These include 1) solid wedge shape without discharge, 2) wedge with centerline slot discharge, 3) wedge with centerline discrete-hole discharge, and 4) wedge with pressure-side cutback slot discharge. These analytical solutions for metal temperature provide fundamental insight into the relevant characteristics affecting the design of trailing-edge configurations for high-pressure turbine airfoils.

Based on the information gained from the analysis, the relevant design features can be summarized as follows: 1) size of the cooling passage, 2) internal cooling features inside the cooling passage, 3) trailing-edge thickness, 4) pressure-side lip thickness, 5) roughness on pressure-side land, and 6) slot film coverage. From this set of features, only two, namely 1 and 2, can be used effectively for centerline discharge configurationswhereas all features, 1–6, can be used effectively for the pressure-side cutback configurations. For the cutback design, there are also the added benefits of improved aerodynamic performance as a result of thinner trailing edges. The life capability to resist thermal-mechanical fatigue and creep is also improved as metal temperature distributions are more evenly distributed over the entire trailing-edge region for the cutback designs.

Overall design parameters were introduced to illustrate how these parameters can be selected during the optimization cycle for the trailing-edge designs. This process leads to the consideration of detailed design features and geometrical attributes, which can be judiciously selected in concert with the available design space. In this regard, cooling effectiveness becomes a strong function of the convective efficiency, and internal cooling features are used to temper step-wise pressure and coolant Mach-number distributions inside the trailing-edge passages. The film-cooling effectiveness and corresponding film coverage, defined in terms of geometrical lip-to-slot ratio t/s and blowing ratio, are also important design parameters. These are influential in maintaining desired film-cooling effectiveness while minimizing trailing-edge cooling flow for specified trailing-edge cutback distances.

In the experimental development, the heat transfer in both the internal cooling chamber and cutback land of an airfoil trailing edge with internal pedestals or pin-fins is presented. For internal cooling, the last row, row 5, where the oblong-shaped teardrop begins, has a relatively low level of heat-transfer enhancement, compared with the previous (fourth) row of pedestals, but still comparable with row 3. Compared with a smooth channel, the magnitude of the enhancement for the entire heat-transfer domain ranges from 2.4 to 3.3 for $3.5 \times 10^3 < Re < 9.5 \times 10^3$. The enhancement is at least 10% higher than that of the uniform pin-fin for $Re < 9.5 \times 10^3$. As for the heat transfer over the cutback land, when blowing ratio $M > 1.32$, the protection is very effective for $x/s < 7.5$. The effect of the nonuniform coolant distribution on the cooling performance is observed when the blowing ratio is relatively small.

REFERENCES

[1] Kuhne, C. M., "Reduced Shock Transonic Airfoil," U.S. Patent No. 2003/0072649A1, April 2003.

[2] Eisemann, K. M., "Uncooled Data Reduction," Pratt & Whitney Internal Documentation, 2003.

[3] Karamcheti, K., *Principles of Ideal-Fluid Aerodynamics*, Krieger, Malabar, FL, 1980, Chap. 13.

[4] Maclachlan, D. W., and Knowles, D. M., "The Effect of Material on the Analysis of Single Crystal Turbine Blades: Part I – Material Model," *Fatigue and Fracture Engineering Material Science 25*, Blackwell Science, Ltd., Boston, MA, 2002, pp. 385–398.

[5] Maclachlan, D. W., and Knowles, D. M., "The Effect of Material on the Analysis of Single Crystal Turbine Blades: Part II – Component Analysis," *Fatigue and Fracture Engineering Material Science 25*, Blackwell Science, Ltd., Boston, MA, 2002, pp. 399–409.

[6] Taslim, M. E., Spring, S. D., and Mehlman, B. P., "Experimental Investigation of Film Cooling Effectiveness for Slots of Various Exits Geometries," *Journal of Thermophysics and Heat Transfer*, Vol. 6, No. 2, 1992, pp. 302–307.

[7] Sivasegaram, S., and Whitelaw, J. H., "Film Cooling Slots: The Importance of Lip Thickness and Injection Angle," *Journal of Mechanical Engineering Science*, Vol. 11, No. 1, 1969, pp. 22–27.

[8] Burns, W., and Stollery, J. L., "The Influence of Foreign Gas Injection and Slot Geometry on Film Cooling Effectiveness," *International Journal of Heat Mass Transfer*, Vol. 12, No. 8, 1969, pp. 935–951.

[9] Pai, B. R., and Whitelaw, J. H., "The Prediction of Wall Temperature in the Presence of Film Cooling," *International Journal of Heat Mass Transfer*, Vol. 14, No. 3, 1971, pp. 409–426.

[10] Kacker, S. C., and Whitelaw, J. H., "An Experimental Investigation of Slot Lip Thickness on the Impervious Wall Effectiveness of the Uniform Density, Two-Dimensional Wall Jet," *International Journal of Mass Heat Transfer*, Vol. 12, No. 9, 1969, pp. 1196–1201.

[11] Uzol, O., and Camci, C., "Aerodynamic Loss Characteristics of a Turbine Blade with Trailing Edge Coolant Ejection: Part 2- External Aerodynamics, Total Pressure Losses, and Predictions," *Journal of Turbomachinery*, Vol. 123, No. 2, 2001, pp. 249–257.

[12] Uzol, O., Camci, C., and Glezer, B., "Aerodynamic Loss Characteristics of a Turbine Blade with Trailing Edge Coolant Ejection: Part 1- Effect of Cutback Length, Spanwise Rib Spacing, Free-Stream Reynolds Number, and Chordwise Rib Length on Discharge Coefficients," *Journal of Turbomachinery*, Vol. 123, No. 2, 2001, pp. 238–248.

[13] Holloway, D. S., Leylek, J. H., and Buck, F. A., "Pressure Side Bleed Film Cooling: Part 1-Steady Framework for Experimental and Computational Results," Paper GT-2002-30471, Vol. 3, American Society of Mechanical Engineers, the Netherlands, 2002, pp. 835–843.

[14] Holloway, D. S., Leylek, J. H., and Buck, F. A., "Pressure Side Bleed Film Cooling: Part 1-Unsteady Framework for Experimental and Computational Results," American Society of Mechanical Engineers, Paper GT-2002-30472, 2002.

[15] Chyu, M. K., Uysal, U., and Li, P.-W., "Convective Heat Transfer in a Triple-Cavity Structure near Turbine Blade Trailing Edge," *Proceedings of IMECE'02, International Mechanical Engineers Congress*, Vol. 4, Paper IMESE 2002-32405, LA, 2002, pp. 265–272.

[16] Martini, P., and Schulz, A., "Experimental and Numerical Investigation of Trailing Edge Film Cooling by Circular Coolant Wall Jets Ejected from a Slot with Internal Rib Arrays," *Journal of Turbomachinery*, Vol. 126, No. 2, 2004, pp. 229–236.

[17] Kim, Y. W., Chad, C., and Moon, H. K., "Film Cooling Characteristics of Pressure Side Discharge Slots in an Accelerating Mainstream Flow," *Proceedings of GT2005, ASME Turbo Expo 2005: Power for Land, Sea and Air*, Vol. 3, GT2005-69061, pp. 889–897.

[18] Martini, P., Schulz, A., and Bauer, H. J., "Film Cooling Effectiveness and Heat Transfer on the Trailing Edge Cutback of Gas Turbine Airfoils with Various Internal Cooling Designs," *Proceedings of GT2005, ASME Turbo Expo 2005: Power for Land, Sea and Air*, Vol. 3, GT2005-68083, pp. 87–96.

[19] Cunha, F. J., Dahmer, M. T., and Chyu, M. K., "Analysis of Airfoil Trailing Edge Heat Transfer and Its Significance in Thermal-Mechanical Design and Durability," *Proceedings of GT2005, ASME Turbo Expo 2005: Power for Land, Sea and Air*, Vol. 3, GT2005-68108, pp. 165–176.

[20] Vedula, R. J., and Metzger, D. E., "A Method for Simultaneously Determination of Local Effectiveness and Heat Transfer Distribution in Three-Temperature Convection Situations," American Society of Mechanical Engineers, Paper 91-GT-345, 1991.

[21] Ekkad, S. V., Zapata, D., and Han, J. C., "Heat Transfer Coefficients over a Flat Surface with Air and CO2 Injection Through a Compound Angle Holes Using a Transient Liquid Crystal Image Method," *Journal of Turbomachinery*, Vol. 119, 1997, pp. 580–586.

[22] Ekkad, S. V., Zapata, D., and Han, J. C., "Film Effectiveness over a Flat Surface with Air and CO_2 Injection Through Compound Angle Holes Using a Transient Liquid Crystal Image Method," *Journal of Turbomachinary*, Vol. 119, No. 3, 1997, pp. 587–593.

[23] Yu, Y., Yen, C.-H., Shih, T. I-P., Chyu, M. K., and Gogineni, S., "Film Cooling Effectiveness and Heat Transfer Coefficient Distributions Around Diffusion Shaped Holes," *Transaction of ASME*, Vol. 124, Oct. 2002, pp. 820–827.

[24] Chyu, M. K., and Ding, H., "Heat Transfer in a Cooling Channel with Vortex Generators," *Journal of Heat Transfer*, Vol. 119, No. 2, 1997, pp. 206.

[25] Chyu, M. K., "Heat Transfer and Pressure Drop for Short Pin-Fin Arrays with Pin-Endwall Fillet," *Journal of Heat Transfer*, Vol. 112, Nov. 1990, pp. 926–932.

[26] Kline, S. J., and McClintock, F. A., "Describing Uncertainties in Single Sample Experiments," *Mechanical Engineering (American Society Mechanical Engineering)*, Vol. 75, No. 1, 1953, pp. 3–8.

[27] Goldstein, R. J., "Film Cooling," *Advanced Heat Transfer*, Vol. 7, No. 1, 1971, pp. 321–379.

[28] Metzger, D. E., Carper, H. J., and Swank, L. R., "Heat Transfer with Film Cooling near Nontangential Injection Slots," *Journal of Engineering for Power*, Vol. 90, 1968, pp. 157–163.

[29] Sen, B., Schmidt, D. L., and Bogard, D. G., "Film Cooling with Compound Angle Holes: Heat Transfer," *Journal of Turbomachinery*, Vol. 118, No. 4, 1996, pp. 800–806.

[30] Langston, L., "A Year of Turbulence," *Power and Energy, ASME Magazine*, June 2004.

[31] Cunha, F. J., "Integrated Steam/Gas Cooling System for Gas Turbines," U.S. Patent No. 5,340,274, Aug. 1994.

[32] Lee, C.-P., "Turbine Blade Trailing Edge Cooling Openings and Slots," U.S. Patent No. 6,174,135B1, Jan. 16, 2001.

[33] Jacala, A., Davis, R. M., Sullivan, M. A., Chyu, R. P., and Staub, F., "Closed Circuit Steam Cooled Bucket," U.S. Patent No. 5,536,143, July 1996.

[34] Cunha, F. J., DeAngelis, D. A., Brown, T. A., Chopra, S., Correia, V. H. S., and Predmore, D. R., "Turbine Stator Vane Segments Having Combined Air and Steam Cooling Circuits," U.S. Patent No. 5,634,766, June 1997.

[35] Cunha, F. J., Dahmer, M. T., and Chyu, M. K., "Thermal-Mechanical Life Prediction System for Anisotropic Turbine Components," IGTI/ASME GT2005-68107, Vol. 3, 2005, pp. 151–164.

[36] Braddy, B. T., "Film Cooled Airfoil Body," U.S. Patent No. 4,303,374, Dec. 1981.

[37] Starkweather, J. H., "Turbine Blade," U.S. Patent No. 5,813,836, Sept. 1998.

[38] Manning, R. F., Acquaviva, P. J., and Demers, E., "Series Impingement Cooled Airfoil," U.S. Patent No. 6,036,441, March 2000.

[39] Cunha, F. J., and DeAngelis, D. A., "Cooling Circuits for Trailing Edge Cavities in Airfoils," U.S. Patent No. 6,056,505, May 2000.

[40] Hill, E. C., Liang, G. P., and Auxier, T., "Airfoil Trailing Edge Cooling Arrangement," U.S. Patent No. 4,601,638, July 1986.

[41] Zukauskas, A. A., "Heat Transfer from Tubes in Cross Flow," *Advances in Heat Transfer*, Vol. 8, No. 1, 1972, pp. 116–133.

[42] Ishida, K., and Hamabe, K., "Effect of Pin-Fin Aspect Ratio and Arrangement on Heat Transfer and Pressure Drop of Pin Fin Duct for Airfoil Internal Cooling Passage," American Society of Mechanical Engineers, 85-WA/HT-62, 1985.

[43] Han, J. C., Dutta, S., and Ekkad, S., *Gas Turbine Heat Transfer and Cooling Technology*, 1st ed., Taylor, and Francis, Philadelphia, 2000, Chap. 1.

[44] Eckert, E. R. G., "Analogies to Heat Transfer Processes," *Measurements in Heat Transfer*, edited by E. R. G. Eckert and R. J. Goldstein, Hemisphere, New York, 1976, pp. 397–423.

CHAPTER 8

Blade Tip Aerodynamics and Heat Transfer

Ronald S. Bunker*

GE Aviation, Cincinnati, Ohio

NOMENCLATURE

C_L	total pressure loss coefficient
C_p	specific heat at constant pressure
h	heat-transfer coefficient
L	characteristic length
P/S	airfoil pressure side
Re	Reynolds number ($\rho VL/\mu$) based on either the passage or tip gap conditions
S/S	airfoil suction side
St	Stanton number ($h/\rho C_p V$)
V	velocity
η	adiabatic film effectiveness
μ	viscosity
ρ	density

ACRONYMS

ACC	active clearance control
BC	boundary condition
CFD	computational fluid dynamics
CMC	ceramic matrix composite
COE	cost of electricity
EGT	exhaust gas temperature
HCF	high-cycle fatigue
HPT	high-pressure turbine
LCF	low-cycle fatigue
LPT	low-pressure turbine

*Principal Engineer.

MCPH	maintenance cost per hour
TBC	thermal barrier coating
TRIT	turbine rotor inlet temperature

Turbine blade tips have been, and continue to be, not only one of the major causes for loss of efficiency in a turbine engine, but also a primary contributing factor in the operational degradation of turbines leading to periodic removal from service for repairs. As with all components of the turbine hot-gas path, blade tips must perform multiple functions while being subject to many design and operational constraints. This is true of both the cooled blade tips in high-pressure turbine stages and also the uncooled blade tips of low-pressure turbine stages. The uniqueness associated with all turbine blade tips comes in the multifaceted complexity of the rotational-stationary interface between the work extraction fluid and the turbine casing.

Turbine blade tips are distinctly different from compressor and fan blade tips in at least two respects. Turbine blade tips are subject to higher temperature gases, in some cases exceeding 1400°C in the first high-pressure turbine stages, and these tips are also subject to far higher aerothermal loading (pressure ratios). Aircraft engine high-pressure blades can see as much as 25 atm pressure whereas large power turbine blades can see about 12 atm, and both can experience blade row pressure ratios up to 2.

A perfectly functioning blade tip will not allow any leakage of valuable working fluid over the tip, from pressure side to suction side, which would bypass or short-circuit the extraction of work by the turbine. A perfect blade tip will also require no cooling, thereby presenting no thermodynamic losses from the use of chargeable flows, and no mixing losses from injection of these flows into the main working fluid. A perfect blade tip will, in addition, generate no secondary flows to reduce stage efficiency or to contribute to losses in downstream airfoil stages. These are the major goals that all designs seek to approach, but none attain. Instead, the more realistic goal is to minimize the impact of the imperfections while also satisfying several other system operational requirements.

This chapter will review the competing requirements and constraints placed on turbine blade tips, describe several approaches used in actual designs to satisfy these conditions, provide in-depth summaries of the aerodynamics and heat transfer associated with the major blade tip designs, and highlight aspects of durability that must be included in any successful turbine design. A more detailed treatment of these topics can be found in the recent von Kármán Institute lecture series of Glezer et al. [1].

I. FUNCTION OF A TURBINE BLADE TIP

The most fundamental question to ask about a blade tip concerns its function as part of the entire blade. The turbine blade is a work extraction device balancing

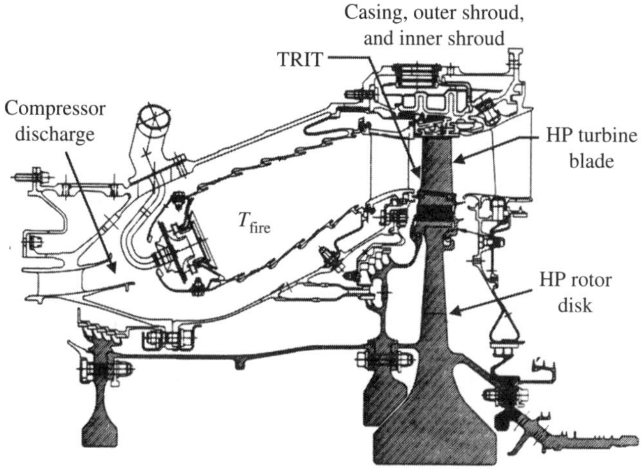

Fig. 1 Gas-turbine engine schematic showing combustor and HP turbine (from [2]).

the requirements of aerodynamics, structural life, material properties, and thermal efficiency. An individual blade should not be considered a singular element. As depicted in Fig. 1 [2], the blade must function in concert with the rotor, the upstream and downstream airfoils, the shrouds and secondary systems, and, of course, the hot-gas-path fluid. As the turbine rotor inlet temperature (TRIT) is raised to improve system thermal efficiency (for a given aerodynamic efficiency), the blade life generally decreases drastically. When blades exceed a safe temperature limit for the materials, they must be cooled, most commonly using the compressor discharge air, to retain adequate material properties and structure. Such cooling represents a short-circuiting or bypassing of the ideal engine main flowpath and decreases the cycle efficiency. A portion of the useful work contained in the cooling air (pressure) is used up in the act of cooling the blades, leading to the term "chargeable air" as this pressure loss is charged as a deficit to the cycle. The remainder of the energy in the cooling air (work and heat) is recovered in the main flowpath though with some mixing losses. Up to a certain point, the economical use of blade cooling allows greater overall thermal efficiency via increased TRIT.

These general considerations apply both to the specific region of the blade tip and to the blade that carries it. The entire blade is designed for maximum work extraction within other system limitations, and the tip is no exception. The blade tip is the required free "end" of the blade, forming the interface between the rotor and the outer diameter flowpath stator (or casing or shroud). The unavoidable physical tip clearance gap is the rotational-to-stationary leakage flow path that detracts from the aerodynamic efficiency of the blade row. As depicted in Fig. 2 for differing types of blade tip designs, stage efficiency is very sensitive to increased relative blade tip clearance with sensitivities from 1:1 to 2:1. This is a far greater issue with smaller turbines, in which the effective tip clearance represents a larger relative percentage of the total annular flowpath height. It is also true that this efficiency derivative decreases as one moves from the high-pressure stages to the low-pressure stages of the turbine. An increasing tip leakage, sometimes called over-tip-flow, decreases the amount of work extracted from the hot gases, which, in turn, increases the engine exhaust gas temperature (EGT) for a given TRIT. The increase in exit temperature of an engine with time is a direct indication of the operational degradation of the engine. At a predetermined EGT setpoint for any engine, also known as the EGT margin relative to the initial new or serviced condition, the engine is either turned down in power or removed from service for maintenance. The losses associated with turbine blade tip degradation, aerodynamic and thermodynamic, typically account for as much as one-third of this EGT margin, hence the high attention paid to blade tips. The increase in aerodynamic loss and decrease in EGT margin is roughly linear with the increase in physical blade tip clearance.

As will be shown in subsequent sections, the aerodynamics surrounding the blade tip region are highly three dimensional, turbulent, and unsteady. Figure 3 shows a generic time-averaged static-pressure distribution around an airfoil blade tip section with and without a tip clearance present. The pressure distribution, although similar to that of the airfoil without a tip clearance, is modified by the general tip leakage flows depicted here. The overall pressure profile is the main driver for tip leakage flows and will change as the clearance changes. Even in the simplest case of a base loaded turbine at essentially constant cycle conditions, the aerodynamic loading, and consequently the thermal loading, will change as a function of time on all tip surfaces. This change is manifested in the gradual loss of the blade tip material due to oxidation and erosion in the absence of other more accelerated loss mechanisms such as tip rubbing. The aerothermal characteristics will change as the tip material is lost and the tip clearance opens. In general, increases in blade tip clearance will lead to higher tip region heat loading, which, in turn, tends to accelerate the tip loss rate with time. This

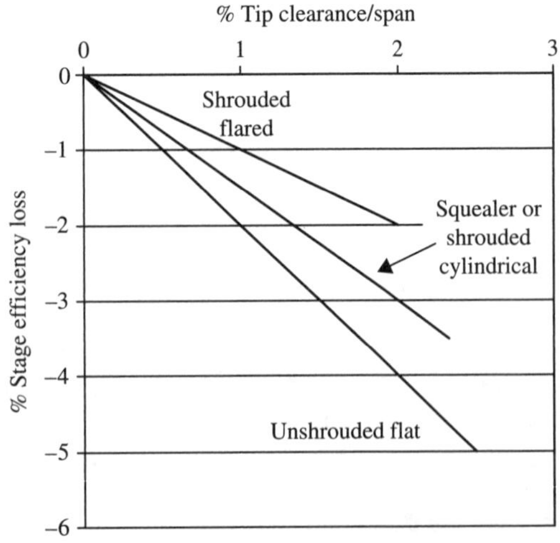

Fig. 2 Effect of blade tip clearance on stage efficiency.

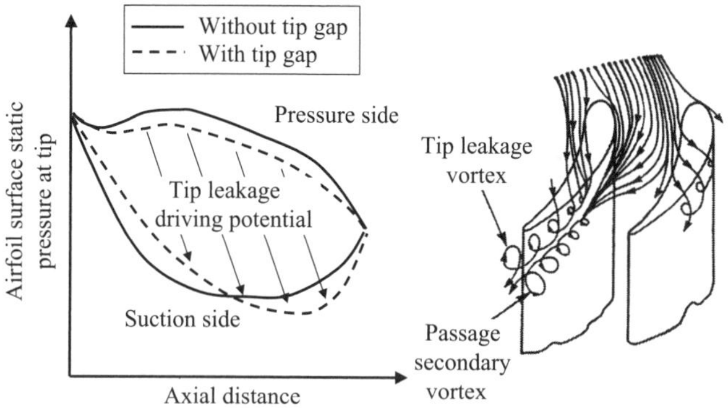

Fig. 3 Generic tip leakage pressure field and flow schematic.

degradation and "failure" of blade tips constitutes about one-third of total high-pressure turbine (HPT) failures, where failure is defined as the loss of the part from service inventory (unrepairable), or the accelerated degradation of the efficiency/output in service. In low-pressure turbines (LPT), which are usually uncooled, failure may more commonly be manifested in terms of simple oxidation and/or creep deformation without much change in the aerothermal characterisitics. The bottom line here is the cost in terms of maintenance cost per hour (MCPH) or cost of electricity (COE).

The HPT turbine blade is typically also a highly complex heat exchanger. The blade tip is an aerodynamically shaped, structural, and most often internally and externally cooled complex fin heat exchanger. The degree of complexity depends on the turbine design functional requirements (such as mission, environment, and cost). In highly cooled designs, a mixture of internal convective and/or impingement cooling with external film cooling is used to ensure adequate tip life between repairs. The cooling techniques employed are dependent on the aerodynamic shape of the tip, the internal cooling circuit of the blade, and the choice of overall tip design (unshrouded or shrouded). The function of blade tip cooling is to maintain the aerodynamic efficiency of the tip (shape and sealing) with the least possible usage of cooling fluid. Because of the tip profile conditions shown in Fig. 3, blade tip cooling must be distributed, act with the complex flowfield, and if possible not degrade as conditions change. This is a substantial challenge, and, as subsequent sections will show, one that has yet to be fully realized.

One further function of the blade tip must be made clear. As the tip provides the rotating-to-stationary interface for the contained working fluid, it must also be able to withstand the conditions associated with tip rub against the stationary shroud/casing. A tight tip clearance provides higher aerodynamic efficiency, but also leads to a higher probability of rubbing. In the event of a tip rub, blade tip material can be lost or deformed. The blade tip region should be designed to maintain its other functions with minimal efficiency loss after a rub has occurred, and it should also be repairable after some degree of material loss or damage.

II. BLADE TIP SYSTEM DESIGN ASPECTS

At a first glance, the turbine blade tip regions might appear to be isolated locations in the overall design, residing at the ends of the blades and only requiring local design attention. In fact, the blade tip is actually a very highly sensitive region that reacts to the entire turbine system design. It is one of two meeting points of the rotating system and the stationary system, which should coexist as closely as possible without actually coming into contact (the other such point being the bearings). Effective and efficient turbine blade tip design has a distinct and important place in the overall turbine engine operation. Blade tip survival offers high impact and high payoff because it affects so many key or critical parameters. In designing blade tips, both cooled and uncooled, for proper operation within the larger turbine system, one must consider several major factors (in no particular order).

1. Stage and turbine aerodynamic efficiency are greatly affected by the blade tip design in terms of the resulting effective leakage clearance. The effective clearance, which can also be thought of as an effective overall tip discharge coefficient, is determined not only by the tip geometry, but also by the tip aerodynamic distribution, injected cooling flows, tip sealing arrangement, rotational speed, shroud surface treatments, and much more. As a first estimate, each stage can be thought of as having an aerodynamically isolated tip region, but the reality of multistage turbines is that all stages must be designed together to obtain maximum benefit. Another important aspect of the aerodynamic efficiency directly tied to blade tips is the mixing loss associated with the tip leakage flows as they combine with the high momentum suction-side passage flow.
2. Stage thermal efficiency and also overall turbine efficiency are strongly affected by the amount of chargeable cooling air used to maintain blade tip integrity and life. In highly cooled HPT blades, the tip region alone may account for as much as 20% of the total blade cooling flow.
3. Bulk material temperature limits must be considered for the entire blade structure. Although the tip region is generally not subject to the same limitations as the rest of the blade in this respect, the tip design does influence the resulting bulk temperatures of the lower blade sections through the overall cooling design. The tip can also present enough weight to require lower bulk temperatures in the main blade sections to avoid creep rupture issues.
4. Maximum local material temperatures are typically a major concern for blade tips, as these regions are the most difficult to cool. Temperature limits will be placed on the metal substrate, the bondcoat, and the thermal

barrier coating (TBC) to avoid, for example, excessive oxidation, high coating strains, and melt infiltration of surface deposits, respectively. This can be thought of as the material system design.

5. Tip sealing methods vary widely, as will be shown later, but all methods attempt to reduce the effective tip clearance. The type of sealing arrangement is intimately tied to the other system design aspects. In many ways, the sealing design is the result of which system design parameters are given the most emphasis.
6. Casing out-of-roundness (noncylindrical shape) will be transmitted through the structure response to the hot-gas flowpath roundness bounding the blade tips. This leads to nonuniform tip gaps around the circumference and potential tip rubs.
7. Shroud segment variation, such as bowing, can result from the thermal gradients present in the design, again leading to nonuniform tip gaps either radially and/or axially.
8. Approaching and leaving disturbances in the flow around blade tips can affect both the aerodynamics and the cooling. Approaching disturbances are most notably associated with the wakes and shocks being shed from the upstream vane row, which to some degree must influence the tip flow and heat transfer by the introduction of unsteady effects. Approaching and leaving disturbances can be encountered in tip designs that involve shroud recesses and axial flow gaps between the stationary shrouds and attached tip shrouds.
9. Gas temperature profiles are the result of the particular combustion system design, the operational point, and mixing through the subsequent stages. The radial gas temperature profile may have severe impact on the blade tip, both in respect to the temperature field itself and the pressure distribution. Stronger radial flows can bring hotter gases to the blade tip than desired whereas gas temperatures can drive strong material thermal gradients and cause lower cooling effectiveness.
10. Aeromechanics must be considered in the overall blade structural design, and the tip region must be included in this response.
11. Stresses, both mechanical and thermal, are key in turbine blade survival. Blade tips must typically deal with very high local thermal stresses. Higher cooling effectiveness in the tip can alleviate thermal stresses, but must be weighed against the cost to the cycle efficiency. As noted earlier, the blade tip design will influence the weight distribution in the entire blade, which must then be dealt with in the allowable stresses, as well as the low-cycle-fatigue (LCF) and high-cycle-fatigue (HCF) responses. This effect will also be transmitted into stress requirements for the blade shank, dovetail, rotor disk posts, and the rotor disk.

12. Operating conditions must be considered at various limiting points in the engine cycle because these change the gas and coolant flow rates, temperatures, and pressures. A blade tip design focused solely on steady-state takeoff conditions might not be well suited for cruise conditions; an optimized or balanced cycle design must be sought.
13. Transients play a major role in the durability and life of any effective blade tip design. The relative displacements, radial and axial, of the rotor and stator systems during various transients will determine the ultimate steady-state operating clearances, as well as the potential for detrimental interference.
14. Durability is desired for both the blade tip and the opposing shroud as a system. In the long term, durability can be associated with oxidation whereas in the short term durability is a matter of survival in the face of tip rubs (intentional or unintentional), plugged cooling holes, and thermal stresses.
15. Materials and material loss must be planned in blade tip design. Blades and blade tips are not automatically designed with the highest temperature capability material, nor the highest strength material. The tip material can be different from the rest of the blade. The compatibility of the tip and shroud materials must be considered (for example, a highly abrasive shroud can damage a relatively weak tip material).
16. Cumulative damage to blade tips is typically experienced in characteristic locations in each design type. Uniform damage or material loss is not the general rule. The change in tip geometry with characteristic damage and loss will alter the aerodynamics and heat transfer, usually leading to accelerated loss.
17. EGT is directly and strongly affected by blade tip clearance. Any improvement in effective tip sealing will preserve valuable EGT margin.
18. Cost of new parts and cost of repair depend on the complexity of tip design.
19. Blade weight impacts the blade root stresses, LCF life, and blade creep. This is not limited to a simple matter of centrifugal stresses, but can also have severe effects on the overall aerodynamic design, changing the reaction and work of the stage.
20. Thrust bearing location and bearing housing distortion affect axial motion and disk sag, which, in turn, are transmitted through the rotor to the blade tip, potentially creating larger clearances on one side of the turbine and rubs on the other side.
21. Rotor and stator systems should be thermally matched to minimize variations in blade tip clearances during transients. Active clearance control systems can aid in this goal by providing fast thermal response of the shroud radial location.

22. Blade tips are commonly at least partially damaged or worn in the course of operation. The ability or inability to repair blade tips becomes an important factor in lifetime cost. The complexity of a blade tip design impacts the decision to provide more or less cooling to balance the cost of repairs. Unrepairable blade tips result in scrapped blades.

Although this summary of system design aspects might appear quite detailed and daunting for such a relatively small region of the turbine, there is one requirement that exceeds all others—the blade tip system design must never cause such severe damage as to liberate blades or pieces of blades in operation. As in the other interacting system relationships within the turbine, prior design and operational experience must guide and temper designs.

III. TRANSIENT OPERATIONAL REQUIREMENTS

Transient cycle conditions represent the most severe test for turbine blade tips in most respects, structurally, aerodynamically, and thermally. Figure 4 presents an example of a typical aircraft engine HPT turbine transient from a cold start through takeoff and climb to steady-state condition at cruise. Also shown is the possible transient condition during a trip, or engine shutdown, from steady state. In Fig. 4, the ordinate represents both the turbine speed and the radial growth ΔR of the rotor and stator portions bounding the blade tip. As the speed of the turbine spools up very quickly, immediate response is seen because of the centrifugal growth of the rotor disk and the blade. During this short time of initial rotor growth, the stator lags in response because its growth is only

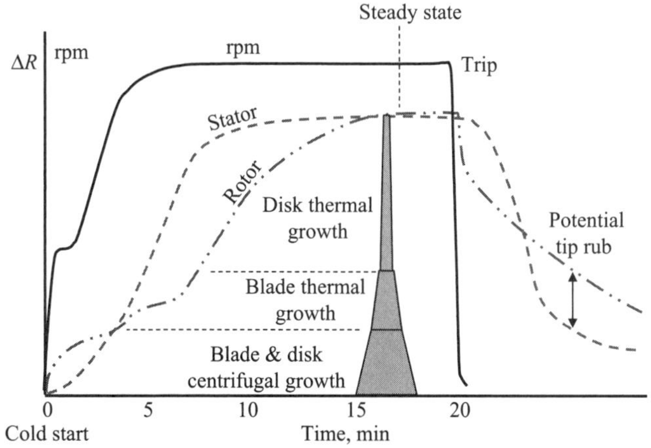

Fig. 4 Transient stator and rotor system radial growth curves.

caused by thermal expansion. The rate of thermal growth of the stator then begins to overtake that of the rotor. The stator is capable of maintaining higher bulk material temperatures, and so it is cooled to a lesser degree than the blade and hence thermally grows outward faster. The blade is highly cooled, so that its thermal growth is less than that of the stator. In the last phase of overall radial growth, the slowly responding thermal mass of the rotor disk expands to its steady-state condition. The overall radial growth of the rotor disk and blade is intended to slightly exceed that of the stator, such that the tip clearance gap closes to an efficient tight position. From this transient history it can easily be seen that off-design conditions, such as overspeed or reduced cooling flow, can quickly result in blade tip rubs.

The postulated trip scenario shown in Fig. 4 highlights a key operational danger in engines. When the rotor speed is suddenly dropped, the centrifugal growth factors of the rotor and blade are reversed quickly while the thermal growth of the stator only slowly responds at first. It is the large thermal growth of the rotor, however, that lags the most in the system. The stator thermally contracts inward before the rotor and blade can move away, resulting in potentially severe and very damaging tip rubs. This is the critical time period for restart of the engine to avoid such harsh interference. A safety factor can be built into the design to allow for larger clearances at cruise conditions such that rub events are less likely, but this has obvious negative efficiency implications. Another alternative might be the design of stator systems with faster thermal response or controlled response.

From the perspective of the blade tip clearance response during transients, Fig. 5 shows the rotor and stator growths relative to the initial cold clearance of the blade tip. Progressing through a similar takeoff, climb, and cruise transient,

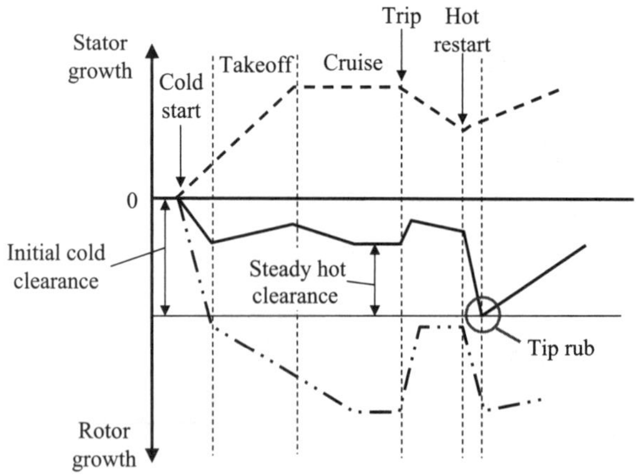

Fig. 5 Blade tip clearance variation with operating condition.

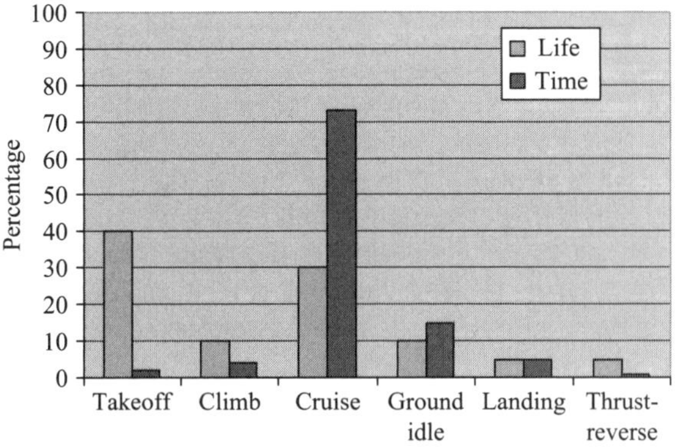

Fig. 6 Example of mission mix effect on blade tip life.

the tip clearance is seen to initially close, then open slightly again, and then close up to the hot steady-state magnitude. Upon a trip condition, the tip clearance first opens somewhat and then closes as the stator response catches up. If a hot restart is then initiated, the blade thermal growth and rotor centrifugal growth rapidly respond again, as the rotor has not had time to cool down. Here again, because the stator cannot respond as quickly, a potential exists for a blade tip rub event. In this case, the rub is generally slight as the entire system moves back to its steady condition.

The basic blade thermal design is based primarily on the conditions present during the "hot" cycle point of the engine, such as hot day takeoff conditions for an aircraft engine. For the balance of peak thermal efficiency with minimal usage of valuable compressed air, however, the blade tip cooling will be designed such that a nominal bulk material (metal or ceramic) temperature is maintained under the most common running condition, while also limiting the maximum local material temperatures experienced under the peak load conditions. For aircraft engines, the nominal operating condition is cruise, and the peak condition is takeoff or thrust reverse. For land-based power turbines, the nominal and peak conditions are the same at the base load operating point of 100% output.

Blade tip life can be treated in design as the cumulative effect of varying thermal loads and cooling from the total time spent under the different transient and steady load conditions. Figure 6 shows this concept in terms of a simple pareto diagram using the major elements of aircraft engine operational time. Keep in mind that the time exposure under takeoff conditions is the smallest, yet the effects on blade tip heat transfer and life can be the most severe. This is analogous to the use of a cumulative damage model in mechanical fatigue of materials. Figure 6 is an illustrative example only as the specific pareto varies widely among system designs.

IV. BLADE TIP DESIGN PHILOSOPHIES

Considering the design and operational requirements discussed in the prior sections, it should come as no surprise that several different forms of blade tips have been devised and used successfully. There is no single solution that satisfies all requirements in an optimized manner, just as there is no single design of turbine engine, nor a single mission definition for all engines. Each blade tip design has its advantages and disadvantages. There are three major design philosophies in practice today: 1) flat cylindrical blade tips, 2) cylindrical tips with squealer sealing rims, and 3) attached tip shrouds used with either cylindrical or flared blade tips.

Flat cylindrical turbine blade tips are no longer very commonplace, but are still in use, even in some modern designs. Figure 7 shows a photo of an older vintage heavy frame turbine first-stage blade tip. The tip is flat and cylindrical (common outer radius along axial direction). This older style blade uses straight cooling passages from hub to tip, discharging the coolant out the tip radially. In one respect, this manner of coolant discharge provides a sort of fluid resistance to hot-gas tip leakage flows, but this was not the intent of the design. By the time the coolant exits the tip in such a design, its thermal potential for cooling purposes has been nearly used up. In modern flat-blade tip designs, complex internal cooling passages deliver tip cooling through film holes near the tip on the pressure side of the airfoil as well as on the tip surface. It should be recognized that at least one tip hole is required per cooling circuit within a blade to allow for dust purge by centrifugal action. Such flat blade tips rely on good thermal matching of the rotor and stator systems to keep tip clearances tight. Because no physical leakage resistance sealing mechanisms are employed in flat tips, these designs typically result in the lowest tip aerodynamic efficiency because of relatively high leakages. The higher leakage flows can also lead to higher heat loads on the tip; sufficient film cooling is crucial in such designs. Flat blade tips might also be more susceptible to damage if and when they do rub as there is no sacrificial buffer material present. One positive aspect of flat tips is that there are no extended surfaces to be cooled, and hence the cooling design can be very simple. As long as the flat tip integrity can be maintained, its performance at a given cycle point condition remains unchanged with time.

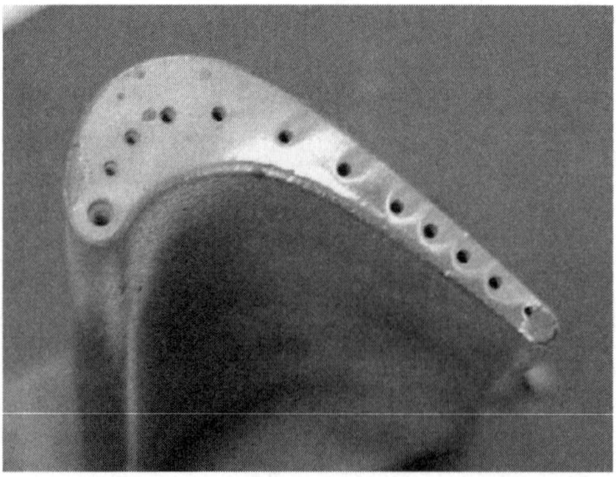

Fig. 7 Power turbine flat blade tip example.

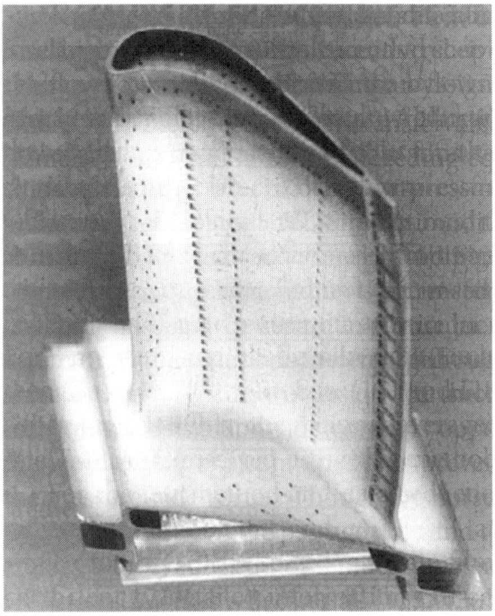

Fig. 8 GE Aircraft Engines HPT blade with squealer tip.

A cylindrical blade tip with a perimeter seal strip, known as the squealer rim, is the design most commonly used today in HPT turbines. Figure 8 shows an example of an advanced, highly cooled HPT turbine blade with squealer tip surrounding the tip cavity. The squealer rim is a natural radial extension of the aerodynamic surface of the airfoil. In most designs, this squealer rim extends as far aft on the tip as possible until the trailing-edge thickness requires closure of the perimeter. In some designs, the trailing-edge portion of the rim is left open on the pressure side, allowing tip cavity cooling air to exit over the suction-side rim as a sort of film cooling.

The function of the squealer rim is as a simple two-tooth labyrinth seal. Tip leakage gas is forced to contract between the pressure-side rim and the shroud, then expand into the tip cavity defined by the entire squealer rim, and then contract again to pass the suction-side rim restriction before expanding into the main flow. This is a very simplistic picture of the tip seal, but captures the intended function. A squealer tip running with a very tight clearance can be a very effective seal, aerodynamically. Likewise, a squealer tip running with a very open clearance can be a very poor seal. The rim thickness is minimized to reduce weight and actually has a negligible role with respect to the aerodynamics. The rim material is allowed to rub against the shroud in transient situations because this material is sacrificial. Minimal, moderate, or even heavy tip rubs in such designs will not compromise the integrity of the remainder of the blade. The difficulty in a squealer rim design lies in cooling of the rim to prevent loss by oxidation and erosion. The rim represents an extended fin that sometimes requires many pressure-side film-cooling holes, and also interior tip cavity cooling holes, to provide adequate cooling coverage, as shown in Fig. 8.

Turbine blade tips using attached tip shrouds are found in some HPT blade designs, but are more common in LPT blades. Figure 9 shows an example of an HPT blade with a highly cooled, attached tip shroud. The tip shroud is in essence a shroud, or bounding outer radius flowpath, that moves with the blade tip. There is a stationary shroud casing outside of this tip shroud. The attached tip shroud allows the hot-gas tip leakage path to be decreased from the full tip profile to only seal gaps between the tip shroud and the casing shroud. The primary gaps are located at the forward and aft circumferential edges of the shroud and radially at each seal tooth on the top of the tip shroud. The blade tip of Fig. 9 employs two circumferential and two axial seal teeth on the upper surface. This design acts as a multitooth labyrinth seal arrangement with complex flow restrictions and interactions. Of all of the current blade tip designs in use today, the tip shroud has the lowest aerodynamic loss when properly implemented. Figure 10 shows an example of a simple LPT blade tip shroud. The function of this design is the same as that of the HPT tip shroud, but the complexity is far less, using only a single seal tooth.

A few other aspects of these designs are worth noting here. The HPT tip shroud is a heavier blade tip than the squealer or flat tip, and so great attention must be paid to stress. This then requires a much more complex cooling system, not only because of the geometry, but also to maintain material temperatures for acceptable stresses. The tip shroud of Fig. 9 is a flared tip, meaning that the mean radius of the tip increases along the axial direction. This requires the casing shroud to do the same, which then allows the axial gaps to be used as sealing faces. This practice has the advantage of greater sealing, but also means that the tip and shroud systems are more sensitive to axial movements of the turbine rotor and stator. The tip shroud of Fig. 10 is a cylindrical tip, but still recessed into a casing shroud. The weight penalty on stress (creep) can be severe enough to require scalloping of the tip shroud. Scalloping is the practice of removing tip shroud material to reduce weight while still attempting to maintain the aerodynamic sealing function. The tip shroud of Fig. 10 is moderately scalloped as shown by the curved edges. This material removal will decrease aerodynamic sealing capability and reduce the airfoil lift coefficient.

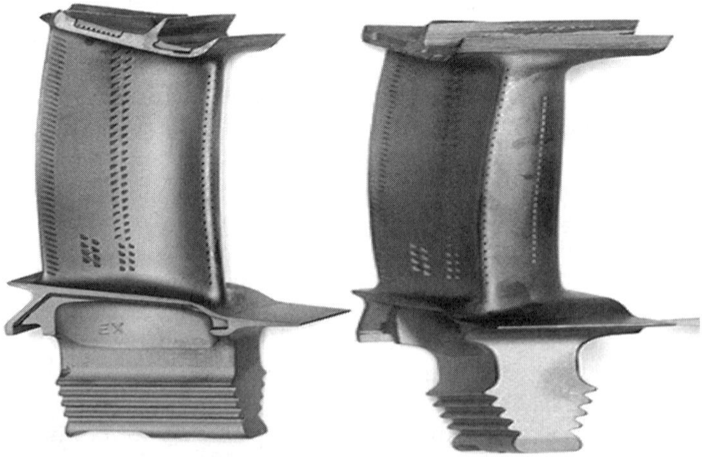

Fig. 9 Rolls-Royce Trent 800 HPT blade after 8299 hours (cleaned) (reproduced by permission of Rolls-Royce).

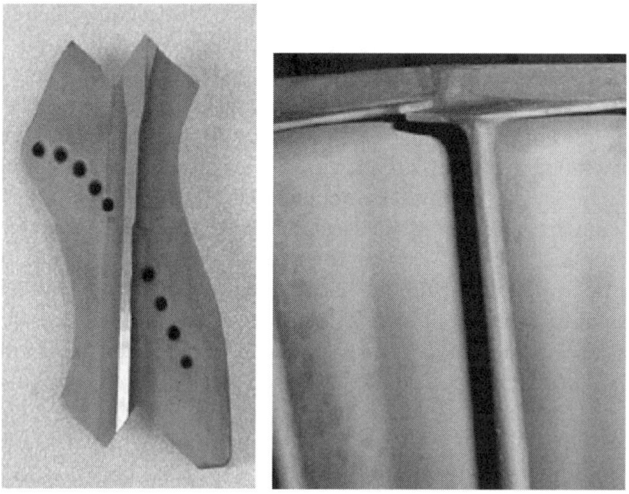

Fig. 10 Scalloped attached tip shroud for power turbine.

A final note concerning the function of attached tip shrouds is that they physically abut one another along the circumferential outer radius. This connection, though not sealed, provides some damping of blade vibrational characteristics. This feature of tip shrouds can be an advantage for the aeromechanics of the blade row. The tip shroud of Fig. 9 shows a straight shroud-to-shroud interface whereas that of Fig. 10 has a so-called Z-lock shape to assist with axial displacements.

Although the three major designs of turbine blade tips have been introduced here, there are many variations on each of these designs, including hybrid designs that attempt to capture the desirable characteristics of each. In every specific blade tip design, tradeoffs must be made to determine the most feasible approach. Here again, the most important single factor that will determine the direction of design is operational experience. In some missions and engines the choice will be obvious. For example, blade tips in a short mission expendable engine will be flat and cylindrical for simplicity and cost reasons. Uncooled blade tips in LPTs, and also those in steam turbines, will be designed with tip shrouds for best aerodynamic performance. Turbine engines requiring long life and high firing temperatures involve a number of blade tip design philosophies.

V. AERODYNAMICS

To gain an appreciation for turbine blade tip aerodynamics, it is instructive to examine the sources of losses within a turbine stage. Figure 11 shows a summary of the breakdown of loss components for several experimental turbines [3].

Included in the overall losses are the vane (stator) profile, endwall, and secondary flow losses, which account for about 25% of the total stage loss. The remaining 75% of stage losses are associated with the blade (rotor). Of these losses, the most significant is that due to tip clearance, which on average accounts for roughly one-third of the total stage loss. Looking at the turbine stage aerodynamic efficiencies noted at the top of Fig. 11, the tip clearance loss represents about 4% of stage efficiency, a very substantial amount.

To explain the specifics of tip clearance loss, the detailed tip region flows must be described. Figure 3 depicts the general pressure distribution around a flat blade tip with and without clearance, showing the driving pressure potentials for the leakage flow. Using a two-dimensional linear blade tip cascade (flat tip), Bindon [4] measured the detailed blade tip and shroud pressure distributions for several tip clearance gap magnitudes. Figure 12 shows the contours of static-pressure coefficient for a tip gap of 2.5% of blade span. These contours show the clear presence of an entry separation region on the tip pressure side in the mid to aft chord location. This feature is also reflected in the shroud contours. The zone of highest tip leakage flows emanates around this region, driven by the pressure-to-suction side overall aerodynamic profile. The leading-edge region of the blade tip is relatively calm, but still has lower magnitude leakages.

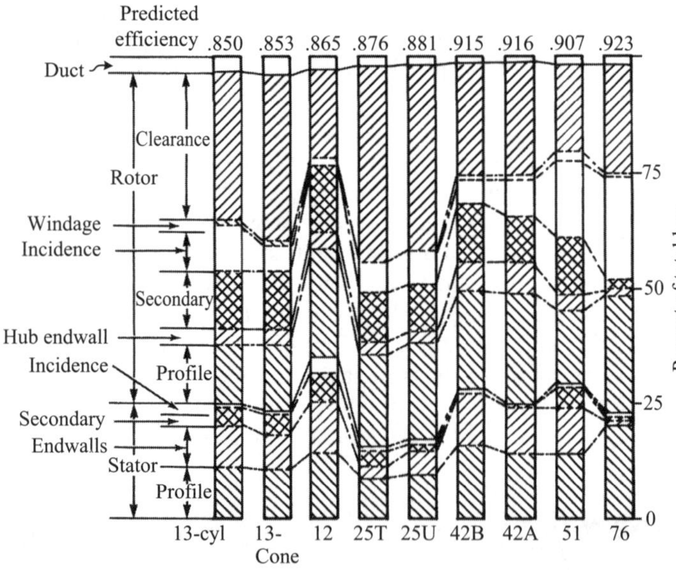

Fig. 11 Efficiency loss contributions for several single-stage turbine rigs (reproduced by permission of von Kármán Institute).

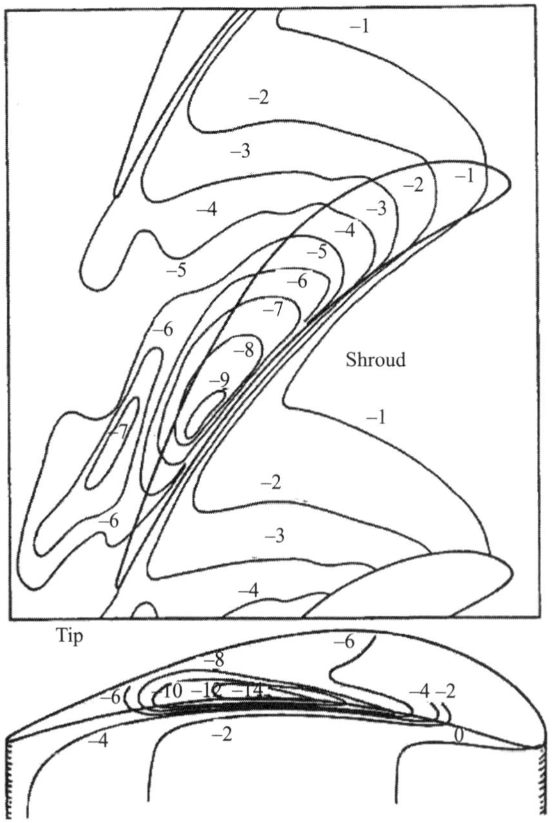

Fig. 12 Flat blade tip and shroud pressure distribution contours [reproduced by permission of American Society of Mechanical Engineers (ASME)].

Yamamoto [5] performed five-hole pitot probe measurements in a linear blade cascade for a plane through the tip clearance and passage region, as well as on a plane downstream of the trailing edge. Figure 13 shows the typical flow vectors and streaklines for the nominal incidence angle of 7.2 deg and a tip gap of 2.1% blade span. The blade tip is flat. The flow vectors exhibit low velocities of leakages in the tip leading-edge region with higher magnitudes in the midchord to aft regions. Leakage flows are typically angled from the regions of highest source pressure to lowest sink pressure. The secondary endwall flows in the passage show direction from the pressure side to the suction side. The tip leakage vortex created by the interaction of the gap flows with the mainstream passage flow is initiated in the region of highest airfoil curvature, where leakage flows first enter the mainstream, and grows from there downstream. The downstream radial plane shows both flow vectors and contour lines of loss coefficients.

The tip vortex aft of the trailing edge is intense and contained within the upper 10% of the span. The vortex has induced a secondary counter-rotating vortex in the mainstream flow, which is larger and penetrates quite far into the passage both radially and circumferentially. Yamamoto also studied the effects of off-incidence angles (negative angles of attack) that simulate conditions at other cycle points. It is common that blade tips might experience very wide variations in flow incidence angles within the cycle, especially in aircraft engines. The tip leakage vortex is roughly unchanged for such conditions, but flows over the tip and induced secondary flows in the passage can be greatly altered. The studies [4, 5] were set in stationary cascades without relative motion effects of the shroud surface. In a rotating turbine rig test, Morphis and Bindon [6] found that such relative motion effects did not strongly influence the resulting pressure distributions.

Given the general flow characteristics surrounding a flat blade tip, an overall picture of the basic tip aerodynamics can be established, as in Fig. 14. This general view shows the leading edge, midchord, and aft region flows as described from the studies just noted. Also shown are two cross-sectional views through differing portions of the blade tip, one in midchord and the other aft.

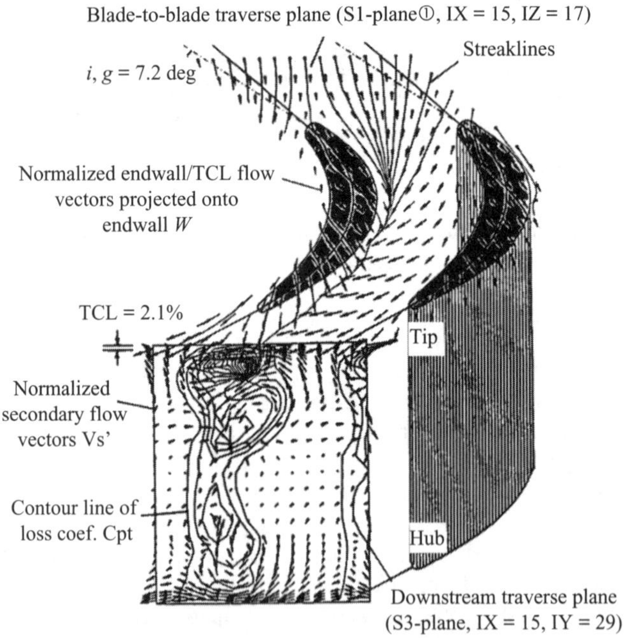

Fig. 13 Blade cascade flow vectors and exit loss coefficient contours (reproduced by permission of ASME).

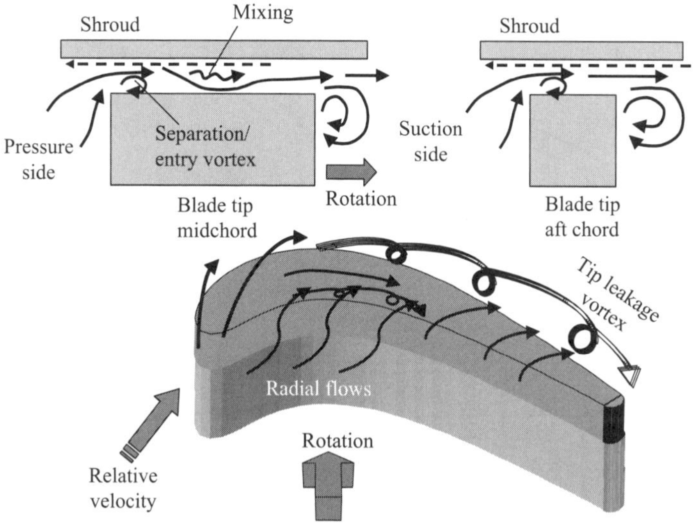

Fig. 14 Schematic of flat blade tip leakage flow characteristics.

These depictions demonstrate that the local blade tip thickness, as well as the local flow conditions, will alter the existence of such effects as the separation vortex, the extent of mixing within the tip gap, and the jetting of leakage flow into the mainstream.

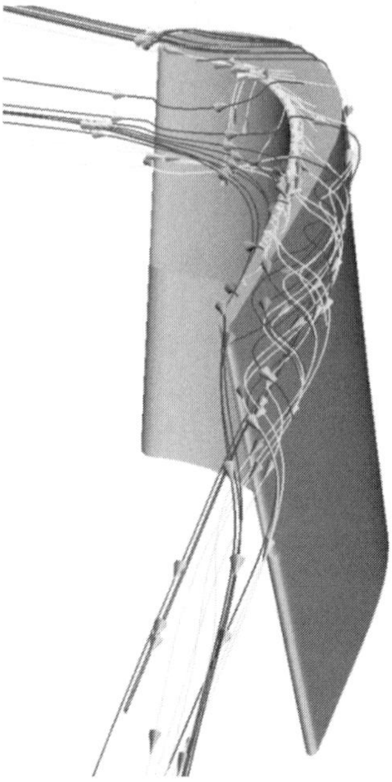

The space shuttle main engine CFD prediction of Ameri and Steinthorrson [7], shown in the particle traces of Fig. 15, clearly illustrates both the tip entry separation and the tip leakage flow vortices, the former being composed primarily of radial flows entering the gap (light lines) and the latter of bulk gap flows (dark lines) mixed with the postseparation flows. Here again, the streamwise growth of the leakage vortex is apparent from the high curvature portion of the airfoil suction side. Bindon [8] captured this vortex growth quite well in his linear blade cascade tests. Figure 16 shows the initiation of the vortex just past 50% axial chord location with subsequent growth and increase of loss coefficients at 80 and 100%

Fig. 15 Computed prediction of flowfield streaklines for flat blade tip (reproduced by permission of ASME).

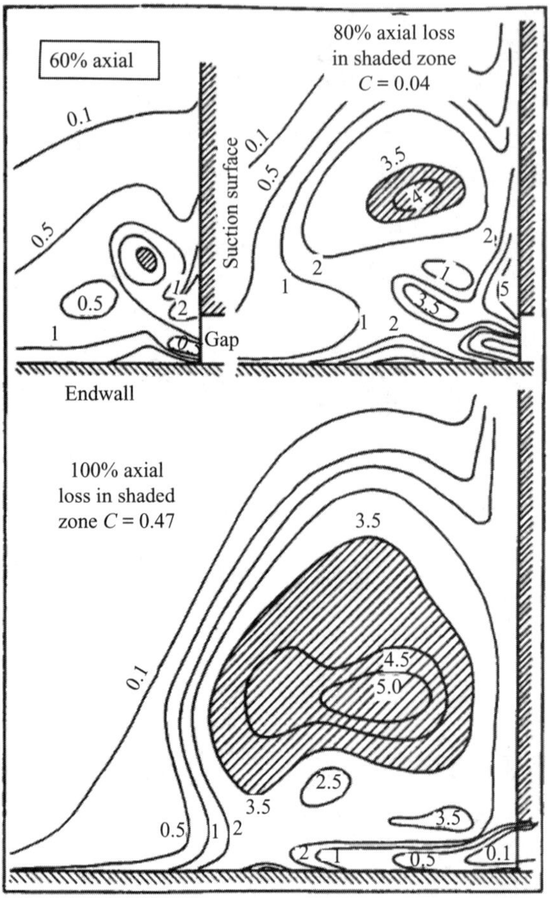

Fig. 16 Growth of suction-side tip leakage vortex strength (reproduced by permission of ASME).

axial chord. Bindon further used such data to quantify the contributions to loss due to endwall secondary flows, internal tip gap mixing, and leakage vortex mixing. Figure 17 shows the results for his airfoil shape and loading. The endwall and secondary flow losses are those present with no tip gap clearance, contributing along the entire axial chord of the passage. The internal gap losses caused by separation, shear flows, and mixing do not begin to contribute until about 40% axial chord, but make up almost half of the total losses. The tip vortex (mixing) losses are seen to begin at about 60% axial chord and rise significantly as the internal gap losses diminish (due to the thin trailing edge). This view of losses is quite generic, but applies well to most flat tip cases, with some

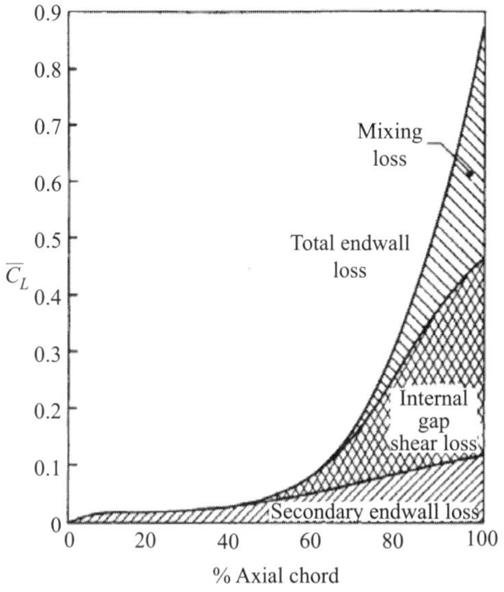

Fig. 17 Contributions to total endwall loss development as function of axial location on tip (reproduced by permission of ASME).

adjustment for specific aerodynamic shapes and loading. The same generic view also holds fairly well for squealer and tip shroud designs. It is apparent that the total blade tip loss for any design can be reduced by attacking its two main contributors, internal gap loss and vortex mixing loss.

Most blade tip designs do not seek to reduce these main aerodynamic losses through detailed manipulations of the interacting flows and structures, but instead take the more straightforward approach of simply reducing the total tip leakage flow. This is accomplished through sealing geometries and tight running clearances. Figure 18 shows the general flow features for three sections of a squealer tip. With the additional flow restrictions, there are separated flows (or vena contracta) on both squealer rims. The typical picture of tip cavity flow is one in which the flow expands into the cavity, reattaches on the cavity floor, and then contracts again to exit the cavity. In the process, recirculation regions are formed in the cavity corners. Figure 18 shows that this is not the case at every section because cavity aspect ratio is not constant along the tip. In the trailing edge, the tip leakage flows can stream across the cavity with little or no impediment. The flowfield in a squealer tip cavity is actually much more complex than this simple view. As the CFD predictions of Ameri et al. [9] show in Fig. 19, the three-dimensional nature of the blade tip and airfoil pressure distribution will

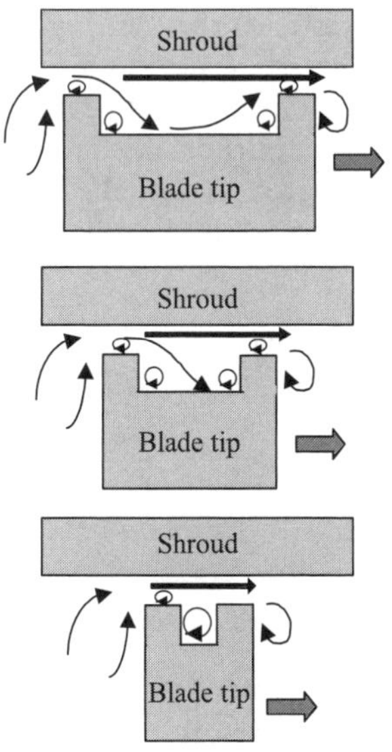

Fig. 18 Flow characteristic changes at three locations in squealer tip.

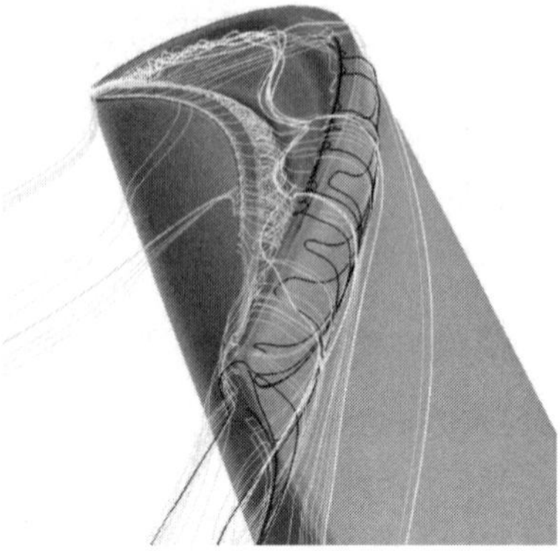

Fig. 19 Computed prediction of flowfield streaklines for squealer blade tip (reproduced by permission of ASME).

cause chordwise tip cavity flows that interact with the main pressure-to-suction side flows. Depending on the specific shape and design, such flows can either help or hinder the function of tip sealing.

In the case of attached tip shrouds, the number and complexity of flow restrictions increases. The simplest geometry using a single tip seal tooth is depicted in Fig. 20. In proper operation, the inboard surface of the tip shroud should line up with the casing shroud surface before and after the casing recess to form a continuous hot-gas flowpath. The flow must then negotiate a forward interface slot to enter the gap region, a seal tooth restriction, and the aft interface slot to reenter the mainstream. Typically, the casing shroud inside the recess will also be treated to increase flow resistance, through the use of, for example, a honeycomb structure. Not shown in this sketch is the fact that the rotating blade tip will cause a significant circumferential flow within the recess gap, especially in the forward section, which will further aid in the reduction of leakage. Figure 21 shows the much more complex design for an HPT tip shroud with axial flaring. Here, two circumferential seal teeth (fins) engaged with a stepped recess in the casing are used in the forward region of the tip shroud. In addition, two chordwise seal teeth (fences) are used in the aft region to help seal the region of highest overall pressure potential across the tip. Together, these devices form a formidable tip leakage resistance network. The aft tip fences will also act to extract useful work (blade lift), as will some portion of a squealer tip rim. Tip shrouds have at least one negative impact on aerodynamics, however, in that they are much more sensitive to variations in alignment or positioning. It is easily seen that a tip shroud which protrudes into the hot-gas path will provide a mainstream blockage that not only reduces the overall blade efficiency, but can also exacerbate tip leakage. Likewise, a recessed forward tip shroud can lead to a protruding aft casing surface, which again can reduce stage efficiency.

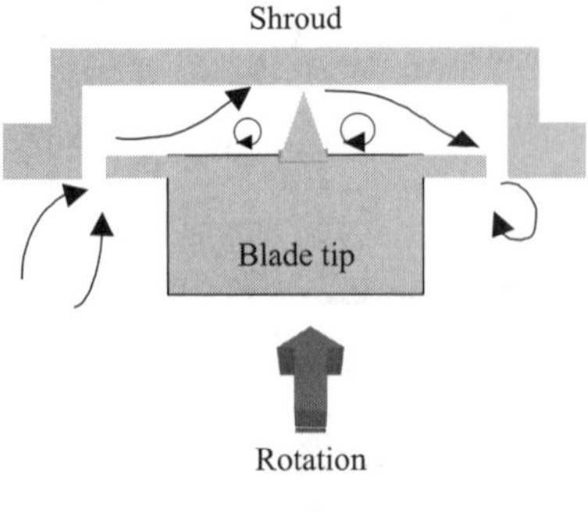

Fig. 20 Recessed tip shroud flow schematic.

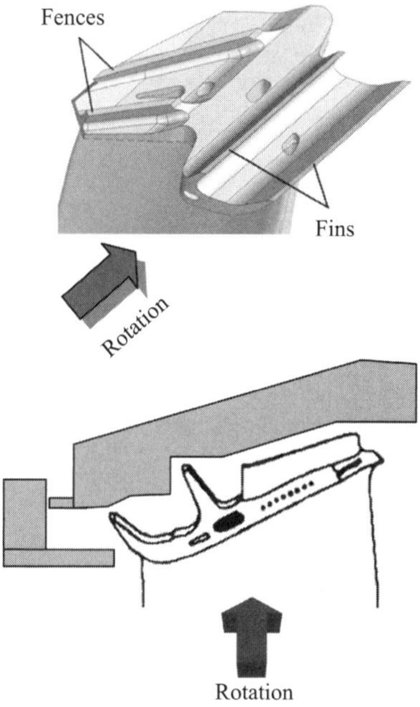

Fig. 21 Complex sealing structure for HPT tip shroud (reproduced by permission of Rolls-Royce).

The noted blade tip designs have common features in aerodynamic efficiency. As in Fig. 2, all designs increase in efficiency as the effective tip clearance is reduced. As the tip clearance is reduced to less than 1% of blade span, the squealer tip and tip shroud designs become roughly equivalent in efficiency. Differences in performance come as blade tips degrade under service conditions and depend on the design's ability to resist such degradation.

VI. HEAT TRANSFER AND COOLING

While the function of a turbine blade tip is to reduce as far as possible the aerodynamic losses at the rotational-to-stationary interface, this function cannot be maintained in cooled blade tips without careful attention to the thermal boundary conditions and the cooling design. Thermal boundary conditions of concern to the blade tip include the following:

1. Mainstream hot-gas temperatures
2. Tip surface heat-transfer coefficient distributions
3. Tip surface film effectiveness distributions
4. Pressure-side sink flow heat-transfer coefficients and film effectiveness
5. Suction-side source flow heat-transfer coefficients and film effectiveness
6. Seal surface heat transfer (rims, seal teeth, etc.)
7. Internal blade tip heat-transfer coefficients
8. Internal coolant temperatures
9. Blade tip substrate and protective coating thermal conductivity

A pareto of the relative impact and uncertainty of each of these boundary conditions upon the resulting blade tip temperatures is shown in Fig. 22. This pareto is only one example of many, but it indicates the typical relative importance of the thermal boundary conditions for most HPT blade tips. A key aspect of blade

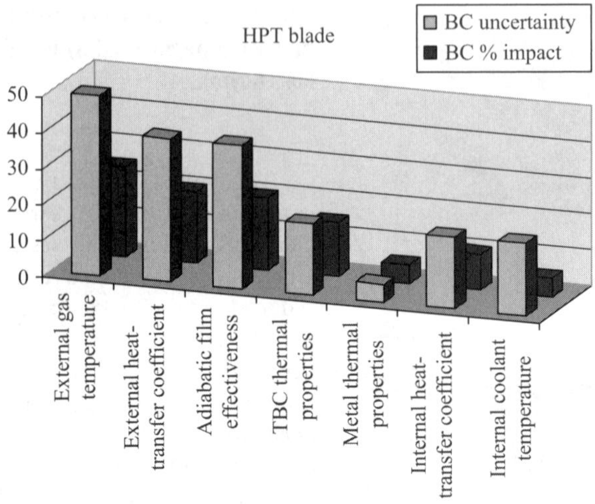

Fig. 22 Pareto of thermal boundary condition impact and uncertainty.

tip heat loading and cooling is that no two aerodynamic or system designs are completely alike. As will be seen later, there are similarities in all blade tip heat-transfer characteristics, but the differences between designs can be subtle or striking. There is consequently no single best blade tip cooling design that can be applied to all engines, only design philosophies, which can be broadly grouped by function and form. Within any particular general design approach there will necessarily be engine-to-engine variations, as well as blade-to-blade variations, which can cause significant deviations from nominal heat loading and cooling conditions. This aspect of blade tip cooling can be thought of as design for reliability applied to the blade or turbine and applies to both the overall cooling design and the individual details of local design features.

The basic heat-transfer characteristics for a flat blade tip are illustrated in Fig. 23, which shows the distribution of heat-transfer coefficients measured by Bunker et al. [10] for a first-stage blade tip in an industrial heavy frame turbine design. This distribution was measured in a linear, high-speed cascade with sharp edged tip, an inlet Mach number of 0.4, overall pressure ratio of 1.43 (atmospheric discharge), and clearance gap of about 1% of blade span. For comparison, Fig. 24 provides the local pressure ratio distributions on the tip surface (total pressure to local static pressure) for the same clearance gap, as well as increased and decreased clearances. Figure 23 is annotated to show the major flow regions and effects for this blade tip. The most striking feature in the heat-transfer coefficient distribution is the "sweet spot" of lowest coefficients in the midchord region. This region is also highlighted in the pressure distributions of Fig. 24, where the pressures across the tip all come into approximate agreement, meaning that there is little pressure difference to drive leakage flows

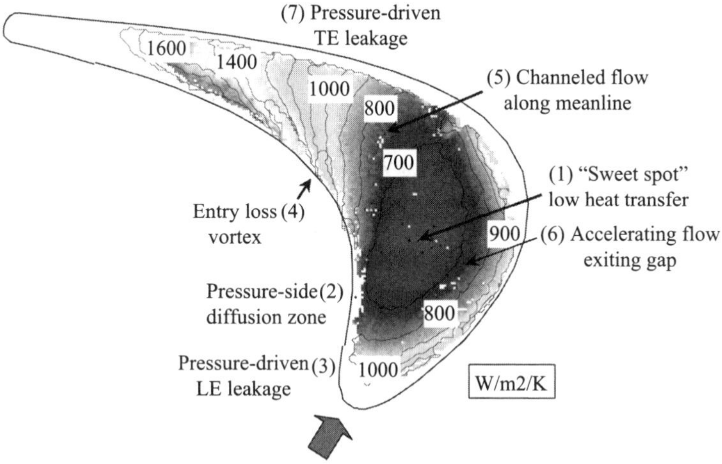

Fig. 23 Flat blade tip heat-transfer coefficient distribution (reproduced by permission of ASME).

here. The leading edge of the airfoil tip ahead of the pressure-side diffusion zone indicates leakage flow along the forward suction-side edge region with higher heat-transfer coefficients. The pressure-side entry loss vortex is clearly observed in the concentration and turning of the heat-transfer coefficient contours. From the region of 40 to 80% axial chord, the pressure differential across the tip is highest, leakage flow is high, and consequently the heat-transfer coefficients are also highest in direct correspondence with the pressure distribution. Given the broad nature of this tip profile, there is also some channeling of leakage flow along the mean chordline from the sweet spot aft towards the suction side where the flow exits.

To emphasize and contrast such heat-transfer characteristics, Fig. 25 shows a second distribution of

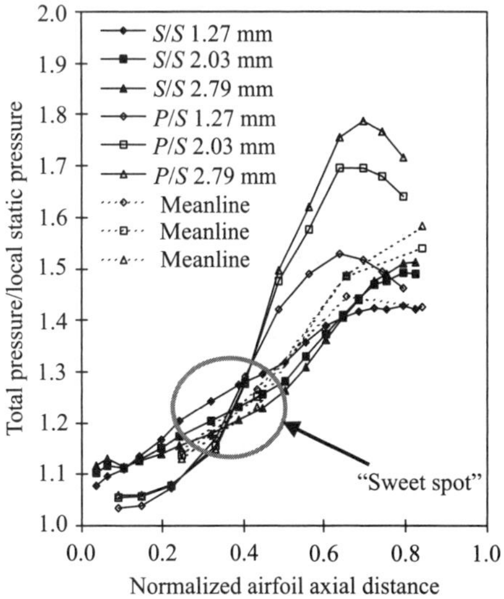

Fig. 24 Flat blade tip pressure field corresponding to Fig. 23: *P/S*, pressure side; *S/S*, suction side (reproduced by permission of ASME).

coefficients measured by Bunker and Bailey [11] for a different, much narrower tip shape under virtually identical cascade conditions. All of the same features can be identified here, but some are extended and others muted by the different aerodynamic profile. The sweet spot is relatively enhanced; however, the entry vortex is more compressed, as compared to Fig. 23.

With the addition of almost any sealing features to a flat blade tip, the driving pressure distribution around the tip is altered, local leakage flows are redistributed, and so too the heat-transfer coefficient distribution on the tip can be substantially changed in both magnitude and form. Figure 26 displays tip heat-transfer coefficients for several variants of simple tip seal features, as obtained by Kwak et al. [12]. In this study, a linear blade tip cascade was again used with a mild overall pressure ratio of 1.2 and a tip clearance of 1% blade span. The flat tip heat-transfer coefficients are shown at the bottom of Fig. 26 (note the slightly higher scale for this case). Immediately apparent is that without a seal it is peaked on the pressure-side entry region; the addition of any tip seal changes the heat-transfer coefficients. This is the case even when the sealing rim is not on the pressure side.

Although the sweet spot on the flat tip is located near the leading edge for this airfoil shape and pressure ratio, this is not the case for most seal geometries. As shown in Fig. 26, the forced redistribution of flow by the seal rims can actually increase the leading-edge region heat transfer, a characteristic commonly not observed in computational predictions alone. The full perimeter squealer rim tip is seen to lower tip cavity heat-transfer coefficients over most of the tip surface, most especially in the midchord and trailing-edge regions. Seal rims placed only on the pressure side, mean chordline, or in combination can tend to act as flow disruptors, or turbulators, leading heat-transfer coefficients that are locally higher than over those of the flat tip.

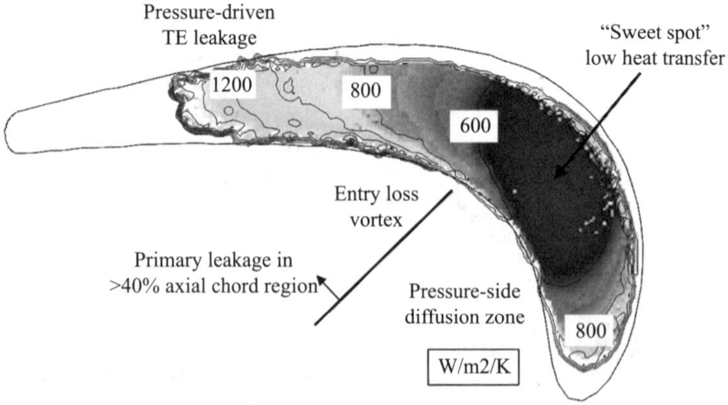

Fig. 25 Blade tip heat-transfer coefficient distribution for narrow aero profile shape (reproduced by permission of ASME).

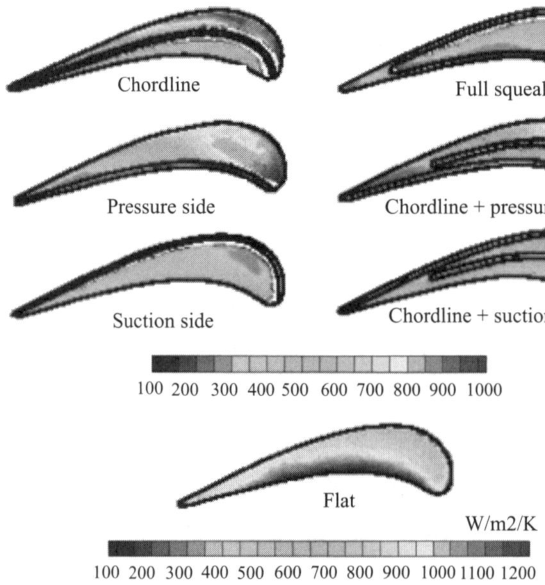

Fig. 26 Effect of various tip seal rim locations on heat-transfer distributions (reproduced by permission of ASME).

The very simple seal geometry using only a suction-side rim provides the lowest overall heat transfer in this example. This positive effect is the combination of two features, effective tip leakage flow reduction by the suction-side seal rim plus the elimination of any tip entry flow disruption from a pressure-side seal rim. This tip sealing geometry is not, however, common in practice as the use of a full squealer rim is generally thought to provide additional leakage reduction (better aerodynamic efficiency), especially in cases of material loss due to tip rubs.

While the use of blade tip seal rims provides a labyrinth seal effect against hot gas leakage flow, the resulting heat-transfer coefficients only resemble those of labyrinth seals in a limited portion of the tip. Three-dimensional flows dictate that only the midchord to aft region of a tip cavity can be treated approximately as a labyrinth seal cavity oriented transverse to the flow. Fortunately, this is also the region of highest leakage flows. In lieu of actual blade tip experimental measurements, data such as that of Metzger et al. [13] for simulated transverse tip grooves can be used to estimate heat-transfer coefficients in these high leakage regions.

The thermal benefit of the squealer tip geometry, noted to be as much as 50% lower heat transfer than the flat tip, is dependent on the survival of the extended seal rims. Depending on the highly three-dimensional flows inside the squealer tip cavity, the heat-transfer coefficients on the rim interior surfaces can be

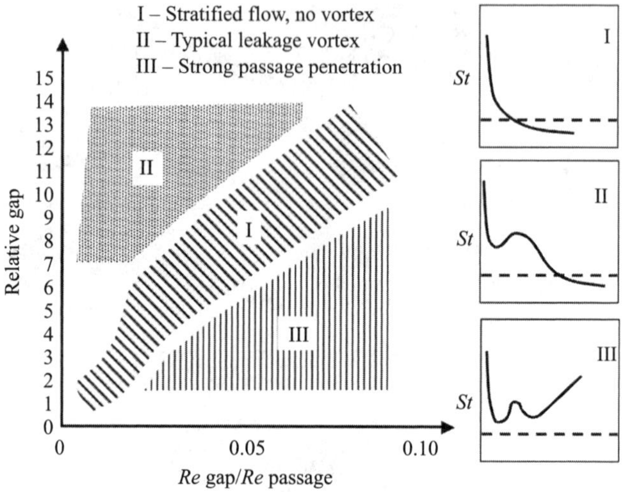

Fig. 27 Heat-transfer characteristics for near-tip sink and source flows (reproduced by permission of ASME).

either very high, typical of the suction-side rim, or very low, typical of the pressure-side rim. The higher heat-transfer coefficients are very similar to those experienced by the exterior seal rim surfaces, both the upper surface and the outer surfaces in the hot-gas flowpath. These latter surfaces can be characterized by the flow and heat transfer for sink and source type flows. The pressure-side entry region can be modeled as a sink flow from the mainstream passage into the tip gap. The suction-side exit region can be modeled as a source flow from the tip gap into the crossing mainstream passage.

Metzger and Rued [14] and Rued and Metzger [15] studied such sink and source flow effects on heat transfer for a range of relative flow strengths and gap magnitudes. Sink flow leads to accelerated flow into the tip gap with increasing heat-transfer coefficients as one approaches the top of the rim seal. Increases of 200 to 300% over heat-transfer coefficients in the mainstream passage below this region can be experienced. Source flow effects are associated directly with the suction-side tip vortex noted earlier and can be quite variable with relative gap size and local leakage-to-mainstream momentum ratio. Figure 27 shows the ranges of resulting heat-transfer enhancements obtained by Rued and Metzger [15] (shown as Stanton-number ratios). The relative gap of the ordinate refers to the magnitude of the tip clearance, where a relative gap size of 7 to 8 should be thought of as nominal operating clearance. The abscissa shows the ratio of source flow Reynolds number based on gap size to the mainstream passage Reynolds number, which can be interpreted as the ratio of mass velocities locally (or a blowing ratio).

Three general behavior characteristics are depicted for the source flow heat-transfer enhancements in the plots to the right of Fig. 27, where the dashed lines represent the level of the heat transfer due to the mainstream passage flow alone, and the abscissa is the radial distance from the tip. In Region I, a stratified flow exists, and the tip rollup vortex is absent. This behavior occurs when the relative gap size and the leakage blowing strength are well matched, such that mixing momentum transfer is minimized only to that required for turning the leakage flow. Such behavior would be desirable by comparison with the other characteristics, but is not generally obtained. Region II behavior is most common, showing the typical tip vortex effect increasing surface heat transfer as the vortex scrubs the near tip area of the suction side (see Figs. 15 and 19). Here the relative gap size is nominal or greater, and the leakage blowing ratio is weak, leading to the formation of the vortex as the flow is mixed and turned. Region III behavior is associated with blade tips that have lost sealing effectiveness, leading to a high leakage momentum that penetrates the mainstream flow deeply, causes major secondary flow losses, and increases near tip heat transfer.

Considering these high heat-transfer coefficients in the hot-gas leakage stream, film cooling of HPT blade tips is common. Figure 28 shows a schematic of application of film-cooling holes to a flat blade tip with many holes aligned in the radial direction near the tip entry on the pressure side and several more holes on the tip surface. Holes on the tip surface are not always film-cooling holes but may be required dust purge holes for the internal cooling circuits of the blade. The pressure-side film holes are angled steeply for best film adherence to the surface, as well as to allow drilling into the internal cooling passages. Similar film hole placement strategies are used on squealer tips as shown in Fig. 8 although the holes in the tip cavity must be located with a good knowledge of the more three-dimensional flows here. As shown in Fig. 28, tip film-cooling flows do not present major alterations to the overall tip leakage flow patterns.

Efficient tip film cooling using minimal coolant amounts is intended to reduce heat flux to the surfaces to enhance survival, not to block leakage flow from entering the tip gap, though this latter effect might be present to a small degree. Most blade tip film-cooling designs are achieved through experience, not through pretest analysis, as there are little or no

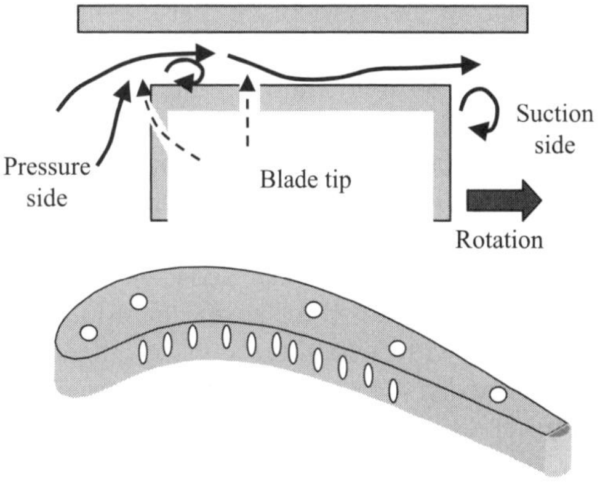

Fig. 28 Schematic of film-cooled blade tip.

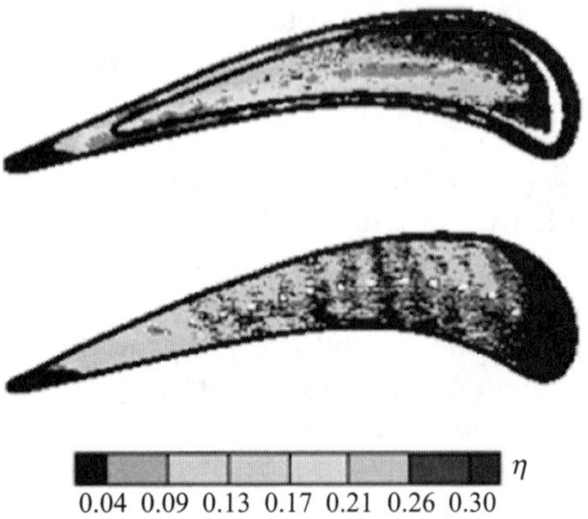

Fig. 29 Film-cooling effectiveness distributions for flat and squealer tip blade cascade (reproduced by permission of ASME).

data available on film cooling of tips comparable to that existing for airfoils and endwalls.

One indication of tip film effectiveness is shown in Fig. 29, from the cascade measurements of Kwak and Han [16]. Using the same blade tip cascade [12] as in Fig. 26, the local adiabatic film effectiveness is shown with 13 radial pressure-side film holes and 13 normal tip surface holes, the group having an average blowing ratio of 2. Effectiveness η is here defined as $(T_{\text{gas inlet}} - T_{\text{adiabatic surface}})/(T_{\text{gas inlet}} - T_{\text{coolant}})$ and shown for both a flat tip and a squealer tip. With a tip clearance of 1.5% blade span, the film-cooling effectiveness magnitudes are fairly low. A majority of the pressure-side film does not survive the entry into the tip. Film injected on the tip follows the leakage flow directly and exhibits a high degree of mixing. Film inside the tip cavity tends to be disrupted by reattaching flows but can collect in the more protected regions to provide higher cooling effectiveness, as seen in the apparent film cooling directed along the chordline. It is extremely difficult to cool the suction-side seal rim of a blade tip unless dedicated film holes are placed nearby. This simple example illustrates the complexity of cooling any blade tip region.

In reality, each tip film-cooling hole will have a different blowing rate and unique mixing with the leakage flow. This fact is driven not only by the external aerodynamics but also by the internal cooling of the blade tip, which, in turn, is directly linked to the overall cooling design of the entire blade. Yet this internal cooling plays an important role in maintaining the tip integrity, and so the internal heat-transfer coefficients must also be well known. Figure 30 shows two internal blade cooling designs, one with a flat tip and the other with a squealer tip. These are but two designs of an almost infinite variety, but depict common cooling circuit features. The flat blade tip utilizes a dedicated cooling passage under the tip surface that contracts from forward to aft to compensate for the extraction of coolant through tip film and purge holes. Such dedicated channel cooling provides a well-characterized internal heat-transfer mechanism. The squealer tip is employed in this example with serpentine cooling channels. The internal tip cooling here is performed by the serpentine turning regions, which typically involve high heat-transfer coefficient enhancements as a result of impingement and secondary flows.

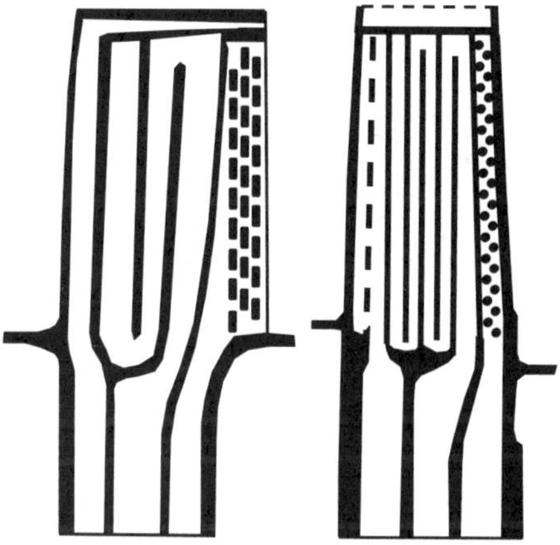

Fig. 30 Example blade internal cooling circuits for flat and squealer tip designs.

Heat transfer and cooling for the various forms of attached tip shrouds differ substantially from that of flat tips and squealer tips. Some similarities are present, such as the sink and source flow effects at the upstream and downstream extensions of the tip shrouds adjacent to the stationary casing, as well as the converging and expanding flows associated with seal teeth. The heat-transfer characteristics of these features are similar to those of the other blade tip designs but modified by the existence of the casing recess volume. For example, a tip shroud will generate a suction-side vortex as the leakage flow interacts with the passage flow, but any increased heat transfer will affect the tip shroud surfaces more than the airfoil. The immediate hot-gas side of the tip shroud acts as an endwall to the airfoil, and so heat transfer will be affected by a leading-edge horseshoe vortex, passage vortex, and other secondary flows, in much the same way as the blade hub region.

For the most part, similar heat-transfer coefficient characteristics can be applied on hub and tip endwalls with a tip shroud present; cooling of the region between the tip shroud and casing is more complex. Figure 31 shows two predictions of tip shroud heat-flux distributions from the study of

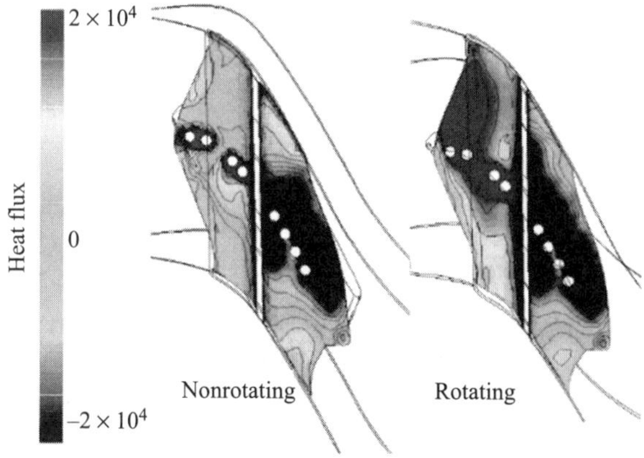

Fig. 31 Predicted tip shroud heat-flux distributions (reproduced by permission of ASME).

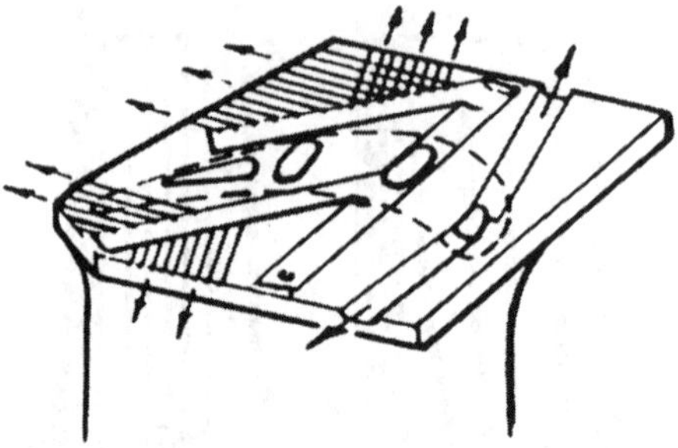

Fig. 32 Distributed internal cooling for a complex tip shroud (reproduced by permission of Rolls-Royce).

Nirmalan et al., one with rotational effects and one stationary, both including coolant injection out the tip holes [17]. The geometry is essentially that of the tip shroud in Fig. 10, but the periodic computational domain is selected based on the underlying airfoil shape. The stationary prediction agrees fairly well with stationary tip shroud cascade data. In the nonrotating case, heat flux is neutral forward of the seal tooth, neither into or out of the tip shroud surface, except the high heat transfer at the cooling hole exits. When rotation is added, a circumferential flow is set up in the forward tip shroud recess volume that drastically enhances heat flux to the tip shroud. The coolant here does not attach to the tip shroud, but instead mixes into the leakage flow and carries over the seal tooth.

Both rotating and nonrotating cases exhibit similar effects aft of the seal tooth. The aft region is flooded by cooling flow to the extent that heat flux is actually out of the tip shroud surface. Such disparities in tip shroud heat flux are undesirable, but not necessarily detrimental. More distributed cooling is required to maintain greater uniformity.

The cooling scheme for a higher temperature tip shroud is shown in Fig. 32 [18]. In this schematic, the upper seal strips (see Figs. 9 and 21) have been removed to expose the internal cooling passages. Cooling channels spread the air to the extended portions of the tip shroud. Coolant is then discharged out of perimeter holes to provide purge sealing against further hot-gas leakage, as well as a degree of film cooling on the mainstream side of the tip shroud. This very complex cooling network is required to maintain the tip shroud integrity against the effects of oxidation and creep. Accompanying this complexity is a higher uncertainty in external heat-transfer coefficients and film effectiveness, which are best determined through operational experience.

VII. IN-SERVICE CONDITIONS AND CHANGES

The main characteristics of blade tip aerodynamics, heat transfer, and cooling have been discussed in the foregoing sections. There are, however, other factors that have impact on operation and durability of blade tips beyond those considered in the basic design. Nominal design conditions do not remain constant either because the blade tips themselves degrade with time and/or because the combustor and surrounding turbine systems change or degrade with time. The degree and rate of modified in-service conditions depend on the mission, the operator, and the local environment of the installation. At least two aspects of blade tip service durability will be present in virtually all designs—blade tips will lose material as a result of oxidation, and blade tips will rub against the shrouds.

Tip material loss caused primarily by oxidation effects is shown in the squealer tip of Fig. 33. This example is for an HPT blade tip with unspecified exposure time (cycles) and temperature history, but depicts the real life distress that such blade tips can experience. The depth of the tip cavity is less than that of the new part though both the leading-edge region and the aftmost part of the tip cavity are fairly intact. Most of the material loss has occurred in the midchord section, especially as noted on the suction-side rim. The pressure-side rim has been oxidized down very close to the exits of the film holes. In this particular example, the blade tip section has not been coated with TBC. One or two deep cracks are visible; this type of cracking is usually associated with local regions of high thermal gradients or stress concentrations caused by cooling holes or plugged core support holes. It is important to note that even with such distress, this blade tip has not failed. The squealer tip has performed its function. The progression of conditions from the new to the degraded involves a complex mix of aerodynamics, heat transfer, and film-cooling changes leading to locally altered material temperatures, all of which can snowball from some starting nucleation region of initial distress.

Fig. 33 Sample squealer blade tip after unspecified hot exposure time in service.

Tip changes as a result of rub events with the stationary shroud might involve the loss of material or in some cases the redistribution of material. Depending on the materials involved, the temperature levels, and the degree of transient severity of the rub event, blade tip and/or shroud material can be worn away and lost or can be redeposited on the tip or shroud. Figure 34, from Corman et al. [19], shows an example of a blade tip rub against a ceramic-matrix-composite (CMC) shroud as evidenced by the smeared local surface region. If the material is simply borne away by the flow, the tip loss is manifested as a uniform increase in clearance. If the material is removed from tip or shroud and deposited on the other component, then a local buildup of material can occur, which, in turn, leads to more progressive rubbing and damage. This situation is possible when the casing and shroud ring are not perfectly round, such that each blade tip in the row rubs in one or more repeated locations. In fact, circumferentially uniform rubs are not found in practice, but instead rubs are localized by the eccentricities in the casing. This mode of tip material loss can become quite severe leading to excessive vibration of the airfoils and in extreme cases to rotor "freeze."

In both types of tip service changes noted here, the aerothermal boundary conditions become moving targets with time. The local or uniform loss of material will alter the tip flowfield and the heat-transfer coefficients. Work extraction and efficiency will change as a result of clearance increases and redistribution of pressures. Film-cooling flow rates will be modified, and in some worst cases cooling holes might even be blocked by oxidation, debris, or rub material. As an example of the degree of change possible, Fig. 35 shows the regional averages of blade tip heat-transfer coefficients obtained by Bunker and Bailey [20] as modified by the progressive uniform loss of a squealer tip rim. The blade tip configuration of Fig. 25 was used in this study with a constant tip clearance gap. The unaffected squealer rim has an average tip cavity heat-transfer coefficient level some 50% lower than that of the flat tip. As the rim material is removed to model a tip rub, average tip cavity heat transfer is increased. When the rim is one-third of its original height, the heat transfer is only 10% below that of the flat tip. Using other data from the same study, this loss of seal rim material is equivalent to about a 40% increase in tip clearance gap. Thus, tip seal material loss leads not only to greater clearance gaps, but also to less effective seal mechanisms. These two effects combine to accelerate tip degradation in service.

Service conditions for blade tips should also include consideration of repair. Because blade tips are typically degraded in service, repair methods have been devised to remove and rebuild all or portions of the tips, thereby avoiding the scrapping of valuable hardware. Repair operations can take place three to five times on a single blade tip before that part is permanently removed from service. Flat blade tips, although simple in design,

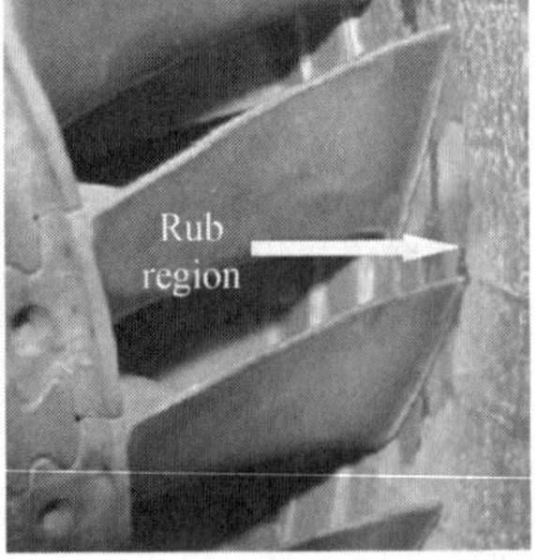

Fig. 34 Example of blade tip rub on stationary shroud.

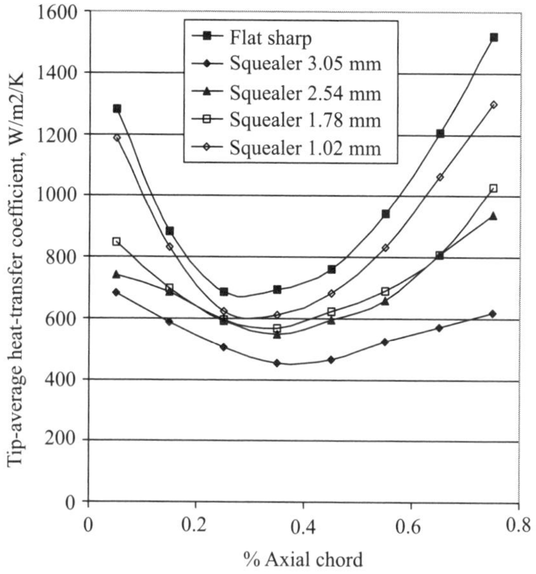

Fig. 35 Effect of squealer tip cavity depth on heat-transfer coefficient distribution (reproduced by permission of ASME).

are perhaps the most difficult for repair. When material must be removed and rebuilt to repair cracks or recover wall thickness, it is virtually impossible to obtain the original new part material properties. Because flat tip repairs occur on wall sections that directly affect the integrity of the cooled airfoil, these are particularly sensitive to changes in properties. Squealer tips have an advantage in this respect, in that the rim material can be removed and rebuilt without intrusion into the structural portion of the cooled airfoil. In fact, a squealer rim might be repaired using a different material from that of the original, a material more suitable to weld operations, for example. Attached tip shrouds have aspects common to both flat and squealer tips. Referring back to Fig. 9, the seal teeth might be easily rebuilt, but any repair required to the cooled sections of the tip shroud will be more involved.

VIII. STATIONARY SHROUD/CASING DESIGN

A few words are in order concerning the function and design of the stationary shroud or casing opposite the blade tip. In many ways, the shroud and the tip are an integral system. This is primarily true for two aspects of their function, first during transient operation and second when they touch. Shrouds are subject to many of the same design constraints and requirements as blade tips, excepting of course the rotational effects. Shrouds in HPTs are usually cooled structures whereas those in LPTs are uncooled. Where shroud designs differ are in their circumferential extent, axial recessing, and materials. Shroud systems can be designed to allow an equal number of shroud segments to blades, one shroud segment for every two blades and sometimes a complete 360-deg shroud ring for the blade row. The advantage of reducing the number of shroud segments comes in decreasing the leakage flows at the interfaces. The disadvantage of a single shroud ring is that it does not allow for ease of repair and replacement. Individual shroud segments, sometimes called shoes, are easier to manufacture and can be refaced or rebuilt easily. Continuous shroud rings are more amenable to roundness control and consistency of transient response. Because of the sheer size of such rings, these types of systems are generally only found on aircraft engines or small turbines. In practice, most shroud systems comprise segmented shrouds with fairly elaborate perimeter sealing.

Cooling of shrouds can usually be accomplished through a combination of impingement and convective methods though some film cooling is sometimes required. Materials are usually similar to those of the partner blade, including the use of TBC. Some advanced materials are now finding their way into service for shrouds, such as the CMC shown in Fig. 34. These materials have higher temperature capability, thereby allowing less cooling, but also present challenges with respect to stresses and system integration. In most casing systems there are actually two sets of shrouds for each blade row, an outer shroud or casing, which holds the inner shrouds that form the hot-gas flowpath. This inner-outer system isolates the hot shroud elements from the colder structure and allows for the introduction of air for both cooling and transient operational control. This latter aspect of shroud systems is referred to as active clearance control (ACC). In concept very simple, ACC uses cool air or heated air at the appropriate times during transients to control the radial thermal displacement of the inner shroud to better match the radial movement of the blade tip, thereby maintaining an efficient tip clearance over a greater range of operating conditions. In practice, ACC is more easily implemented on power turbine installations than in aircraft engines.

A great variety of recessing geometries opposite the blade tip are used in shroud designs. Flat tips and squealer tips are generally opposed by a shroud of constant cylindrical arc sector or radius (except at the shroud-to-shroud interfaces). Sometimes these shrouds are located at slightly larger radius than the upstream vane endwall to avoid the possibility of undesired flowpath disturbances. Tip shrouds will always be recessed into the stationary shroud, as discussed earlier. Such recessed casings can be cylindrical, flared, or even stepped radially outward. These recessed shroud surfaces frequently employ additional sealing methods, such as honeycomb materials, or labyrinth seal teeth acting in concert with the tip shroud seal teeth. In every case, the stationary shroud surfaces must bear the brunt of any tip rubs. It is a requirement that shrouds be sacrificial in this respect, sparing the life and operability of the rotating blade.

IX. CONCLUSION

The present discussion has highlighted the basic functional and operational requirements associated with axial turbine blade tips. These requirements, in conjunction with the many other competing system aspects of turbine engines, lead to several design solutions found in practice today. Conceptually, the hot-gas-path interface between the rotor and the casing is quite simple, but, in fact, the extreme environmental conditions along with cascaded systems issues make this interface sensitive to small changes. In modern turbines, the blade tip regions are responsible for about one-third of the total aerodynamic losses, require a significant amount of cooling, and thus pose one of the major remaining challenges to efficiency gains. These same blade tips are also one of the major causes for limiting the service times of turbine engines, impacting blade life, operational costs, and maintenance costs. The proposed and applied detailed design solutions for efficient and durable blade tips are many. All designs offer tradeoffs with respect to the blade tips and the turbine system. In the end, each design must balance aerodynamic performance, consequent thermal loading and cooling requirements, blade and rotor stresses, transient operational characteristics, and durability issues, to produce a successful outcome.

REFERENCES

[1] Glezer, B., Harvey, N., Camci, C., Bunker, R., and Ameri, A., *Turbine Blade Tip Design and Tip Clearance Treatment*, edited by T. Arts and von Kármán Inst. Lecture Series, VLI LS 2004-02, Belgium, 2004.

[2] Fowler, T. W. (ed.), *Jet Engines and Propulsion Systems for Engineers*, General Electric Aircraft Engines, Cincinnati, OH, 1989.

[3] Booth, T. C., "Importance of Tip Clearance Flows in Turbine Design," *Tip Clearance Effects in Axial Turbomachines*, von Kármán Inst. Lecture Series, VKI LS 1985-05, Belgium, 1985.

[4] Bindon, J. P., "Pressure Distributions in the Tip Clearance Region of an Unshrouded Axial Turbine as Affecting the Problem of Tip Burnout," International Gas Turbine Conference, American Society of Mechanical Engineers, Paper 87-GT-230, June 1987.

[5] Yamamoto, A., "Endwall Flow/Loss Mechanisms in a Linear Turbine Cascade with Blade Tip Clearance," *Journal of Turbomachinery*, Vol. 111, 1989, pp. 264-275.

[6] Morphis, G., and Bindon, J. P., "The Effects of Relative Motion, Blade Edge Radius and Gap Size on the Blade Tip Pressure Distribution in an Annular Turbine Cascade," International Gas Turbine Conference, American Society of Mechanical Engineers, Paper 88-GT-256, June 1988.

[7] Ameri, A. A., and Steinthorrson, E., "Prediction of Unshrouded Rotor Blade Tip Heat Transfer," International Gas Turbine Conference, American Society of Mechanical Engineers, Paper 95-GT-142, June 1995.

[8] Bindon, J. P., "The Measurement and Formation of Tip Clearance Loss," *Journal of Turbomachinery*, Vol. 111, 1989, pp. 257-263.

[9] Ameri, A. A., Steinthorsson, E., and Rigby, D. L., "Effect of Squealer Tip on Rotor Heat Transfer and Efficiency," *Journal of Turbomachinery*, Vol. 120, 1997, pp. 753–759.

[10] Bunker, R. S., Bailey, J. C., and Ameri, A. A., "Heat Transfer and Flow on the First Stage Blade Tip of a Power Generation Gas Turbine Part 1: Experimental Results," *Journal of Turbomachinery*, Vol. 122, 1999, pp. 263–271.

[11] Bunker, R. S., and Bailey, J. C., "Blade Tip Heat Transfer and Flow with Chordwise Sealing Strips," *Proceedings of the 8th International Symposium on Transport Phenomena and Dynamics of Rotating Machinery*, edited by J. C. Han, Pacific Center of Thermal Fluids Engineering, Maui, HI, March 2000, pp. 548–555.

[12] Kwak, J. S., Ahn, J., Han, J. C., Bunker, R. S., Lee, C. P., Boyle, R., and Gaugler, R., "Heat Transfer Coefficients on the Squealer Tip and Near-Tip Regions of a Gas Turbine Blade with Single or Double Squealer," American Society of Mechanical Engineers, Paper GT2003–38907, June 2003.

[13] Metzger, D. E., Bunker, R. S., and Chyu, M. K., "Cavity Heat Transfer on a Transverse Grooved Wall in a Narrow Flow Channel," *Journal of Heat Transfer*, Vol. 111, 1989, pp. 73–79.

[14] Metzger, D. E., and Rued, K., "The Influence of Turbine Clearance Gap Leakage on Passage Velocity and Heat Transfer near Blade Tips: Part I - Sink Flow Effects on Blade Pressure Side," *Journal of Turbomachinery*, Vol. 111, 1989, pp. 284–292.

[15] Rued, K., and Metzger, D. E., "The Influence of Turbine Clearance Gap Leakage on Passage Velocity and Heat Transfer near Blade Tips: Part II - Source Flow Effects on Blade Suction Sides," *Journal of Turbomachinery*, Vol. 111, 1989, pp. 293–300.

[16] Kwak, J. S., and Han, J. C., "Heat Transfer Coefficient and Film Cooling Effectiveness on a Gas Turbine Blade Tip," American Society of Mechanical Engineers, Paper 2002-GT-30194, June 2002.

[17] Nirmalan, N. V., Bailey, J. C., and Braaten, M. E., "Experimental and Computational Investigation of Heat Transfer Effectiveness and Pressure Distribution of a Shrouded Blade Tip Section," American Society of Mechanical Engineers, Paper GT2004–53279, June 2004.

[18] Hartley, R., "High Pressure Turbine Tip Clearance Performance Investigation," M.Sc. Thesis, Cranfield Univ., Bedford, England, U.K., March 1996.

[19] Corman, G. S., Dean, A. J., Brabetz, S., Brun, M. K., Luthra, K. L., Tognarelli, L., and Pecchioli, M., "Rig and Engine Testing of Melt Infiltrated Ceramic Composites for Combustor and Shroud Applications," American Society of Mechanical Engineers, Paper 2000-GT-638, May 2002.

[20] Bunker, R. S., and Bailey, J. C., "Effect of Squealer Cavity Depth and Oxidation on Turbine Blade Tip Heat Transfer," American Society of Mechanical Engineers, Paper 2001-GT-155, June 2001.

CHAPTER 9

Modeling and Simulation of Turbine Cooling

Tom I-P. Shih*
Purdue University, West Lafayette, Indiana

Paul A. Durbin†
Iowa State University, Ames, Iowa

The efficiency and service life of a gas-turbine engine is strongly affected by its turbine component where the thermal energy contained in the high-pressure and high-temperature gas is converted into mechanical energy [1]. The most effective way to improve the efficiency of the turbine component is to raise the temperature of the gas that enters it, which could be as high as the adiabatic flame temperature from the combustion of fuel and oxidizer [1, 2]. Advances in two areas have enabled the steady increase in the turbine inlet temperature over the past few decades. One is the development of high-temperature-resistant materials, including thermal and environmental barrier coatings [3, 4]. The other is the development of innovative cooling technologies [1, 2]. With cooling, the material's temperature can be maintained below the maximum allowable even though the temperature of the gas in the turbine's hot-gas path far exceeds it. Because cooling requires work, efficiency demands effective cooling with the minimum amount of cooling flow.

Considerable efforts have been made for decades to quantify heat transfer for a wide variety of geometries and to develop efficient and effective cooling designs [5–11]. With these and other efforts, great progress has been made as evidenced by the impressive advances in the efficiency and the service life of gas turbines over the years. Every advance has required a significant leap in understanding and insight. Computational fluid dynamics and heat transfer (CFD/HT) offers the potential to provide the leap in understanding and insight that are needed to make the next advance in cooling [12]. There are, however, a few issues that affect the reliability of CFD/HT in predicting the flow and the heat transfer with the accuracy that is needed to make the next advance.

Verification, validation, and uncertainty quantification are aimed at addressing CFD/HT's reliability [13, 14]. Verification is concerned with the accuracy with which the governing equations – the continuity, momentum, and energy

*Professor and Head, School of Aeronautics and Astronautics.
†Martin C. Jischke Professor, Department of Aerospace Engineering.

equations, along with the turbulence model – are solved by the numerical method of solution. If there are no errors from the verification analysis, then the numerical solution generated is identical to the exact solution of the governing equations. The main challenges in verification are understanding the effects of grid spacing and time-step size on the computed solution and quantifying the errors from inadequate resolution. Validation is concerned with the accuracy of the governing equations in describing the physics of the problem being studied. If there are no errors from a validation analysis, then the governing equations used to represent the problem are taken to be perfect. In validation, the principle difficulties are the modeling of turbulence and ensuring that CFD/HT is truly solving the same problem as the experimentalist when comparing CFD/HT predictions with experimental data. Thus, validation is about solving the "right" governing equations whereas verification is about solving the governing equations "right." Uncertainty quantification (UQ) is concerned with the fact that the inputs into CFD/HT simulations, such as material properties, geometry, and operating conditions, contain uncertainties because of limitations in materials processing, manufacturing, and control. UQ seeks to understand the consequences of these uncertainties in the inputs on CFD/HT predictions and how these uncertainties can be accounted for to guide risk analysis and design decisions.

This chapter addresses aspects of verification, validation, and uncertainty quantification that are important in the modeling and simulation of turbine cooling by CFD/HT. This chapter also gives an overview of turbulence models that are widely used in CFD/HT with focus on understanding the fundamental basis of the models.

I. VERIFICATION: ESTIMATING GRID-INDUCED ERRORS

Most CFD/HT methods generate solutions to the governing partial differential equations by invoking two approximations. The first is the discretization of the domain with the time domain replaced by a finite number of time levels and the space domain replaced by a finite number of points or cells, referred to as a grid or a mesh. The second is the discretization of the governing equations on that discretized domain by using a finite-difference (FD), finite-volume (FV), or finite-expansion method (FE). These two approximations produce two types of errors in the numerical method of solution, and they are connected. Errors from the discretized domain arise when the cells and the time-step sizes are not small enough and/or numerous enough for the discretized governing equations to resolve the geometry and/or the relevant flow features. Errors also arise from having poor quality cells such as lack of smoothness when the size of the cells changes from one location to another and lack of alignment of a high-aspect-ratio cell with the flow direction. Errors from the discretized governing equations arise from not suppressing spurious modes such as numerical diffusion and dispersion and from not correctly representing physical phenomena such as advection, diffusion, pressure waves, shock waves, positivity, and monotonicity.

Many CFD/HT software packages used by researchers and engineers have built-in schemes that are constructed to control spurious modes, to enable high-fidelity simulations via high-resolution algorithms, and to generate high-quality grids or meshes. Grids or meshes are at the heart of the verification problem. For CFD/HT analysis of turbine cooling, the number of cells needed to resolve the geometry and the flow features is huge—easily in the hundreds of millions of cells for just one blade with its internal cooling passages and film-cooling holes. Because one level of grid refinement in three dimensions, which involves halving the grid spacing in every direction, could increase the total number of cells by a factor of eight, it is impractical to do a comprehensive grid-sensitivity study. Thus, methods are needed that can provide an estimate of errors from the grid without a comprehensive grid-sensitivity study.

Roache [13] classified methods for estimating grid-induced errors into two categories: 1) methods based on multiple grids and 2) methods based on a single grid. Methods based on multiple grids [15–18] require solutions to be generated on at least three increasingly finer grids. This class of methods can be prohibitively expensive for turbine cooling. Also, if Richardson's extrapolation is used to understand grid convergence, then the grid must be sufficiently fine for the truncated Taylor series to be bounded before this method can yield meaningful results. Single-grid error-estimation methods can be classified as those based on algebraic equations and those based on partial-differential equations (PDEs). Methods based on algebraic equations assume that the error at a cell is a function of the cell, the cells that are adjacent to it, and the solutions at those cells. Much work has been done on such methods, mostly in connection to solution-adaptive mesh refinement. Most algebraic error-estimators consider only gradients of scalars such as the second-derivative of pressure or velocity [19]. Shih et al. [20] and Gu and colleagues [21–23] have proposed measures that account for the vectorial and tensorial nature of the solution and linked them to the geometry and size of each cell in a grid. Methods based on PDEs, first proposed by Babuska and colleagues [24, 25], recognize that errors once generated can be transported by advection and diffusion to other parts of the flowfield. With PDE methods, a transport equation is needed to account for the generation of errors and their propagation. PDE methods offer a general framework to quantify grid-induced errors and are reviewed in the next two subsections.

A. ERROR-TRANSPORT EQUATION

As noted, a method for deriving transport equations for errors was first presented by Babuska and colleagues [24, 25], and it was developed for finite element methods. Ferziger [26], Van Straalen et al. [27], and Zhang et al. [28, 29]

applied it to finite difference and finite volume methods. The method proceeds as follows. Consider the following PDE whose solution is being sought

$$L(U) = f \tag{1}$$

where L is the differential operator operating on the dependent variable U and f contains the nonhomogeneous terms. If U_a is an approximate solution of Eq. (1), then its substitution into Eq. (1) will produce a residual R because it will not satisfy Eq. (1), that is,

$$L(U_a) - f = R \tag{2}$$

If Eqs. (1) and (2) are linear, then subtracting Eq. (2) from Eq. (1) yields the following transport equation:

$$L(e) = -R \quad \text{or} \quad Le = -R \tag{3}$$

where the error in the solution, denoted by e, is given by

$$e = U - U_a \tag{4a}$$

and a more accurate solution denoted by U can be obtained by adding e to U_a, that is,

$$U = U_a + e \tag{4b}$$

If Eqs. (1) and (2) are not linear, then they will need to be linearized before subtracting.

Equation (3) is used to compute e, which provides an estimate of the error in the CFD/HT solution. Also, by adding e to the original CFD/HT solution as shown in Eq. (4b), then a more accurate CFD solution is generated. Note that because the error in Eq. (4a) is defined with respect to the exact solution of Eq. (1), the error-transport equation given by Eqs. (3) and (4) can account for errors from discretizing the domain and errors from discretizing the governing equations.

Qin and Shih [30] noted that Eqs. (3) and (4) are only valid for FE methods such as finite element, spectral, and spectral element, but not for collocation-type methods such as FD and FV. This is because for an FD and an FV method, the differential operator L in Eq. (2) is replaced by a discrete operator, so that subtracting Eq. (2) from Eq. (1) will not yield Eqs. (3) and (4). This inconsistency was also noted by Roache [13], Ferziger [26], and Van Straalen et al. [27]. To rectify this inconsistency, these investigators suggested that FD and FV solutions at grid points or cells be made into a continuous or a piecewise continuous function. However, Roache [13] recognized that the resulting function is not unique and will depend on the interpolation formula used.

Because FD, FV, and FE methods can be unified in integral form via the method of weighted residuals through appropriate weighting functions [31], the most general error-transport equation should be in integral form. The differential form given by Eqs. (3) and (4) can be recovered from the integral form only if the weighting function is continuous (finite element, but not FD or FV) and if the solution generated is genuine (no weak solutions). Giles and Pierce [32], Giles [33], Venditti and Darmofal [34], Park [35], and Hicken and Zingg [36] employed the integral approach for finite element and FV methods via the adjoint-variable formulation to estimate errors of integral quantities such as lift and drag and to determine optimal grid-point distributions that minimize the errors in those integral quantities.

Qin and Shih [30] took a different approach, which can also be applied to FD and FV methods as well as FE methods though they only applied it to FD and FV methods. They construct what are referred to as discrete error-transport equations (DETEs) directly from the FD/FV equations without any reference to the original PDEs. By disregarding the original PDEs, their approach only provides errors induced by the grid and the time-step size, but their method does provide an estimate of the error at every cell for FV methods and every grid point for FD methods at any time level. The error estimated can be added back to the solution to provide a more accurate solution.

B. DISCRETE-ERROR TRANSPORT EQUATION

The DETE concept presented by Qin and Shih [30, 37], Qin et al. [38], and Shih [39] proceeds as follows. Suppose for a FD/FV method, the differential operator L in Eq. (1) is replaced by the discrete operator L_D on a discrete domain so that the FD/FV equation corresponding to Eq. (1) is

$$L_D\left(U_i^n, h, k\right) = 0 \tag{5}$$

where U_i^n is the solution at grid point or cell i and time level n; h denotes the grid spacing or cell size; and k denotes the time-step size. As h and k are refined, the solution U_i^n approaches the grid-independent solution $U_{g,i}$, which according to the Lax equivalence theorem [40] is also the exact solution of the PDE approximated by the FD/FV equations if the PDE is well posed and linear and if the FD/FV equation is consistent and stable. However, because PDEs of interest in CFD/HT are not linear, the less restrictive term of grid-independent solution is used. With Eq. (5), the grid-independent solution is given by

$$L_D(U_{i,g}^n, h_g, k_g) = 0 \tag{6}$$

where h_g is the grid spacing or cell size and k_g is the time-step size needed to obtain the grid-independent solution. Note that the number of grid-independent solutions is infinite because the cell size and the time-step size could always be refined even though the solution is already well converged (satisfies the convergence criteria). This ambiguity should not be an issue. This is because the difference between grid-independent solutions should be in the last few digits of the total number of digits used to represent a real number.

From Eqs. (5) and (6), two different DETEs can be derived. One is based on the coarse grid spacing h and coarse time step k, and the other is based on h_g and k_g. Qin and Shih [30] showed that only the one based on h and k has meaning. This is because interpolating $U^n_{i,g}$ onto h and k is well posed (that is, all interpolants give similar results) whereas interpolating the coarse grid solution U_i onto h_g and k_g is ill posed (that is, different interpolants can give very different results). Accordingly, Qin and Shih [30] inserted the grid-independent solution $U_{g,i}$ into Eq. (5) with the coarse mesh and time-step size (h, k) to get

$$L_D(h, k)U^n_{i,g} = R^n_{i,g} \tag{7}$$

If L_D is linear, then subtracting Eq. (5) from Eq. (7) gives the discrete-error-transport equation (DETE), which is

$$L_D(h, k)e^n_i = R^n_{i,g} \tag{8}$$

where

$$e^n_i = U^n_{i,g} - U^n_i \tag{9a}$$

or

$$U^n_{i,g} = U^n_i + e^n_i \tag{9b}$$

If L_D is not linear, then L_D needs to be linearized. Suppose $L_{D,L}$ is the linearized operator, then

$$L_{D,L}(h, k)U^n_i = R^n_{i,c} \tag{10}$$

Thus, though $L_D(U^n_i, h_g, k_g) = 0$, $L_{D,L}(h, k)U^n_i$ is in general nonzero because the linearization changed L_D. Because

$$L_{D,L}(h, k)U^n_{i,g} = R^n_{i,g} \tag{11}$$

Subtracting Eq. (11) from Eq. (10) yields the following DETE for PDEs that are not linear:

$$L_{D,L}(h, k)e^n_i = R^n_{i,g} - R^n_{i,c} \tag{12}$$

where e^n_i is still given by Eq. (9a).

From Eqs. (7) to (9) for linear PDEs and Eqs. (10) to (12) for PDEs that are not linear, the following observations can be made. The first is that the residual can be computed exactly via Eq. (7) or Eq. (11) if the grid-independent solution is known. Knowing the exact residual is useful in guiding the modeling of the residual and its validation. The second is that once the residual is known (e.g., through modeling), then Eq. (8) or Eq. (12) can be used to estimate the grid-induced errors e^n_i at every cell or grid point at any time level. The third is that the residual $R^n_{i,g}$ defined by Eq. (7) or Eq. (11) is a local quantity whereas the solution error e^n_i given as a solution to Eq. (8) or Eq. (12) is a global function. Error is generated or destroyed at a grid point or a cell if the residual at the point or cell is nonzero. This indicates that grid adaptation should be made where the residual is high and not where the error is high.

Qin and Shih [30, 37] showed that if the "actual" residual defined by Eq. (7) or Eq. (11) is used, that is,

$$R_{i,g}^{n} = L_D(h, k)U_{i,g}^{n} \quad \text{or} \quad R_{i,g}^{n} = L_{D,L}(h, k)U_{i,g}^{n} \tag{13}$$

then the DETE given by Eq. (8) or Eq. (12) can predict the grid-induced error for solutions generated on an "imperfect" grid perfectly for FD/FV equations that are linear, quasilinear, steady, or unsteady through the following one-dimensional model equations: advection-diffusion equation, linear wave equation, and inviscid Burger equation. They also found that the linearization procedure used to derive the DETEs for nonlinear FD/FV equations need not be conservative when estimating grid-induced errors involving weak solutions. This is because the error-propagation speed and jump conditions are captured by the FD/FV equations, which are in conservative form.

C. MODELING THE RESIDUAL IN THE DISCRETE-ERROR TRANSPORT EQUATION

The residual in Eq. (8) or Eq. (12) must be modeled before it can be used to estimate grid-induced errors in the CFD/HT solution generated on a grid that is not sufficiently fine and/or might be of poor quality. Modeling the residual is the main challenge when implementing both the continuous Eq. (3) and the discrete Eq. (8) or Eq. (12). The usefulness of the error-transport equations depends on how well the residual is modeled.

1. SINGLE-GRID METHODS

One way to model the residual is by using the modified equation of Warming and Hyett [41], which is a PDE that is truly represented by the FD/FV equation. Zhang et al. [28, 29] and Qin and Shih [30, 37] showed that modeling the residual by using the leading terms of the truncation error in the modified equation can give excellent results if the grid spacing or cell sizes are sufficiently small. However, when the grid spacing or cell size is too large, then the estimated error predictions can be poor [30, 37]. Celik et al. [42] modeled the residual in Eq. (12) by expanding terms in the FD/FV equations about the cell center and keeping only the leading terms. This modeling approach is essentially the same as using the modified equation, except that the time derivatives are not replaced by spatial derivatives.

As an alternative to the modified equation, Qin and Shih [43] proposed a method based on data mining. This approach involves two steps. The first is to study the nature of the residual by evaluating the "actual" residual through Eq. (7) or Eq. (11) by using solutions generated on a variety of poor quality meshes in a systematic way. The second step is to model the residual based on the understanding gained on the nature of the residual through statistical analysis and curve fitting.

Note that generating two solutions on the same grid, using a higher-order method and a lower-order method, to get a residual—commonly used in finite element methods—does not work for DETE. This is because L_D changes when the order of the method changes.

2. MULTIPLE-GRID METHODS

Though the DETE approach belongs to the single-grid-error-estimator category, Qin et al. [44], Shih and Qin [45], and Shih and Williams [46] also examined a method for estimating the residual from solutions generated on multiple grids. One approach for steady-state solutions is as follows. Suppose U_h is the CFD/HT solution obtained on a baseline grid generated by using best practices, U_{h+} is a slightly refined grid, and U_{h-} is a slightly coarsened grid. The coarsening and refining are implemented in regions where the geometry and flow features are expected to have the greatest influence on the overall solution, such as about the stagnation region and about points/lines where separation and reattachment occur. From these three solutions, two estimates of the residual can be computed by using Eq. (7) or Eq. (11). The first estimate is computed by using U_{h+} and U_h, which gives

$$R_{h,h+} = L_D(h, k)U_{h+} \quad \text{or} \quad R_{h,h+} = L_{D,L}(h, k)U_{h+} \tag{14}$$

The second estimate is computed by using U_h and U_{h-}, which gives

$$R_{h-,h} = L_D(h-, k)U_h \quad \text{or} \quad R_{h-,h} = L_{D,L}(h-, k)U_h \tag{15}$$

Note that the residual is always computed on the coarser grid. Thus, in Eq. (14), the residual is computed on the baseline grid (denoted as h), and in Eq. (15), the residual is computed on the coarsest grid (denoted as $h-$). However, one would like to estimate the residual on the finest grid (denoted as $h+$).

The residual on the finest grid can be estimated by using one of the following two equations:

$$R_{h+} = R_{h,h+}(V/V_h) \tag{16}$$

$$R_{h+} = bV + cV^2 \tag{17}$$

where V is the volume of the cell in the finest mesh denoted by $h+$. Equations (16) and (17) are curve fits of the residuals computed by using Eqs. (14) and (15) plus the recognition that $R = 0$ if the grid spacing is zero.

II. VALIDATION

To validate a CFD/HT solution, three criteria must be satisfied. The first is that the CFD/HT solution being validated must be a grid-independent solution. This is because once the solution is grid independent, then one can assume that all remaining errors in the CFD/HT solution are caused by errors in the governing equations, such as the modeling of turbulence. This assumption is based on the Lax equivalence theorem [40], which states that for a well-posed linear PDE, if a numerical method is consistent and stable, then it is convergent (that is, the numerical solution approaches the exact solution of the PDE as the grid spacing or cell size and time-step size approach zero). Though the governing equations used in CFD/HT are not linear, the Lax-equivalent theorem is still a necessary condition, albeit an insufficient one.

The second criterion is that the CFD/HT must be solving the same problem as the experimental study that generated the data to validate the CFD/HT solution. The second criterion is a difficult one to satisfy for two reasons. First, CFD/HT needs boundary conditions (BC). More specifically, it needs profiles of velocity, temperature, and turbulence quantities at the entire inflow boundary and the pressure at the outflow boundary. It also needs the temperature or heat flux on all walls unless enough information is provided to enable a conjugate analysis. Experimentalists generally do not provide enough information for CFD/HT to specify the boundary conditions in the detail needed because it is very difficult to measure what is needed. For internal cooling, typically only a "nominal" Reynolds number and an operating pressure are provided. For external flows, typically only the boundary-layer thickness and the freestream temperature and velocity are provided. Though the mean velocity profile is sometimes provided, one also needs the turbulence quantities sustaining that profile, which is almost never provided. The turbulence quantities are needed because boundary layers in experimental studies are often made turbulent by trips or turbulators. If the distance from the trip to the location where the velocity profile is measured is not sufficiently long, then the velocity profile will be a strong function of the turbulent structures induced by the trip.

The second reason why it is difficult for CFD/HT to solve the same problem as the experimental study is that the flow must not reverse at the inflow and outflow boundaries of the CFD/HT computational domain; that is, the flow at the inflow boundary must all enter with none exiting, and the flow at the outflow boundary must all exit with none entering. Many experimental configurations do not meet this requirement.

The third criterion for validating a CFD/HT solution is the need to understand how the experimental study measured and postprocessed the data. For example, how was the heat-transfer coefficient measured or calculated? How were the surface heat flux, surface temperature, and bulk temperatures measured, approximated, or calculated? Like CFD/HT, experimental methods have errors and make assumptions. These errors and assumptions must be understood during the validation process [47]. Also, the bulk temperature used to compute the heat-transfer coefficient is rarely given when presenting Nusselt-number correlations [48, 49] so that this is another source of uncertainty in the validation process.

Of the aforementioned three criteria, the CFD/HT user has full control of only one—the generation of a grid-independent solution. Even that criterion can be difficult to achieve, depending upon the complexity of the problem and the number of cells needed to resolve all of the relevant features of the flow and heat transfer. Of the other two criteria, CFD/HT can only approximate the experimental study as closely as possible based on the information provided. Thus, validating a CFD/HT solution is not an easy task.

Experimental studies whose goal is to generate data for validating CFD/HT studies should attempt to satisfy the following criteria: 1) design the experimental setup so that a reasonable computational domain can be defined, where the flowfield at the inflow and outflow boundaries do not have reverse flows and are relatively easy to reproduce computationally; 2) provide enough information for the CFD/HT study to repeat the experiment so that all boundary conditions can be specified with reasonable fidelity to the experiment; and 3) provide sufficent information on how the data are measured and postprocessed.

III. UNCERTAINTY QUANTIFICATION

Oberkampf et al. [50] classified uncertainties in computations and measurements into two categories: aleatory and epistemic. Aleatory uncertainties are caused by variabilities in the system that cannot be controlled. In CFD/HT modeling and simulation of turbine cooling, aleatory uncertainties include variability in material properties and in the geometry due to tolerances and limitations in material processing and manufacturing processes. Variability in operating conditions produced by the uncontrollable parts of the environment and the engine are also aleatory. The epistemic uncertainties are because of humankind's as well as the engineer's/scientist's lack of knowledge about the inputs to the problem being studied and the tools used to study the problem. These uncertainties could include not knowing the thermomechanical and transport properties of the materials, not knowing all of the operating conditions, and not knowing the limitations or the capabilities of the turbulence models used in CFD/HT. Epistemic uncertainties also include discrepancies between the CFD/HT models and the experimental setup in validation studies.

There are a number of methods that can be employed to address aleatory uncertainties and their consequences. Because aleatory errors tend to be random, they could be described by probability theory. The variability in the input parameters about the intended values is often described by Gaussian distributions. To assess the effects of aleatory uncertainty in each input parameter, one could use a sensitivity analysis from any gradient-based optimization method such as finite difference [51] or adjoint-variable [52]. The finite difference method is straightforward but tedious because a CFD/HT simulation must be performed for each deviation about the intended value for each of the input parameters that has uncertainty. For steady-state flows, the adjoint-variable method is highly efficient, being able to provide sensitivity on all input parameters simultaneously. For unsteady flows, the adjoint-variable method could be costly. This is because it requires information for the entire duration of interest [36]. The costliness of some unsteady problems could be addressed by the time-spectral adjoint-variable method [53].

To assess the net effect of aleatory uncertainties from all input parameters is difficult. This is because the worst-case scenario might never take place, and assuming that it does imposes limits on the design that are unnecessarily restrictive. Thus, one needs to understand the relationship among the aleatory uncertainties among the input parameters. The aleatory uncertainties in geometry and operating conditions could be obtained via in situ measurements. With such data, a number of mathematical frameworks rooted in Kalman filtering and Bayesian statistics could be used to estimate the net effects of aleatory uncertainties. These include Monte Carlo [54] and stochastic expansion methods such as polynomial chaos [55]. The effects of aleatory uncertainties on optimization of objective functions with constraints need special attention. This is because an optimal design that does not take into consideration uncertainties in the input parameters could violate the imposed constraints when uncertainties are accounted for [56].

Epistemic uncertainty is connected to a lack of knowledge of things that are deterministic. Epistemic errors will reduce as our understanding and knowledge increase. Because lack of knowledge is not a random process, it should not be described by probability theory. One major source of epistemic error in CFD/HT solutions is failure to correctly model the problem of interest in terms of geometry, materials, and operating conditions. As just mentioned, epistemic error is a major source of error in validation of CFD/HT solutions because the source of the benchmark data for validation often does not provide enough information to accurately reproduce the experiment. Another major source of epistemic errors in CFD/HT studies is turbulence modeling. This is the topic of the next section.

IV. MODELING OF TURBULENCE

One of the major issues in CFD/HT modeling and simulation of turbine cooling is how to address turbulence and its effects on the flow and surface heat transfer. Most mathematical models of turbulent flows are based on the continuity, momentum (Navier–Stokes), and total energy equations. These conservation/balance equations, henceforth referred to as the N-S equations, are used because the smallest length and time scales of most turbulent flows, including those encountered in turbine cooling, are still in the continuum regime. These mathematical models can be classified into the following four categories: 1) those that analyze the unsteady N-S equations in three dimensions (3-D) without any averaging or filtering—known as direct numerical simulation (DNS); 2) those that analyze the unsteady N-S equations in 3-D that have been spatially filtered (i.e., removing all length scales smaller than a certain dimension)—known as large-eddy simulation (LES); 3) those that analyze the ensemble- or Reynolds-averaged N-S (RANS) equations; and 4) those that use different approaches in different parts of the turbulent flowfield (hybrid). A popular hybrid model is detached eddy simulation (DES), which uses the unsteady form of the RANS equations in the near-wall region and LES in regions away from walls.

DNS is analogous to experiments if initial and boundary conditions (such as the nature of the turbulent flow that enters the turbine component from the combustor) are sufficiently well known to correctly initiate the simulation. With current computing capabilities, DNS is not feasible for routine analysis of turbine-cooling designs. This is because DNS requires the resolution of every length and timescale of the turbulent flow, and for turbine-cooling flows (with its high Reynolds numbers), this would require a prohibitively large number of grid points/cells and time steps. In addition, long run times are needed to produce meaningful statistics, which adds further to its cost. Presently, DNS is used to generate high-quality benchmark data to understand the underlying mechanisms of turbulent flows at low Reynolds numbers and to develop and validate models for LES, RANS, and hybrid.

LES, like DNS, also can be informative and be a source of benchmark data. They too are analogous to experiments and require a good deal of computational resources to produce statistics. However, by simulating only the larger scales of the turbulent flowfield with the physics from the smaller scales modeled, they fall short of first principles simulation. Though LES requires less computational cost than DNS, LES remains quite costly and requires considerable attention to grid dependency. LES is reliable in separated flow and away from boundaries. Near the wall, it suffers from inaccuracy unless highly refined grids are used. This is because small eddies next to walls must be captured to obtain accurate simulations. For heat-transfer applications, regions near walls are critical. The expense of simulating eddy structures near walls remains far too high for regular use in turbine cooling. For this reason, hybrid methods such as DES were developed, where LES is used further away from walls and RANS in regions next to walls. Though DES is computationally less costly than LES, it is still computationally intensive, and the coupling between RANS and LES at their interface is still a research area.

It is because of the high cost of DNS, LES, and DES that the vast majority of CFD/HT studies on turbine cooling still use the RANS formulation with closure models to represent the effects of turbulent mixing. Though RANS is not a first principles approach, it is a practical engineering method that relies on empirical data to determine coefficients in the model. Conversely, RANS is not brute-force simulation of turbulent eddy structures. The mean flow is computed from the ensemble-averaged equations without simulating the seemingly random turbulence and accumulating statistics. Instead, it models the turbulent transport by eddy diffusivity, guided by insights and data from experiments and

DNS/LES simulations. The remainder of this chapter reviews some issues in RANS computation.

A. REYNOLDS-AVERAGED NAVIER–STOKES EQUATIONS

The ensemble-averaged continuity, momentum, and thermal-energy equations for a thermally perfect gas with temperature-dependent properties are given by [57]

$$\frac{\partial \rho}{\partial t} + \nabla \cdot \rho U = 0$$

$$\frac{\partial \rho U}{\partial t} + \nabla \cdot \rho UU = -\nabla P + \nabla \cdot \Sigma + \nabla \cdot \tau_T \tag{18a}$$

$$\frac{\partial \rho C_v T}{\partial t} + \nabla \cdot (\rho C_v T + P)U = \nabla \cdot \kappa \nabla T + \nabla \cdot Q_T + \Phi$$

where the viscous stress tensor is

$$\Sigma_{ij} = \mu\left(\frac{\partial U_i}{\partial x_j} + \frac{\partial U_j}{\partial x_i}\right) - \frac{2}{3}\mu \frac{\partial U_k}{\partial x_k}\delta_{ij} \tag{18b}$$

In the preceding equations, ρ, U, P, and T are respectively the averaged density, velocity, pressure, and temperature; C_v is the specific heat capacity; μ is the dynamic viscosity; and κ is the thermal conductivity. The terms τ and Q represent turbulent stress and heat-flux tensors, the details of which can be found in Pletcher et al. [57]. For incompressible flows with temperature-independent properties, $\tau_{ij} = -\rho\overline{u_i u_j}$ and $Q_i = -\rho C_v \overline{u_i \theta}$, where u_i and θ represent turbulent fluctuations of the velocity and temperature, respectively. The closure problem involves devising equations that relate these turbulent fluctuations to the mean flow.

Equation (18) is often referred to as the Reynolds-Averaged Navier–Stokes (RANS) equations. It is derived by time or ensemble averaging if the mean flow is steady and by ensemble averaging if the mean flow is unsteady. Even though turbulent flows are always 3-D and unsteady, the RANS equations given by Eq. (18) can be 1-D, 2-D, or 3-D, and steady or unsteady depending on how the mean quantities vary. Thus, RANS is computationally less intensive and hence more practical. The problem with RANS equations, however, is in modeling the turbulent fluctuations such as Reynolds stresses that arise from the averaging process. Basically, a universally valid model is impossible because turbulent fluctuations such as Reynolds stresses have all scales (largest to smallest) embedded in them, and larger scales depend on geometry and boundary conditions, which are problem dependent. The turbulence modeling community, however, believes that it is possible to construct reasonably general models for different classes of flows (see for example Wilcox [58], Pope [59], and Durbin and Pettersson-Reif [60]). As mentioned, DNS and LES, along with detailed experimental measurements, are needed to guide the development of such RANS models for turbine cooling.

Models for the turbulent fluctuations such as Reynolds stresses can be classified into the following categories: Reynolds-stress models (RSMs), algebraic-stress models (ASMs), nonlinear constitutive relations (NCRs), and eddy-diffusivity models (EDMs). Of these models, RSMs represent the highest level of closure. RSM models (also known as second moment models) require the solution of a transport equation for each component of the symmetric second-order tensor τ and for the first-order tensor Q, in addition to a scale equation, which results in 10 transport equations in addition to the ensemble-averaged continuity, momentum, and energy equations, plus species if there are chemical reactions. Closure at this tensoral level captures more phenomena than simpler, scalar closures. This is in part because the production tensor in the Reynolds-stress equation is retained in its exact form. However, the added expense, numerical stiffness, and uncertainties in the near-wall behavior limit the utility of tensor closures. Extensive description of Reynolds transport closures can be found in various texts (for example, see Wilcox [58], Pope [59], and Durbin and Pettersson-Reif [60]) and review articles (for example, see Launder [61], Speziale [62], and Durbin and Shih [63]).

The more practical forms of closure—ASMs, NCRs, and EDMs—invoke a scalar eddy viscosity. Such models also utilize transport equations, but the transported quantities are scalars—such as the turbulent kinetic energy k and its dissipation rate ε. Though scalar models have undeniable inaccuracies, they can be used to predict turbine heat transfer and are widely used in industry and academia. In the remainder of this section, only closure models based on scalar variables are reviewed with focus on the basic methodology of eddy-viscosity modeling and some of its shortcomings that are relevant to turbine cooling.

B. EDDY VISCOSITY

Eddy-viscosity models (EVMs) tackle the Reynolds-stress closure problem by assuming that the Reynolds stresses are aligned with the mean rate of strain (invoking the molecular analogy) and by introducing the concept of eddy viscosity as a proportionality factor. Thus, the turbulent stress and heat flux in Eq. (18a) are modeled by

$$\begin{aligned} \tau_{Tij} &= 2\rho\nu_T\left(S_{ij} - \frac{1}{3}S_{kk}\delta_{ij}\right) - \frac{2}{3}k\,\delta_{ij} \\ Q_{Ti} &= \kappa_T\frac{\partial T}{\partial x_i} \end{aligned} \tag{19a}$$

where the strain tensor is given by

$$S_{ij} = \frac{1}{2}\left(\frac{\partial U_i}{\partial x_j} + \frac{\partial U_j}{\partial x_i}\right) \tag{19b}$$

With Eq. (19), the eddy viscosity and the eddy thermal conductivity are simply added to the molecular viscosity and the molecular thermal conductivity in Eq. (18). The eddy thermal conductivity is connected to the eddy viscosity through the turbulent Prandtl number.

There are a number of ways to derive an equation for the eddy viscosity, including dimensional analysis and kinetic theory of gases. Here, a result from turbulent dispersion theory that states the transport coefficient for random convection in homogeneous turbulence can be expressed as the product of a Lagrangian correlation timescale and a velocity variance (for example, see Durbin and Pettersson-Reif [60]) is invoked, which gives

$$\nu_T = \overline{u^2} T_L \tag{20}$$

Various models of the eddy viscosity are distinguished by how they provide the time and velocity-fluctuation scales. In most two-equation models, the velocity variance is taken to be the turbulent kinetic energy per unit mass k. In the k-ω model, a transport equation is solved for a reciprocal timescale, $\omega = 1/T_L$. Thus, the eddy viscosity is

$$\nu_T = k/\omega \tag{21}$$

In the k-ε model, the timescale is related to the rate of the dissipation of the turbulent kinetic energy ε so that $T_L = C_\mu k/\varepsilon$, where C_μ is a coefficient of proportionality. Thus, the eddy viscosity is

$$\nu_T = C_\mu k^2/\varepsilon \tag{22}$$

In the v^2-f model, it is argued that a velocity scale, which behaves like the wall-normal velocity, is more appropriate near the wall. This wall-normal velocity scale is denoted as $\overline{v^2}$. If the timescale remains as $C_\mu k/\varepsilon$, then the eddy viscosity is given by

$$\nu_T = C_\mu \overline{v^2} k/\varepsilon \tag{23}$$

For the aforementioned three eddy-viscosity models, a transport equation for turbulent kinetic energy is needed in addition to a transport equation for ω, ε, and/or $\overline{v^2}$. Transport equations for k and the other variables provide the spatial distribution of eddy viscosity within the flowfield via the preceding formulas.

The transport equation for the turbulent kinetic energy per unit mass k is given by (for example, see Pope [59])

$$\frac{Dk}{Dt} = \mathcal{P} - \varepsilon + \frac{1}{\rho}\nabla \cdot \left[\left(\mu + \frac{\rho \nu_T}{\sigma_k}\right)\nabla k\right] \tag{24}$$

The preceding equation describes the imbalance between production and dissipation of k along with its transport in space. Production represents the rate of transfer from the mean flow to the turbulence. With the eddy-viscosity approximation, it is

$$\mathcal{P} = \frac{\tau_{ij}}{\rho}\frac{\partial U_i}{\partial x_j} = 2\nu_T\left(|S|^2 - \frac{1}{3}S_{kk}^2\right) - \frac{2}{3}kS_{kk} \tag{25}$$

where $|S^2| = S_{ij}S_{ji}$ with S_{ij} being the rate of strain tensor, and $S_{kk} = \partial_k U_k$ the velocity divergence, which vanishes in incompressible flow.

To close Eq. (24), ε is needed. In the k-ω model, ε is given by

$$\varepsilon = C_\mu k\omega \tag{26}$$

and a transport equation for ω is provided, while in the k-ε model, a transport equation for ε is provided. The transport equations for ω and ε are dimensionally consistent analogies to the k equation, Eq. (24). For ε, it is

$$\frac{D\varepsilon}{Dt} = \frac{C_{\varepsilon 1}\mathcal{P} - C_{\varepsilon 2}\varepsilon}{T_L} + \frac{1}{\rho}\nabla \cdot \left[\left(\mu + \frac{\rho \nu_T}{\sigma_\varepsilon}\right)\nabla \varepsilon\right] \tag{27}$$

Standard model constants are $C_\mu = 0.09$, $C_{\varepsilon 1} = 1.44$, $C_{\varepsilon 2} = 1.92$, $\sigma_\varepsilon = 1.3$, $\sigma_k = 1.0$. The ω transport equation is

$$\frac{D\omega}{Dt} = 2C_\mu C_{\omega 1}|S|^2 - C_{\omega 2}\omega^2 + \frac{1}{\rho}\nabla \cdot [(\mu + \rho \nu_T \sigma_\omega)\nabla \varepsilon] \tag{28}$$

The standard constants are $C_{\omega 1} = 5/9$, $C_{\omega 2} = 5/6$, $\sigma_\omega = \sigma_k = 2$, and $C_\mu = 0.09$. In the case of the v^2-f model, a transport equation is needed for $\overline{v^2}$. In the v^2-f model, the transport equation for $\overline{v^2}$ is paired with a modified Helmholtz equation for the intermediate variable f. Thus, $\overline{v^2}$ is predicted by the following pair of equations:

$$\begin{aligned} \frac{D\overline{v^2}}{Dt} &= kf - \varepsilon\frac{\overline{v^2}}{k} + \frac{1}{\rho}\nabla \cdot \left[\left(\mu + \frac{\rho \nu_T}{\sigma_k}\right)\nabla v^2\right] \\ L^2\nabla^2 f - f &= -C_2\frac{P}{k} + \frac{C_1}{T}\left(\frac{\overline{v^2}}{k} - \frac{2}{3}\right) \end{aligned} \tag{29}$$

The variance $\overline{v^2}$ can be considered as the wall-normal turbulence intensity or as a mixing intensity. Here, $\overline{v^2}/\mathrm{k}$ can be understood as an anisotropy parameter that tends to zero as it approaches the wall. Hanjalic et al. [64] adopted $\zeta \equiv \overline{v^2}/k$ as the dependent variable in their ζ-f model.

C. MODIFICATIONS

The k-ε and k-ω models have served as workhorses of applied CFD for at least 20 years. In the course of time, a number of modifications have been proposed to improve their predictive capabilities. This section discusses a few of these modifications and their motives.

The k-ε model is especially problematic in the near-wall region. The ε equation given by Eq. (27) has an irregular singular point at a no-slip wall because $T_L = k/\varepsilon$ vanishes there. That irregularity can be removed by placing a lower bound on T_L. The Kolmogoroff timescale could be used for this purpose as follows:

$$T_L = \max\left[\frac{k}{\varepsilon}, C_T\sqrt{\frac{\nu}{\varepsilon}}\right] \tag{30}$$

where C_T is a constant (typically 6) and prevents the singularity. However, the failure of k-ε is more fundamental than this. The velocity variance k produces too much mixing. The notion of a damping function f_μ has therefore been introduced in the literature. The eddy viscosity is revised to

$$\nu_T = f_\mu C_\mu k^2/\epsilon \tag{31}$$

with f_μ increasing from zero at the wall to unity further away from the wall. This approach has not proved satisfactory. Instead of f_μ, the two-layer models and wall functions are commonly used (Chen and Jaw [65] and Durbin and Pettersson-Reif [60]).

These modifications introduce undesirable ambiguities. For heat-transfer problems, where transport to the surface is critical, the k-ε model has been found to be less accurate and generally less satisfactory than k-ω variants.

1. LARGE RATES OF STRAIN

The most common, scalar eddy-viscosity models are subject to a spurious overproduction of turbulent kinetic energy in highly strained flows. This occurs generally under high rates of strain and can be seen in turbine and compressor passages. It is manifested by turbulent intensities growing to highly unphysical levels. Physically correct and anomalous turbulent-kinetic energy levels are illustrated in Fig. 1.

The source of this anomaly can be explained as follows. The production is given by

$$\mathcal{P} = -\overline{u_i u_j} S_{ij} \tag{32}$$

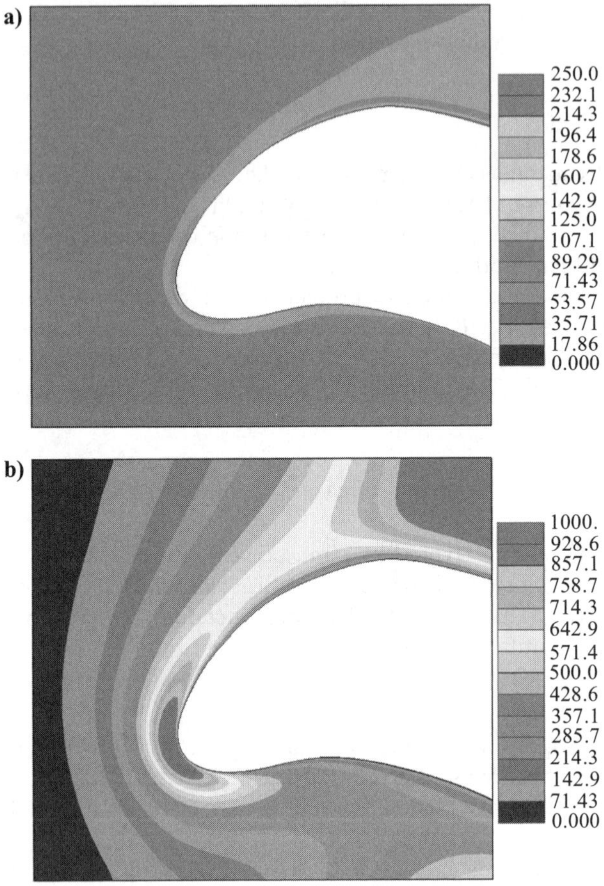

Fig. 1 Turbulent kinetic energy *k* about the leading edge of a turbine airfoil. Scale of *k* is arbitrary, but is the same for both plots of *k*: a) physically correct *k* and b) anomalous *k* predicted by *k*-ε.

In the principal axes of the rate of strain tensor,

$$\mathcal{P} = -\overline{u_1u_1}S_{11} - \overline{u_2u_2}S_{22} - \overline{u_3u_3}S_{33} \tag{33}$$

Each component of the normal stress satisfies

$$\overline{u_\alpha u_\alpha} \le 2k$$

In incompressible flow, $S_{jj} = 0$. Thus, at least one $S_{\alpha\alpha}$ is negative, so that

$$\mathcal{P} \le 2k \max_\alpha(-SS_{\alpha\alpha})$$

It can be shown that

$$|S_{\alpha\alpha}| \leq \sqrt{2|S|^2/3}$$

in three dimensions. Hence, on purely mathematical grounds,

$$\mathcal{P} \leq 2k\sqrt{2|S|^2/3} \tag{34}$$

In particular, production should increase linearly with $|S|$ at large rates of strain. With the eddy-viscosity model given by Eq. (19), however, the production as given in Eq. (32) is

$$\mathcal{P} = 2\nu_t|S|^2 \tag{35}$$

for incompressible flow, which appears to increase quadratically with rate of strain. Inserting Eq. (35) into Eq. (34) suggests the bound

$$\nu_t \leq \sqrt{\frac{2}{3}}\frac{k}{|S|} \tag{36}$$

Alternatively, instead of phrasing Eq. (36) as a bound on eddy viscosity, it can be stated as the timscale bound

$$T_L \leq \frac{2}{\sqrt{6}C_\mu|S|} \tag{37}$$

This result, with a different numerical factor, can also be derived as a realizability condition on the Reynolds-stress tensor. This is not a convincing argument when the mean flow is computed from the eddy viscosity. The real issue is to avoid excessive production of turbulent kinetic energy.

The condition given by Eq. (37) can be implemented for the k-ω model by

$$T_L = \min\left[\frac{1}{\omega}, \frac{2\alpha}{\sqrt{6}|S|}\right] \tag{38}$$

α is an adjustable constant that must be less than or equal to one. Medic and Durbin [66] set $\alpha = 0.6$.

Many investigators have proposed to make model constants functions of the rate of strain—either explicitly or implicitly. For instance, the shear stress limiter of Menter [67] can be expressed in that form. Zhu and Shih [68] invoke a strain-dependent formula for C_μ based on realizability in two dimensions of Eq. (19a) with Eq. (22). Thus, Eq. (36) can alternatively be phrased as

$$C_\mu \leq \sqrt{\frac{2}{3}}\frac{\varepsilon}{|S|k}$$

Rather than a limit per se, Zhu and Shih [68] propose the formula

$$C_\mu = \frac{2/3}{5.5 + |S|k/\varepsilon}$$

to effectively create a bound.

An early method to constrain the production of turbulent kinetic energy is that of Launder and Kato [69]. They replaced the formula $\mathcal{P} = 2\nu_T|S|^2$ by $\mathcal{P} = 2\nu_T|S||\Omega|$. This is not consistent with Eq. (33). Rather, it is designed to make production vanish in pure straining flow. However, turbulence production is actually energy transfer from the mean flow to the turbulence. Hence, the Launder and Kato [69] replacement is not consistent with energy conservation.

Figure 2 shows how limiters affect turbulent intensity on the stagnation line of a blunt body. Figure 2a shows k to be produced excessively by the standard k-ε model. With the bound given by Eq. (37), excessive production is prevented, and k is closer to data as shown in Fig. 2b—albeit still not perfect. Though not shown, the k-ω model produces similar results.

An example of the effect of timescale limiting on the mean heat transfer was provided by Medic and Durbin [66] and is shown in Fig. 3. Figure 3a shows the mean heat-transfer coefficient to a blade in a linear turbine cascade predicted by the k-ω model with and without the limiter given by Eq. (38). Figures 3b and 3c show the turbulent intensity, defined locally by $\sqrt{k}/U(x)$. Both numerator and denominator increase in strongly accelerated flow, but it is known from experiment and theory that their ratio decreases. Without limiting the rate of turbulent-energy production, the k-ω model predicts a substantial increase of

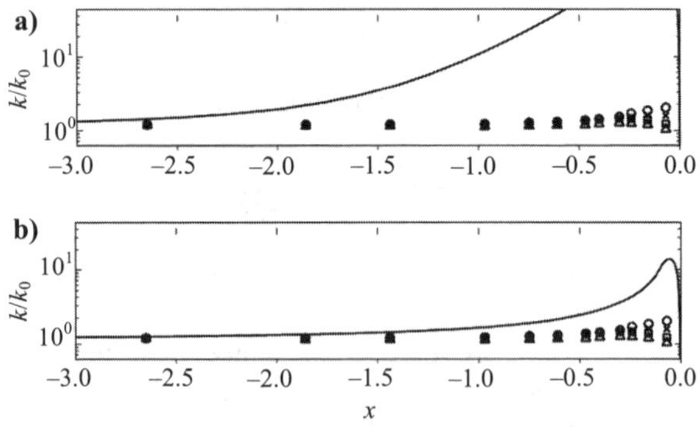

Fig. 2 Turbulent kinetic energy *k* normalized by *k* in the freestream k_0 about the stagnation line of a bluff body predicted by the *k*-ε model with and without timescale limiters. $x = 0$ is the stagnation point on the bluff body. Lines represent simulations. Circles, squares, and triangles are experimental measurements: a) without limiters and b) with limiters.

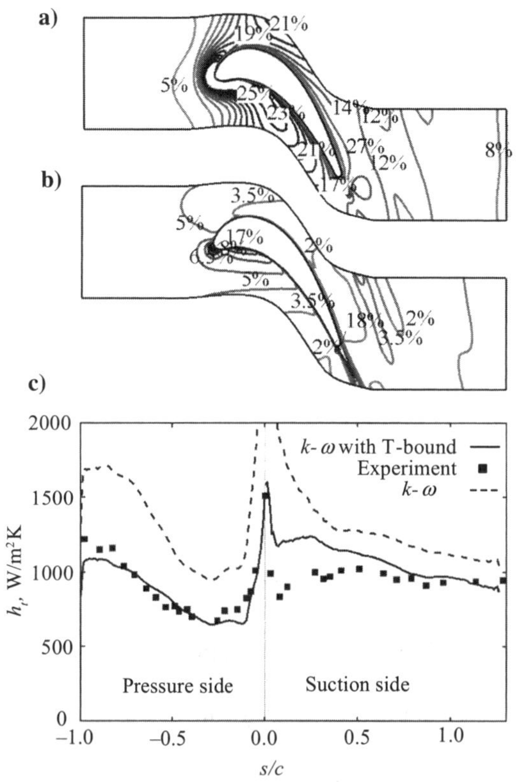

Fig. 3 Turbulent intensity ($I = \sqrt{2k/3}/U$, where U is the freestream velocity) about a turbine airfoil and heat-transfer coefficient h_t on the pressure and suction surfaces of that airfoil predicted by the k-ω model with and without limiters: a) turbulent intensity without limiters, b) turbulent intensity with limiters, and c) heat-transfer coefficient h_t with and without limiters along with experimental data.

the ratio as shown in Fig. 3b, which is an error. When the timescale is bounded by Eq. (38), a qualitatively correct behavior is obtained, as shown in Fig. 3c. Figure 3a illustrates the effect of this on the mean heat-transfer coefficient. It illustrates how excessive turbulent-energy production leads to excessive heat transfer. With the timescale bound imposed, agreement to data is improved.

2. NEAR-WALL ANISOTROPY

Experiments show that wall jets spread more rapidly in the lateral direction than in the wall-normal direction. Lateral spreading might be on the order of four to five times faster. One rationale that has been offered is that the turbulent stress is greater in the z direction than in the y direction for a jet flowing in the x direction. Profiles of $\overline{v^2}$ and $\overline{w^2}$ for turbulent flow between parallel plates are shown in Fig. 4. From this figure, $\overline{w^2}$ and k can be seen to be larger than $\overline{v^2}$ as the wall is approached, and $\overline{v^2}$ is suppressed more than $\overline{w^2}$. The eddy viscosity follows the behavior of $\overline{v^2}$, which is the motivation for Eq. (23).

Because the disparity between lateral and normal spreading is caused by anisotropic eddy viscosity, it can be argued that $C_\mu \overline{w^2} T$ is a lateral viscosity and $C_\mu \overline{v^2} T$ is a normal viscosity. In a tensorally consistent form,

$$\nu_{Tij} = C_\mu \overline{u_i u_j} T$$

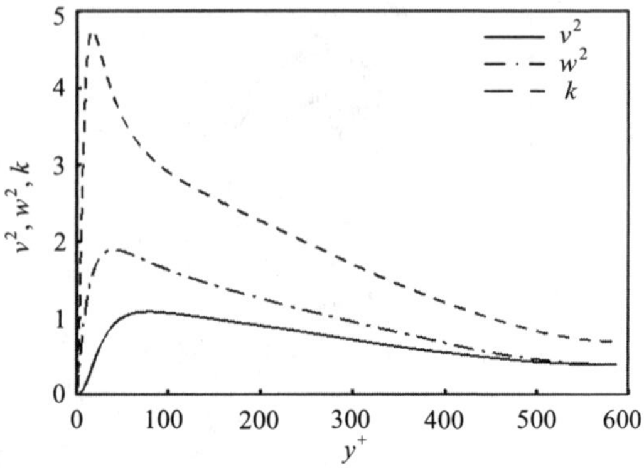

Fig. 4 Measured Reynolds stresses – $\overline{v^2}$ and $\overline{w^2}$ denoted as v^2 and w^2—and the turbulent kinetic energy k in a plane channel as a function of distance from the wall (y^+).

The preceding formula is of no predictive value because the objective of the eddy viscosity is to provide a closure to the unknown Reynolds stresses $\overline{u_i u_j}$. It is just a rationale for lateral vs normal spreading. In a model that predicts a velocity scale $\overline{v^2}$ and k, one could create a tensor such as

$$\nu_{Tij} = C_\mu T\left[k\delta_{ij} + \left(\overline{v^2} - k\right)\hat{n}_i\hat{n}_j\right]$$

where $\hat{n}$ is the wall normal vector although this would have to be replaced by a vector that is defined throughout the flow.

Anisotropic eddy viscosities have been explored by various investigators as a means to increase lateral spreading. Lakehal et al. [70] fit curves through DNS data on $\overline{v^2}/k$ and $\overline{w^2}/k$ vs y and applied a tensorally inconsistent formula

$$-\overline{u_i u_j} = 2\nu_{Tj}S_{ij} - \frac{2}{3}k\delta_{ij}$$

(with summation on j suspended). They produced higher lateral spreading and noted that this improved agreement to experiments.

Craft and Launder [71] presented evidence that enhanced lateral spreading is not caused by anisotropic eddy viscosity. They attributed it to streamwise vorticity generated by normal stress anisotropy. Rather than being a transport mechanism, normal-stress anisotropy acts as a rotational body force. In the absence of baroclinic torque, streamwise vorticity is created by the first term on the right of

$$\frac{D\Omega_x}{Dt} = \frac{\partial}{\partial y}\frac{\partial}{\partial z}\left(\overline{v^2} - \overline{w^2}\right) - \frac{\partial^2}{\partial y^2}\partial^2\overline{vw} + \frac{\partial^2}{\partial z^2}\overline{vw} + \nu\nabla^2\Omega_x \tag{39}$$

where Ω_x is the streamwise vorticity. Their study of three-dimensional, spreading wall jets concluded with the observations that anisotropic diffusion plays a small role and that the only significant mechanism driving the high lateral rate of spreading is the turbulent body force in the vorticity equation, Eq. (39).

D. CURVATURE AND ROTATION

Consider a boundary layer flowing in the tangential direction on a convex, curved wall. The radius of curvature is R, and r is the direction normal to the wall. The components of the Reynolds stress production tensor are

$$\begin{aligned}
\mathcal{P}_{11} &= -2\overline{uv}\frac{\partial U}{\partial r} \\
\mathcal{P}_{22} &= 2\overline{uv}\frac{U}{R} \\
\mathcal{P}_{33} &= 0 \\
\mathcal{P}_{12} &= -\overline{v^2}\frac{\partial U}{\partial r}U + \overline{u^2}\frac{U}{R}
\end{aligned} \tag{40}$$

In shear flow, $\overline{uv} < \overline{0}$, and so the first equation in Eq. (40) is the usual shear production of $\overline{u^2}$. The second equation in Eq. (40), $\mathcal{P}_{22}$, is negative in shear flow, which causes suppression of $\overline{v^2}$. Thus, curvature exerts a stabilizing influence on that component. As a result, the last equation in Eq. (40), the dominant production, $-\overline{v^2}\partial_r U$, of shear stress is reduced. In turn, reduction of the Reynolds shear stress reduces $\mathcal{P}_{11}$ and hence the turbulent kinetic energy. Thus, the stabilizing influence of convex curvature emerges by this circuitous rationale. Concave curvature can be represented by letting R be negative, reversing the previous argument to conclude that production of turbulent kinetic energy is increased by concave curvature.

It is crucial that these effects start with a particular component $\mathcal{P}_{22}$. This is the component in the direction of the radius of curvature. This direction must be distinguished in order to capture the influence of curvature. Models based on total turbulent energy k are unable to capture the influence of curvature, because only the rate of turbulent kinetic energy production $\mathcal{P} = 2V_t|S|^2$ is used.

Rotation has an effect analogous to that of curvature. The component of Reynolds normal stress directed toward the center of rotation can be suppressed or enhanced. Rotation in the direction of shear is stabilizing whereas rotation against the shear is destabilizing. In a rotating channel, for instance, the shear near one wall corotates with the channel (the flow next to the leading face of the rotating duct) while along the other it counter-rotates (the flow next to trailing face). Figure 5 shows the Reynolds shear stress and mean flow at four rotation rates for a rotating duct [72, 73]. From this figure, it can be seen that the turbulence is reduced near the leading wall $(y/h = 2)$ and enhanced at the trailing wall $(y/h = 0)$. The mean velocity is symmetric in the nonrotating case. The flow passes more easily on the side where turbulent stress is reduced by rotation. Again, this effect is not captured by scalar-eddy-viscosity models. They predict symmetric flow with no effect of rotation.

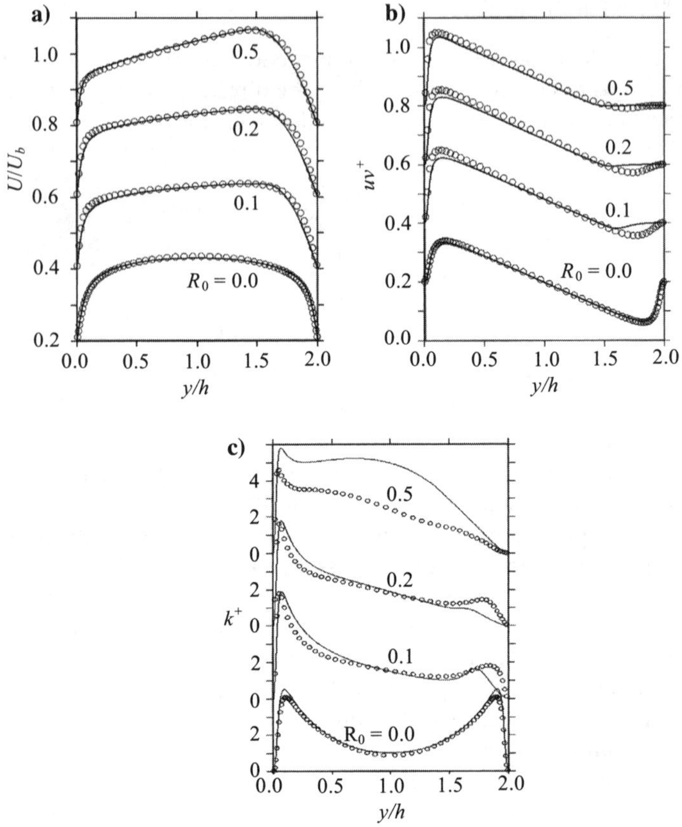

Fig. 5 Incompressible radially outward flow in a rotating channel, where $y/h = 0$ is the trailing face, and $y/h = 2$ is the leading face. Lines are from RANS [72], and circles are from DNS [73]: a) mean velocity profile U/U_b, where U_b is the mean flow; b) Reynolds stress, $uv^+ = \overline{uv}/u_\tau^2$, where u_τ is the friction velocity; and c) turbulent kinetic energy, $k^+ = k/u_\tau^2$.

Reynolds-stress transport models can capture effects of curvature and rotation because they contain the full production tensor. It can be shown analytically [60, 74] that equilibrium branches bifurcate at a critical rotation rate, corresponding to suppression of turbulence. The analysis can be characterized as deriving an effective C_μ that depends on shear and rotation—actually on rate of strain and absolute rate of rotation. Hence, it is possible to create a scalar that bifurcates by constructing a C_μ that depends on the rate of strain and rotation of the mean flow [72]. The effect of curvature and rotation arises through their influence on turbulence. For instance, the velocity profiles in Fig. 5 would remain symmetric, independent of rotation, in laminar flow.

The flow in internal-cooling ducts located inside a rotating turbine blade experiences Coriolis and buoyancy forces. Stephens and Shih [75] and Lin et al. [76] showed that for radially outward flow in a rotating duct, the Coriolis force creates a pair of counter-rotating secondary flows that transport the cooler air from near the center of the duct to the trailing face first as shown in Fig. 6a. Because the thermal boundary layer starts on the trailing face, Fig. 6a shows the temperature of the cooling air next to the leading face to be higher than the temperature of the air next to the trailing face. With higher temperature and hence lower density next to the leading face, centrifugal buoyancy along the radial direction decelerates the flow next to the leading face as shown in Fig. 6b. This deceleration can lead to flow separation on the leading face. When centrifugal buoyancy causes flow separation on the leading face, the Coriolis-induced secondary flows can cause the formation of additional pairs of vortices next to the leading face [75]. When there are inclined ribs on the leading and trailing faces, the secondary flows induced by the ribs enhance those that rotate in the same direction and retard those that do not [76]. To capture centrifugal buoyancy correctly in turbine cooling, density variations that arise from temperature variations and centrifugal forces must be accounted for in the governing equations because they are so large (that is, invoking the Boussinesq approximation is inadequate, the flow cannot be treated as incompressible, even though the Mach number of the flow is much less than 0.3).

Buoyancy fluctuations also alter the turbulence. They produce an unclosed contribution to the turbulent kinetic energy equation. The influence of density fluctuations upon turbulence is usually considered a secondary effect in turbine applications because the Mach number of the flow is not large. Temperature

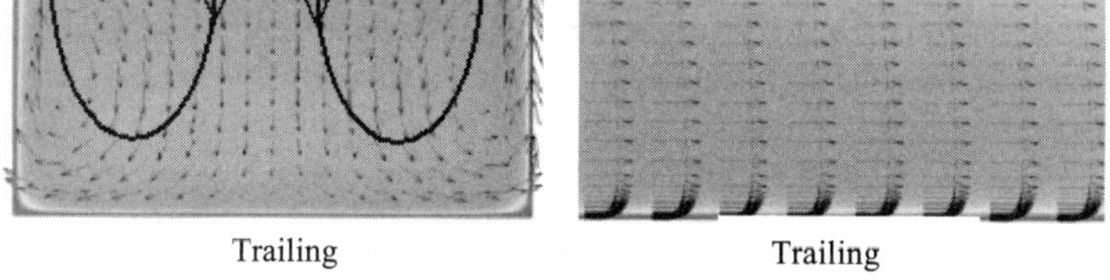

Fig. 6 Flow induced by rotation and buoyancy for radially outward flow in a duct: a) Coriolis induced secondary flow in the duct's cross section and b) centrifugal buoyancy induced flow deceleration along the leading face.

fluctuations could also produce density fluctuations, however. In film cooling, the ratio of the density of the film-cooling air to that of the hot gas could be 1.5 or higher from the temperature differences. Also, the temperature gradients in the thermal boundary layers in the hot-gas path and in the internal-cooling passages can be extremely high, so that the density gradient is also appreciable. Thus, temperature-fluctuation induced buoyancy fluctuations might be important in rotating components, and it might need to be modeled.

E. HEAT FLUX

In conjunction with eddy-viscosity models, heat transport usually is represented by gradient diffusion with a turbulent Prandtl number given by

$$Q_i = \frac{K}{Pr_T}\frac{\partial T}{\partial x_i} \tag{41}$$

where $Pr_T = 0.9$ is the standard value for boundary layers and tends to provide satisfactory predictions in attached boundary layers. A single value of Pr_T is not, however, adequate for all applications; a significantly lower value is needed to represent mixing in a combustor.

In complex geometries Eq. (41) might be replaced by the generalized gradient diffusion hypothesis (GGDH) in which the diffusion coeffient is a second-order tensor:

$$Q_i = K_{ij}\frac{\partial T}{\partial x_j} \tag{42}$$

The diffusion tensor $\boldsymbol{K}$ is not symmetric. For instance, boundary-layer transport in the x direction due to a temperature gradient in y does not have the same diffusivity as that for transport in y due to a gradient in x. In parallel shear flow, an expression of the form

$$K_{ij} = \frac{C_\mu T_L}{Pr_T}\left[\overline{u_i u_j} - \beta T_L \overline{u_i u_j}\frac{\partial U_i}{\partial x_k}\right] \tag{43}$$

has some justification, where β is a constant. Asymmetry ($K_{ij} \neq K_{ji}$) is caused by the mean shear through the second term inside the brackets of Eq. (43). Although $\boldsymbol{K}$ can be quite asymmetric, in a nearly parallel shear flow, with small streamwise temperature gradients, cross-stream transport is dominant. For instance, if the largest gradients are in the y and z directions, then Eq. (42) gives

$$Q_2 = K_{22}\frac{\partial T}{\partial x_2} + K_{23}\frac{\partial T}{\partial x_3}$$

$$Q_3 = K_{32}\frac{\partial T}{\partial x_2} + K_{33}\frac{\partial T}{\partial x_3}$$

The last term of Eq. (43) only contributes to K_{1j} so that asymmetry is not relevant. Based on the eddy-viscosity formula Eq. (19a), $K_{23} = 0$ if the flow is in the x direction. Perhaps the most important conclusion is that transport perpendicular to a wall at $y = 0$ should be represented by the eddy diffusivity given by

$$K_{22} = \frac{C_\mu}{Pr_t}\overline{v^2}T_L$$

Thus, it should be based on the wall normal component of turbulent intensity.

Another issue that receives considerable attention in heat-transfer analysis is whether to use a wall function or to integrate to the wall. This question arises because of the turbulence model. Integrating to the wall requires that the turbulence model be accurate and well posed down to $y = 0$. The wall function replaces the region next to the wall with an effective boundary condition above the wall. Numerous studies have shown that accurate prediction of heat transfer argues for integration to the wall [12, 75–77]. As mentioned earlier, the k-ε model is problematic if integrated to the wall. The k-ω, or its shear stress transport (SST) variant, is preferred among the two-equation models for integrating to the wall. The v^2-f, or its variant, is also designed for integrating to the wall.

In the wall function approach, the region between the wall and the log layer is not meshed or gridded. If the surface is ribbed, as is common in internal passages of blades, then these will not be resolved with a wall function method [77]. Similarly, film-cooling jets will not be properly represented. Even without these arguments against wall functions, studies have found that integrating to the wall produces more accurate predictions of heat transfer than with wall functions [12, 78]. The deficiency of eddy-viscosity models for film cooling is related to the anisotropy of the turbulence connected to wall jets.

F. UNSTEADY RANS

Turbulence models represent ensemble-averaged effects of turbulence. Ensemble averaging should not be confused with time averaging. Some flows in gas turbines could have unsteadiness that is periodic in time. When this occurs, the frequency spectrum of turbulence contains spikes at the coherent frequency and its harmonics. These rise above the broadband spectrum. The turbulence model only seeks to represent turbulence associated with the broadband component. The spikes are part of the mean flow and must be resolved temporally. Ignoring coherent unsteadiness, that is including only the incoherent component via a turbulence model, generally underpredicts mixing. Steady-state calculations of turbulent flows with unsteady mean produce mistakes in the mean flow. Such errors should not be attributed to the turbulence model.

In many cases, improved predictions have been obtained by unsteady RANS computations [79, 80]. A notable example is the phenomenon of hot streak migration [81, 82]. It is observed that the pressure side of a turbine blade is hotter than the suction side. Streaks of hot fluid exiting the combustor and flowing through the first stator vanes impinge on the first rotor. If the pressure is nearly uniform around the circumference between the stator and rotor, the Mach number is also. For a given Mach number, hot fluid has a higher velocity. Consider the rotor relative frame. If the velocity vector of the colder gas is directed through the blade passage, the velocity triangle for the hot gas will be deflected toward a blade surface. Because the velocity is higher, the direction of deflection is toward the pressure side. Thus, as a hot streak passes the interblade passage, the flow will be deflected. The temperature field is thereby rectified to produce a higher time-averaged temperature on the pressure side. This phenomenon is called the Kerrebrock–Mikolajczak effect. It is seen in unsteady computations, even with Euler equations [81], although a full simulation of the hot streak evolution requires a turbulence model.

V. SUMMARY

Computational fluid dynamics and heat transfer (CFD/HT) has advanced to the point of being able to provide a fairly deep understanding of the effects of geometric and operating parameters on turbine cooling designs, as well as insight which may lead to new design paradigms. Despite this progress, however, attention is still needed in verification, validation, and uncertainty quantification. Verification can be problematic because the sheer magnitude of the problem might render grid-sensitivity studies impractical. Thus, new methods are required, to provide a posteriori estimates of grid-induced error in the computed solutions. One such method based on the discrete-error-transport equation (DETE) was presented. For this method, the challenge is in modeling the residual in the DETE, especially for unsteady flows. Validation poses challenges in that it is important for the CFD/HT study to truly simulate the experimental study. It is equally important for the experimental study to provide enough details so that the CFD/HT study can replicate the experiment as closely as possible. In addition, the experimental study needs to explain how measurements were made and how data were postprocessed. For the heat-transfer coefficient, it is important to provide information on the bulk temperatures and the wall temperatures used to compute the heat-transfer coefficients. Uncertainty quantification requires a dual approach. Aleatory errors in CFD/HT solutions could be examined using statistical methods and probability theory because such errors are random. Epistemic errors in CFD/HT solutions, however, are much more complex. Epistemic errors connected to the validation process are outlined. Epistemic errors connected to turbulence models for steady and unsteady RANS are described by reviewing three popular two-equation models—k-ε, k-ω, and v^2-f—with focus on the thinking process that created them and efforts to address their deficiencies.

REFERENCES

[1] Suo, M., "Turbine Cooling," *Aerothermodynamics of Aircraft Engine Components*, edited by G. C. Oates, AIAA, New York, 1985, pp. 275–328.

[2] Shih, T. I.-P., and Chyu, M. (eds.), "Special Section on Turbine Science and Technology," *Journal of Propulsion and Power*, Vol. 22, No. 2, 2006, pp. 225–396.

[3] Pollock, T. M., and Tin, S., "Nickel-Based Superalloys for Advanced Turbine Engines: Chemistry, Microstructure, and Properties," *Journal of Propulsion and Power*, Vol. 22, No. 2, 2006, pp. 361–374.

[4] Gleeson, B., "Thermal Barrier Coatings for Aerospace Applications," *Journal of Propulsion and Power*, Vol. 22, No. 2, 2006, pp. 375–383.

[5] Metzger, D. E., "Cooling Techniques for Gas Turbine Airfoils," AGARD CP 390, NATO, 1985, pp. 1–12.

[6] Moffat, R. J., "Turbine Blade Heat Transfer," *Heat Transfer and Fluid Flow in Rotating Machinery*, edited by W. J. Yang, Hemisphere, Washington, D.C., 1987, pp. 1–24.

[7] Goldstein, R. J., "Film Cooling," *Advances in Heat Transfer*, Academic Press, New York, 1971, Vol. 7, pp. 321–379.

[8] Han, J. C., Dutta, S., and Ekkad, S. V., *Gas Turbine Heat Transfer and Cooling Technology*, Taylor & Francis, New York, 2000.

[9] Goldstein, R. (ed.), *Heat Transfer in Gas Turbine Systems*, Annals of the New York Academy of Sciences, Vol. 934, May 2001.

[10] Sundén, B., and Faghri, M. (eds.), *Heat Transfer in Gas Turbines*, WIT Press, Ashurst, Southhampton, England, U.K., 2001.

[11] Bunker, R. S., "Gas Turbine Cooling: Moving from Macro to Micro Cooling," American Society of Mechanical Engineers, Paper GT2013-94277, June 2013.

[12] Shih, T. I.-P., and Sultanian, B., "Computations of Internal and Film Cooling," *Heat Transfer in Gas Turbines*, edited by B. Sundén and M. Faghri, WIT Press, Ashurst, Southhampton, England, U.K., 2001, pp. 175–225.

[13] Roache, P. J., *Verification and Validation in Computational Science and Engineering*, Hermosa Publishers, New Mexico, 1998.

[14] Oberkampf, W. L., and Roy, C. J., *Verification and Validation in Scientific Computing*, Cambridge Univ. Press, Cambridge, England, U.K., 2010.

[15] Celik, I., Chen, C. J., Roache, P. J., and Scheuer, G. (eds.), *Symposium on Quantification of Uncertainty in Computational Fluid Dynamics*, FED-Vol. 158, American Society of Mechanical Engineers, New York, 1993.

[16] Celik, I., and Zhang, W.-M., "Calculation of Numerical Uncertainty Using Richardson Extrapolation: Application to Some Simple Turbulent Flow Calculations," *ASME Journal of Fluids Engineering*, Vol. 117, 1995, pp. 439–445.

[17] Wilson, R. V., and Stern, F., "Verification and Validation for RANS Simulation of a Naval Surface Combatant," AIAA Paper 2002-0904, 2002.

[18] Roy, C. J., "Review of Discretization Error Estimators in Scientific Computing," AIAA Paper 2010-126, 2010.

[19] Carey, G. F., *Computational Grids: Generation, Adaptation, and Solution Strategies*, Taylor & Francis, Washington, D.C., 1997.

[20] Shih, T. I.-P., Gu, X., and Chu, D., "Grid-Quality Measures and Error Estimates," *Numerical Grid Generation in Computational Field Simulations*, edited by B. K. Soni, J. F. Thompson, J. Hauser, and P. R. Eiseman, ISSG, Mississippi State Univ., MS, 2000, pp. 799–808.

[21] Gu, X., Hernandez, E., Chu, D., Sun, R., Keller, P., and Shih, T. I.-P., "Grid-Quality Measures for Error Estimation of CFD Solutions," *Computational Fluid and Solid Mechanics*, Vol. 1, edited by K. J. Bathe, Elsevier, New York, 2001, pp. 843–846.

[22] Gu, X., Schock, H. J., Shih, T. I.-P., Hernandez, E. C., Chu, D., Keller, P. S., and Sun, R. L., "Grid-Quality Measures for Structured and Unstructured Meshes," AIAA Paper 2001-0652, 2001.

[23] Gu, X., and Shih, T. I.-P., "Differentiating Between Error Source and Error Location in Solution-Adaptive Mesh Refinement," AIAA Paper 2001-2660, 2001.

[24] Babuska, I., and Rheinboldt, W., "Error Estimates for Adaptive Finite Element Computations," *SIAM Journal on Numerical Analysis*, Vol. 15, No. 4, 1978, pp. 736–754.

[25] Babuska, I., Strouboulis, T., Gangaraj, S. K., and Upadhyay, C. S., "Pollution Error in the h-Version of the Finite Element Method and Local Quality of the Recovered Derivatives," *Computational Methods in Applied Mechanics and Engineering*, Vol. 140, 1997, pp. 1–37.

[26] Ferziger, J. H., "Estimation and Reduction of Numerical Error," *Symposium on Quantification of Uncertainty in Computational Fluid Dynamics*, FED-Vol. 158, American Society of Mechanical Engineers, New York, 1993, pp. 1–8.

[27] van Straalen, B. P., Simpson, R. B., and Stubley, G. D., "A Posteriori Error Estimation for Finite-Volume Simulations of Fluid Flow Transport," *Third Annual Conference of the CFD Society of Canada*, Vol. 1, Baniff, Alberta, 1995.

[28] Zhang, X. D., Trépanier, J.-Y., and Camarero, R., "A Posteriori Error Estimation for Finite-Volume Solutions of Hyperbolic Conservation Laws," *Computational Methods in Applied Mechanics and Engineering*, Vol. 185, 2000, pp. 1–19.

[29] Zhang, X. D., Pelletier, D., Trépanier, J.-Y., and Camarero, R., "Numerical Assessment of Error Estimators for Euler Equations," *AIAA Journal*, Vol. 39, No. 9, 2001, pp. 1706–1715.

[30] Qin, Y., and Shih, T. I.-P., "A Discrete Transport Equation for Error Estimation in CFD," AIAA Paper 2002-0906, 2002.

[31] Finlayson, B. A., *Method of Weighted Residuals and Variational Principles*, Elsevier Science & Technology Books, New York, 1972.

[32] Giles, M. B., and Pierce, N. A., "On the Properties of Solutions of the Adjoint Euler Equations," *Sixth ICFD Conference on Numerical Methods for Fluid Dynamics*, Oxford, England, U.K., 1998.

[33] Giles, M. B., "On Adjoint Equations for Error Analysis and Optimal Grid Adaptation," *Frontiers of Computational Fluid Dynamics*, edited by D. A. Caughey and M. M. Hafez, World Scientific, Singapore, 1998, pp. 155–170.

[34] Venditti, D. A., and Darmofal, D. L., "Anisotropic Grid Adaptation for Functional Outputs: Application to Two-Dimensional Viscous Flows," *Journal of Computational Physics*, Vol. 187, 2003, pp. 22–46.

[35] Park, M. A., "Adjoint-Based, Three-Dimensional Error Prediction and Grid Adaptation," *AIAA Journal*, Vol. 42, No. 9, 2004, pp. 1854–1862.

[36] Hicken, J. E., and Zingg, D. W., "The Role of Dual Consistency in Functional Accuracy: Error Estimation and Superconvergence," AIAA Paper 2011-3855, 2011.

[37] Qin, Y., and Shih, T. I.-P., "A Method for Estimating Grid-Induced Errors in Finite-Difference and Finite-Volume Methods," AIAA Paper 2003-0845, 2003.

[38] Qin, Y., Chi, X., and Shih, T. I.-P., "Estimating Grid-Induced Errors in Navier-Stokes Solutions by Euler Discrete-Error-Transport Equations," AIAA Paper 2005-0567, 2005.

[39] Shih, T. I.-P., "Estimating Grid-Induced Errors in CFD Solutions," *Frontiers of Computational Fluid Dynamics*, edited by D. A. Caughey and M. M. Hafez, World Scientific, Singapore, 2005, Chap. 9, pp. 183–198.

[40] Richtmyer, R. D., and Morton, K. W., *Difference Methods for Initial-Value Problems*, 2nd ed., Interscience Publisher, New York, 1967.

[41] Warming, R. F., and Hyett, B. J., "The Modified Equation Approach to the Stability and Accuracy Analysis of Finite-Difference Methods," *Journal of Computational Physics*, Vol. 14, 1974, pp. 159–179.

[42] Celik, I., Hu, G., and Badeau, A., "Further Refinement and Bench Marking of a Single-Grid Error Estimation Technique," AIAA Paper 2003-0626, 2003.

[43] Qin, Y., and Shih, T. I.-P., "Analysis and Modeling of the Residual in the Discrete Error Transport Equation," AIAA Paper 2003-3850, 2003.

[44] Qin, Y., Chi, X., and Shih, T. I.-P., "Modelling the Residual in the Error-Transport Equation for Estimating Grid-Induced Errors in CFD Solutions," AIAA Paper 2006-892, 2006.

[45] Shih, T. I.-P., and Qin, Y., "A Posteriori Method for Estimating and Correcting Grid-Induced Errors in CFD Solutions - Part 1: Theory and Method," AIAA Paper 2007-100, 2007.

[46] Shih, T. I.-P., and Williams, B., "Development and Evaluation of an à Posteriori Method for Estimating and Correcting Grid-Induced Errors in Solutions of the Navier-Stokes Equations," AIAA Paper 2009-1499, 2009.

[47] Shih, T. I.-P., Gomatam-Ramachandran, S., and Chyu, M. K., "Time-Accurate CFD Conjugate Analysis of Transient Measurements of the Heat-Transfer Coefficient in a Channel with Pin Fins," *Journal of Propulsion and Power Research*, Vol. 2, No. 1, 2013, pp. 10–19; doi: http://dx.doi.org: 10.1016/j.jppr.2013.01.003

[48] Chi, X., and Shih, T. I.-P., "Bulk Temperature, Heat-Transfer Coefficient, and Nusselt Number - Revisited," AIAA Paper 2012-0807, 2012.

[49] Shih, T. I.-P., and Sathyanarayanan, S. K., "A New Nusselt Number for Complicated Configurations," American Society of Mechanical Engineers, Paper HT-2013-17114, 2013.

[50] Oberkampf, W. L., Helton, J. C., and Sentz, K., "Mathematical Representation of Uncertainty," AIAA Paper 2001-1645, 2001.

[51] Nocedal, J., and Wright, S. J., *Numerical Optimization*, Springer-Verlag, New York, 1999.

[52] Jameson, A., "Aerodynamic Design via Control Theory," *Journal of Scientific Computing*, Vol. 3, No. 3, 1988, pp. 233–260.

[53] Choi, S., and Datta, A., "Prediction of Helicopter Rotor Loads Using Time-Spectral Computational Fluid Dynamics and an Exact Fluid-Structure Interface," AIAA Paper 08-7325, 2008.

[54] Caisch, R. E., "Monte Carlo and Quasi-Monte Carlo Methods," *Acta Numerica*, Vol. 7, 1998, pp. 1–49.

[55] Najm, H. N., "Uncertainty Quantification and Polynomial Chaos Techniques in Computational Fluid Dynamics," *Annual Review of Fluid Mechanics*, Vol. 41, 2009, pp. 35–52.

[56] Padulo, M., Campobasso, M. S., and Guenov, M. D., "Novel Uncertainty Propagation Methods for Robust Aerodynamics Design," *AIAA Journal*, Vol. 49, No. 3, 2011, pp. 530–543.

[57] Pletcher, R. H., Tannehill, J. C., and Anderson, D. A., *Computational Fluid Mechanics and Heat Transfer*, 3rd ed., Taylor & Francis, Washington, D.C., 2012.

[58] Wilcox, D. C., *Turbulence Modeling for CFD*, DCW Industries, La Canada, CA, 1993.

[59] Pope, S., *Turbulent Flows*, Cambridge Univ. Press, Cambridge, England, U.K., 2000.

[60] Durbin, P., and Pettersson-Reif, B. A., *Statistical Theory and Modeling for Turbulent Flow*, 2nd ed., Wiley, Chichester, England, U.K., 2010.

[61] Launder, B. E., "Second-Moment Closure: Present . . . and Future?," *International Journal of Heat and Fluid Flow*, Vol. 10, 1989, pp. 282–300.

[62] Speziale, C. G., "Analytic Methods for the Development of Reynolds Stress Closures in Turbulence," *Annual Review of Fluid Mechanics*, Vol. 23, 1991, pp. 107–157.

[63] Durbin, P. A., and Shih, T. I.-P., "An Overview of Turbulence Modeling," *Modelling and Simulation of Turbulent Heat Transfer*, edited by B. Sundén and M. Faghri, WIT Press, Southhampton, England, U.K., 2005, Chap. 1, pp. 3–31.

[64] Hanjalic, K., Popovac, M., and Hadziabdic, M., "A Robust Near-Wall Elliptic Relaxation Eddy-Viscosity Turbulence Model for CFD," *International Journal of Heat and Fluid Flow*, Vol. 25, 2005, pp. 1047–1051.

[65] Chen, C. J., and Jaw, S. Y., *Fundamentals of Turbulence Modeling*, Taylor and Francis, Philadelphia, 1998.

[66] Medic, G., and Durbin, P. A., "Toward Improved Prediction of Heat Transfer on Turbine Blades," *Journal of Turbomachinery*, Vol. 124, 2002, pp. 187–192.

[67] Menter, F. R., "Two-Equation Eddy-Viscosity Turbulence Models for Engineering Applications," *AIAA Journal*, Vol. 32, No. 8, 1994, pp. 1598–1605.

[68] Zhu, J., and Shih, T. H., "Calculations of Turbulent Separated Flows," NASA Technical Memorandum TM-106721, 1993.

[69] Launder, B. E., and Kato, M., "Modelling Flow-Induced Oscillations in Turbulent Flow around a Square Cylinder," American Society of Mechanical Engineers, FED, Vol. 157, 1993, pp. 189–199.

[70] Lakehal, D., Theordoridis, G., and Rodi, W., "Three Dimensional Flow and Heat Transfer Calculations at the Leading Edge of a Symmetrical Turbine Blade," *International Journal Heat and Fluid Flow*, Vol. 22, 2001, pp. 113–122.

[71] Craft, T. J., and Launder, B. E., "On the Spreading Mechanism of the Three-Dimensional Wall Jet," *Journal of Fluid Mechanics*, Vol. 435, 2001, pp. 305–326.

[72] Pettersson-Reif, B. A., Durbin, P. A., and Ooi, A., "Modeling Rotational Effects in Eddy-Viscosity Closures," *International Journal of Heat and Fluid Flow*, Vol. 20, 1999, pp. 563–573.

[73] Kristoffersen, R., and Andersson, H. I., "Direct Simulations of Low-Reynolds-Number Turbulent Flow in a Rotating Channel," *Journal of Fluid Mechanics*, Vol. 256, 1993, pp. 163–197.

[74] Speziale, C. G., and Mhuiris, N. M. G., "On the Prediction of Equilibrium States in Homogeneous Turbulence," *Journal of Fluid Mechanics*, Vol. 209, 1989, pp. 591–615.

[75] Stephens, M. A., and Shih, T. I.-P., "Computations of Flow and Heat Transfer in a Smooth U-Shaped Square Duct with and Without Rotation," *Journal of Propulsion and Power*, Vol. 15, No. 2, 1999, pp. 272–279.

[76] Lin, Y.-L., Shih, T. I.-P., Stephens, M. A., and Chyu, M. K., "A Numerical Study of Flow and Heat Transfer in a Smooth and a Ribbed U-Duct with and Without Rotation," *ASME Journal of Heat Transfer*, Vol. 123, 2001, pp. 219–232.

[77] Gu, X., Wu, H.-W., Shen, S., and Shih, T. I.-P., "Numerical Simulation of Flow and Heat Transfer in a Duct with Square Ribs and Bleed Holes," AIAA Paper 2004-0658, 2004.

[78] Lin, Y.-L., and Shih, T. I.-P., "Film Cooling of a Semi-Cylindrical Leading Edge with Injection through Rows of Compound-Angled Holes," *ASME Journal of Heat Transfer*, Vol. 123, 2001, pp. 645–654.

[79] Iaccarino, G., Ooi, A., Durbin, P. A., and Behnia, M., "Reynolds Averaged Simulation of Unsteady Separated Flow," *International Journal of Heat and Fluid Flow*, Vol. 24, 2003, pp. 147–156.

[80] Medic, G., and Durbin, P. A., "Unsteady Effects on Trailing Edge Cooling," *ASME Journal of Heat Transfer*, Vol. 127, 2005, pp. 388–392.

[81] Shang, T., and Epstein, A. H., "Analysis of Hot Streak Effects on Turbine Rotor Heat Load," *ASME Journal of Turbomachinery*, Vol. 119, 1997, pp. 544–553.

[82] Abhari, R. S., "Unsteady Fluid Dynamics of Turbines," *Workshop on Unsteady Flows in Turbomachinery*, edited by V. Minnowbrook, NASA CP-2006-214484, 2006, pp. 3–24.

CHAPTER 10

Nickel-Based Superalloys

Sammy Tin*
Illinois Institute of Technology, Chicago, Illinois

Tresa M. Pollock†
University of California, Santa Barbara, Santa Barbara, California

NOMENCLATURE

a	lattice parameter
a_γ	lattice parameter of γ phase
$a_{\gamma'}$	lattice parameter of γ' phase
b	Burgers vector
C_l	local composition of liquid
C_o	nominal alloy composition
C_S	local composition of solid
D_T	thermal diffusivity
G	thermal gradient
g	gravitational acceleration
h	height of mushy zone
K	average permeability
k	distribution coefficient
R	solidification or withdrawal rate; stress ratio
Ra_a	modified Rayleigh number for freckle formation
S	Schmid factor
T_M	homologous melting temperature
w	width of channel
α	Fourier number
ΔK_{th}	threshold stress intensity
$\Delta\rho/\rho_o$	density gradient in liquid
δ	lattice misfit
μ	shear modulus
ν	kinematic viscosity
σ_{OR}	uniaxial stress required to glide dislocation through narrow channel

*Associate Professor, Department of Mechanical, Materials, and Aerospace Engineering.
†Professor and Chair, Materials Department.

ACRONYMS

ASTM	American Society of Testing and Materials
BCT	body-centered tetragonal
EBCHR	electron-beam cold hearth remelting
ESR	electroslag remelting
fcc	face-centered cubic
HIP	hot isostatic pressing
LMC	liquid metal cooling
PDAS	primary dendrite arm spacing
PM	powder metallurgy
TCP	topological closed packed
TMF	thermomechanical fatigue
VAR	vacuum arc remelting
VIM	vacuum induction melting

The continuing drive for improvements in the performance, operating temperatures, and emissions in advanced aircraft engines has driven the development of nickel-based superalloys for decades. Nickel-based superalloys are an unusual class of metallic materials with an exceptional combination of toughness, strength at high temperatures, and resistance to degradation in corrosive or oxidizing environments. Their performance in materials for severe combustion and chemical environments has led to their widespread use in aircraft and power generation turbines, rocket engines, nuclear powerplant components, marine engines, and a variety of chemical processing operations. Intensive process development activities over the past few decades have resulted in alloys that can tolerate average temperatures of 1050°C with occasional excursions (as in local hot spots near airfoil tips) to temperatures as high as 1200°C, which is approximately 90% of the melting point of the material [1]. The fundamental aspects of the composition and structure that result in these exceptional properties are briefly reviewed here. Major classes of superalloys that are utilized in gas turbine engines and the complex thermomechanical processes for their production are outlined, along with characteristic mechanical and physical properties.

I. SUPERALLOYS IN GAS TURBINE ENGINES

Nickel-based superalloys are a critical class of engineering materials for aircraft engines, typically comprising 40–50% of the total weight of an aircraft engine [1]. Components fabricated from this class of materials (Fig. 1a) include disks, vanes, blades, combustor liners, bolts, seals, shafts, and structural housings. Broadly speaking, superalloys can be divided into two classes based on the

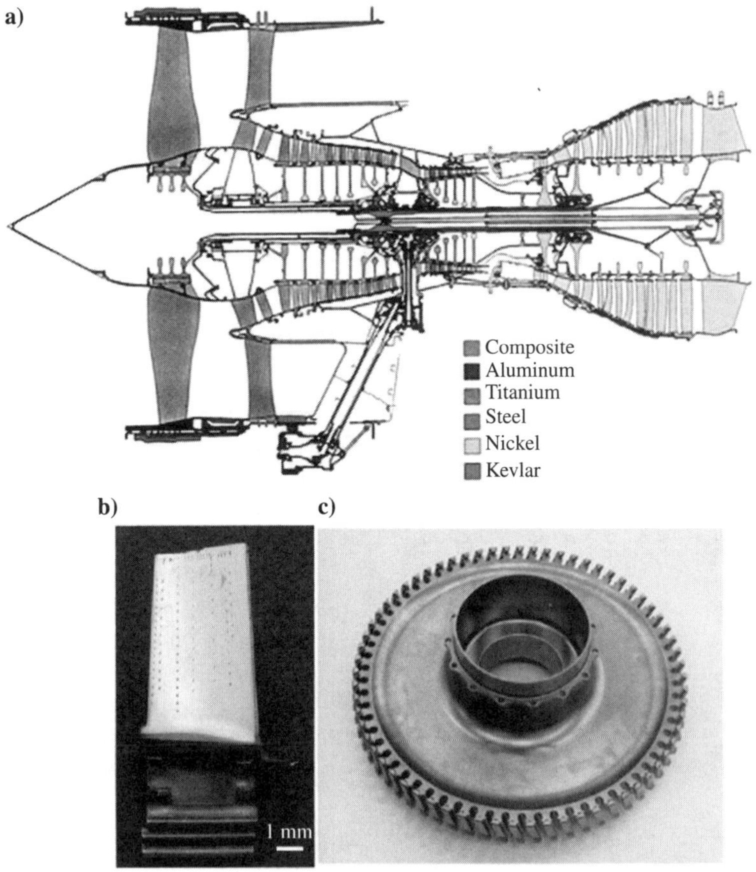

Fig. 1 Schematic of a) a turbine engine (Rolls Royce Trent 800 [3]) with major classes of materials, b) superalloy turbine blade with cooling holes and ceramic thermal barrier coating, c) and turbine disk.

processing approach required: 1) wrought or 2) cast. Table 1 lists the nominal composition of several common cast and wrought commercial superalloys utilized in gas turbine engines.

As will be discussed in subsequent sections, individual alloys have been developed in view of the constraints and unique features of the various cast or wrought processing approaches. Because of the high thermal and mechanical loads imposed on components made of these materials and the associated performance and safety issues, continued development of superalloys for turbine disks and turbine airfoils has been a priority, resulting in successive generations of alloys

TABLE 1 COMPOSITION OF COMMERCIAL Ni-BASED SUPERALLOYS (WT. %, BAL. Ni)

Alloy	Cr	Co	Mo	W	Ta	Re	Nb	Al	Ti	Hf	C	B	Y	Zr	Other
Conventionally Cast Alloys															
Mar-M246	8.3	10.0	0.7	10.0	3.0	—	—	5.5	1.0	1.50	0.14	0.02	—	0.05	—
René 80	14.0	9.5	4.0	4.0	—	—	—	3.0	5.0	—	0.17	0.02	—	0.03	—
IN-713LC	12.0	—	4.5	—	—	—	2.0	5.9	0.6	—	0.05	0.01	—	0.10	—
C1023	15.5	10.0	8.5	—	—	—	—	4.2	3.6	—	0.16	0.01	—	—	—
Directionally Solidified Alloys															
IN792	12.6	9.0	1.9	4.3	4.3	—	—	3.4	4.0	1.00	0.09	0.02	—	0.06	—
GTD111	14.0	9.5	1.5	3.8	2.8	—	—	3.0	4.9	—	0.10	0.01	—	—	—
1st-Generation Single-Crystal Alloys															
PWA 1480	10.0	5.0	—	4.0	12.0	—	—	5.0	1.5	—	—	—	—	—	—
René N4	9.8	7.5	1.5	6.0	4.8	—	0.5	4.2	3.5	0.15	0.05	0.00	—	—	—
CMSX-3	8.0	5.0	0.6	8.0	6.0	—	—	5.6	1.0	0.10	—	—	—	—	—
2nd-Generation Single-Crystal Alloys															
PWA 1484	5.0	10.0	2.0	6.0	9.0	3.0	—	5.6	—	0.10	—	—	—	—	—
René N5	7.0	7.5	1.5	5.0	6.5	3.0	—	6.2	—	0.15	0.05	0.00	0.01	—	—
CMSX-4	6.5	9.0	0.6	6.0	6.5	3.0	—	5.6	1.0	0.10	—	—	—	—	—
3rd-Generation Single-Crystal Alloys															
René N6	4.2	12.5	1.4	6.0	7.2	5.4	—	5.8	—	0.15	0.05	0.00	0.01	—	—
CMSX-10	2.0	3.0	0.4	5.0	8.0	6.0	0.1	5.7	0.2	0.03	—	—	—	—	—

(Continued)

TABLE 1 COMPOSITION OF COMMERCIAL Ni-BASED SUPERALLOYS (WT. %, BAL. Ni) (*CONTINUED*)

Alloy	Cr	Co	Mo	W	Ta	Re	Nb	Al	Ti	Hf	C	B	Y	Zr	Other
Wrought Superalloys															
IN 718	19.0	—	3.0	—	—	—	5.1	0.5	0.9	—	—	0.02	—	—	18.5Fe
René 41	19.0	11.0	10.0	—	—	—	—	1.5	3.1	—	0.09	0.005	—	—	—
Nimonic 80A	19.5	—	—	—	—	—	—	1.4	2.4	—	0.06	0.003	—	0.06	—
Waspaloy	19.5	13.5	4.3	—	—	—	—	1.3	3.0	—	0.08	0.006	—	—	—
Udimet 720	17.9	14.7	3.0	1.3	—	—	—	2.5	5.0	—	0.03	0.03	—	0.03	—
Powder-Processed Superalloys															
René 95	13.0	8.0	3.5	3.5	—	—	3.5	3.5	2.5	—	0.065	0.013	—	0.05	—
René 88 DT	16.0	13.0	4.0	4.0	—	—	0.7	2.1	3.7	—	0.03	0.015	—	—	—
N18	11.2	15.6	6.5	—	—	—	—	4.4	4.4	0.5	0.02	0.015	—	0.03	—
IN100	12.4	18.4	3.2	—	—	—	—	4.9	4.3	—	0.07	0.02		0.07	

with increasing capabilities (Table 1). Accordingly, turbine inlet temperatures in successive generations of aircraft engines have scaled upward with these improvements in turbine material temperature capability [2, 3]. The fundamental aspects of alloy design that have permitted these improvements are discussed in Sec. II. While superalloys are currently operating near their limits in terms of operating temperatures, alternate materials systems that could replace superalloys in jet engines remain elusive, driving continued alloy, process, and coating innovations.

Aircraft engines typically contain several hundred cast superalloy turbine blades. In virtually all advanced aircraft engines, hot section blades are geometrically intricate to accommodate elaborate airfoil cooling schemes (Fig. 1b). These components are investment cast, with internal ceramic cores that define internal cooling features. The final geometry of the blade is achieved after a series of post-casting operations, including core leaching, grinding, laser drilling and often coating, along with a series of inspection operations. Depending on the casting approach employed, turbine airfoils can be composed of equiaxed grains or columnar grains or can be cast as single crystals, completely eliminating all high-angle grain boundaries. Because grain boundaries are sites for damage accumulation at high temperatures, the blades in the early stages of the turbine are typically single crystals while the blades in the later (cooler) stages of the turbine are fabricated from equiaxed alloys. Structural components such as engine cases are also fabricated by investment casting processes with equiaxed alloys. The investment casting process and the cast microstructure of superalloys is discussed in detail in Sec. III.

Turbine disks serve to transmit the energy harnessed from the airfoils to drive the compressor and generate useful work energy that is converted to thrust or electrical power. These critical rotating structures (Fig. 1c) are fabricated via wrought processing approaches that begin with either cast ingots or consolidated superalloy powder preforms. A complex series of forging, heat treatment, and machining operations are typically employed to ensure high levels of strength and damage tolerance. Close control of the structure of the material at each stage of processing is critical; defects that must be avoided in the final product are discussed in Sec. III. As a result, exceptional combinations of strength, toughness, and fatigue and crack growth resistance can be achieved in equiaxed disk materials, as highlighted in Sec. IV.

II. CONSTITUTION OF SUPERALLOYS

As apparent from Table 1, face-centered cubic (fcc) γ-nickel is the major elemental constituent of superalloys. Pure nickel has a melting point of 1455°C and a density of 8.9 g/cm^3. In terms of cost, pure nickel is more expensive than aluminum and stainless steel, but less costly than titanium and cobalt. Beyond nickel, superalloys contain up to 40 wt% of a combination of 5 to 10 other elements, with chromium present in nearly all superalloys. The elements typically alloyed with nickel to form a superalloy are highlighted in Fig. 2a.

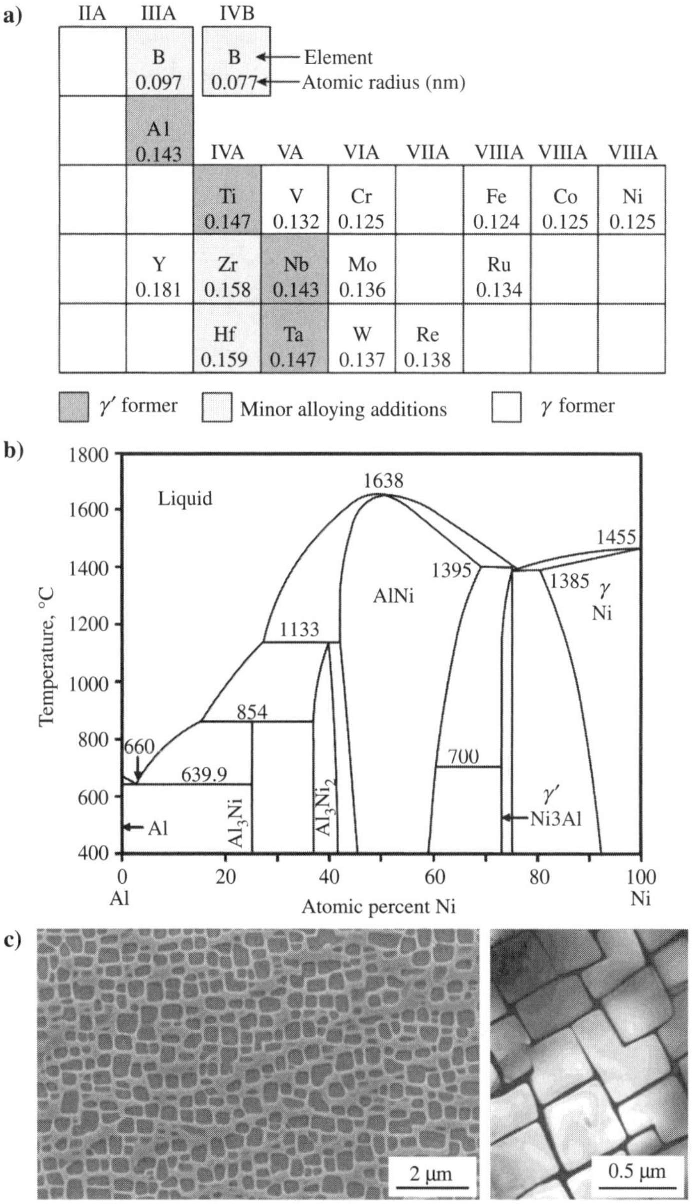

Fig. 2 a) Alloying elements present in Ni-base superalloys (adapted from [2]). b) Binary Ni-Al phase diagram with the γ-Ni and γ'–Ni_3Al phases at the nickel-rich end of the phase diagram. c) Microstructures of a Ni-based superalloy single crystal revealing a high volume fraction of γ' precipitates, which have a cuboidal morphology.

The nickel-aluminum system is the binary alloy system that forms the basis for higher-order superalloy compositions (Fig. 2b). As the level of aluminum added to fcc γ-nickel increases, the solubility of Ni for Al is eventually exceeded, and a second phase forms by a precipitation process. This second phase has a nominal composition of Ni_3Al, is designated the γ' phase, and has an ordered intermetallic $L1_2$ crystal structure. Formation of the γ' phase occurs in the solid state as the supersaturated solid solution of γ-nickel is cooled below its equilibrium solvus temperature. Solvus temperatures are highly variable from alloy to alloy and are generally lower in turbine disk alloys, varying from 900°C up to the melting temperature. The precipitation and growth kinetics of the γ' phase are highly sensitive to the rate at which the alloy is cooled through the solvus temperature. A unimodal distribution of fine γ' precipitates (300–500 nm) is typically associated with cooling rates in excess of $\sim$40 K/min whereas slower cooling rates tend to promote the formation of multiple populations of γ' precipitates consisting primarily of large (>500 nm) and small (<50 nm) precipitates. The presence of a high volume fraction of the γ' phase is key to strengthening as will be discussed in more detail in Sec. IV. These two phases remain the major constituents of the superalloy microstructure, even with the addition of 6–10 additional elements. Figure 2c shows a typical two-phase $\gamma - \gamma'$ microstructure for a single crystal turbine blade alloy. In general, refractory alloying elements such as Mo, W, Nb, and Re, which have large differences in electronic structure and atomic radii compared to Ni, are added for solid-solution strengthening of the γ phase. Additions of Ti, Ta, and Nb contribute to the formation and strengthening of the Ni_3(Al, Ti, Ta, Nb) γ' phase [2, 3]. Additions of Cr, Y, and La typically improve oxidation and/or corrosion behavior, which is optimized via formation of an adherent slow-growing alumina scale.

Additions of iron form the basis of an important group of Ni-Fe superalloys with the most common commercially available alloy being IN 718 (Table 1) [4]. This unique subclass of superalloys typically relies on additions of Nb for high temperature strength and contains significant levels of iron, which are intentionally added to reduce the overall levels of nickel and cobalt. Because this results in a significant reduction in the cost of the alloy, many of the Ni-Fe superalloys are used in structures toward the rear of the turbine where extreme high temperature strength is not required. In terms of volume, a large majority of the commercial superalloys market is composed of these iron-containing Ni-base superalloys.

The microstructures of Ni-Fe-Nb alloys are highly complex, and multiple intermetallic phases can exist within the microstructure of the alloy. The combination of Fe and Nb present within the fcc Ni-matrix leads to the precipitation of both γ' and Ni_3Nb. Interestingly, these Ni_3Nb precipitates can exist as two distinct phases. At relatively high temperatures, precipitation occurs in the form of orthorhombic δ phase precipitates at the grain boundaries. Lower temperatures promote the formation of meta-stable, coherent disk-shaped precipitates γ'' with a body-centered-tetragonal (BCT) $D0_{22}$ crystal structure within the γ matrix.

Despite the low overall volume fraction, the low γ'' coarsening rates at service temperatures, combined with large interfacial misfit strains, provide an unusually high degree of strengthening in Ni-Fe superalloys. Because of the phase instabilities associated with the high levels of Fe present within these alloys, useful service temperatures are limited to below 650°C. Above this temperature, the structural properties rapidly degrade as TCP phases form and microstructural changes occur.

Several elements are added in small quantities for control of grain structure and mechanical properties that are strongly influenced by grain boundaries. Minor additions of B, C, Hf, and Zr tend to result in the formation of borides or carbides, often located at the grain boundaries. These elements are important for control of grain size during wrought processing and for minimization of damage accumulation at grain boundaries in service [5]. Carbon exhibits a high affinity for elements such as Hf, Zr, Ta, Ti, Nb, W, Mo, V, and Cr, and tends to form primary MC (where M = metal atom) carbides directly from the liquid during solidification of Ni-based superalloys. Some general characteristics of selected MC carbide phases are listed in Table 2. Depending on the composition of the primary MC carbide and the constituent elements present in the alloy, subsequent solid-state transformation can decompose the MC carbide into a variety of $M_{23}C_6$, M_6C and M_7C_3 carbides during heat treatment or in service [6]. For example, Cr from the matrix might react with the less stable TiC and NbC carbides to form a series of $M_{23}C_6$, M_6C and M_7C_3 carbides. As the primary MC carbides are consumed, the depletion of Cr from the matrix surrounding the carbides results in the formation of a layer of γ' at the carbide interface. Because these carbides occupy a significantly larger volume along the grain boundaries, and are often interconnected, particularly if they form with a script morphology (Fig. 3),

TABLE 2 CHARACTERISTIC PROPERTIES OF SELECTED REFRACTORY CARBIDES

Carbide	Crystal Structure	ΔH at 298 K, eV/atom	Density, g/cm^3	Lattice Parameter, Å
TiC	B1 (NaCl)	1.91	4.91	4.328
ZrC	B1 (NaCl)	2.04	6.59	4.698
HfC	B1 (NaCl)	2.17	12.67	4.640
VC	B1 (NaCl)	1.06	5.05	4.166
NbC	B1 (NaCl)	1.46	7.79	4.470
TaC	B1 (NaCl)	1.48	14.50	4.456
CrC	B1 (NaCl)	− 0.01	–	–
MoC	Hexagonal	0.13	9.06	a: 2.932 c: 10.97
WC	Hexagonal	0.42	15.80	a: 2.906 c: 2.837

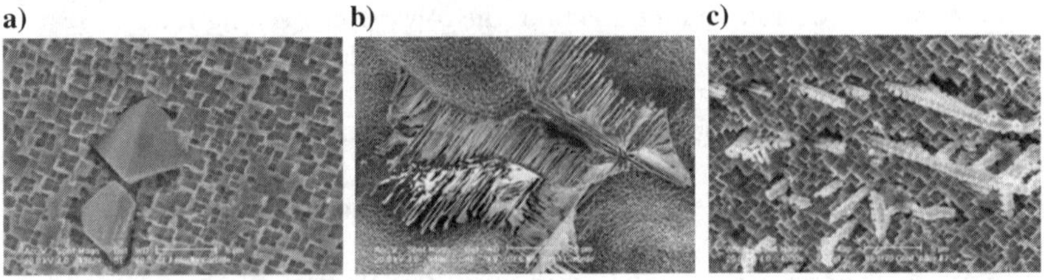

Fig. 3 SEM micrographs of a) blocky, b) script, and c) nodular MC carbides contained within the interdendritic regions of carbon-containing Ni-based superalloys.

potential degradation of the mechanical properties can occur when the MC carbides undergo this phase transformation. Although carbides may serve as crack initiation sites during fatigue (in the absence of other more major types of defects) [7], the presence of discrete carbides at the grain boundaries inhibits sliding and damage accumulation during high-temperature creep [5].

Other minor elements also play an important role in superalloy metallurgy. Impurities present in elemental additions or revert material in trace amounts are often found in the final product of melting of Ni-based superalloys. The influence of alloy cleanliness on structural properties has been the subject of many detailed investigations [8, 9]. Advances in vacuum melting technology have minimized the levels of undesirable low-melting-point elements [10] such as Pb, Bi, Se, Ag, As, Sb, Cu, Te, and S. Sulfur, which is often the most difficult to control, is a deleterious impurity that adversely affects mechanical properties and oxidation resistance. The content of S in the alloy can be minimized by melting in MgO-Al_2O_3-lined crucibles to form MgS, which can then be removed from the melt.

Thermodynamic driving forces typically result in a preferential partitioning of individual alloying elements to either the γ or γ' phases. Table 3 shows the composition of the constituent phases in the turbine blade alloy René N5 [11] and the turbine disk alloy IN100 [12]. From these data, it is apparent that Re, Mo, Cr, and Co preferentially partition to the matrix gamma phase while Ti, Ta, and Al partition to the precipitate phase. Certain elements, such as W and Ru, are soluble within both the γ and γ' phases and are typically much more evenly distributed between the two phases [13]. Element partitioning is an important alloy design consideration as the composition of the constituent phases will impact both the mechanical and the environmental characteristics of the alloy. Because the precipitate and matrix phases are crystallographically coherent, the compositions of the phases influence their lattice parameters and precipitate-matrix misfit δ:

$$\delta = \frac{a_{\gamma'} - a_{\gamma}}{\frac{1}{2}(a_{\gamma'} + a_{\gamma})} \tag{1}$$

TABLE 3 COMPOSITION OF THE CONSTITUTENT PHASES (WT%)

Alloy/Phase	Al	Co	Cr	Mo	W	Ta	Ti	Re	V	Ni
René N5/γ Phase	2.3	12.1	15.0	2.7	5.3	0.4	—	7.6	—	Bal
René N5/γ' Phase	7.5	4.5	2.4	0.8	4.0	3.1	—	0.5	—	Bal
IN100/γ Phase	4.5	20.6	14.8	3.2	—	—	2.7	—	0.7	Bal
IN100/γ' Phase	6.8	11.8	4.1	1.6	—	—	7.6	—	0.8	Bal

where $a_{\gamma'}$ and a_γ are the lattice parameters of the γ' and γ phases, respectively. The misfit will result in initial internal stresses, which, in concert with the high degree of elastic anisotropy, strongly influence precipitate shapes and resultant mechanical properties [14–23]. Precipitates are typically spherical in alloys with near-zero misfit and become cuboidal in shape as the magnitude of the misfit approaches $|\delta| \approx 0.3$ [20].

A major concern in the design of new alloys and in the definition of specification limits for the acceptable range of individual alloying elements is the avoidance of a class of phases known as topologically close-packed phases (TCPs). These phases are typically rich in refractory alloying elements and possess complex crystal structures characterized by close-packed layers of atoms (atomic coordination number >12). Examples of phases typically considered as TCPs include the orthorhombic P phase, the tetragonal σ phase, the rhombohedral R, and rhombohedral μ phases [24–28]. The TCPs often form "basket weave" sheets that are aligned with the octahedral planes in the FCC nickel matrix (Fig. 4). Similarities in the composition and crystallography of the various TCPs allow these precipitates to develop as mixed structures consisting of a number of different phases [28]. The TCPs are detrimental because they deplete strengthening elements from the microstructure and/or serve as crack initiation sites during cyclic loading [24–26]. Precipitation kinetics for these phases are often very sluggish, resulting in precipitation only after extended times in service. New alloy design tools based on the Calphad method [29–32] are increasingly utilized in the design of new alloys and to establish or modify specification ranges for existing alloys to avoid such deleterious phases. The ability to predict phase compositions and their ranges of stability is dependent on the development of thermodynamic models for these complex intermetallic phases and on the availability of databases to validate the modeling.

III. PROCESSING OF SUPERALLOYS

Superalloy processing begins with the fabrication of large ingots that are subsequently utilized for one of three major processing routes: 1) remelting and subsequent investment casting; 2) remelting followed by wrought processing; or

3) remelting to form superalloy powder, which is subsequently consolidated and subjected to wrought processing operations. Ingots are fabricated by vacuum induction melting (VIM) in a refractory crucible to consolidate elemental and/or revert materials to form a base alloy. Although selected alloys can be melted in air/slag environments using electric arc furnaces, VIM melting of superalloys is much more effective in the removal of low melting point trace contaminants. Following the vaporization of the contaminants, the carbon boil reaction is utilized to deoxidize the melt prior to the addition of the reactive γ' forming elements such as Ti, Al, and Hf. Once the desired alloy composition of the VIM ingot is attained, the solidified ingot is subjected to additional melting or consolidation processes that are selected based on the final application of the material. Charge weights of VIM ingots may range from $\sim$2,500 kg to in excess of 27,500 kg [33].

Considering the stringent requirements for minimizing defects in turbine engine components, a detailed understanding of structure evolution in each of the just-mentioned processing paths is essential. In the following sections, we briefly review the processing approaches and aspects of superalloy structure that influence properties. Mechanical properties are discussed in more detail in Sec. IV.

A. CAST SUPERALLOYS

Investment casting is the primary casting process for fabrication of superalloy components with complex shapes, including blades and vanes. Ceramic molds containing alumina, silica, and/or zirconia are utilized in this process (Fig. 5). The molds are fabricated by progressive buildup of ceramic layers around a

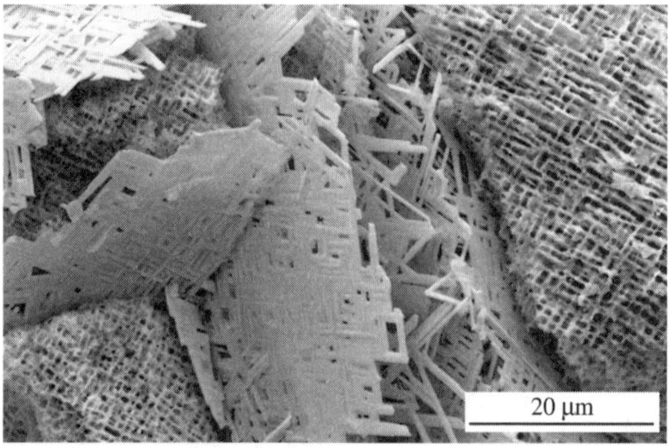

Fig. 4 Ni-based superalloys containing elevated levels of refractory elements are prone to the precipitation of various TCP phases when exposed to elevated temperatures. Refractory-rich TCPs (bright contrast) appear as interwoven sheets of TCPs in a partially extracted sample.

wax pattern of the cast component. Ceramic cores can be embedded in the wax to produce complex internal cooling structures. Prior to casting, a thermal cycle is applied to remove the wax and sinter the ceramic investment mold. Following this, the mold is filled with remelted superalloy in a preheated vacuum chamber to obtain a shaped casting. The single-use mold is removed once the alloy has cooled to room temperature.

Castings can be equiaxed, columnar-grained, or single crystal. Equiaxed castings solidify fairly uniformly throughout their volume while columnar and single crystal castings are withdrawn from a hot zone in the furnace to a cold zone at a controlled rate. Following initial solidification, castings are subjected to a series of subsequent heat treatment cycles that serve to reduce segregation [34], establish one or more size populations of γ' precipitates [35], modify the structure of grain boundary phases (particularly carbides) [36], and/or assist in the application of environmental coatings [37].

In all casting processes, the final structure (and therefore properties) of the material is sensitive to the thermal conditions present during solidification of the casting. Solidification is dendritic in character, and the primary and secondary dendrite arm spacings are dependent on cooling rate, $G*R$ (Fig. 6). Associated with the dendritic solidification is segregation of the constituent alloying elements. The extent of segregation is quantified by the distribution coefficient k, where

$$k = \frac{C_s}{C_l} \tag{2}$$

with C_s the local composition of the solid and C_l the local composition of the liquid. Considering the requirement for mass balance plus some degree of back-diffusion in the solid during solidification, the variation in solid composition as a function of fraction solid f_s from the beginning of solidification (at the dendrite core) to the end of solidification (in the interdendritic region) can be described by the modified Scheil equation [38]:

$$C_s = kC_o[1 - (1 - 2\alpha k) f_s]^{\frac{k-1}{1-2\alpha k}} \tag{3}$$

where C_o is the nominal alloy composition and α is the Fourier number.

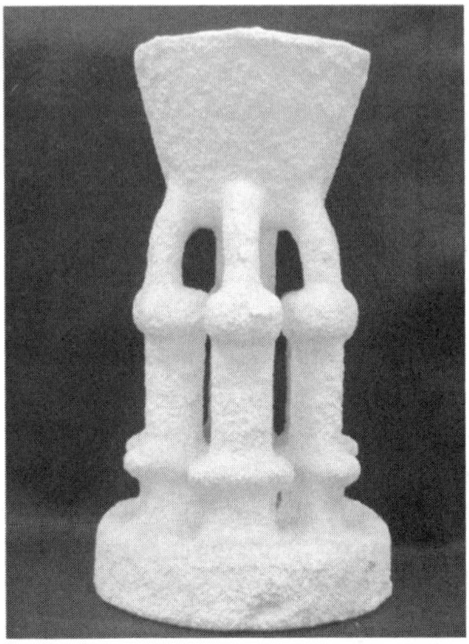

Fig. 5 Ceramic investment casting mold with single crystal starter at the bottom of the plate and single crystal plate following directional solidification and removal of ceramic mold (courtesy of A.J. Elliott).

Fig. 6 Variation in dendrite morphology and primary dendrite arm spacing (PDAS) with cooling rate (G * R) during solidification.

In recent years, segregation in multicomponent superalloys has received significant attention because of adverse effects of segregating elements on grain defect formation during solidification of advanced alloys [38–41]. An example of the variation of tungsten content across the dendritic structure of two different single crystal alloys [42] is shown in Fig. 7. Note that the modified Scheil equation provides a good estimate of the segregation tendencies of W ($k = 1.54$, $\alpha = 0.01$ for Alloy SX-2), except at the lowest and highest solid fractions, where the sampling error of the local chemistry is more significant. Table 4 shows the range of values of distribution coefficients of individual elements measured in a large set of alloys characteristic of generation 2 and 3 single crystal alloys [39]. Note that values of the distribution coefficient closer to 1.0 are desirable for minimizing segregation and that Ta, W, and Re tend to most strongly segregate in the as-cast microstructure. Again, prior to use in service, the as-cast components are often subjected to complex heat treatments designed to reduce or eliminate

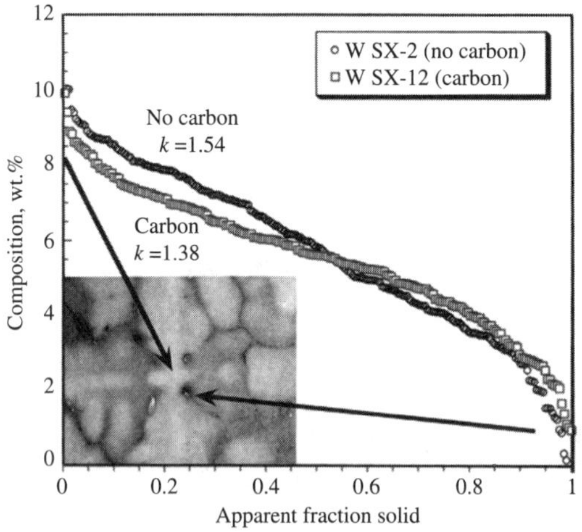

Fig. 7 Dendritic microsegregation leads to the formation of significant compositional gradients within the as-cast structure of alloy SX-3 (with overall composition of Ni-5.7Al-4.0Cr-11.5Co-5.0Re-6.0Ta-5.0 W wt.%). Elevated levels of Ta correspond to the interdendritic regions of the microstructure. The dendrite core contains elevated levels of Re.

these solidification-induced compositional gradients and to establish a controlled size and distribution of precipitates.

B. DIRECTIONALLY SOLIDIFIED ALLOYS

Although cast Ni-based superalloys have inherently good high-temperature properties, these properties can be improved through processing. The creep rupture resistance of Ni-based superalloys can be enhanced by orienting the grain boundaries parallel to the applied stress direction or by removing the grain boundaries entirely (Fig. 8). Nickel-based alloys are directionally solidified into columnar-grained or single crystal forms by withdrawing an investment mold downward through a radiation baffle from a high temperature furnace at moderate thermal gradients (G) and withdrawal rates (R), in the range of 10 to 100°C/cm and 5 to 40 cm/h, respectively. A unidirectional thermal gradient produces preferential, oriented grain growth. Because Ni-based superalloys exhibit cubic symmetry, these processing conditions encourage solidification along the orthogonal <001> crystallographic orientations. The <001> dendrites aligned most

TABLE 4 RANGES OF DISTRIBUTION COEFFICIENTS FOR GENERATION 2 AND 3 ALLOYS AND CORRESPONDING DENSITIES OF PURE ELEMENTS

	Al	Cr	Co	Ta	W	Re	Mo
Distribution coefficient, *k*	0.81–0.95	1.05–1.17	1.03–1.13	0.67–0.80	1.28–1.58	1.23–1.60	1.13–1.46
Density @ 20°C*	2.7	7.2	8.8	16.7	19.3	21.0	10.2

*Density of pure Ni = 8.9 gm/cm³.

Fig. 8 Grain structures of single crystal, directionally solidified, and conventionally cast turbine blades.

favorably with the thermal gradient tend to grow rapidly while the superheated bulk liquid prevents the formation of equi-axed grains ahead of the solidifying interface. Consequently, directionally solidified components typically consist of a number of columnar <001> grains aligned parallel to the solidification direction (Fig. 8). In addition to favorably affecting the creep rupture properties, selection of the <001> grain orientation during solidification results in an elastic anisotropy that enhances resistance to thermal fatigue. Because the <001> orientation is considered to be elastically "soft" (it exhibits a significantly lower modulus as compared to the other primary directions), the cyclic thermal stresses associated with expansion and contraction of the component are minimized. As a result, orientation control of directionally solidified castings, particularly single crystal components, is extremely important [43–44]. Castings exhibiting misorientation angles measuring greater than 10 deg between the primary stress axis and the <001> crystal orientation are typically rejected.

With directional solidification processing, several types of chemistry-sensitive grain defects can develop [33]. The two most common grain defects that cause rejection of directionally solidified production components are freckle chains and misoriented grains (Fig. 9). Freckle-type defects, first studied in superalloys by Giamei and Kear [45] and Copley et al. [46], arise due to convective instabilities in the mushy zone that develop because of density inversions created by progressive segregation of individual alloying elements during solidification. The fluid flow within "channels" that develop as a result of these instabilities results in fragmentation of dendrite arms, producing a small chain of equiaxed grains aligned

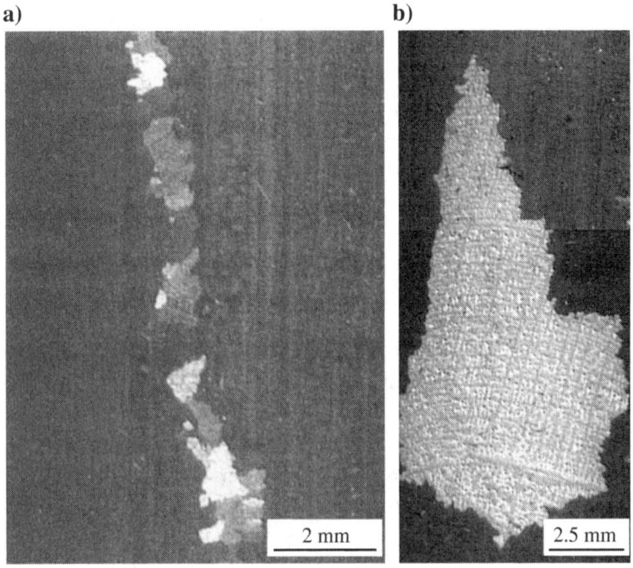

Fig. 9 Macroscopic chemistry-sensitive grain defects present on the surface of single crystal Ni-based superalloy castings, including a) freckles and b) a misoriented grain.

approximately parallel to the solidification direction [47, 48]. Freckles are enriched in elements that segregate to the interdendritic region during solidification, and thus differ in composition from the base alloy [42]. Freckle formation is promoted by low cooling rates (low thermal gradients) and corresponding large dendrite arm spacings [33, 42] for a fixed alloy composition. Misoriented grains differ from freckles in that they have the same nominal composition as the base alloy, but are typically larger and elongated along the solidification direction. Misoriented grains possessing high angle grain boundaries with respect to the parent crystal form under the same alloy and process conditions as freckles, which suggests that thermosolutal convection and fragmentation also contribute to their formation [42]. The high angle boundaries associated with these defects serve as crack initiation sites, degrade mechanical properties, and must be avoided.

Because freckles and misoriented grains develop as a result of thermosolutal convection, it has been suggested that they will form when a critical Rayleigh number is exceeded [40, 42, 49, 50]. The Rayleigh number Ra_s is a measure of the ratio of the buoyancy force to the retarding frictional force in the mushy zone:

$$Ra_s = \frac{(\Delta\rho/\rho_o)gKh}{\alpha\nu} \tag{4}$$

This mean value of the Rayleigh number over the height h of the mushy zone [40] is dependent on the density gradient in the liquid of $\Delta\rho/\rho_o$, average permeability

K, gravitational acceleration g, thermal diffusivity D_T, and kinematic viscosity ν. In this form of the expression, the magnitude of the Rayleigh number is proportional to the height of the mushy zone, which, in turn, varies with the square of the dendrite arm spacing. From this criterion it is apparent that defect occurrence can be avoided by reducing h, which is accomplished by increasing the thermal gradient in the process, or by reducing the density gradient in the mushy zone. During solidification, Re and W are progressively depleted in the mushy zone, increasing the Rayleigh number, while Ta is progressively enriched (Table 5). These elements all have a strong influence on liquid density, and so achieving a balance of Re and W vs Ta or reducing the overall levels of Re and W will reduce the driving force for convective instabilities [42]. Unfortunately, these elements are also the most important for strengthening, and so there has recently been greater effort aimed at increasing thermal gradients during solidification.

Higher thermal gradients during solidification can permit the use of more complex alloys in physically larger blades. Higher thermal gradients during solidification also permit implementation of more advanced cooling schemes; a number of "high gradient" processes have been under development in recent years. One approach involves the use of liquid metal coolants (LMC) during solidification [51]. Figure 10a shows a schematic of a Bridgman system modified to use liquid tin as a cooling medium. The LMC process using aluminum as the cooling medium has been utilized in the former Soviet Union for the regular production of aero-engine blades [51] and is a proven process for smaller aircraft-engine castings. Recent investigations of the process show promising results for single crystal/columnar-grained castings with substantially larger cross sections, of the type needed for large aircraft engines or industrial gas turbines [52, 53]. Substantial increases in cooling rate and elimination of freckle defects have recently been demonstrated with a liquid tin LMC process involving directionally solidified castings with cross-sectional areas measuring up to 5×9.5 cm (Fig. 10b) [54]. Another new approach involves the use of high-velocity inert gas that is injected from a baffle located below the hot zone of the furnace and directed at the mold surface at the location of the solidification front [55]. A third approach involves the use of a fluidized bed located below the furnace and isolated with a rigid baffle [56].

TABLE 5 COMPOSITIONS OF FRECKLES AND WHITE SPOT DEFECTS IN IN718

	Al	Ti	Cr	Fe	Ni	Nb	Mo	Si
Nominal IN718	0.5	0.9	19.0	18.5	Bal	5.1	3.0	0.2
Freckle	0.43	1.33	17.4	15.2	Bal	9.43	3.51	0.16
White Spot	0.41	0.62	17.7	19.2	Bal	2.96	3.2	0.19

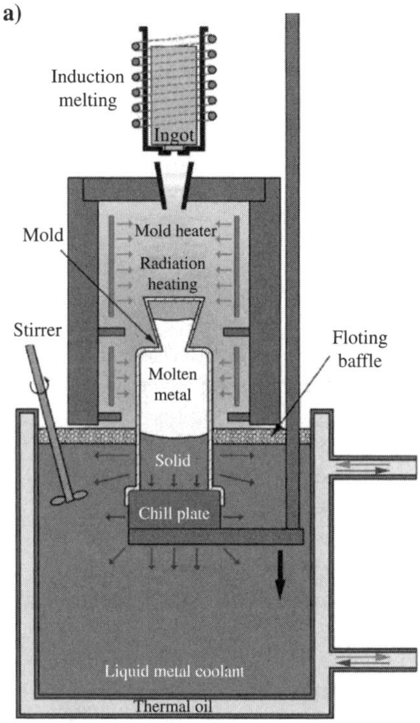

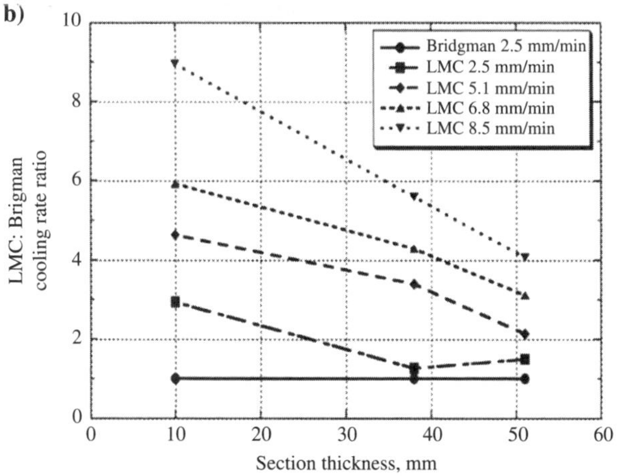

Fig. 10 a) Schematic of the Bridgman process modified to utilize liquid metal cooling (LMC) during unidirectional solidification of Ni-based superalloy components. b) Note the substantial cooling rate benefit of the LMC process, compared to the conventional Bridgman process.

There are additional defects of concern in cast alloys that are sensitive to the details of the casting geometry and casting procedures and less sensitive to alloy chemistry. These defects include dimensional noncompliance because of core shift, porosity, hot tearing, slivers, and low- and high-angle boundaries (in single crystals) [2]. These defects can limit mechanical properties and are carefully controlled in specifications and monitored by nondestructive inspection approaches.

C. WROUGHT ALLOYS

As mentioned earlier, wrought alloys are typically fabricated by remelting of VIM ingots to form a secondary ingot or powder for subsequent deformation processing. A secondary melting process is necessary for wrought alloys because the high-temperature structural properties of Ni-based superalloys are highly sensitive to microstructural variations, chemical inhomogeneities, and inclusions. As ingot sizes increase, VIM melting often results in macrosegregation or the formation of large shrinkage cavities during solidification. The formation of these solidification defects is caused by large-scale solute segregation associated with dendritic solidification under low thermal gradients. Because heat transfer during solidification of VIM ingots is limited by the low intrinsic thermal conductivity of the solidifying mass, large ingots are highly prone to the formation of these features. Thus other secondary melting processes are utilized, including vacuum arc remelting (VAR), electroslag remelting (ESR), and electron-beam cold hearth refining (EBCHR) [2]. Here, only the more common ESR and VAR processes are discussed in the context of avoiding property-reducing defects.

1. VACUUM ARC REMELTING

For the production of critical rotating components, such as turbine disks, vacuum arc remelting (VAR) is used to refine the ingot and eliminate macrosegregation. Consumable electrodes (30 to 50 cm in diameter) cast from the VIM charge are remelted into a water-cooled copper crucible. Unlike the VIM process, in which the entire charge of the alloy is molten and allowed to solidify, VAR only involves localized melting of the electrode tip (Fig. 11). Melt rates of VAR are on the order of $\sim$0.5 to 1 kg/s. Defect features, such as macrosegregation and shrinkage, are effectively minimized as high thermal gradients are maintained during solidification of the comparatively smaller melt pool. Processing parameters are selected such that the melt pool exhibits a steady-state size and shape [57].

Although VAR can effectively eliminate the undesirable features of the VIM process, this remelt process can introduce inclusions into the finished ingot. Inclusions in the VAR process can be classified into two groups: extrinsic and intrinsic. Extrinsic inclusions can come from a variety of sources. Incomplete removal of refractory ceramic particles originating from the refractory liner and agglomerates of oxides and nitrides in the revert material used during VIM

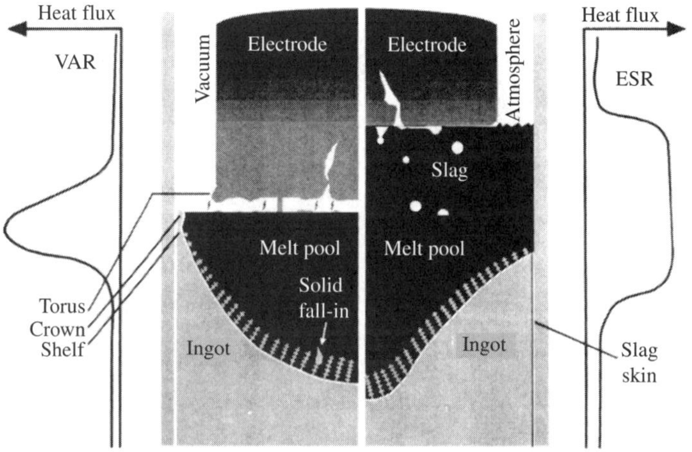

Fig. 11 Schematic comparing the melt pool and heat flux during vacuum arc remelting (VAR) and electroslag remelting (ESR) process, using a VIM ingot electrode as the input.

melting can enable these inclusions to be present in the remelted ingot. As the surface of the VIM ingot is machined to form a consumable electrode, fragments of tungsten carbide cutting tools can be embedded within the ingot. Steel shot used to clean the copper crucible and splash from the previous melt in the VIM crucible might also potentially serve as extrinsic inclusions. With clean melting practices and stringent quality control measures, many of these extrinsic inclusions can be minimized. Intrinsic inclusions, however, are much more difficult to control during processing and are often dependent upon the chemistry of the alloy.

Thermal and compositional perturbations in the mushy zone during solidification lead to the formation of intrinsic microstructural defects, such as tree rings, freckles, and white spots (Fig. 12). In VAR ingots, freckle defects consist of chains of equiaxed grains aligned parallel to the melt pool profile or solidification direction [58–59]. These microstructural features are highly enriched with solute and are compositionally different from the bulk alloy. Similar to the freckle defects found in single crystal and columnar grain investment castings, freckles in VAR products also form as a result of thermosolutal convection [49]. In many multicomponent superalloys, as solute accumulates within the mushy zone during dendritic solidification, the subsequent density imbalance between the solute and bulk liquid serves as a driving force for the onset of convective fluid flow. Upon cooling, the solute-enriched convective instabilities solidify as isolated regions of equiaxed grains. The geometry of the solidification front is more complex in the VAR process than in the directional solidification process. For this reason, development of predictive models for the occurrence of freckling is more challenging. White

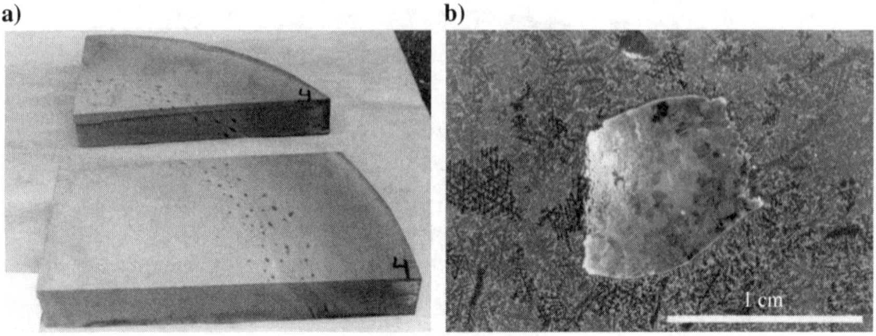

Fig. 12 Freckle and white spot defects in IN718.

spots are discrete features in the superalloy billet that are observed after chemical etching [60]. Although compositionally similar, these features are typically less heavily alloyed than the superalloy matrix. Compositions of white spots and freckles in IN718 are listed in Table 6. White spot formation is commonly attributed to the entrapment of fragments from the melting electrode or crown of the solidifying shelf.

2. ELECTROSLAG REMELTING

For the production of high-grade superalloy billets with minimal sulfur levels and inclusion content, the VIM ingot can be further refined using an electroslag remelting (ESR) process [61, 62] prior to VAR melting. As in the VAR process, consumable electrodes, measuring 60 to 80 cm in diameter, are cast from the VIM melt. Instead of electrical arcing, however, resistance heating from the molten slag is used to remelt the electrode. Macrosegregation and chemical heterogeneities are minimized when solidification is restricted to a comparatively small volume of molten metal. During ESR, droplets of molten metal are passed through a layer of CaO-MgO-CaF_2-Al_2O_3 slag and then accumulate in the melt pool (Fig. 11). The slag resides on the surface of the melt pool and effectively removes the residual sulfur and traps ceramic inclusions that are drawn out

TABLE 6 COMPOSITION OF FRECKLE CHAINS IN HIGH-REFRACTORY SINGLE-CRYSTAL SUPERALLOYS

	Al	Cr	Co	Hf	Re	Ta	W	Ni
Nominal	6.0	4.5	12.5	0.16	6.3	7.0	5.8	Bal
Freckle	8.2	3.6	11.0	0.18	2.3	10.0	2.9	Bal
Interdendritic	7.6	4.4	12.2	0.13	3.4	8.2	3.9	Bal

to the surface. Although VAR processes are restricted to round ingot geometries, however, ESR processes can be adapted to yield shaped ingots, such as rectangular slabs for sheet production. ESR can also be used to refine alloy ingots melted in air.

In addition to the potential for the entrapment of slag within the ingot, a number of other limitations are associated with ESR processing. The protective layer of slag resident on the surface of the melt pool enables the melting process to occur in atmosphere and a vacuum environment is unnecessary. Although protective, however, the layer of hot molten slag is also thermally insulating. Consequently, when compared to VAR, solidification within the melt pool occurs under lower thermal gradients and results in a larger melt pool volume increase [63]. As the size of the melt pool increases, the ESR ingot becomes increasingly susceptible to the onset of thermosolutal convection and the formation of freckle defects. In practice, ESR is limited to the production of moderately sized ingots or slabs for the manufacture of bar or sheet products. The associated changes in the melt pool shape also induce the formation of textured columnar grains within the resultant microstructure of the ESR ingot. The anisotropy associated with the aligned microstructure causes ESR ingots to be less compliant during forging and other hot working processes. Moreover, compositional changes also occur during ESR melting, and volatile elements, such as Ti, Al, and Hf, can react with the nonvacuum atmosphere or constituents in the slag and result in appreciable losses.

3. POWDER METALLURGY ALLOYS

To increase the strength of polycrystalline Ni-base superalloys, levels of refractory alloying additions and γ' forming elements have gradually been increased to levels that prohibit conventional processing [64]. Elements such as W, Mo, Ti, Ta, and Nb effectively strengthen the alloy, but also result in severe segregation within the ingot upon solidification. Additionally, the limited ductility of the high-strength alloys renders the ingot susceptible to cracking as thermally induced stresses evolve during cooling. Powder-processing routes have been developed to overcome the difficulties associated with melt-related defects and are appropriate for the production of advanced high-strength polycrystalline superalloy components. Listed in Table 1 are the compositions of some commercially available powder-processed Ni-based superalloys. As a result of their superior mechanical properties at elevated temperatures as compared to cast and wrought alloys, powder-processed Ni-based superalloys are used primarily in high-pressure turbine and compressor disk applications.

Powder processing begins with gas atomization of a highly alloyed VIM ingot. Rapid solidification of the fine powders effectively suppresses macrosegregation within the alloy. Because the low ductility associated with the correspondingly high strength causes many of these advanced superalloys to be highly sensitive to initial flaw sizes, the atomized powders are separated based on particle size. Standard 150 or 270 meshes are used to separate the powders into sizes $>100\ \mu m$

and >50 μm, respectively. Powder sizes directly influence the initial potential crack size present in the finished component [65]. Although finer powder sizes are desired to minimize initial defect sizes, costs increase substantially as yields are greatly reduced.

Once powders are collected into steel cans, the cans are evacuated under vacuum and sealed. The cans are then hot isostatically pressed (HIP) or extruded to consolidate the powder. The HIP process consists of heating the alloy to just below the γ' solvus temperature under a hydrostatic pressure of up to 310 MPa. After four to five hours, diffusion bonding and sintering of the powders under pressure yields a fully dense superalloy billet. Billet sizes are limited by the capacity of the HIP furnace, but systems capable of forming billets up to 150 cm diam and 300 cm height are available. Consolidation under hot extrusion is often preferred over HIP because of the ability to produce fine-grained structures (ASTM 12) and reduce effects associated with prior particle boundaries. The evacuated can containing the superalloy powder is hot extruded through a set of dies that greatly reduces the diameter. During this thermomechanical process, the individual powder particles are subjected to deformation, and any oxide films initially present on the surfaces of the powder are broken up. Because substantial plastic deformation and adiabatic heating occurs during this process, hot extrusion temperatures are selected such that temperatures are maintained below the γ' solvus temperature.

4. DEFORMATION PROCESSING

Forging and cogging are common hot working processes by which superalloy ingots are converted into useful structural components. Because of the high intrinsic strength of Ni-based superalloys, forming of these materials generally occurs at high temperatures (~1000°C). Hot working processes are primarily designed to refine the microstructure to yield isotropic properties and attain a near net shape component. Prior to hot deformation, cast ingots are heat treated at elevated temperatures to homogenize the microstructure and minimize chemical heterogeneities resulting from microsegregation during solidification. As a result, the microstructures in the homogenized ingot are typically extremely coarse (grain sizes >10 mm) and often have a residual columnar-grained structure. Ideally, depending on the application, uniform equiaxed grain sizes on the order of ASTM 12 to 6 (5–50 μm diam) are desired in the forged components. Conversion of the microstructure after homogenization into the fine-grained structure is achieved via dynamic and metadynamic recrystallization during and after hot working, respectively [66, 67]. Process variables, such as strain, strain rate, die and workpiece temperature, are carefully controlled, such that complete recrystallization occurs throughout the material and a uniform microstructure is attained (Fig. 13). Superalloy sheets or small diameter billets (up to 13 cm) can be rolled or forged directly from cast slabs or bars. Prior to the forging of large net shape superalloy disks for turbine engine applications, large homogenized ingots

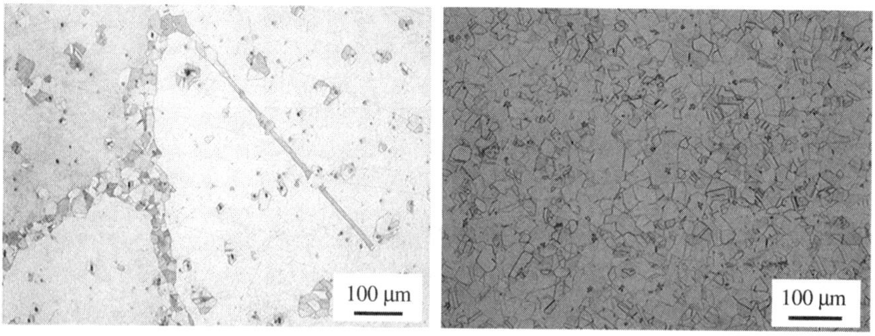

Fig. 13 Micrographs showing recrystallization of IN718 during hot deformation.

(36–60 cm diam) are cogged, or hot worked through a set of open dies, to form billets measuring 15–40 cm in diameter. The cogging process assists in breaking down the initial grain structure such that complete recrystallization occurs in the final forging.

One of the major advantages of powder-processed superalloy billets is the initial starting microstructure. Because of the rapid solidification of the powder particles and the negligible amount of elemental segregation within the microstructure, homogenization heat treatments are not required. Powder consolidation processes also tend to yield a fine equiaxed microstructure in the billet ideal for the direct production of net shape components (Fig. 14). Conventional forging practices, however, are not ideal for powder-processed Ni-based superalloys. Because of the limited ductility associated with the high levels of refractory alloying elements, isothermal forging of these alloys is generally required. As opposed to conventional forging processes where dies are often cold or warm and strain rates are high, dies used for isothermal forging are at the same temperature as the billet [68]. In most instances, isothermal forging temperatures are maintained just below the γ' solvus temperature. This enables the fine initial grain sizes to be retained throughout the forging process. The high temperatures combined with the fine grain size and characteristically low strain rates enable the high strength superalloy to be formed via superplastic flow during isothermal forging.

5. HEAT TREATMENT

Engineering of the grain structure in both cast and wrought powder-processed Ni-base superalloys is extremely important, as this can be used to impart both damage tolerance and strength. Generally, a fine, equiaxed grain structure is desired in most applications where the design lives are governed by strength and damage accumulation during cyclic loading. As service temperatures increase above $\sim$700°C, however, these fine-grained microstructures become susceptible to creep [69, 70]. Hence, coarse-grained microstructures tend to be preferred for elevated temperature (>700°C) applications because of their resistance to grain boundary sliding during creep deformation. Heat treatments are used to modify both the grain structure and intragranular precipitate microstructures following the final forging step. When forged components are heat-treated just below the γ' solvus temperature, fine grain structures tend to be retained as only the intragranular γ' precipitates are returned into the solid solution. This is also commonly referred to as "subsolvus" processing. In this instance, the γ' precipitates residing along the grain boundaries remain within the microstructure and are effective in inhibiting grain growth during heat treatment (Fig. 14b). Coarse-grained microstructures are formed during supersolvus processing, where the heat treatment temperature is performed above the γ' solvus. At these temperatures, complete solutioning of the γ' precipitates occurs thereby freeing the grain boundaries and enabling growth (Fig. 14c). Currently, the large majority of Ni-based superalloy forgings are subsolvus processed after deformation. Supersolvus processing tends to be limited to powder-processed Ni-based superalloy forgings intended for creep limited designs.

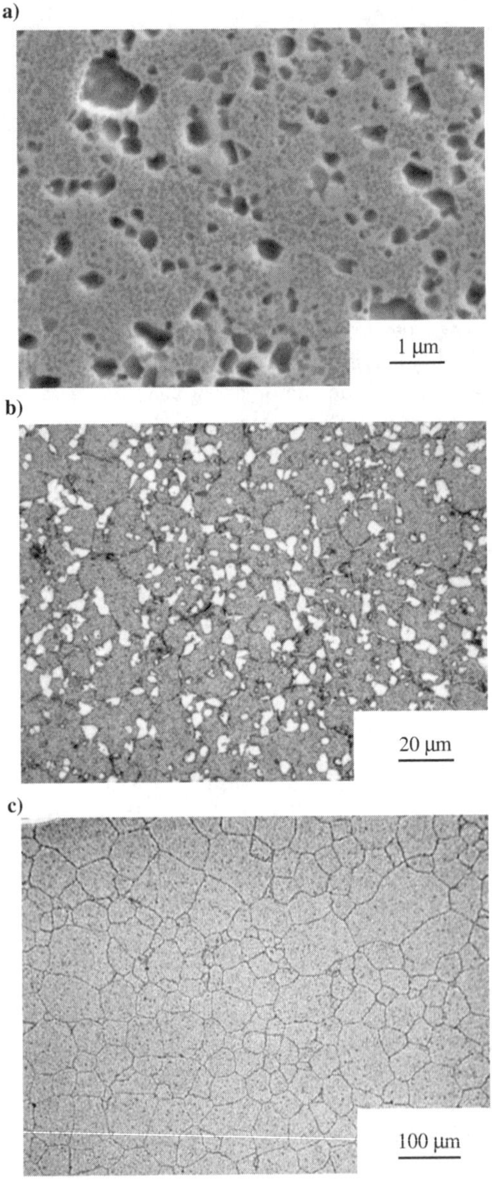

Fig. 14 Micrographs showing the microstructure of power-processed Ni-based superalloys in the a) extruded billet condition, b) subsolvus processed condition, and c) supersolvus processed condition.

Careful control of the cooling rate from the "sub-" or "supersolvus" heat treatment is also extremely important as this will strongly influence the γ' precipitate size, distribution, and morphologies. As will be discussed in Sec. IV, high cooling rates induce a fine uniform dispersion of γ' precipitates within the microstructure and maximize strength. Water fog, oil, and gas fan cooling methods are commonly used to achieve the desired cooling rates. Although accelerated cooling rates from the sub- or supersolvus heat treatment temperatures are desired, physically large forgings experience nonuniform cooling rates because of their low intrinsic thermal conductivities. As a result, highly alloyed superalloy components are prone to the formation of residual stresses that can result in quench cracking or distortion during machining. To relieve these residual stresses and refine the intragranular precipitate morphologies, a secondary aging heat treatment at temperatures ranging between 700–800°C tends to be employed.

Rotating turbine disk components are subjected to large cyclic stresses during service that render these components highly sensitive to the presence of defects and stress concentrations. As a result, these components must be carefully inspected with a variety of nondestructive x-ray and ultrasonic techniques. Moreover, in an effort to ensure microstructural homogeneity within the forged parts, up to 70% of the material by volume can be machined off during the final net-shape forging.

IV. PROPERTIES OF SUPERALLOYS

Superalloys constitute a large fraction of the materials of construction in turbine engines because of their unique combination of physical and mechanical properties. Table 7 lists some typical physical properties of superalloys. In aircraft engines, it is typical to consider density-normalized properties; thus alloy densities, which are typically in the range of 7.7–9.0 g/cm^3, are of specific interest. Optimization of the relevant set of mechanical properties is of paramount importance and is dependent on a high level of control and understanding of the

TABLE 7 TYPICAL PHYSICAL PROPERTIES OF SUPERALLOYS

Property	Typical Ranges
Density	7.7–9.0 g/cm^3
Melting temperature (liquidus)	1320–1450°C
Elastic modulus	Room temp: 210 GPa; 800°C: 160 GPa
Thermal expansion	8–18 × 10^{-6}/°C
Thermal conductivity	Room temp: 11 W/m-K; 800°C: 22 W/m-K

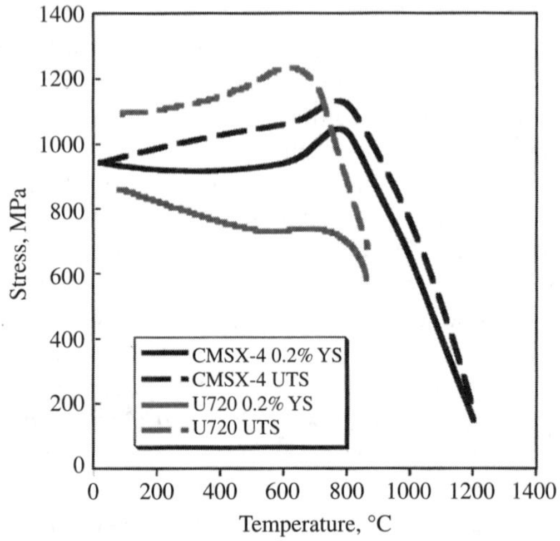

Fig. 15 Tensile strength dependence on temperature for single-crystal Ni-based superalloy CMSX-4 and polycrystalline Ni-based superalloy Udimet 720.

processes summarized in Sec. III because mechanical properties are a strong function of microstructure. Mechanical properties of primary interest include tensile properties, creep, fatigue, and cyclic crack growth. Depending on the details of component design, any one of these four properties can be life-limiting. Each of these properties is briefly discussed in the following sections. It is worth noting that models for prediction of these properties must treat many aspects of microstructure at different length scales. For this reason, predictive models for each of these properties presently remain under development though in recent years there has been increased effort aimed at property modeling [67].

A. TENSILE PROPERTIES

Nickel-based superalloys have relatively high yield and ultimate tensile strengths with yield strengths often in the range of 900–1300 MPa and ultimate tensile strengths from 1200–1600 MPa at room temperature. Figure 15 shows the temperature dependence of the yield strength of a single crystal alloy and a powder disk alloy [71, 72]. Turbine disk alloys are typically developed to have higher strengths for flexibility in the design against burst of the turbine disk in the event of an engine overspeed. Note that the tensile properties do not substantially decay until temperatures pass approximately 850°C. The slight rise in the yield strength of the alloys at intermediate temperatures is because of the unusual flow behavior of the intermetallic Ni_3Al γ' phase. The temperature dependence of yielding in single phase Ni_3Al, with a strong increase at intermediate temperatures, is shown in Fig. 16. Deformation of the precipitates gives a corresponding but weaker rise in the flow stress of superalloys at intermediate temperatures. Note also that the two-phase superalloys are much stronger than either the matrix or precipitate materials in their bulk form (Fig. 16).

Strengthening in two phase superalloys arises from multiple microstructural sources, including solid solution strengthening, grain size strengthening, and the interaction of dislocations with precipitates (Orowan bowing between precipitates or shearing through precipitates in strong or weak coupled modes). Heat

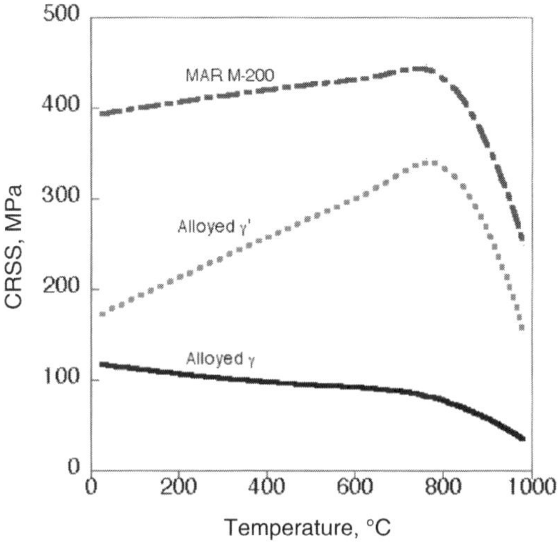

Fig. 16 Comparison of the critical resolved shear stress (CRSS) corresponding to the Ni-based superalloy MAR-M200 and the individual constituent phases (adapted from [4]).

treatments and various processing steps at different temperatures can result in multiple populations of precipitates with substantially varying mean sizes [12]. For example, three distinct populations of γ' precipitates in the microstructure of powder processed RR1000 are apparent in Fig. 17, where the microstructure is imaged at various magnifications. The three populations of γ' precipitates are referred to as "primary," "secondary," and "tertiary" in order of decreasing size and precipitation temperature. The primary precipitates form during powder consolidation and are not solutioned in the late stages of processing and therefore inhibit grain growth and indirectly provide grain size strengthening. Disk alloys with these large ($\approx$1 μm) precipitates are referred to as subsolvus processed. If only one or two populations of submicron-size precipitates are present, because of complete solutioning of the primary γ' then the grains are much larger (30–50 μm); these microstructural conditions result from supersolvus processing. Maximizing strength is a major challenge and requires close attention to all phases of processing. Modeling that considers the various possible strengthening mechanisms for each microstructural constituent in PM IN100 also demonstrates that there is no unique combination of microstructural parameters that results in a high strength alloy (Fig. 18).

B. CREEP PROPERTIES

Because superalloys experience extended periods under stress at high temperature, a high resistance to time-dependent creep deformation is essential. This is very important for cast blade alloys because they will experience temperatures up to 1100°C whereas disk alloys are typically limited to less than 700°C. For a fixed stress and temperature, two-phase superalloys have a much higher creep resistance, compared to their single-phase counterparts (Fig. 19). As with all properties that are governed by plastic deformation processes, creep properties are sensitive to microstructure. Figure 20 shows the creep rupture life as a function of volume

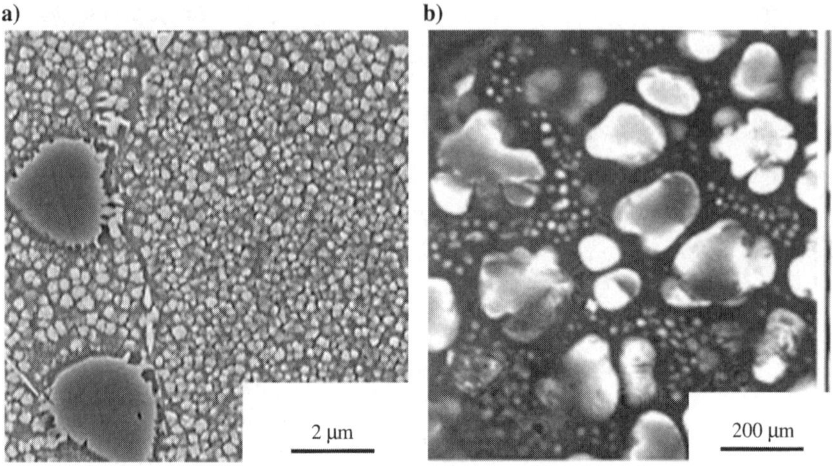

Fig. 17 Micrographs of subsolvus processed Ni-based superalloy RR1000 revealing a) the presence of primary γ' precipitates along the grain boundary and b) a fine dispersion of secondary and tertiary γ' within the grains.

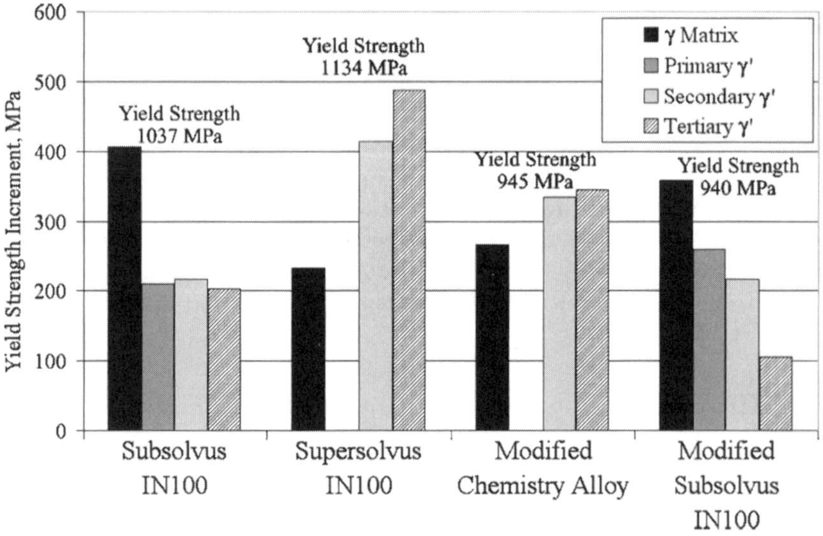

Fig. 18 The dependence of yield strength on microstructure and the strength contributions from the γ matrix and γ' precipitate populations for several variants of IN100. The modified chemistry version of IN100 has somewhat higher levels of Al, Cr, and Co and lower levels of Ti and V, compared to the nominal composition of IN100.

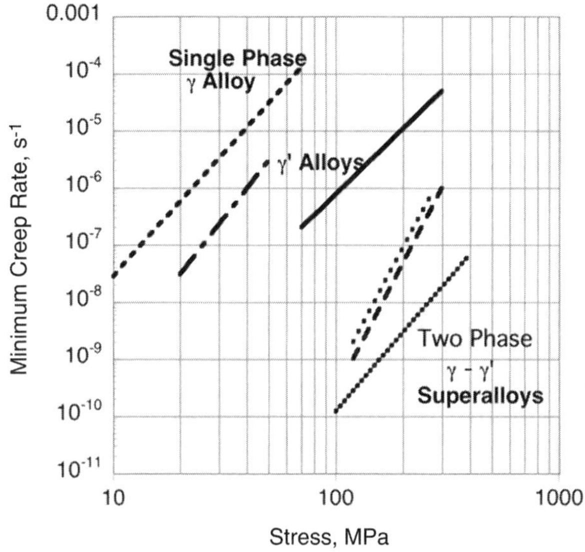

Fig. 19 As reflected by the minimum creep rates, two-phase $\gamma - \gamma'$ superalloys exhibit significantly improved creep resistance when compared to their single-phase constituents.

fraction of precipitates for several cast alloys [73]. Note that the strength peaks when the precipitate volume fractions are in the range of 0.6–0.7. Not surprisingly, many alloys contain volume fractions of precipitates in this peak range.

Alloy chemistry is also important to creep properties. Because the rate-controlling processes in creep are diffusion-controlled, elements that have low interdiffusion coefficients with nickel are generally beneficial to creep. Interdiffusion data for various alloying elements in nickel have recently been studied in detail [74, 75]. Elements most effective at slowing diffusion in Ni-based alloys include Ir, Re, Ru, Pt, W, Rh, and Mo. Advanced, creep-resistant alloys benefit from substantial additions of Re, W, and Mo (see Table 1).

A combination of increasing refractory alloying additions in conjunction with advances in processing has resulted in substantial increases in the maximum temperature capability of superalloys over the past few decades. For example, considering a creep rupture life of 1000 h at a stress of 137 MPa, the most recently developed single-crystal superalloys have a temperature capability of approximately 1100°C [76] while conventionally cast equiaxed alloys developed in the 1970s had a temperature capability of 900°C–950°C [77]. The temperature capabilities have reached 85–90% of the melting point, an usually high fraction of melting, compared to the operating conditions of any other class of structural materials. This indicates a need for the development of a completely new class of materials with higher melting points; unfortunately, this is a major challenge, and there are no obvious replacements for superalloys in the hottest sections of the turbine engine.

The exceptional creep properties Ni-based superalloys result primarily from the high resistance of the precipitates to shearing even at elevated temperatures. The uniaxial stress σ_{OR} required to glide a dislocation through the narrow matrix channels of the superalloy microstructure is

$$\sigma_{OR} = \sqrt{\frac{2}{3}}\left(\frac{\mu b}{wS}\right) \tag{5}$$

where μ is the shear modulus, b is the Burger's vector, h is the width of the channel, and S is the Schmid factor. Typical values of these material properties at 850°C are $\mu = 48.2$ GPa, $b = 0.254$ nm, and $h = 60$ nm [23]. For these parameters, an applied stress of 408 MPa must be exceeded for the onset of dislocation glide through the channels at 850°C. Thus, this resistance accounts for a large fraction, though not all, of the creep resistance of the two phase material. Figure 21 shows dislocations gliding through matrix channels during creep of CMSX-3 at 850°C and 552 MPa.

There have recently been efforts [78] to model the creep behavior of superalloys with the use of continuum damage mechanics approaches, but formulating models that capture the essence of the wide array of complex deformation mechanisms remains a challenge. The details of the deformation processes are very sensitive to temperature and applied stress, and it is most convenient to consider mechanisms of creep deformation [11] at low, intermediate, and high temperatures (and high, intermediate, and low stresses, respectively).

1. LOW-TEMPERATURE CREEP

Superalloys are highly resistant to creep deformation at temperatures below 800°C. In general, creep deformation occurs by deformation on <110> {111} slip systems, with an initial preference for

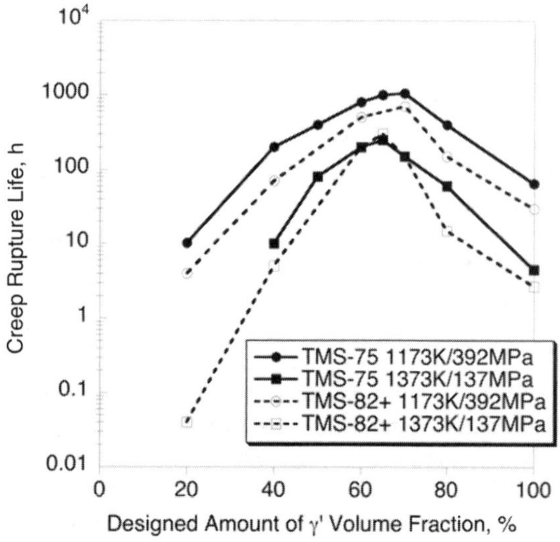

Fig. 20 Variation in creep rupture life as a function of γ' volume fraction for polycrystalline IN713C and single crystal TMS-75 (adapted from [72]).

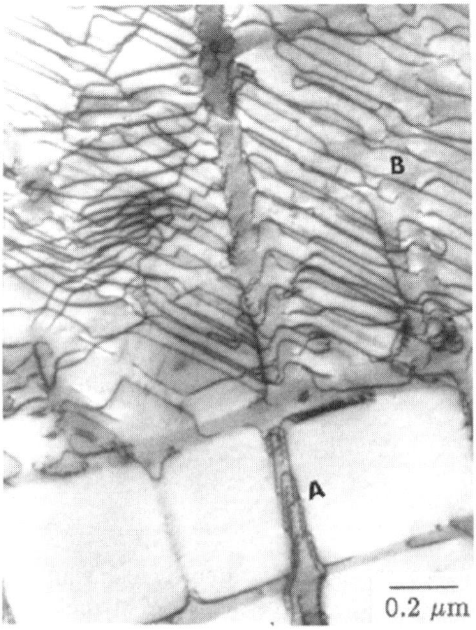

Fig. 21 After an initial incubation period where dislocations fill the horizontal channels, dislocations are forced to bow through the channels because of the high resistance of the precipitates to shearing.

dislocation glide through the continuous matrix. At temperatures $<0.6\ T_M$, however, high uniaxial stresses can result in the activation of $<112>\{111\}$ slip systems [79]. As dislocations accumulate at the $\gamma - \gamma'$ interface, $1/2<110>$ dislocations undergo reactions that result in $<112>$ type dislocations that are able to penetrate into the γ' precipitate. The details of the dislocation reactions that result in such precipitate shearing processes are still under investigation [80]. Nevertheless, as this shearing process occurs, strain is accumulated rapidly as a result of the planar nature of the slip and can result in high strains during the primary creep transient. This mode of deformation is observed in both single-crystal and polycrystalline alloys during creep. For single crystals, $\{110\}$ and $\{111\}$ oriented crystals are less prone to this type of deformation whereas in $\{001\}$ crystals, planar slip along $<112>$ continues until crystal rotations enable the resolved shear stresses to activate slip on other planes. Other factors that influence these shearing processes include alloy composition, γ' size, and volume fraction.

2. INTERMEDIATE-TEMPERATURE CREEP

At intermediate temperatures, stress levels are typically insufficient to result in shearing of the γ' precipitates. Thus, deformation within the microstructure is generally confined to the γ matrix and results in unusual creep curves that contain an initial incubation period and a brief primary transient, followed by an extended period of accelerating creep [23]. In general a "steady-state" creep rate is not achieved.

No macroscopic straining occurs during the incubation period that involves the distribution of grown-in dislocations. These initial dislocations serve as sources from which dislocations in the γ matrix are able to multiply. Single-crystal experiments show that when an external uniaxial stress is applied, the misfit stresses between the γ matrix and γ' precipitate are unbalanced, and the effective stresses enable preferential flow of dislocations within the horizontal channels. When the Orowan stresses are sufficiently high that dislocations are unable to bow between the vertical channels formed because of the aligned precipitates, dislocation glide in the matrix channels continues until the percolation process is complete.

The deformation mechanism associated with the primary creep transient during intermediate temperature creep is distinctly different from that of low-temperature creep. Unlike dislocation shearing of the γ' precipitates along the $<112>$ direction at high stresses and low temperatures, the primary creep transient at intermediate temperatures can be attributed to the relief of coherency stresses at the $\gamma - \gamma'$ interface as dislocations are accommodated. At the end of primary creep, a three-dimensional network of dislocations is formed around the precipitates. Despite the lack of a steady-state strain rate, these dislocation networks surrounding the precipitates are extremely stable and contribute to the gradual progression into tertiary creep.

3. HIGH-TEMPERATURE CREEP

The enhanced diffusivity associated with deformation at extremely high temperatures results in morphological changes within the microstructure. With the application of an external stress to alloys with significant precipitate-matrix misfit, the discrete cuboidal γ' precipitates coalescence into rafts or rods aligned perpendicular or parallel to the applied stress direction. The kinetics of directional coarsening is strongly influenced by the temperature as well as the stresses associated with the coherency strains at the $\gamma - \gamma'$ interface. Alignment of the rafts or rods, however, is dependent upon whether the external and misfit stresses are compressive or tensile. For example, uniaxial tensile stresses cause alloys with negative misfit to form rafts perpendicular to the applied stress direction whereas compressive stresses applied to the same alloy will result in the formation of rods aligned parallel to the direction of applied stress. Because most commercial directionally solidified and single-crystal alloys exhibit a negative misfit and are used to sustain tensile loads, rafts are generally formed perpendicular to the applied stress direction (Fig. 22). Because the rafts extend for many microns, dislocation climb and /or cross-slip assisted bypassing these coarse structures cannot occur. As a result, for creep to continue, the rafts must be sheared by dislocations. Thus, the chemical composition of the rafts and the structure of the $\gamma - \gamma'$ interface become important. Two different rate-limiting mechanisms of creep have been proposed for the directional coarsening regime [81, 82]. The first model [81] assumes that a diffusion-assisted motion of $a<010>$ dislocations with noncompact cores across the γ' raft limits creep deformation. The second model, by Carroll et al. [82], postulates that deformation is controlled by diffusion-limited climb of ordinary $a/2<110>$ dislocations along the $\gamma - \gamma'$ interfaces into paired configurations with interfacial dislocations that subsequently shear of the precipitates as superdislocations.

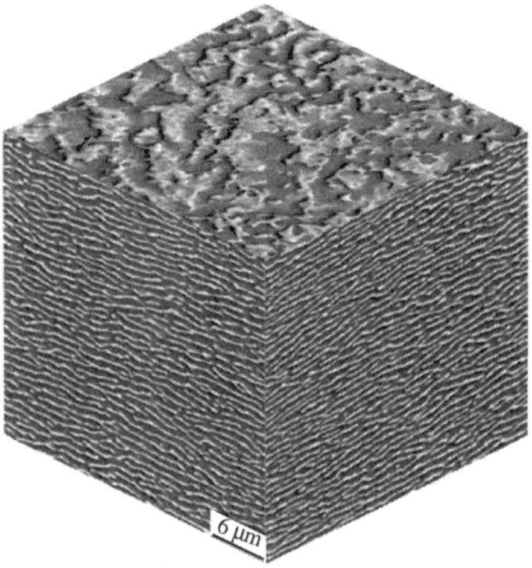

Fig. 22 Directional coarsening at elevated temperatures results in the formation of γ rafts aligned perpendicular to the direction of the applied stress. The matrix inverts as the γ rafts are contained within a γ' matrix.

With single-crystal and directionally solidified Ni-based superalloys containing in excess of 60–65% γ' by volume, directional coarsening can also result in an inversion of the microstructure. Once rafting is complete, rafts of γ are contained within an intermetallic γ' matrix. As the rafts of γ are discrete, a continuous path for dislocation motion along the matrix channels no longer exists. Provided that the rafted structure remains stable, rafted structures are highly resistant to deformation at low stresses, where dislocations are unable to repeatedly shear the plates of each phase. When stresses are sufficient to cause shearing of the rafts, microstructural damage is able to accumulate rapidly, and tertiary creep rates are accelerated.

C. FATIGUE AND FATIGUE CRACK GROWTH

Turbine engine components experience significant fluctuations in stress and temperature during their repeated takeoff-cruise-landing cycles. These cycles can result in localized, small plastic strains. Thus low-cycle, low-frequency fatigue is of interest to engine designers. Engine vibrations and airflow between the stages of the turbine can also result in high cycle fatigue, with rapid cycle accumulation in airfoils at much higher frequencies, in the kHz range.

Figure 23 shows fatigue properties for the powder disk alloy René 88DT at room temperature and 593°C [83]. Such polycrystalline disk alloys exhibit outstanding fatigue properties with fatigue strengths that are a high fraction of the monotonic yield strength. Under strain-controlled testing conditions in the low cycle regime, extensive cyclic hardening typically occurs. At lower temperatures, cyclic deformation tends to occur fairly inhomogeneously through shearing of precipitates along $\{111\}$ planes, with dislocations isolated in planar slip bands [84, 85]. As the ease of cross-slip increases with temperature, deformation occurs more homogeneously through the microstructure [84].

Fatigue properties are sensitive to mean stress, particularly at elevated temperatures. A Goodman approach to accounting for mean stresses is often used for component design [86]. Variability in fatigue lives tends to increase at lower

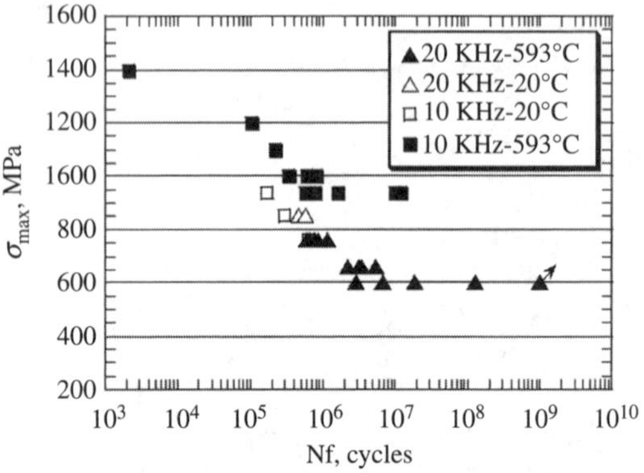

Fig. 23 The S-N behavior of René 88DT at 20 and 593°C at a load-ratio of 0.05 at frequencies of 10 Hz and 20 kHz. An arrow indicates a specimen that ran out to 10^9 cycles.

stresses and longer lives in superalloys [83, 87]. An example of this is shown in Fig. 23 for René 88DT, where fatigue lives at 593°C vary between approximately 10^6 cycles and 10^9 cycles at $R = 0.05$ and $\sigma_{max} = 600$ MPa. Such variability might be caused by intrinsic variations in microstructure and/or caused by the infrequent appearance of extrinsic defects. Crack initiation during fatigue of disk alloys can occur at extrinsic inclusions that are introduced during processing or at specific microstructural features such as larger grains [83, 87]. In cast alloys, during fatigue at higher temperatures, a high density of cracks initiates at the surface, either in the brittle oxide or in the aluminide coating, with only a small fraction of these cracks propagating inward [88–91]. There is a general tendency for initiation to shift from sample surfaces to subsurface regions at lower temperatures and in the high cycle regime, for both equiaxed and single-crystal alloys [86, 90, 92, 93]. Subsurface initiation sites include porosity (Fig. 24), carbides, or eutectic.

Hold times in compression or tension also have an important influence on fatigue. For single crystals, prolonged hold times in tension cause less reduction in fatigue life than hold times in compression at temperatures

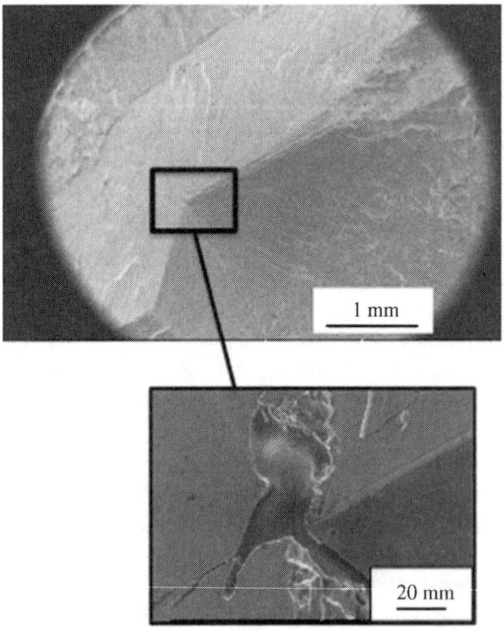

Fig. 24 Fatigue crack propagation curves at conventional (20 Hz) and ultrasonic (20 kHz) frequencies for René 88DT in air and vacuum. Long crack behavior has also been included at a frequency of 20 Hz.

above 950°C [94]. At the peak compressive load, where the hold time is imposed, bulk creep deformation occurs along with rafting [95]. This results in residual tension upon unloading. The tension permits opening of cracks that develop on the superalloy surface or in the aluminide coating and oxidation of material at the crack tip. It is thought that upon recompression the extension along the direction of the crack plane (as a result of the Poisson effect) induces tension in the oxide that further pushes it into the creeping substrate [95]. Thus, crack tip oxidation kinetics in the coating and/or the superalloy and the creep resistance of the superalloy must be optimized to minimize degradation under compressive hold conditions. For more complex thermomechanical fatigue (TMF) cycles, a general correlation between TMF life and the stress amplitude in the tensile half-cycle has been noted [96].

Because of stringent safety and associated lifing requirements, fatigue crack propagation is also an important aspect of material behavior, particularly for turbine disk materials. Paris law crack growth behavior is displayed over a wide range of stress intensities for most superalloys. Cyclic threshold stress intensities are relatively high, with ΔK_{th} often in the range of 8–20 MPa-m$^{1/2}$ [92]. Crack growth rates are sensitive to microstructural features, including grain and precipitate sizes and volume fractions [92]. At temperatures above approximately 500°C, environmental effects and cyclic frequency become significant factors, with higher crack growth rates in air, as compared to vacuum. Figure 25 shows an example of fatigue crack growth behavior for the powder disk alloy René 88DT. Note that crack growth rates are sensitive to test environment and test frequency at 593°C. In single-crystal alloys, cracks can grow crystallographically along {111} planes, particularly in the early stages of growth (stage I) [7, 92]. Depending on testing conditions, as cracks grow longer (stage II), they can advance in a less crystallographic manner [97, 98] with a greater tendency toward mode 1 behavior.

Although fatigue and fatigue crack growth are often limiting properties, comprehensive models for life prediction that account for complex loading, crack initiation, and growth behavior, as well as microstructure and environment, continue to be developed. Integration of physics-based models with advanced sensors that can diagnose the

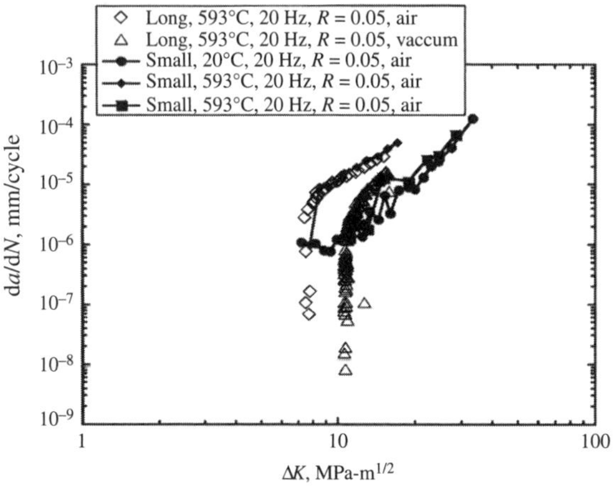

Fig. 25 Fatigue crack initiation at a near-surface pore in a single crystal superalloy (photo courtesy of C. Brundidge).

current "damage state" remains a promising approach for component life prediction [99, 100].

V. CONCLUSIONS

The challenging environment of the aircraft engine has led to the development of an exceptional class of high-temperature materials: nickel-based superalloys. Continued improvements in the properties of these materials have been possible through close control of chemistry and microstructure and with continued innovation in the processing of both cast and wrought classes of materials. Further improvements are likely with the development, implementation, and integration of tools for alloy design and models for microstructure-process evolution and mechanical properties. Integration of the materials models with aero, thermal, and structural engine design tools remains an interesting challenge and opportunity.

ACKNOWLEDGMENTS

The authors are grateful for the contributions of a number of their collaborators to this manuscript, including R. J. Mitchell, A. C. Yeh, L. J. Carroll, A. J. Elliott, Q. Feng, A. Shyam, C. Brundidge, A. Suzuki, J. Yi, C. J. Torbet, and J. W. Jones.

REFERENCES

[1] Schafrik, R., and Sprague, R., "Gas Turbine Materials Part II," *Advanced Materials & Processes*, Vol. 162, No. 4, 2004, pp. 27–30.

[2] Sims, C. T., Stoloff, N. S., and Hagel, W. C., *Superalloys II*, Wiley, New York, 1986.

[3] Reed, R. C., *The Superalloys: Fundamentals and Applications*, Cambridge Univ. Press, Cambridge, England, U.K., 2006.

[4] Loria, E. A., *Superalloy 718: Metallurgy and Applications*, TMS, Warrendale, PA, 1989.

[5] VerSnyder, F. L., and Shank, M. E., "The Development of Columnar Grain and Single Crystal High Temperature Materials Through Directional Solidification," *Materials Science Engineering*, Vol. 6, 1970, pp. 213–247.

[6] Collins, H. E., "Relative Stability of Carbide and Intermetallic Phases in Nickel-Base Superalloys," *Superalloys*, ASM, Metals Park, OH, 1968, pp. 171–198.

[7] Crompton, J. S., and Martin, J. W., "Crack Growth in a Single Crystal Superalloy at Elevated Temperature," *Metallurgical Transactions A*, Vol. 15A, 1984, pp. 1711–1719.

[8] Meetham, G. W., "Trace Elements in Superalloys–An Overview," *Metals Technology*, Vol. 11, 1984, pp. 414–418.

[9] McLean, M., and Strang, A., "Effects of Trace Elements on Mechanical Properties of Superalloys," *Metals Technology*, Vol. 11, 1984, pp. 454–464.

[10] Machlin, E. S., "Kinetics of Vacuum Induction Theory," *Transactions of the AIME*, Vol. 218, 1961, pp. 314–326.

[11] Pollock, T. M., and Field, R. D., *Dislocations and High Temperature Plastic Deformation of Superalloy Single Crystals*, in *Dislocations in Solids*, edited by F. R. N. Nabarro, and M. S. Duesbery, Elsevier, Amsterdam, 2002.

[12] Wusatowska-Sarnek, A. M., Ghosh, G., Olson, G. B., Blackburn, M. J., and Aindow, M., "Characterization of the Microstructure and Phase Equilibria Calculations for the Powder Metallurgy Superalloy IN100," *Journal of Materials Research*, Vol. 18, 2003, pp. 2653–2663.

[13] Reed, R. C., Yeh, A. C., Tin, S., Babu, S. S., and Miller, M. K., "Identification of the Partitioning Characteristics of Ruthenium in Single Crystal Superalloys Using Atom Probe Tomography," *Scripta Mater.*, Vol. 51, 2004, pp. 327–331.

[14] Tien, J. K., and Copley, S. M., "The Effect of Uniaxial Stress on the Periodic Morphology of Coherent Gamma Prime Precipitates in Nickel-Base Superalloy Crystals," *Metallurgical Transactions*, Vol. 2, 1971, pp. 215–219.

[15] Grose, D. A., and Ansell, G. S., "The Influence of Coherency Strain on the Elevated Temperature Tensile Behavior of Ni-15Cr-AI-Ti-Mo Alloys," *Metallurgical Transactions A*, Vol. 12A, 1981, pp. 1631–1645.

[16] Pineau, A., "Influence of Uniaxial Stress on the Morphology of Coherent Precipitates During Coarsening–Elastic Energy Considerations," *Acta Metallurgica*, Vol. 24, 1976, pp. 559–564.

[17] Ricks, R. A., Porter, A. J., and Ecob, R. C., "The Growth of γ' Precipitates in Nickel-Base Superalloys," *Acta Metallurgica*, Vol. 31, 1983, pp. 43–53.

[18] Doi, M., and Miyazaki, T., "Microstructural Development Under the Influence of Elastic Energy in Nickel-Base Alloys Containing Gamma Precipitates," *Superalloys 1988*, TMS, Warrendale, PA, 1988, pp. 663–672.

[19] Giamei, A. F., and Anton, D. L., "Rhenium Additions to a Ni-Base Superalloy: Effects on Microstructure," *Metallurgical Transactions*, Vol. 16A, 1985, pp. 1997–2005.

[20] Fahrmann, M., Hermann, W., Fahrmann, E., Boegli, A., Pollock, T. M., and Sockel, H. G., "Determination of Matrix and Precipitate Elastic Constants in (g-g') Ni-Base Model Alloys and Their Relevance to Rafting," *Materials Science and Engineering A*, Vol. 60, 1999, pp. 212–221.

[21] Nathal, M. V., MacKay, R. A., and Miner, R. V., "Influence of Precipitate Morphology on Intermediate Temperature Creep Properties of a Nickel-Base Superalloy Single Crystal," *Metallurgical Transactions A*, Vol. 20A, 1989, pp. 133–141.

[22] Muller, L., Glatzel, U., and Feller-Kniepmeier, M., "Modelling Thermal Misfit Stresses in Nickel-Base Superalloys Containing High Volume Fraction of γ' Phase," *Acta Metallurgica et Materials*, Vol. 41, 1992, pp. 3401–3411.

[23] Pollock, T. M., and Argon, A. S., "Creep Resistance of CMSX-3 Nickel Base Superalloy Single Crystals," *Acta Metallurgica Materials*, Vol. 40, 1992, pp. 1–30.

[24] Ross, E. W., "Rene 100- A Sigma-Free Turbine Blade Alloy (Rene 100 Alloy High Temperature Rupture Strength, Stability and Resistance to Formation of Embrittling Intermetallic Phase)," *Journal of Metals*, Vol. x, 1967, pp. 12–14.

[25] Nystrom, J. D., Pollock, T. M., Murphy, W. H., and Garg, A., "Discontinuous Cellular Precipitation in a High-Refractory Nickel-Base Superalloy," *Metallurgical Materials Transactions*, Vol. 28A, 1997, pp. 2443–2452.

[26] Wlodek, S. T., "The Structure of in-100 (Structure in Two Heats of in-100 Nickel-Base Alloy Before and After Heating, and in Presence of Stress)," *Transactions of the ASME*, Vol. 57, 1964, pp. 110–119.

[27] Rae, C. M. F., and Reed, R. C., "The Precipitation of Topologically Close-Packed Phases in Rhenium-Containing Superalloys," *Acta Materialia*, Vol. 49, 2001, pp. 4113–4125.

[28] Darolia, R., Lahrman, D. F., and Field, R. D., "Formation of Topologically Closed Packed Phases in Nickel-Base Single Crystal Superalloys," *Superalloys 1988*, TMS, Warrendale, PA, 1988, pp. 255–264.

[29] Agren, J., "Calculation of Phase Diagrams: Calphad," *Current Opinion in Solid State & Materials Science*, Vol. 1, 1996, pp. 355–360.

[30] Kattner, U. R., "The Thermodynamic Modeling of Multicomponent Phase Equilibria," *Journal of Metals*, Vol. 49, 1997, pp. 14–19.

[31] Saunders, N., Fahrmann, M., and Small, C. J., "The Application of Calphad Calculations to Ni-Based Superalloys,"*Superalloys 2000*, TMS, Warrendale, PA, 2000, pp. 803–811.

[32] Wu, K., Chang, Y. A., and Wang, Y., "Simulating Interdiffusion Microstructures in Ni-Al-Cr Diffusion Couples: A Phase Field Approach Coupled with CALPHAD Database," *Scripta Materialia*, 2004, Vol. 50, pp. 1145–1150.

[33] Tien, J. K., and Caulfield, T., *Superalloys, Supercomposites and Superceramics.* Academic Press, New York, 1988.

[34] Ma, D., and Grafe, U., "Microsegregation in Directionally Solidified Dendritic-Cellar Structure of Superalloys CMSX-4," *Materials Science Engineering A*, Vol. 270, No. 2, 1999, pp. 339–342.

[35] Nathal, M. V., MacKay, R. A., and Miner, R. V., "Influence of Precipitate Morphology on Intermediate Temperature Creep Properties of a Nickel-Base Superalloy Single Crystal," *Metallurgical Transactions*, Vol. 20, No. 1, 1989, pp. 133–141.

[36] Sabol, G. P., and Stickler, R., "Microstructure of Nickel-Based Superalloys," *Physica Status Solidi (b)*, Vol. 35, No. 1, 1969, pp. 11–52.

[37] Angenete, J., and Stiller, K., "Comparison of Inward and Outward Grown Pt Modified Aluminide Diffusion Coatings on a Ni Based Single Crystal Superalloy," *Surface Coatings Tech.*, Vol. 150, No. 2–3, 2002, pp. 107–118.

[38] Tin, S., Pollock, T. M., and Murphy, W., "Stabilization of Thermosolutal Convective Instabilities in Ni-Based Single-Crystal Superalloys: Carbon Additions and Freckle Formation," *Metallurgical Material Transactions*, 2001, Vol. 32A, pp. 1743–1753.

[39] Tin, S., and Pollock, T. M., "Predicting Freckle Formation in Single Crystal Ni-Base Superalloys," *Journal of Materials Science*, Vol. 39, 2004, pp. 7199–7205.

[40] Beckermann, C., Gu, J. P., and Boettinger, W. J., "Modeling of Micro- and Macrosegregation and Freckle Formation in a Single-Crystal Nickel-Base Superalloy Directional Solidification," *Metallurgical Material Transactions*, Vol. 31A, 2000, pp. 2545–2557.

[41] Wang, W., Lee, P. D., and McLean, M., "A Model of Solidification Microstructures in Nickel-Based Superalloys: Predicting Primary Dendrite Spacing Selection," *Acta Materialia*, Vol. 51, 2003, pp. 2971–2987.

[42] Pollock, T. M., and Murphy, W. H., "The Breakdown of Single-Crystal Solidification in High Refractory Nickel-Base Alloys," *Metallurgical Materials Transactions*, Vol. 27A, 1996, pp. 1081–1094.

[43] MacKay, R. A., and Maier, R. D., "The Influence of Orientation on the Stress Rupture Properties of Nickel-Base Superalloy Single Crystals," *Metallurgical Transactions*, Vol. 13, No. 10, 1982, pp. 1747–1754.

[44] Sass, V., Schneider, W., and Mughrabi, H., "On the Orientation Dependence of the Intermediate-Temperature Creep Behavior of a Monocrystalline Nickel-Base Superalloy," *Scripta Metallurgica et. Materials*, Vol. 31, No. 7, 1994, pp. 885–890.

[45] Giamei, A. F., and Kear, B. H., "On the Nature of Freckles in Nickel Base Superalloys," *Metallurgical Transactions*, Vol. 1, 1970, pp. 2185–2192.

[46] Copley, S. M., Giamei, A. F., Johnson, S. M., and Hornbecker, M. F., "The Origin of Freckles in Unidirectionally Solidified Castings," *Metallurgical Transactions*, Vol. 1, 1970, pp. 2193–2204.

[47] Sample, A. K., and Hellawell, A., "The Mechanisms of Formation and Prevention of Channel Segregation During Alloy Solidification," *Metallurgical Transactions*, Vol. 15A, 1984, pp. 2163–2173.

[48] Schneider, M. C., Gu, J. P., Beckermann, C., Boettinger, W. J., and Kattner, U. R., "Modeling of Micro- and Macrosegregation and Freckle Formation in Single-Crystal Nickel-Base Superalloy Directional Solidification," *Metallurgical Transactions*, Vol. 28A, 1997, pp. 1517–1531.

[49] Auburtin, P., Cockcroft, S. L., and Mitchell, A., "Liquid Density Inversions During the Solidification of Superalloys and Their Relationship to Freckle Formation in Castings," *Superalloys 1996*, 1996, TMS, Warrendale, PA, pp. 443–450.

[50] Sarazin, J. R., and Hellawell, A., "Channel Formation in Pb-Sn, Pb-Sb, and Pb-Sn-Sb Alloy Ingots and Comparison with the System NH4CI-H2O," *Metallurgical Transactions*, Vol. 19A, 1988, pp. 1861–1871.

[51] Giamei, A. F., and Tschinkel, J. G., "Liquid Metal Cooling: a New Solidification Technique," *Metallurgical Transactions*, Vol. 7A, 1976, pp. 1427–1434.

[52] Hugo, F., Betz, U., Ren, J., Huang, S.-C., Bondarenko, J. A., and Gerasimov, V., "Casting of Directionally Solidified and Single Crystal Components Using Liquid Metal Cooling (LMC)," *International Symposium on Liquid Metal Processing and Casting*, 1999, AVS, pp. 16–30.

[53] Lohmuller, A., Esser, W., Gossmann, Hordler, M., Preuhs, J., and Singer, R. F., "Improved Quality and Economics of Investment Castings by Liquid Metal Cooling–The Selection of Cooling Media," *Superalloys 2000*, TMS, Warrendale, PA, 2000, pp. 181–188.

[54] Elliott, A. J., Tin, S., King, W. T., Huang, S.-C., Gigliotti, M. F. X., and Pollock, T. M., "Directional Solidification of Large Superalloy Castings with Radiation and Liquid-Metal Cooling: A Comparative Assessment," *Metallurgical and Materials Transactions A*, Vol. 35A, 2004, pp. 3221–3231.

[55] Konter, M., Kats, E., and Hofman, N., "A Novel Casting Process for Single Crystal Gas Turbine Components," *Superalloys 2000*, TMS, Warrendale, PA, 2000, pp. 189–200.

[56] Graham, L., "Fluidized Bed with Baffle," U.S. Patent 6,889,747, 2005

[57] Kermanapur, A., Evans, D. G., Siddall, R. J., Lee, P. D., and McLean, M., "Effect of Process Parameters on Grain Structure Formation During VAR of INCONEL Alloy 718," *Journal of Materials Science*, Vol. 39, 2004, pp. 7175–7182.

[58] Auburtin, P., Wang, T., Cockcroft, S. L., and Mitchell, A., "Freckle Formation and Freckle Criterion in Superalloy Castings," *Metallurgical Materials Transactions B*, 2000, Vol. 31, No. 4, pp. 801–811.

[59] VanDenAvyle, J. A., Brooks, J. A., and Powell, A. C., "Reducing Defects in Remelting Processes for High-Performance Alloys," *Journal of Metals*, Vol. 50, No. 3, 1998, pp. 22–25.

[60] Jackman, L. A., Maurer, G. E., and Widge, S., "New Knowledge About 'White Spots' in Superalloys," *Advanced Materials and Processes*, Vol. 5, 1993, pp. 18–25.

[61] Durber, G. L. R., Jones, C. L., and Dykes, A. J., "VIM + ESR Alloy 718–An Assessment of Chemistry Control, Alloy Cleanliness and Mechanical Properties (Vacuum Induction Melting and Electroslag Refining)," *Superalloys 1984*, TMS, Warrendale, PA, 1984, pp. 433–442.

[62] Shamblen, C. E., Chang, D. R., and Corrado, J. R., "Superalloy Melting and Cleanliness Evaluation," *Superalloys 1984*, TMS, Warrendale, PA, 1984, pp. 509–520.

[63] Choudhurdy, A., "State of the Art of Superalloy Production for Aerospace and Other Application Using VIM/VAR or VIM/ESR," *ISIJ International*, Vol. 32, No. 5, 1992, pp. 563–574.

[64] Maurer, G. E., Castledine, W., Schweizer, F. A., and Mancuso, S., "Development of HIP Consolidated P/M Superalloys for Conventional Forging to Gas Turbine Engine Components," *Superalloys 1996*, Warrendale, PA, TMS, 1996, pp. 645–652.

[65] Chang, D. R., Krueger, D. D., and Sprague, R. A., "Superalloy Powder Processing Properties and Turbine Disk Applications," *Superalloys 1984*, TMS, Warrendale, PA, 1984, pp. 245–273.

[66] Luton, M. J., and Sellars, C. M., "Dynamic Recrystallization in Nickel and Nickel-Iron Alloys During High Temperature Deformation," *Acta Metallurgica*, Vol. 17, 1969, pp. 1033–1043.

[67] Shen, G., Semiatin, S. L., and Shivpuri, R., "Modeling Microstructural Development During the Forging of Waspaloy," *Metallurgy and Material Transactions*, Vol. 26A, 1995, p. 1795.

[68] Shen, G., and Furrer, D., "Manufacturing of Aerospace Forgings," *Journal of Material Processing Technology*, Vol. 98, No. 2, 2000, pp. 189–195.

[69] Bain, K. R., Gambone, M. L., Hyzak, J. M., and Thomas, M. C., "Development of Damage Tolerant Microstructures in Udimet 720," *Superalloys 1988*, TMS, Warrendale, PA, 1988, pp. 13–22.

[70] Miner, R. V., Gayda, J., and Maier, R. D., "Fatigue and Creep-Fatigue Deformation of Several Nickel-Base Superalloys at 650°C," *Metallurgical Transactions*, Vol. 13A, 1982, pp. 1755–1765.

[71] Backman, D. G., Mourer, D. P., Bain, K. R., and Walston, W. S., "AIM-A New Methodology for Developing Disk Materials," *Advanced Materials and Processes for Gas Turbines*, TMS, Warrendale, PA, 2003, pp. 255–264.

[72] Huron, E. S., "Serrated Yielding in a Nickel-Base Superalloy," *Superalloys 1992*, TMS, Warrendale, PA, 1992, pp. 675–684.

[73] Murakumo, T., Kobayashi, T., Koizumi, Y., and Harada, H., "Creep Behaviour of Ni-Base Single-Crystal Superalloys with Various Volume γ' Fraction," *Acta Materialia*, Vol. 52, 2004, pp. 3737–3744.

[74] Karunarante, M. S. A., and Reed, R. C., "Interdiffusion of the Platinum-Group Metals in Nickel at Elevated Temperatures," *Acta Materiala*, Vol. 51, 2003, pp. 2905–2914.

[75] Reed, R. C., and Karunarantne, M. S. A., "Interdiffusion in the Face-Centred Cubic Phase of the Ni-Re, Ni-Ta and Ni-W Systems Between 900 and 1300°C," *Materials Science and Engineering A*, Vol. 281, 2000, pp. 229–233.

[76] Walston, W. S., Cetel, A., MacKay, R., O'Hara, K., Duhl, D., and Dreshfield, R., "Joint Development of a Fourth Generation Single Crystal Superalloy," *Superalloys 2004*, TMS, Warrendale, PA, 2004, pp. 15–24.

[77] Tanaka, R., "Research and Development of Ultra-High Temperature Materials in Japan," *Materials a High Temperatures*, Vol. 17, 2000, pp. 457–464.

[78] McLean, M., and Dyson, B. F., "Modeling the Effects of Damage and Microstructural Evolution on the Creep Behavior of Engineering Alloys," *Journal of Engineering Materials and Technology*, Vol. 122, 2000, pp. 273–278.

[79] Rae, C. M. F., Cox, D. C., Rist, M. A., Reed, R. C., and Matan, N. C., "On the Primary Creep of CMSX-4 Superalloy Single Crystals," *Metallurgy and Materials Transactions A*, Vol. 31A, 2000, pp. 2219–2228.

[80] Rae, C. M. F., and Reed, R. C., "Primary Creep in Single Crystal Superalloys: Origins Mechanisms and Effects," *Acta Materialia*, 2007, Vol. 55, No. 3, pp. 1067–1081.

[81] Srinivasan, R., Eggeler, G. F., and Mills, M. J., "γ'-Cutting as Rate-Controlling Recovery Process During High-Temperature and Low-Stress Creep of Superalloy Single Crystals," *Acta Materialia*, Vol. 48, 2000, pp. 4867–4878.

[82] Carroll, L. J., Feng, Q., and Pollock, T. M., "Interfacial Dislocation Networks and Creep in Directional Coarsened Ru-Containing Nickel-Base Single-Crystal Superalloys," *Metallurgical and Materials Transactions*, Vol. 39A, 2008, pp. 1647–1657.

[83] Shyam, A., Torbet, C. J., Jha, S. K., Larsen, J. M., Caton, M. J., Szczepanski, C. J., Pollock, T. M., and Jones, J. W., "Development of Ultrasonic Fatigue for Rapid, High Temperature Fatigue Studies in Turbine Engine Materials," *Superalloys 2004*, Warrendale, PA, TMS, 2004, pp. 259–267.

[84] Tong, J., Dalby, S., Byrne, J., Henderson, M. B., and Hardy, M. C., "Creep, Fatigue and Oxidation in Crack Growth in Advanced Nickel Base Superalloys," *International Journal of Fatigue*, Vol. 23, No. 10, 2001, pp. 897–902.

[85] Clavel, M., and Pineau, A., "Fatigue Behaviour of Two Nickel-Base Alloys I: Experimental Results on Low Cycle Fatigue, Fatigue Crack Propagation and Substructures," *Material Science Engineering*, Vol. 55, 1982, pp. 157–171.

[86] Chan, K. S., Hack, J. E., and Leverant, G. R., "Fatigue Crack Growth in MAR-M200 Single Crystals," *Metallurgical Transactions A*, Vol. 18A, 1987, pp. 581–591.

[87] Jha, S. K., Caton, M. J., and Larsen, J. M., "A New Paradigm of Fatigue Variability Behavior and Implications for Life Prediction," *Material Science Engineering A*, 2007, pp. 23–32, 468–470.

[88] Totemeier, T. C., Gale, W. F., and King, J. E., "Fracture Behaviour of an Aluminide Coating on a Single Crystal Nickel Base Superalloy," *Metallurgical Materials Transactions A*, Vol. 25A, 1994, pp. 2837–2840

[89] Ohtani, R., Tada, N., Shibata, M., and Taniyama, "High Temperature Fatigue of the Nickel- Base Single- Crystal Superalloy CMSX- 10," *Fatigue Fract. Eng. Mater. Structure*, Vol. 24, 2001, pp. 867–876.

[90] MacLachlan, D. W., and Knowles, D. M., "Fatigue Behaviour and Lifing of Two Single Crystal Superalloys," *Fatigue Fract. Eng. Mater. Structure*, Vol. 24, 2001, pp. 503–521.

[91] Fleury, E., and Rémy, L., "Low Cycle Fatigue Damage in Nickel-Base Superalloy Single Crystals at Elevated Temperature," *Materials Science Engineering*, Vol. A167, 1993, pp. 23–30.

[92] Antolovich, S. D., and Lerch, B., "Cyclic Deformation, Fatigue and Fatigue Crack Propagation in Ni-Base Alloys," *Superalloys, Supercomposites and Superceramics*, edited by J. K. Tien, and T. Caulfield, Academic Press, New York, 1989, pp. 363–412.

[93] Yi, J. Z., Torbet, C. J., Feng, Q., Pollock, T. M., and Jones, J. W., "Ultrasonic Fatigue of a Single Crystal Ni-Base Superalloy at 1000°C," *Materials Science Engineering A*, Vol. 443, No. 1–2, 2007, pp. 142–149.

[94] Frenz, H., Kinder, J., Klingelhoffer, H., and Portella, P. D., "Behavior of Single Crystal Superalloys Under Cyclic Loading at High Temperatures," *Superalloys 1996*, TMS, Warrendale, PA, 1996, pp. 305–312.

[95] Evans, A. G., He, M. Y., Suzuki, A., Gigliotti, M., Hazel, B., and Pollock, T. M., "A Mechanism Governing Oxidation-Assisted Low-Cycle Fatigue of Superalloys," *Acta Materialia*, Vol. 57, No. 10, 2009, pp. 2969–2983.

[96] Kraft, S., Zauter, R., and Mughrabi, H., "Aspects of High-Temperature Low- Cycle Thermomechanical Fatigue of a Single Crystal Nickel- Base Superalloy," *Fatigue & Fracture of Engineering Materials & Structures*, Vol. 16, No. 2, 1993, pp. 237–253.

[97] Wright, P. K., Jain, M., and Cameron, D., "High Cycle Fatigue in a Single Crystal Superalloy: Time Dependence at Elevated Temperature," *Superalloys 2004*, TMS, Warrendale, PA, 2004, pp. 657–666.

[98] Crompton, J. S., and Martin, J. W., "Crack Growth in a Single Crystal Superalloy at Elevated Temperature," *Metallurgical Transactions A*, Vol. 15A, 1984, pp. 1711–1718.

[99] Larsen, J. M., and Christodoulou, L., "Using Materials Prognosis to Maximize the Utilization Potential of Complex Mechanical Systems," *Journal of Metals*, Vol. 56, 2004, p. 14.

[100] *Integrated Computational Materials Engineering*, National Research Council Report, National Academies Press, Washington, D.C., 2008.

CHAPTER 11

Thermal Barrier Coatings

Rodney Trice* and Kevin Trumble*
Purdue University, West Lafayette, Indiana

NOMENCLATURE

k_{th}	thermal conductivity
Q	absorbed heat
T_H	high-temperature reservoir
T_L	low-temperature reservoir
W	useful work for propulsion
η	engine efficiency

ACRONYMS

APS	air plasma spray
CMAS	calcium magnesium aluminosilicate
CTE	coefficient of thermal expansion
EB-PVD	electron-beam physical vapor deposition
$Gd_2Zr_2O_7$	gadolinium zirconate
LPPS	low-pressure plasma spray
MCrAlY	bondcoat material where M is typically Ni or Co
Na_2SO_4	sodium sulfate
(Ni,Pt)Al	platinum modified nickel aluminide bondcoat
PS	plasma spray
PS-PVD	plasma spray physical vapor deposition
SPPS	solution precursor plasma spray
SPS	suspension plasma spray
TBC	thermal barrier coating
TEM	transmission electron microscopy
TET	turbine entry temperature
TGO	thermally grown oxide
VLPPS	very low-pressure plasma spray

*Professor of Materials Engineering, School of Materials Engineering.

V_2O_5	vanadium oxide
Y_2O_3	yttria
ZrO_2	zirconia
7YSZ	7 wt% Y_2O_3-93 wt% ZrO_2, the baseline TBC composition

A gas turbine can be thermodynamically modeled as a Carnot reversible heat engine, with the maximum efficiency given by [1]

$$\eta = \frac{W}{Q} = 1 - \frac{T_L}{T_H} \tag{1}$$

T_H represents the turbine entry temperature or TET, and an increase in this value will increase the overall efficiency of the gas turbine. The desire to increase T_H has driven research on superalloys and thermal barrier coating systems for the past five decades.

Increased creep resistance has been obtained through careful and innovative processing of Ni-based superalloy microstructures. These microstructures have changed from equiaxed (1950–1980) and directionally cast (1970–1990) grains to, most recently, single crystal. Microstructural refinement has improved creep properties dramatically in nickel-based superalloys, but with a melting temperature of 1455°C(1728 K), nickel has its operational limits. For example, a TET of 1200°C(1473 K) would require nickel to operate at 85% of its melting temperature. The temperature limits of nickel-based superalloys have been overcome by internal cooling in the first-stage turbine blades and since the 1980s via the use of thermal barrier coatings. Although the development of single-crystal Ni-based superalloys has increased operating temperature limits by ~50°C over earlier directionally cast parts [2], this is significantly less than the 160–200°C improvement in operating temperature that comes from a 200-μm-thick thermal barrier coating or TBC [3]. TBC systems are now applied to static parts such as combustors as well as rotating turbine blades.

Thermal barrier coating systems, defined here as the three-layer structure composed of the ceramic topcoat, thermally grown oxide (TGO), and bondcoat, are applied to the surface of the superalloy. A schematic representation is shown in Fig. 1. The industry standard ceramic topcoat is 7–8 wt% Y_2O_3-stabilized ZrO_2 (7YSZ) and is applied using either electron-beam physical vapor deposition (EB-PVD) or plasma-spray (PS) processes, depending on the location of the component in the engine. The TGO is not directly applied, but rather is a stable oxidation product of aluminum (from the bondcoat) and oxygen diffusing through the 7YSZ. The bondcoat is typically a MCrAlY, where M is Ni or Co, or platinum-modified nickel aluminide (Ni,Pt)Al. Because of the desire to lower transportation costs, the expectations for TBCs have continued to increase. Whereas TBCs were once beneficial add-ons that provided additional protection for a superalloy structure that could survive without the coating, the goal is now for TBCs to become prime-reliant systems. Thus, the superalloy is

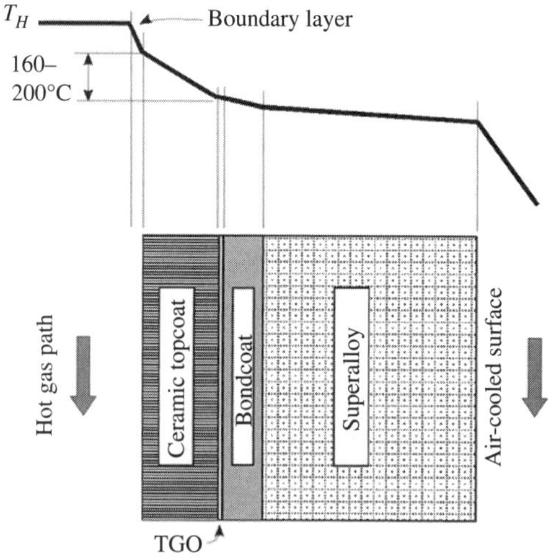

Fig. 1 Schematic representation of thermal barrier coating, including low thermal conductivity ceramic topcoat, thermally grown oxide layer, and bondcoat (adapted from [4]).

operated at its temperature limit, and the coating surface offers protection to several hundred degrees higher. The end result is that the turbine can be operated at a greater T_H, and thus a gain in efficiency is achieved. In this chapter, the essential components of this system will be presented and discussed with reference to their high-temperature performance and interactions with their environment.

I. SERVICE REQUIREMENTS FOR THERMAL BARRIER COATINGS

The gas turbine for aviation or industrial purposes is an open cycle described best by an idealized Brayton cycle. Air is compressed at ratios of 25:1 and higher, mixed with fuel in the combustor chamber, ignited, and allowed to expand through the turbine section. In a turbojet, the work of expansion is used to spin the shaft that compresses the air and to provide aircraft thrust. In a turbofan, additional thrust comes from a fan added to the front of the engine. The highest temperatures occur in the combustor and in the first turbine section; these are not static temperatures but depend upon the flight patterns of the aircraft [1]. For example, the TET is greatest during takeoff and landing, but is still quite high during high-altitude cruising.

High temperatures drive many basic physical phenomena, as noted by Padture et al. [4], including diffusion, oxidation, thermal expansion and conduction, radiation, and microstructural phase transformations and sintering. The high surface temperatures on turbine parts create large thermal gradients through the thickness of the TBC, which produce stresses within the plane of the coating. These stresses vary from compressive during engine startup to tensile on cooldown. High-temperature compressive relaxation of the coating causes the TBC to permanently shrink, and the subsequent reversal of stress state during cooldown can cause through-thickness cracking and delamination. In addition to temperature extremes and gradients, erosion of the TBC on the leading edge of individual turbine blades is common because of high rotation speeds and the interaction of the TBC with micron- to millimeter-sized debris from parts of the coating system upstream, as well as particulates in the air, such as calcium magnesium aluminosilicates (CMAS) below 1240°C, or ash. Impurities in the fuel or air that are ingested by the engine (e.g., sulfur, sodium, vanadium, CMAS above 1240°C) can also corrode and/or infiltrate the ceramic topcoat causing substantial damage to the TBCs. The development of a nonplanar TGO at the interface between the ceramic topcoat and the bondcoat can cause in-plane cracks to develop and eventually link up, causing delamination of the thermally protective layer.

II. CERAMIC TOPCOAT: 7 WT% Y_2O_3-STABILIZED ZrO_2

The outermost layer of the TBC, the ceramic topcoat, is subject to the highest temperatures and interacts directly with the environment. Thus, the ideal ceramic topcoat should 1) possess a low thermal conductivity to maximize the temperature drop between it and the metallic structure, 2) be microstructurally stable at high temperatures, 3) exhibit thermal shock resistance sufficient to withstand numerous heating/cooling cycles associated with gas turbine usage, 4) have a CTE matched to the superalloy to minimize thermally generated stresses, 5) be tough enough to resist erosion and impact with small particulates, and 6) resist reaction and infiltration with foreign corrosive species. Since the development of thermal barrier coatings in the 1970s, the ceramic topcoat has most often been composed of zirconia (ZrO_2) with its crystal structure stabilized by yttria (Y_2O_3).

A phase diagram for the ZrO_2-Y_2O_3 system is shown in Fig. 2, reproduced from [5]. Zirconia exists in three crystal structures: monoclinic *m*, tetragonal *t*, and cubic *c*. While pure zirconia adopts the monoclinic crystal structure at room temperature, the addition of a stabilizer like Y_2O_3 can change its

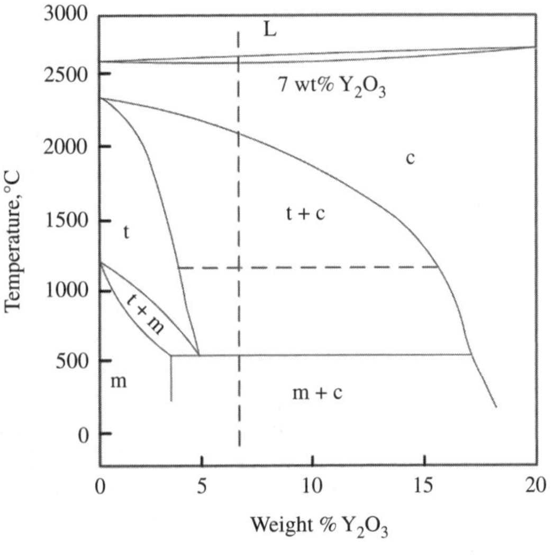

Fig. 2 Phase diagram for Y_2O_3-ZrO_2. The standard ceramic topcoat composition used in thermal barriers is 7 wt% Y_2O_3/ 93 wt% ZrO_2 (reproduced from [5]).

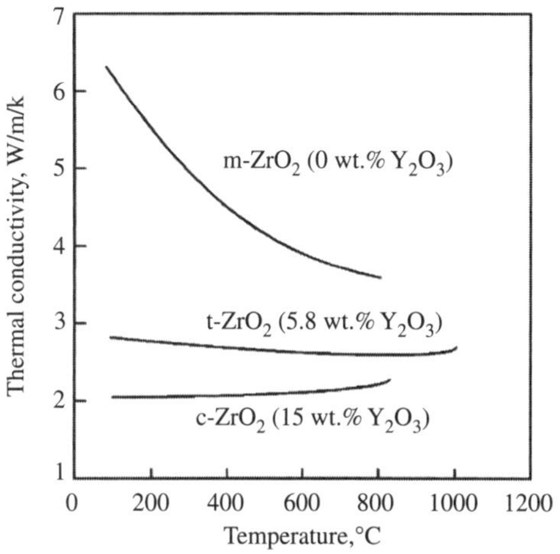

Fig. 3 Plot of thermal conductivity vs temperature for three ZrO_2 polymorphs. Crystal structure was controlled by the amount of yttria added to the zirconia. (This figure was adapted from [10]). Samples ranged from 96 to 100% dense. Values are subject to $\sim\pm 6\%$ error.

equilibrium crystal structure as the 22% larger Y^{3+} ion replaces the smaller Zr^{4+} ion on its lattice site. As Y^{3+} is added, it first stabilizes zirconia in the t-ZrO_2 crystal structure, and with further additions, c-ZrO_2 is formed.

The equilibrium phase diagram for 7YSZ would predict both m-ZrO_2 (containing 3 wt% Y_2O_3) and c-ZrO_2 (with 16–18 wt% Y_2O_3) at room temperature. In fact, this is not the case because of nonequilibrium processing conditions (that is, high quench rates) associated with either PS or EB-PVD application of 7YSZ. Rather than the equilibrium phases predicted, a nonequilibrium single phase material, often identified in the literature as t'-ZrO_2, forms [6, 7]. Unlike t-ZrO_2, which will transform to m-ZrO_2 upon cooling, t'-ZrO_2 is nontransformable. This is desired, as the transformation of t- to m-ZrO_2 is associated with a 3.5% volume increase, which can lead to cracking and delamination of the coating [8]. The large amount of yttria in the t'-ZrO_2, 7–8 wt%, contributes to crystal stability. Because of similar lattice parameters, differentiation of t-, t'-, and c-ZrO_2 by x-ray diffraction requires careful analysis of 2Θ peaks between 72–76 deg with respect to published lattice parameters [9].

In addition to stabilizing zirconia, Y^{3+} provides another benefit. Because of its valence difference, for every 2 Y^{3+} ions that substitute on Zr^{4+} atom sites, an oxygen vacancy is created to preserve charge neutrality. This has an important consequence for the thermal properties of 7YSZ. The t-ZrO_2 with 7–8 wt% Y_2O_3 stabilizer has a significant number of oxygen vacancies, which interact with thermal energy (phonons) to reduce thermal transport. The net effect is to lower the thermal conductivity (k_{th}) of the material as shown in Fig. 3 [10]. Phonon scattering is not observed in m-ZrO_2 because of the limited solubility of yttria in the crystal; thus, m-ZrO_2 demonstrates a higher k_{th}. The lower the k_{th}, and thus the greater thermal gradient across the thin ceramic topcoat, the more protection afforded the superalloy. Because 7YSZ has a high concentration of oxygen vacancies, it has a very high oxygen diffusivity of 10^{-11}–10^{-10} m^2/s [11]. This allows for quick oxygen transport through the ceramic topcoat,

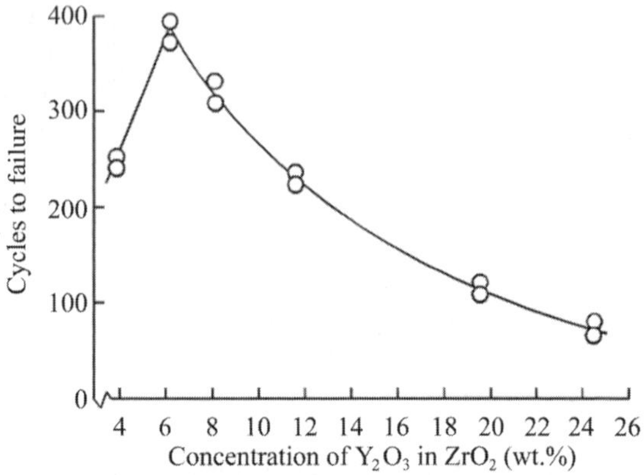

Fig. 4 Effect of yttria content on lifetime of a ZrO_2-Y_2O_3/NiCrAlY bondcoat subject to 1 h/1100°C thermal cycles, as adapted from [13].

enabling growth of the TGO. Additional oxygen transport takes place as a result of the porous nature of PS or EB-PVD coatings.

The effect of the high combustor temperatures T_H is to partition nonequilibrium t'-ZrO_2 (with 7–8 wt% Y_2O_3) into equilibrium t-ZrO_2 (3–4 wt% Y_2O_3) and c-ZrO_2 (14–16 wt% Y_2O_3) phases via diffusion of Y^{3+} ions (according to the phase diagram). This transformation will not be observed unless the sample is 1) held above ~1250°C for 50+ h and 2) cooled down below the transformation temperature. As can be seen in Fig. 2, this is slightly greater than 500°C. There is evidence that this transformation changes the mechanical properties of plasma-sprayed coatings. For example, Wallace et al. [12] observed that the elastic modulus measured at room temperature dramatically increased after hold times less than 10 h at 1400°C, but dramatically decreased after hold times greater than 10 h at the same temperature. The initial increase in modulus is because of sintering of the microstructure and the decrease in modulus after the 10 h hold at 1400°C is correlated with the t-ZrO_2 to m-ZrO_2 transformation.

It would be a reasonable question to ask, "Why not fully stabilize the zirconia to the cubic phase to lower k_{th} and provide the most stable crystal structure?" The answer is that studies performed by Stecura [13, 14] revealed limited cycles to failure for fully stabilized zirconia, as shown in Fig. 4. In fact, the 6–8 wt% Y_2O_3 baseline composition used for most ceramic topcoats is based on the peak in cycles necessary to fail the topcoat as well as increased erosion and toughness compared to other candidate coating materials. Related to the compatibility with the superalloy structure is the relatively high coefficient of thermal expansion (for a ceramic) of 9–11 × 10^{-6}/°C for 7YSZ. Finally, the density of 7YSZ is 6.08 g/cm^3.

III. APPLICATION METHODS FOR THE CERAMIC TOPCOAT

The ceramic topcoat is generally applied one of two ways, either through plasma spray (PS) or electron-beam physical vapor deposition (EB-PVD). Nonrotating parts, like a combustor chamber or guide vanes, are typically plasma sprayed.

Turbine blades, with integrated cooling holes, generally use vapor-deposited coatings. PS TBCs are generally applied more thickly than EB-PVD TBCs at 300 and 120 μm, respectively.

A. PLASMA SPRAY

Plasma spray involves injecting >10-μm-diam powders into a plasma jet plume, where they are melted and propelled toward the gas-turbine structure. Melted droplets flatten upon impact and quickly cool (10^6 K/s) [15]. Coatings are formed by rastering the plasma gun over the part to be coated, and deposition rates can be 10–50 μm/pass. The plasma-sprayed microstructure is extremely complex and defective as shown in cross-sectional view in Fig. 5 [16, 17]. Small interlamellar pores (<0.1 μm) exist between the flattened powders or lamellae; the actual contact between adjacent lamellae has been suggested to be as low as 20% [15]. In addition, intralamellar microcracks are numerous as a result of the thermal stresses that develop during cooldown. The orientation of the pores and microcracks within the coating has been extensively studied using small-angle neutron scattering (SANS) by Ilavsky and co-authors [18–20]. Results indicate that the long axis of the oblate interlamellar pores and the intralamellar cracks are orthogonal to one another, with the pores existing within the plane of the coating and the microcracks oriented parallel to the spray direction. Furthermore,

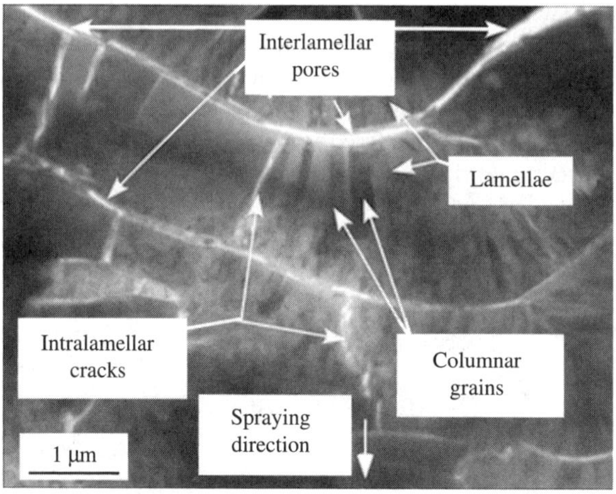

Fig. 5 TEM image showing microstructural features of plasma-sprayed 7YSZ from [17]. Columnar grains observed in each lamella result from heat flow into substrate during processing. Interlamellar porosity exists between lamella, and intralamellar cracks tend to follow the columnar grain boundaries. To a first approximation, interlamellar porosity and intralamellar cracks are orthogonal and parallel, respectively, to the spray direction.

the microstructure is strongly influenced by the plasma spray parameters that influence the velocity [21–25] and temperature [25, 26] of the powder. Overall porosity in PS coatings is typically 5–20%, depending on the spray parameters.

B. EB-PVD

Electron-beam physical vapor deposition involves directing a high-energy electron beam at the ingot to be vaporized (in this case, 7YSZ). The vapor deposits as solid in the form of a coating on multiple rotating turbine blades (preheated to ~1000°C) within the line of sight of the vapor. Deposition rates are usually no greater than 1 μm/min and can be as low as several nanometers per minute. The microstructure of EB-PVD coatings is composed of columnar grains, as shown in Fig. 6 (adapted from [27]) with porosity between the columns and very fine porosity within the columns [28]. The columnar structure enhances the compliance of the coating, and this can contribute to increased coating lifetimes. Despite having few cracks compared to PS coatings, pathways do exist between the columnar grains (along the porosity), allowing infiltration of corrosive species. EB-PVD coatings are typically much smoother than PS applied coatings.

A comparison and discussion of the differential costs associated with each process is presented in [29]. Costs are broken down into three categories to include one-time costs such as capital equipment, direct coating-related costs involving coating materials and substrate preparation costs, and indirect cost factors as associated with equipment maintenance and tooling replacement. Although the capital equipment costs required for a PS system are ~$500,000, costs can be in the millions of dollars for EB-PVD systems. Labor costs, however, are analyzed to be higher for PS parts. Thus, all three costs, along with part volume, must be considered in any economic analysis.

The microstructural differences of 7YSZ coatings made from PS and EB-PVD result in different thermal and mechanical properties for each coating. EB-PVD coatings generally possess a higher as-fabricated k_{th} than PS coatings,

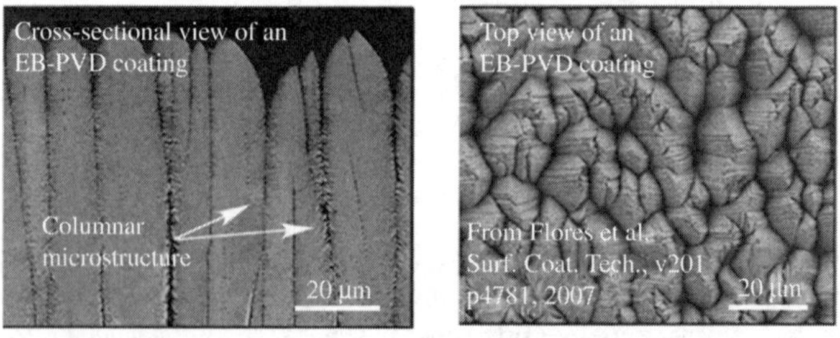

Fig. 6 Typical microstructure of EB-PVD coating of 7YSZ from cross-sectional and top views, showing the columnar grains (from [27]).

~1.5 W/m/K vs ~1–1.2 W/m/K. This difference is caused by lower amounts of porosity and the shape of the porosity found in EB-PVD coatings. EB-PVD coatings are, however, much more compliant than PS coatings because of the columnar microstructure, which can accommodate thermally applied strains. For this reason, EB-PVD coatings are often used on critical parts, such as turbine blades.

IV. INTERACTION OF THE 7YSZ CERAMIC TOPCOAT AND THE ENVIRONMENT

A. EFFECT OF TEMPERATURE ON MICROSTRUCTURE AND PROPERTIES OF PS AND EB-PVD 7YSZ COATINGS

Dramatic changes occur in the microstructure of plasma-sprayed coatings in service. As shown by small-angle neutron scattering [30] and TEM [31, 32], the microstructure in PS coatings will sinter at high service temperatures (800–1300°C). Intralamellar cracks tend to heal between 800 and 1000°C, and interlamellar pores tend to sinter above 1000°C [30]. Thus, the two predominant defects disappear in two temperature regimes. Using Mullins' thermal grooving theories, Erk et al. [33] showed that at 1200°C and above, volume diffusion is the dominant mass transport mechanism. This is significant because at this representative service temperature, interlamellar pores, which act to lower the k_{th}, can completely close during sintering. EB-PVD coatings also sinter with exposure to high temperatures. Lughi et al. [34] noted that the feathery morphology of the 7YSZ columns first smoothes upon high-temperature exposure and then develops low-amplitude undulations. Next, some of the undulations on the columns grow in amplitude and bridge adjacent columns, forming rows of local sintering necks. Because the cracks and pores provide a pathway for gases and liquids, changes in the morphology that occur as a result of high-temperature service will change infiltration of potentially damaging corrosive species.

Closure of the microcracks and pores, and the m-ZrO_2 phase change, affect the coating properties. For example, elastic modulus doubles or triples as a result of sintering of the coating after exposure to simulated use temperatures. In one study, researchers [35] observed an increase in modulus from 12 to 33 GPa after 12 h at 1200°C. This increase results from a change in the morphology of the PS grains from lamellae comprised of columnar grains (as-sprayed) to equiaxed grains with necking between adjacent grains [16]. The effect of the increase in modulus is to increase the stress transferred from the substrate to the coating, which can lead to either higher spallation stresses when the coating is under compression or more severe through-thickness coating cracks when the coating is under tension.

The k_{th} of PS 7YSZ is not constant, but dramatically changes with extended hold times at temperatures above 800°C [36]. For example, the as-plasma-sprayed k_{th} of 7YSZ can more than double (from 1.2 to 2.7 W/m/K) after 50 h at 1200°C

[32]. The observed changes in k_{th} have been linked to the changes in microstructure described earlier. After 50 h at 1400°C, a continued increase in k_{th} is believed to be a result of the t-ZrO_2 to m-ZrO_2 transformation, increasing the volume fraction of the high conductivity phase. Petorak et al. showed that the application of stress and temperature can increase k_{th} of 7YSZ above that expected for only exposure to temperature [37]. For comparison, Renteria et al. [28] found the average thermal conductivity of an EB-PVD coating increased from 1.2 W/m/K in the as-deposited condition to 1.4 W/m/K following 100 h at 1100°C. Also, Zhu et al. [38] observed an increase in k_{th} from 1.5 to 1.8 W/m·K following a 200-h total time test where the surface temperature reached 1280°C. Increases in k_{th} were attributed to sintering of the microporosity in the EB-PVD prepared coatings.

B. EFFECT OF CORROSIVE SPECIES ON 7YSZ

Hot corrosion is defined as the accelerated degradation resulting from the deposition, infiltration, and/or reaction of a corrosive species with the ceramic topcoat [39]. Penetration of the molten species is greatly enhanced by the porous and/or microcracked microstructure intrinsic to PS or EB-PVD coatings. The primary contaminants causing hot corrosion in a gas turbine include sodium, sulfur, vanadium, glass-forming silicates (such as Ca-Mg-Al silicates or CMAS), and volcanic ash. [40]. Because of the large quantity of hot gas passing through the turbine, significant accumulation of impurities is possible even though their concentration might only be parts per million. For example, an estimate on General Electric gas turbines indicated 1 ppm of contaminant is equivalent to 0.45 kg of contaminant entering the turbine section every 50 hours [41].

Sodium is a universal contaminant in fuel, but might also contaminate the fuel during marine transportation. In aviation gas-turbine fuels, the sodium and sulfur contents are relatively low, approximately 0.05 and 0.24 wt%, respectively. The sulfur amount is about 1 wt% in typical industrial fuel and reaches 4 wt% in heavy oil [42]. As high as the sulfur content seems to be in many fuels, sulfur is generally not harmful without sodium. In fact, turbine components can tolerate up to 1 wt% sulfur in the absence of trace metal contaminants [43]. When combined with sodium, however, sulfur forms corrosive deposits that are detrimental to the life of turbine parts. Sodium sulfate, which melts at 884°C and decomposes at 1100°C, is the predominant constituent of salt deposition inside gas turbines [44]. In addition to the Na_2SO_4 already present as an impurity of sea salt, ingested NaCl is converted to Na_2SO_4 by oxides of sulfur in the combustion environment.

Vanadium impurities come almost exclusively from the fuel and are oxidized during combustion. Vanadium oxide forms a liquid in the turbine as the service temperature exceeds its melting temperature of 690°C [45]. Their concentrations are negligible in natural gas, but are of concern in less-refined liquid fuels, as well as syngas produced from gasification of coal and liquid fuel. Compared to sodium sulfate hot corrosion, the chemical reaction between vanadium contamination and the zirconia-based coatings is much faster and therefore more deleterious. Chen et al. [46, 47] showed that at temperatures as low as 700°C the reaction can proceed rapidly. Simultaneous reaction of V_2O_5 with ZrO_2 and Y_2O_3 (in solid solution with the ZrO_2) forms YVO_4 and ZrV_2O_7. More importantly, one of the final reaction products is m-ZrO_2 [48, 49]. Vanadium compounds combine with sodium sulfate to form a eutectic liquid with an even lower melting temperature [50] thus expanding the temperature range of hot corrosion. In addition, the incorporation of vanadium compounds into the salt deposit will increase the solubility of the protective oxides, thus accelerating acidic fluxing [51].

With the drive to operate gas turbines at hotter temperatures, the surface of TBCs has exceeded 1250°C. This has resulted in a new threat to TBCs from airborne impurities that include CMAS [52] and volcanic ash [53, 54] that are pulled into the gas turbine. At lower operating temperatures, these particulates remain solid and only increase the surface erosion on the leading edge of the blade as they pass through the turbine. At higher engine operating temperatures, they become molten and infiltrate the coating. Several researchers have noted that this sand particulate leads to premature spallation of the coating from the substrate because of thermal shock via deterioration of thermal strain tolerance [55, 56]. While the composition of CMAS is variable depending on geographic location [52], a composition of 33CaO-9MgO-13$AlO_{1.5}$-45SiO_2 (in mol% oxide) is often studied [57, 58]. With the small addition of Fe or Ni (from alloys upstream of the deposit), it is believed that the melting points over a broad range of CMAS compositions would be ~1200°C [58].

Volcanic ash is formed as a result of the violent nature of eruptions, with one mechanism of formation a consequence of the interaction of steam and magma. This interaction can propel ash high up into the air. For example, The Anatahan Volcano of the Commonwealth of Northern Mariana Islands created an ash cloud 13.4 km (44,000 ft) high, easily above the cruising range of jet aircraft [59]. According to an *Oxford Economics* report, global aviation losses associated with the 2010 eruption of Eyjafjallajökull in Iceland in 2010 were tallied at 2.2 billion dollars [60]. Productivity losses were valued at 490 million dollars from stranded passengers (~7 million) and loss of perishable items. The volcanic ash cloud caused 80% of European flights to be cancelled at the peak of eruption. Although large pieces of tephra (>50 μm) will fall back to the Earth in a matter of hours [59], the smaller ash can remain suspended for days and be subsequently spread by wind. It is not the highly visible ash clouds that are critical as aircraft can easily detect these and maneuver around them [61]. Rather, it is the diffuse ash clouds, like the one that a NASA DC-8 aircraft flew through in February of 2000 [62, 63], that seriously degrade an engine. Ash composed of small particles (<10 μm) has low reflectance and might not be detected by weather radar [59]. Complicating strategies to address the interaction of the ash and gas-turbine components is the fact that the composition of the ash depends on where it is formed. For example, ash from Mount St. Helens, which erupted in May of 1980, was composed of 65 mol.% SiO_2, 18 mol.% Al_2O_3, 5 mol.% Fe_2O_3, 2 mol.% MgO, 4 mol.% of CaO and Na_2O, 0.1 mol.% S along with 37 trace metals [64].

Interaction of jet engines with diffuse ash clouds can cause engine flameout [62], an extremely dangerous condition in which improper mixing of air and fuel extinguishes the flame in the combustion chamber. Longer-term problems include drastically shortened lifetimes for key engine components, such as the turbine blades, as a result of ash/7YSZ coating interactions [61]. Just as important as the chemical reaction between the ash and the coating is the erosive effect of the particulate as it interacts with moving engine hardware. Cooling holes in the turbine blade, essential for reducing the overall temperature of the structure, can become filled with ash. Plugged cooling holes cause the entire thermal barrier system to overheat and can hasten reaction kinetics between the 7YSZ and the ash and/or other corrosive species.

C. MICROSTRUCTURAL INFILTRATION OF THE CORROSIVE SPECIES

PS and EB-PVD coatings are permeable to infiltration of molten corrosive phases because of their porous and/or cracked microstructures. McKee et al. [65] examined the hot corrosion behavior of PS 7YSZ coatings in burner rig tests using NaCl and sulfur-doped fuel. After tests at 871°C, Na_2SO_4 was found in the pores of the zirconia topcoat. Moreover, sulfide formation was detected at the bond coat/substrate interface, which indicates that molten sulfate (or possibly a gaseous S_2 phase) had penetrated through both the zirconia layer and PS NiCrAlY bond coat. This finding demonstrates that the penetration of the molten salt can endanger the bondcoat and superalloy.

When Chen et al. [66] studied the infiltration behavior of V_2O_5 compounds into PS 7YSZ, they observed three distinct regions. The first region, a planar reaction zone, was caused by the V_2O_5 melt reacting extensively with the surface 7YSZ. This dissolution reaction front proceeded in a planar fashion. The second region was a melt infiltration reaction zone (MIRZ), where the liquid V_2O_5 infiltrated deep into the coating via the porosity and microcracked pathways. Reaction of the V_2O_5 with the PS 7YSZ coating was only noted near the infiltrated pores and cracks. For example, a previously open pore was filled with the liquid V_2O_5 during infiltration, which, as determined by a combination of X-ray diffraction and TEM, transformed the t'-ZrO_2 into m-ZrO_2 and YVO_4 [67]. The third region was below the MIRZ and was unaffected by the corrosive liquid.

Krämer et al. [58] noted that CMAS just above its melting temperature ($\sim$1240°C) completely infiltrates a 200-μm EB-PVD coating in less than 4 h, dissolving the t'-ZrO_2 phase and reprecipitating a phase determined by the local composition. TEM micrographs revealed the formation of a thin reaction layer in the CMAS-infiltrated pores. The reaction layer was determined to be an yttria-depleted zone. Reaction of the corrosive liquid with the TBC also can affect the pore geometry. For example, the reaction products of $Gd_2Zr_2O_7$ (an alternate TBC composition) and CMAS tend to close the pores to further infiltration [68] while V_2O_5 can open the pores or cracks by dissolving the 7YSZ [66].

D. EROSION OF TBCs

Prior studies on TBC erosion have investigated the effect of coating application method (APS or EB-PVD), chemical composition, erodent particulate size and velocity, and temperature [69–73]. Beginning with microstructure studies, Wellman and Nicholls [69] compared the erosion behavior of as-sprayed APS and as-deposited EB-PVD 7YSZ coatings, noting that EB-PVD coatings possess ~10 times the erosion resistance of APS. EB-PVD coatings impacted by hard particles tend to locally damage the tops of the columnar grains, which then can break off in small fragments. Under similar testing parameters, APS coatings tend to fracture at the interface between lamella, coming off in larger pieces. Thus, APS coatings demonstrated much higher erosion rates (~200 g of coating per kg of erodent) than their EB-PVD counterparts (~20 g/kg). As testing is conducted at realistic higher operating temperatures, more erosion of the coating is observed [70]. For example, erosion of APS 7YSZ was measured to be ~200 g/kg at 25°C and ~300 g/kg at 910°C, when impacted with 40-μm alumina normal to its surface.

Janos et al. [71] investigated the effect of thermal aging on erosion resistance and found that 7YSZ APS coatings aged for times/temperatures as short as 16 h at 1260°C exhibited 50% less erosion than the as-sprayed coatings. This is presumably because of sintering of the interlamellar pores and cracks. Particulate size was shown to affect the erosion rate of APS coatings with a marked increase noted for particles ~20 μm and larger [72]. The composition of the coating plays a role in erosion behavior, as observed by Steenbakker et al. [73], who showed that Gd-doped zirconia TBCs possess a lower erosion resistance than 7YSZ at elevated temperatures.

Particle velocity and size, along with impacting temperature, also play a key role in the erosion failure mechanism observed. Wellman and Nicholls [69] categorize the erosion mechanisms for EB-PVD coatings into three damage modes. Mode I is erosion near the surface, caused by surface cracking and lateral cracking. This is typically observed for small particulate striking the surface with lateral cracks forming 20 μm below the surface. Subsequent impacts will tend to link these cracks until parts of the coating delaminate and the process is repeated again. The Mode II deformation mechanism is compaction below the impact site that is attributed to closure of the nano-sized porosity between columns. This type of damage occurs for intermediate-sized particulate and would be favored to occur at elevated temperatures. Mode III deformation is associated with larger particulate sizes or foreign object damage. Gross plasticity under the impact site is observed, consistent with high-temperature deformation. Chen et al. [74] published a modeling study with a strong correlation to the experimental observations of Wellman and Nicholls [69].

V. ALTERNATE COMPOSITION CERAMIC TOPCOATS

Zhu and Miller [75] have developed low-conductivity and sintering-resistant coatings using a multicomponent oxide defect-clustering approach. Their approach was to design 4.5 mol% Y_2O_3-ZrO_2 (equivalent to 7YSZ) coatings, but with the addition of multicomponent, paired-cluster, rare-Earth oxide dopants. These additional oxide dopants were selected to promote the creation of thermodynamically stable, highly defective lattice structures with immobile defect clusters [76–78]. The clusters were designed to attenuate and scatter lattice phonon waves and also scatter radiative photon waves, lowering k_{th} of the coating as compared to 7YSZ-only coatings. As evidence, consider the k_{th} data presented in Fig. 7 for two conventionally prepared plasma-sprayed coatings [77]. Using a laser heat flux to heat the surface to 1316°C for 1–20 h revealed very significant differences in thermal conductivity between baseline 7YSZ (no dopants added) and a 7YSZ coating with an additional 9 mol% Nd_2O_3-Yb_2O_3 dopant. Both the k_{th} and the rate at which k_{th} was increasing in the 9 mol% $(Nd_2O_3/Yb_2O_3)/4.5$ mol% Y_2O_3-ZrO_2 coatings were significantly reduced compared to 4.5 mol% Y_2O_3-ZrO_2 coatings. Other dopant pairs, such as Gd_2O_3-Yb_2O_3 and Gd_2O_3-Nd_2O_3, added to 7YSZ also demonstrated a lower k_{th} than

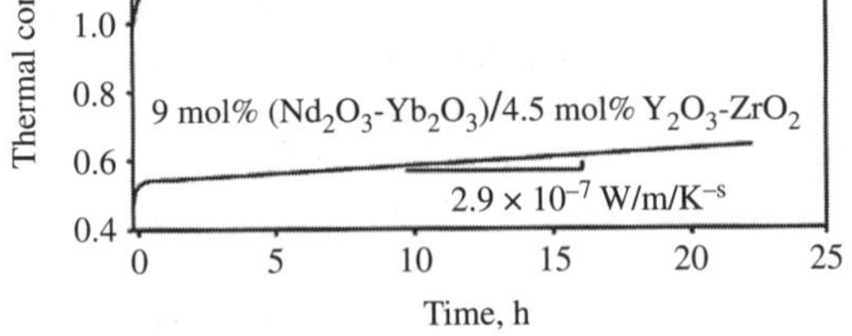

Fig. 7 Plot of thermal conductivity vs time at 1316°C comparing the response of standard 7YSZ coatings and coatings engineered to have a defect oxide structure using the approach of Zhu and Miller. The coating with an additional 9 mol% Nd_2O_3-Yb_2O_3 dopants added to the 4.5 mol% Y_2O_3 found in the ZrO_2 demonstrated a lower overall thermal conductivity, and less effect of sintering. (The figure was adapted from [77].)

coatings of only 7YSZ. High-resolution transmission electron microscopy of the plasma-sprayed microstructures revealed that extensive nano-scale rare-Earth dopant segregation exists within the zirconia lattice, causing large lattice distortion. It was suggested that the presence of the 5–100-nm clusters was the cause of the low k_{th} observed in these types of coatings [78].

An excellent review of alternate TBCs has been published with an overview of key physical properties of alternate TBC materials [79]. More recently, $Gd_2Zr_2O_7$ [80, 81] has come to be of interest because of its low thermal conductivity (approximately 30% lower than 7YSZ). Furthermore, there is evidence that it resists infiltration of CMAS [67, 82] via formation of crystalline products that fill the pores [67]. Garnet ceramics, such as $Y_3Al_5O_{12}$ or YAG, have also been suggested, as they possess ~10 orders of magnitude lower oxygen diffusivity than 7YSZ, limiting TGO growth [83]. Layered structures of 7YSZ with alternating coating materials might eventually provide the best combination of thermal, mechanical, and physical properties.

VI. OTHER COATING FABRICATION METHODS

There is currently a drive to fabricate coatings with smaller, even nanometer-sized microstructural features. In PS this can be challenging as powders smaller than 10 μm cannot be fed using conventional methods as a result of electrostratic attractions. Two approaches have emerged to overcome this limitation: solution precursor plasma spray (SPPS) [84–87] and suspension plasma spray (SPS) [88–92]. In SPPS, a liquid ceramic precursor is sprayed into a plasma plume where it fragments into small droplets. Depending on the processing variables, each droplet then experiences some or all of the following: evaporation of the solvent, precipitation, pyrolysis, and melting. Alternatively, SPS involves first suspending typically <200 nm-size powders in a liquid, generally ethanol, and then injecting the suspension into the plasma plume. In SPPS, composition of the final coating is controlled by the ceramic precursor(s). The precursor used for forming 7YSZ is an aqueous solution containing zirconium and yttrium salts [84]. In SPS the composition is controlled by the ceramic powder although it was recently shown that the composition of the final coating can be controlled via the addition of soluble nitrate rare-Earth compounds into the ethanol-based suspension [93]. Both SPPS and SPS are injection technologies with only limited modification of standard PS equipment required. Microstructures for coatings produced by these two methods can range from a dense collection of mostly spherical nanometer-sized features to stacked lamella as in typical PS coatings.

Low-pressure plasma spray (LPPS) is the general name designating plasma spray that occurs in a vacuum with the plasma gun located within the vacuum chamber. Plasma spraying at pressures below 5 torr is referred to as very low-pressure plasma spray or VLPPS. In an important discovery, Refke et al. [94] of

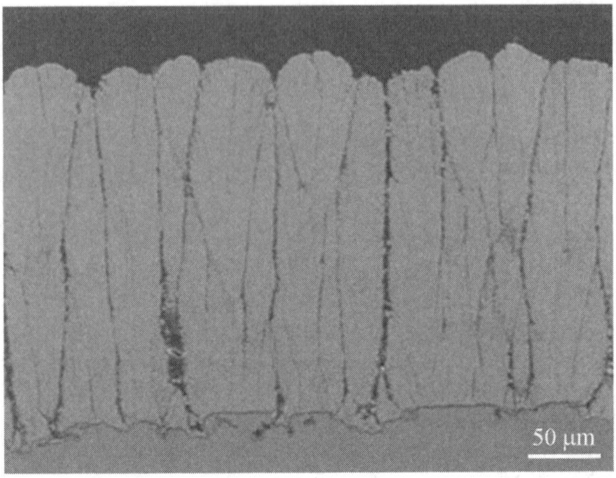

Fig. 8 SEM image of the microstructure created by very low-pressure plasma spray. The microstructure is similar to that produced using EB-PVD. (Micrograph is courtesy of Sulzer Metco.)

Sulzer Metco have demonstrated recently that 7YSZ coatings can be made from a vapor, rather than droplet buildup, by carefully controlling the processing conditions. In the Refke et al. [94] study, large 15-μm-diam spray-dried 7YSZ powders were fed into a plasma gun operated at ~90 kW, a power sufficient to vaporize the powders. The standoff distance between the gun and the substrate varied between 900–1300 mm, and the chamber pressure was maintained at ~1 torr. By carefully controlling the temperature of the superalloy substrate, it was possible to keep the microstructure of the 7YSZ coating similar to that formed by EB-PVD methods, as shown in Fig. 8. Deposition rates can be as high as 10–20 μm/min depending on processing conditions and substrate configuration. This new process is referred to as plasma spray/physical vapor deposition, or PS-PVD. The newly discovered PS-PVD process fills the gap between the conventional PVD technologies and standard thermal spray processes.

VII. BONDCOATS

Between the ceramic topcoat and the superalloy substrate (see Fig. 1), a metallic bondcoat layer is employed. The bondcoat is typically 50–100 μm thick, depending on the topcoat, superalloy substrate, and application. Although ceramic topcoat layers can adhere directly to the superalloy component surfaces, their permeability to gases as a result of microcrack or porosity networks leads to strongly oxidizing conditions at the interface. Superalloys have good oxidation resistance, but not good enough to prevent significant scale formation and associated volume increase at the interface, leading to decohesion. A primary role of the bondcoat is to provide enhanced oxidation resistance to limit oxide growth at the interface. This effect is achieved through the presence of higher Cr and Al concentrations in the bondcoat than in the superalloys.

The two main classes of bondcoats currently in use are MCrAlY, applied by plasma spray and used mainly with plasma-sprayed topcoats, and (Pt,Ni)Al, applied by aluminization and diffusion-reaction with electroplated Pt and used exclusively with the EB-PVD topcoats. These systems are reviewed as follows.

A. MCrAlY BONDCOATS

The original bondcoat compositions for 7YSZ topcoats emerged from the overlay coatings based on NiCrAlY applied to superalloy components for oxidation/corrosion resistance prior to the development of TBC systems [95]. Cobalt can be substituted for some or all of the Ni, so that the composition is commonly designated MCrAlY. Although the base metals (Ni and Co) form relatively nonprotective oxide scales, in the presence of sufficient Cr and Al selective oxidation leads to the formation of continuous, adherent scales of the more stable oxides, Cr_2O_3 and Al_2O_3. Furthermore, the much lower diffusivity in these oxides compared to base metal oxides leads to slow scale growth, which is vital to the life of the system. MCrAlY compositions typically contain more Cr ($\sim$20–35 wt%) than Al ($\sim$5–15 wt%), and it is possible to tune the selective oxidation behavior to produce continuous Cr_2O_3 or Al_2O_3 scales, depending on the concentrations of Cr and Al in the coating [96]. Alumina scales are usually preferred, although chromia imparts better hot corrosion (sulfidation) resistance. Aluminum oxide exhibits a large number of metastable polymorphs (crystal structures), referred to as transition aluminas. With increasing temperature and certain alloy chemistry effects, which are not well understood, the stable α-Al_2O_3 structure (corundum) forms. A significant feature of the microstructure of α-Al_2O_3 scales is their much larger grain size compared to the transition aluminas.

Another important compositional feature is the addition of Y (typically $<$0.5 wt%) in MCrAlY. Yttrium and other so-called reactive elements (e.g., Zr and Hf) impart dramatic benefit to cyclic oxidation resistance. Without reactive element additions, alumina scales spall during cooling because of thermal expansion mismatch with the substrate. Spallation exposes subscale, which reoxidizes on the next heating cycle. The process repeats with thermal cycling leading to rapid weight loss. The mechanism by which reactive element additions improve scale adhesion and delay the onset of weight loss in cyclic oxidation has been widely studied over many years [97]. It is now well established that the main effect of Y is to get sulfur, which, when present even in low concentrations as impurity in the bulk, segregates strongly to the scale-metal interface, weakening the chemical bonding. Other minor additions to MCrAlY coatings (such as Si) have offered improvements in performance, but these details are beyond the scope of this chapter.

Although MCrAlY coatings were originally produced by diffusion coating though pack cementation processes, plasma spray quickly became the preferred method of application, primarily because of better control of coating composition and thickness. The powders for spraying are typically produced to precise compositions by inert gas atomization. Although air plasma spray can be used, MCrAlY bondcoats are typically applied by low-pressure (vacuum) plasma spray (LPPS) or inert shrouded plasma spray in order to minimize oxidation during the coating process.

The MCrAlY bondcoat typically consists of a two-phase microstructure of β-NiAl and the γ-Ni solid solution. The β-NiAl intermetallic compound phase is brittle whereas the γ-Ni solid solution phase is relatively ductile. At low temperatures, the two-phase bondcoat structure lacks ductility relative to the superalloy substrate and has limited strain tolerance. At high temperatures, on the other hand, the low strength of the γ-Ni is considered detrimental under cyclic deformation condition. Improving MCrAlY bondcoat mechanical properties is a challenge as a result of the tradeoffs in properties with the high Al and Cr concentrations necessary to provide sufficient oxidation resistance.

B. β-(Ni,Pt)Al Bondcoats

This second class of bondcoats evolved from the need for higher performance capabilities of the EB-PVD 7YSZ topcoats employed on single-crystal, high-pressure turbine blades. Improved oxidation behavior is achieved by incorporating Pt in a β-NiAl aluminide coating. The application of these bondcoats involves a two-step process. A platinum layer up to $\sim$10 μm is electroplated on the superalloy surface and then aluminized and alloyed by interdiffusion (hence the name "diffusion" coatings). The resulting structure consists of a single-phase β-(Ni,Pt)Al outer layer up to $\sim$50 μm thick with a multiphase "interdiffusion" layer between it and the superalloy substrate. In some cases, the aluminization step is omitted and the electroplated Pt layer alloys with aluminum from the superalloy substrate.

Although the beneficial effects of Pt on oxidation behavior are well established, many possible mechanism(s) have been investigated. Gleeson [98] has reviewed the proposed mechanisms, which include mechanical keying of the oxide to the bondcoat, enhanced diffusion of aluminum for rapid selective oxidation, improved interface adherence, promotion of the stable α-Al_2O_3 over the transition alumina structures, and suppressing interface void formation. Platinum does not oxidize in air, and so platinum aluminide forms pure alumina scales. Perhaps the most compelling evidence for the Pt effect is that its presence in the nickel aluminide promotes selective oxidation of aluminum so that continuous alumina scales form without the occurrence of less stable nickel oxides [99]. Platinum can physically limit the availability of Ni during the early stages of oxidation [100]. Hayashi and coworkers [101] have also shown that Pt lowers the chemical activity of Al in nickel aluminide coatings, driving interdiffusion fluxes that enrich Al toward the oxide coating interface. These observations led to a new bondcoat composition based on Pt-modified γ-Ni + γ'-Ni_3Al in which the lower aluminum concentration compared to β-(Ni,Pt) coatings reduces the depletion of aluminum from the bondcoat by diffusion into the superalloy substrate [102]. Work continues to optimize oxidation behavior of the Pt-enhanced diffusion coatings and to reduce the amount of Pt incorporated. Indeed, at current prices, the material cost alone for a 10-μm-thick Pt layer is on the order of ~\$1/cm^2.

VIII. TBC SYSTEM FAILURE

Given the complexity of their microstructure and the service conditions under which they function, it should not be surprising that there are a variety of failure modes and mechanisms in TBC systems. Two main extrinsic mechanisms of TBC topcoat degradation have been discussed already in Sec. IV. Hot corrosion occurs as a result of fuel contaminants, as well as ingested salts and mineral dusts that form molten species that infiltrate and react with the topcoat. In addition, hard particle debris ingested or released from coated components within the engine can cause erosion by high velocity impact. Whereas substantial efforts have been leveled at characterizing hot corrosion and developing mitigation strategies, less research has been conducted on erosion.

A variety of intrinsic mechanisms also limit the performance and life of TBC systems. Sintering of porosity, especially in the columnar EB-PVD topcoats (also detailed in Sec. IV), reduces strain compliance and increases k_{th}. By far the main focus of research, however, has been on the process of spallation that ultimately limits coating life. Evans et al. [103] provided a clear framework for understanding the microstructural and micromechanics mechanisms that govern spallation of TBCs. Although there are differences in the details depending on coating system microstructure and service conditions, which are beyond the scope of this chapter, spallation originates from stresses associated with the thermally grown oxide (TGO). Very large residual compressive stresses develop in the TGO upon cooling because of the large thermal expansion mismatch with the superalloy substrate. Large compressive stresses also develop from the growth of the oxide layer itself, their magnitude increasing with increasing TGO thickness (typically ~1 to 10 μm). In one common mode of failure, the interface develops undulations as the TGO relaxes by creep during thermal cycling. Cracks nucleate at defects and grow in the topcoat near the TGO in response to induced normal tensile stress and eventually link together over sufficient area that the topcoat can undergo large-scale buckling. In other system configurations, cracking can occur at the TGO/bondcoat interface or the TGO/topcoat interface. In either case, losing most or all of the topcoat exposes the superalloy to higher temperatures and thus more aggressive oxidation.

In light of the TGO-induced failure mechanisms, slowing the growth rate of the TGO should benefit coating life. At the same time, however, the bondcoat must maintain sufficient aluminum in order to prevent base metal oxides from forming. Because the bondcoat typically contains higher Al concentration (activity) than the superalloy substrate, Al tends to diffuse from the bondcoat into the superalloy substrate. In principle, if TGO growth slowed enough, the bondcoat at the TGO interface could become depleted of Al by backdiffusion into the substrate. The chance for this type of failure may play a role in optimizing bondcoat thickness, although the need for a certain bondcoat thickness may be governed more by the risk of not having sufficient Al reserve in case of spallation failure.

Finally, through a series of systematic experiments, Wu and coworkers [104] have drawn attention to the role of superalloy substrate differences in determining TBC performance. These authors applied three different commercial coating systems on five different single-crystal alloys and studied the cyclic oxidation resistance. The number of cycles to failure trended consistently with coating type within each alloy type whereas significant differences were observed for a given coating type across the different alloys. The initial interface fracture toughness should depend only on the bondcoat chemistry and structure, which was constant (for a given coating type) on the different substrate alloys. Because the test specimen geometry and thermal cycling conditions were fixed, the authors concluded that the different degrees of interface fracture toughness degradation could only be explained by differences in the superalloy substrate chemistry. Although the particular alloying elements responsible for the differences were not identified, these results do provide strong evidence that the particular superalloy substrate (chemistry) plays as important a role in coating performance as the coating system itself.

IX. CONCLUSIONS

TBC systems represent a high level of materials selection and processing sophistication. Clearly, the development and use of TBCs has driven many of the increases in gas-turbine operating temperature, making these complex systems more fuel efficient. Balancing thermal and mechanical properties is no easy task, but is being addressed by the large team of engineers and scientists in academia, national laboratories, and industry working to solve these problems. The desire for ever-increasing engine efficiency, and therefore increases in T_H, will continue to push the envelope of TBC usage. Future operating conditions, including demand for continued increases in operating temperature combined with exposure to aggressive environments, will present a new set of scientific and engineering problems. Whether the solutions to future technical requirements will be found in existing, alternative, or even new TBC materials remains to be seen.

REFERENCES

[1] Reed, R. C., *The Superalloys: Fundamentals and Applications*, Cambridge Univ. Press, Cambridge, England, U.K., 2006, pp. 8–11.

[2] Clarke, D. R., and Phillpot, S. R., "Thermal Barrier Coating Materials," *Materials Today*, Vol. 8, No. 6, 2005, pp. 22–29.

[3] Meier, S. M., Gupta, D. K., and Sheffler, K. D., "Ceramic Thermal Barrier Coatings for Commercial Gas Turbine Engines," *Journal of the Minerals Metals & Materials Society*, Vol. 43, No. 3, 1991, pp. 50–53.

[4] Feuerstein, A., Knapp, J., Taylor, T., Ashary, A., Bolcavage, A., and Hitchman, N., "Technical and Economical Aspects of Current Thermal Barrier Coating Systems for Gas Turbines by Thermal Spray and EB-PVD: A Review," *Journal of Thermal Spray Technology*, Vol. 17, No. 2, 2008, pp. 199–213.

[5] Padture, N. P., Gell, M., and Jordan, E., "Thermal Barrier Coatings for Gas Turbine Engine Applications," *Science*, Vol. 296, No. 5566, 2002, pp. 280–284.

[6] Stevens, R., *Zirconia and Zirconia Ceramics*, Magnesium Elektron, Manchester, England, U.K., 1986.

[7] Miller, R. A., Smialek, J. L., and Garlick, R. G., "Phase Stability in Plasma-Sprayed, Partially Stabilized Zirconia-Yttria," *Advances in Ceramics, Science and Technology of Zirconia*, Vol. 3, edited by A. H. Heuer and L. W. Hobbs, The American Ceramic Society, Westerville, OH, 1981, pp. 241–253.

[8] Lanteri, V., Chaim, R., and Heuer, A. H., "On the Microstructures Resulting from the Diffusionless Cubic to Tetragonal Transformation in ZrO_2–Y_2O_3 Alloys," *Journal of the American Ceramic Society*, Vol. 69, No. 10, 1986, pp. C251–C261.

[9] Heuer, A. H., "Transformation Toughening in ZrO_2–Containing Ceramics," *Journal of the American Ceramic Society*, Vol. 70, No. 10, 1987, pp. 689–698.

[10] Muraleedharan, K., Subrahmanyam, J., and Bhaduri, S. B., "Identification of t′ Phase in ZrO_2–7.5 wt% Y_2O_3 Thermal Barrier Coatings," *Journal of the American Ceramic Society*, Vol. 71, No. 5, 1988, pp. C226–C227.

[11] Raghavan, S., Wang, H., Dinwiddie, R. B., Porter, W. D., and Mayo, M. J., "The Effect of Grain Size, Porosity, and Yttria Content on the Thermal Conductivity of Nanocrystalline Zirconia," *Scripta Materialia*, Vol. 39, No. 8, 1998, pp. 1119–1125.

[12] Oishi, Y., and Ando, K., "Oxygen Self-Diffusion in Cubic ZrO_2 Solid Solutions," *Transport in Nonstoichiometric Compounds*, edited by G. Simkovich and V. S. Stubican, Plenum Press, New York, 1985, pp. 189–202.

[13] Wallace, J., Stalick, J., Ilavsky, J., and Allen, A., "Response of ZrO_2 TBC's to Annealing," American Ceramic Society Annual Meeting, April 1999.

[14] Stecura, S., "Effects of Compositional Changes on the Performance of a Thermal Barrier Coating System," American Ceramic Society, 3rd Annual Conference on Composites and Advanced Materials, Jan. 1979.

[15] Stecura, S., "Optimization of the NiCrAl-Y/ZrO_2-Y_2O_3 Thermal Barrier System," NASA TM 86905, 1985.

[16] McPherson, R., "A Review of Microstructure and Properties of Plasma Sprayed Ceramic Coatings," *Surface and Coatings Technology*, Vol. 39/40, No. 1–3, 1989, pp. 173–181.

[17] Trice, R. W., and Faber, K. T., "Deformation Mechanisms in Compression Loaded Stand–Alone Plasma–Sprayed Alumina Coatings," *Journal of the American Ceramic Society*, Vol. 83, No. 12, 2000, pp. 3057–3064.

[18] Deschaseaux, C., "A Sintering Study of Plasma-Sprayed Yttria-Stabilized Zirconia Thermal Barrier Coatings Using Stand-Alone Coating Tests," M.Sc. Thesis, Materials Science and Engineering, Purdue Univ., IN, 2002.

[19] Ilavsky, J., Allen, A., Long, G. G., Krueger, S., Berndt, C. C., and Herman, H., "Influence of Spray Angle on the Pore and Crack Microstructure of Plasma–Sprayed Deposits," *Journal of the American Ceramic Society*, Vol. 80, No. 3, 1997, pp. 733–742.

[20] Ilavsky, J., Allen, A., Long, G. G., Herman, H., and Berndt, C. C., "Characterization of the Closed Porosity in Plasma–Sprayed Alumina," *Journal of Materials Science*, Vol. 32, No. 13, 1997, pp. 3407–3410.

[21] Ilavsky, J., Long, G. G., Allen, A. J., Leblanc, L., Prystay, M., and Moreau, C., "Anisotropic Microstructure of Plasma–Sprayed Deposits," *Thermal Spray: Meeting the Challenge of the 21st Century, Proceedings of the ITSC*, edited by C. Coddet, ASM International, Materials Park, OH, 1998, pp. 1577–1582.

[22] Bianchi, L., Grimaud, A., Blein, F., Lucchese, P., and Fauchais, P., "Comparison of Plasma–Sprayed Alumina Coatings by RF and DC Plasma Spraying," *Journal of Thermal Spray Technology*, ol. 4, No. 1, 1995, pp. 59–65.

[23] Fantassi, S., Vardelle, M., Vardelle, A., and Fauchais, P., "Influence of the Velocity of Plasma Sprayed Particles on the Splat Formation," *Thermal Spray Coatings: Research, Design and Applications*, edited by C. C. Berndt and T. F. Bernecki, ASM International, Materials Park, OH, 1993, pp. 1–6.

[24] Bianchi, L., Lucchese, P., Denoirjean, A., and Fauchais, P., "Zirconia Splat Formation and Resulting Coating Properties," *Advances in Thermal Spray Science & Technology*, edited by C. C. Berndt and S. Sampath, ASM International, Materials Park, OH, 1995, pp. 261–266.

[25] Vardelle, M., Vardelle, A., Leger, A. C., Fauchais, P., and Gobin, D., "Influence of Particle Parameters at Impact on Splat Formation and Solidification in Plasma–Spraying Process," *Journal of Thermal Spray Technology*, Vol. 4, No. 1, 1995, pp. 50–58.

[26] Li, C., Ohmori, A., and Harada, Y., "Experimental Investigation of the Morphology of Plasma–Sprayed Copper Splats," *Proceedings of the International Thermal Spray Conference*, 1995, pp. 333–339.

[27] Yankee, S. J., and Pletka, B. J., "Effect of Plasma–Spray Processing Variations on Particle Melting and Splat Spreading of Hydroxylapatite and Alumina," *Journal of Thermal Spray Technology*, Vol. 2, No. 3, 1993, pp. 271–281.

[28] Flores Renteria, A., Saruhan, B., Ilavsky, J., and Allen, A. J., "Application of USAXS Analysis and Non-Interacting Approximation to Determine the Influence of Process Parameters and Ageing on the Thermal Conductivity of Electron-Beam Physical Vapor Deposited Thermal Barrier Coatings," *Surface & Coatings Technology*, Vol. 201, No. 8, 2007, pp. 4781–4788.

[29] Flores Renteria, A., Saruhan, B., Schulz, U., Raetzer-Scheibe, H.-J., Haug, J., and Wiedenmann, W., "Effect of Morphology on Thermal Conductivity of EB-PVD YSZ TBCs," *Surface & Coatings Technology*, Vol. 201, No. 6, 2006, pp. 2611–2620.

[30] Ilavsky, J., Long, G. G., Allen, A. J., and Berndt, C. C., "Evolution of the Void Structure in Plasma-Sprayed YSZ Deposits During Heating," *Materials Science and Engineering: A*, Vol. A272, No. 1, 1999, pp. 215–221.

[31] Dutton, R., Wheeler, R., Ravichandran, K. S., and An, K., "Effect of Heat Treatment on the Thermal Conductivity of Plasma Sprayed Thermal Barrier Coatings," *Journal of Thermal Spray Technology*, Vol. 9, No. 2, 2000, pp. 204–209.

[32] Trice, R. W., Su, Y. J., Mawdsley, J., Faber, K. T., de Arellano-Lopez, A. R., Wang, H., and Porter, W., "Effect of Heat-Treatment on Phase Stability, Lamella and Grain Morphology, and Thermal Conductivity of Plasma-Sprayed YSZ," *Journal of Materials Science*, Vol. 37, No. 11, 2002, pp. 2359–2365.

[33] Erk, K., Deshcaseaux, C., and Trice, R. W., "Grain-Boundary Grooving of Plasma-Sprayed Yttria-Stabilized Zirconia Thermal Barrier Coatings," *Journal of the American Ceramic Society*, Vol. 89, No. 5, 2006, pp. 1673–1678.

[34] Lughi, V., Tolpygo, V. K., and Clarke, D. R., "Microstructural Aspects of the Sintering of Thermal Barrier Coatings," *Materials Science and Engineering: A*, Vol. A368, No. 1-2, 2004, pp. 212–221.

[35] Thurn, G., Schneider, G., and Aldinger, F., "High–Temperature Deformation of Plasma–Sprayed ZrO_2 Thermal Barrier Coatings," *Materials Science and Engineering: A*, Vol. A233, No. 1-2, 1997, pp. 176–182.

[36] Zhu, D., and Miller, R. A., "Thermal Conductivity and Elastic Modulus Evolution of Thermal Barrier Coatings Under High Heat Flux Conditions," *Journal of Thermal Spray Technology*, Vol. 9, No. 2, 2000, pp. 175–180.

[37] Petorak, C., Ilavsky, J., Wang, H., Porter, W., and Trice, R., "Microstructural Evolution of 7 wt% Y_2O_3-ZrO_2 Thermal Barrier Coatings due to Stress Relaxation at Elevated Temperatures and the Concomitant Changes in Thermal Conductivity," *Surface & Coatings Technology*, Vol. 205, No. 1, 2010, pp. 57–65.

[38] Zhu, D., Miller, R. A., Nagaraj, B., and Bruce, R. W., "Thermal Conductivity of EB-PVD Thermal Barrier Coatings Evaluated by a Steady-State Laser Heat Flux Technique," *Surface & Coatings Technology*, Vol. 138, No. 1, 2001, pp. 1–8.

[39] Eliaz, N., Shemesh, G., and Latanision, R. M., "Hot Corrosion in Gas Turbine Components," *Engineering Failure Analysis*, Vol. 9, No. 1, 2002, pp. 31–43.

[40] Jones, R. L., "Some Aspects of the Hot Corrosion of Thermal Barrier Coatings," *Journal of Thermal Spray Technology*, Vol. 6, No. 1, 1997, pp. 77–84.

[41] Foster, A. D., von Doering, H. E., and Hilt, M. B., "Fuels Flexibility in Heavy-Duty Gas Turbines," General Electric, Reference Document, GER-3428a, 1983.

[42] Marple, B. R., Voyer, J., Moreau, C., and Nagy, D. R., "Corrosion of Thermal Barrier Coating by Vanadium and Sulfur Compounds," *Materials at High Temperatures*, Vol. 17, No. 3, 2000, pp. 397–412.

[43] *Specification for Fuel Gases for Combustion in Heavy-Duty Gas Turbines*, GEI 41040G, Rev. Jan. 2002.

[44] Bornstein, N. S., "Reviewing Sulfidation Corrosion – Yesterday and Today," *JOM: the Journal of the Mineral, Metals & Materials Society*, Vol. 48, No. 11, 1996, pp. 37–39.

[45] Susnitzky, D. W., Hertl, W., and Carter, C. B., "Destabilization of Zirconia Thermal Barriers in the Presence of V_2O_5," *Journal of the American Ceramic Society*, Vol. 71, No. 11, 1998, pp. 992–1004.

[46] Chen, Z., "Hot Corrosion and High Temperature Deformation of Yttria-Stabilized Zirconia Thermal Barrier Coatings," Ph.D. Dissertation, Purdue Univ., West Lafayette, IN, 2006.

[47] Chen, Z., Speakman, S., Howe, J., Wang, H., Porter, W., and Trice, R., "Investigation of Reactions Between Vanadium Oxide and Plasma-Sprayed Yttria Stabilized Zirconia Coatings," *Journal of the European Ceramic Society*, Vol. 29, 2009, pp. 1403–1411.

[48] Jones, R. L., Williams, C., and Jones, S., "Reaction of Vanadium Compounds with Ceramic Oxides," *Journal of the Electrochemical Society: Solid-State Science and Technology*, Vol. 133, No. 1, 1986, pp. 227–230.

[49] *Phase Diagrams for Zirconium and Zirconia Systems*, American Ceramic Society, Westerville, OH, 1998.

[50] Hancock, P., "Vanadic and Chloride Attack of Superalloys," *Materials Science and Technology*, Vol. 3, No. 7, 1987, pp. 536–544.

[51] Zhang, Y., and Rapp, R., "Solubilities of CeO_2, HfO_2 and Y_2O_3 in Fused Na_2SO_4-30 Mol% $NaVO_3$ and CeO_2 in Pure Na_2SO_4 at 900°C," *Corrosion*, Vol. 43, No. 6, 1987, pp. 348–352.

[52] Smialek, J., Archer, F. A., and Garlick, R. G., "Turbine Airfoil Degradation in the Persian Gulf War," *JOM: the Journal of the Mineral, Metals & Materials Society*, Vol. 12, No. 12, 1994, pp. 39–41.

[53] Grindle, T., and Burcham, F. W., Jr., "Engine Damage to a NASA DC-8-72 Airplane from a High-Altitude Encounter with a Diffuse Volcanic Ash Cloud," NASA/TM-2003-212030, 2003.

[54] Kim, J., Dunn, M. G., Baran, A. J., Wade, D. P., and Tremba, E. L., "Deposition of Volcanic Materials in the Hot Sections of Two Gas Turbine Engines," *Journal of Engineering for Gas Turbines and Power*, Vol. 115, No. 3, 1993, pp. 641–651.

[55] Chen, X., "Calcium-Magnesium-Alumina-Silicate (CMAS) Delamination Mechanisms in EB-PVD Thermal Barrier Coatings," *Surface & Coatings Technology*, Vol. 200, No. 11, 2006, pp. 3418–3427.

[56] Mercer, C., Faulhaber, S., Evans, A. G., and Darolia, R., "A Delamination Mechanism for Thermal Barrier Coatings Subject to Calcium-Magnesium-Alumino-Silicate (CMAS) Infiltration," *Acta Materialia*, Vol. 53, No. 4, 2005, pp. 1029–1039.

[57] Borom, M. P., Johnson, C. A., and Peluso, L. A., "Role of Environmental Deposits and Operating Surface Temperature in Spallation of Air Plasma-Sprayed Thermal Barrier Coatings," *Surface & Coatings Technology*, Vol. 86–87, No. 1–3, 1996, pp. 116–126.

[58] Krämer, S., Yang, J., Levi, C. G., and Johnson, C. A., "Thermomechanical Interaction of Thermal Barrier Coatings with Molten CaO-MgO-Al_2O_3-SiO_2 (CMAS) Deposits," *Journal of the American Ceramic Society*, Vol. 89, No. 10, 2006, pp. 3167–3175.

[59] Guffanti, M., Ewert, J. W., Gallina, G. M., Bluth, G., and Swanson, G., "Volcanic-Ash Hazard to Aviation During the 2003–2004 Eruptive Activity of Anatahan Volcano, Commonwealth of the Northern Mariana Islands," *Journal of Volcanology and Geothermal Research*, Vol. 146, No. 1–3, 2005, pp. 241–255.

[60] "The Economic Impacts of Air Travel Restrictions due to Volcanic Ash," *Oxford Economics* (report prepared for Airbus).

[61] Prata, A. J., "Observations of Volcanic Ash Clouds in the 10-12 μm Window Using AVHRR/2 Data," *International Journal of Remote Sensing*, Vol. 10, No. 4–5, 1989, pp. 751–761.

[62] Grindle, T. J., and Burcham, F. W., Jr., "Engine Damage to a NASA DC-8-72 Airplane from a High-Altitude Encounter with Diffuse Volcanic Ash," NASA/TM-2003-212030, Aug. 2003.

[63] Pieri, D., Ma, C., Simpson, J. J., Hufford, G., Grindle, T., and Grove, C., "Analyses of in-situ Airborne Volcanic Ash from the February 2000 Eruption of Hekla Volcano, Iceland," *Geophysical Research Letters*, Vol. 29, No. 16, 2002, pp. 19-1–19-4.

[64] Taylor, H. E., and Lichte, F. E., "Chemical Composition of Mount St. Helens Volcanic Ash," *Geophysical Research Letters*, Vol. 7, No. 11, 1980, pp. 949–952.

[65] McKee, D. W., Luthra, K. L., Siemers, P., and Palko, J. E., "Resistance of Thermal Barrier Ceramic Coatings to Hot Salt Corrosion," *Proceedings of the 1st Conference on Advanced Materials for Alternative Fuel Capable Directly Fired Heat Engines*, CONF-790749, edited by J. W. Fairbands and J. Stringer, U.S. Dept. of Energy, 1979, pp. 258–269.

[66] Chen, Z., Mabon, J., Wen, J.-G., and Trice, R., "Degradation of Plasma-Sprayed Yttria-Stabilized Zirconia Coatings via Ingress of Vanadium Oxide," *Journal of the European Ceramic Society*, Vol. 29, 2009, pp. 1647–1656.

[67] Lanteri, V., Chaim, R., and Heuer, A. H., "On the Microstructures Resulting from the Diffusionless Cubic to Tetragonal Transformation in ZrO_2–Y_2O_3 Alloys," *Journal of the American Ceramic Society*, Vol. 69, No. 10, 1986, pp. C251–C261.

[68] Krämer, S., Yang, J., and Levi, C. G., "Infiltration-Inhibiting Reaction of Gadolinium Zirconate Thermal Barrier Coatings with CMAS Melts," *Journal of the American Ceramic Society*, Vol. 91, No. 2, 2008, pp. 576–583.

[69] Wellman, R. G., and Nicholls, J. R., "A Review of the Erosion of Thermal Barrier Coatings," *Journal of Physics D - Applied Physics*, Vol. 40, No. 16, 2007, pp. R293–R305.

[70] Nicholls, J. R., Wellman, R. G., and Deakin, M. J., "Erosion of Thermal Barrier Coatings," *Materials at High Temperatures*, Vol. 20, No. 2, 2003, pp. 207–218.

[71] Janos, B. Z., Lugscheider, E., and Remer, P., "Effect of Thermal Aging on the Erosion Resistance of Air Plasma Sprayed Zirconia Thermal Barrier Coatings," *Surface & Coatings Technology*, Vol. 113, No. 3, 1999, pp. 278–285.

[72] Nicholls, J. R., "Laboratory Studies of Erosion-Corrosion Processes Under Oxidising and Oxidising/Sulphidising Conditions," *Materials at High Temperatures*, Vol. 14, No. 3, 1997, pp. 289–306.

[73] Steenbakker, R. J. L., Wellman, R. G., and Nicholls, J. R., "Erosion of Gadolinia Doped EB-PVD TBCs," *Surface & Coatings Technology*, Vol. 201, No. 6, 2006, p. 2140.

[74] Chen, X., He, M. Y., Spitsberg, I., Fleck, N. A., Hutchinson, J. W., and Evans, A. G., "Mechanisms Governing the High Temperature Erosion of Thermal Barrier Coatings," *Wear*, Vol. 256, No. 7–8, 2004, pp. 735–746.

[75] Zhu, D., and Miller, R. A., "Low Conductivity and Sintering-Resistant Thermal Barrier Coatings," U.S. Patent 7,186,166, 6 March 2007.

[76] Zhu, D., and Miller, R. A., "Development of Advanced Low Conductivity Thermal Barrier Coatings," *International Journal of Applied Ceramic Technology*, Vol. 1, No. 1, 2004, pp. 86–94.

[77] Zhu, D., and Miller, R. A., "Thermal Conductivity and Sintering Behavior of Advanced Thermal Barrier Coatings," *Ceramic Engiineering Science Proceedings*, Vol. 23, 2002, pp. 457–468.

[78] Zhu, D., Chen, Y. L., and Miller, R. A., "Defect Clustering and Nano-Phase Structure Characterization of Multicomponent Rare-Earth Oxide Doped Zirconia-Yttria Thermal Barrier Coatings," *Ceramic Engineering Science Proceedings*, Vol. 24, No. 3, 2003, pp. 525–534.

[79] Cao, X. Q., Vassen, R., and Stoever, D., "Ceramic Materials for Thermal Barrier Coatings," *Journal of the European Ceramic Society*, Vol. 24, No. 1, 2004, pp. 1–10.

[80] Lehmann, H., Pitzer, D., Pracht, G., Vassen, R., and Stover, D., "Thermal Conductivity and Thermal Expansion Coefficients of the Lanthanum Rare-Earth-Element Zirconate System," *Journal of the American Ceramic Society*, Vol. 86, No. 8, 2003, pp. 1338–1344.

[81] Wu, J., Wei, X., Padture, N., Klemens, P., Gell, M., Garcia, E., Miranzo, P., and Osedni, M., "Low-Thermal-Conductivity Rare-Earth Zirconates for Potential Thermal Barrier Coating Applications," *Journal of the American Ceramic Society*, Vol. 85, No. 12, 2002, pp. 3031–3035.

[82] Krämer, S., Leckie, R. M., Grant, K. M., and Levi, C., "Molten Deposits on Coated Gas Turbine Components: The Price Paid for Higher Temperature Operation," *MS&T* 2006, Cincinnati, OH, Oct. 2006.

[83] Padture, N. P., and Klemens, P. G., "Low Thermal Conductivity in Garnets," *Journal of the American Ceramic Society*, Vol. 80, No. 4, 1997, pp. 1018–1020.

[84] Padture, N., Schlichting, K., Bhatia, T., Ozturk, A., Jordan, E., Gell, M., Jiang, S., Xiao, T., Strutt, P., Garcia, E., Miranzo, P., and Osendi, M., "Towards Durable Thermal Barrier Coatings with Novel Microstructures Deposited by Solution Precursor Plasma Spray," *Acta Materialia*, Vol. 49, No. 12, 2001, pp. 2251–2257.

[85] Bhatia, T., Ozturk, A., Xie, L., Jordan, E., Cetegen, B., Gell, M., Ma, X., and Padture, N., "Mechanisms of Ceramic Coating Deposition in Solution Precursor Plasma Spray," *Journal of Materials Research*, Vol. 17, No. 9, 2002, pp. 2363–2372.

[86] Xie, L., Chen, D., Jordan, E., Ozturk, A., Wu, F., Ma, X., Cetegen, B., and Gell, M., "Formation of Vertical Cracks in Solution-Precursor Plasma-Sprayed Thermal Barrier Coatings," *Surface & Coatings Technology*, Vol. 201, No. 3–4, 2006, pp. 1058–1064.

[87] Jadhav, A., Padture, N., Jordan, E., Gell, M., Miranzo, P., and Fuller, E, Jr., "Low-Thermal Conductivity Plasma-Sprayed Thermal Barrier Coatings with Engineered Microstructures," *Acta Materialia*, Vol. 54, No. 12, 2006, pp. 3343–3349.

[88] Fauchais, P., Etchart-Salas, R., Rat, V., Coudert, J., Caron, N., and Wittmann-Teneza, K., "Parameters Controlling Liquid Plasma Spraying: Solutions, Sols, or Suspensions," *Journal of Thermal Spray Technology*, Vol. 17, No. 1, 2008, pp. 31–59.

[89] Fauchais, P., Rat, V., Coudert, J., Etchart-Salas, R., and Montavon, G., "Operating Parameters for Suspension and Solution Plasma-Spray Coatings," *Surface & Coatings Technology*, Vol. 202, No. 18, 2008, pp. 4309–4317.

[90] Fazilleau, J., Delbox, C., Rat, V., Coudert, J., Fauchais, P., and Pateyron, B., "Phenomena Involved in Suspension Plasma Spraying Part 1: Suspension Injection and Behavior," *Plasma Chemical Plasma Process*, Vol. 26, No. 4, 2006, pp. 371–391.

[91] Kassner, H., Siegert, R., Hathiramani, D., Vassen, R., and Stoever, D., "Application of Suspension Plasma Spraying (SPS) for Manufacture of Ceramic Coatings," *Journal of Thermal Spray Technology*, Vol. 17, No. 1, 2008, pp. 115–123.

[92] Chen, Z., Trice, R., Besser, M., Yang, X., and Sordelet, D., "Air-Plasma Spraying Colloidal Solutions of Nanosized Powders," *Journal of Materials Research*, Vol. 39, No. 13, 2004, pp. 4171–4178.

[93] VanEvery, K., Krane, M., Trice, R. W., Porter, W., Wang, H., Besser, M., Sordelet, D., Ilavsky, J., and Almer, J., "In-Flight Alloying of Nanocrystalline Ytrria-Stabilized Zirconia Using Suspension Plasma Spray to Produce Ultra-Low Thermal Conductivity Thermal Barriers," *International Journal of Applied Ceramic Technology*, Vol. 8, No. 6, 2011, pp. 1382–1392.

[94] Refke, A., Hawley, D., Doesburg, J., and Schmid, R. K., "LPPS Thin Film Technology for the Application of TBC Systems," *International Thermal Spray Conference*, ASM, Materials Park, OH, 2005, pp. 438–443.

[95] Miller, R. A., "Thermal Barrier Coatings for Aircraft Engines: History and Directions," *Journal of Thermal Spray Technology*, Vol. 6, No. 1, 1997, pp. 35–42.

[96] Pettit, F. S., and Goward, G. W., "High Temperature Corrosion and Use of Coatings for Protection," *Metallurgical Treatises, The Metallurgical Society of AIME*, 1981, pp. 603–619.

[97] Birk, N., Meier, G. H., and Pettit, F. S., *Introduction to the High-Temperature Oxidation of Metals*, 2nd ed., Cambridge Univ. Press, Cambridge, England, U.K., 2006, pp. 144–148.

[98] Gleeson, B., "Thermal Barrier Coatings for Aeroengine Applications," *Journal of Propulsion and Power*, Vol. 22, No. 2, 2006, pp. 375–383.

[99] Pint, B. A., "The Role of Chemical Composition on Oxidation Performance of Aluminide Coatings," *Surface & Coatings Technology*, Vol. 188–189, No. 11–12, 2004, pp. 71–78.

[100] Hayashi, S., Narita, T., and Gleeson, B., "Early-Stage Oxidation Behavior of γ'-Ni_3Al-based Alloys with and Without Pt Addition," *Materials Science Forum*, Vol. 522–523, 2006, pp. 229–238.

[101] Hayashi, S., Wang, W., Sordelet, D. J., and Gleeson, B., "Interdiffusion Behavior of Pt-Modified γ-Ni + γ'-Ni_3Al Alloys Coupled to Ni-Al-Based Alloys," *Metallurgical and Materials Transactions*, Vol. 36A, No. 7, 2005, pp. 1769–1775.

[102] Gleeson, B., Sordelet, D. J., and Wang, W., "High-Temperature Coatings with Pt Metal Modified γ-Ni + γ'-Ni3Al Alloy Composition," U.S. Patent No. 7,273,662 B2, 25 Sept. 2007.

[103] Evans, A. G., Mumm, D. R., Hutchinson, J. W., Meier, G. H., and Pettit, F. S., "Mechanisms Controlling the Durability of Thermal Barrier Coatings," *Progress in Materials Science*, Vol. 45, No. 5, 2001, pp. 505–553.

[104] Wu, R. T., Kawagishi, K., Harada, H., and Reed, R. C., "Retention of Thermal Barrier Coating Systems on Single-Crystal Superalloys: Effects of Substrate Composition," *Acta Materialia*, Vol. 56, No. 14, 2008, pp. 3622–3629.

CHAPTER 12

Turbine Materials and Mechanics

Anthony G. Evans*
University of California, Santa Barbara, Santa Barbara, California

David R. Clarke†
Harvard University, Cambridge, Massachusetts

Carlos G. Levi‡
University of California, Santa Barbara, Santa Barbara, California

Multilayer oxides and Ni-alloys have been successfully incorporated into turbines for aeropropulsion and for power generation. By enabling elevated operating temperatures, they exert a crucial influence on the fuel efficiency and performance. In this chapter, the roles of these materials within the overall system are described, and their most important properties are outlined. The mechanisms that govern their properties are presented, and approaches for adjusting them in desirable directions are discussed. The focus is on gas turbines for propulsion, but the same issues pertain to land-based turbines; indeed, many of the more advanced land-based turbines for electrical generation today are derived from aeroturbines. Where there are major differences, these are explicitly discussed. Opportunities for new materials with potential for further improvements in fuel efficiency are assessed.

I. BACKGROUND

A. MATERIALS STRATEGY

Oxides and Ni-alloys are present in the hot section of turbines used for propulsion and power generation. By designing and using these materials in optimal combinations, it has been possible to systematically increase the combustion temperature (Fig. 1) (Schafrik, R., "Fuel Efficiency of Aircraft Engines," private communication to T. Evans, 2004). In turn, this temperature increase has enhanced the fuel efficiency (Fig. 2) (Schafrik, R., "Fuel Efficiency of Aircraft Engines," private communication to T. Evans, 2004). The utility of these types of materials is greatest in the high-pressure turbine, especially the airfoil (Schafrik,

*Deceased, formerly Professor of Materials and Professor of Mechanical Engineering.
†Extended Tarr Family Professor of Materials, School of Engineering and Applied Sciences.
‡Professor of Materials and Mechanical Engineering, College of Engineering.

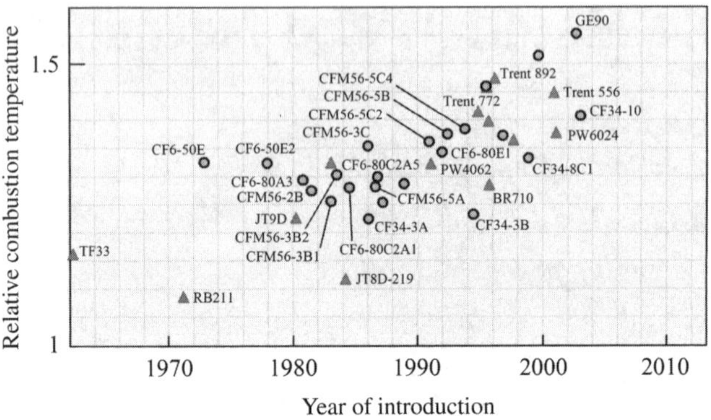

Fig. 1 The trend in combustion temperature in commercial aeroengines over the last five decades (Schafrik, R., "Fuel Efficiency of Aircraft Engines," private communication to T. Evans, 2004).

R., "Fuel Efficiency of Aircraft Engines," private communication to T. Evans, 2004) [1–8]. The technology has demonstrated how oxides can be used to protect load-bearing, Ni-basedstructural members that experience environmental extremes. It involves choices of materials and spatial configurations, as well as survivability upon extreme temperature cycling, without loss of functionality. The overall system is depicted in Fig. 3. This chapter examines the materials, the configurations, and the mechanisms affecting performance and durability. The materials and the basic mechanisms governing performance and durability are

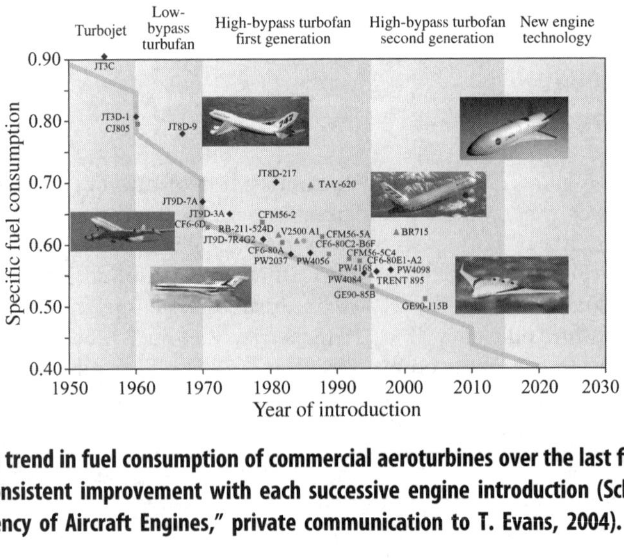

Fig. 2 The trend in fuel consumption of commercial aeroturbines over the last five decades. Note the consistent improvement with each successive engine introduction (Schafrik, R., "Fuel Efficiency of Aircraft Engines," private communication to T. Evans, 2004).

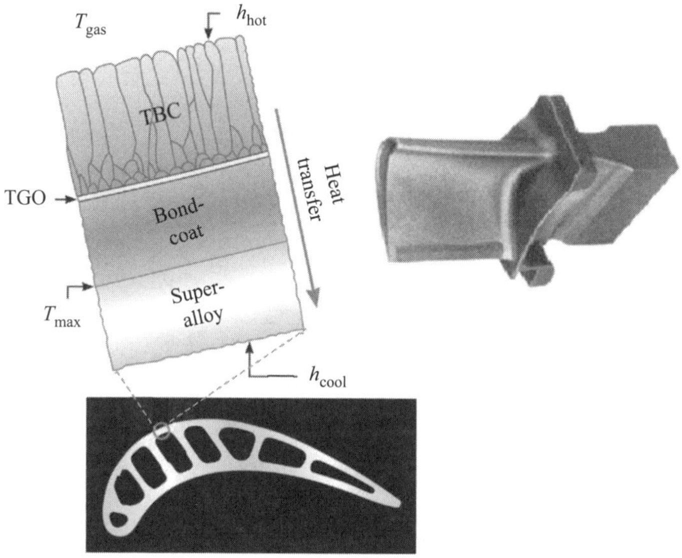

Fig. 3 A schematic of an airfoil and a magnified view of a surface zone with the TBC and bondcoat layers identified. The thermal conditions are defined.

summarized in this section, embellishing a review presented elsewhere [9]. Subsequently, the key roles played by the constituents are detailed. In Sec. II, the interdependent influences of the bondcoat alloy and the thermally grown oxide are elaborated. In Sec. III, the properties of the thermally insulating oxide are described and assessed, and important phenomena that limit performance are described. In Sec. IV, the temperature limits imposed by the infiltration of the coating by CMAS (calcium-magnesium-alumino-silicate) deposited from the environment are described, as well as the limits imposed by sintering and phase transformations. In Sec. V, the mechanics governing the spalling of the coating are reviewed with implications for limits associated with thermal gradients through the coating. In Sec. VI, the status of life prediction schemes is examined, including the role of nondestructive probes. Finally, in Sec. VII, opportunities for future enhancement in fuel efficiency through new materials and improved models are presented.

1. MATERIALS AND CONFIGURATIONS

The following considerations have motivated the choice of materials and their spatial configurations. The thermal requirements are established by directing air through internal channels in the structural Ni-alloy (Fig. 3) with heat-transfer coefficient determined by the flow rate and the channel geometry. Subject to a

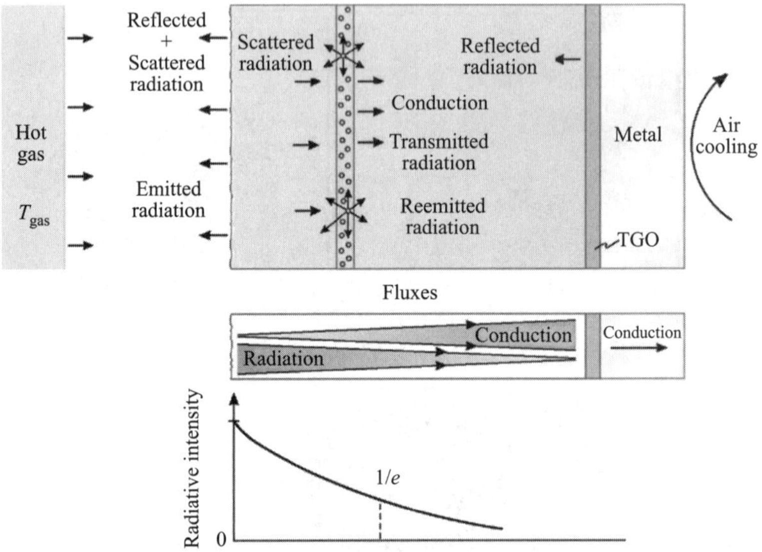

Fig. 4 The combination of conduction and radiation that occur through the thickness of a TBC.

combustion temperature T_{gas} and an external heat-transfer coefficient, by superposing an external insulating oxide, T_{gas} can be raised while the alloy is kept at an allowable maximum temperature. The conduction and radiation effects involved are summarized in Fig. 4. Remarkably, insulating oxides deposited onto geometrically complex structural components, such as airfoils, remain attached for extended periods despite cycling through an enormous temperature range (in excess of 1200°C) within an oxidizing environment despite large thermal expansions mismatch. A single material would be incapable of satisfying these requirements. The viable solution is an oxide/metal multilayer (Fig. 5). The outer oxide imparts thermal protection while the metallic layer (bondcoat) not only affords oxidation protection through the formation of a second oxide, but also facilitates plastic accommodation of misfit strains (Schafrik, R., "Fuel Efficiency of Aircraft Engines," private communication to T. Evans, 2004) [1–7].

When the technology was originally developed, the preferred insulating oxide was determined to be yttria-stabilized zirconia (YSZ), chosen because of its low, temperature-invariant, thermal conductivity (Fig. 6) [9]. The most desirable phase was ascertained by conducting laboratory-based thermal cycle tests to seek the composition affording greatest durability, that is, the largest number of cycles before the coating spalls (Fig. 7) [10]. The outcome was 7wt% yttria-stabilized-zirconia (7YSZ). This composition is still used, despite the discovery of lower thermal conductivity options (Fig. 6b) [4, 5, 11–14]. It remains the material of choice because other properties (especially toughness [15–20]) are also crucial. Moreover, YSZ is amenable to processing with minimal difficulty by air plasma spray (APS) and electron-beam physical vapor deposition (EB-PVD) as well as other, less commonly used techniques [21]. The preferred thickness of the oxide layer is component dependent. On rotating components, such as airfoils, the layer must be thick enough to achieve the desired temperature drop, yet thin enough to avert excessive inertial loads as a result of the extra mass. The outcome is thickness in the range $100 \leq H_{tbc} \leq 250$ μm. On stationary components, such as shrouds and combustors, the mass is less critical, and much thicker layers can be used. The choice is typically 500 μm $\leq H_{tbc} \leq 1$ mm.

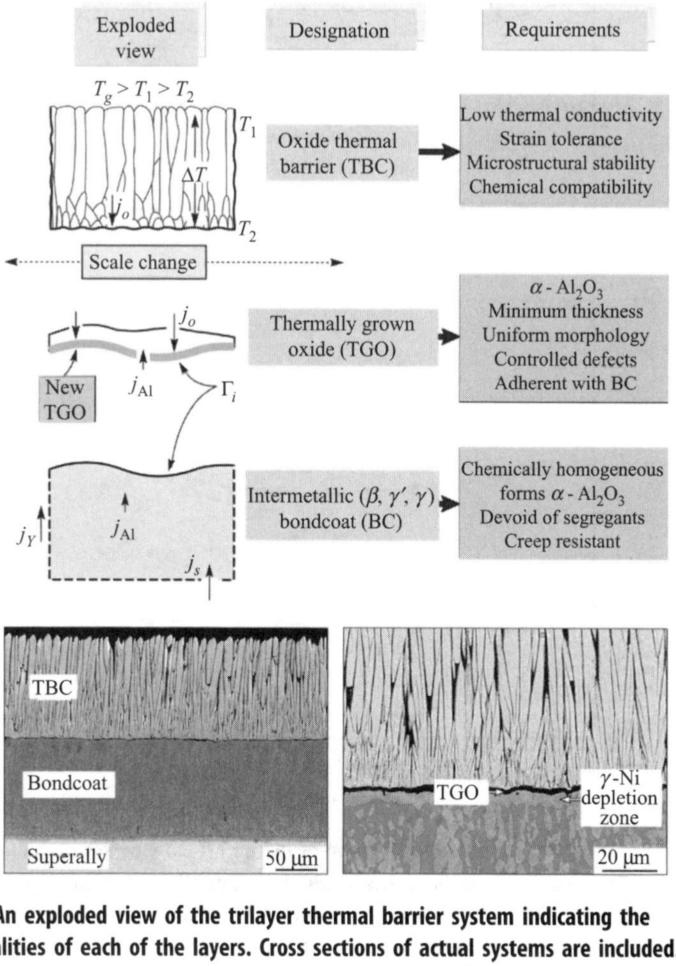

Fig. 5 An exploded view of the trilayer thermal barrier system indicating the functionalities of each of the layers. Cross sections of actual systems are included.

The principles governing the choice of materials for oxidation protection are notionally straightforward. The basic requirements are as follows. 1) A thermally grown oxide (TGO) forms at the bondcoat surface by reaction with the

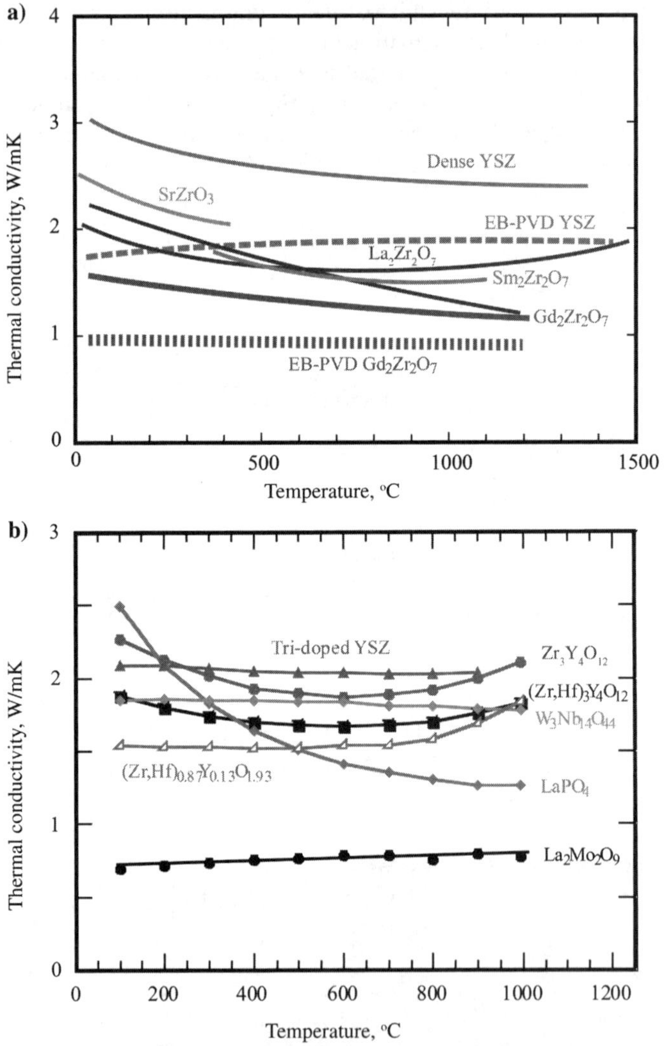

Fig. 6 a) The thermal conductivity of several insulating, ternary oxides as a function of temperature. The solid lines are for dense solids. The dotted lines are for vapor deposited coatings incorporating spatially configured porosity. b) The thermal conductivity of new compositions (all dense solids) identified since 2000.

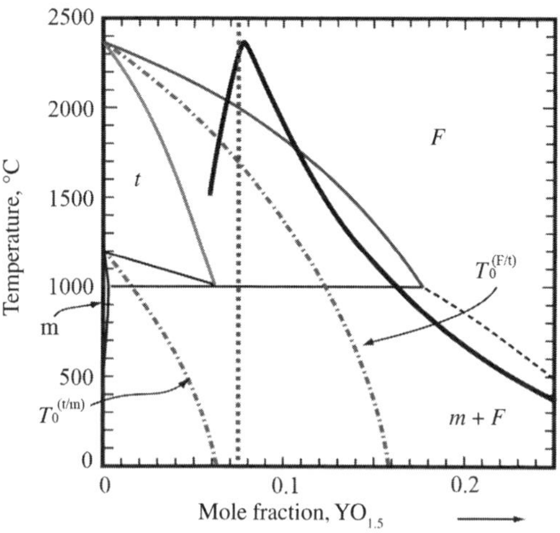

Fig. 7 A binary phase diagram for the $ZrO_2(YO_{1.5})$ system, showing the phases expected. A solid line representative of the cyclic durability is superposed. The ellipse denotes the maximum range of pertinent to TBC operation [10].

combustion gas. The preferred TGO should have the lowest possible oxygen ingress at the temperatures of interest (900–1150°C) with correspondingly small counterdiffusion of the metallic elements. 2) The bondcoat should have sufficient thermochemical compatibility with the structural alloy that the basic composition, microstructure, and properties are retained for the expected life of the system. The singular solution is an alloy that forms $\alpha-Al_2O_3$ upon oxidation. To achieve this, near its surface, the alloy must contain sufficient Al that the primary oxidation product is, indeed, $\alpha-Al_2O_3$ and, moreover, should act as a reservoir for reformation of $\alpha-Al_2O_3$ should spallation occur. The common choices are alloys based on Ni(Al) with alloy additions (such as Cr, Co, Pt, Y, Hf). Other requirements are more nuanced. They dictate competitive advantage through key aspects of system performance and durability. In practice, three categories of bondcoat have been implemented, differentiated by the phases present and the alloy additions: 1) one consists of a single β-phase usually made by interdiffusing Al and Pt with Ni adjacent to the surface of the superalloy [22–24]; 2) a second consists of a two-phase, $\gamma+\beta$-alloy, usually deposited onto the substrate by low-pressure plasma spraying (LPPS) or EB-PVD [25–27]; and 3) the third is a two-phase $\gamma+\gamma'$ alloy made by infusing Pt (and Hf) into the substrate [28, 29]. Systems made using these bondcoats perform differently with durability governed by different mechanisms.

2. PERFORMANCE AND DURABILITY

For the TBC to survive extreme thermal cycling, the misfit strains between the layers must be understood and managed [3, 30–35]. These strains arise because of differences in thermal expansion coefficient, as well as phase transformations [36] and interdiffusion. With temperature cycling, they cause residual stresses, which activate inelastic mechanisms that, in turn, limit durability. The importance of the misfit differs for each of the layers. It is least important for the external oxide (7YSZ) because this layer need not be dense: it serves only to insulate the underlying alloy and does not provide oxidation protection. It is designed with a microstructure having spatially configured porosity that affords low in-plane stiffness and high strain tolerance [26, 37–41]. This strategy cannot be used for either the TGO or the bondcoat; to serve their functions, both need to be dense (minimal porosity). The TGO misfit cannot be independently controlled, but its adverse consequences can be managed by limiting its thickness. The misfits between the bondcoat and substrate are more nuanced; they result not only from thermal expansion, but also from phase transformations [35] and swelling [42]. Understanding these misfits, ascertaining their importance to system durability, and finding means to control them, has been an important research focus.

Ultimately, durability is governed by spalling of the external insulating oxide (Fig. 8) [2]. Small diameter spalls can be tolerated because backside cooling and boundary-layer effects still allow the exposed surface to be protected by the (surrounding) intact oxide. Degradation only becomes a concern after an appreciable area fraction of the coating has been removed. Actual spall formation is preceded by smaller cracks that extend and coalesce along delamination planes located either within the oxide layer or at the interface between the TGO and the bondcoat [43, 44].

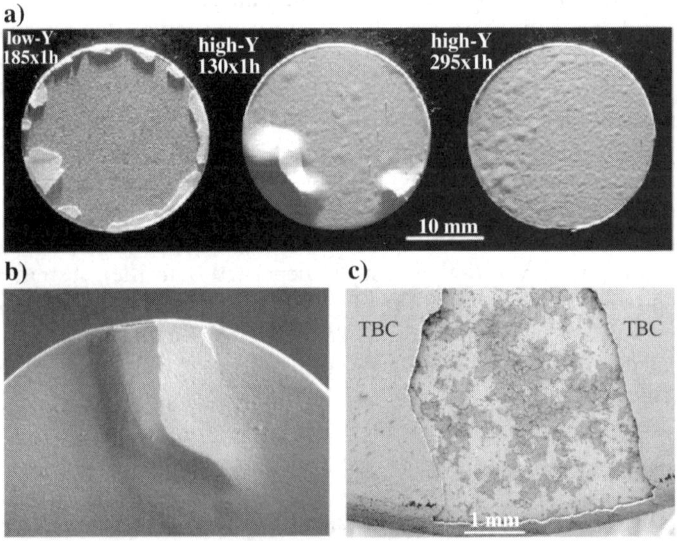

Fig. 8 a) Three examples of spalling found after furnace cycling tests. b) A large-scale buckle in a TBC coating with a ridge crack along the top. This buckle initiated at the upper free edge. c) The surface exposed when the buckled region spalls. In this case, because of rumpling, the exposed surface is a mix of YSZ and alumina.

B. CONSTITUENTS AND THEIR THERMOMECHANICAL PROPERTIES

1. INSULATING OXIDE

The thermal expansion coefficient of this layer α_{tbc} is appreciably lower than that of the substrate α_{sub}. The difference is about $\alpha_{tbc} - \alpha_{sub} \equiv \Delta\alpha_{tbc} \approx -3\text{ppm}/^{\circ}\text{C}$ although it is dependent on the specific compositions of the substrate alloy and the oxide. To prevent spontaneous delamination as a result of this misfit, the in-plane modulus of the layer E_{tbc} must be controlled as illustrated by the following simple argument. At the highest temperature, creep in the YSZ causes it to become stress-free. Subsequent cooling induces residual stress through the thermal expansion misfit. If the YSZ were fully dense ($E_{tbc} = 200$ GPa, $\nu_{tbc} \approx 0.2$), for a typical value of the average temperature drop ($\Delta T \approx 1000^{\circ}$C), the residual stress at ambient would be $\sigma_R \approx E_{tbc}\Delta\alpha_{tbc}\Delta T/(1 - \nu_{tbc}) \approx -0.8$ GPa. For thickness, $H_{tbc} \geq 100$ μm, the stored energy/area, $U_{tbc} \equiv \sigma_R^2 H_{tbc}/2E_{tbc} \geq 160\,\text{J/m}^2$, would exceed the mode I toughness ($\Gamma_{tbc} \approx 45\,\text{J/m}^2$ for dense 7-YSZ [17]), rendering the system prone to spontaneous delamination [16]. To obviate this problem, deposition methods have been developed that create a non-dense microstructure with appreciably lower in-plane modulus, $E_{tbc} \leq 50$ GPa [37–39]. In this modulus range, the stored energy becomes $U_{tbc} < 45\,\text{J/m}^2$ for $H_{tbc} = 150$ μm (namely, of order the toughness), thereby enabling implementation. The columnar structure developed by EB-PVD is an especially effective way of attaining low E_{tbc} [26] (Fig. 9a). In air plasma systems the development of a pattern of dense vertical cracks (DVC) during manufacturing [40] (Fig. 9c) serves a similar purpose.

2. BONDCOAT

The relationships between the properties of the bondcoat and system durability are much more complex because of the highly nonlinear interplay with the substrate and the TGO [32, 33]. The ideal bondcoat would have the following attributes: 1) resistance to interdiffusion with the substrate, 2) minimal strain misfit with the substrate (based on thermal expansion, phase transformations, and interdiffusion induced swelling), and 3) high creep strength with adequate ductility. These preferences cannot be realized simultaneously. The challenge has been to identify those attributes having the greatest importance.

3. THERMALLY GROWN OXIDE

The characteristics of the TGO are controlled largely by the bondcoat microstructure and microchemistry, modulated by impurities, water vapor, and dopants. Upon initial oxidation, transient phases of alumina generally form. Later, these convert into α–Al_2O_3 [36, 45–51]. The bondcoats used in practice develop this phase at a relatively early stage within the cyclic life, minimizing adverse influences of the phase transformation on durability. As the α–Al_2O_3 layer grows, it

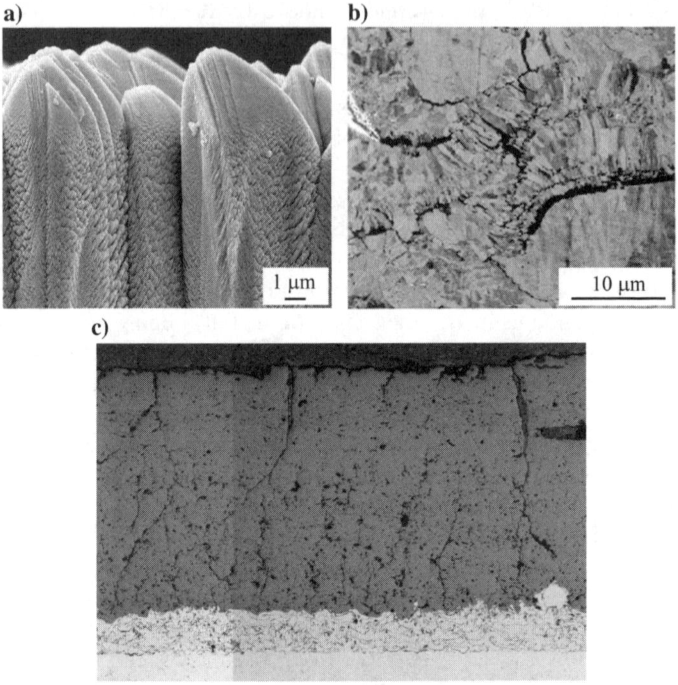

Fig. 9 Typical microstructures of thermal barrier coatings: a) columnar structure of material deposited by electron-beam physical vapor deposition incorporating planar nanoscale porosity created by rotation during deposition; b) splat microstructure of material deposited by atmospheric plasma spray; and c) dense vertical cracks in APS coatings that alleviate the residual stress and enhance the delamination resistance.

develops a small (but significant) compressive stress [52, 53]. These values are consistent with growth stresses formed on other alumina-forming metals [54]. Upon cooling, the compression increases dramatically as a result of thermal expansion misfit with the substrate: $\alpha_{tgo} - \alpha_{sub} \equiv \Delta\alpha_{tgo} \approx -7$ ppm/°C, such that $\sigma_{tgo} \approx -4$ GPa at ambient [3, 55–57]. Consequently, even though the TGO might be relatively thin at the end of the cyclic life ($h_{tgo} \approx$ 3–6 μm), the energy stored/area is quite large $U_{tgo} = \sigma_{tgo}^2 h_{tgo}/2E_{tgo} \approx 80\ \mathrm{Jm^{-2}}$ and contributes substantially to the potential for delamination at the TGO/bondcoat interface (Fig. 10).

4. INTERFACES

Although interfaces between metals and oxides involve fundamentally strong (covalent and ionic) bonds [58–60], their adhesion can be compromised by minor impurities in the alloys (S is especially detrimental) [60]. To inhibit such degradation, there has been a long history in the industry of systematically lowering the S level in superalloys and using selected alloy additions (Y, Pt, Hf, etc.) to tie up remnant S.

C. MECHANISMS LIMITING THE DURABILITY OF HOT SECTION COMPONENTS

An early challenge in the implementation of thermal barrier systems was the difficulty in realizing laboratory tests that reproduced the conditions that arise in an operating turbine. Furnace cycle and burner rig tests were widely used, but the spalling mechanisms were not always representative of those found in airfoils, shrouds, or combustors removed from actual engine service. As the body of information on components accumulated, this concern became less problematic. A remaining issue is the merit of purported failure mechanisms presented in the literature, based on laboratory test results. Each of the mechanisms presented next has been carefully scrutinized and correlated with engine experience. That is, these mechanisms are those that the authors deem reproducible and verifiable on the basis of engine experience. They fall into two basic categories: intrinsic and extrinsic (Fig. 11). Those in the intrinsic category are not especially sensitive to the presence of a thermal gradient in the component whereas those in the extrinsic category are sensitive to thermal gradients.

The intrinsic category is characterized by a group of mechanisms that arise because of the strain misfits associated with the constituent materials. These mechanisms can often be reproduced in well-executed furnace cycle and burner rig tests. The failures are ultimately manifest as spalls, usually occurring in the hottest areas of the TBC. When button-shaped laboratory specimens are subjected to thermal cycling (furnace cycle tests), buckled areas develop, followed by spalling, as indicated on Fig. 8. In these tests, the large-scale buckles usually initiate at the perimeter and are

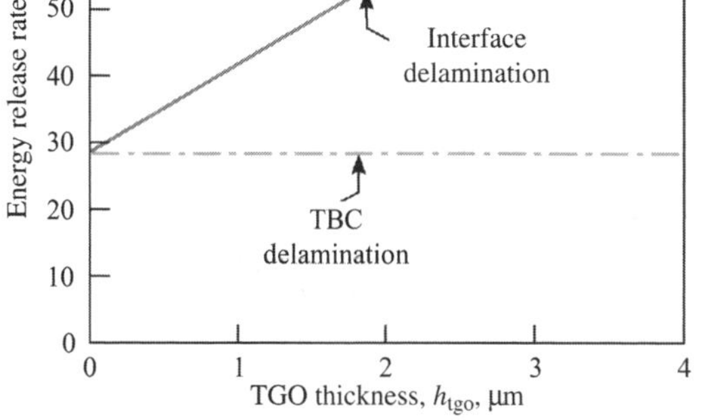

Fig. 10 The energy release rates for delamination along either the TGO/bondcoat interface, as a function of TGO thickness (solid line), or internally, within the TBC (dash-dot line). Also shown is an estimate of the mode II toughness of the interface.

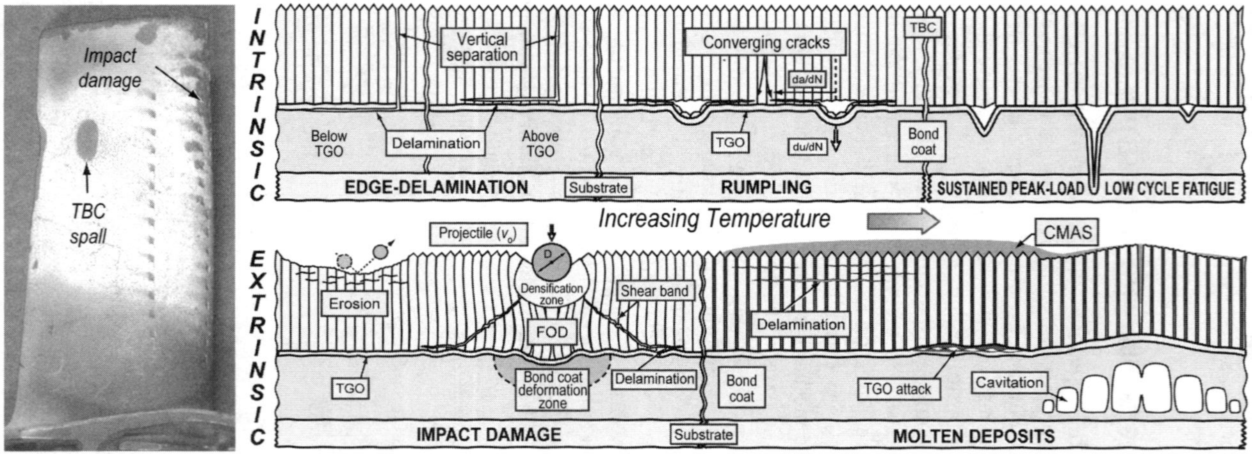

Fig. 11 A summary of the various mechanisms that can cause spalling of the TBC on turbine airfoils. The intrinsic mechanisms are governed by strain misfits between the constituent layers upon thermal cycling. The extrinsic mechanisms are determined by external factors. Also shown at the bottom is an airfoil removed from engine service that contains several spalled regions.

accompanied by ridge cracks. The end of cyclic life is usually defined as when these buckles cover a predetermined proportion of the coating area. Because the TBC is in residual compression at ambient, the buckles arise once a separation zone has developed close to the interface and large enough to satisfy the critical-size requirement for elastic buckling of a film/coating on a stiff substrate [2], typically several times the coating thickness. Accordingly, the time- and cycle-dependent development of this separation zone governs the number of cycles that can be sustained prior to buckling/spalling. Removal of the buckled TBC identifies the underlying plane of separation.

In systems with EB-PVD coatings, three different failure pathways have been identified, differentiated through the surface exposed by the spall. 1) One exposes YSZ and some alumina on both delamination surfaces (Fig. 8). Cross sectioning indicates that this separation mode is accompanied by rumpling (or ratcheting) of the TGO, manifest as undulations that, locally, penetrate into the bondcoat (Fig. 12) [31, 61, 62]. This mechanism arises primarily (but not exclusively) in systems with β-phase bondcoats. 2) A second exposes the bondcoat, sometimes with periodic islands of TGO and entrained zirconia (Fig. 13). The bondcoat exhibits imprints of the grains in the TGO, suggesting brittle failure by loss of adhesion at the metal/oxide interface. Cross sections affirm that the failure occurs primarily by delamination along the interface, sometimes with local extension through thickness heterogeneities in the TGO (Fig. 13) [27]. This mechanism is prevalent for bondcoats comprising $\gamma + \beta$- and $\gamma + \gamma'$-phases. It also occurs with β-phase bondcoats when rumpling is suppressed or does not occur. This

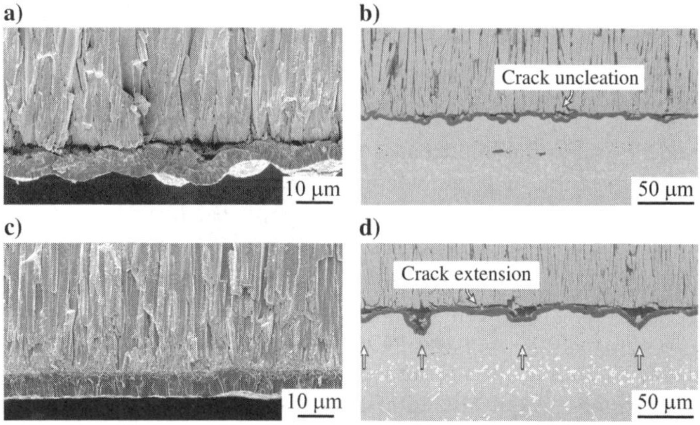

Fig. 12 Scanning electron images of TBC-coated furnace cycle specimens with β-phase bondcoats. TBC and TGO with substrate appearing black: a) grit blast surface that experiences rumpling and c) one that resists rumpling and does not fail. Substrate and bondcoat present b) at 20% and d) at 80% of cyclic life. The coating in c) is intact, but the alloy below appears black.

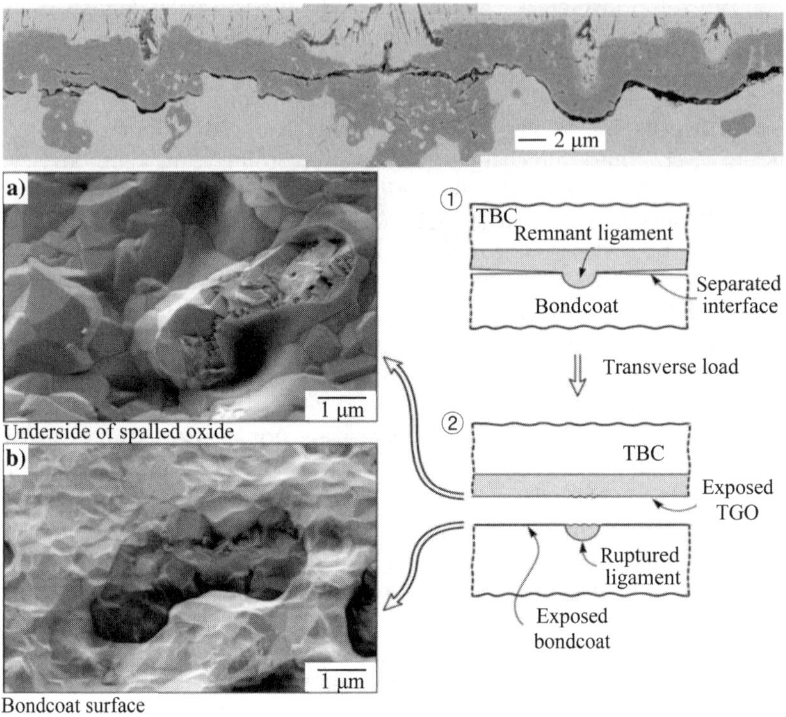

Fig. 13 Images of systems with a β bondcoat: a) a delamination at the interface and b) a separated interface imaged at matching locations.

typically occurs when the bondcoat is pre-oxidized before the TBC is deposited (Fig. 12), or when the maximum temperature attained in the thermal cycle is below 1100°C [63, 64]. 3) A third exposes the bondcoat, but now with superposed features indicative of voids formed at longer times [65]. All intrinsic mechanisms have a characteristic TGO thickness h_{crit} at the incidence of spalling. However, h_{crit} depends on the bondcoat composition and microstructure, as well as the thermal cycling history. In itself, it is not a useful metric for characterizing failure across a range of bondcoats and cycling scenarios. However, for many land-based turbines where operation typically consists of relatively long cycle times at lower temperatures, the use of a critical thickness has some merit but not at short cycle times [66].

The extrinsic category of failure mechanisms cannot be reproduced in furnace cycling or conventional burner rig tests. This category includes damage induced by particle impact (erosion and foreign object damage) [67–72] and delaminations enabled by the penetration of molten deposits of calcium-magnesium-alumino-silicate (CMAS) formed from the ingress into the engine of siliceous debris (sands, dust, volcanic ash, etc.) in the atmosphere [73–77]. This damage is dominated by the microstructure and properties of the insulating oxide and the composition of the molten deposits. The manifestations in turbine hardware are as follows. Foreign object damage (FOD) is apparent as spalls at the leading edges of airfoils. Less severe particle impacts cause the gradual thinning of the TBC by erosion in the vicinity of the leading edges. CMAS damage is found in the hottest sections of airfoils, especially along the pressure surface, and in shrouds. The presence of CMAS is evident through a yellow coloration associated with transition elements (such as Fe) in the deposits. The mechanism operates in the presence of a thermal gradient that plays a dual role [16, 77, 78]: 1) it enables the CMAS (once molten) to penetrate to a specified depth into the oxide, and 2) it causes the surface to experience residual tensile stress upon cooling [76]. In turn, these stresses provide the energy release rate that favors the formation of vertical stress cracks, followed by internal delamination. Much of the mechanistic understanding has been based on observations and measurements made on components removed from engines, but recent developments in testing facilities offer promise for duplicating the failure modes in laboratory tests [77, 79, 80].

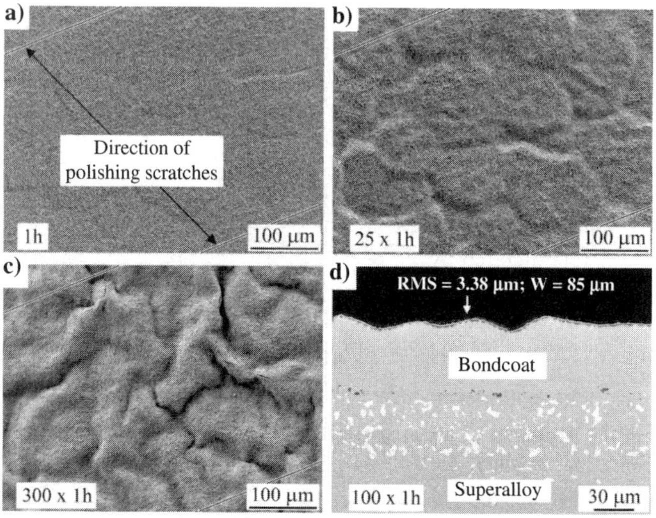

Fig. 14 Surface rumpling of the bondcoated samples during cyclic oxidation in air at 1150°C: a) 1 cycle; b) 25 cycles; c) 300 cycles); and d) cross section after 100 cycles showing roughness parameters of the bondcoat surface. The direction of polishing scratches on the bondcoat surface prior to oxidation is indicated on a), b), and c).

II. INTERDEPENDENT ROLES OF ALUMINA AND BONDCOAT

A. RUMPLING

As already mentioned, rumpling is commonly found in systems with β-phase bondcoats (Fig. 14), but is uncommon in two-phase systems. Its occurrence in the former has been characterized through critical experiments and a model [30–34, 63, 64]. Together, these elucidate the basic trends. The rate at which the rumpling undulations grow in amplitude depends in a highly nonlinear manner on the properties of the constituent materials. The amplitude growth is much greater upon thermal cycling than under isothermal conditions and increases with peak temperature and the number of cycles [63, 64] and so is less pronounced for the hottest sections of land-based turbines. In all practical systems, initial roughness exists, and so trends focused on minimizing this are not practical. Among the many different material properties involved, only three can be modified through compositional changes in the constituent materials: the creep strength of the bondcoat, the strain misfit between the bondcoat and substrate, and the thickening rate of the TGO. Accordingly, a mechanism map has been constructed that uses these as coordinates (Fig. 15). The larger the misfit or the thickening rate, or the lower the creep strength, the more rapid the growth of undulations.

The basic features are as follows. 1) During the cooling half-cycle, the thermal expansion misfit between the TGO and the substrate induces in-plane compression, as well as a small (but crucial) downward pressure (due to the surface curvature), as depicted in Fig. 16. This pressure induces creep deformations in the bondcoat that accommodate amplitude growth of the TGO. As the temperature continues to drop, the pressure exerted by the TGO increases, but so does the creep strength of the bondcoat, usually nonlinearly. Because of this interplay, most of the undulation growth occurs just below the peak temperature T_{max}. 2) Upon reheating to T_{max}, oxidation resumes, and the TGO thickens and experiences a lateral strain-rate dictated by the counterflux of O and Al. This lateral strain reinitializes

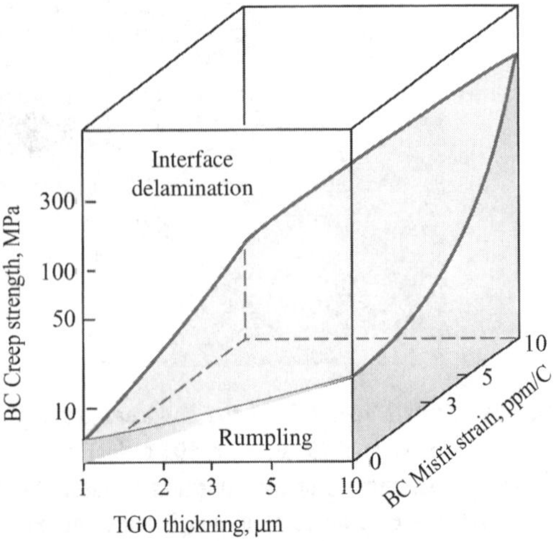

Fig. 15 A mechanism map expresses the parameters that affect the onset of rumpling. The combination of parameters that reside in the domain beneath the dividing curved surface result in rumpling. Moreover, the further beneath this surface the more rapid the undulation growth rate.

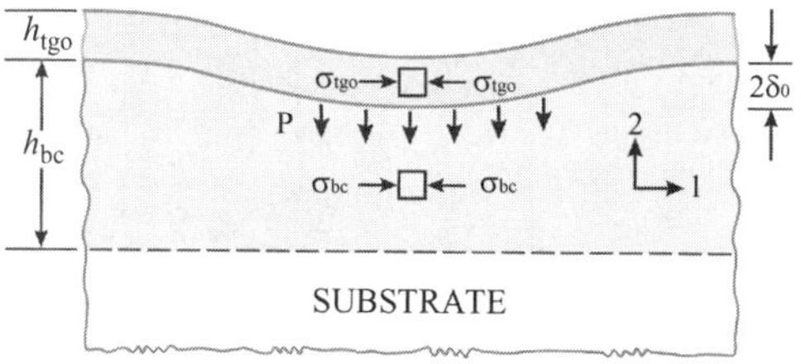

Fig. 16 A schematic showing the stresses created during thermal cycling that facilitate rumpling. The TGO exerts a downward pressure *p* on the bondcoat, which causes it to creep and accommodate the TGO amplitude growth.

the TGO after each cycle allowing the process to proceed on a cycle-by-cycle basis. 3) The strain misfit between the bondcoat and substrate is important because it induces stress in the bondcoat upon cooling. This stress accelerates the creep allowing more extensive accommodation. 4) Typically, the thicker the TGO, the greater the undulation growth per cycle because of the larger pressure it exerts on the bondcoat (more than compensating for the higher bending stiffness). Consequently, more rapid thickening leads to greater amplitude growth. Subject to the foregoing guidelines, the ensuing discussion addresses current understanding of the salient material properties and how they might be influenced.

1. THICKENING AND ELONGATION OF THE TGO

Over the relevant thickness range, the salient phenomena are dictated by columnar $\alpha-Al_2O_3$ (Fig. 17). At the temperatures of interest (1050–1150°C), the

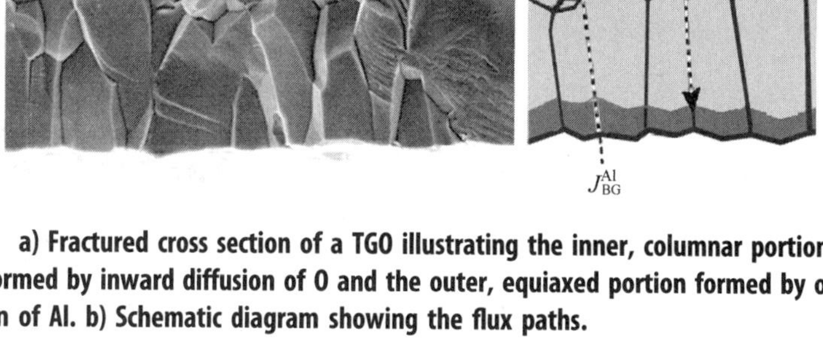

Fig. 17 a) Fractured cross section of a TGO illustrating the inner, columnar portion of the oxide formed by inward diffusion of O and the outer, equiaxed portion formed by outward diffusion of Al. b) Schematic diagram showing the flux paths.

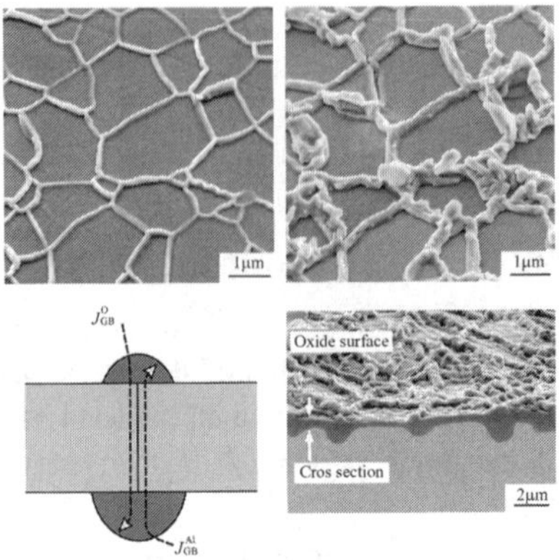

Fig. 18 Illustration of the same area of an alumina TGO on reoxidation after smooth polishing of the TGO formed in the first oxidation step. New oxide forms along the grain boundaries of the initially formed TGO, and the amount increases with further oxidation. This is a schematic of the counterdiffusion of O and Al along the TGO grain boundaries leading to a thickening of the oxide above and below the oxide on either side of the boundary. In between grain boundaries, the oxide is significantly thinner, as indicated by the arrows.

growth involves counterdiffusion of oxygen and aluminum along the $\alpha-Al_2O_3$ grain boundaries (Figs. 17 and 18) [81]. The diffusion flux and hence the parabolic rate constant are dictated by the grain size of the $\alpha-Al_2O_3$ as well as by impurities and by dopants that diffuse to the TGO and become entrained. Despite a wealth of empirical data, the fundamentals remain elusive although it is clear that the following two issues are especially critical.

Phase transformation from amorphous alumina (formed during initial stages of heating) through various transient polymorphs to the stable $\alpha-Al_2O_3$ determines the nucleation rate of the stable phase, as well as its lateral growth rate. In turn, this determines the grain size of the $\alpha-Al_2O_3$, thereby controlling the subsequent thickening rate (oxidation proceeds by grain boundary diffusion). The $\alpha-Al_2O_3$ forms by means of a nucleation and growth process, affected by certain dopants, such as Cr (always incorporated for corrosion resistance) and the rare earths (Hf and Y used to control interface adhesion). Doping with Hf and Y retards the growth rate of the $\alpha-Al_2O_3$, once nucleated [82]. This benefit happens because these cations are soluble in gamma and theta but not in alpha, causing them to be rejected from the growing α. Consequently, the local driving force is insufficient for the interface to advance into the metastable phase until the excess solute precipitates [83]. Conversely, Cr and Fe (cations soluble in α) accelerate the growth rate [84, 85]. Given these counteracting influences of the dopants, a compromise in the relative dopant levels is desired . Yet, because a quantitative model is lacking, reliance has been placed on empiricism, slowing progress toward optimal doping strategies.

The relative inward and outward diffusive fluxes along the grain boundaries in the ensuing $\alpha-Al_2O_3$ influence both thickening and elongation. The grain boundaries governing these effects are clean and devoid of amorphous interphases (although sub-mono-layers of Hf or Zr can be entrained). The conventional picture is that, once $\alpha-Al_2O_3$ is formed, oxide thickening is dominated by inward diffusion of O. In practice, there is a counterflux of cations [81, 82] demonstrated by the following protocol. The alloy is oxidized to form a continuous TGO, which is then polished at an angle, and re-oxidized. Some of the new oxide forms as ridges along locations where the oxide grain boundaries intersect the surface (Fig. 18). There are corresponding ridges where the grain boundaries intersect the interface. Measurement of the volume of these ridges allows assessment of the relative anion and cation fluxes (evidently comparable in Fig. 18). Such measurements reveal that the counterfluxes depend sensitively on dopants such as Hf and Y entrained in the grain boundaries. The unresolved questions are 1) how to think about the atomic mechanisms of counterdiffusion and 2) how elongation strain is determined by this process.

2. STRESSES IN THE TGO

If all of the new $\alpha-Al_2O_3$ formed at the interface with the bondcoat, the ensuing volume increase would be accommodated by upward (rigid body) motion of the prior TGO, obviating any growth stress. Instead, the outward counterflux of Al causes some new $\alpha-Al_2O_3$ to form at dislocations/ledges along the transverse grain boundaries [22, 86], as well as that formed on the surface of the TGO (Fig. 18). The new oxide formed at the boundaries must be accommodated by lateral deformation of the neighboring grains, causing a compressive growth stress (Fig. 19). Because $\alpha-Al_2O_3$ is susceptible to plastic deformation at the growth temperature, the stress attains a steady state, wherein the strain-rate induced by the growth is balanced by the creep rate. The magnitude of this stress has been measured in situ for the TGO formed on several different bondcoats and is of order $\sigma_{growth} \approx -300$ MPa [50] (Fig. 20). While the stress level is consistent with deformation mechanism maps for $\alpha-Al_2O_3$ and stress relaxation rates in a typical TGO, it remains to develop a quantitative model. Moreover, the influences of role of dopants on growth strains and relaxations are poorly understood.

Upon cooling, because of its low thermal expansion coefficient (relative to the substrate), a large in-plane compression develops (Fig. 20). At ambient, whenever the TGO remains planar (no rumpling), the compressive stress is in the range $-3.5 < \sigma_{tgo} < -6$ GPa, depending on the thermal expansion coefficient for the substrate [57]. When rumpling occurs, the stress diminishes because of bending and elongation of the TGO [33, 56].

3. BONDCOAT STRAIN MISFIT

This strain is affected by the thermal expansion difference with the substrate, martensitic transformation (at Al compositions where it occurs) and swelling

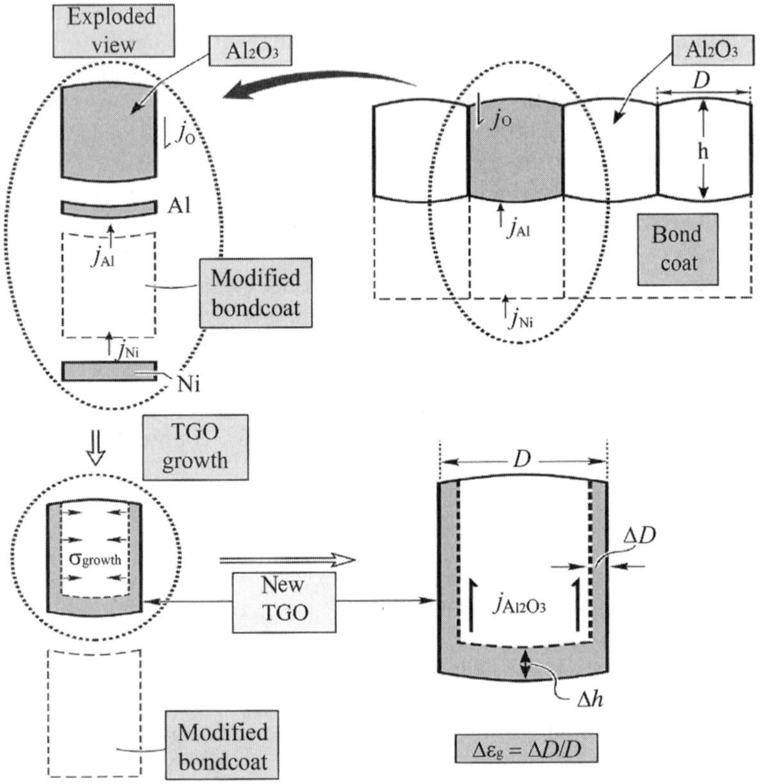

Fig. 19 Schematic indicates the elastic compression of the TGO grains needed to accommodate the formation of new alumina at the internal grain boundaries.

caused by interdiffusion with the substrate (Fig. 21). It induces an extra stress upon thermal cycling that accelerates the creep deformations in the bondcoat in response to the pressure exerted by the TGO. All of

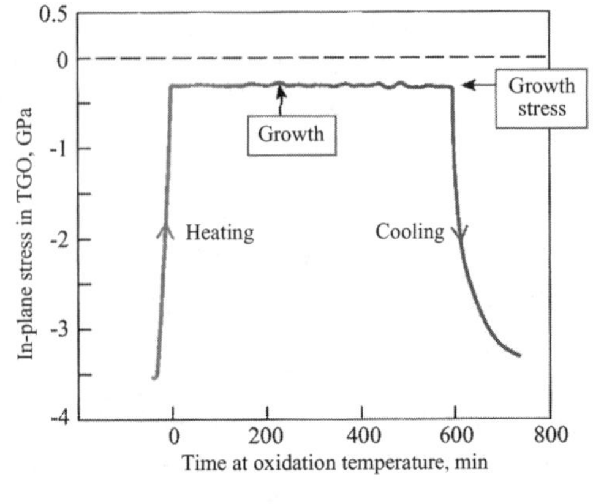

Fig. 20 Illustration of the stress in the TGO measured in situ in the synchotron during a single thermal cycle from ambient to 1125°C and back to ambient [47, 48].

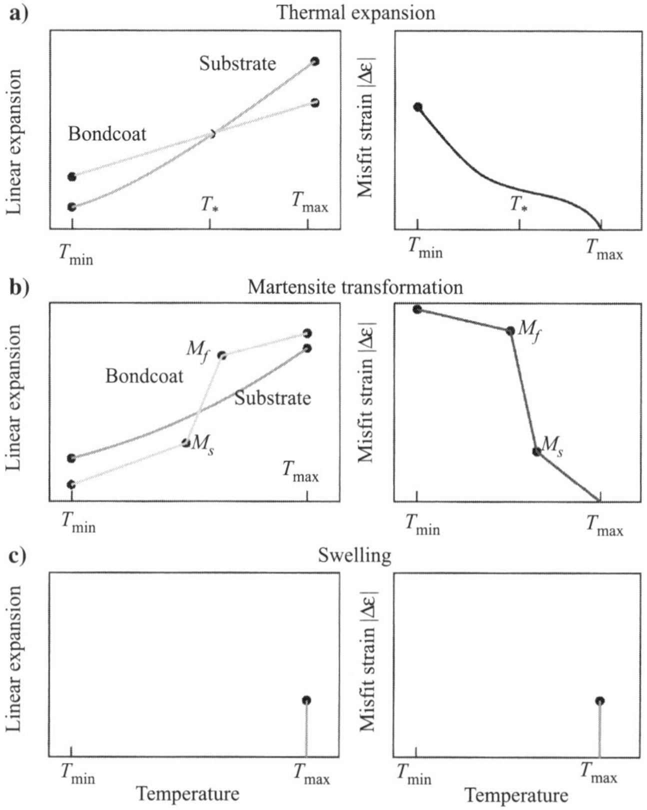

Fig. 21 The three sources of misfit strain between the bondcoat and substrate (schematic graphs): thermal expansion, martensite transformation, and swelling. On the left are expansions, and on the right are misfit strains.

these misfits can be included in the model and can be examined experimentally. Some of the key features, summarized in Fig. 21, are as follows. Initially, the Al content is relatively high (within the β-field), and there is no martensite transformation. In this circumstance, the thermal expansion is the primary misfit governing undulation growth. Note that it does not matter whether the bondcoat has a higher or lower thermal expansion than the substrate; only the absolute difference is relevant (Figs. 21a and 21b). The rate of change of the misfit is largest close to T_{max} (Fig. 22). Consequently, creep accelerates as soon as the temperature starts to drop, resulting in a maximum undulation amplitude when the lowest temperature in the thermal cycle T_{min} is about 100–150°C below T_{max} (Fig. 22). There is also a large misfit at low temperatures (Fig. 21b), but this is not effective because the associated high creep strength inhibits plastic accommodation of the TGO.

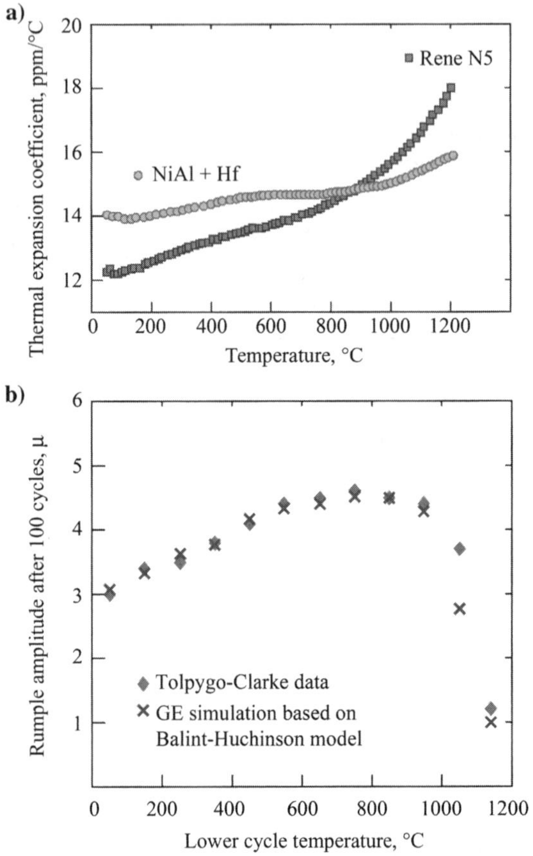

Fig. 22 a) The thermal expansion coefficients of a representative Rene N5 substrate and a β-phase (NiAl/Hf) bondcoat. b) The rumpling amplitude after 100 cycles as a function of the lowest temperature in the cycle: comparison of measurements and predictions.

As the Al content in the bondcoat decreases because of interdiffusion and formation of the TGO, a γ'-phase forms, and the Al content in the β-phase also diminishes. Subsequently, at some stage, the β-phase experiences a martensitic transformation ($B2 \rightarrow L1_0$) on cooling and reheating, with accompanying volume change (Figs. 21c and 12d). The Martensite start temperature M_s depends on composition [87]. When $M_s \leq 550°C$, the creep strength is sufficiently high that the associated strain does not lead to TGO accommodation. When larger, accommodation is possible. Moreover, the higher M_s, the greater the accommodation is. Eventually, the undulation growth associated with the transformation is predicted to exceed that caused by the thermal expansion misfit.

B. INTERFACE ADHESION

When rumpling is suppressed, the intrinsic durability is limited by delamination along the interface between the TGO and the bondcoat. The energy release rate enabling this mechanism (Fig. 10) is communicated to the interface as a mode II (shear) delamination. In principle, equating the energy release rate to the mode II toughness of the interface predicts a lower bound on the TGO thickness that causes delamination (Fig. 10). In practice, this potential for predicting the critical thickness has only been partially realized because of the difficulty in measuring the mode II toughness. Even measurements of the mode I and mixed mode interface toughness have been challenging. The difficulties are as follows. 1) Loading scenarios that extend cracks solely along the interface with defined mode mixity

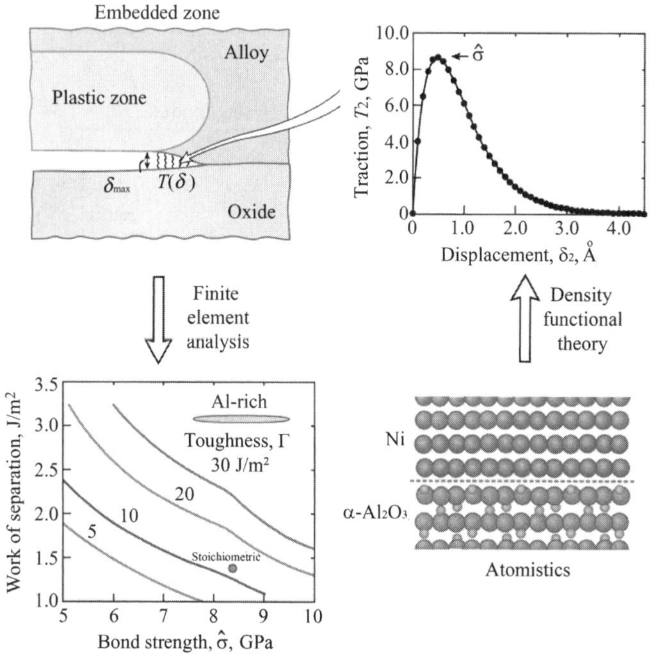

Fig. 23 Schematic showing how traction/separation results obtained from first principles calculations become an embedded process for a crack extension simulation that captures the multiplicative influence of plastic dissipation on the interface toughness. Bottom left shows contours of interface toughness in space defined by the work of separation and bond strength determined from first principles calculations. The results are for plasticity parameters measured for an NiCoCrAlY bondcoat (plasticity length scale $\ell = 50$ nm). Superimposed are the locations for stoichiometric and Al-rich interfaces: toughness ~ 10 J/m^2 and >30 J/m^2 respectively.

are scarce. Conical and wedge indentation approaches have been used as well as buckle delaminations [2]. 2) Extracting the interface toughness from the measurements is nontrivial partly because of the (difficult to quantify) contributions to the energy release rate from the release of the residual stresses in the TBC and TGO layers and partly because of uncertainties in measuring the precise location of the delamination front. Tests that have attempted to deconvolute the measurements suggest that the interface toughness (at a TGO thickness of order $h_{tgo} \approx 3\ \mu m$) is in the range $40 \rightarrow 60\ Jm^{-2}$, at mode mixity 60 deg → 90 deg [2].

Because the measurements are sparse and the toughness depends on the presence of segregants and dopants, however, a simulation scheme that distinguishes the factors dominating the adhesion (Fig. 23) has been pursued [88–90]. It has two basic ingredients. 1) The traction/separation characteristics during bond rupture at the interface are ascertained using a first principles approach based on density functional theory. 2) These results are input to an embedded process zone (EPZ) simulation of interface crack extension that captures the

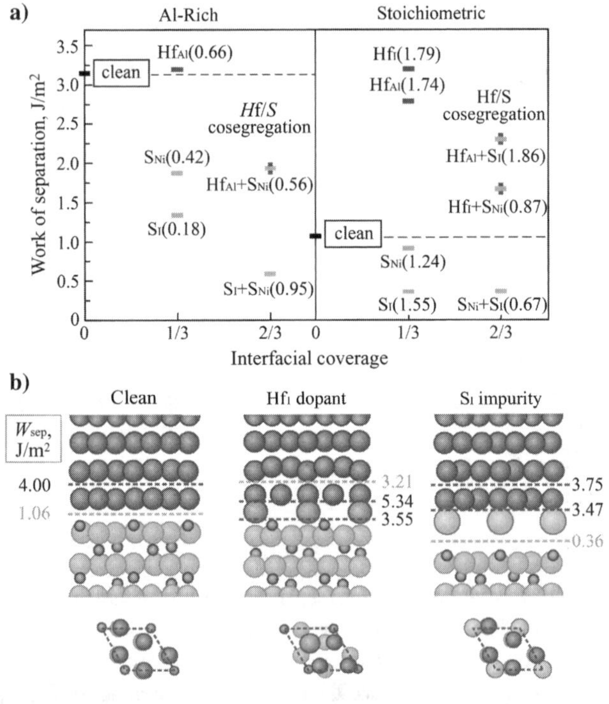

Fig. 24 a) Trends in the work of separation calculated using density functional theory for both stoichiometric and Al-rich interfaces. b) The atomic arrangements at a stoichiometric interface, with and without dopants and segregants. The numbers refer to the work of separation on the indicated plane.

multiplicative influence on the toughness of the plastic dissipation occurring in the bondcoat. In general, the traction/separation curves are found to depend on the termination plane (stoichiometry) of the $\alpha-Al_2O_3$, as well as the presence of dopants and impurities. For γ-Ni(Al) alloys, the termination has been ascertained to be a mix of stoichiometric and Al-rich. The former is the least adherent (Fig. 24). Moreover, when present, S segregates to this interface and further decreases the adhesion (by up to 70%) because the interfacial covalent-ionic Ni-O bonds are replaced with weaker ionic-covalent S-Al bonds. Doping with Hf obviates the detriment, especially when it segregates on interstitial sites (Hf_I) (Fig. 24) because Hf-Ni and Hf-O bonds effectively knit the surfaces together. When integrated (Fig. 23), these results establish a rationale for designing tough interfaces. Recall that Hf has the additional benefit that it affects the creep strength of the TGO when it segregates to the grain boundaries.

III. THERMALLY INSULATING OXIDE

The performance of the system and the operative degradation mechanisms are intimately connected to the properties of the insulating layer as affected by its composition and microstructure. The salient properties include 1) thermal conductivity, 2) toughness, 3) yield strength (at high temperature), 4) in-plane Young's modulus, and 5) phase stability. (The thermal expansion coefficient is less critical because the misfit stresses are accommodated by ensuring that the modulus is sufficiently low). To fully exploit materials with lower thermal conductivity than 7YSZ, the surface temperature must be increased, which enhances concerns about such issues as thermal gradient delamination and phase stability. Therefore, it is imprudent to examine changes in one property without considering the consequences for the others. For example, incorporating materials with reduced toughness would be problematic because all of the delamination mechanisms scale directly with this property. Indeed, a crucial attribute of 7YSZ is its relatively high toughness compared to almost all other oxides.

A. THERMAL CONDUCTIVITY

At the time when YSZ was first used in thermal barrier coatings, it was the oxide with the lowest known thermal conductivity, apart from UO_2 and ThO_2. In seeking an alternative, lower conductivity oxide, the most successful approach is guided by the concept of a minimum thermal conductivity. This is the thermal conductivity when the the phonon mean free path is equal to the interatomic distance [91]. This value of the conductivity can be expressed by [4]

$$\kappa_{\min} = 0.87 k_B \overline{\Omega}^{-2/3} (E/\rho)^{1/2} \tag{1}$$

where $\overline{\Omega}_a$ is an effective atomic volume, $\overline{\Omega}_a = M/m\rho N_A$, with M being the mean atomic mass of the ions in the unit cell, m the number of ions in the unit cell, ρ the density, and E Young's modulus. The k_B and N_A are the Boltzmann constant and Avogadro's number, respectively. In many cases, the elastic properties of complex oxides are unknown, and $\kappa_{\min}$ cannot be estimated very accurately, that is, better than a factor of 50%. Nevertheless, materials with large atomic mass, a large number of atoms per unit cell, and flexible atomic bonding are likely to have low modulus and, hence, low thermal conductivity [91]. On this basis, low conductivity has been demonstrated in a variety of oxides with previously unknown thermal and elastic properties (Fig. 6b). Several of these, notably $La_2Mo_2O_9$ and oxy-apatites such as $Ca_2Gd_8(SiO_4)_6O_2$, are anionic defect structures that also exhibit unusually high oxygen conductivity. Others, such as $Bi_4Ti_3O_{12}$, have highly anisotropic crystal structures although the important factor is not the anisotropy per se, but the mass difference between the block structures. The results shown in Fig. 6b are encouraging because they suggest considerable potential for the discovery of oxides with still lower thermal conductivity although viable compositions within this range are restricted by other requirements, such as thermochemical compatibilities and toughness.

B. TOUGHNESS

Among dense (nonporous) oxides, remarkably, the toughness range realizable among all materials is fully encompassed by YSZ across the composition range between cubic and tetragonal (Fig. 25) [17]. The cubic materials (c-ZrO_2, e.g., 20-YSZ) are exceptionally brittle (toughness, $\Gamma \approx 6\ \mathrm{Jm}^{-2}$) while partially stabilized tetragonal materials (t-ZrO_2, e.g., 3-YSZ), which exhibit a martensitic transformation to the monoclinic phase (m-ZrO_2), are among the toughest ($\Gamma > 300\ \mathrm{Jm}^{-2}$). The transformation mechanism, however, is inapplicable for two related reasons: 1) It is thermodynamically forbidden at elevated temperatures, specifically those above $T_0(t/m)$ (Fig. 26a) because there is no driving

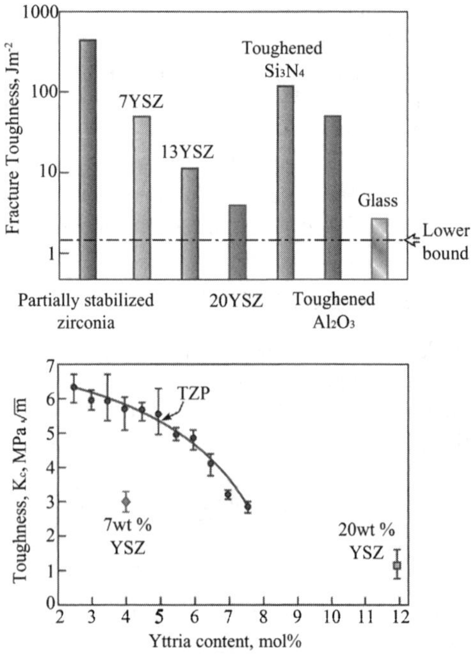

Fig. 25 The toughness of monolithic ceramics showing the wide range achievable in the YSZ system [19].

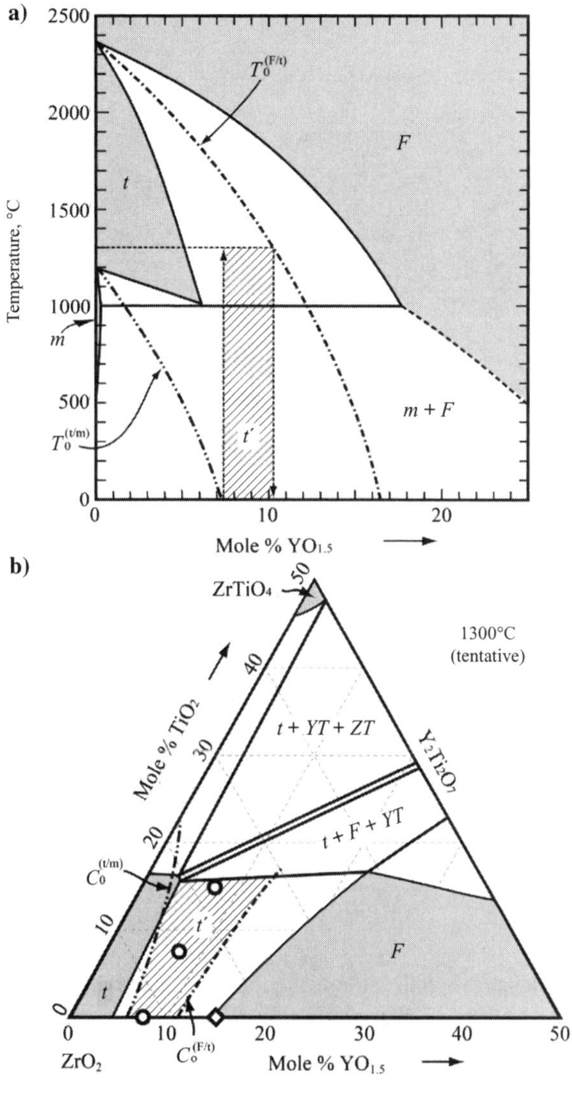

Fig. 26 a) ZrO_2-$YO_{1.5}$ phase diagram showing the domain wherein non-transformable-tetragonal solid solutions are thermodynamically feasible and favored over the cubic form of the same composition up to 1300°C (hatched area). b) Extension of that domain into the ZrO_2-$YO_{1.5}$-TiO_2 system. $C_0(t/m)$ is the trace of the $T_0(t/m)$ surface at ambient temperature, and $C_0(F/t)$ is the trace of the $T_0(F/t)$ surface on the 1300°C isotherm [19].

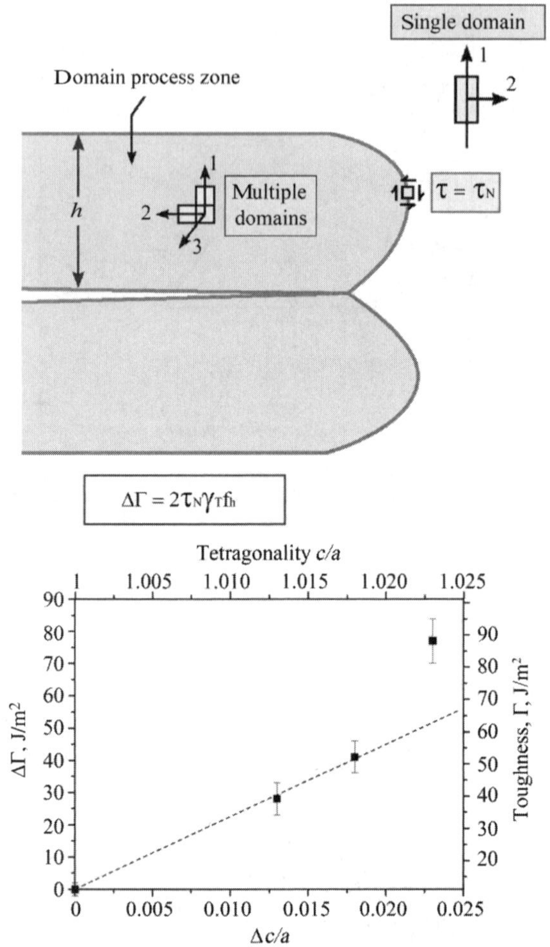

Fig. 27 Trend in toughness with tetragonality for the compositions marked in Fig. 26b. From left: $15.2YO_{1.5}$, $7.6YO_{1.5}$, $7.6YO_{1.5}$-$7.6TiO_2$, $7.6YO_{1.5}$-$15.2TiO_2$.

force for the partitionless $t \rightarrow m$ transformation; and 2) repeated cycling across the $T_0(t/m)$ results in disruptive volume changes every time the t-ZrO_2 transforms to m-ZrO_2 on cooling and regenerates upon heating with concomitant microcracking. The ensuing loss of mechanical integrity increases susceptibility to other forms of damage, such as erosion. Compositions within the nontransformable tetragonal (t') phase field, bound by the compositions for which $T_0(t/m)$ is below ambient and $T_0(c/t)$ is above the maximum operating temperature, provide the best performance [18]. Because tetragonality typically decreases with increasing dopant content [92], the preferred compositions within this range are those at the lower end, preeminently exemplified by 7YSZ. Typically, the mode I toughness is in the range, $40 < \Gamma < 50$ Jm^{-2} [17–19, 93], which is sufficient to prevent spalling after manufacturing and to suppress large-scale delamination when exposed to thermal gradients. Such compositions are ferro-elastic [94–96]. Upon crack extension, dissipation occurs through the formation (or switching) of nanoscale domains, resulting in toughening that scales with the tetragonality, c/a, and the coercive stress [17, 18, 94–96]. The concept that the tetragonality governs the toughness of tetragonal oxides has led to the development of new compositions with appreciably higher toughness than 7YSZ [18–20, 97]. The most prominent example resides within the ZrO_2-$YO_{1.5}$-TiO_2 ternary phase field (Fig. 27), having toughness (measured by microindentation) of order $\Gamma \approx 90$ Jm^{-2} [18].

The toughness of actual TBCs is quite different although the measurements are sparse. Surprisingly, despite the porosity (which normally reduces toughness), TBCs can have higher toughness than the dense materials for the same composition (sometimes considerably). One documented example is for 7YSZ generated by air plasma spray, which has toughness $300 \leq \Gamma \leq 400$ J/m^2 [98]. This elevated toughness suggests important effects of the microstructure although the underlying reasons remain to be explained.

C. MODULUS

Given that the in-plane modulus of the layer is so important, it is remarkable that published measurements are still sparse. The few comprehensive studies in the literature [37–39] reveal values in the range $20 < E_{tbc} < 40$ GPa (compared with 200 GPa for dense 7-YSZ), remarkably consistent for both APS and

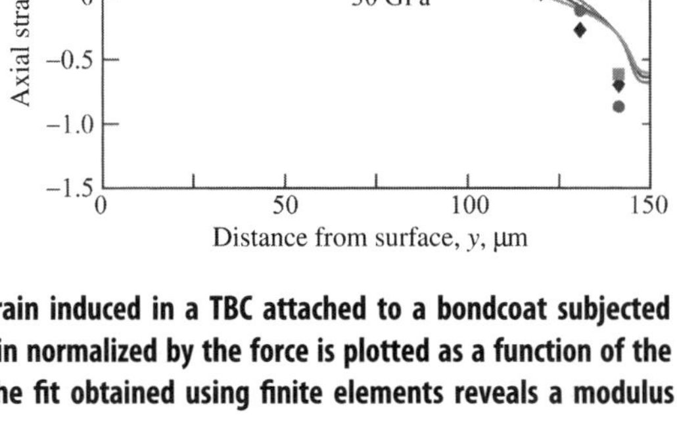

Fig. 28 Axial strain induced in a TBC attached to a bondcoat subjected to three-point bending. The strain normalized by the force is plotted as a function of the distance from the tensile surface. The fit obtained using finite elements reveals a modulus of 20 GPa.

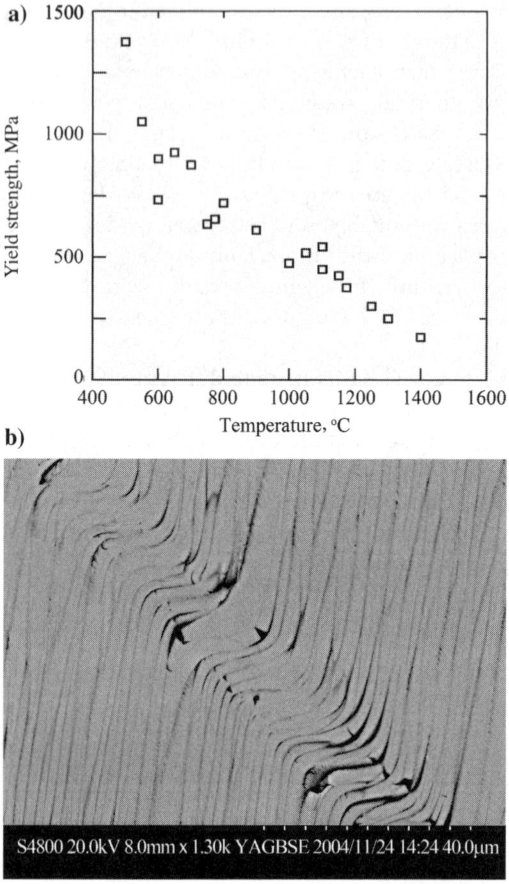

Fig. 29 a) The yield strength of 7-YSZ as a function of temperature. b) A kink band induced in a columnar material by impressing at 1100°C. Note the plastic bending across the band.

EB-PVD materials. Recent measurements conducted on a columnar material as a function of an imposed tensile strain (Fig. 28) [99] indicate that, at strains up to the onset of fracture, the modulus is strain invariant with $E_{tbc} \approx 20$ GPa. Values in this range, in conjunction with the strain invariance, are consistent with a microstructure comprising columns bonded over a small fraction of their total transverse area.

D. YIELDING

The yield strength plays a role through the ability of plasticity to dissipate the kinetic energy from particles circulating in the turbine that impinge onto the rapidly rotating airfoils. The ability of the layer to yield when impacted by foreign objects in the engine is a significant attribute. The plastic response upon impact at high temperature is reflected in the relatively low yield strength of 7YSZ above 900°C [96] (Fig. 29a). The associated plastic dissipation serves to absorb much of the kinetic energy from the impact and thereby diminish the amplitude of the elastic waves that propagate through the layer [15]. Also of interest is the role of yielding in the development of kink bands in systems having columnar microstructure [71, 72] (Fig. 29b). The plastic bending of the columns is apparent, as well as the incidence of cracks wherever the bending induces large tensile strains [69, 93].

IV. TEMPERATURE LIMITS

Engine performance is linked to the temperature capability of the thermal barrier oxide, which for currently available 7YSZ materials is limited by three phenomena that degrade the strain tolerance and insulating function. Two of these arise from intrinsic changes in the oxide microstructure over time, namely, sintering [100–103] and destabilization of the t′-phase [104–110]. The third is associated with the penetration of molten deposits, generically known as CMAS (calcium-magnesium-alumino-silicates), into the void spaces with consequent stiffening and increased susceptibility to delamination upon cooling [73, 75, 77] (see Sec. V). For aeroturbines, CMAS degradation is arguably the most critical of these phenomena because the surface temperature is bound by its incipient melting point, which can be as low as 1200°C. Increasingly, it is also a problem for land-based turbines located near sources of dust or those being operated with alternative fuels. Absent CMAS degradation, sintering effects arise before t' destabilization, generally manifest as increased thermal conductivity [111, 112] and subsequent loss of compliance [39, 113]. Destabilization of t' represents the ultimate limit. Challenges to and opportunities for the development of new compositions with higher temperature capabilities are briefly discussed in this section.

A. PHASE EVOLUTION

This problem is rooted in the metastable nature of the nontransformable t'-phase required for attainment of a desirable toughness over the range of temperatures of interest (Fig. 26a). For most viable stabilizers, t' is supersaturated and partitions into the equilibrium (depleted tetragonal and cubic) phases upon sufficient high temperature exposure [4]. Thereafter, the depleted tetragonal phase is susceptible to the disruptive transformation to monoclinic upon cooling below $T_0(t/m)$ [104–110, 114]. The transformation can be suppressed upon cooling and then occurs gradually over time, with a rate dependent on temperature and the presence of moisture [39, 115, 116]. The basic behavior is common to all systems stabilized with oversized trivalent cations. The kinetics depend strongly, however, on dopant size; along the lanthanide series, the onset of monoclinic is fastest for

La and slowest for Yb [4]. The problem is obviated when the total stabilizer content moves the composition toward the cubic range, often with significant reductions in thermal conductivity, but at the expense of the toughness because of the concomitant decrease in tetragonality.

Because the most common stabilization approach is based on the introduction of anion vacancies that remove the underlying reason for the tetragonal distortion, none of these systems offers viable avenues to enhanced phase stability while maintaining tetragonality and toughness. The challenge requires an alternate strategy, based on smaller cations with valence $\geq 4+$ such as Ti^{4+} and Ta^{5+} that induce tetragonality by exacerbating the mismatch between the average cation size and the surrounding O cage. These dopants cannot be used alone because there is no viable nontransformable regime in the binary diagrams. Successful examples of compositions exhibiting desirable combinations of toughness and phase stability can thus be found in ZrO_2 co-doped with $Y^{3+} + Ti^{4+}$ [18], $Y^{3+} + Ta^{5+}$ [19], $Yb^{3+} + Ta^{5+}$ [117], or $Ce^{4+} + Ti^{4+}$ [20]. All of these systems exhibit regions within the single-phase tetragonal field that are nontransformable to monoclinic and, by extension, associated supersaturated t' compositions that would also be nontransformable even if they were to partition into the equilibrium phases at high temperature, for example, Fig. 26b. Stable tetragonal phases are thus available at temperatures as high as 1500°C in some of these systems, but the range of nontransformable compositions tends to shrink as the temperature decreases, for example, [118], motivating the use of graded or layered architectures to satisfy the phase stability requirements over the entire thickness of the coating. Moreover, these compositions present more significant processing challenges than YSZ owing in part to differences in vapor pressures [119] and the susceptibility of Ti, Ta, and Ce to change oxidation states during deposition.

B. SINTERING

Sintering refers to time/temperature evolutions of the pore architecture that might affect the desired levels of compliance and thermal resistivity of the TBC. Sintering adversely affects thermal resistance [111, 112]. The consequences for compliance and hence, delamination resistance, are not entirely clear; the weight of the evidence indicates that sintering has minimal adverse effects. The occurrence of sintering is evident as an increase in the thermal conductivity/diffusivity (Fig. 30) at temperatures as low as 900°C for both EB-PVD [101] and APS [39] 7YSZ. The rate of increase is highest upon first exposure, followed by a much slower evolution, with eventual saturation (Fig. 30). The trend is apparent for both isothermal conditions [39, 101] as well as under high thermal gradients [111], and the details are specific to the material: the initial microstructure [120], the thermal history, and, most importantly, whether the coating is attached to the substrate or free-standing. The first stage represents the rapid evolution of highly anisotropic features, such as the spheroidization of the feathery pores within the columns of EB-PVD coatings [121] or bonding across intersplat

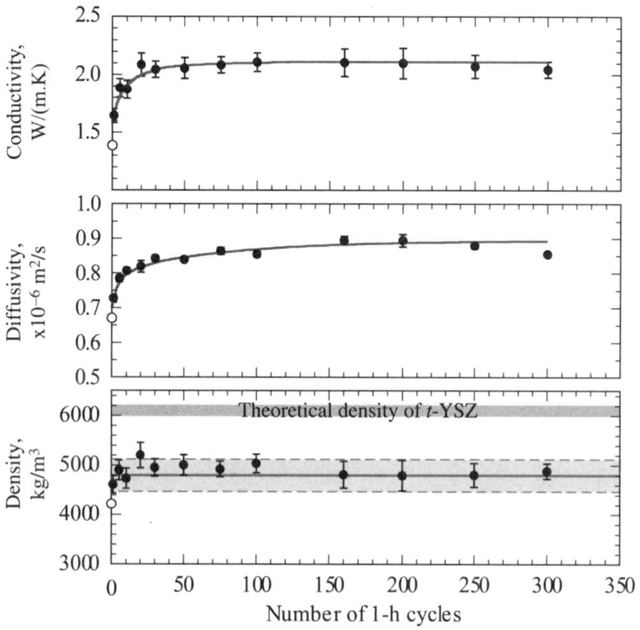

Fig. 30 Effect of thermal cycling from ambient to 1150°C on the thermal diffusivity, thermal conductivity, density, and optical penetration depth of an EB-PVD coating: 1-h cycles.

boundaries in APS [113]. The second stage involves bonding across intercolumnar gaps in EB-PVD [100] and transverse microcracks in APS. The processes are largely controlled by surface diffusion, although additional mechanisms can rise above 1100°C [101].

Although the effects of pore reconfiguration in blocking heat flow [122], and inducing optical scattering, are clear, the influences on the modulus are elusive. Indentation tests do not measure the macroscopic in-plane compliance, relevant to delamination. Bending tests are more appropriate, but most measurements have been performed on free-standing coatings, with results quite different from those for attached coatings. When unconstrained, the TBC shrinks and densifies, causing the modulus to increase by a factor of 5–6 in APS materials (after 200 h/1250°C) [39] and 100 h/1300°C [120]). Conversely, on a similar TBC aged for 100 h/1300°C while still attached, but then detached for measurement, the modulus increases by a factor of only 1.5 [120]. The latter is attributed to the sintering stresses induced by the substrate constraint and their subsequent relief by the formation of a pattern of "mud cracks" that recover the in-plane compliance over the scale of the coating [102]. The phenomenon is exploited in dense, vertically cracked (DVC) APS materials (Fig. 9c), which exhibit superior delamination resistance, and emerges naturally in EB-PVD coatings, where the intercolumnar gaps provide ready nucleation sites. There have been suggestions that the mud cracks can induce stresses into the TGO and at interfaces, which might cause preferential TGO cracking and delamination, but no experimental evidence has been presented.

Increasing the stability of the coating against morphological evolution should preserve the initially low conductivity, with attendant benefits to the evolution of the TGO. The effect is illustrated in Fig. 31, which shows 7YSZ and Gd zirconate ($Gd_2Zr_2O_7$, GZO) in the as-deposited (EB-PVD) condition and after 100 h at 1200°C [123]. The 7YSZ exhibits smoothing of the outer column surfaces, concurrent with the spheroidization of the feathery pores [101, 121]. In contrast, the GZO columns retain much of their intercolumnar porosity. Grain boundary grooving experiments conducted on these materials provide insight about the associated surface diffusivities. The example presented in Fig. 32a contrasts the rate of grooving (and hence surface diffusivity) for three zirconias with different combinations of Y and Gd (Gandhi, A. S., "Thermal Grooving Experiments in Polycrystalline Yttria Stabilized Zirconia," unpublished work, 2004). The contrast is remarkable and consistent with the microstructural observation (from Fig. 31) that the GZO has much lower sintering susceptibility than YSZ. Insight into the stresses induced upon sintering can be provided by curvature measurements conducted on heat-treated coatings

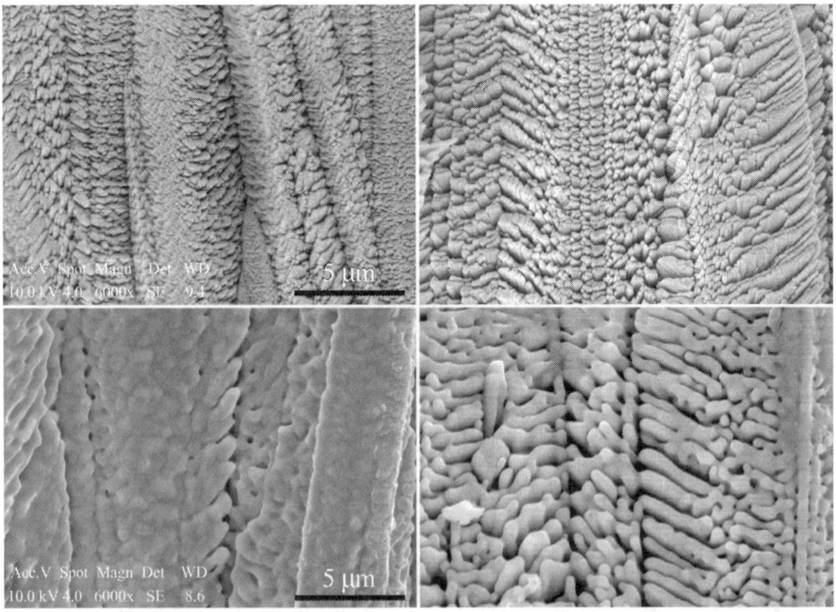

Fig. 31 Comparison of the morphological evolution of the pore structure in 7YSZ (left) and GZO (right) EB-PVD coatings. Top: as-deposited condition; bottom: after 100 h at 1200°C.

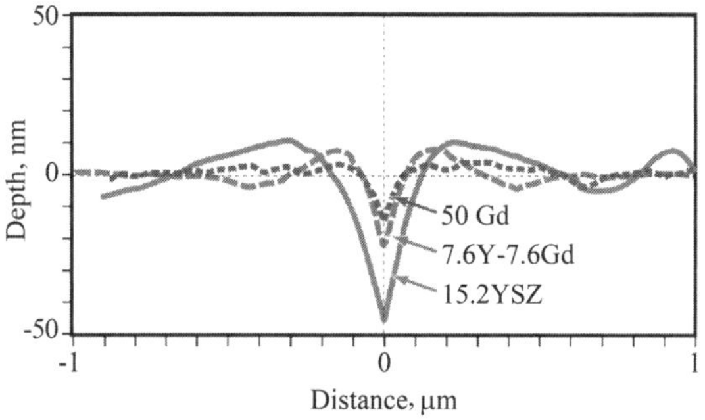

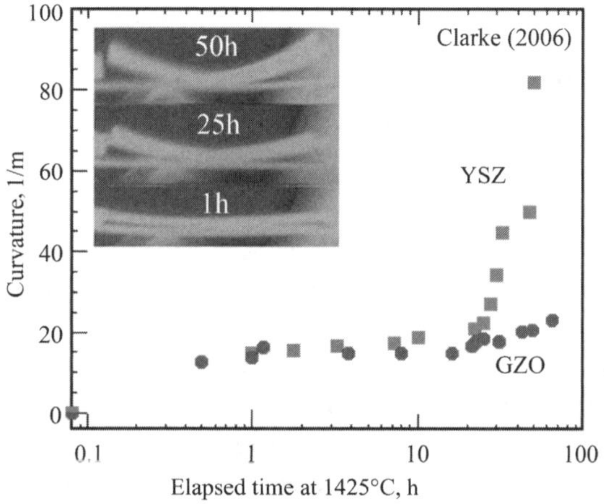

Fig. 32 a) Grain boundary grooving profiles for initially flat surfaces on three different zirconia-based materials, showing the substantial reduction in the surface diffusion kinetics with increasing the Gd content in the solid solution. b) Curvature induced by the intercolumnar sintering stresses in EB-PVD coatings deposited on thin sapphire substrates. Note the pronounced acceleration of the kinetics in 7YSZ relative to that of GZO after the first few hours.

deposited on thin alumina substrates (Lughi, V., and Clarke, D. R., "In-Situ Measurements of TBC Sintering," unpublished work, 2002) (Fig. 32b). The curvature results from the accommodation of the sintering stresses by inelastic (creep) bending of the thin substrate. The diminished bending confirms that the sintering resistance of GZO is greatly superior to that of 7YSZ.

C. CMAS DEGRADATION

The CMAS problem is eminently thermomechanical, as illustrated in Fig. 33 and elaborated in the following section [74, 76, 77]. Nevertheless, issues related to the

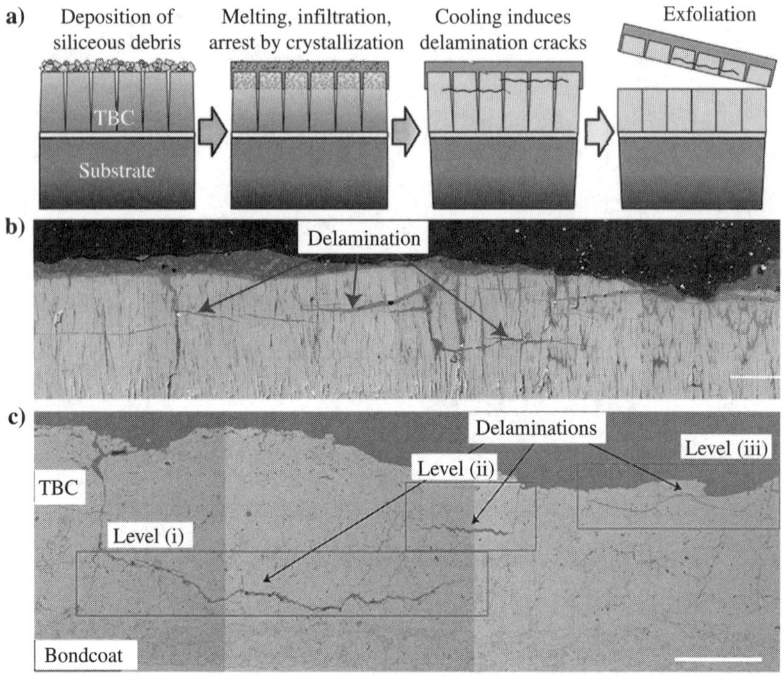

Fig. 33 Schematic of a) the thermomechanical degradation mechanism in TBCs by penetration of molten silicate deposits. The ensuing damage is illustrated by b) field specimens of an EB-PVD coated airfoil and c) an APS-DVC coated shroud.

thermochemical aspects of the infiltration mechanism deserve special consideration. The first relates to the manner in which the penetrating liquid is arrested as it flows down the coating capillary passages within the thermal gradient (Fig. 34). (Absent a gradient, the melt penetrates the full thickness in minutes, at flow velocities of order ~3 μm/s). Because of kinetic barriers to nucleation, the melt does not crystallize when reaching the depth within the coating corresponding to the liquidus isotherm, but continues down the thermal gradient until sufficient supercooling is generated to trigger crystallization. This behavior has been demonstrated by DSC experiments, wherein the melting point is found to be ~1240°C and the crystallization exotherm upon cooling peaks at ~1100°C (Fig. 34). It is consistent with the profile of residual stresses measured on engine hardware (Fig. 35), with TEM confirmation that the CMAS located between the melting and crystallization isotherms is crystalline at ambient, whereas that located within the outer region, where temperatures are above the melting isotherm, remains amorphous (Fig. 34). The region where the CMAS remains molten during high temperature exposure experiences dissolution of the 7YSZ and reprecipitation as a modified tetragonal phase, transformable to monoclinic [75]. The amounts of Y and Zr incorporated into the melt as it flows down the coating capillaries do not seem to be sufficient to modify its crystallization behavior [75].

This emerging understanding of the CMAS arrest process suggests that melt penetration can be mitigated by dynamically modifying the melt through reaction with the TBC or a suitable sacrificial oxide. Notable examples are rare Earth zirconates, such as GZO, which rapidly react with CMAS to precipitate a crystalline mixture of an apatite silicate based on $Ca_2Gd_8(SiO_4)_6O_2$. Isothermal experiments show that penetration is arrested within ~30 μm, even upon isothermal exposure at 1300°C, well above the melting point of CMAS [124]. Reaction occurs first within the pore spaces, which are rapidly sealed against further penetration. Attack of the TBC surface occurs much more slowly (Fig. 36a). The thermal scenario demonstrated in Fig. 36b indicates that the mitigation approach works, despite the increase in surface temperature caused by the higher thermal

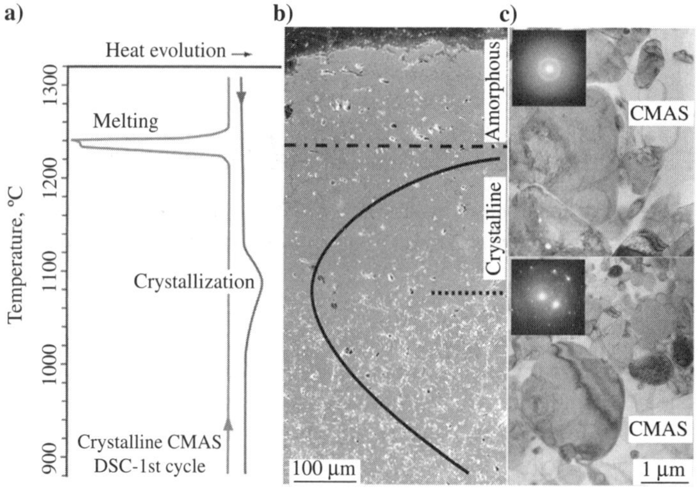

Fig. 34 These figures show a) melting and crystallization behavior for a model CMAS mixture ($Ca_{33}Mg_9Al_{13}Si_{45}$ in cation percent) determined by differential scanning calorimetry using a presintered mixture of the crystalline oxides and heating/cooling rates of ~3 k/min and b) a low magnification image of the penetrated region in an APS-DVC coating, where the silicate within the void space c) in the top part is amorphous and d) in the bottom part is crystalline. The latter is assumed to result from the undercooling of the silicate melt as it penetrated below its melting isotherm. Temperatures estimated from residual stress measurements (Fig. 35) coupled with thermomechanical models [111] suggest that the isotherm corresponding to the lower bound of the penetrated region is consistent with the peak of the crystallization exotherm detected by DSC.

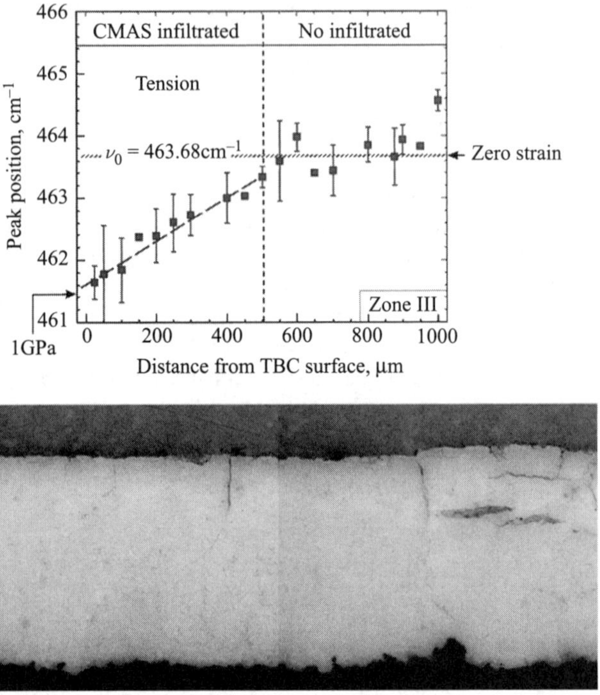

Fig. 35 Residual stresses induced upon cooling within the penetrated region of the APS-DVC coating in Fig. 33 and the transverse cracks resulting within the tensile region near the surface.

resistance of the zirconate (Fig. 6a). There is a concomitant reduction in the bondcoat temperature, leading to slower growth of the TGO. Note, however, that the zirconate requires an intermediate 7YSZ layer to preclude deleterious thermochemical interaction with the TGO [125] and to mitigate delamination [4, 77, 126]. Approaches to enhance the intrinsic toughness of the zirconates remain elusive.

V. DELAMINATION AND SPALLING PHENOMENA

Spalling is preceded by the formation and extension of delaminations along planes parallel to the substrate, at the TGO/bondcoat interface, or in the TBC above the TGO, or (rarely) within the TGO itself. It occurs in accordance with the intrinsic and extrinsic mechanisms discussed in Sec. I. A full predictive capability remains elusive because of the numerous scenarios, as well as substantive measurement challenges, and nuances in energy release rate and mode mixity. Basic differences in the mechanics arise for circumstances that activate intrinsic and extrinsic mechanisms. Consequently, they are examined separately.

A. INTRINSIC MECHANISMS IN COMPRESSED COATINGS

When the intrinsic mechanisms prevail, the TGO and TBC are both in net-residual-compression throughout a thermal cycle although (importantly) localized tensile stress can arise. The following synopsis highlights issues that are understood for compressed coatings (but not always appreciated) and others that remain unresolved [16, 127–135].

1. An internal delamination on a flat surface has zero energy release rate [16]. Such a delamination cannot propagate unless it is either connected to a free edge or large enough to cause buckling (Fig. 37) [128].
2. When free edges exist, pure mode II conditions prevail [16, 128]. That is, delaminations extend subject to a mode II toughness. Buckle delaminations are mixed mode [131, 133].
3. Edge-initiated buckles and spalls (Fig. 8) are rarely distinguished from those that initiate internally, even though the mechanics are completely different, complicating interpretation of the cyclic life.

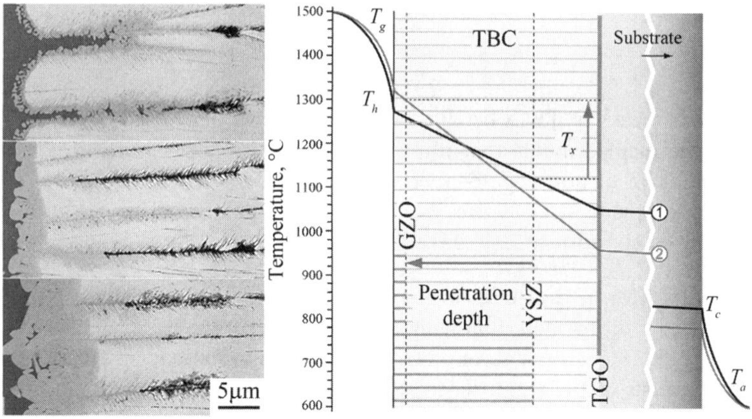

Fig. 36 Micrographs illustrating the sealing of the intercolumnar gaps in a $Gd_2Zr_2O_7$ TBC by interaction between the zirconate and the molten CMAS at 1300°C (top). The channels are sealed within fractions of a minute by rapid crystallization of a mixture of apatite and fluorite phases. Afterward the CMAS reacts with the tip of the columns to form similar products, but in a much longer timescale (hours). The schematic thermal profiles on the right reflect the dual benefits of the zirconate relative to 7YSZ, both in arresting CMAS penetration near the surface by a dynamic modification of the crystallization temperature and by reducing the temperature at the bondcoat/thermal barrier interface, owing to the lower thermal conductivity of the oxide.

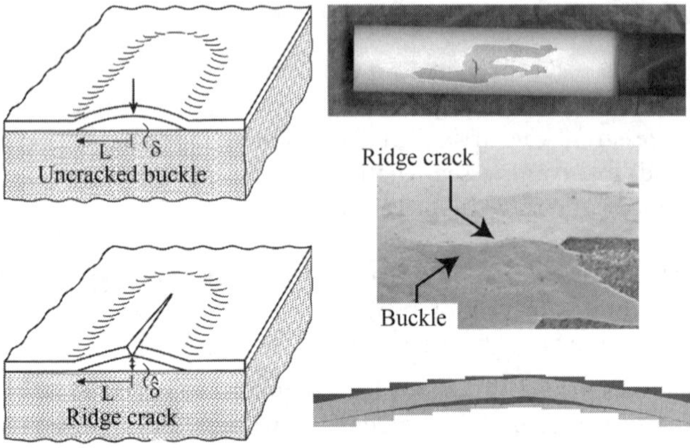

Fig. 37 Synopsis of buckling phenomena found in TBCs. On the left, the schematics show edge buckles with and without a ridge crack. On the right are images of spalls and buckles. On the top is a large axial spall created by introducing a wedge indentation through an intact TBC, demonstrating the supercritical nature of the delamination. In the middle is the end region of one of the spalls from the upper image, showing the buckle with ridge crack that preceded spalling. (On the bottom is a cross section through the buckle revealing that the crack does not extend through the entire coating, but arrests at the TGO.

4. Vertical cracks in TBCs behave as internal edges [27, 132]. They accommodate the displacements that impart the energy release rate required for the propagation of non-edge-connected delaminations. Such cracks form when in-plane tensile stresses are induced, either upon sintering or with rapid cooling of the surface (Fig. 35).

For a spall to form, the state of compression in the coating must be relieved locally. This can happen following buckling, augmented by ridge cracking (Fig. 37) [133, 136]. In the absence of buckling, delaminations arrest before converging because, as the residual stress relaxes, the associated energy release rate drops below the toughness [137, 138]. In such circumstances, the coating remains attached. Spalling requires application of a subsequent mechanical force, such as an impact.

B. IMPERFECTIONS AND SUPERCRITICALITY

Imperfections become critical when they impart an energy release rate that approaches the toughness $G \rightarrow \Gamma$. Free-edges are enablers, as are distributed cracks in the TBC (above the TGO) when rumpling occurs (Fig. 12). Absent rumpling and when free-edges are excluded, the connections between imperfections, energy release rates, toughness, and interface delamination remain ill-defined. Many noncritical imperfections can exist and evolve with cycling, and yet the TBC retains its integrity. In such circumstances, supercriticality can arise in the sense that the TBC spalls catastrophically if a large-scale edge imperfection is introduced (Fig. 37) that causes $G > \Gamma$ [27, 133]. Accordingly, in standard tests and in airfoils, the coating remains intact until a critical imperfection is introduced, at which point spalling occurs abruptly and unstably. This situation resembles that for monolithic ceramics exposed to thermo-chemo-mechanical environments that introduce flaws during service. In both cases, locating the critical flaws and characterizing them has been a daunting challenge.

Because the interface delaminates and most of the energy release rate derive from relaxation of the residual stress in the TGO (Fig. 10) [128], *to become critical, an imperfection must traverse the TGO and extend along the interface.* Moreover, the TGO must be free to displace so that its energy density is fully transmitted to the delamination. Among the imperfections postulated in the literature, the only one capable of providing an energy release rate sufficient for large-scale delamination ($G \to \Gamma$ for $a \gg h_{tgo}$) is the internal edge, manifest as vertical separations in the TBC (Fig. 35). Even then, the separation must induce a tensile strain concentration in the TGO that allows penetration to the interface. Although plausible, this mechanism has not been observed, and the mechanics have not been substantiated. Pegs at rare Earth oxide entrainments induce local tensile stress sufficiently large to form small cracks, but they arrest without causing delamination [139]. Inclined slip bands in the TGO, should they occur (never observed), could initiate interface cracks, but these are also stable and can only extend a short distance. The reality is that many different flaws could be created with time at temperature: local ternary oxides (spinels), interface segregants, and so on. Given the absence of a viable mechanism, supported by observation, the pragmatic approach is to use weakest link statistics, replicating the scheme used for characterizing the failure of monolithic ceramics. The approach is elaborated in Sec. VI.

C. INFLUENCE OF THERMAL GRADIENTS

During operation, a thermal gradient exists across the TBC. It is believed that the TBC is largely stress-free under these conditions because of the relatively low creep strength of a typical YSZ above 1000°C. With the exception of scheduled, controlled cooling of land-based turbines, the cooling rates are sufficiently rapid that stresses develop, largely through the thermoelasticity of the material, because any creep relaxation is minimal. When these circumstances apply, at ambient, a residual stress gradient will develop in the TBC with (in some cases) the surface being in tension (Fig. 34 [16, 140, 141]). Because of the gradient, prospective delaminations experience a bending moment in addition to an axial force. The moment affects not only the energy release rate but also the mode

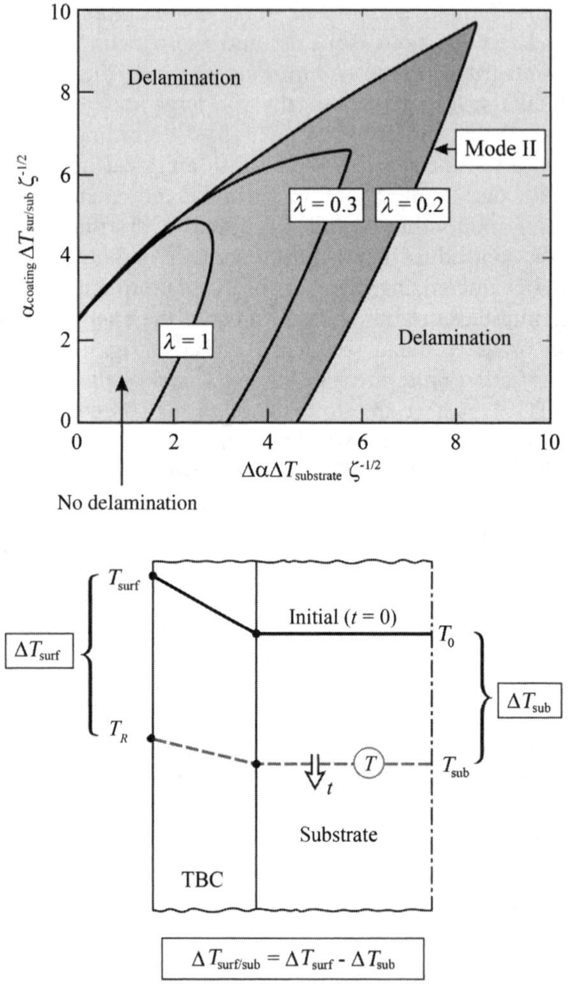

Fig. 38 Delamination map for cooling of a coating subject to a through-thickness thermal gradient.

mixity, resulting in situations wherein delaminations in the TBC can experience mode I as the system cools. The mechanics involved is straightforward, but unwieldy. For simplicity of visualization, the results can be condensed into a delamination map (Fig. 38) with coordinates (X, Y), given by [16]:

$$
\begin{aligned}
X &= \Delta\alpha\Delta T_{\text{substrate}}\,\varsigma^{-1/2} \\
Y &= \alpha_{\text{tbc}}\Delta T_{\text{sur/sub}}\,\varsigma^{-1/2}
\end{aligned}
\tag{2}
$$

where

$$\varsigma = \left[\frac{(1-\nu_{\rm tbc})}{(1+\nu_{\rm tbc})}\frac{\Gamma_{\rm IC}}{E_{\rm tbc}h_{\rm tbc}}\right]$$

$$\Delta T_{\rm sur/sub} = \Delta T_{\rm surface} - \Delta T_{\rm substrate}$$

$$\Delta\alpha = \alpha_{\rm sub} - \alpha_{\rm tbc}$$

with $\Gamma_{\rm IC}$ the mode I fracture toughness on a plane parallel to the substrate. On the map, λ is a parameter that captures the influence of mode mixity on the toughness [16, 127]. Implicit in the map is that free edges or holes exist and that the

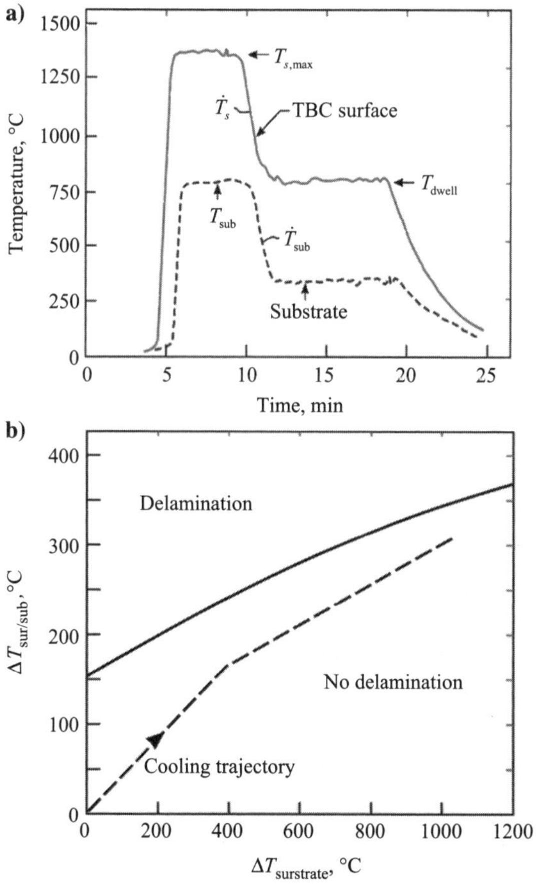

Fig. 39 a) Representative thermal scenario and b) its superposition on a delamination map for a 1-mm-thick APS coating.

delaminations start from these edges. Inserting the material properties for a 1-mm-thick APS 7YSZ coating gives the map presented in Fig. 39b. To implement the maps, a cooling trajectory must be superposed. If that trajectory remains within the blue region, delamination is prohibited. Conversely, if it enters a brown region, delamination is likely. A representative thermal excursion for an aircraft takeoff is shown in Fig. 39a and sketched onto the delamination map. For this example, the TBC surface experiences 1400°C during takeoff while the bondcoat is at 800°C. During cruise, the TBC surface temperature decreases to 800°C while the bondcoat temperature drops to 350°C. This drop accounts for the initial segment of the cooling trajectory. Upon landing, the temperatures drop to ambient. During this stage, even though the slope of the trajectory increases, it stops before crossing into the "delamination" region, and the TBC survives the thermal excursion. The situation changes dramatically when the TBC has been partially infiltrated with CMAS [16, 77].

VI. LIFE-PREDICTION SCHEMES

A. MODELS

Progress toward a life-prediction capability has been sporadic and is still unsatisfactory. The difficulties are embedded in the multiplicity of mechanisms involved (Fig. 11) and the highly nonlinear nature of many of the phenomena. A universal code is not feasible, given the different mechanisms. Attention has been devoted to schemes applicable when the intrinsic mechanisms pertain, even though the extrinsic mechanisms are generally more limiting. Moreover, the schemes have been envisioned for bondcoat systems that result in delamination at the TGO/bondcoat interface. They would not be expected to apply for delamination in the TBC induced by rumpling, but might be modified to incorporate failures of this type.

Approaches have been based on strength, energy densities, or energy release rates. A fourth approach is based on probabilistic methods with the potential for incorporating weakest link statistics. The approaches and their limitations are described, and then pathways toward predictive, mechanism-inspired, models are discussed.

1. STRENGTH-BASED ENGINEERING MODELS

The strength-based engineering models are inspired by two basic ideas: 1) the influence of the time at temperature can be captured by the thickness of the TGO, relative to a critical value h_{crit} with the premise that the remnant cycles to failure decrease as $h_{tgo} \rightarrow h_{crit}$; and 2) upon thermal cycling, the strength/toughness degrades along the separation plane, through a time/cycle dependent "strength" parameter S_{tbc}. The result is expressed as a number of cycles to failure [66, 140],

$$N_f \sim \frac{S_{\text{tbc}}}{\Delta\sigma}\left[1 - \left(\frac{h_{\text{tgo}}}{h_{\text{crit}}}\right)^b\right] \tag{3}$$

where $\Delta\sigma$ is the residual stress in the TBC because of its thermal expansion misfit with the substrate and b is an empirical exponent. With three fitting parameters S_{tbc}, b, h_{crit}, Eq. (3) has the flexibility to correlate data. Utility is restricted to interpolation between measurements because the parameters differ for different systems and depend on cycle frequency, as well as coating thickness, and other factors. Moreover, the cyclic life is often lower than the model suggests because phenomena unrelated to the TGO thickness intervene obviating any dependence on h_{crit}.

2. ENERGY DENSITY APPROACHES

Energy density approaches ascertain the strain energies stored in the TBC and TGO and compare the total energy with a measure of the toughness along the separation plane, as indicated in Sec. I. The method incorporates the thermoelastic properties of both the TBC and TGO and their thickness, but it is deficient in three major ways: 1) not all of the stored energy is available for delamination (G is not the same as the energy density); 2) no allowance is made for whether the separations extend in opening or shear (or intermediate) mode; and 3) the role of imperfections is discounted. The method is useful for obtaining bounds (as in Sec. I) but is overly conservative predicting longer lives than observed.

3. STRAIN ENERGY RELEASE RATES

As described in the preceding section, delamination mechanics is based on determinations of the energy release rate and mode mixity and comparisons with the (mode-dependent) toughness. The method is rigorous when failure is controlled by the propagation of delaminations and has been successfully applied to a wide range of problems involving films and coatings [16, 127–133, 137, 138], including critical thickness determination, predictions of buckle-driven delamination, and interfacial cracking. It is less suitable when failure is dominated by imperfections. Basic delamination mechanics has been applied to TBCs, but important gaps remain. To develop a rigorous life-prediction scheme, there is an opportunity to address these limitations. A simple example illustrates the possibilities. The likelihood of interface delamination must increase as the excess (supercritical) energy release rate increases, defined as (Fig. 10):

$$\Delta G = \sigma_{\text{tgo}}^2 h_{\text{tgo}}/2E_{\text{tgo}}(1-\nu_{\text{tgo}}) + \sigma_{\text{tbc}}^2 h_{\text{tbc}}/2E_{\text{tbc}}(1-\nu_{\text{tbc}}) - \Gamma_{\text{int}} \tag{4}$$

Within this parameter, the hot time and temperature are embedded in h_{tgo} and possibly E_{tbc} (if sintering occurs) and Γ_{int} (if segregant weakening occurs). In this formulation there is no cyclic effect unless the interface is subject to fatigue crack growth or rumples [142]. By assigning a relationship between the probability of spalling and ΔG, an expression for the TBC life can be determined, as described next.

4. PROBABILISTIC METHODS

The simplest approach (and the only one used thus far) asserts that the probability of survival is related to the number of thermal cycles (at specified peak temperature) N by a Weibull distribution:

$$1 - \Phi = \exp\{-[(N - N_{th})/N_0]^m\} \tag{5}$$

where Φ is the failure probability, m is the shape parameter, N_0 the scale parameter, and N_{th} the threshold. This formula is simply used to fit to measurements of the incidence of spalling and used for interpolation. It has the flexibility to fit data (given three fitting parameters) but does not provide any insight into the imperfections or the influence of variables (material properties, TGO thickness, temperature). To incorporate the variables, an alternative based on the excess energy release rate [Eq. (4)] could be explored: $1 - \Phi = \exp\{-[\Delta G/\Gamma_0]^m\}$, where Γ_0 is the scale parameter.

B. NONDESTRUCTIVE PROBES

The deficiencies of these models suggest that higher-fidelity life-prediction schemes might be devised by integrating a monitoring tool with an adaptive model. An illustration is presented for cases wherein performance is limited by rumpling; it combines the use of a piezospectroscopy stress probe of the TGO with neural networks. It is possible to monitor the stresses in the buried TGO by piezospectroscopy because YSZ is optically translucent. A laser probe penetrates the TBC and excites the characteristic R-line luminescence from trace Cr^{3+} impurities incorporated into the TGO during oxidation [143]. From a spectral analysis of the luminescence, the average biaxial stresses in the TGO, σ_B, as well as the distribution, can be determined [144]. The biaxial stress is related to the frequency shift of the R2 line by the piezospectroscopy relation [145]: $\overline{\Delta \nu} = 5.08\sigma_B$, and so the piezospectroscopic frequency shift is proportional to the square root of the strain energy density. A typical spectrum change upon cycling is shown in Fig. 40, together with the frequency shift from the unstressed state. Also shown is the deconvolution of the spectra into two sets of R1 and R2 lines: one indicative of an intact TGO and the other of portions of the TGO damaged by thermal cycling. The frequency of the R2 lines monotonically decreases as the rumpling amplitude increases (Fig. 41a) because of the local curvatures induced in the TGO as it conforms to the bondcoat deformation. Moreover, the frequency decreases as local delaminations develop (Fig. 41b). Specifically, the decrease in frequency $\Delta\nu$ can be related to the area fraction f of the TBC that separates above the rumples as $f = f_0 + b\Delta\nu + c\Delta\nu^2$, where f_0 is the initial separated area after processing, with b and c coefficients that depend on the system [43].

Direct information concerning the evolution of stresses in the TGO has the potential to be incorporated into the fracture-mechanics-based models, as well as being used in conjunction with a neural network model "trained" with piezo-spectroscopy results. To be effective in turbine operation, implementation would require periodic measurements of coated blades during service, as well as the development of large area piezospectroscopy imaging techniques [146].

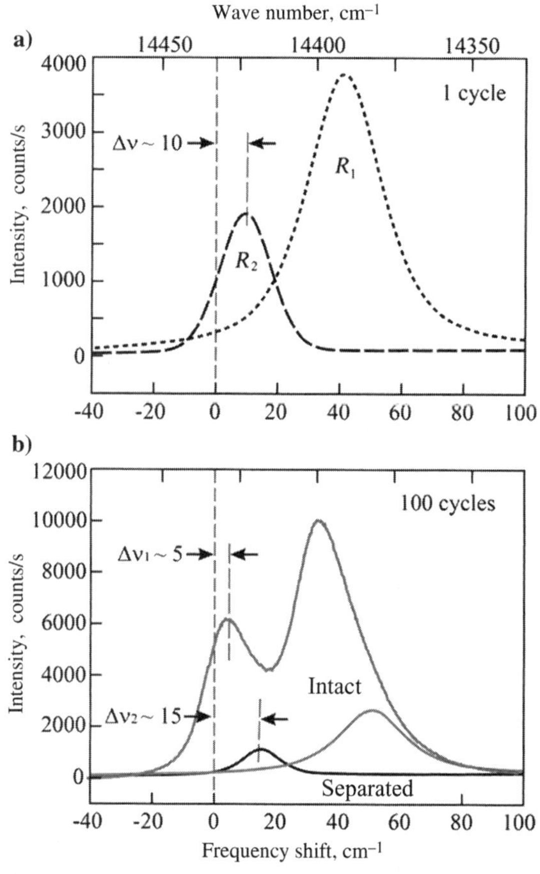

Fig. 40 Photo-stimulated luminescence spectra after thermal cycling from room temperature to 1150°C: a) one cycle and b) 100 cycles, each 1 h. The lower spectra has been deconvoluted into two sets of R1 and R2 lines to indicate the spectra from the intact and damaged TGO.

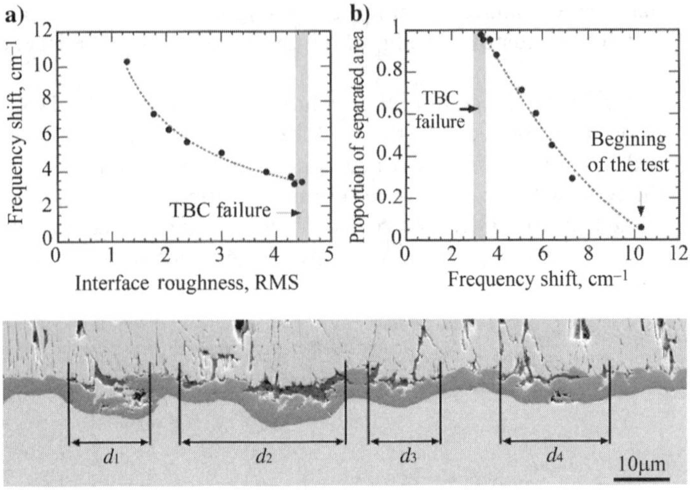

Fig. 41 a) Monotonic decrease in luminescence frequency shift with increased rumpling amplitude. b) Correlation between the proportion of separated area and frequency shift.

VII. RESEARCH OPPORTUNITIES

The foregoing synopsis summarizes the opportunities for research on alumina and zirconia (and related oxides) that might further enhance the performance of hot section components in turbines.

1. Despite decades of research, the thickening rate of the $\alpha-Al_2O_3$ (once formed) remains to be understood. There are dramatic effects of impurities (detrimental) and dopants (beneficial). The grain size is also important because the diffusional fluxes occur primarily along the grain boundaries. The counterfluxes of oxygen and Al have yet to be accounted for in the context of a viable atomistic model. This is of key importance because the latter are directly responsible for the growth stress, which, while measurable, has not been adequately rationalized or predicted.
2. A basic understanding of the adhesion of the $\alpha-Al_2O_3/\gamma-Ni$ interface is progressing. First principles computations based on density functional theory have demonstrated trends in the work of separation of the interface Γ_0 upon segregation of impurities and dopants, consistent with experimental findings. A scheme for linking Γ_0 to the toughness through a crack extension simulation is in place, but results are not yet formalized because of ambiguities regarding the plasticity length scale for the bondcoat adjacent to the interface.
3. The toughness of tetragonal oxides has been attributed to their ferroelastic properties. The dissipation that causes toughening occurs through nanoscale

domain formation during crack extension, which suggests a dependence of toughness on the tetragonality c/a and coercive stress. An ensuing focus on compositions in the ZrO_2-$YO_{1.5}$-TiO_2 system having larger c/a has, indeed, led to a doubling of the toughness. While consistent with the ferroelastic mechanism, direct evidence has yet to be provided. Moreover, such enhancements suggest that systems with yet larger toughness compositions remain to be discovered.

4. There are still too few direct measurements of the in-plane modulus of the YSZ coatings generated by EB-PVD. Approaches, such as vibrating flexural beams that provide further measurements while attached to the substrate and characterize trends with intercolumnar connectivity, are merited.
5. Yielding of YSZ contributes significantly to its ability to withstand impact by projectiles at high temperature. Plastic deformation also participates in kink band development during impact. A systematic assessment of the deformation mechanisms and of trends in yield strength with composition would facilitate design for impact tolerance.

ACKNOWLEDGMENTS

This chapter incorporates ideas developed under research programs sponsored by the Office of Naval Research (MURI/N00014-00-1-0438, N00014-08-1-0522, N00014-09-1-1068) and the National Science Foundation (DMR-0099695, DMR-0605700, DMR-1105672). The authors are grateful to numerous students, postdoctoral associates, and colleagues who have contributed to the research in this area.

REFERENCES

[1] Miller, R. A., "Thermal Barrier Coatings for Aircraft Engines: History and Directions," *Journal of Thermal Spray Technology*, Vol. 6, No. 1, 1997, pp. 35–42.

[2] Evans, A. G., Mumm, D. R., Hutchinson, J. W., Meier, G. H., and Pettit, F. S., "Mechanisms Controlling the Durability of Thermal Barrier Coatings," *Progress in Materials Science*, Vol. 46, No. 5, 2001, pp. 505–553.

[3] Clarke, D. R., "Stress Generation During High-Temperature Oxidation of Metallic Alloys," *Current Opinion in Solid State and Materials Science*, Vol. 6, No. 3, 2002, pp. 237–244.

[4] Levi, C. G., "Emerging Materials and Processes for Thermal Barrier Systems," *Current Opinion in Solid State and Materials Science*, Vol. 8, No. 1, 2004, pp. 77–91.

[5] Clarke, D. R., and Levi, C. G., "Materials Design for the Next Generation Thermal Barrier Coatings," *Annual Reviews of Materials Research*, Vol. 33, 2003, pp. 383–417.

[6] Padture, N. P., Gell, M., and Jordan, E. H., "Thermal Barrier Coatings for Gas Turbine Engine Applications," *Science*, Vol. 296, No. 5566, 2002, pp. 280–284.

[7] Stiger, M. J., Yanar, N. M., Pettit, F. S., and Meier, G. H., "Mechanisms for the Failure of Electron Beam Physical Vapor Deposited Thermal Barrier Coatings Induced by High Temperature Oxidation," *Elevated Temperature Coatings: Science and Technology III*, edited by J. M. Hampikian and N. B. Dahotre, The Minerals, Metals and Materials Society: Warrendale, PA, 1999, pp. 51–65.

[8] Clarke, D. R., Oechsner, M., and Padture, N. P., "Thermal Barrier Coatings for More Efficient Gas-Turbine Engines," *Materials Research Society Bulletin*, Vol. 37, No. 10, 2012, pp. 891–941.

[9] Evans, A. G., Clarke, D. R., and Levi, C. G., "The Influence of Oxides on the Performance of Advanced Gas Turbines," *Journal of the European Ceramic Society*, Vol. 28, No. 7, 2008, pp. 1405–1419.

[10] Stecura, S., "Optimization of the NiCrAl-Y/ZrO_2-Y_2O_3 Thermal Barrier System," NASA TM-86905, 1985.

[11] Mévrel, R., Laizet, J. C., Azzopardi, A., Leclercq, B., Poulain, M., Lavigne, O., and Demange, D., "Thermal Diffusivity and Conductivity of $Zr_{1-x}Y_xO_{2-x/2}$ ($x = 0$, 0.084 and 0.179) Single Crystals," *Journal of the European Ceramic Society*, Vol. 24, No. 10-11, 2004, pp. 3081–3089.

[12] Winter, M. R., and Clarke, D. R., "Oxide Materials with Low Thermal Conductivity," *Journal of the American Ceramic Society*, Vol. 90, No. 2, 2007, pp. 533–540.

[13] Nicholls, J. R., Lawson, K. J., Johnstone, A., and Rickerby, D. S., "Methods to Reduce the Thermal Conductivity of EB-PVD TBCs," *Surface and Coatings Technology*, Vol. 151–152, March 2002, pp. 383–391.

[14] Lehmann, H., Pitzer, D., Pracht, G., Vassen, R., and Stöver, D., "Thermal Conductivity and Thermal Expansion Coefficient of the Lanthanum Rare-Earth Element Zirconate System," *Journal of the American Ceramic Society*, Vol. 86, No. 8, 2003, pp. 1338–1344.

[15] Evans, A. G., Fleck, N. A., Faulhaber, S., Vermaak, N., Maloney, M., and Darolia, R., "Scaling Laws Governing the Erosion and Impact Resistance of Thermal Barrier Coatings," *Wear*, Vol. 260, No. 7–8, 2006, pp. 886–894.

[16] Evans, A. G., and Hutchinson, J. W., "The Mechanics of Coating Delamination in Thermal Gradients," *Surface and Coatings Technology*, Vol. 201, No. 18, 2007, pp. 7905–7916.

[17] Mercer, C., Williams, J. R., Clarke, D. R., and Evans, A. G., "On a Ferroelastic Mechanism Governing the Toughness of Metastable Tetragonal-Prime (T′) Yttria Stabilized Zirconia," *Proceedings of the Royal Society A-Mathematical Physical and Engineering Sciences*, Vol. 463, No. 2081, 2007, pp. 1393–1408.

[18] Schaedler, T. A., Leckie, R. M., Krämer, S., Evans, A. G., and Levi, C. G., "Toughening of Non-Transformable t′-YSZ by Addition of Titania," *Journal of the American Ceramic Society*, Vol. 90, No. 12, 2007, pp. 3896–3901.

[19] Pitek, F. M., and Levi, C. G., "Opportunities for TBCs in the ZrO_2-$YO_{1.5}$-$TaO_{2.5}$ System," *Surface and Coatings Technology*, Vol. 201, No. 12, 2007, pp. 6044–6050.

[20] Krogstad, J. A., Lepple, M., and Levi, C. G., "Opportunities for Improved TBC Durability in the CeO_2-TiO_2-ZrO_2 System," *Surface and Coatings Technology*, Vol. 221, April 2013, pp. 44–52.

[21] Sampath, S., Schulz, U., Jarligo, M. O., and Kuroda, S., "Processing Science of Advanced Thermal-Barrier Systems," *MRS Bulletin*, Vol. 37, No. 10, 2012, pp. 903–910.

[22] Clarke, D. R., "The Lateral Growth Strain Accompanying the Formation of a Thermally Grown Oxide," *Acta Materialia*, Vol. 51, No. 5, 2003, pp. 1393–1407.

[23] Nicholls, J. R., "Advances in Coating Design for High Performance Gas Turbines," *MRS Bulletin*, Vol. 28, No. 9, 2003, pp. 659–670.

[24] Warnes, B. M., and Punola, D. C., "Clean Diffusion Coatings by Chemical Vapor Deposition," *Surface and Coatings Technology*, Vol. 94–95, Oct. 1997, pp. 1–6.

[25] Goward, G. W., "Progress in Coatings for Gas Turbine Airfoils," *Surface and Coatings Technology*, Vol. 108–109, Oct. 1998, pp. 73–79.

[26] Strangman, T. E., "Thermal Barrier Coatings for Turbine Airfoils," *Thin Solid Films*, Vol. 127, No. 1-2, 1985, pp. 93–105.

[27] Xu, T., Faulhaber, S., Mercer, C., Maloney, M., and Evans, A. G., "Observations and Analyses of Failure Mechanisms in Thermal Barrier Systems with Two-Phase Bond Coats Based on NiCoCrAlY," *Acta Materialia*, Vol. 52, No. 6, 2004, pp. 1349–1350.

[28] Rickerby, D. S., and Wing, R. G., *Thermal Barrier Coating for a Superalloy Article and a Method of Application Thereof*, edited by U.S.P. Office, Rolls-Royce, Chromalloy, England, U.K, 1999.

[29] Gleeson, B. M., Sordelet, D. J., and Wang, W., *High-Temperature Coatings with Pt Metal Modified. Gamma.-Ni + .Gamma.'-Ni_3Al Alloy Compositions*, edited by U.S.P. Office, Iowa State Univ. Research Foundation, Des Moines, IA, 2012.

[30] Tolpygo, V. K., and Clarke, D. R., "Surface Rumpling of a (Ni,Pt)Al Bond Coat Induced by Cyclic Oxidation," *Acta Materialia*, Vol. 48, No. 13, 2000, pp. 3283–3293.

[31] Tolpygo, V. K., and Clarke, D. R., "Rumpling Induced by Thermal Cycling of an Overlay Coating: The Effect of Coating Thickness," *Acta Materialia*, Vol. 52, No. 3, 2004, pp. 615–621.

[32] Balint, D. S., and Hutchinson, J. W., "An Analytical Model of Rumpling in Thermal Barrier Coatings," *Journal of Mechanics and Physics of Solids*, Vol. 53, No. 4, 2005, pp. 949–973.

[33] Davis, A. W., and Evans, A. G., "A Protocol for Validating Models of the Cyclic Undulation of Thermally Grown Oxides," *Acta Materialia*, Vol. 53, No. 7, 2005, pp. 1895–1905.

[34] Karlsson, A. M., Hutchinson, J. W., and Evans, A. G., "A Fundamental Model of Cyclic Instabilities in Thermal Barrier Systems," *Journal of the Mechanics and Physics of Solids*, Vol. 50, No. 8, 2002, pp. 1565–1589.

[35] Chen, M. W., Glynn, M. L., Ott, R. T., Hufnagel, T. C., and Hemker, K. J., "Characterization and Modeling of a Martensitic Transformation in a Platinum Modified Diffusion Aluminide Bond Coat for Thermal Barrier Coatings," *Acta Materialia*, Vol. 51, No. 14, 2003, pp. 4279–4294.

[36] Tolpygo, V. K., and Clarke, D. R., "Microstructural Study of the Theta-to-Alpha Transformation in Alumina Scales Formed on Nickel-Aluminides," *Materials at High Temperatures*, Vol. 17, No. 1, 2000, pp. 59–70.

[37] Johnson, C. A., Ruud, J. A., Bruce, R., and Wortman, D. J., "Relationships Between Residual Stress, Microstructure and Mechanical Properties of Electron Beam-Physical Vapor Deposition Thermal Barrier Coatings," *Surface and Coatings Technology*, Vol. 108–109, Oct. 1998, pp. 80–85.

[38] Gregori, G., Li, L., Nychka, J. A., and Clarke, D. R., "Vibration Damping of Superalloys and Thermal Barrier Coatings at High-Temperatures," *Materials Science and Engineering A*, Vol. 466, No. 1-2, 2007, pp. 256–264.

[39] Cernuschi, F., Bison, P. G., Marinetti, S., and Scardi, P., "Thermophysical, Mechanical and Microstructural Characterization of Aged Free-Standing Plasma-Sprayed Zirconia Coatings," *Acta Materialia*, Vol. 56, No. 16, 2008, pp. 4477–4488.

[40] Taylor, T. A., "Thermal Barrier Coating for Substrates and Process for Producing It," U.S. Patent 5,073,433, Technology Corp., Danbury, CT,. 1991.

[41] Bruce, R. W., and Schaeffer, J. C., "Sprayed ZrO_2 Thermal Barrier Coating with Vertical Cracks," European Patent EP1281788, General Electric Co., 2003.

[42] Tolpygo, V. K., and Clarke, D. R., "On the Rumpling Mechanism in Nickel-Aluminide Coatings. Part II: Characterization of Surface Undulations and Bond Coat Swelling," *Acta Materialia*, Vol. 52, No. 17, 2004, pp. 5129–5141.

[43] Tolpygo, V. K., Clarke, D. R., and Murphy, K. S., "Evaluation of Interface Degradation During Cyclic Oxidation of EB-PVD Thermal Barrier Coatings and Correlation with Luminescence. Surface and Coatings Technology," Vol. 188–189, Nov.-Dec. 2004, pp. 62–70.

[44] Heeg, B., Tolpygo, V. K., and Clarke, D. R., "Damage Evolution in Thermal Barrier Coatings with Thermal Cycling," *Journal of the American Ceramic Society*, Vol. 94, No. 6, 2011, pp. S112–S119.

[45] Mennicke, C., Mumm, D. R., and Clarke, D. R., "Transient Phase Evolution During Oxidation of a Two-Phase Nicocraly Bond Coat," *Zeitschrift für Metallkunde*, Vol. 90, No. 12, 1999, pp. 1079–1084.

[46] Levi, C. G., Sommer, E., Terry, S. G., Catanoiu, A., and Rühle, M., "Alumina Grown During Deposition of Thermal Barrier Coatings on NiCrAlY," *Journal of the American Ceramic Society*, Vol. 86, No. 4, 2002, pp. 676–685.

[47] Murphy, K. S., More, K. L., and Lance, M. J., "As-Deposited Mixed Zone in Thermally Grown Oxide Beneath a Thermal Barrier Coating," *Surface and Coatings Technology*, Vol. 146–147, Sept.-Oct. 2001, pp. 152–161.

[48] Pint, B. A., Martin, J. R., and Hobbs, L. W., "The Oxidation Mechanism of θ-Al_2O_3 Scales," *Solid State Ionics*, Vol. 78, No. 1-2, 1995, pp. 99–107.

[49] Brumm, M. W., and Grabke, H. J., "The Oxidation Behaviour of NiAl-I. Phase Transformations in the Alumina Scale During Oxidation of NiAl and NiAl-Cr Alloys," *Corrosion*, Vol. 33, No. 11, 1992, pp. 1677–1690.

[50] Doychak, J., Smialek, J. L., and Mitchell, T. E., "Transient Oxidation of Single-Crystal Beta-NiAl," *Metallurgical Transactions A*, Vol. 20, No. 3, 1989, pp. 499–518.

[51] Rybicki, G. C., and Smialek, J. L., "Effect of the q-a-Al_20_3 Transformation on the Oxidation Behavior of b-NiAI + Zr. Oxidation of Metals," Vol. 31, No. 3/4, 1989, pp. 275–304.

[52] Reddy, A., Hovis, D. G., Veal, B., Paulikas, A., and Heuer, A. H., "In-Situ Study of Oxidation-Induced Growth Strains in a Model Nicraly Bond Coat Alloy," *Oxidation of Metals*, Vol. 67, No. 3–4, 2007, pp. 153–177.

[53] Heuer, A. H., Reddy, A., Hovis, D. G., Veal, B., Paulikas, A., Vlad, A., and Ruhle, M., "The Effect of Surface Orientation on Oxidation-Induced Growth Strains in Single

Crystal Nial: an in-Situ Synchrotron Study," *Scripta Materialia*, Vol. 54, No. 11, 2006, pp. 1907–1912.

[54] Tolpygo, V. K., Dryden, J., and Clarke, D. R., "Determination of the Growth Stresses Generated During the Oxidation of Fe-22Cr-4.8Al-0.3Y Alloy," *Acta Materialia*, Vol. 46, No. 3, 1998, pp. 927–937.

[55] Clarke, D. R., and Adar, F., "Measurement of the Crystallographically Transformed Zone Produced by Fracture in Ceramics Containing Tetragonal Zirconia," *Journal of the American Ceramic Society*, Vol. 65, No. 6, 1982, pp. 284–288.

[56] Gong, X., and Clarke, D. R., "On the Measurement of Strain in Coatings Formed on a Wrinkled Elastic Substrate," *Oxidation of Metals*, Vol. 50, No. 5–6, 1998, pp. 355–376.

[57] Lipkin, D. M., and Clarke, D. R., "Measurement of the Stress in Oxide Scales Formed by Oxidation of Alumina-Forming Alloys," *Oxidation of Metals*, Vol. 45, No. 3/4, 1996, pp. 267–280.

[58] Zhang, W., Smith, J. R., Wang, X-G., and Evans, A. G., "Influence of Sulfur on the Adhesion of the Nickel/Alumina Interface," *Physical Review B*, Vol. 67, No. 245414, 2003, pp. 1–12.

[59] Zhang, W., Smith, J. R., and Evans, A. G., "The Connection Between ab Initio Calculations and Interface Adhesion Measurements on Metal/Oxide Systems: Ni/Al_2O_3 and Cu/Al_2O_3," *Acta Materialia*, Vol. 15, No. 15, 2002, pp. 3803–3816.

[60] Smialek, J. L., "Effect of Sulfur Removal on Al_2O_3 Scale Adhesion," *Metallurgical Transactions A*, Vol. 22A, No. 3, 1991, pp. 739–752.

[61] Ruud, J. A., Bartz, A., Borom, M. P., and Johnson, C. A., "Strength Degradation and Failure Mechanisms of Electron-Beam Physical-Vapor-Deposited Thermal Barrier Coatings," *Journal of the American Ceramic Society*, Vol. 84, No. 7, 2001, pp. 1545–1552.

[62] Mumm, D. R., Evans, A. G., and Spitsberg, I. T., "Characterization of a Cyclic Displacement Instability for a Thermally Grown Oxide in a Thermal Barrier System," *Acta Materialia*, Vol. 49, No. 12, 2001, pp. 2329–2340.

[63] Tolpygo, V. K., and Clarke, D. R., "Temperature and Cycle-Time Dependence of Rumpling in Platinum-Modified Diffusion Aluminide Coatings," *Scripta Materialia*, Vol. 57, No. 7, 2007, pp. 563–566.

[64] Tolpygo, V. K., and Clarke, D. R., "Rumpling of CVD (Ni,Pt)Al Diffusion Coatings Under Intermediate Temperature Cycling," *Surface and Coatings Technology*, Vol. 203, No. 20-21, 2009, pp. 3278–3285.

[65] Tolpygo, V. K., "Development of Internal Cavities in Platinum-Aluminide Coatings During Cyclic Oxidation," *Surface and Coatings Technology*, Vol. 202, No. 4-7, 2007, pp. 617–622.

[66] Chan, K., Cheruvu, S., and Viswanathan, R., "Development of a Thermal Barrier Coating Life Model," American Society of Mechanical Engineers, Paper GT2003-38171, June 2003, pp. 591–595; doi: 10.1115/GT2003-38171.

[67] Chen, X., Wang, R., Yao, N., Evans, A. G., and Darolia, R., "Foreign Object Damage in a Thermal Barrier System: Mechanisms and Simulations," *Materials Science and Engineering*, Vol. A352, 2003, No. 1–2, pp. 221–231.

[68] Nicholls, J. R., Deakin, M. J., and Rickerby, D. S., "A Comparison Between the Erosion Behavior of Thermal Spray and Electron-Beam Physical Vapour

Deposition Thermal Barrier Coatings," *Wear*, Vol. 233–235, Dec. 1999, pp. 352–361.

[69] Wellman, R. G., and Nicholls, J. R., "Some Observations on Erosion Mechanisms of EB-PVD TBCs," *Wear*, 2000, Vol. 242, pp. 89–96.

[70] Bruce, R. W., "Development of 1232 degrees C (2250 degrees F) Erosion and Impact Tests for Thermal Barrier Coatings," *Tribology Transactions*, Vol. 41, No. 4, 1998, pp. 399–410.

[71] Crowell, M. W., Wang, J., McMeeking, R. M., and Evans, A. G., "Dynamics of Kink Band Formation in Columnar Thermal Barrier Oxides," *Acta Materialia*, Vol. 56, No. 16, 2008, pp. 4150–4159.

[72] Fleck, N. A., and Thanasis, Z., "The Erosion of EB-PVD Thermal Barrier Coatings: The Competition Between Mechanisms," *Wear*, Vol. 268, Nos. 11–12, 2010, pp. 1214–1224.

[73] Borom, M. P., Johnson, C. A., and Peluso, L. A., "Role of Environmental Deposits and Operating Surface Temperature in Spallation of Air Plasma Sprayed Thermal Barrier Coatings," *Surface and Coatings Technology*, Vol. 86–87, Dec. 1996, pp. 116–126.

[74] Mercer, C., Faulhaber, S., Evans, A. G., and Darolia, R., "A Delamination Mechanism for Thermal Barrier Coatings Subject to Calcium-Magnesium-Alumino-Silicate (CMAS) Infiltration," *Acta Materialia*, Vol. 53, No. 4, 2005, pp. 1029–1039.

[75] Krämer, S., Yang, J. Y., Johnson, C. A., and Levi, C. G., "Thermochemical Interactions of Thermal Barrier Coatings with Molten CaO-MgO-Al_2O_3-SiO_2 (CMAS) Deposits," *Journal of the American Ceramic Society*, Vol. 89, No. 10, 2006, pp. 3167–3175.

[76] Krämer, S., Faulhaber, S., Chambers, M., Clarke, D. R., Levi, C. G., Hutchinson, J. W., and Evans, A. G., "Mechanisms of Cracking and Delamination Within Thermal Barrier Systems in Aero-Engines Subject to Calcium-Magnesium-Alumino-Silicate (CMAS) Penetration," *Materials Science and Engineering A*, Vol. 490, No. 1-2, 2008, pp. 26–35.

[77] Levi, C. G., Hutchinson, J. W., Vidal-Setif, M-H., and Johnson, C. A., "Environmental Degradation of TBCs by Molten Deposits," *MRS Bulletin*, Vol. 37, No. 10, 2012, pp. 932–941.

[78] Sundaram, S., Lipkin, D. M., Johnson, C. A., and Hutchinson, J. W., "The Influence of Transient Thermal Gradients and Substrate Constraint on Delamination of Thermal Barrier Coatings," *Journal of Applied Mechanics*, Vol. 80, No. 011002, 2013, pp. 1–13.

[79] Steinke, T., Sebold, D., Mack, D. E., Vaßen, R., and Stöver, D., "A Novel Test Approach for Plasma-Sprayed Coatings Tested Simultaneously Under CMAS and Thermal Gradient Cycling Conditions," *Surface and Coatings Technology*, Vol. 205, No. 7, 2010, pp. 2287–2295.

[80] Drexler, J. M., Aygun, A., Li, D., Vaßen, R., Steinke, T., and Padture, N. P., "Thermal-Gradient Testing of Thermal Barrier Coatings Under Simultaneous Attack by Molten Glassy Deposits and Its Mitigation. Surface and Coatings Technology," Vol. 204, Nos. 16–17, 2010, pp. 2683–2688.

[81] Nychka, J. A., and Clarke, D. R., "Quantification of Aluminum Outward Diffusion During Oxidation of FeCrAl Alloys," *Oxidation of Metals*, Vol. 63, Nos. 5–6, 2005, pp. 325–352.

[82] Tolpygo, V. K., and Clarke, D. R., "Microstructural Evidence for Counter-Diffusion of Aluminum and Oxygen During the Growth of Alumina Scales Materials at High Temperatures," Vol. 20, No. 3, 2003, pp. 261–271.

[83] Clarke, D. R., "Epitaxial Phase Transformation of Alumina," *Physics Status Solidi, (a)*, Vol. 166, No. 1, 1998, pp. 183–196.

[84] Simpson, T. W., Wen, Q., Yu, N., and Clarke, D. R., "Kinetics of the Amorphous $\rightarrow \gamma \rightarrow \alpha$ Transformations in Aluminum Oxide: Effect of Crystallographic Orientation," *Journal of the American Ceramic Society*, Vol. 81, No. 1, 1998, pp. 61–66.

[85] Ragan, D. D., Mates, T., and Clarke, D. R., "Effect of Yttrium and Erbium Ions on Phase Transformations in Epitaxial Alumina," *Journal of the American Ceramic Society*, Vol. 86, No. 4, 2003, pp. 541–545.

[86] Rhines, F. N., and Wolf, J. S., "The Role of Oxide Microstructure and Growth Stresses in the High-Temperature Scaling of Nickel," *Metallurgical Transactions*, Vol. 1, June 1970, pp. 1701–1710.

[87] Smialek, J. L., and Hehemann, R. F., "Transformation Temperature of Martensite in β-phase Nickel Aluminide," *Metallurgical Transactions*, Vol. 4, No. 6, 1973, pp. 1571–1575.

[88] Wei, Y., and Hutchinson, J. W., "Toughness of Ni/Al_2O_3 Interfaces as Dependent on Micron-Scale Plasticity and Atomistic-Scale Separation," *Philosophical Magazine*, Vol. 88, Nos. 30–32, 2008, pp. 3841–3859.

[89] Smith, J. R., Jiang, Y., and Evans, A. G., "Adhesion of the Gamma-Ni(Al)/Alpha-Al_2O_3 Interface: a First-Principles Assessment," *International Journal of Materials Research*, Vol. 98, No. 12, 2007, pp. 1214–1221.

[90] Jiang, Y., Smith, J. R., and Evans, A. G., "First Principles Assessment of Metal/Oxide Interface Adhesion," *Applied Physics Letters*, Vol. 92, No. 14, 2008, p. 141918.

[91] Clarke, D. R., "Materials Selection Guidelines for Low Thermal Conductivity Thermal Barrier Coatings," *Surface and Coatings Technology*, Vol. 163–164, Jan. 2003, pp. 67–74.

[92] Yoshimura, M., Yashima, M., Noma, T., and Somiya, S., "Formation of Diffusionlessly Transformed Tetragonal Phases by Rapid Quenching of Melts in ZrO_2-$RO_{1.5}$ Systems (R = Rare Earths)," *Journal of Materials Science*, Vol. 25, No. 8, 1990, pp. 2011–2016.

[93] Watanabe, M., et al., "A Probe for the High Temperature Deformation of Oxides Used for Thermal Barrier Systems," *Acta Materialia*, Vol. 52, No. 6, 2004, pp. 1479–1487.

[94] Virkar, A. V., "Role of Ferroelasticity in Toughening of Zirconia Ceramics," *Key Engineering Materials*, Vol. 153–154, No. 4, 1998, pp. 183–210.

[95] Chan, C. J., Lange, F. F., Rühle, M., Jue, J. F., and Virkar, A. V., "Ferroelastic Domain Switching in Tetragonal Zirconia Single Crystals-Microstructural Aspects," *Journal of the American Ceramic Society*, Vol. 74, No. 4, 1991, pp. 807–813.

[96] Baither, D., Bartsch, M., Baufeld, B., Tikhonovsky, A., Ruhle, M., and Messerschmidt, U., "Ferroelastic and Plastic Deformation of t′-Zirconia Single Crystals," *Journal of the American Ceramic Society*, Vol. 84, No. 8, 2001, pp. 1755–1762.

[97] Spitsberg, I., and Boutwell, B. A., "Thermal Barrier Coatings with Improved Impact and Erosion Resistance," U.S. Patent, General Electric Co., Schenectady, NY, 2005.

[98] Donohue, E. M., Philips, N. R., Begley, M. R., and Levi, C. G., "Thermal Barrier Coating Toughness: Measurement and Identification of a Bridging Mechanism Enabled by Segmented Microstructure," *Materials Science and Engineering A*, Vol. 564, March 2013, pp. 324–330.

[99] Eberl, C., Gianola, D. S., Wang, X., He, M. Y., Evans, A. G., and Hemker, K. J., "A Method for in situ Measurement of the Elastic Behavior of a Columnar Thermal Barrier Coating," *Acta Materialia*, Vol. 59, No. 9, 2011, pp. 3612–3620.

[100] Lughi, V., Tolpygo, V. K., and Clarke, D. R., "Microstructural Aspects of the Sintering of Thermal Barrier Coatings," *Materials Science and Engineering*, Vol. 368, Nos. 1–2, 2004, pp. 212–221.

[101] Flores-Renteria, A., and Saruhan, B., "Effect of Ageing on Microstructure Changes in EB-PVD Manufactured Standard PYSZ Top Coat of Thermal Barrier Coatings," *Journal of the European Ceramic Society*, Vol. 26, No. 12, 2006, pp. 2249–2255.

[102] Hutchinson, R. G., Fleck, N. A., and Cocks, A. C. F., "A Sintering Model for Thermal Barrier Coatings," *Acta Materialia*, Vol. 54, No. 5, 2006, pp. 1297–1306.

[103] Cipitria, A., Golosnoy, I. O., and Clyne, T. W., "Sintering Kinetics of Plasma Sprayed Zirconia TBCs," *Journal of Thermal Spray Technology*, Vol. 16, Nos. 5–6, 2007, pp. 809–815.

[104] Miller, R. A., Smialek, J. L., and Garlick, R. G., "Phase Stability in Plasma-Sprayed, Partially Stabilized Zirconia-Yttria," *Science and Technology of Zirconia*, edited by A. H. Heuer and L. W. Hobbs, The American Ceramic Society, Columbus, OH, 1981, pp. 241–253.

[105] Schulz, U., "Phase Transformation in EB-PVD Yttria Partially Stabilized Zirconia Thermal Barrier Coatings During Annealing," *Journal of the American Ceramic Society*, Vol. 83, No. 4, 2000, pp. 904–910.

[106] Lughi, V., and Clarke, D. R., "Transformation of Electron-Beam Physical Vapor Deposited 8wt.% Yttria-Stabilized Zirconia Thermal Barrier Coatings," *Journal of the American Ceramic Society*, Vol. 88, No. 9, 2005, pp. 2552–2558.

[107] Cairney, J. M., Rebollo, N. R., Ruhle, M., and Levi, C. G., "Phase Stability of Thermal Barrier Oxides: a Comparative Study of Y and Yb Additions," *International Journal of Materials Research*, Vol. 98, No. 12, 2007, pp. 1177–187.

[108] Krogstad, J. A., Krämer, S., Lipkin, D. M., Johnson, C. A., Mitchell, D. R. G., Cairney, J. M., and Levi, C. G., "Phase Stablity of t′-Zirconia Based Thermal Barrier Coatings: Mechanistic Insights," *Journal of the American Ceramic Society*, Vol. 94, No. S1, 2011, pp. S168–S177.

[109] Lipkin, D. M., Krogstad, J. A., Gao, Y., Johnson, C. A., Nelson, W. A., and Levi, C. G., "Phase Evolution upon Aging of Air Plasma Sprayed t′-Zirconia Coatings: I-Synchrotron X-Ray Diffraction," *Journal of the American Ceramic Society*, Vol. 96, No. 1, 2013, pp. 290–298.

[110] Krogstad, J. A., Leckie, R. M., Krämer, S., Cairney, J. M., Lipkin, D. M., Johnson, C. A., and Levi, C. G., "Phase Evolution upon Aging of Air Plasma Sprayed t′-Zirconia Coatings: II–Microstructure Evolution," *Journal of the American Ceramic Society*, Vol. 96, No. 1, 2013, pp. 299–307.

[111] Zhu, D., and Miller, R. A., "Thermal Conductivity and Elastic Modulus Evolution of Thermal Barrier Coatings Under High Heat Flux Conditions," *Journal of Thermal Spray Technology*, Vol. 9, No. 2, 2000, pp. 175–180.

[112] Kakuda, T. R., Limarga, A. M., Bennett, T. D., and Clarke, D. R., "Evolution of Thermal Properties of EB-PVD 7YSZ Thermal Barrier Coatings with Thermal Cycling," *Acta Materialia*, Vol. 57, No. 8, 2009, pp. 2583–2591.

[113] Thompson, J. A., and Clyne, T. W., "The Effect of Heat Treatment on the Stiffness of Zirconia Top Coats in Plasma Sprayed TBCs," *Acta Materialia*, Vol. 49, No. 9, 2001, pp. 1565–1575.

[114] Witz, G., Shklover, V., Steurer, W., Bachegowda, S., and Bossmann, H. P., "Phase Evolution in Yttria-Stabilized Zirconia Thermal Barrier Coatings Studied by Rietveld Refinement of X-Ray Powder Diffraction Patterns," *Journal of the American Ceramic Society*, Vol. 90, No. 9, 2007, pp. 2935–2940.

[115] Lughi, V., and Clarke, D. R., "Low-Temperature Transformation Kinetics of Electron-Beam Deposited 5wt.% Yttria-Stabilized Zirconia," *Acta Materialia*, Vol. 55, No. 6, 2007, pp. 2049–2055.

[116] Chevalier, J., Gremillard, L., and Deville, S., "Low-Temperature Degradation of Zirconia and Implications for Biomedical Implants," *Annual Reviews of Materials Research*, Vol. 37, 2007, pp. 1–32.

[117] Shen, Y., Leckie, R. M., Levi, C. G., and Clarke, D. R., "Low Thermal Conductivity Without Oxygen Vacancies in Equimolar $YO_{1.5}$-$TaO_{2.5}$- and $YbO_{1.5}$-$TaO_{2.5}$-Stabilized Tetragonal Zirconia Ceramics," *Acta Materialia*, Vol. 58, No. 13, 2010, pp. 4424–4431.

[118] Schaedler, T. A., Fabrichnaya, O., and Levi, C. G., "Phase Equilibria in the TiO_2-$YO_{1.5}$-ZrO_2 System," *Journal of the European Ceramic Society*, Vol. 28, No. 13, 2008, pp. 2509–2520.

[119] Schulz, U., Saruhan, B., Fritscher, K., and Leyens, C., "Review on Advanced EB-PVD Ceramic Topcoats for TBC Applications," *International Journal of Applied Ceramic Technology*, Vol. 1, No. 4, 2004, pp. 302–315.

[120] Schulz, U., Rätzer-Scheibe, H-J., Saruhan, B., and Renteria, A. F., "Thermal Conductivity Issues of EB-PVD Thermal Barrier Coatings," *Materialwissenschaft und Werkstofftechnik*, Vol. 38, No. 9, 2007, pp. 659–666.

[121] Rätzer-Scheibe, H.-J., and Schulz, U., "The Effects of Heat Treatment and Gas Atmosphere on the Thermal Conductivity of APS and EB-PVD Thermal Barrier Coatings," *Surface and Coatings Technology*, Vol. 201, No. 18, 2007, pp. 7880–7888.

[122] Lu, T. J., Levi, C. G., Wadley, H. N. G., and Evans, A. G., "Distributed Porosity as a Control Parameter for Oxide Thermal Barriers Made by Physical Vapor Deposition," *Journal of the American Ceramic Society*, Vol. 84, No. 12, 2001, pp. 2937–2946.

[123] Leckie, R. M. R., Rebollo, N. R., Yang, J. Y., and Levi, C. G., "Microstructure Stability Issues in Emerging TBC Materials," *International Symposium on Thermal Barrier Coatings and Titanium Aluminides for Gas Turbines (TurbOMat)*, DLR, Bonn, Germany, 2002.

[124] Krämer, S., Yang, J. Y., and Levi, C. G., "Infiltration-Inhibiting Reaction of Gadolinium Zirconate Thermal Barrier Coatings with CMAS Melts," *Journal of the American Ceramic Society*, Vol. 91, No. 2, 2008, pp. 576–583.

[125] Leckie, R. M., Krämer, S., Ruhle, M., and Levi, C. G., "Thermochemical Compatibility Between Alumina and ZrO_2-$GdO_{3/2}$ Thermal Barrier Coatings," *Acta Materialia*, 2005, Vol. 53, No. 11, pp. 3281–3292.

[126] Vaßen, R., Traeger, F., and Stöver, D., "New Thermal Barrier Coatings Based on Pyrochlore/YSZ Double-Layer Systems," *International Journal of Applied Ceramic Technology*, Vol. 1, No. 4, 2004, pp. 351–361.

[127] Hutchinson, J. W., and Suo, Z., "Mixed Mode Cracking in Layered Materials," *Advances in Applied Mechanics*, Vol. 39, 1992, pp. 63–191.

[128] Choi, S. R., Hutchinson, J. W., and Evans, A. G., "Delamination of Multilayer Thermal Barrier Coatings," *Mechanics of Materials*, Vol. 31, No. 7, 1999, pp. 431–447.

[129] Begley, M. R., Mumm, D. R., Evans, A. G., and Hutchinson, J. W., "Analysis of a Wedge Impression Test for Measuring the Interface Toughness Between Films/Coatings and Ductile Substrates," *Acta Materialia*, Vol. 48, No. 12, 2000, pp. 3211–3220.

[130] Hutchinson, J. W., and Evans, A. G., "On the Delamination of Thermal Barrier Coatings in a Thermal Gradient," *Surface and Coatings Technology*, Vol. 149, Nos. 2–3, 2002, pp. 179–184.

[131] Moon, M. W., Jensen, H. M., Hutchinson, J. W., Oh, K. H., and Evans, A. G., "The Characterization of Telephone Cord Buckling of Compressed Thin Films on Substrates," *Journal of the Mechanics and Physics of Solids*, Vol. 50, No. 11, 2002, pp. 2355–2377.

[132] Yu, H. H., He, M. Y., and Hutchinson, J. W., "Edge Effects in Thin Film Delamination," *Acta Materialia*, Vol. 49, No. 1, 2001, pp. 93–107.

[133] Faulhaber, S., Mercer, C., Moon, M. W., Hutchinson, J. W., and Evans, A. G., "Buckling Delamination in Compressed Multilayers on Curved Substrates with Accompanying Ridge Cracks," *Journal of Mechanics and Physics of Solids*, Vol. 54, No. 5, 2006, pp. 1004–1028.

[134] Tolpygo, V. K., and Clarke, D. R., "Spalling of $\alpha-$Alumina Films Grown by Oxidation. I. Dependence on Cooling Rate and Metal Thickness," *Materials Science and Engineering*, Vol. A 278, No. 1-2, 2000, pp. 142–150.

[135] Tolpygo, V. K., and Clarke, D. R., "Spalling of α-Alumina Films Grown by Oxidation. II. Decohesion Nucleation and Growth," *Materials Science and Engineering*, Vol. A 278, No. 1-2, 2000, pp. 151–161.

[136] Sergo, V., and Clarke, D. R., "Observation of Subcritical Spall Propagation of a Thermal Barrier Coating," *Journal of the American Ceramic Society*, Vol. 81, No. 12, 1998, pp. 3237–3242.

[137] Chen, X., Hutchinson, J. W., He, M. Y., and Evans, A. G., "On the Propagation and Coalescence of Delamination Cracks in Compressed Coatings: with Application to Thermal Barrier Systems," *Acta Materialia*, Vol. 51, No. 7, 2003, pp. 2017–2030.

[138] He, M. Y., Evans, A. G., and Hutchinson, J. W., "Convergent Debonding of Films and Fibers," *Acta Materialia*, Vol. 45, No. 8, 1997, pp. 3481–3489.

[139] Zhu, H. X., Fleck, N. A., Cocks, A. C. F., and Evans, A. G., "Numerical Simulations of Crack Formation from Pegs in Thermal Barrier Systems with NiCoCrAlY Bond Coats," *Materials Science and Engineering*, Vol. A404, Nos. 1–2, 2005, pp. 26–32.

[140] Cruse, T. A., Stewart, S. E., and Ortiz, M., "Thermal Barrier Coating Life Prediction Model Development," *Journal of Engineering for Gas Turbines and Power*, Vol. 110, No. 4, 1988, pp. 610–616.

[141] Limarga, A. M., Vassen, R., and Clarke, D. R., “Stress Distributions in Plasma Sprayed Thermal Barrier Coatings Under Thermal Cycling in a Temperature Gradient,” *Journal of Applied Mechanics*, Vol. 78, No. 0118003-1/9, 2011.

[142] Busso, E. P., Lin, J., and Sakuraf, S., “A Mechanistic Study of Oxidation-Induced Degradation in a Plasma-Sprayed Thermal Barrier Coating System: Part II. Life Prediction Model, *Acta Materialia*, Vol. 49, No. 9, 2001, pp. 1529–1536.

[143] Christensen, R. J., Lipkin, D. M., Clarke, D. R., and Murphy, K. S., “Nondestructive Evaluation of the Oxidation Stresses Through Thermal Barrier Coatings Using Cr^{3+} Piezospectroscopy,” *Applied Physics Letters*, Vol. 69, No. 24, 1996, pp. 3754–3756.

[144] Peng, X., and Clarke, D. R., “Piezospectroscopic Analysis of Interface Debonding in Thermal Barrier Coatings,” *Journal of the American Ceramic Society*, Vol. 83, No. 5, 2000, pp. 1165–1170.

[145] Clarke, D. R., and Gardiner, D. J., “Recent Advances in Piezospectroscopy,” *International Journal of Materials Research*, Vol. 98, No. 8, 2007, pp. 756–762.

[146] Heeg, B., and Abbiss, J. B., “Piezospectroscopic Imaging with a Tunable Fabry-Perot Filter and Tikhonov Reconstruction,” *Optics Letters*, Vol. 32, No. 7, 2007, pp. 859–861.

CHAPTER 13

Nondestructive Evaluation

R. Bruce Thompson* **and Lisa J. H. Brasche**†
Iowa State University, Ames, Iowa

As the source of aircraft propulsion, the turbine engine is a critical component. Its failure can have a variety of deleterious consequences, a number of which can cause the loss of vehicle and life, and its efficient operation plays a key role in aircraft performance and economics. From the safety perspective, the consequences of total loss of power depend on the number of engines on an aircraft and on its mission, but can include emergency landings, loss of vehicle over the ocean due to inability to reach destination, crashes because of inability to maintain altitude or to climb to clear hazards such as mountains, or loss of vehicle in military combat because of reduced maneuverability. Uncontained failures, in which high-energy rotating components escape the casing and penetrate the fuselage, can lead to loss of life if individuals are struck by the penetrating parts, or loss of vehicle if critical parts of the aircraft are damaged. Figure 1 shows the damaged engine components that were the cause in two such events. Other forms of engine degradation can lead to loss of efficiency and hence have implications that are primarily related to economic, as opposed to safety, concerns.

Various engineering tools and strategies are employed to prevent such events; this chapter deals with the tools of nondestructive evaluation (NDE) that are currently being applied, on a component-by-component basis, to detect flaws or changes in material condition that could lead to failure or loss of efficiency. Primary emphasis is placed on the control of fatigue with more limited attention given to other degradation mechanisms.

I. DESIGN/LIFE MANAGEMENT PRACTICES

To fully appreciate the use of NDE, one must first understand the design/life management strategies that are currently in use and set the context for the noninvasive measurements.

*Former Director, Center of Nondestructive Evaluation and Distinguished Professor of Materials Science and Aerospace Engineering; Dr. Thompson passed away in 2011 after a celebrated career in NDE research and education.

†Associate Director, Center for Nondestructive Evaluation; currently Principal Engineer, Pratt & Whitney, East Hartford, Connecticut 06108.

Fig. 1 Examples of uncontained aircraft engine failures. Left: disk failure from Sioux City incident of 1989. Failure was attributed to the presence of hard alpha inclusions, which led to inservice fatigue crack propagation and eventual disk rupture. Consequence was a crash landing and loss of 111 lives. Right: fan hub failure and associated engine and airframe damage from Pensacola incident of 1996. Failure resulted from presence of anomalous machining condition, which led to crack nucleation, propagation, and eventual disk rupture. A consequence was the loss of two lives when passengers were struck by broken pieces.

A. SAFE-LIFE MANAGEMENT OF FATIGUE

The operation of aircraft turbine engines is carefully managed to avoid problems such as those illustrated in Fig. 1, and the approach used depends on the design process initially employed. Fatigue is perhaps the most widely encountered damage mechanism, and over time, two major design philosophies have been employed. The safe-life approach is based on either stress vs life-cycle curves or strain vs life-cycle curves. In this probabilistic strategy, components are removed from service when the probability of failure becomes unacceptably high. Thompson and Thompson offer the following overview [1]. "Before 1970, structures whose life was limited by fatigue were designed in accordance with the 'safe life' philosophy" [2]. Therein, it is assumed that the structure is initially unflawed, and the life is considered to be limited by fatigue, a cumulative damage process in which microscopic cracks are nucleated from minute irregularities as the material undergoes cyclic loading. These cracks gradually grow and coalesce until a macrocrack develops which would ultimately extend to failure. In this "safe life" approach it is intended that a given structure should not develop macroscopic fatigue cracks during its service life. In practice, this means that a component is considered to have "failed" as soon as a crack of some finite size has formed. For example, 1/32 in (0.79 mm) was the size criterion used in certain military jet engine applications [3]. If such a crack is found, the part is removed from service. This is known as an initiation criterion, since no attempt is made to utilize the remaining life during which the finite fatigue crack extends to a larger size which would fail abruptly."

Safe-life procedures can be quite conservative. As Thompson and Thompson point out [1], "all parts are not identical on a microscopic scale and hence the initiation of fatigue damage must be described on a statistical basis." Suppose all parts are used for an identical design life, selected on the basis of a test program, to ensure that no more than a specified percent "fail" ("failure" being defined as the development of observable damage). Then the life of many of the parts, which would exceed this design life, will be wasted. In the example shown in Fig. 2, typical of a military aircraft turbine disk [4], the design life is set by the requirement of less than 0.1 percent "failure." This design life is only 25 percent of the mean life, with the result that only 33 percent of potential disk life is utilized. In some practical situations, this fraction has been estimated to be less than 10 percent."

To put this in a more recent context, a brief description is provided of how the safe-life approach is currently employed by the U.S. Air Force in managing fatigue in the closely related problem of metal aircraft structures [5]: "The so-called safe-life approach is a probabilistic based method." The safe-life of a structure is that usage period in flight hours when there is a low probability that the strength will degrade below its design ultimate value due to fatigue cracking. The determination of the safe-life of an aircraft depends primarily on the results of a full-scale fatigue test of the structure. The number of simulated flight hours of operational service successfully completed in the laboratory is the "test life" of the structure. The safe-life also depends on the expected distribution of failures. The distribution of failures provides the basis for factoring the test life. The factor is called the "scatter factor." The distribution of failures may be derived from past experience from similar aircraft or from the results of design development testing preceding the full-scale fatigue test. The test life is divided by the scatter factor to determine the safe life. The scatter factor (usually in the interval from two

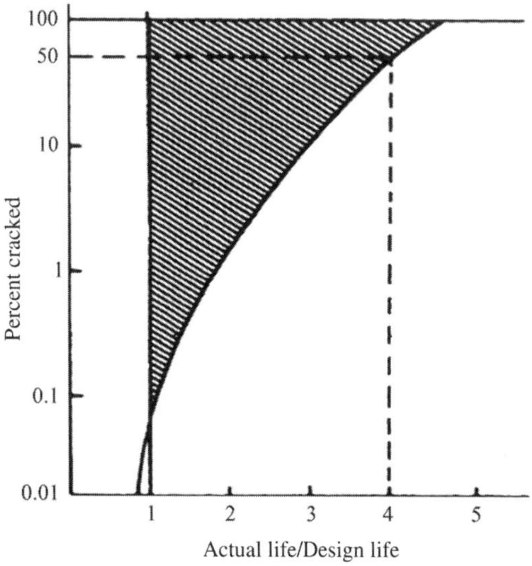

Fig. 2 Schematic representation of the distribution of actual failure lives of aircraft turbine disks as compared to the nominal design life, which is established to ensure that less than one in 1000 components develop a crack (from [4]).

to four) is supposed to account for material property and fabrication variations in the population of aircraft. As noted by Cameron and Hoeppner [6], situations in which inspection is "not a regularly employed practice, impractical, unfeasible, or occasionally physically impossible . . . are prime candidates for the applications of safe-life techniques when coupled with the appropriate technologies to demonstrate the likelihood of failure to be sufficiently remote."

When an inspection is performed on a component being managed by safe-life procedures, any indication of a defect is generally considered to be grounds for removal of the part from service. There are infrequent occasions in which design does not take into account all possible failure mechanisms, leading to unexpected failures. A field campaign to inspect the remaining fleet and ensure that similar defects are detected and repaired or replaced to prevent future failures will generally ensue.

B. DAMAGE TOLERANCE MANAGEMENT OF FATIGUE

The advent of fracture mechanics and other procedures to predict the evolution of damage laid the foundation for a second major design method, the damage tolerance approach. Damage tolerance is the attribute of a structure that permits it to retain its required residual strength for a period of usage without repair after the structure has sustained specified levels of fatigue, corrosion, and accidental or discrete source damage. In this section, attention will be restricted to the example of fatigue, described by the most elementary model. It should be recognized, however, that many refinements have been or are being developed.

We again follow Thompson [5]. The damage tolerance approach to components whose lives are controlled by fatigue is based on the physical understanding of damage evolution, as summarized in Fig. 3. The growth and/or fracture of a

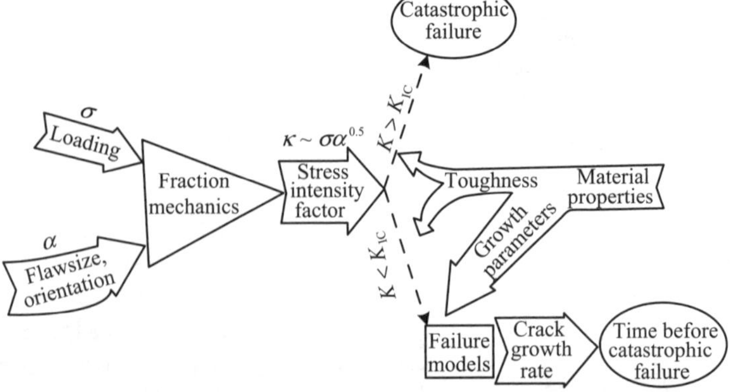

Fig. 3 Fracture mechanics concepts governing the prediction of failure under conditions of cyclic fatigue.

crack are assumed to be controlled by the stress intensity factor K, which defines the singularity in stress at the tip of the crack and is a function of the applied loads and the crack size. If the stress intensity factor exceeds a material property known as the facture toughness K_{IC}, failure is imminent. If the load is cyclic and the maximum stress intensity factor is less than the fracture toughness, the cracks will grow at a rate that is controlled by the excursions in the cyclically varying stress intensity factor, until the fracture criteria is reached. Anderson [7], Suresh [8], and Kanninon and Popelar [9] provide extensive discussions of this linear elastic fracture mechanics approach, as well as many refinements.

An excellent overview of the safe-life and damage tolerant design concepts, and the role of NDE in their implementation, can be found in the text by Grandt [10].

C. ROLE OF NONDESTRUCTIVE EVALUATION IN THE FATIGUE MANAGEMENT

With a safe-life design, the initial expectation is that the component will not require inspection during its useful life; for damage tolerance designs, inspection is part of the life management strategy for inservice components. Before discussing this latter case, it is worth noting that, even for safe-life components, inspection often plays a key role in continued airworthiness and safe operation. Field durability issues can arise for a number of reasons [11] including the following:

- Metallurgical defects not detected at manufacture
- Design errors from missed stresses, missed temperatures, etc.
- Manufacturing errors, such as improper heat treatment
- Maintenance errors, such as faulty repairs or improper assembly
- Operational errors, such as overtemps or overspeeds
- Other accidental, mechanical, or physical damage

Therefore, although the safe-life philosophy does not incorporate inspection considerations into the design calculations, inservice inspection plays a role when unexpected events occur. In 1999, the Federal Aviation Administration issued an advisory circular that includes instructions on the role of inspection to ensure continued airworthiness of the commercial fleet [12]. Advisory circular AC 33.4-2 states that "when the engine is in the shop and the engine parts and components are exposed, the parts and components should be subjected to appropriate inspections to determine their eligibility for reinstallation in an engine for continued service." With this change in FAA policy, inservice inspection came to play a greater role in commercial engine maintenance, even for engines designed in accordance with the safe-life philosophy. Referred to as "inspections of opportunity," the FAA tracked the impact of the policy from 1998-2007, with 70 cracked critical parts removed from service in that period [13].

In parts designed for damage tolerance, the prediction of failure during fatigue depends on three inputs: stress levels, flaw properties (e.g., size) and material properties (e.g., fracture toughness and crack growth rate parameters). Advent of damage tolerance design opened a new role for inspection, providing information about the flaw properties as input to the fracture mechanics calculations. Figure 4 conceptually illustrates how NDE techniques, providing this information, fit into the damage tolerance approach. It is assumed that the service-induced damage in a part will increase with use over time. In the case of fatigue, the measure of damage is crack length. It is also assumed that failure will occur when the crack size reaches a critical value A_F.

Suppose that there is a distribution of initial crack sizes (or crack nucleating conditions) within the material, as illustrated by the range of flaw sizes at zero time, and that the inspection has perfect differentiation, that is, that it can identify all flaw sizes greater than some value A_I for removal from service. The curves in Fig. 4 (neglecting for the moment the difference between the solid and dashed segments) show the expected evolution of crack size with time, based on deterministic fracture mechanics. The time T_{F1} is taken to be the time at which the first failure (or an acceptably low probability of failure) would occur. Then inspection would be conducted at an earlier time T_{I1} (dictated by the desired safety margin). The hypothesized perfect inspection would then allow all components containing cracks of size greater than A_I to be removed from service. The damage evolution that was avoided by the inspection is denoted by the dashed segments of the curves. The expected life of the remaining components will now be greater. As indicated, the process could be repeated additional times, extending the life of remaining components even further.

The details of this general strategy depend on the application. For example, as discussed by Grandt [10], in some Air Force applications T_{F1} is taken as the time that the flaw intended to be rejected by NDE would grow to failure. T_{I1} is taken to be 50% of this time to provide a safety factor.

By way of comparison, in the safe-life approach, one simply defines the service life to be sufficiently short that the probability of failure due to this natural evolution of crack size is at an acceptably low level. As noted earlier, the distribution of component lives implies that much useful life is lost in this scenario. In the damage tolerance approach, the use of inspection allows a reduction of the conservatism of the safe-life approach.

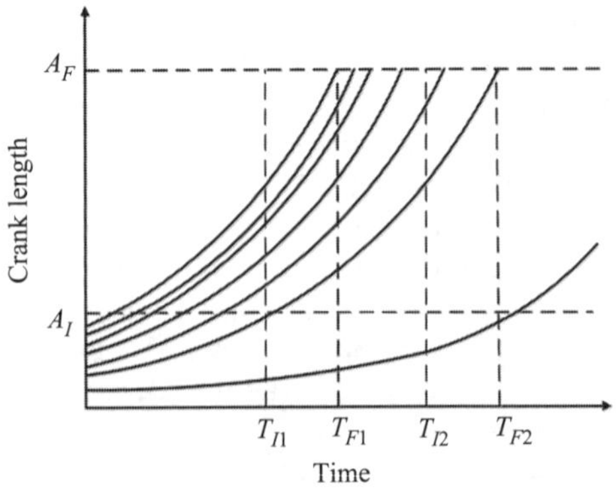

Fig. 4 Conceptual illustration of the damage tolerance approach to life assessment.

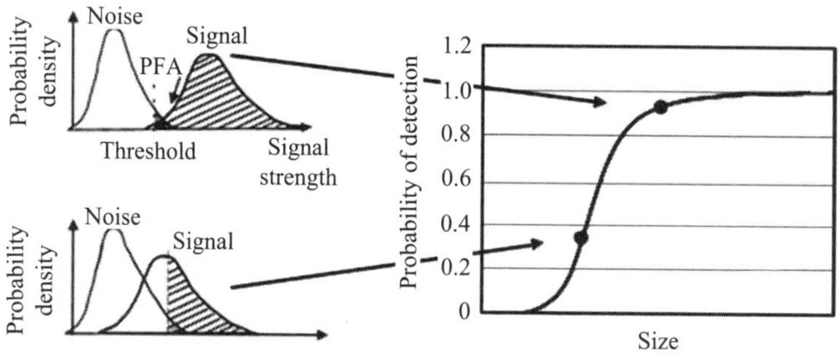

Fig. 5 Probability-of-detection concepts. POD and PFA are determined by fractions of signal and noise distributions above a threshold. Signal distribution generally shifts to higher levels as crack size increases, leading to a POD curve of the general shape shown.

A means is provided to utilize the life remaining in individual components, which, for the majority of the population, could be much longer than the safe life limit. It is recognized that defects will be present, and NDE techniques are expected to remove those components from service that contain defects whose size is greater than the inspectable flaw size A_I.

More sophisticated life management approaches recognize that no NDE technique is perfect. Given a family of cracks of nominally identical size, there will be different NDE responses as a result of such effects as variation in the morphology of the flaw, influences of the nearby microstructure, and variations in measurement systems and human operators.

A metric of an inspection is the degree to which it can correctly identify for removal from service those components whose flaw size exceeds the target size, for example, A_I. Formally, this metric is a subset of a more general metric known as the probability of detection (POD). As sketched in Fig. 5, the POD is based on the concept that, for a given flaw size, there will be a distribution of flaw signals as a result of variabilities such as those noted in the preceding paragraph. The POD is the fraction of the signal distribution above a threshold based on an accept/reject criterion. As shown to the left, POD usually increases with flaw size because the signal distribution shifts to larger values. In addition, it must be recognized that, in the absence of any flaw, there will also be a distribution of noise signals from benign discontinuities (part geometry, grain boundaries, etc.), and that these signals could be misinterpreted as flaw signals and hence lead to false alarms.

Lowering the threshold will increase the POD for a given flaw size, but it will also increase the probability of false alarms (PFA) because it will be more likely that noise signals from geometrical or microstructural features produce signals exceeding the threshold. Thus, the POD and PFA are defined, respectively, by the distribution of signals that would be produced by flaws of the same nominal size and the distribution of noise in the absence of a flaw, each as compared to the threshold, or accept/reject criteria, as shown in Fig. 5 [14]. The threshold is selected to balance achieving a sufficiently high POD (thereby avoiding failed parts and their consequences) and keeping the PFA sufficiently low (avoiding removal of an unacceptably high number of good parts from service and the associated cost). The accept/reject level must be determined within the context of the overall life management program for a particular component with a key ingredient being the relative consequences of false accepts (possible part failure) and false rejects (cost of lost parts).

The POD, or any other measure of the effectiveness of a technique, will obviously depend on the criterion for acceptance or rejection of a part. For example, in the case of a pulse-echo ultrasonic inspection, a received electrical signal at a time T after the wave was launched indicates that there is a discontinuity at a distance $v/2T$ from the transducer, where v is the velocity of sound in the material under inspection. Such a signal is generally interpreted as being caused by a flaw if its amplitude exceeds a predetermined threshold, often deduced from the strength of signals reflected from a known calibration reflector such as a flat-bottomed or side-drilled hole.

These conceptual ideas are implemented in detailed ways that are industry specific. Utilization of damage tolerance design philosophy has long been a standard in military jet engines and is the recommended approach for future commercial engine designs [15]. With the initiation of the Engine Structural Integrity Program (ENSIP) in the mid-1980s, the U.S. Air Force has defined damage tolerance as "the ability of the engine to resist failure due to the presence of flaws, cracks, or other damage, for a specified period of unrepaired usage." Guidance is provided by the Air Force in the ENSIP handbook [16], which suggests an organized and disciplined approach to structural design, analysis, qualification, production and life management of gas turbine engines, and has as its goal to ensure engine structural safety, durability, reduced life-cycle costs and increase service readiness. The Mil–HNBK 1783B [16] includes typical flaw sizes that have been demonstrated as part of probability of detection (POD) studies. The conduct and analysis of POD studies is covered in a companion document Mil–HNBK 1823 [17].

The current U.S. Air Force approach to damage tolerance is primarily a deterministic method. Many aspects of it, however, are based on probabilistic concepts, the most notable of which, perhaps, is the "inspectable" flaw size. The inspectable flaw size is often taken as that size which will be detected with a specific probability and confidence. The USAF often uses 90% probability and 95% confidence for its inspectable-flaw criterion when the inspection is performed manually and 90% probability and 50% confidence for an automatic inspection. As was generally illustrated in Fig. 4, a fracture mechanics analysis is used to determine the time for the growth of the inspectable crack to critical length, at which time the crack would be expected to become unstable and propagate rapidly, possibly resulting in catastrophic loss of the aircraft. The inspection is typically performed at one-half the time for the inspectable crack to grow to critical length. When a

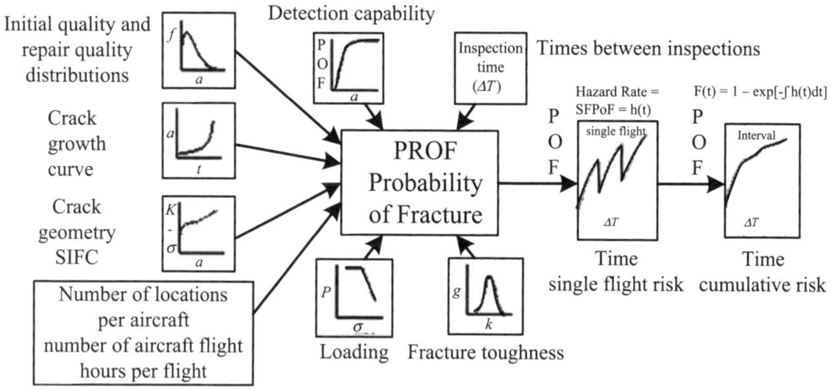

Fig. 6 Risk analysis input parameters and single flight probability of failure (from [18]).

crack is found by NDE, the structure is repaired, replaced, or modified to eliminate the defect.

As another example, fully probabilistic approaches provide the foundation for advisory guidance from the FAA regarding the design and life management of commercial aircraft engines [15], and probabilistic risk assessment techniques are being adopted by the Air Force [18]. Figure 6 illustrates the inputs to those strategies [18].

A key step in the implementation of damage tolerant life management is the NDE reliability or POD demonstration, which determines the POD vs flaw size relationship defining the capability of an NDE system. As indicated in Mil-HNBK 1823 [17], variation in NDE system response is caused by both the physical attributes of a flaw and the NDE process variables or parameters. Such variability can come about because of variation in the measurement system (such as the probe, the electronic noise for electronic methods, etc.), in the test piece (such as geometry, material effects, etc.), the flaw (size, shape, orientation), or the operator. Through proper experimental design and execution, these factors can be accounted for in performance of empirical POD studies. Recent efforts to supplement empirical data with physics-based model calculations have been reported by Thompson [19]. A comprehensive collection of POD curves has been compiled under FAA funding and is available from the Nondestructive Testing Information Analysis Center (NTIAC) [20].

D. ROLE OF NDE IN MANAGING DAMAGE MECHANISMS OTHER THAN FATIGUE

1. The majority of propulsion life management has concentrated on the engine disk and the detection of fatigue damage, but other components and issues are also of interest when considering the life and performance of a modern

day jet engine. Much of the design challenge for propulsion is directed toward achieving higher thrust and efficiency, which is to a great extent dictated by the engine operating temperature. In high-temperature operation, the airfoils must be protected from corrosion, erosion, creep, and other forms of high-temperature degradation. In the past two decades or so, thermal barrier coatings (TBCs) have become the method of choice for also protecting airfoils [21]. TBCs were first used to protect stationary parts such as the augmenter or afterburner; because of the poor durability of the coatings in high heat-flux environments, such as those in which airfoils operate, it was only much later that TBCs began to be used on blades and vanes. Modern engines rely on directionally solidified cast blades or single-crystal designs. To these blades, a yttria-stabilized zirconia coating is typically applied using plasma spray coating methods. In many cases, a bondcoat will be used to minimize the mismatch between the TBC layers and the base alloy used in the blade. A top coat is also included to provide the insulation properties needed to protect the blade. Further engineering details of TBC blade design can be found in a number of sources, including current websites (data available online at http://www.mtu.de/en/technologies/engineering_news/conventional_and_advanced_coatings.pdf).

The demanding environment and operational parameters [22] for airfoils of the high-pressure turbine section include temperatures of about 1000°C, with excursions above 1100°C, which is close to 90% of the alloys' melting points. The necessity of close control of the materials' surface temperatures comes from the fact that blade life as affected by creep is halved for every 10–15 deg increase in temperature [23]. Today's engine designs are inching toward temperatures in excess of 1400°C. In addition to efforts to improve the TBC performance, inspection methods to verify coating thickness, coating integrity, and defect detection are of interest. Defects of concern include spallation of the coating, which is often preceded by cracking and micropores in one or more layers of the TBC [24].

As with other segments of the aerospace industry, propulsion components are increasingly made of composites. With the GE90 certification in 1995, composites were used successfully in blade applications for the first time in commercial jet transportation. The blade design includes a titanium shank to which the composite airfoil is formed. As with all critical parts, the composite blades are thoroughly inspected.

II. PRIMARY NDE TECHNIQUES FOR DETECTION OF DISCRETE DAMAGE

The majority of inspection attention is focused on the disk or drum rotor components, whose failure is likely to do significant damage. Different inspection methods are utilized at various points in the engine life cycle to screen for fatigue cracks or other forms of discrete damage. Selection of the inspection method is based on the capability of a given method, often as characterized by

POD, and the flaw size requirements for the component design. Inspection cost and turn time are also considered when selecting an inspection strategy. Examples of the principles and applications of several of the major inspection techniques are provided next. For more detail, the interested reader can consult a number of texts [10, 25–28] and handbooks [29, 30].

A. FLUORESCENT PENETRANT INSPECTION

Fluorescent penetrant inspection (FPI) is the most widely used inspection method for engine components with over 90% of propulsion components inspected with this method at least once in their lifetime. During FPI, the part is exposed to a fluid containing fluorescent particles. Capillary action draws the fluid into the crack, excess penetrant is removed from the surface of the part, some of the fluid is drawn from the crack by the application of a developer, and the indication is viewed under a black light. The widespread use of FPI is driven by two major factors: the ability to detect the surface breaking cracks that are responsible for the majority of fatigue failures, and the ability to provide a whole-field, or

Fig. 7 General steps of fluorescent penetrant inspection: a) the process begins with a clean, dry part to which the pernetrant is applied, and aerospace applications use the fluorescent penetrant method typically in a dip tank; b) after the specified dwell time, excess penetrant is removed typically using a spray prerinse of acceptable temperature and pressure; c) if the postemulsifiable process is being used, the part is then dipped in the emulsification bath to make the oil-based penetrant water washable; d) the emulsification step is followed by a postrinse step; e) after a drying step, developer is applied, typically using either a spray application or developer chamber as shown here; and f) upon completion of adequate developer dwell time, the component is inspected under blacklight in a darkened room or booth.

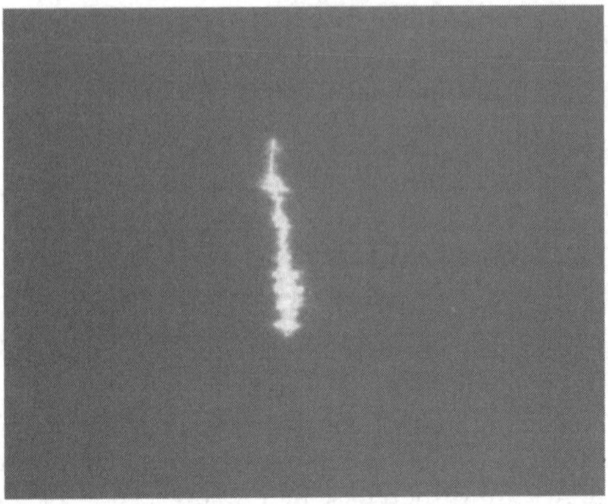

Fig. 8 A typical FPI image of a fatigue crack (55 mil length) is shown. This image was taken using a 40 x UVa microscope.

global inspection of a part of complex geometry for those surfaces that have line-of-sight access.

The general principles of this mature technology are well established, as illustrated in Fig. 7. The surface of the part is first cleaned and dried. The penetrant (containing fluorescent dye) is then applied, by spraying, brushing, or immersing the component in a penetrant bath. After a dwell time in which capillary action draws the penetrant into the flaw, excess penetrant is then removed from the surface of the component with the constraint that the removal of penetrant from flaws must be minimized. Depending on the type of penetrant used, this step might involve cleaning with a solvent, direct rinsing with water, or treatment with an emulsifier followed by rinsing. After drying to remove the rinse water, developer is then applied to the surface of the part to draw the penetrant trapped in flaws back to the surface, where it can be seen. The developer can be applied by dusting or dipping or spraying of either dry powder or wet solutions containing developer. The component is then examined under UV light, with the flaw being indicated by the change in color of the region where the penetrant has been drawn to the surface. Figure 8 shows an example of a flaw indication. After completion of the inspection, the component is either cleaned and returned to service or scrapped, depending on the outcome.

There are a great many process details that can influence the effectiveness of the FPI procedure. A significant issue is the reliance on the acuity of the inspector, who must remain attentive throughout a repetitive task in which flaws are only rarely found. Beyond this, a number of physical factors can play a major role in influencing the contrast of the indication and hence the ease with which the inspector can see it. These include the manner in which the part is cleaned and dried, chemical and physical properties of the penetrant at the temperature of inspection, the dwell time, the manner in which the excess penetrant is removed, and the manner and effectiveness of the developer application. Optimizing these processes is difficult, and procedures currently vary between OEMs. Continuing changes in the available penetrants in response to environmental concerns has also complicated the issue.

2. Nevertheless, FPI plays a crucial role in military and commercial applications, and in the timeframe 1996–2000, the FAA issued nearly 200 airworthiness

directives that called for the use of FPI. Because of the important contributions that FPI makes to engine safety, several research projects are underway to develop a better understanding of the effectiveness of various procedures. Results of the program have not yet been documented archivally but can be found on the internet (data available online at http://www.cnde.iastate.edu/faa-casr/fpi).

B. EDDY CURRENT INSPECTION

Despite its widespread use and many successes, FPI is limited by its dependence on operator acuity and sensitivity to a number of factors that can obscure or compete with indications. An alternative that has steadily received increasing usage over the last few decades is eddy current (EC) inspection, involving scanning a small electrical coil over the part and monitoring the flaw-induced change in its electrical impedance. EC techniques generally have a higher POD than FPI techniques, but suffer from the extra complication that the probe must be scanned over the entire area to be inspected. This can lead to significant inspection times and can be challenging in the presence of complex geometries. EC inspection is a widely used fatigue crack detection method, however, with manual, semi-automated, and automated inspections routinely used.

Figure 9 illustrates the general principles (see also http://www.ndt_ed.org for background information). The probe is a coil of wire (often wound on a ferrite core), through which alternating current is passed at a frequency that generally ranges from a few kilohertz to a few megahertz. When the coil is placed next to a conducting component, the dynamic magnetic fields of the coil will induce eddy currents in the component. These will produce dynamic magnetic fields, some of which will pass through the coil, leading to a change in its electrical impedance. When a crack is present and interrupts the flow of the eddy currents, there will be a further change in the impedance. The characteristics of this change in impedance as the coil is scanned from an unflawed region to a flawed region indicate the presence of the flaw and provide information about its characteristics.

EC inspection is a near-surface technique because the eddy currents decay exponentially with depth as a result of the

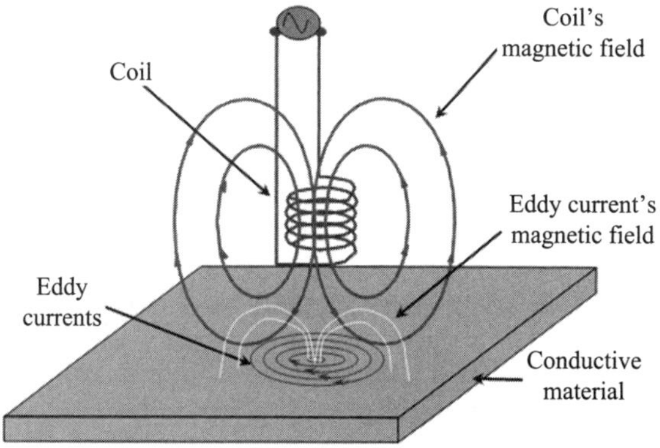

Fig. 9 General principles of eddy current testing (from http://www.ndt_ed.org)

electromagnetic skin effect. The depth of penetration varies inversely with the square root of frequency, electrical conductivity, and magnetic permeability (for ferrous materials). Hence, the skin depth is relatively small for high-frequency inspection of aluminum and large for low-frequency inspection of titanium or nickel-based superalloys. For example, for titanium, the skin depth has values of 147.7, 46.7, and 14.8 mils (3.75, 1.18, and 0.38 mm) at 10 kHz, 100 kHz, and 1 MHz, respectively [26].

Figure 10 schematically indicates the nature of the change in the electrical impedance as influenced by a number of experimental factors (data available online at http://www.ndt_ed.org). When the coil is placed in air, it will have an impedance that can often be approximated by that of a resistor in series with an inductor. If the coil is placed on a nonmagnetic, conducting part, the resistance will increase (because additional energy is lost to the heat generated by the eddy currents in the conductive material), and the inductance will drop (because the stored magnetic energy is reduced by the shielding of the interior of the conductive material by the eddy currents). The impedance shifts along the arc marked "conductivity," with the magnitude of the shift controlled by the product of frequency at which the probe is driven and the electrical conductivity of the test piece. If the probe is then placed over a crack, the current flow and fields will be further perturbed, and there will be a further change in impedance, as shown.

Changes in liftoff (distance from the probe to the test piece) will also change the coil impedance (which must approach the air value for large liftoff). This is important because, in practical inspections, it is often hard to control the exact distance of the probe above the surface. One of the important challenges in eddy current inspection is to differentiate signals from flaws from impedance changes as a result of liftoff variations. Approaches include careful fixturing to control liftoff and discrimination of flaw and liftoff signal based on their different phase angles. This discrimination depends on the inspection frequency and the electrical conductivity of the component.

On magnetic parts, the eddy current inspection is different because the inductance often increases rather than decreases when the coil is moved closer to the component. The relative effects of changes in liftoff,

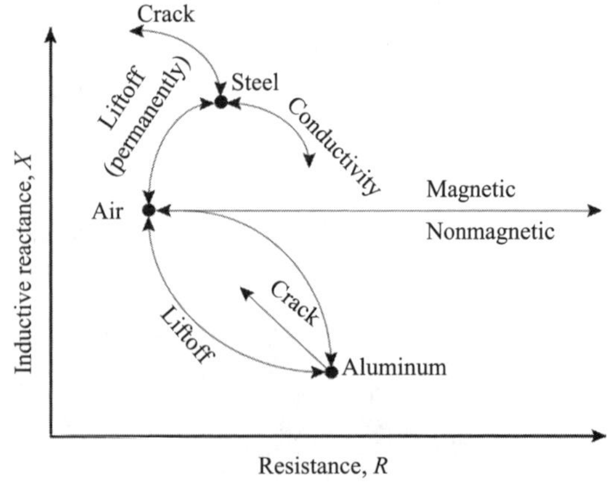

Fig. 10 Overview of the effects of many factors on the impedance plane plot. A plot of this nature is called an impedance plane plot of the eddy current response (from http://www.ndt_ed.org).

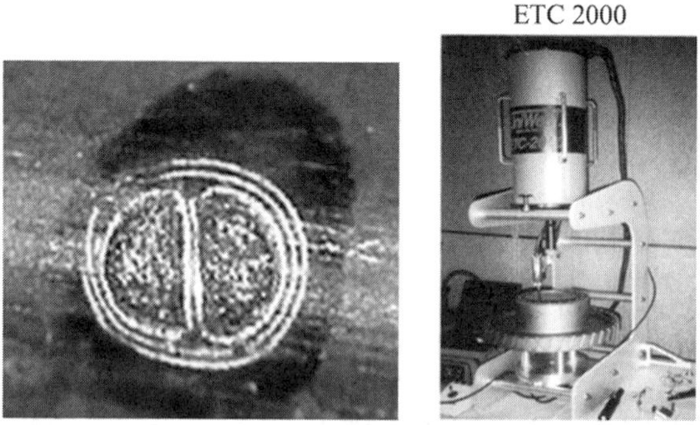

Fig. 11 Photograph of a split-D coil. Hybrid coil consists of an encircling driver and differential pickups (from http://www.ndt_ed.org).

conductivity, and flaw size are also shown schematically in Fig. 10. One of the major challenges in inspecting ferromagnetic components with eddy currents is the fact that normally occurring fluctuations in the local magnetic permeability can produce changes in signals that might be difficult to distinguish from those caused by flaws.

There are a great number of practical variations on this basic theme. The coil exciting the eddy currents might be different from the coil picking up the fields and hence detecting the changes induced by the presence of a flaw (pitch-catch or reflection probe); the difference in the impedance of two similar coils can be measured (differential probe), a technique that discriminates against the influence of liftoff (which will change the impedance of both coils in the same manner). One example that combines a separate excitation coil with a differential pickup is the Split-D coil, as shown in Fig. 11 (data available online at http://www.ndt_ed.org). This configuration is widely used in the inspection of bolt holes of rotating components. Figure 12 shows a widely used scanning system that allows the high-speed scanning of the holes.

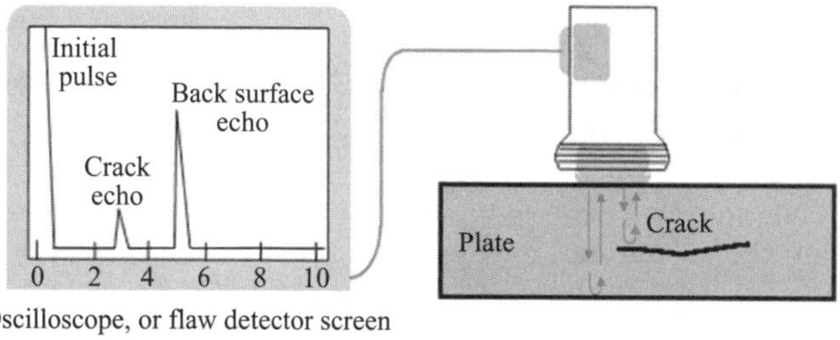

Fig. 12 Photograph of high-speed scanning system used in the inspection of bolt holes in turbine disks.

C. ULTRASONIC INSPECTION

In contrast to FPI and eddy current inspections, ultrasonic inspection provides the opportunity to gain information about flaws in the interior of a component away from any surface. One of its major applications is in searching for inclusions in rotating components that might nucleate fatigue cracks. Additional in-service applications include searching for fatigue cracks in geometrically inaccessible regions.

The basic concept of an ultrasonic inspection in the pulse-echo, contact mode is illustrated in Fig. 13 (data available online at http://www.ndt_ed.org). A transducer, whose active element is a piezoelectric material, excites short pulses of high frequency (often in the range of 1 – 10 MHz) elastic wave energy. The elastic energy leaves the transducer and propagates through a thin, fluid, coupling layer into the component under inspection. Any energy reflected from discontinuities returns to the transducer and is detected by reciprocal processes in the piezoelectric element. In the most elementary form of inspection, the signals ("echoes") are displayed as a function of time (A-scan), as illustrated on the left-hand side of Fig. 13. In a rough sense, the amplitude of the echo is related to the size of the flaw (although many uncontrolled factors can make this a very weak relationship), and the time of the echo gives the distance of the flaw from the surface. In Fig. 13, the transducer is shown in contact with the part, a configuration that is generally used in manual inspections, for example, in response to field problems. In a variant of the geometry shown in Fig. 13, the probe can be mounted on a wedge, so that the wave propagates at an angle to the part surface.

Among the challenges of the contact mode of ultrasonic inspection are the facts that 1) the signals observed depend on the thickness of the coupling layer, with best results requiring this to be a small fraction of the ultrasonic wavelength ($\lambda = 150$ μm at 10 MHz in water); and 2) it is hard to maintain uniform coupling, particularly if rapid scans are desired. Hence an alternate, immersion mode is used when possible, and this is the approach of choice during

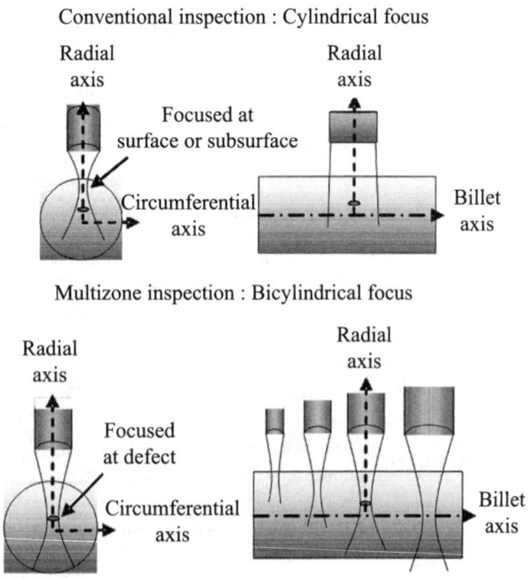

Fig. 13 Schematic of ultrasonic inspection in the pulse-echo contact mode (from [13]).

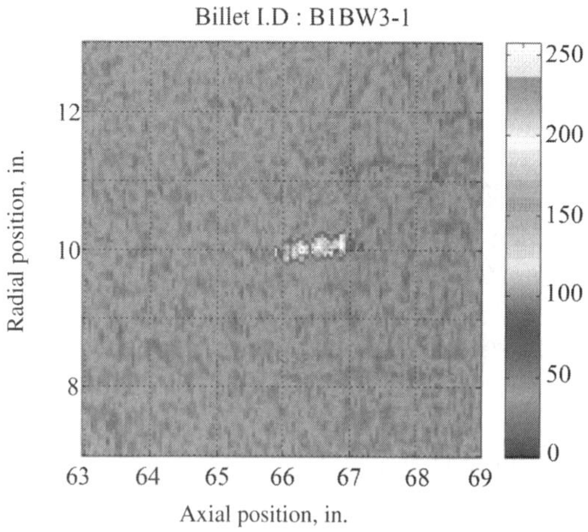

Fig. 14 Schematic inspection system for aircraft engine billets: top—single transducer conventional inspection; bottom—multiple transducer multizone inspection.

manufacturing inspections. In immersion inspections, the part and the transducer are placed in a water tank with the ultrasonic pulse propagating from the transducer, through the water, and into the part through the liquid–solid interface. After reflecting from defects or other discontinuities, the energy returns to the same or a different transducer for detection. Advantages of this approach include reproducible signals, amenability to high-speed scans, the ability to scan complex component geometries, and the presentation of the data in image format.

Figure 14 schematically shows two variants of immersion inspection, as practiced during the inspection of billet intended for disk manufacture. For this geometry, the billet is rotated θ about its axis, and the transducer is scanned parallel to the axis z. A time gate is used to isolate the response of defects at a given range of depths, and the peak amplitude of the signal is recorded. In the conventional approach, as shown at the top, a single transducer is used, often focused near the surface, and the gate is adjusted such that the signal from a flaw at any depth between the surface (other than a "near"-surface dead zone) and the centerline is accepted. In the multizone approach, shown at the bottom, a set of transducers is used, each focused at a different depth [31]. The gate is adjusted to select a region of material near the focal plane of each transducer for inspection. In current implementations, this depth is often 1 in. (2.54 cm). By plotting the amplitude of the peak signal in the gate as a function of θ and z, an image of the flaw, known as a C-scan, is captured. As an example, Fig. 15 shows an image of a hard-alpha inclusion in a titanium billet. More recently, the use of a single, phased array ultrasonic transducer has come into use.

Ultrasonic inspections are performed in immersion during manufacturing on billets (as before) and forgings (machined into a "sonic shape" that has simpler surface geometry and hence is more readily inspectable than the final shape). Additional inspections can be performed during service. Contact inspections are sometimes employed in-service when part geometry or on-wing location render immersion impracticable. Both rotating parts and blades can be inspected.

Fig. 15 C-Scan image of hard-alpha. Amplitude of peak signal in gate is plotted vs axial (abscissa) and circumferential (ordinate) position.

A technology that is currently gaining widespread attention is phased-array ultrasonic inspection. By replacement of the single piezoelectric element found in traditional transducers with multiple elements, each provided with separate electronic transmitter and receiver circuitry, it is possible to electronically focus and scan the beam. The result can be a faster and/or more efficient inspection. Figure 16 shows an experimental demonstration of a phased-array inspection of a sonic shape forging. The positioning of the phased-array probe with respect to the forging is shown in the left insert, and the layout of the phased array elements is shown to the right. This system was designed to detect planar defects 1/128 in. (0.2 mm) in diameter throughout the forging.

D. X-RAY INSPECTION

X-ray inspection also sees extensive use as part of production qualification, particularly in complex shaped parts, such as blades, and the static cases that

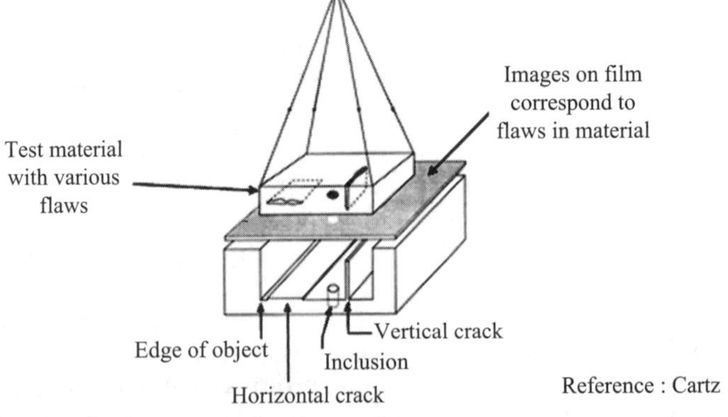

Fig. 16 Phased-array inspection with positioning of probe with respect to forging and layout of phased-array elements shown in inserts.

house the engine. Many of the components that are inspected with radiography are cast structures. Internal defects are of primary concern and include pores, inclusions, and dross. X-ray radiography is one of the oldest forms of inspection, with the basic principles being illustrated in Fig. 17. X-rays are emitted from a source, often a tungsten target, against which electrons are driven by an accelerating voltage in the 20–400 kV range. After propagating to and through the target, the x-ray intensity is recorded on a detector. Flaws are imaged because of the differences in their x-ray absorption coefficient with respect to the host materials.

The sketch in Fig. 17 illustrates how the geometry of the flaw influences the contrast in the image [26]. The contrast in the image is controlled by relative absorption of the x-rays in the flaw and surrounding material. Because cracks or voids have a lower absorption than engine materials, there will be an increased intensity for x-rays passing through them. For the case of a crack, the contrast (determined by the differences in these intensities) will be much greater if the x-rays propagate parallel to the plane of the defects, so that the differential absorption is present over a greater path length. On the other hand, it is very difficult to detect tight cracks when x-rays propagate perpendicular to their faces. Inclusions can have either higher or lower absorptions than the matrix, depending roughly on their relative atomic masses. Hence, they can be seen as either an increase or a decrease in x-ray intensity.

Traditionally, film has been used as the detector. However, advances in technology now make digital recording media an attractive alternative. Among the advantages are the speed at which an image can be obtained (no developing required) and the ease with which the information can be exchanged, stored, and discarded. A topic of recent discussion is the relative quality of the images

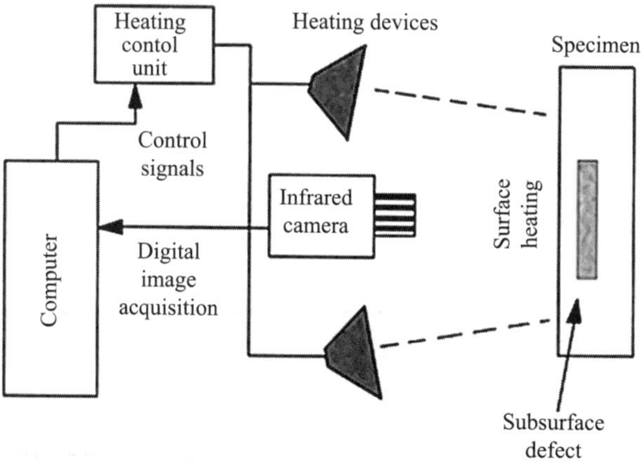

Fig. 17 Schematic of x-ray measurement (from [26]).

and hence the relative ability to detect flaws using the two detection media. It has historically been the case that, under ideal conditions, higher contrast images could be obtained with film-based techniques, but digital images now have sufficient quality for the purposes of detecting flaws, and their other advantages make them an increasingly preferred technique.

There are a number of technical factors that must be considered in setting up an x-ray. One of the most important is the value of the accelerating voltage because the x-ray absorption coefficient is highly dependent on the energy of the photons (which are characterized by a Brehmstrahlung spectrum related to the accelerating voltage). In general, higher voltages must be used to penetrate through thicker sections and produce a useful image in a reasonable time. The higher energy of the photons has the effect of reducing the absorption coefficients and the contrast in the image (because higher absorption coefficients generally imply greater differences between the x-ray intensity passing through comparable thicknesses of base material and flaw). X-ray inspection can involve multiple shots for a given component with some applications requiring in excess of 200 shots. As has been noted, the technology is moving toward more extensive use of digital radiography, particularly with advances that are being made in detectors and imaging software. As further improvements are made in sensitivity and resolution, use of digital radiography will see further increase. Use of radiography techniques for in-service applications is predominantly in verification of repairs. Limited applications of isotope source use have occurred for on-wing applications. Selection of isotope methods is made when crack detection is needed for inaccessible or hidden structures and for wide cracks where crack orientation effects will not deter detectabilty. Typically, if other methods are feasible, they will be utilized.

The preceding discussion has dealt with projection radiography. As illustrated schematically in Fig. 17, the image formed on the detector is controlled by the absorption processes along each ray path from source to detector. The goal in

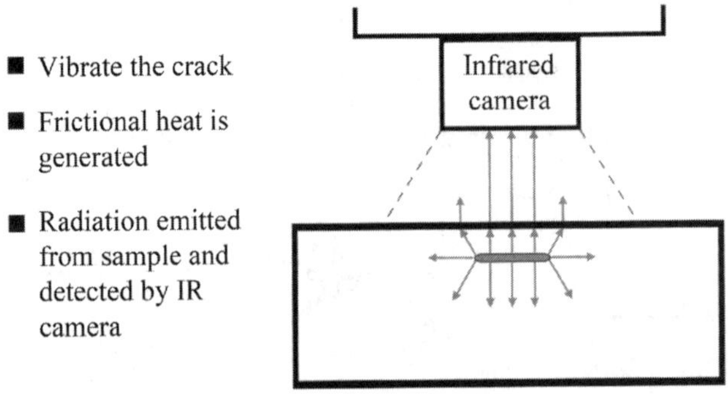

Fig. 18 Pulsed thermography imaging system (from [32]).

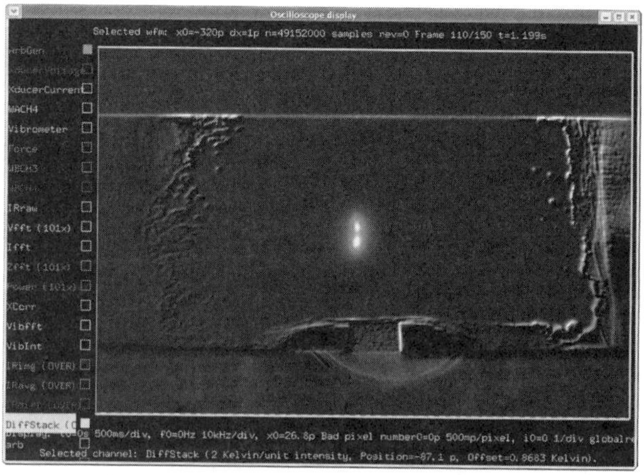

Fig. 19 Principles of vibrothermography.

computerized tomography (CT) is to obtain a three-dimensional image of the defect structure. Clearly, this cannot be accomplished with a single measurement. At an intuitive level, the approach is simple; one makes multiple x-ray measurements (with different angles with respect to the component). The component is considered to be discretized into a series of voxels, and the x-ray absorption within each voxel is inferred from the data set. In practice, this can be a challenging task because of the large datasets that must be handled, and hence a number of approximate approaches have been developed, such as filtered back projection and spiral CT. CT can produce exquisite images of flaw structures, but this is realized with increased data acquisition and processing time.

E. THERMOGRAPHIC INSPECTION

Flash thermography, as illustrated in Fig. 18, is increasingly used for NDE. Heating devices, such as xenon flash lamps, are pulsed to produce a transient heat pulse. An infrared camera is used to monitor the temperature evolution at the surface of the sample. The presence of a subsurface defect, as shown, will modify the flow of heat and hence the temperature image. Time is an important parameter in flash thermography because the difference in temperature between the flawed and unflawed region will depend on the time after the flash.

F. VIBROTHERMOGRAPHY

Vibrothermography, also known as sonic infrared imaging, is an emerging technique that combines vibrational excitation and infrared detection [33]. Figure 19 schematically illustrates the physical idea. The component under inspection is vibrated, often in the 20–40-kHz frequency range, by a narrowband ultrasonic welding device, a broadband piezoelectric transducer whose frequency can be swept, or some other excitation source. Heat is generated at cracks, generally attributed to frictional sources, and the resulting temperature rise is detected by an IR camera.

One of the major advantages of this approach is that the crack is seen in a dark field. When observed in real time, the crack "lights up" when the excitation is applied. This is a relatively new technique, and research to tie down a number of physical principles (including the mechanism of the heating) and procedural details such as the optimum means of holding the sample (without unduly modifying its vibration), applying the excitation, and being sure that the excitation is truly nondestructive was completed.

III. CHARACTERIZATION OF DISTRIBUTED DAMAGE AND STRESS

All of the techniques just cited have been designed to detect and characterize discrete defects and are used in industrial practice. Interest is rapidly increasing, however, in techniques to characterize material properties and conditions that are precursors of discrete damage or influence the rate of initiation and/or growth of defects. These are generally still under development, and a full discussion of their status goes beyond the scope of this article. Techniques designed to control fatigue in disks include the use of ultrasonic harmonic generation to monitor the evolution of dislocation structures and/or microcracks associated with fatigue, ultrasonic backscattering and attenuation to monitor grain size, and eddy current resistivity and high-energy x-ray measurements to monitor shot-peening-induced stresses in nickel-based superalloys. To provide a flavor of this work, three examples will be discussed in more detail.

Surface treatments such as shot peening have long been utilized to impart compressive residual stresses in critical components such as jet engine disks. Compressive residual stresses have demonstrated benefits in improved fatigue life, retardation of crack growth, and resistance to foreign object damage [34, 35]. Given the value of surface treatments such as shot peening, laser peening, and low plasticity burnishing to impart beneficial residual surface stresses, these are also employed in the manufacture of turbine engine components. Although the benefits of these surface treatments are known, in most cases, the extent to which they are retained after a period of service, including the depth profile, is unknown and cannot be taken into account in life calculations. The Air Force, through efforts in the Engine Reliability and Life Extension program, has estimated that quantitative assessment of residual stress would enable significant life extension with costs savings estimated in the millions [36]. The traditional method of measuring residual stress is low-energy x-ray diffraction. To generate the depth profile needed to make decisions about remaining residual stress, a sequence of x-ray diffraction measurements followed by controlled chemical etching is used. Because of the need to remove material to generate the depth profile, this technique is not truly nondestructive.

Significant research efforts have been made to nondestructively quantify residual stress as a function of depth [37]. This has included electromagnetic methods [38–45] as well as x-ray-based methods [46]. Prior research has shown that for nickel alloys, electrical conductivity measurements can be correlated with residual stress. Current work has identified the need to effectively separate the effects of near-surface compressive residual stress from the competing effects of cold work and surface roughness on the eddy current response. The measurement of eddy current conductivity through multiple frequency inspection is expected to allow the nondestructive evaluation of subsurface residual stress profiles in shotpeened and otherwise surface-treated specimens. Microstructural complexities in titanium alloys have made use of electromagnetic measurements more problematic, but high-energy x-ray techniques are being developed and show significant promise of addressing this problem in titanium alloys, as well as providing an alternative approach for nickel-based superalloys [46].

Several characterization and inspection approaches have been considered for use on TBCs (for example, see [47]). Inspection efforts have included electrochemical impedance spectroscopy and laser fluorescence. In the laser fluorescence method, stresses were measured in laboratory specimens, which were cycled to failure and correlated with remaining TBC life. Good correlations were also found between laser measurements and remaining life in lab tests and with field run hardware [48]. Work at the University of Connecticut and University of California—Santa Barbara included development of prototype instrumentation to extend this approach to field measurements. Damage progression has been assessed using thermal methods [49] where it was found that the thermal wave amplitude reached saturation near the end of useful coating life. Electromagnetic methods, including the meandering winding magnetometer [50], developed by JENTEK, also show promise.

Techniques used in the inspection of composites include the use of through-transmission ultrasonics to verify flowpath surfaces to ensure full utilization of the 3D aerodynamic design and computed tomography including the 17-in. (43.2-cm) dovetails. A 6-MeV CT inspection system is used for the fan blade inspection. Higher-temperature composites include the bismaleimide (BMI) components, which can operate at temperatures up to 245°C and are seeing increasing use in propulsion systems and offer new challenges for inspection.

IV. BROADER ISSUES

Degradation of components through the growth of cracks can have safety and/or economic implications. As implied by previous discussions, the rupture of a high-energy rotating component is clearly in the former category because the engine casing is generally not designed to contain such violent but infrequent events and loss of the aircraft or life is a possible outcome. Blade or blade coating failures will generally not lead to such catastrophic consequences, but they will lead to loss of efficiency and possibly to engine damage and loss of power. Regulatory bodies tend to be most concerned with rotating components whereas operators/OEMs take a broader perspective.

In the commercial arena, the safety of the flying public is contingent on the symbiotic relationship of the regulator, the manufacturer, and the operator. During the design and fabrication of aviation components, the manufacturer must comply with FAA regulations put in place to ensure the safety of the commercial fleet. The OEM will submit design and manufacturing data to the FAA for approval and issuance of a type certificate. As part of the data submittal, the OEM must also define long-term maintenance requirements. With the sale of the aircraft to an operator, the OEM provides guidance on the maintenance, inspection, and repair for a given aircraft, component or propulsion system, which must be managed and adhered to by the operator. The FAA also has regulations in place to govern the operation and maintenance of the aircraft to which the operator must ensure compliance.

For many years, major carriers performed the majority of their maintenance work in-house. More recently, economic pressures have increased the use of third-party organizations for maintenance, repair, and overhaul (MRO). The commercial jet transport MRO market was valued at $37B in 2004 [51]. A major component of the "low cost carrier model" is the extensive use of outsourcing for all MRO activities. The "legacy" carriers are searching for the most profitable balance between the use of internal labor for MRO functions and outsourcing. Legacy carriers with internal MRO capabilities are most often also service providers, that is, they not only perform maintenance for their own fleets but offer this as a service for fee to other airlines. The OEMs are also significant players in the MRO marketplace, with the engine manufacturers all having MRO operations within their companies, and there are examples of partnerships between air carriers and OEMs to provide MRO services to others. These relationships are further complicated by an evolving business model, which causes the OEM to assume an increasing responsibility for the reliable operation of the engine throughout its life, a practice sometimes referred to a "power-by-the-hour."

The U.S. Air Force operates in a different manner, specifying performance and operational criteria to which the OEM will design and build. In most cases, the Air Force performs its own maintenance, inspection, and repair functions using procedures that have often been developed in cooperation with the OEM.

The complexity of these varied business relationships brings with it complexity for the inspection community. Understanding the implications of a given technology requires interaction with each of the stakeholders, from the regulator, to the manufacturer, to the operator—whether military or commercial—to the service provider.

In both the military and commercial sectors, the push is toward the use of image-based technologies. Additional emphasis is being placed on automated data analysis with trends toward digital data acquisition and the associated storage issues.

Implementation of damage tolerance design/life management strategies by using NDE rather than safe-life strategies and removing from service those components whose individual lifetimes are nearing their end has been very effective at improving safety while lowering operational costs. For example, the U.S. Air Force has saved hundreds of millions of dollars in unnecessary replacement components through the Retirement for Cause (RFC) program, in which eddy current inspections were used to remove service turbine engine disks that were developing fatigue cracks [52]. Despite its improvements over the previous system, this approach still entails significant cost because it requires periodic inspection on the individual components of disassembled engines at a fixed interval, regardless of the condition of the component. There is currently a vigorous research and development effort underway to develop improved, more cost-effective approaches for identifying components and systems near the end of their lives, a goal described as condition-based maintenance. These approaches will not be reviewed in detail in this chapter because the field is relatively young and a number of major programs are in progress [53, 54].

V. CONCLUSIONS

The reliable operation of the aircraft turbine engine is crucial to the successful completion of both commercial and military missions, and failures can have safety and/or economic implications. In this chapter, the design/life management practices that provide a context for NDE are described. Current NDE practices to support this life management are described in the context of the economic and safety factors that must be considered. New challenges and the future research directions that they imply are reviewed, including efforts to find both smaller defects and material anomalies associated with new failure modes.

NDE of aircraft engines is a complex field, which is driven by many factors including the operational needs of the users (commercial airlines or military), the needs and regulatory requirements imposed by the government, the manufacturing and inspection capabilities of the OEMs, and new ideas provided by the research community. Economic considerations play an important role in this field, and there are inevitably tensions among the requirements of safety, reliability, performance, and economics. These often drive the solutions being developed. The chapter provides the authors' view of the current status of the field, as influenced by all of these factors.

It is tempting to speculate on future directions, but the crystal ball is rendered somewhat cloudy by an uncertainty regarding how the current economic challenges of the commercial aviation industry will be resolved, as well as how the military balances operational costs, performance requirements, and changes in mission. It is safe to say, however, that there will continue to be a drive for innovative approaches that combine improved economics and reliability, factors that will be at a premium as new materials are introduced to increase performance and as design and cost margins tighten.

REFERENCES

[1] Thompson, R. B., and Thompson, D. O., "Ultrasonics in Nondestructive Evaluation," *Proceedings of the IEEE*, Vol. 73, 1985, pp. 1716–1754.

[2] Coffin, M. D., and Tiffany, C. F., "New Air Force Requirements for Structural Safety, Durability and Life Management," *Journal Aircraft Safety*, Vol. 13, 1976, pp. 93–98.

[3] Satter, S. A., and Sundt, C. V., "Gas Turbine Engine Disk Cyclic Life Prediction," *Journal of Aircraft*, Vol. 12, 1975, pp. 360–365.

[4] Rao, C. A., Jr., "The Impact of Inspection and Analysis Uncertainty on Reliability Predictions and Life Extension Strategy," *Proceedings of the ARPA/AFML, Review of Progress in Quantitative Nondestructive Evaluation* (Tech Rept. AFML-TR-78-205), Air Force Materials Lab., Dayton, OH, 1979, pp. 150–161.

[5] Thompson, R. B., "Nondestructive Evaluation and Life Assessment," *ASM Handbook*, Vol. 11: *Failure Analysis and Prevention*, edited by W. T. Becker and R. J. Shipley, ASM International, Materials Park, OH, 2002, pp. 269–275.

[6] Cameron, D. W., and Hoeppner, D. W., "Fatigue Properties in Engineering," *Nondestructive Evaluation and Quality Control*, Vol. 17: ASM Handbook, ASM International, Materials Park, OH, 1989, pp. 15–26.

[7] Anderson, T. L., *Fracture Mechanics: Fundamentals and Applications*, CRC Press, Boca Raton, FL, 1991.

[8] Suresh, S., *Fatigue of Materials*, Cambridge Univ. Press, Cambridge, England, U.K., 1998.

[9] Kanninen, M. F., and Popelar, C. H., *Advanced Fracture Mechanics*, Oxford Univ. Press, New York, 1985.

[10] Grandt, A. F., *Fundamentals of Structural Integrity*, Wiley, New York, 2004.

[11] "Titanium Rotating Components Review Team Report," Federal Aviation Administration, Aircraft Certification Service Engine and Propeller Directorate, Burlington, MA, Dec. 1990.

[12] "Instructions for Continued Airworthiness," Federal Aviation Administration, Advisory Circular 33.4–1, Burlington, MA, Aug. 1999; www.faa.gov.

[13] Broz, A., "An Inspection Success Story," ATA NDT Forum, Aug. 2007.

[14] Rummel, W. D., Hardy, G. L., and Cooper, T. D., "Applications of NDE Reliability to Systems," *Nondestructive Evaluation and Quality Control,* Vol. 17: ASM Handbook, ASM International, Materials Park, OH, 1989, pp. 674–688.

[15] "Damage Tolerance for High Energy Turbine Engine Rotors," Federal Aviation Administration, Advisory Circular AC 33.14-1, Engine and Propellor Directorate, Burlington, MA.

[16] *Engine Structural Integrity Program (ENSIP)*, U.S. Dept. of Defense Military Handbook, MIL-HDBK-1783B, Feb. 2002; http://assist.daps.mil/.

[17] *Nondestructive Evaluation System Reliability Assessment*, U.S. Dept. of Defense Military Handbook, MIL-HDBK-1823, April 1999; http://assist.daps.mil/.

[18] Gallagher, J. P., Babish, C. A., and Malas, J. C., "Damage Tolerant Risk Analysis Techniques for Evaluating the Structural Integrity of Aircraft Structures," *Proceedings of the 11th International Conference on Fracture*, Turin Italy, March 2005.

[19] Thompson, R. B., "Using Physical Models of the Testing Process in the Determination of Probability of Detection," *Materials Evaluation*, Vol. 59, 2001, pp. 861–865.

[20] Matzkanin, G. A., and Rummel, W. D., *NDE Capabilities Data Book*, NTIAC-DB-97-02, 1997.

[21] Soechting, F., "A Design Perspective on Thermal Barrier Coatings," *Journal of Thermal Spray Technology*, Vol. 8, No. 4, Dec. 1999, pp. 505–511.

[22] Schulz, U., Fritscher, K., Leyens, C., Peters, M., and Kaysser, W. A., "The Themocyclic Behaviour of Differently Stablized and Structured EB_PVD TBCs," *Journal of Materials*, Vol. 10, No. 10, Oct. 1997.

[23] Harrison, G. F., *Proceedings of the European Propulsion Forum*, Royal Aeronautical Society, London, 1993, pp. 3.1–3.16.

[24] Padture, N. P., Gell, M., and Jordan, E. H., "Thermal Barrier Coatings for Gas-Turbine Engine Applications," *Science*, Vol. 296, No. 5566, April 2002, p. 280.

[25] Halmshaw, R., *Non-Destructive Testing*, 2nd ed., Edward Arnold, London, 1991.

[26] Cartz, L., *Nondestructive Testing*, ASM International, Materials Park, OH, 1995.

[27] Bray, D. G., and Stanley, R. K., *Nondestructive Evaluation: A Tool in Design Manufacturing and Service*, CRC Press, Boca Raton, FL, 1997.

[28] Jiles, D., *Introduction to the Principles of Materials Evaluation*, Taylor and Francis, Philadelphia, 2007.

[29] *Nondestructive Testing Handbook Series*, American Society for Nondestructive Testing, Columbus, OH (multiple years).

[30] *Nondestructive Evaluation and Quality Control*, Vol. 17: ASM Handbook ASM International, Materials Park, OH, 1989.

[31] Nieters, E. J., Gilmore, R. S., Trzaskos, R. C., Young, J. D., Copley, D. C., Howard, P. J., Keller, M. E., and Leach, W. J., "A Multizone Technique for Billet Inspection," *Review of Progress in Quantitative Nondestructive Evaluation*, edited by D. O. Thompson and D. E. Chimenti, Plenum Press, NY, 1995, pp. 2137–2144.

[32] Sun, G., Wang, X., Feng, Z. J., Jin, H., Siu, H., Quyang, Z., Han, X., Favro, L. D., Thomas, R. L., and Bomback, J. L., "Imaging and Quantitative Measurement of Corrosion in Painted Automotive and Aircraft Structures," *Review of Progress in Quantitative Nondestructive Evaluation*, Vol. 19, edited by D. O. Thompson and D. E. Chimenti, AIP, NY, 2000, pp. 603–607.

[33] Thomas, R. L., "Thermal NDE Techniques—from Photoacoustics to Thermosonics," *Review of Progress in Quantitative Nondestructive Evaluation*, Vol. 21, edited by D. O. Thompson and D. E. Chimenti, AIP, NY, 2002, pp. 3–13.

[34] De Los Rios, E. R., Mercier, P., and El-Sehily, B. M., "Short Crack Growth Behaviour Under Variable Amplitude Loading of Shot Peened Surfaces,"

Fatigue and Fracture of Engineering Materials and Structures, Vol. 19, No. 2/3, 1996, pp. 175–184.

[35] Prevey, P. S., "The Effect of Cold Work on the Thermal Stability of Residual Compression in Surface Enhanced IN718," *Proceedings of the 20th ASM*, 2000.

[36] John, R., Larsen, J. M., Buchanan, D. J., and Ashbaugh, N. E., "Incorporating Residual Stresses in Life Prediction of Turbine Engine Disks," *Proceedings of NATO RTO (AVT) Symposium on Monitoring and Management of Gas Turbine Fleets for Extended Life and Reduced Costs*, Oct. 2001.

[37] Blodgett, M., and Moran, T., "Residual Stress Program Plan," *Proceedings of the Residual Stress Workshop*, April 2005.

[38] Kalyanasundaram, K., and Nagy, P. B., "A Simple Numerical Model for Calculating the Apparent Loss of Eddy Current Conductivity due to Surface Roughness," *NDT&E International*, Vol. 37, 2004, pp. 47–56.

[39] Blodgett, M. P., and Nagy, P. B., "Eddy Current Assessment of Near-Surface Residual Stress in Shot-Peened Nickel-Base Superalloys," *Journal of Nondestructive Evaluation*, Vol. 23, 2004, pp. 107–123.

[40] Yu, F., and Nagy, P. B., "Numerical Method for Calculating the Apparent Eddy Current Conductivity Loss on Randomly Rough Surfaces," *Journal of Applied Physics*, Vol. 95, 2004, pp. 8340–8351.

[41] Yu, F., and Nagy, P. B., "Simple Analytical Approximations for Eddy Current Profiling of the near-Surface Residual Stress in Shot-Peened Metals," *Journal of Applied Physics*, Vol. 96, 2004, pp. 1257–1266.

[42] Blodgett, M. P., and Nagy, P. B., "Eddy Current Assessment of near-Surface Residual Stress in Shot-Peened Nickel-Base Superalloys," *Journal of Nondestructive Evaluation*, Vol. 23, 2004, pp. 107–123.

[43] Nakagawa, N., Shen, Y., and Frishman, A. M., "A Study of Correlation Between Conductivity Measurement and Surface Residual Stress," *Review of Progress in Quantitative Nondestructive Evaluation*, Vol. 24, edited by D. O. Thompson and D. E. Chimenti, American Inst. of Physics, Melville, NY, 2005, pp. 1363–1370.

[44] Lee, C., Shen, Y., and Nakagawa, N., "A High-Frequency Eddy Current Inspection System and Its Application to the Residual Stress Characterization," Presentation, Review of Progress in Quantitative Nondestructive Evaluation, Brunswick, Aug. 2005.

[45] Nakagawa, N., Shen, Y., and Frishman, A. M., "A Study of the Relation Between Surface Residual Stress and Conductivity Profiles," Presentation, Review of Progress in Quantitative Nondestructive Evaluation, Brunswick, Aug. 2005.

[46] Gray, J., "Residual Stress Measurements Using High Energy X-Rays," *Proceedings of the Residual Stress Workshop*, April 2005.

[47] Final Rept. Univ. Turbine Systems Research Program (USDOE); http://www.osti.gov/bridge/servlets/purl/836410-HYaNFS/native/836410.PDF

[48] Wen, M., Jordan, E. H., and Gell, M., "Remaining Life Prediction of Thermal Barrier Coatings Based on Photoluminescence Piezospectroscopy Measurements," *Transactions of ASME*, Vol. 128, July 2006, p. 610.

[49] Newaz, G., and Chen, X., "Progressive Damage Assessment in Thermal Barrier Coatings Using Thermal Wave Imaging Technique," *Science and Coatings Technology*, Vol. 190, No. 3, Jan. 2005, p. 7.

[50] Ziberstein, V., Shay, I., Goldfine, N., Malow, T., and Reiche, R., "Validation of Multi-Frequency Eddy Current MWM Sensors and MWM-Arrays for Coating Production Quality and Refurbishment," ASME/EGTI Turbo Expo, American Society of Mechanical Engineers, June 2003.

[51] "Ripe for Recovery," *Aviation Week & Space Technology*, 19 April 2004.

[52] Annis, C. G., Jr., VanWanderham, M. D., Harris, J. A., Jr., and Sims, D. L., "Gas Turbine Retirement for Cause: A Nondestructive Evaluation (NDE) and Fracture Mechanics Based Maintenance Concept," *Nondestructive Evaluation: Microstructural Characterization and Reliability Strategies*, edited by O. Buck and S. M. Wolf, Metallurgical Society of AI/ME, Warrendale, PA, 1981, pp. 53–63.

[53] Christodoulou, L., and Larsen, J. M., "Using Materials Progress to Maximize the Potential of Complex Mechanical Systems," *Materials Damage Prognosis*, edited by J. M. Larsen, L. Christodoulou, J. R. Calcaterra, M. L. Dent, M. M. Derriso, J. W. Jones, and S. M. Russ, TMS, 2005.

[54] Christodoulou, L., and Larsen, J. M., "Materials Damage Prognosis: A Revolution in Asset Management," *Materials Damage Prognosis*, edited by J. M. Larsen, L. Christodoulou, J. R. Calcaterra, M. L. Dent, M. M. Derriso, J. W. Jones, and S. M. Russ, TMS, 2005.

CHAPTER 14

Erosion, Deposition, and Their Effect on Performance

Awatef A. Hamed* **and Widen Tabakoff**†

University of Cincinnati, Cincinnati, Ohio

Richard Wenglarz‡

South Carolina Institute for Energy Studies, Clemson, South Carolina

Suspended solid particles are often encountered in the turbomachinery operating environment because of several mechanisms that contribute to particle ingestion in gas turbine engines. Solid and molten particles can be produced during the combustion process from burning heavy oils or synthetic fuels, and aircraft engines can encounter particles transported by sandstorms to several thousand feet altitude [1]. Thrust reverser efflux at low airplane speeds, as well as engine inlet to ground vortex during high power setting with the aircraft standing or moving on the runway, can blow sand, dust, ice, and other particles into the engine. Helicopter engines are especially susceptible to large amounts of dust and sand ingestion during hover, takeoff, and landing. Dust erosion proved so severe during the Vietnam field operations that some engines had to be removed from service after fewer than 100 h of operation [2]. Even after two decades of technological advances, the loss in power and surge margins as a result of compressor blade erosion caused some helicopter units to be removed after fewer than 20 h during the Gulf War field operations [3].

Particulate clouds from the eruption of volcanoes present one of the most dangerous environments for aircraft engines. Several incidents have been related to engine operation in volcanic ash cloud environments. Two incidents are a British Airways Boeing 747 powered by four Rolls Royce RB11 engines flying over the Mt. Galunggung volcano on 23 June 1982, and a Singapore Airlines Boeing 747–400 aircraft powered by Pratt and Whitney engines approaching Anchorage, Alaska, on 15 December 1989 that entered the volcanic ash clouds from the Mt. Redoubt volcano. Tests performed at the University of Cincinnati showed that volcanic ash is four times more erosive than quartz sand [4]. Commercial aircraft engines encountering volcanic ash clouds indicated more severe

*Professor and Director, Center for Intelligent Propulsion and Advanced Life Management of Systems; hameda@ucmail.uc.edu.

†Emeritus Professor.

‡Consultant, Seneca, South Carolina.

problems than those identified in routine Arizona road dust tests, according to military specification MIL E 5007D. Compressor blade and rotor path erosion, deposition of material on hot-section components, and blockage of cooling passages are some of the phenomena experienced in volcanic ash cloud encounters [5–7]. Dunn et al. [8] observed that it is possible to consume the surge margin very quickly when the engine operates in a dust cloud. They attributed this phenomenon to the dust and volcanic ash deposition on the high-pressure turbine vanes (Fig. 1) and the associated rapid increase in burner and compressor discharge static pressure.

In land-based engines, experience with the early coal-burning gas turbine projects of the Locomotive Development Committee [9] between 1944 and 1963 and subsequently with the Australian coal-burning gas turbine project [10] produced a great deal of information on deposition and erosion. The type of coal used and the engine flowpath affected both deposition and erosion, but rotor tip speed had the greatest impact on erosion. Dussourd [11] proposed a simple one-dimensional model for predicting material loss of turbine blades caused by erosion based on adaptation of bluff-body relations for particle impact velocity and capture. Using existing databases to calibrate the one constant and the velocity exponent in the derived equation, Dussourd projected 1 x blade life improvement with a 50–60% reduction in flow velocity. Debond or spallation of thermal barrier coatings is an additional degradation mechanism in air- and ground-based turbines. In a recent surface characterization study of nearly 100 land-based turbine components, Bons et al. [12] found surface roughness levels four to eight times greater than the level for production line hardware and observed that film-cooling sites are particularly prone to surface degradation. Even small particles of 1 to 30 μm in size have been known to cause severe damage to the exposed components of gas turbines [13]. The associated degradation of blades and flowpath through erosion and deposition degrades performance thermal protection and

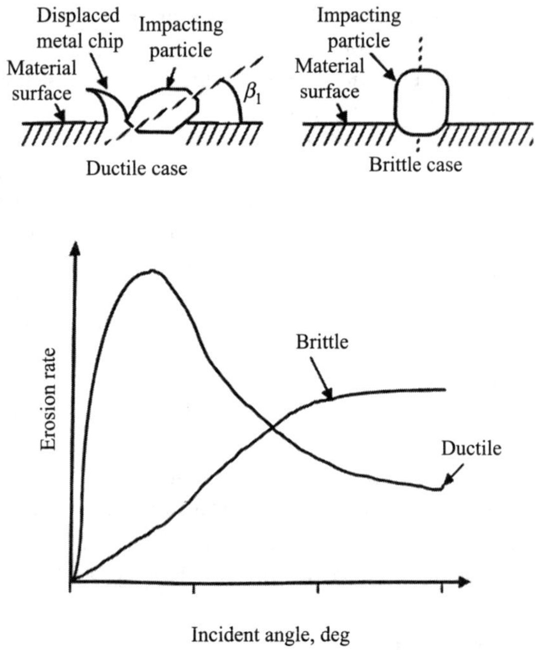

Fig. 1 Erosion rate variation with impact angle.

cooling effectiveness, thus reducing stability and life, and can even lead to a complete loss of power.

This chapter presents a review of the experimental and numerical investigations of the various phenomena associated with particle ingestion, erosion, and deposition in turbomachines and their effects on performance. It includes a summary of results, descriptions of test facilities, and discussion of the methodologies developed for erosion and deposition predictions.

I. EROSION

Turbomachinery erosion is affected by many factors, including ingested particle characteristics, gas flowpath, blade geometry, operating conditions, and blade material. Both experimental and numerical studies have been conducted to determine the pattern and intensity of compressor and turbine blade erosion. Grant and Tabakoff [14] and Balan and Tabakoff [15] conducted experimental studies of single-stage axial-flow compressor erosion. Examination of disassembled rotor blades after sand ingestion revealed blunted leading edges, sharpened trailing edges, reduced blade chords, and increased pressure surface roughness. Sugano et al. [16] reported similar observations regarding the changes produced by erosion in axial induced draft fans of coal-fired boilers. They also determined that blade chord reduction and material removal from the pressure surface increased with particle size.

Richardson et al. [17] presented the results of a JT9D high-pressure compressor diagnostic study in which they documented the changes in airfoil roughness, blade airfoil, and tip clearance with service. The study indicated that in general the changes correlated well with engine cycles and not with hours of engine service. Rotor blade erosion was observed mainly in the outer 50% of the span, where significant reductions in the blade chord and thickness and changes in the leading- and trailing-edge geometries were observed. Surface roughness measurements indicated quick buildup with no trends observed beyond 2000 cycles. Tip clearances increased as a result of both blade shortening and rubstrip erosion. Dunn et al. [18] measured tip clearance that exceeded specifications by a factor of three with operation in dust-laden environments and reported surge occurrences when the engine was run in this deteriorated state.

A. COATING AND BLADE MATERIAL EROSION STUDIES AND FACILITIES

Theoretical studies of material loss by solid particle erosion are predominantly empirical, involving basic assumptions as to the process governing material removal. Different combinations of cutting, fatigue, brittle fracture, and melting mechanisms have been proposed and supported by experimental data from erosion tests. Experimental studies of particle surface impacts are necessary to provide blade material erosion and particle rebound characteristics over the range of impinging conditions encountered in turbomachines. New blade

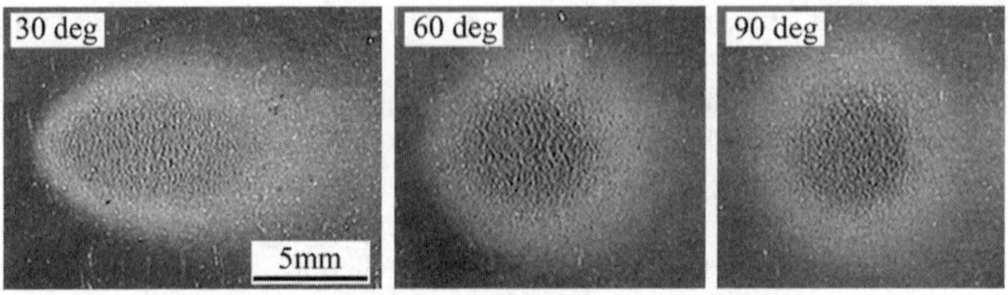

Fig. 2 Samples tested in jet-blasted facility [22].

coating materials are often tested for erosion at specified temperatures and particle impact velocities and impingement angles. These tests are carried out in facilities that control particle-laden flow around the target to achieve the desired impact conditions over the tested coupons.

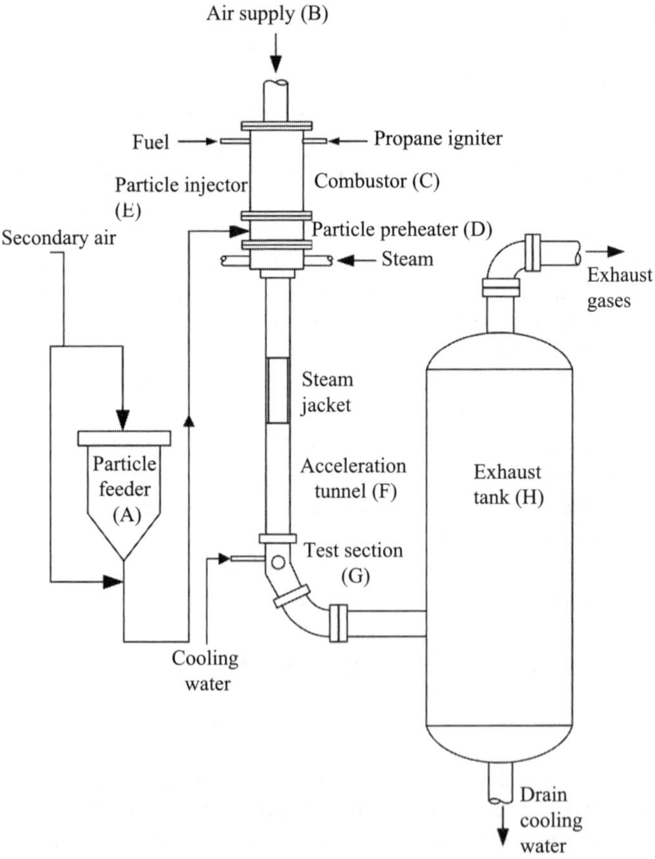

Fig. 3 High-temperature erosion tunnel.

A testing method utilizing a small jet of particle-laden air impacting a stationary specimen was used by Finnie [19] and later by other investigators [20–23] for measuring the erosion characteristics of materials. Photos of samples tested using this type of blast facility at different inclination angles are shown in Fig. 2 [23]. Large variations in the depth and roughness of the tested surface can be seen in the figure. Dosanjh and Humphry [24] performed a computational study of a particle-laden jet impinging normally on a flat wall. The results indicated significant radial variations in particle concentration, impact velocities, and impingement angle at the target surface. The computed variations were strongly dependent on particle size and on the temperature and level of turbulence of the jet. Erosion wind tunnels control the particles' distribution and velocities in the test section and provide uniform particle impact conditions over the tested surface [25]. In addition, the hot erosion tunnel developed by Tabakoff and Wakeman [26] and shown schematically in Fig. 3, provides uniform high test-section temperatures for testing turbine blade materials and coating. Erosion tunnels also enable testing of actual vanes [27–29].

The results of erosion studies often express the ratio of surface mass or volume removal to impinging particle mass. In general, the erosion rate of a given material is affected by the particles' impact velocity and impingement angle. The variation of erosion rate with impingement angle is characteristically different for ductile and brittle material, as shown schematically in Fig. 1. This is attributed to the predominantly different mechanisms of cutting and brittle fracture. Typical test results of blade material [30] and thermal barrier coating [31] erosion rates are shown in Figs. 4 and 5.

Erosion rate is also affected by particle composition [32] and shape. Figure 6 shows magnified scanning electron micrographs of fly ash, silica sand, and aluminum oxide particles. The latter are most erosive because of their angular shapes and very sharp corners. Erosion test results obtained by Grant and Tabakoff [14] using aluminum oxide

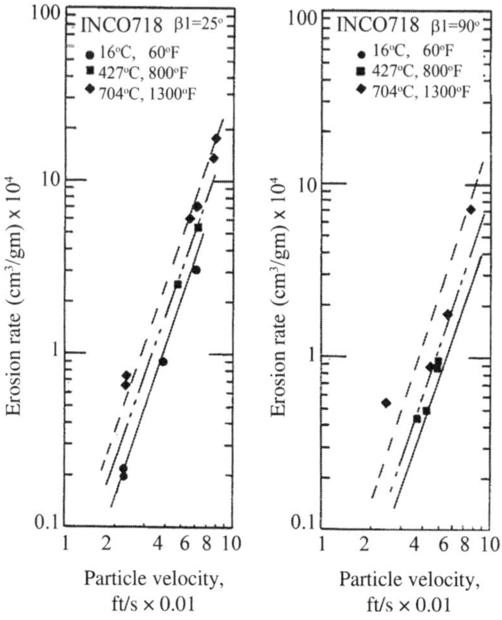

Fig. 4 Erosion test results showing effects of temperature and impact velocity [30].

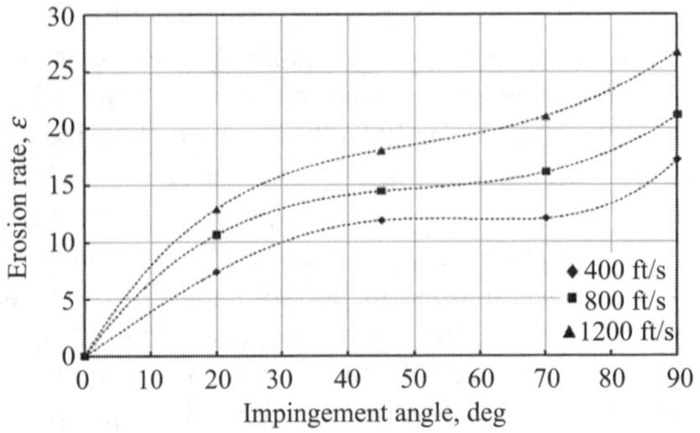

Fig. 5 Measure erosion rate of TBC at 2000°F [31].

particles and by Kotwal and Tabakoff [32] using alumina and silica particles of different sizes indicate that larger particles produced higher erosion rates, but that the effect of particle size on erosion rate diminished as impact velocity decreased. Table 1 and 2 give lists of some erosion tests with the blade/coating materials and particles used in the tests, the test conditions, and the reference where the results were reported.

Surface roughness characteristics were measured after erosion tests in some investigations [27, 29] and were found to correlate closely with the erosion rate in terms of variation with the impact angle, velocity, and particle size. The eroded surface roughness did not change beyond a certain limit even with additional mass removal by erosion [29]. Richardson et al. [17] also reported that compression system airfoil surface roughness did not change beyond 2000 cycles. Figure 7 [29] clearly shows the difference in roughness between the exposed and protected vane surfaces after testing in the erosion tunnel.

In general, particles encounter repeated impacts with the turbine and compressor surface, and their trajectories are affected by the rebound conditions after each impact. Experimental studies have been conducted to measure the magnitude and direction of particle rebound velocity. Finnie [33] developed a system to measure particle velocities by tracking double-exposed pictures using a stroboscopic light source. Hussein and Tabakoff [34] used high-speed photography to investigate the rebound characteristics of particles from flat targets and to track actual particle trajectories in turbine cascades. Subsequently, Grant and Tabakoff [14], Tabakoff et al. [25], and Wakeman and Tabakoff [30]

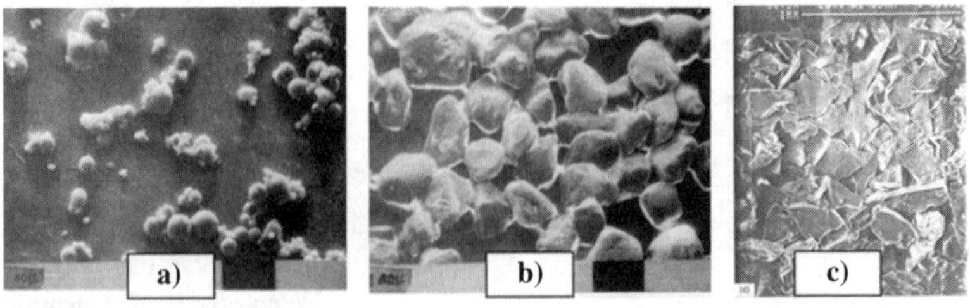

Fig. 6 Electron micrographs of a) fly ash, b) silica sand, and c) aluminum-oxide particles.

TABLE 1 WIND-TUNNEL TESTS OF BLADE MATERIAL EROSION

Substrate Material	Impact α, deg	Temperature	Velocity, m/s	Particles	Reference
MAR 246	0–90	815°C	366	Fly ash	5
X-40	0–90	538°C	305	Fly ash	33
Inco 738	0–90	538°C	305	Fly ash	
Cobalt	0–90	538°C	145–259	Fly ash	7
Rene 41	0–90	649°C	182–305	Fly ash	34
AM 355	0–90	316°C	122–305	Fly ash	
Al 2024	0–90	Ambient	65–137	Fly ash	8
St St 304	0–90	Ambient	65–137	Fly ash	35
St St 304	0–90	Ambient	128	Quartz	
St St 304	0–90	Ambient	122	Alumina	
Ti 6A1–4 V	0–90	Ambient	65–137	Ash	
Al 2024	0–90	Ambient	92–152	Quartz	9, 36
Ti-6–4	0–90	16–704°C	152	Al oxide	13
Inco 718	0–90	16–704°C	152	Al oxide	38
Steel 304	0–90	30–650°C	18–305	Al oxide	
Inco 600	0–90	371–493°C	100–300	Runway sand	14, 39
	0–90	538°C	305	Chromite	40
Inco 738	0–90	482°C	183–305		16
FSX-414	0–90	482°C	183–305	Fly ash	41
X-40	0–90	482°C	183–305		
St Steel 304	0–90	30–650°C	200–330		17
Rene 41	0–90	30–650°C	200–330	Fly ash	42
A286	0–90	30–650°C	200–330		
St St 304	0–90	316, 650°C	120–300	Fly ash	19, 44
Inco 718	0–90	16–704°C	65–244	Quartz	21
Ti-6–4	0–90	16–704°C	65–244	Quartz	46
St St 355	0–90	Ambient-538°C	99–152	Silica	23
			122–305	Fly ash	47

(*Continued*)

TABLE 1 WIND-TUNNEL TESTS OF BLADE MATERIAL EROSION (*CONTINUED*)

Substrate Material	Impact α, deg	Temperature	Velocity, m/s	Particles	Reference
St St 304	0–90	315–650°C	183–305		24
Rene 41	0–90	315–650°C	183–305	Ash	48
Ti-6-4	0–90	16–704°C	152		
Inco 600	0–90	370–577°C	120–240	Quartz	25, 49

used high-speed cameras for particle restitution measurements in the erosion wind tunnel, which was equipped with optical access through the test section.

Because photographic methods were limited to particle sizes greater than 30 μ, Tabakoff and Sugiyama [35] developed a method to use laser Doppler velocimetry (LDV) to measure fly ash restitution characteristics. LDV was subsequently used in other investigations [36–40] to measure the restitution characteristics for different particle material combinations. Figure 8 [37] shows typical LDV results for the velocity and directional restitution ratios. The restitution ratios exhibit variance around a mean value, which depends on the impact angle. The variance is probably associated with the orientation of nonrounded particles at the time of impact and with the erosion-produced target surface irregularities. Particle rebound characteristics were found to be unaffected by the gas or target temperature [26].

B. NUMERICAL SIMULATIONS OF PARTICLE TRAJECTORIES IN TURBOMACHINES

Trajectory simulations are based on the numerical integration of the particles' equations of motion through the turbomachinery blade passages. Because of their higher inertia, the particles lag the gas in turning and acceleration or deceleration, and this causes them to impact the surfaces constituting the various blade passage boundaries. Trajectory simulations therefore require the turbomachines' flowfield and flowpath definition through the blade passage as input and a model for particle restitution conditions following each surface impact. Hussein and Tabakoff [41] pioneered the methodology for particle trajectory simulations through axial turbine and compressor stages and the use of experiment-based particle restitution models. They presented sample particle trajectories based on the velocity field at the mean diameter and the mean value of the experimental restitution ratios. Their simulations indicated that the frequency of blade pressure surface impacts increase with increased particle size and with their initial velocity at the stage inlet. They also demonstrated that particles gain large circumferential velocities from rotor blade impacts, which

TABLE 2 WIND-TUNNEL TESTS OF BLADE COATING EROSION

Substrate Material	Coating	Impact α, deg	Temperature	Velocity, m/s	Particles	Reference
Cemented	Al_2O_3	0–90	260°C	140	Aluminum	2
Tungsten	T_iC	0–90	260°C	140	oxide	32
Carbide	T_iN	0–90	260°C	140		
MAR 246	T_iC	0–90	815°C	366	Fly ash	10
MAR 246	RT22	0–90	815°C	366	Fly ash	37
MAR 246	T_iC	0–90	815°C	305	Chromite	
Waspaloy	T_iC	0–90	538°C	305	Chromite	40
St St 410	T_iC	0–90	540°C	305	Chromite	18
INCO 718	Iron nitride	0–90	540°C	305	Chromite	43
Ti-6A1–4 V	Various multilayer coatings	0–90	Ambient	185	Aluminum oxide	20, 45, 49
St St 410	SDG-2207	0–90	565°C	305	Chromite	26
Inco 718	T_iC	0–90	538°C	305	Chromite	51
Waspaloy	T_iC	0–90	538°C	305	Chromite	
WC	T_iC	0–90	Amb-650°C	140	Chromite	
WC	Al_2O_3	0–90	Amb-650°C	140	Chromite	
WC	T_iN	0–90	Amb-650°C	140	Chromite	
WC	Uncoated	0–90	Amb-650°C	140	Chromite	
Inco 718	7YSZ EP PVD	0–90	1600–1800°F	400–1200	Aluminum oxide	31

Fig. 7 Vane surface roughness due to erosion [29].

causes them to centrifuge toward the outer casing. In the case of axial flow turbines, many particles bounce back and forth between the blunt leading edges of the rotor blade and the nozzle vane trailing edge before finally going through the rotor. The computed trajectories through the turbine nozzle were consistent with their earlier experimental visualization in a turbine cascade tunnel using high-speed photography [42].

Particle trajectory simulations have progressed to include viscous and three-dimensional flow effects following the general advances in the turbomachinery flowfield solutions. The basis for trajectory simulations in turbomachines continues to be Eulerian–Lagrangian, with one-way coupling between particles and flow. Flowfield representations in axial flow machine trajectory simulations progressed from mean streamline combined with spanwise and cross-passage velocities from secondary-flow theory for turbines [43] to inviscid flow on a number of the blade-to-blade stream surfaces for multistage compressors [44]. This was combined with secondary flow and experimentally based streamwise and crossflow velocity gradients near the end walls [45] for multistage turbines. On the other hand, early trajectory simulations in radial flow machines [46, 47] were based on flow solutions on the meridinal planes and panel methods. Currently, three-dimensional flowfield solutions of the Reynolds-averaged Navier–Stokes equations for turbulent flow through blade passages are frequently used in turbomachinery trajectory simulations [29, 48].

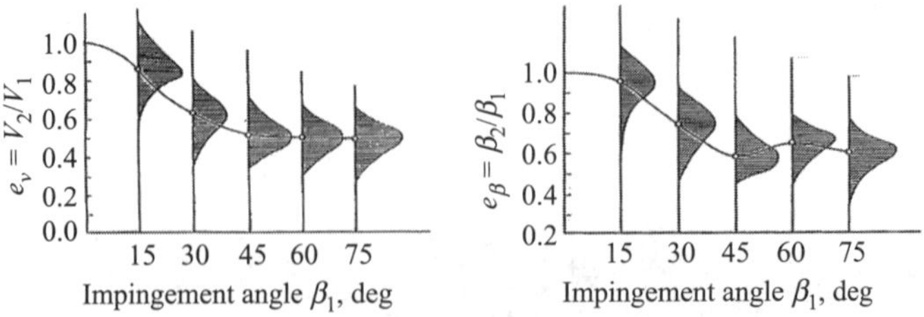

Fig. 8 Typical LDV results for velocity and restitution ratios [37].

In general, particle trajectories are determined from the numerical integration of the equations of motion in the blade rotating reference frame:

$$\frac{\mathrm{d}^2 r_p}{\mathrm{d}\tau^2} = F_r + r_p\left(\frac{\mathrm{d}\theta_p}{\mathrm{d}\tau} + \omega\right)^2 \tag{1}$$

$$r_p\frac{\mathrm{d}^2 \theta_p}{\mathrm{d}\tau^2} = F_\theta - 2\frac{\mathrm{d}r_p}{\mathrm{d}\tau}\left(\frac{\mathrm{d}\theta_p}{\mathrm{d}\tau} + \omega\right) \tag{2}$$

$$\frac{\mathrm{d}^2 Z_p}{\mathrm{d}\tau^2} = F_z \tag{3}$$

In the preceding equations, r_p, θ_p, and z_p define the particle location in cylindrical coordinates and ω the blade angular velocity. The last terms in the right-hand side (RHS) of the first two equations represent the centrifugal force and Coriolis acceleration. The first term in the RHS of Eqs. (1–3) represents the components of the aerodynamic force of interaction between the two phases. The drag caused by the relative velocity is considered as the primary aerodynamic force on the suspended particles because the forces caused by gravity and buoyancy are negligible compared to the aerodynamic and centrifugal forces. Forces caused by interparticle interactions and pressure gradient are also negligible for the small particles and low concentrations encountered in turbomachines. The aerodynamic force of interaction is expressed in terms of the drag coefficient and the particle slip velocity as follows:

$$\overline{F} = \frac{3}{8}\frac{C_D}{r_p}|\overline{V} - \overline{V}_p| \cdot (\overline{V} - \overline{V}_P) \tag{4}$$

where $\overline{V}$ and $\overline{V}_P$ are the gas and the droplet velocity vectors, respectively. The drag coefficient C_D is computed from empirical correlations involving the Reynolds number based on the relative velocities of the particle and the gas.

According to Eq. (4), the aerodynamic force increases with the square of particle slip velocity. On the other hand, the lag in particle velocity at a given blade row exit will cause incidence angle to deviate from that of the gas in the following blade row frame of reference. The loss of momentum as a result of stator pressure surface impacts will produce positive incidence in the following compressor rotor but negative incidence in the following turbine rotor. Rotor pressure surface impacts, on the other hand, increase the absolute momentum of the particles.

Sample trajectories through a turbine rotor are shown in Fig. 9 to demonstrate the strong influence of particle size on the location of particle impacts with the rotor blade surface [29]. In general, the particles lag behind the gas when they leave the proceeding stator blade because of the loss in their momentum caused by pressure surface impacts. The smaller particles lag less because they are accelerated after they rebound by the surrounding high-speed flow. Viewed in the rotor relative frame of reference, the lag in absolute velocity produces

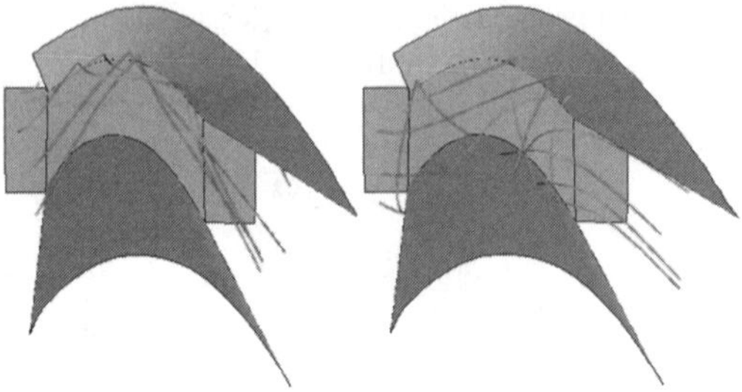

Fig. 9 Effect and size on particle trajectories in turbine rotor: a)10-μ particles; b) 50-μ particles.

negative incidence [31]; hence, the larger particles tend to impact the rotor suction surface as seen in Fig. 9.

C. TURBOMACHINERY BLADE EROSION

Hamed and Tabakoff [49] and Hamed et al. [29] developed a methodology to predict turbomachinery blade surface erosion patterns using the computed blade surface statistical impact data from particle trajectory simulations in combination with correlations of erosion test results for blade and coating materials. Tabakoff [44] presented computational results for blade erosion through a T700 five-stage axial compressor. The computed blade leading edge and pressure-surface erosion along the first rotor was followed by rotor tip and stator root erosion in subsequent stages as a result of particle radial migration following initial rotor impact. This is consistent with Mann and Warnes [1] observations of multistage compressor blade erosion pattern and with Richardson et al.'s [17] documentation of the changes in compressor blade airfoils and surface roughness with service. Diagnostic measurements of the particles' size variation through helicopter engine [1] compression systems using isokinetic sampling indicated that it became nearly independent of the original size after the low-pressure (LP) compression system.

A picture of a T-53 G compressor after erosion tests conducted with runway sand at the University of Cincinnati is shown in Fig. 10. The light colors indicate the eroded rotor and stator blade surfaces. The leading edge and pressure surface erosion is visible along the first rotor and toward the rotor blade tips and stator blade roots in subsequent stages.

Hamed and Tabakoff [49] and Elfeki and Tabakoff [46] presented computational results for particle trajectories and erosion in a supercharger centrifugal compressor impeller with one and two splitter blades; the results indicated that particle size strongly influences both blade erosion pattern and intensity. Blade pressure surfaces' erosion was predicted near the casing and increased towards the tip especially for larger particles. The predictions, which were verified in lab tests of centrifugal compressor erosion, are consistent with the mechanical damage pattern of a number of impeller blades following helicopter engine air cleaner (AC) coarse dust ingestion reported by Mann and Warnes [1].

In modern gas turbines, the loss of thermal barrier coating protection through erosion can be detrimental to blade life. Experimental evaluation of thermal barrier coating erosion resistance requires experimental erosion rate

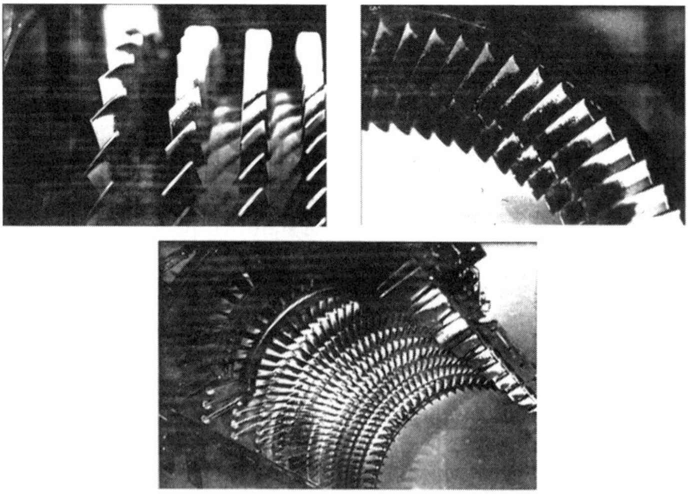

Fig. 10 Multistage compressor erosion.

measurements test over a range of elevated temperatures and high velocities [31]. Figure 11 demonstrates the predicted erosion pattern of TBC based on the simulated trajectories of Fig. 9 for different size particles [29].

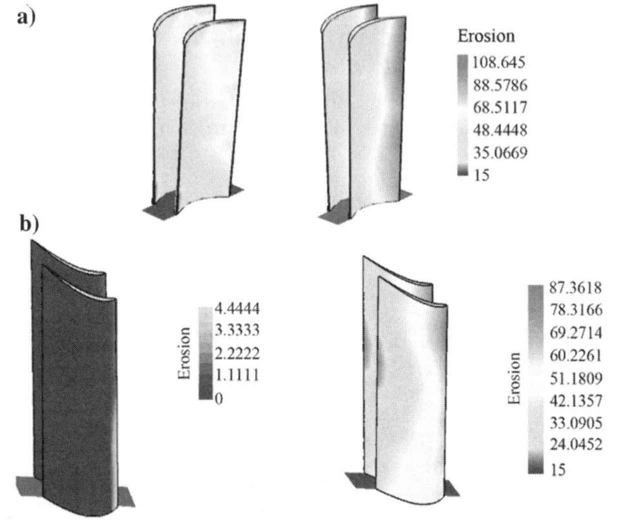

Fig. 11 Predicted erosion rate on rotor pressure and suction surface Mg/g/m^2: a) 10-μ particles; b) 50-μ particles.

The erosion rate is expressed in terms of mass removal per unit surface area per total mass ingestion into the turbine. According to Fig. 11, the smaller 10-μ particles cause erosion over the second half of rotor blade pressure surface and no suction surface erosion. On the other hand, the larger 50-μ particles cause TBC erosion over the first half of the rotor suction surface and less erosion over the pressure surface. Some of the large particles reenter the stator blade passages after they rebound from rotor leading-edge suction surface impacts. These particles rebound after impacting the stator blade suction side and might bounce back and forth, thus weakening the blades' thermal protection in these critical areas before they finally leave the turbine stage [31]. The three-dimensional computational results in a single-stage commercial and NASA-designed compressor drive turbine indicate that the erosion patterns were similar in the two turbines with high TBC erosion rates at the casing towards the trailing edge of the stator blade suction side and at the shroud towards the leading edge of the rotor blade suction side.

Figure 12 demonstrates that flight cycles and not flight hours determine compressor blade erosion. By contrast, Fig. 13 shows how noncyclic volcanic ingestion can deposit and cause immediate damage in turbine vanes [50].

Beacher and Tabakoff [45] performed particle trajectory analyses through a multistage, coal-fired gas turbine. Although no erosion predictions were available, the high concentration of particles near the casing, past the first rotor, correlated well with the observed leading edge and pressure surface wear pattern reported by Smith et al. [9]. Metwally et al. [51] conducted a computational study to investigate the effect of blade coating on the erosion of an automotive gas turbine. Their blade surface erosion predictions indicated substantial reduction associated with rhodium platinum aluminide (CRT22B) coating compared to the base MAR-M246 alloy blade.

Tabakoff and Hamed's study [52] of radial inflow turbines indicated that the highest erosion rate was at the rotor pressure surface near the outer corner of the exit. Experimental and analytical studies of the performance of aircraft auxiliary power turbines with silicon dioxide particle ingestion [53] indicated a unique phenomenon in which particles became trapped in the vortex region below the nozzles and rotor. This phenomenon, which is caused by the

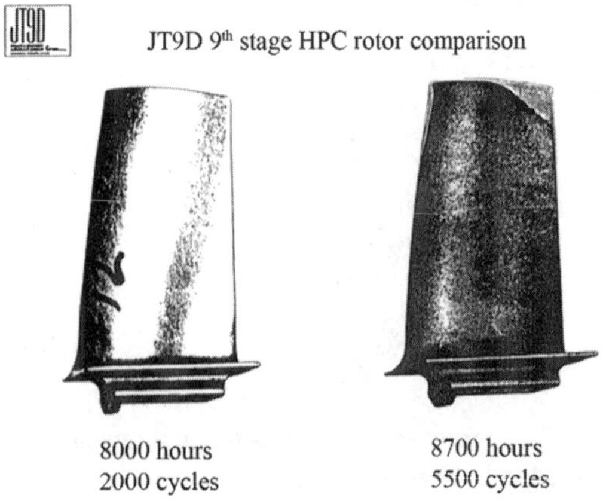

Fig. 12 Effect of flight cycles on compressor blade erosion.

Fig. 13 Volcanic ash deposition on turbine vanes.

balance between the radial components of the aerodynamic drag and the centrifugal forces acting on the particles, was recorded on film and showed accumulation and eventual blocking of the flow passage. A large reduction in the wheel speed was measured within a few seconds and continued even after discontinuation of particle ingestion.

The motion of suspended particles through turbomachines is essentially a stochastic process because individual particle's point of entry into the machine, its initial velocity vector, and its size and shape are all subject to statistical variation. A number of studies have been conducted to investigate the influence of modeling various aspects of these variances on the computed particle trajectories and blade erosion predictions. Tabakoff et al. [54] compared the computed particle trajectories through a two-stage turbine and the associated blade erosion for Cincinnati Gas & Electric Company's fly ash particle size distribution to those based on the mean particle diameter. The results indicated that nonuniform particle impacts were spread over more of the blade surface, resulting in lower peak erosion values.

Different methodologies have been considered for modelling the effects of the experimentally observed variance in particle rebound characteristics on the erosion of a two-stage turbine. Initially, Hamed [55] used the fast probability integration method (FPIM) to model the influence of the measured variance in particle rebounds on the particle trajectories through an axial flow turbine and the associated blade erosion. The FPI-based model resulted in lower estimates of both peak and mean blade surface erosion as compared to those computed based on the mean value of the experimentally measured restitution ratios. Subsequently, Hamed and Kuhn [56] developed stochastic particle trajectory simulations based on direct sampling of the actual experimentally measured variance in particle rebound characteristics [30]. These results confirmed that the deterministic bounce model overestimates the blade pressure surface erosion. Another important difference was in the stator blade suction surface erosion near the trailing edge, which was predicted with the direct sampling of the experimental rebound statistics, but not with FPI. This was found to be associated with particles reentering the nozzle passage after rebounding from the following rotor.

D. EFFECTS OF TURBOMACHINERY EROSION ON ENGINE PERFORMANCE AND LIFE

In addition to safety considerations, the damage resulting from turbomachinery erosion has a series of consequences from both engineering and economic standpoints. According to Kleinert [57], erosion is the primary cause of fuel consumption increase in modern turbofan engines. Measurements of isolated compressor and cascade performance following erosion cycles [14] have indicated reduction in compressor adiabatic efficiency and stage loading and an increase in cascade total pressure losses. Blade loading reduction, which was noticeable in the outer 50% of the rotor span, increased with erosion cycles as more erosive silica particles passed through the compressor. Sugano et al. [16] presented the measured change in axial induced draft fan performance with the laps of running time in a coal ash environment. The most significant effect was the drop in stall point by 5% when the eroded blade chord reduction reached 10%. Similar lowering of the surge limit in eroded fans was reported by Ghenaiet et al. [48], who also characterized the increased tip clearance and measured the drop in efficiency caused by sand erosion of a single-stage ventilation fan with C4 rotor blades made from cast aluminum.

Tabakoff and Simpson [58] recently conducted an exhaustive experimental study of the erosion characteristics of various compressors and turbine blade materials and coatings. In addition to erosion weight loss, they characterized the corresponding change in chord and thickness of compressor cascades with and without coatings. Subsequently, Kline and Simpson [28] conducted a full engine sand ingestion test demonstration of a T64 RB01 "rainbow" compressor with alternate bare and coated blades. They reported 25% loss in horsepower after ingestion of 35 kg f , and they had to stop the engine because of surging. They confirmed the cascade erosion results [58] and determined that virtually 100% of the engine performance loss was attributable to erosion of the bare blades. Edwards and Rouse [59] explained how the gas-generator power and surge margins are affected by the eroded compressor performance both through the drop in surge line caused by erosion and through the rise in operating line caused by the increased turbine inlet temperature required to maintain the power level with the loss in compressor efficiency. They also discussed how turbine efficiency loss caused by erosion reduces gas-generator efficiency and requires operation at increased temperatures, which also causes the operating line to rise above normal and contributes to the reduction in surge margin.

Schmucker and Schaffer [60] conducted an experimental study to determine the effects of reworked blades for the most common defects associated with erosion on axial compressor performance, namely, damaged leading and trailing edges, rounded tips, and rubbed coatings. A high-pressure five-stage research compressor was tested with reworked leading- and trailing-edge blades mixed with new blades, with 1.5-mm tip rounded rotor blades at the leading and trailing edges and with 1–3% equivalent radial tip clearance. The largest losses in surge margin and in efficiency (7.5 and 2% respectively) were associated with a 1%

increase in tip clearance. The rounded tip rotor blades resulted in 4% loss in surge margin and 0.4% loss in efficiency. The losses in performance for reworked blades were 2% in surge margin and less than 0.5% in efficiency and mass flow rate.

Several investigators have developed models to simulate the effects on performance of surface roughness and various aspects of increased tip clearance changes in compressor blade airfoils shapes. Richardson et al. [17] developed a model for the associated high-pressure compressor performance deterioration based on measurements of in-service engine parts. They reported data on each stage tip clearance change as a result of blade and flowpath erosion and on rotor airfoil changes at six radial locations for each stage. They used the data in a performance model to estimate the loss in efficiency and flow capacity associated with changes in tip clearance and in airfoil leading- and trailing-edge angles, chord, and thickness. Their estimates of compressor efficiency loss and engine TSFC rise agreed with fleet prerepair engine performance average above 1500 cycles. The authors also reported that cold-section refurbishment through restoration of tip clearances, cleaning of airfoils, and replacement of those with chord lengths out of a recommended limit were credited with 1.3% restoration in TSFC. Batcho et al. [61] developed a model for compressor stage performance deterioration that incorporated tip clearance and secondary flow loss models and thin airfoil theory lift and drag changes associated with airfoil mean camberline. They used the model to examine the response of an eroded compressor and estimated 51% reduction in surge margins and 45% reduction in surge pressure ratio with compressor erosion. Tabakoff et al. [62] and Hamed et al. [63] developed a stage-stacking model for the loss in performance as a result of compressor erosion; it was validated using the single stage data of Balan and Tabakoff [15]. Subsequently, Tabakoff et al. [64] used the same analysis combined with a thermodynamic model to study the restoration of performance through water injection.

Nagy et al. [65] developed an erosion-resistant-coating life model and applied it to coated compressor blade erosion by quartz particles. The life model, which is based on 1.8% reduction in the chord length, was used to calculate the mass of erodent for coated airfoil life for various coating thicknesses. Naik et al. [66] presented the results of a detailed investigation on the erosion resistance and durability of polymer matrix composite coating on Rolls-Royce AE 3007 bypass vanes. The rainbow (coated/uncoated) vane erosion tests in the erosion tunnel demonstrated two to eight times improvement relative to the bare metal under conditions simulating 5000 flight hours. In addition, both structural laboratory vibratory tests and engine durability tests demonstrated the capabilities of the coatings for propulsion applications.

II. DEPOSITION

There are two types of mechanisms involved in turbine deposition and effects on performance: delivery of impurities to turbine surfaces and attachment and (buildup) of impurities delivered to surfaces.

A. PARTICLE DELIVERY

Impurities from inlet air or fuel can enter the turbine flow passages as particles (in solid or liquid form) and, often for the hot section, as gaseous species that had been vaporized in upstream combustion or gasification processes. Vaporized impurities that enter the hot section can condense as liquids on cooled turbine surfaces or in the gas stream as the temperature and pressure drop through the turbine stages. Dominant mechanisms of delivery of particles to turbine flowpath surfaces are inertial impaction, turbulent diffusion/eddy impaction, Brownian diffusion, and thermophoresis.

In inertial impaction, the particles have sufficient mass to deviate from turning gas flow streamlines, penetrate airfoil boundary layers, and essentially crash onto airfoil surfaces. Smaller particles can be entrained in turbulent eddies in the surface boundary layers to be swept toward airfoils and end walls (turbulent diffusion). Even though eddies dissipate near surfaces, the particles have sufficient inertia to coast to the surfaces (eddy impaction). Yet smaller particles with insufficient mass to be delivered by inertial effects can be transported to surfaces by impacts with the thermally agitated gas molecules in surface boundary layers. For extremely small particles, the random impacts can produce "random walk" Brownian diffusion delivery to the surfaces. If the surface is cooled (as for airfoils of upstream hot-section stages), the energy of the random impacts on particles from thermally agitated gas molecules in the thermal boundary layer is higher at the hot side of the particle farther from the cooled surface than at the cooler side of the particle. This produces a net average impact force from gas molecules in the direction toward the surface that transports these particles to cooled components (thermophoresis).

Perhaps the earliest work that applied existing theories of particle transport to turbines was a series of analyses to predict deposition on airfoils due to inertial impaction, vapor diffusion, and Brownian diffusion. The inertial impaction relations used by Smith [67] resulted from prior work by Taylor [68], who had studied the impingement of water droplets on aircraft wings. In these and subsequent analyses of inertial impaction deposition in turbines, Newtonian equations of motion for particles subject to drag forces from the fluid were integrated, and their trajectories and impact rates on airfoil surfaces were calculated. McCreath [69] integrated equations of motion for 15-μ particles in Tyne turbine stator vane and rotor blade passages and found reasonable agreement with deposition buildup measured over pressure surfaces in experiments. Dring et al. [70] showed excellent agreement between calculated trajectories and photographs of trajectories over a range of particle diameters (Stokes numbers from $\sim$0.1 to 1.9) for experiments using a symmetric airfoil. At turbine flowpath conditions, integration of particle equations of motion considering only fluid drag forces typically applies to particles larger than a few microns in diameter (Stokes number on the order of 1 or larger), for which particles have sufficient inertia that the other mechanisms just described have a relatively small effect on transport to airfoil nose and pressure (concave) surfaces. Convex (suction) airfoil surfaces are shielded from direct inertial impaction of larger particles, so that the other mechanisms just described cause deposition on those surfaces from particles smaller than a few microns in diameter.

Developments in theory of particle transport and deposition for small particles (Stokes number much less unity) that provided the basis for later applications to turbines include the work of Lin et al. [71], Friedlander and Johnstone [72], Davies [73], and Cleaver and Yates [74]. The developments of Lin et al. were used by Parker and Lee [75] in studies of deposition of submicron particles on turbine blades. Friedlander and Johnstone indicated that, for particles on the order of a micron in diameter, existing Brownian and turbulent theories underpredicted deposition measured on surfaces in experiments, and they proposed that particle transport in that size range near to surfaces is associated with inertial flight to surfaces (eddy impaction) resulting from velocity imparted to particles by turbulent eddies. Davies developed a relation for stopping distance from the surface for which transport is dominated by inertial flight. Moore and Crane [76] incorporated Davies' stopping distance relations in their diffusion analyses of particle transport to turbine blades related to corrosion. They also calculated inertial impaction delivery to turbine airfoils for particles in the diameter range from 1 to 10 μ. Hidy and Heisler [77] published a survey of the state of the art of small particle transport and deposition in the late 1970s.

B. APPLICATION OF PARTICLE DELIVERY MODELS TO TURBINE DEPOSITION

From the late 1970s to mid-1980s, Rosner [78] and associates [79] published extensively on the theories of condensation, turbulent diffusion, and thermophoretic transport of particles in boundary layers. Much of this research addressed delivery of corrosive compounds from turbine flowpaths to bounding surfaces (such as airfoils). Menguturk and Sverdrup [80] incorporated previous particle delivery theory advances for the mechanisms of turbulent and Brownian diffusion into a turbine deposition model and showed that the model predicted deposition rates that agreed reasonably well with experimental deposition data for pipes and a turbine cascade. Wenglarz [81] utilized this model to calculate deposition rates in a 50-MW coal-fired PFBC turbine for alternative particulate cleanup systems. An approach was also developed to estimate turbine power drops as a results of blockage of the stator passage throats (minimum flow area in the expander) and maintenance intervals for deposit removal to restore power. Other later examples of applying particle delivery models to predict turbine deposition are given by Ahluwalia et al. [82] and Frackrell et al. [83]. Ahluwalia et al. combined several mechanistic models for particle and vapor transport to include the simultaneous contributions of Brownian and turbulent diffusion, thermophoresis, eddy impaction, and inertial impaction. Predicted deposition agreed well with deposition measured in pipe flow and reasonably well with measured deposition in a turbine cascade. Particle delivery rates were then calculated on the surfaces of the first-stage stator vane of a large turbine. Frackrell et al. reviewed particle delivery modeling approaches for application to turbines and compared model predictions against experimental deposition data for pipe flow and flow around cylinders, including a probe exposed to deposition in a rig representing combustion products from a coal gasification system. Calculated deposition profiles using an inertial impaction model for particle sizes of about 5 and 15 μ were shown to agree well with deposition measured on the first-stage vanes and blades in a low-speed, two-stage model turbine. Deposition rates over the concave and convex surfaces of the first-stage stator vane of a large utility turbine were then calculated for two ranges of particle sizes.

In the 1990s, natural gas became the fuel of choice for land-based turbines, and most of the attention shifted away from alternate turbine fuels along with concerns about resulting turbine flowpath degradation. By the turn of the century, little research and development was directed to turbine deposition. An exception has been work in Europe by El-Batsh and Haselbacher, described in a number of publications. These authors published evaluations and verification of particle delivery models for applications to turbines in 2000 [84] and calculation of turbine cascade deposition effects in 2002 [85].

C. EFFECT OF PARTICLE SIZE

Figure 14 illustrates the effect of particle diameter on deposition velocities (deposition rates normalized to gas stream particle concentration) as calculated by a particle delivery model for the concave (pressure) surface trailing edge of the first stator vanes in a large utility turbine. Turbulent diffusion dominates at the smaller diameters shown on the plot. For increasing diameters, eddy impaction enhances turbulent diffusion for surface delivery, and then inertial impaction dominates for particles larger than a few microns in diameter. The deposition velocity curve starts to flatten at diameters in the vicinity of 0.1 μ as a result of the effects of Brownian diffusion at smaller sizes. Because Brownian diffusion rates increase with decreasing particle diameter, this results in a minimum in

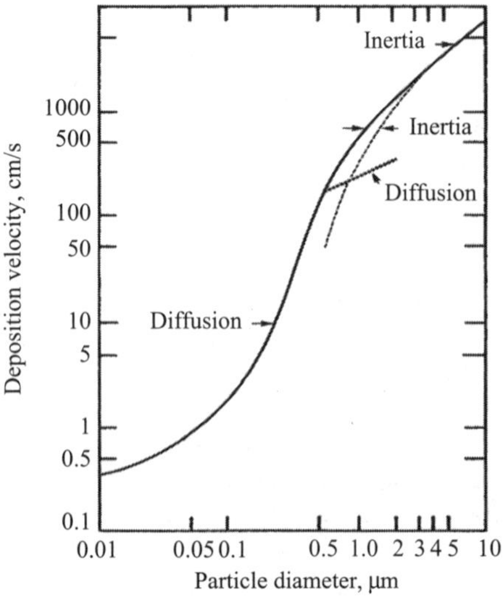

Fig. 14 Turbine vane deposition vs particle diameter.

the deposition velocity curve at a small diameter below 0.01 μ that is not shown on the plot. Although the model used did not include thermophoresis, the depth of the minimum would depend on the degree of airfoil surface cooling and corresponding magnitude of thermophoric effects.

D. BUILDUP OF IMPURITIES DELIVERED TO TURBINE SURFACES

Because of the high mass flow rates for gas turbines, the preceding mechanisms are sufficient to cause significant mass delivery of impurities to turbine surfaces, even for minute concentrations (<1 ppmw) of impurities in the flow stream. For example, a large turbine with mass flow of 1000 lb/s experiences more than 28,000 lb of impurities in an 8000-h operating year for a flowpath particulate concentration of 1 ppmw.

Although significant quantities of impurities can be delivered by the preceding mechanisms to turbine passage airfoils and end walls, whether or not excessive deposition occurs depends or whether or not there is attachment or sticking of the impurities upon arrival at those surfaces. Competing mechanisms of attachment and removal are described by Tabakoff et al. [86] and are not discussed in detail here. One of the principle conclusions, however, was that molten phases need to be delivered to the turbine surfaces to sufficiently attach particles so they are not removed and re-entrained in the flow stream. It was also noted that a small molten mass fraction (a few percent, and sometimes less) of total material delivered to the surfaces can result in excessive rates of deposition and strong deposits. A review of past test results for a number of alternate fuels by Wenglarz and Wright [87] showed a fuel-ash-dependent transition temperature above which deposition (and corrosion) increase drastically, along with the characteristics of these degradations (that is, an increase in gas temperature of 200°F can increase deposition rates by two orders of magnitude [88]). Below the transition temperature, the main contributor to molten phases is vaporized ash species that condense in a small diameter range (~0.01 μ). Above the transition temperature, larger particles in the 1-μ and larger diameter range are molten. As illustrated in Fig. 14, delivery rates to turbine surfaces for the micron diameter range are much higher than rates for the 0.01-μ range, so that much greater levels of molten phases can be delivered to turbine surfaces for gas stream temperatures above the transition temperature. Accordingly, Wenglarz and Wright concluded that the most important factor determining the level of molten phases delivered to the turbine hot-section surfaces, and whether there are extreme rates of deposition, is probably the gas stream temperature relative to the melting point of the flowpath impurities in the larger particle diameter range.

ACKNOWLEDGMENTS

The first two authors wish to acknowledge the support of the U.S. Army, U.S. Air Force, U.S. Department of Energy, NASA, and our numerous national and international partners.

REFERENCES

[1] Mann, D. L., and Warnes, G. D., "Future Direction in Helicopter Engine Protection System Configuration," AGARD-CP-588, Paper No. 4, NEUILLY-SUR-SEINE, France, Nov. 1994.

[2] Mund, M. G., and Guhna, H., "Gas Turbine Dust Air Cleaners," American Society of Mechanical Engineers, Paper 70-GT-104, May 1970.

[3] Sirs, R. C., "The Operation of Gas Turbine Engines in Hot and Sandy Conditions-Royal Air Force Experiences in the Gulf War," AGARD-CP-558, Paper No. 2, NEUILLY-SUR-SEINE, France, Nov. 1994.

[4] Tabakoff, W., "Review of Material Erosion Exposed to Aerodynamic Conditions," U.S. Dept. of Engineering, DOE Report on Program Review, Oak Ridge, TN, April 1986.

[5] Mitchell, H. J., and Gilmore, F. R., "Dust-Cloud Effects on Aircraft Engines: Emerging Issues and New Damage Mechanisms," Defence of Nuclear Agency, RDA-TR-120012-001, Washington, D.C., March 1982.

[6] Chambers, J. C., "The 1982 Encounter of British Airways 747 with the Mt. Galuggung Eruption Cloud," AIAA Paper 85–0097, Jan. 1985.

[7] Smith, W. S., "International Efforts to Avoid Volcanic Ash Clouds," AIAA Paper 85–0101, Jan. 1985.

[8] Dunn, M. G., Baran, A. J., and Miatah, J., "Operation of Gas Turbine Engines in Volcanic Ash Clouds," American Society of Mechanical Engineers, Paper 94-GT-170, Oct. 1994.

[9] Smith, J., Cargill, R. W., Strimbeck, D. C., Nabors, W. M., and McGee, J. P., "Bureau of Mines Coal-Fired Gas Turbine Research Project: Test of New Turbine Blade Design," U.S. Dept. of the Interior, Bureau of Mines, RI, 6920, Washington, D. C., 1967.

[10] Atkin, M. L., and Duke, G. A., "The Operation of a Modified Ruston Hornsby Gas Turbine on N.S.W. Bituminous Coal," Aeronautical Research Lab., Rept. 133, Dept. of Supply, Australian Defense Scientific Service, Canberra, Australia, 1971.

[11] Dussourd, J. L., "A Simple One-Dimensional Model for Primary Turbine Blade Erosion Prediction," American Society of Mechanical Engineers, Paper 83-GT-164, June 1983.

[12] Bons, J. P., Taylor, R. P., McClain, S. T., and Rivir, R. B., "The Many Facets of Turbine Surface Roughness," American Society of Mechanical Engineers, Paper 2001-GT-0163, Oct. 2001.

[13] MacCay, R., "The Gas Turbine as a Source of Continuous Precise Power," American Society of Mechanical Engineers, Paper 69-GT-20, May 1969.

[14] Grant, G., and Tabakoff, W., "Erosion Prediction in Turbomachinery Resulting from Environmental Particles," *Journal of Aircraft*, Vol. 12, No. 5, 1975, pp. 471–478.

[15] Balan, C., and Tabakoff, W., "Axial Compressor Performance Deterioration," AIAA Paper 84–1208, June 1984.

[16] Sugano, H., Yamaguchi, N., and Taguchi, S., "A Study on the Ash Erosion of Axial Induced Draft Fans of Coal-Fired Boilers," Vol. 19, Mitsubishi Heavy Industries, Tokyo, Japan, 1982.

[17] Richardson, J. H., Sallee, G. P., and Smakula, F. K., "Causes of High Pressure Compressor Deterioration in Service," AIAA Paper 79–1234, June 1979.

[18] Dunn, M. G., Padova, C., Moller, J. G., and Adams, R. M., "Performance Deterioration of a Turbofan and a Turbojet Engine upon Exposure to Dust Environment," *Journal of Engineering for Gas Turbine and Power*, Vol. 109, No. 2, 1987, pp. 336–343.

[19] Finnie, I., "Erosion of Surfaces by Solid Particles," *Wear*, Vol. 3, No. 2, 1960, pp. 87–103.

[20] Sheldon, G. L., "Similarities and Differences in the Erosion Behavior of Materials," *Journal of Basic Engineering*, Vol. 89, No. 3, 1970, pp. 619–625.

[21] Tilly, G. P., "A Two Stage Mechanism of Ductile Erosion," *Wear*, Vol. 23, No. 1, 1973, pp. 87–93.

[22] Hutchings, W. H., "A Model for the Erosion of Metals by Spherical Particles at Normal Incidence," *Wear*, Vol. 70, No. 3, Aug. 1983, pp. 269–281.

[23] Oka, Y. I., Nishimura, M., Nagahashi, K., and Matsumura, M., "Control and Evaluation of Particle Impact Conditions in a Sand Erosion Test Facility," *Wear*, Vol. 250, No. 1-12, 2001, pp. 736–743.

[24] Dosanjh, S., and Humphry, J. A., "The Influence of Turbulence on Erosion by a Particle-Laden Fluid Jet," *Wear*, Vol. 102, No. 4, 1985, pp. 309–329.

[25] Tabakoff, W., Grant, G., and Ball, R., "An Experimental Investigation of Certain Aerodynamic Effects on Erosion," AIAA Paper 74-639, July 1974.

[26] Tabakoff, W., and Wakeman, T., "Test Facility for Material Erosion at High Temperature," Special Publication 664, American Society for Testing and Materials, West Conshohocken, PA, 1979, pp. 123–135.

[27] Balan, C., and Tabakoff, W., "A Method for Predicting the Performance Deterioration of a Compressor Cascade due to Sand Erosion," AIAA Paper 84-1208, June 1984.

[28] Kline, M., and Simpson, G., "The Development of Innovative Methods for Erosion Testing a Russian Coating on GE T64 Gas Turbine Engine Compressor Blades," American Society of Mechanical Engineers, Paper GT2004-54336, June 2004.

[29] Hamed, A., Tabakoff, W., Rivir, R. B., Das, K., and Arora, P., "Turbine Blade Surface Deterioration by Erosion," *Journal of Turbomachinery*, Vol. 127, No. 3, 2005, pp. 445–452.

[30] Wakeman, T., and Tabakoff, W., "Measured Particle Rebound Characteristics Useful for Erosion Prediction," American Society of Mechanical Engineers, Paper 82-GT-170, 1982.

[31] Hamed, A., Tabakoff, W., Swar, R., Shin, D., Woggon, N., and Miller, R.. "Combined Experimental and Numerical Simulations of Thermal Barrier Coated Turbine Blades Erosion," National Aeronautics and Space Administrations, NASA/TM-2013-217857, Cleveland, OH, April 2013.

[32] Kotwal, R., and Tabakoff, W., "A New Approach for Erosion Prediction due to Fly Ash," *Journal of Engineering for Power*, Vol. 103, No. 2, 1981, pp. 265–267.

[33] Finnie, I., "An Experimental Study on Erosion," *Proceedings of the Society for Experimental Stress Analysis*, Vol. 17, No. 2, 1960, pp. 65–70.

[34] Hussein, M. F., and Tabakoff, W., "Dynamic Behavior of Solid Particles Suspended by Polluted Flow in Turbine Stage," *Journal of Aircraft*, Vol. 10, No. 7, 1973, pp. 334–340.

[35] Tabakoff, W., and Sugiyama, Y., "Experimental Methods of Determining Particle Restitution Coefficients," *Proceedings of the ASME Symposium on Polyphase Flow*

and Transient Technology, edited by R. A. Bajura, American Society of Mechanical Engineers, New York, 1980, pp. 203–210.

[36] Tabakoff, W., Malak, M., and Hamed, A., "Laser Measurements of Solid-Particle Rebound Parameters Impacting on 2024 Aluminum and 6AI-4 V Titanium Alloys," *AIAA Journal*, Vol. 25, No. 5, 1987, pp. 721–726.

[37] Tabakoff, W., and Malak, M., "Laser Measurements of Fly Ash Rebound Characteristics for Use in Trajectory Calculations," *Journal of Turbomachinery*, Vol. 109, No. 4, 1987, pp. 535–540.

[38] Tabakoff, W., "Measurements of Particles Rebound Characteristics on Materials Used in Gas Turbines," *Journal of Propulsion and Power*, Vol. 7, No. 5, 1991, pp. 805–813.

[39] Tabakoff, W., and Hamed, A., "Experimental Investigation of Particle Surface Interactions for Turomachinery Applications," *Laser Anemometry Advances and Applications*, edited by A. Dybbs and B. Ghorashi, American Society of Mechanical Engineers, New York, 1991, pp. 775–780.

[40] Tabakoff, W., Hamed, A., and Murugan, X., "Effect of Target Materials on Particle Restitution Characteristics for Turbomachinery Applications," *Journal of Propulsion and Power*, Vol. 12, No. 5, 1996, pp. 260–266.

[41] Hussein, M. F., and Tabakoff, W., "Computation and Plotting of Solid Particle Flow in Rotating Cascades," *Computers and Fluids*, Vol. 2, No. 1, 1974, pp. 1–15.

[42] Tabakoff, W., and Hussein, M. F., "Trajectories of Particles Suspended in Flows Through Cascades," *Journal of Aircraft*, Vol. 8, No. 1, 1971, pp. 60–64.

[43] Ulke, A., and Roulean, W. T., "The Effects of Secondary Flows on Turbine Blade Erosion," American Society of Mechanical Engineers, Paper 76-GT-74, March 1976.

[44] Tabakoff, W., "Compressor Erosion and Performance Deterioration," *Journal of Fluids Engineering*, Vol. 109, 1987, pp. 297–306.

[45] Beacher, B., and Tabakoff, W., "Trajectories of Ash Particles Through a Coal-Burning Gas Turbine," American Society of Mechanical Engineers, Paper 84-GT-122, May 1984.

[46] Elfeki, S., and Tabakoff, W., "Erosion Study of Radial Flow Compressor with Splitters," American Society of Mechanical Engineers, Paper 86-GT-240, Jan. 1986.

[47] Tabakoff, W., and Hamed, A., "Effect of Environmental Particles on a Radial Compressor," AIAA Paper 88–0366, Jan. 1988.

[48] Ghenaiet, A., Elder, R. L., and Tan, S. C., "Particle Trajectories Through an Axial Fan and Performance Degradation due to Sand Ingestion," American Society of Mechanical Engineers, Paper 2001-GT-0497, June 2001.

[49] Hamed, A., and Tabakoff, W., "Experimental and Numerical Simulations of the Effects of Ingested Particles in Gas Turbine Engines," *Erosion, Corrosion and Foreign Object Effects in Gas Turbines*, AGARD-CP-558, Paper No. 11, NATO Science and Technology Organization, NEUILLY-SUR-SEINE, France, 1994, pp. (11)1–13.

[50] Swar, R., Hamed, A., Shin, D., Woggon, N., and Miller, R., "Deterioration of Thermal Barrier Coated Turbine Blades by Erosion," *International Journal of Rotating Machinery*, Vol. 2012, Oct. 2012, Article ID 601837, 10 pp.

[51] Metwally, M., Tabakoff, W., and Hamed, A., "Blade Erosion in Automotive Gas Turbine Engine," *Journal of Engineering for Gas Turbines and Power*, Vol. 117, Jan. 1995, pp. 213–219.

[52] Tabakoff, W., and Hamed, A., "Temperature Effect on Particle Dynamics and Erosion in Radial Inflow Turbines," *Journal of Turbomachinery*, Vol. 110, April 1988, pp. 258–246.

[53] Clevenger, W., and Tabakoff, W., "Dust Particle Trajectories in Aircraft Radial Turbine," *Journal of Aircraft*, Vol. 13, No. 10, Oct. 1976, pp. 786–791.

[54] Tabakoff, W., Hamed, A., and Metwally, M., "Effect of Particle Size Distribution on Particle Dynamics and Blade Erosion in Axial Flow Turbines," *Journal of Gas Turbine and Power*, Vol. 113, Oct. 1991, pp. 607–615.

[55] Hamed, A., "An Investigation in the Variance in Particle Surface Interactions and Their Effects in Gas Turbines," *Journal of Engineering for Gas Turbine and Power*, Vol. 114, April 1992, pp. 235–241.

[56] Hamed, A., and Kuhn, T. P., "Effects of Variational Particle Restitution Characteristics on Turbomachinery Erosion," *Journal of Engineering for Gas Turbine and Power*, Vol. 117, July 1995, pp. 432–440.

[57] Kleinert, G., "Turbofan Engine Maintenance for Fuel and Hardware Conservation," *Seramtec Review*, Vol. 33, 1990, pp. 2, 3.

[58] Tabakoff, W., and Simpson, G., "Experimental Study of Deterioration and Retention on Coated and Uncoated Compressor and Turbine Blades," AIAA Paper 2002-0373, Jan. 2002.

[59] Edwards, V. R., and Rouse, P. L., "U.S. Army Rotorcraft Turboshaft Engines–Dust Erosion Considerations," AGARD-CP-558, Paper No. 3, NATO Science and Technology Organization, NEUILLY-SUR-SEINE, France, 1994, pp. (3)1–10.

[60] Schmucker, J., and Schaffer, A., "Performance Deterioration of Axial Compressors due to Blade Defects," AGARD-CP-558, Paper 16, NATO Science and Technology Organization, NEUILLY-SUR-SEINE, France, 1994, pp. (16)1–5.

[61] Batcho, P. F., Moller, J. C., Padova, C., and Dunn, M. G., "Interpretation of Gas Turbine Response due to Dust Ingestion," Vol. 109, No. 3, 1978, pp. 344–352.

[62] Tabakoff, W., Lakshiminarasimha, A. N., and Pasin, M., "Simulation of Compressor Performance Deterioration due to Erosion," *Journal of Turbomachinery*, Vol. 112, No. 1, 1990, pp. 78–112.

[63] Hamed, A., Tabakoff, W., and Singh, D., "Modeling of Compressor Performance Deterioration due to Erosion," *Int'l Journal of Rotating Machinery*, Vol. 4, No. 4, 1998, pp. 243–248.

[64] Tabakoff, W., Kaushik, S., and Lakshiminarasimha, A. N., "Performance Improvement of an Eroded Axial Flow Compressor Using Water Injection," AIAA Paper 90-2016, July 1990.

[65] Nagy, D. R., Parameswaran, V. R., MacLeod, J. D., and Immarigeon, J. P., "Protective Coatings for Compressor Gas Path Components," AGARD-CP-558, Paper 27, NATO Science and Technology Organization, NEUILLY-SUR-SEINE, France, 1994, pp. (27)1–10.

[66] Naik, S. K., Sutter, J. K., Tabakoff, W., Siefker, R. G., Haller, H. S., Cupp, R. J., and Miyoshi, K. K., "Wear Resistant Polymer Matrix Composites for Aerospace Applications," American Society of Mechanical Engineers, Paper GT2004-54330, June 2004.

[67] Smith, H. C. G., "A Theoretical Note upon the Mechanism of Deposition in Turbine Airfoil Fouling," National Gas Turbine Establishment, Memo. M.145, Feb.1952.

[68] Taylor, G. I., "Notes on Possible Equipment and Technique for Experiments on Icing on Aircraft," A.R.C. Technical Report, Issue No. 2024, H. M. Stationery Office, London, 1940, pp. 1–7.

[69] McCreath, C. G., "The Role of Salt Particulate Matter in the Hot Corrosion Phenomenon Exhibited by Gas Turbine Power Plants Operated in a Marine Environment," *Power Industry Research*, Vol. 2, No. 1, 1982, pp. 1–15.

[70] Dring, R. P., Casper, J. R., and Suo, M., "Particle Trajectories in Turbine Cascades," *Journal of Energy*, Vol. 3, No. 3, May–June 1979, pp. 161–166.

[71] Lin, C. S., Moulton, R. W., and Putnam, G. L., "Mass Transfer Between Solid Walls and Fluid Streams," *Industrial and Engineering Chemistry*, Vol. 45, No. 1, 1953, pp. 636–640.

[72] Friedlander, S. K., and Johnstone, H. F., "Deposition of Suspended Particles from Turbulent Gas Streams," *Industrial and Engineering Chemistry*, Vol. 49, No. 7, 1957, pp. 1151–1156.

[73] Davies, C. N., "Deposition from Moving Aerosols," *Aerosol Science*, Academic Press, New York, 1966, pp. 393–446.

[74] Cleaver, J. W., and Yates, B., "A Sub Layer Model for Deposition of Particles from a Turbulent Flow," *Chemical Engineering Science*, Vol. 30, No. 8, 1975, pp. 983–992.

[75] Parker, G. J., and Lee, P., "Studies of the Deposition of Submicron Particles on Turbine Blades," *Proceedings of the Institute of Mechanical Engineers*, Vol. 186, No. 38, 1972, pp. 519–526.

[76] Moore, M. J., and Crane, R. I., "Aerodynamic Aspects of Gas Turbine Blade Corrosion," *Deposition and Corrosion in Gas Turbines*, edited by A. Hart and A. Cutler, Wiley, New York, 1973, pp. 35–57.

[77] Hidy, G. M., and Heisler, S. L., "Transport and Deposition of Flowing Aerosols," edited by D. T. Shaw, *Recent Developments in Aerosol Science*, Wiley, New York, 1978, pp. 135–165.

[78] Rosner, D. E., Chen, G. C., Fryburg, G. C., and Kohl, F. J., "Chemically Frozen Multicomponent Boundary Theory of Salt and/or Ash Deposition Rates for Combustion Gases," *Combustion Science and Technology*, Vol. 20, Nos. 3-4, 1979, pp. 87–106.

[79] Rosner, D. E., and Fernandez de la Mora, J., "Particle Transport Across Turbulent Non-Isothermal Boundary Layers," *Journal of Engineering for Power*, Vol. 104, No. 4, 1982, pp. 885–994.

[80] Menguturk, M., and Sverdrup, E. F., "A Theory for Fine Particle Deposition in 2-D Boundary Layer Flows and Application to Gas Turbines," American Society of Mechanical Engineers, Paper 81-GT-54, Jan. 1981.

[81] Wenglarz, R. A., "An Assessment of Deposition in PFBC Power Plant Gas Turbines," *Journal of Engineering for Power*, Vol. 103, No. 3, July 1981, pp. 552–560.

[82] Ahluwalia, R. K., Im, K. H., Chuang, C. F., and Hajduk, J. C., "Particle and Vapor Deposition in Coal-Fired Gas Turbines," American Society of Mechanical Engineers, Paper 86-GT-239, June 1986.

[83] Frackrell, J., Brown, K., and Young, J., "Modelling Particle Deposition in Gas Turbines Employed in Advanced Coal-Fired Systems," American Society of Mechanical Engineers, Paper 94-GT-467, June 1994.

[84] El-Batsh, H., and Haselbacher, H., "Effect of Turbulence Modeling on Particle Dispersion and Deposition on Compressor and Turbine Blade Surfaces," American Society of Mechanical Engineers, Paper 2000-GT-519, 2000.

[85] El-Batsh, H., and Haselbacher, H., "Numerical Investigation of the Effect of Ash Particle Deposition on the Flow Field Through Turbine Cascades," American Society of Mechanical Engineers, Paper GT-2002-30600, 2002.

[86] Tabakoff, W., Hamed, A., and Wenglarz, R., *Particulate Flows and Blade Erosion*, Lecture Series Publications, von Karman Inst for Fluid Dynamics, Sint-Genesius-Rode, Belgium, 1988.

[87] Wenglarz, R. A., and Wright, I. G., "Alternate Fuels for Land-Based Turbines," *Proceedings of the Workshop on Materials and Practices to Improve Resistance to Fuel Derived Environmental Damage in Land-and Sea-Based Turbines*, edited by D. L. Olson, Colorado School of Mines, Golden, CO, 2003, pp. 786–791.

[88] Wenglarz, R. A., and Fox, R. G., "Physical Aspects of Deposition from Coal-Water Fuels Under Gas Turbine Conditions," *Journal of Engineering for Gas Turbines and Power*, Vol. 112, No. 1, Jan. 1990, pp. 9–14.

CHAPTER 15

Surface Roughness Effects

Jeffrey P. Bons*

Ohio State University, Columbus, Ohio

NOMENCLATURE

A_f	windward frontal surface area of roughness elements on sample
A_s	windward wetted surface area of roughness elements on sample
B	log law constant
b_2	centrifugal impeller discharge height
c	vane or blade true chord
c_f	skin-friction coefficient
c_p	specific heat at constant pressure
c_x	vane or blade axial chord
d_h	hydraulic diameter
h	heat-transfer coefficient
k	roughness height
k^+	dimensionless roughness parameter $(k_s u_\tau/\nu)$
k_s	sand roughness or "equivalent" sand roughness height
$k_{s,\mathrm{adm}}$	admissible sand roughness
M	Mach number or film-cooling blowing ratio, $\rho_c U_c/\rho_\infty U_\infty$
N	number of points in profile record
Nu	Nusselt number
Ra	centerline average roughness, Eq. (1)
Re_c	Reynolds number based on true chord and inlet or exit conditions (for compressor or turbine respectively), unless otherwise noted
Re_{cx}	Reynolds number based on axial chord and inlet or exit conditions (for compressor or turbine respectively), unless otherwise noted
Re_k	roughness Reynolds number $(k_s U_\infty/\nu)$

*Professor of Aerospace Engineering.

Re_x	Reynolds number based on streamwise distance (x)
Rq	rms roughness, Eq. (2)
Rt	maximum peak to valley roughness, Eq. (3)
Rz	average peak-to-valley roughness
S	surface area of sample without roughness
S_f	total frontal surface area of sample
S_w	total wetted surface area of sample
St	Stanton number ($h/\rho c_p U_\infty$)
T	temperature
Tu	freestream turbulence level, %
t	roughness spacing
U, u	velocity
u^+	dimensionless velocity (u/u_τ)
u_τ	shear or friction velocity $(\tau/\rho)^{1/2}$
x	streamwise distance
y	surface height coordinate after removal of polynomial fit to surface curvature
y^+	dimensionless wall distance (yu_τ/ν)
α_f	mean forward-facing angle roughness parameter from Bons [26]
δ	boundary-layer thickness
η	efficiency
η_{film}	film-cooling effectiveness $(T_s - T_\infty)/(T_c - T_\infty)$
θ	boundary-layer momentum thickness
κ	von Kármán constant (0.41)
Λ_s	roughness shape density parameter from Sigal and Danberg [29] [Eq. (5)]
λ	roughness shape density parameter from Kind et al. [52]
μ	viscosity
μ_t	turbulent eddy viscosity
ν	kinematic viscosity
π_c	compressor pressure rise
ρ	fluid density
ω	total pressure loss coefficient

SUBSCRIPTS

c	compressor or coolant
ex	exit conditions
in	inlet conditions
s	surface
tr	boundary-layer transition
∞	freestream conditions
1	inlet conditions
2	exit conditions

This chapter addresses the effects of surface roughness on gas-turbine performance, including both the fundamental effects of roughness on fluid flow and heat transfer and their aggregate effect on gas-turbine power output and fuel efficiency. Though considerable effort has been made to synthesize the major findings from gas-turbine roughness research conducted over the last 60 years, the references employed are by no means comprehensive. Rather, this chapter is intended to provide a sampling of past and current research in a very active and difficult area of study. To briefly summarize, the most significant conclusions of this chapter are as follows:

- Empirical roughness correlations routinely employed for drag and heat-transfer estimates in gas turbines are often inadequate and at times misapplied.
- Roughness affects engine performance by causing earlier boundary-layer transition, increased boundary-layer momentum loss (i.e., thickness), and/or flow separation.
- Roughness effects in the compressor and turbine are dependent on the Reynolds number (*Re*), the ratio of roughness size to airfoil chord, and to a lesser extent Mach number.
  - At low *Re*, roughness can eliminate laminar separation bubbles (thus reducing loss).
  - At high *Re* (when the boundary layer is already turbulent), roughness can thicken the boundary layer to the point of separation (thus increasing loss).
- Roughness has the undesirable effect of increasing the convective heat transfer from the gas stream to metal surfaces in the turbine. In most cases roughness also degrades film-cooling performance.
- Recent advances in roughness modeling for computational-fluid- dynamics (CFD) predictions show significant promise.
- Considerable research is yet necessary to fully understand and properly model the role of roughness in gas turbines.

Though numerous excellent texts have addressed these (and other) roughness-related questions for general fluid flows, the particular focus of this chapter is on roughness as it relates specifically to gas turbines.

I. WHAT IS SURFACE ROUGHNESS?

In the context of the present discussion, surface roughness is any alteration to a solid surface (intended or not) that presents a nonsmooth boundary to the gas path. This alteration could be an obstruction that protrudes out into the flow or a depression into the solid surface. Generally, roughness is characterized either by discrete elements (perhaps ordered arrays of roughness elements) or a continuous but random variation of the surface above and below a smooth-line datum (Fig. 1). Many of the components in gas turbines are designed with complex

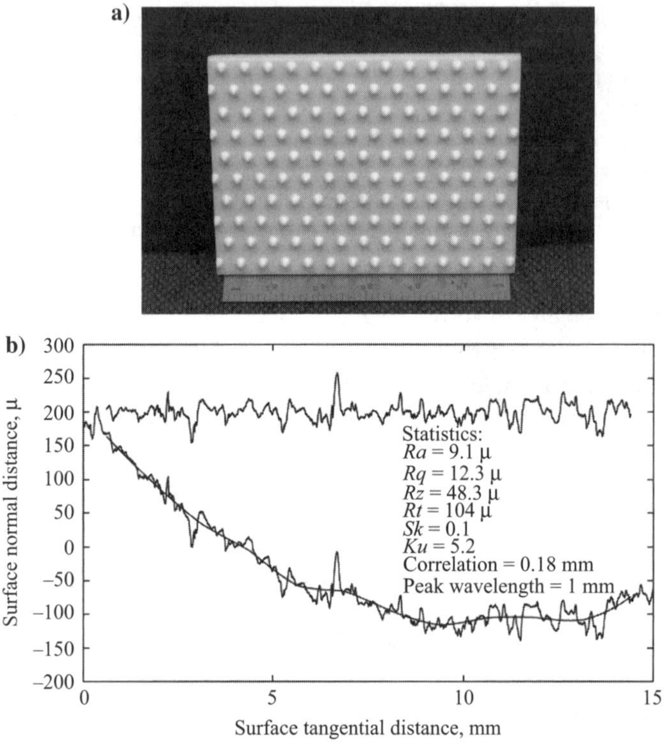

Fig. 1 Examples of discrete/ordered roughness elements and continuous roughness: a) ordered roughness elements; b) random roughness with and without meanline curvature removed.

geometries such as three-dimensional (3-D) gas paths and changes in curvature, which necessarily result in a convoluted flowpath. In most cases, however, these design features would not be characterized as "roughness." To be classified as "roughness," surface aberrations are typically abrupt, with a vertical scale on the same order as the boundary-layer dimensions (or smaller).

II. WHY IS ROUGHNESS IMPORTANT?

It is well understood that surface roughness affects both the aerodynamics and thermodynamics (heat transfer) of fluid flow over a surface. Both of these effects are important in a gas turbine because the output (thrust for aeroengines and shaft power for turboshaft or land-based engines) and efficiency (fuel efficiency) are both implicated.

Fig. 2 Deposition on first-stage vane of utility gas turbine after approximately 8000 h of service.

Foreign particulate that deposits on gas-turbine surfaces can result in modified airfoil shapes (e.g., at the leading edge [1, 2]), increased surface roughness [3–6], and clogged film-cooling holes [5, 7, 8]. Figure 2 shows a dramatic example of deposition on a nozzle guide vane as a result of inadequately cleaned fuel used in a land-based turbine. In extreme cases such as this, the roughness elements created by deposition are large enough to significantly reduce the flowpath area and alter the operating point of the engine. Because many turbines are operated at or near choke conditions, this reduction in flowpath area reduces the operating mass flow rate of the engine and directly results in lower power output. The deposit roughness also increases the heat-transfer rate from the hot gas path to the turbine surface [3, 5, 9], requiring the engine to operate at a lower firing temperature and/or with more cooling flow to the turbine—both of which reduce the cycle (i.e., fuel) efficiency. To make matters worse, turbine deposits have been known to contribute to spallation of ceramic thermal barrier coatings because of the thermal expansion coefficient mismatch of the ceramic with the foreign deposit material [10]. This, in addition to the clogging of film-cooling holes cited earlier, severely curtails the operating limits of the turbine.

Roughness caused by surface erosion (typically caused by large particles such as desert sand impacting and scraping the surface) can be equally damaging. Erosion at the leading edge of an airfoil reduces the effective chord of the blade or vane while erosion at a rotor tip increases the tip gap flow and associated secondary losses [2, 11, 12]. An example of compressor rotor tip wear as a result of sand ingestion is shown in Fig. 3. In cases of severe wear such as this, loss of compressor stall margin can make safe operation of the gas turbine altogether impossible [12–17]. The airfoil leading-edge region is particularly sensitive to

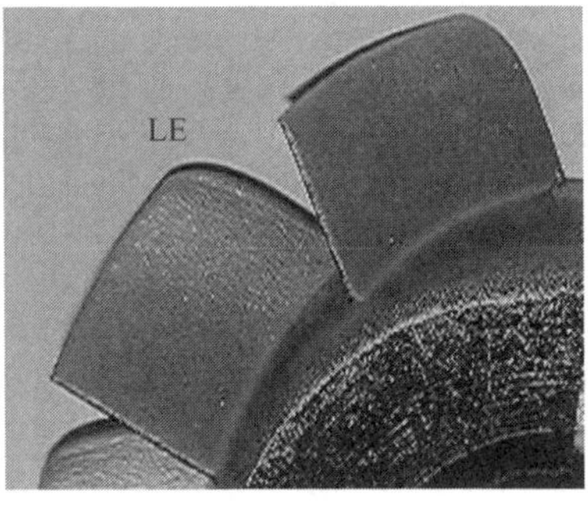

Fig. 3 Test rotor blades eroded as a result of sand ingestion near design point during 9 h (Fig. 9 from Ghenaiet et al. [12]).

roughness effects because the character of the boundary layer (laminar or turbulent) plays a critical role in determining the maximum flow diffusion that can be achieved before blade stall in a compressor [18] and the peak heat-transfer level (and maximum aerodynamic turning) in a turbine [19]. Finally, many of the environmental sources that cause degradation and surface roughness contain corrosive elements that promote rapid deterioration of metal components in the gas path, potentially leading to catastrophic failure.

Because of the importance of surface degradation in gas turbines, there have been a number of excellent review articles addressing its root causes and exploring preventive measures [13, 14]. Technical advancements in the design and manufacture of gas-path turbomachinery components over the last two decades have, however, heightened the significance of understanding the effects of surface roughness on gas-turbine operation. Surface coatings in both the compressor and turbine, more aggressive airfoil shapes, and advanced rotor tip and passage endwall designs, as well as an increased number of bleeds to feed more intricate film-cooling hole geometries, are among the technologies that have created an increased urgency for fundamental research into the root causes and effects of surface roughness. In essence, as computational models of gas-turbine component performance have improved in fidelity and accuracy, designers have pursued more sophisticated (and complex) flowpaths. This, in turn, has driven the need to better understand the performance-limiting effects of service-induced surface roughness.

III. WHAT PREVENTIVE MEASURES ARE AVAILABLE?

Because the primary sources of surface degradation and the attendant surface roughness are from the gas stream and fuel, roughness levels can be controlled by operating with highly filtered inlet flow and stringently cleaned fuels. Because this is not always practical (particularly for aeroengines), frequent boroscopic evaluations are often needed to detect and assess surface damage before it becomes performance limiting. In addition, the prudent application of other preventive measures can mitigate unacceptable losses in performance and/or availability [13]. For example, deposit buildup on compressor blading has traditionally been removed using abrasive chaff (such as nutshells). With the advent of coated compressor blades in the last decade, compressor maintenance has moved to online water washing, sometimes with detergents [16, 20–23]. More stubborn deposits can be removed with off-line (crank) washing of both compressor and turbine blading. Dirty fuels (such as biomass or high ash-bearing heavy fuel oils [24]) can be safely used with adequate cold- and/or hot-gas cleanup, though these auxiliary systems can be expensive and require regular maintenance. Ultimately, the most severe cases of degradation (erosion and corrosion in particular) might necessitate a major engine overhaul with significant downtime. Only then can individual blades or vane sectors be removed for refurbishment and resurfacing to recover lost performance margin.

IV. HOW IS ROUGHNESS CHARACTERIZED?

Unless the surface roughness is assembly related (rivet depressions on an inlet diffuser cowling or steps at a rotor-stator interface in a turbine), local measurements must be taken to assess the extent to which surface "smoothness" has been compromised during service. This is typically done either with a contact stylus or an optical surface measurement system. The contact stylus method is much more commonly used and consists of a diamond-tipped conical stylus (typically a 60-deg conical stylus with 1.5-μm radius tip) that follows the vertical surface features with submicron precision as it traverses the part of interest. Such "profilometers" can be small handheld units for field inspections or computer-controlled, multi-axis machines for benchtop assessments during engine teardown. Laser-based optical methods are significantly more expensive but have the advantage that they are "noncontact" and do not alter the true surface character by dislodging deposits. Both methods are limited by the conical tip dimensions (for contact methods) and the finite laser spot size (for optical methods), as described by Bons et al. [6]; these limitations are especially important in addressing highly irregular surfaces.

A typical contact stylus surface trace is shown in Fig. 1b. (The vertical axis has been magnified to reveal the micron-level roughness over the 15-mm-long trace.) As this trace demonstrates, the stylus picks up both the surface roughness and the surface contour (or "waviness"). For flow that is not separated, the flow streamlines near the surface generally follow the local surface contour. Because the objective is to assess the effect of roughness on flow-related parameters such as Stanton number for heat transfer and skin friction for aerodynamic loss, only surface features that depart from the natural form (and thus the local flow streamlines) are considered to be of interest. Thus, the relevant roughness data must be extracted from the raw data trace by performing a mean-line fit to the natural curvature of the part. This can be done using a spline-fit to the raw data (also indicated in Fig. 1b). With the form removed, relevant statistics can be extracted from the residual roughness data, including the centerline averaged roughness *Ra*, the rms roughness *Rq*, and the maximum peak-to-valley roughness *Rt*, as defined next:

$$Ra = \frac{1}{N}\sum_{i=1}^{N} |y_i - \bar{y}| \tag{1}$$

$$Rq = \sqrt{\frac{1}{N}\sum_{i=1}^{N} (y_i - \bar{y})^2} \tag{2}$$

$$Rt = y_{\max} - y_{\min} \tag{3}$$

The trace can also be subdivided into smaller increments or "cutoffs" (e.g., 10–15 subdivisions per trace) to determine *Rz*, which is defined as the average over the entire trace of the *Rt* values from each subdivision. *Rz* is commonly used as an estimate of the average roughness height *k*.

Though not as commonly used, two-dimensional surface traces can also be evaluated to determine the relative surface angle between two consecutive data points:

$$\alpha_i = \tan^{-1}\left(\frac{y_{i+1} - y_i}{x_{i+1} - x_i}\right) \tag{4}$$

The rms of this angle distribution and the average of all forward-facing angles in a given trace have also been used to characterize roughness [25, 26].

Finally, in some cases, three-dimensional surface roughness data are available for evaluation. This might be from an ordered array of discrete roughness elements of known dimensions and spacing (Fig. 1a), or from a full 3-D map of a randomly rough surface. Three-dimensional roughness data can be used to determine any of the several roughness shape/density parameters offered in the literature [27–29]. For example, Sigal and Danberg [29] developed the shape/density parameter Λ_s to successfully correlate skin-friction data from two- and three-dimensional roughness elements mounted to a smooth surface. It is defined as

$$\Lambda_s = \frac{S}{S_f}\left(\frac{A_f}{A_s}\right)^{-1.6} \tag{5}$$

For the evaluation of randomly rough surfaces (rather than ordered cones or hemispheres), the calculation of A_f, S_f, and A_s must be adapted as described by Bons et al. [9] and van Rij et al. [30].

V. WHAT CORRELATIONS EXIST TO ACCOUNT FOR ROUGHNESS EFFECTS?

Because it is often difficult to extract universal correlations from completely random roughness characterizations, the bulk of roughness research has historically involved well-characterized roughness elements. One of the earliest roughness correlations dates back three-quarters of a century to the turbulent pipe flow study of Nikuradse [31]. His pressure loss data were taken in pipes whose walls had been roughened with sand that had been sieved to discrete size ranges (e.g., 0.1 mm $< k_s <$ 0.3 mm). An adhesive was applied to the internal surface of the pipe, and a single layer of the sand particles created a uniform roughness-to-diameter ratio for each test. Nikuradse found different dependencies of pressure loss on Reynolds number and roughness for different flow regimes (Fig. 4). He defined a dimensionless roughness parameter, $k^+ = k_s u_\tau/\nu$, using the actual sandgrain diameter k_s, the measured friction or shear velocity [$u_\tau = (\tau/\rho)^{1/2}$], and the kinematic viscosity ν. He found that for values of this parameter greater than 70, the pipe loss coefficient was only a function of k_s whereas for $5 < k^+ < 70$, both Re and k_s were important. These regimes are termed "completely" or "fully" rough and "transitionally" rough, respectively. For k^+ values below 5, roughness was found to have no effect on pressure loss

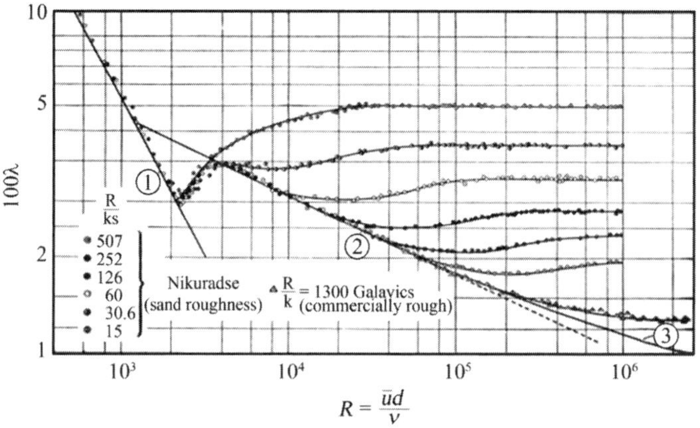

Fig. 4 Nikuradse's original sand-roughened pipe flow data (Fig. 20.18 from Schlichting [35]).

because the roughness peaks were wholly immersed in the laminar sublayer of the turbulent boundary layer. Thus, this regime is declared "hydraulically smooth." Nikuradse observed that the turbulent boundary layer "law of the wall" ($u^+ = 1/\kappa^* \ell n\, y^+ + B$) was still valid for rough surfaces, except that the constant B was shown to be dependent on k^+. For "completely rough" surfaces, this dependency can be approximated as $B = 5.5 - 1/\kappa\{\ell n\,(1 + 0.3k^+)\}$ [32]. Note that the constant in this expression varies between 5 and 5.5 in the literature, and care should be taken to consistently employ the same expression when comparing various models. In this way, k^+ can be used (in a boundary-layer calculation, for example [33]) to calculate skin friction for a rough wall.

Schlichting [34] subsequently used Nikuradse's data to correlate various types of "nonsand" roughness (e.g., rivets, bumps, protuberances). In so doing, Schlichting [35] coined the term "equivalent sandgrain roughness" to connote a roughness feature (and spacing) that has the "equivalent" effect on skin-friction losses as a uniform layer of actual sandgrains of diameter k_s. Though Schlichting's quantitative results have since been disputed by Coleman et al. [36], his equivalent sandgrain methodology has gained universal acceptance. Until recently, practically every roughness correlation for skin friction, convective heat transfer, and even boundary-layer transition utilized k_s [32, 35, 37–40]. Using Nikuradse's pipe flow data, as well as his own channel flow data, Schlichting [35] offered the following correlation for skin friction:

$$c_f = [2.87 + 1.58 \log (x/k_s)]^{-2.5} \tag{6}$$

where x is the streamwise distance. This relation is valid for the "completely rough" flow regime, as evidenced by the lack of Reynolds-number dependency.

TABLE 1 ROUGHNESS CORRELATIONS FOR k_s DETERMINATION OF GAS-TURBINE ROUGHNESS

Year	Reference	k_s Relation	Surface Type
1962	Speidel [41]	$k_s = Rz/5$	Milled surface with grooves parallel (within 10 deg) to flow (Rz = groove height)
		$k_s = Rz/2.56$	Milled surface with grooves greater than 10 deg from flow-aligned
1967	Forster [42]	$k_s \approx 2Ra$	Machined surfaces
		$k_s \approx 7Ra$	Emery papers
1976	Koch and Smith [43]	$k_s \approx 6Ra$	Sand grains
1976	Bammert and Sandstede [44]	$k_s \approx 2.2Ra^{0.88}$	Mechanically produced surface and emery grain surface
1980	Schaffler [45]	$k_s \approx 8.9Ra$	Forged and machined blades
1984	Simon and Bulskamper [46]	$k_s \approx 2Ra$	Machined surfaces
1990	Sigal and Danberg [29] (as used by Boyle and Civinskas [47] Bogard et al. [5])	$0.5 < k_s/k < 5$ as $f(\Lambda_s)$	Isolated 2-D and 3-D roughness elements of height k
1997	Barlow and Kim [48]	$k_s \approx 2.7k$ or $k_s \approx 16Ra$	Ordered array of pedestals of height k
1996	Hoffs et al. [49]	$k_s = Rz$	Liquid crystal surface
1998	Guo et al. [50]	$k_s = Rz$	Liquid crystal surface
1998	Bogard et al. [5]	$k_s \approx 4Ra$	Turbine vane surface roughness
1998	Abuaf et al. [51]	$Ra < k_s < 10Ra$	Cast and polished metal surfaces
1998	Kind et al. [52]	$2.4 < k_s/k < 6.1$ as $f(\lambda)$	Sparsely distributed sand grains of average size k
2001	Boyle et al. [53]	$k_s \approx 2.1Rq$	Research vane surface

(Continued)

TABLE 1 ROUGHNESS CORRELATIONS FOR k_s DETERMINATION OF GAS-TURBINE ROUGHNESS (*CONTINUED*)

Year	Reference	k_s Relation	Surface Type
2003	Boyle and Senyitko [39]	$k_s \approx 4.8Rq$	ZrO spray on roughness particles
2003	Bunker [54]	$k_s \approx 10Ra$	Polished TBC
2004	Shabbir and Turner [55]	$k_s = 8.9Ra$	Turbine roughness
2004	Zhang and Ligrani [56]	$k_s \approx 1.9Rz$ as $f(\Lambda_s)$	20–150-μm Ni particles
2005	Bons [26]	$0.5 < k_s/Rz < 3.5$ as $f(\alpha_f)$	Scaled turbine blade roughness
2005	Syverud et al. [1]	$0.4 < k_s/k < 2$ as $f(\lambda)^{47}$	Salt grains from sea-spray ingestion
2005	Stripf et al. [57]	$2 < k_s/k < 5$ as $f(k, t)$	Ordered arrays of truncated cones of height k and spacing t
2005	Hummel et al. [58]	$k_s \approx 5.2Ra$	Correlation with various surfaces
2006	Yuan and Kind [59]	$k_s \approx 1.8k$	Sparsely distributed sand grains of average size k

Other notable fluid mechanics textbooks offer similar correlations based on the "equivalent sandgrain" parameter:

$$c_f = [1.4 + 3.7\log(x/k_s)]^{-2} \quad \text{from White [32]} \tag{7}$$

$$c_f = 0.168[\ell n(84\delta/k_s)]^{-2} \quad \text{from Kays and Crawford [38]} \tag{8}$$

$$c_f = [3.476 + 0.707\log(x/k_s)]^{-2.46} \quad \text{from Mills [37]} \tag{9}$$

In roughness work related to gas turbines, various correlations have been employed to convert commonly measured surface roughness parameters (*Ra*, *Rq*, or *Rz*) to equivalent sandgrain roughness k_s, following the Schlichting model. Table 1 contains a survey from the open literature.

Table 1 shows a wide variety of proposed correlations. Because many of the correlations vary by up to a factor of five, the uninitiated might have difficulty selecting an appropriate value for a new application [51]. Several of the researchers identified in the table have lamented this morass of data, ultimately defining their own correlation. For example, Hummel et al. [58] used a turbine blade loss model from Traupel [60] to correlate estimated blade row losses vs Ra/c

from several previous experiments (including some of their own data) and arrived at a best fit using $k_s = 5.2Ra$. Boyle and Senyitko [39] provide a detailed evaluation of a number of the leading correlations, showing how estimates for k_s can vary up to an order of magnitude depending on the method used.

Of course, part of the challenge of defining a single universal correlation is the variety of roughness types: machined, sand grain, ordered arrays of identical roughness elements (such as cones), and, finally, actual degraded surface roughness. Recognizing this difficulty, many have proposed correlations based on something other than the standard roughness metrics (*Ra*, *Rq*, or *Rz*). Parameters that account for the individual shape and density of roughness elements (Λ_s, λ, α_f) have a more physical basis but still fall short of being universal [1, 26–29, 52, 55, 57]. Moreover, the vast majority of these correlations are defined using "artificial" or "simulated" roughness rather than "real" roughness. Bons [9] studied scaled models of "real" roughness samples taken from inservice turbine hardware and found a markedly different k_s correlation when compared to ordered arrays of deterministic roughness elements (cones or hemispheres). Figure 5 shows data from Bons' randomly rough surfaces compared with Sigal and Danberg's [29] data from ordered roughness elements on the same plot. The two data sets yield very different roughness correlations for k_s/k vs Λ_s:

$$\log\left(\frac{k_s}{k}\right) = -1.31\log(\Lambda_s) + 2.15 \tag{10}$$

$$\log\left(\frac{k_{\text{sadj}}}{k}\right) = -0.43\log(\Lambda_s) + 0.82 \tag{11}$$

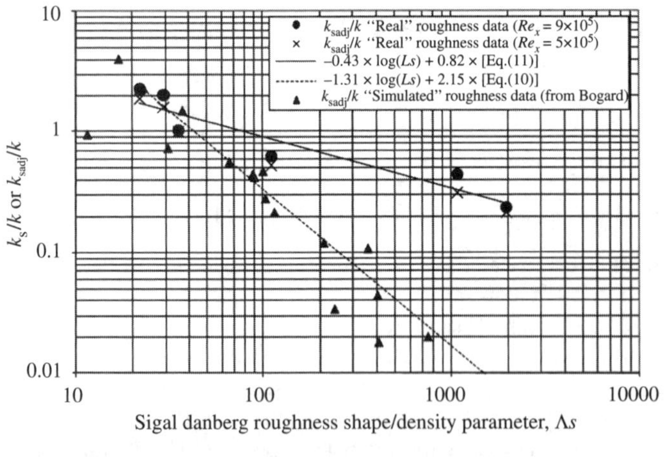

Fig. 5 Equivalent sandgrain correlation data vs fit. Data from Bogard et al. [5] with accompanying fit [Eq. (10)]. Data from "real" roughness surfaces at two Reynolds numbers with accompanying fit [Eq. (11)] (Fig. 9 from Bons [9]).

This discrepancy is significant because the majority of roughness studies have been conducted using sand-roughened walls or ordered arrays of roughness elements. Indeed, sand grains are the most common roughness characterization employed historically for gas-turbine research (as well as roughness work in other industries), despite considerable evidence that degradation-induced gas-turbine surface roughness is not similar to uniform sand-grain roughness. For example, Dunn et al. [7] show pictures of compressor rotor blades subjected to "simulated" volcanic-ash ingestion with varying levels of erosion. Though the images do not allow an up-close assessment of surface character, it is evident that the erosion is not spatially uniform. Erosion is primarily evident in the leading edge and tip regions. Accelerated salt-spray ingestion tests by Kacprzynski et al. [20] and Syverud et al. [1] show similar spatial variation. A picture of second-stage compressor vanes shows 0.5-mm-thick salt deposits at the vane leading edge, with a strong preference for the hub vs the tip annulus [1]. A boroscope image of a first-stage compressor blade shown in [20] has regions with pronounced ridge-lines of salt deposit, as well as regions without any obvious deposit. Finally, a magnified image of salt grains on a first-stage vane pressure surface shows a nonuniform distribution of salt grains with sizes from $3 < k < 30$ μm (mean grain size = 22 μm, mean grain spacing = 88 μm). In 2000, Leipold et al. [61] studied roughness on precision-forged compressor blades and found sparsely distributed roughness elements rather than the close-packed elements typical of sandgrain characterizations. Kind et al. [52] published a similar finding.

For turbines, Taylor [4] measured surface roughness on two sets of first-stage turbine vanes, 30 from a TF-39 and 30 from an F-100 engine. He found that the roughness level varied by an order of magnitude around the blades on average ($1 < Ra < 10$ μm), with degradation favoring the leading-edge suction surface for one engine and the trailing-edge pressure surface for the other. In addition, individual traces showed the roughness to be highly non-Gaussian in many cases (nonzero skewness and kurtosis). Tarada and Suzuki [3] report on a survey of 58 used turbine blades from aero, marine, and industrial engines. Peak roughness levels (Ra) ranged from 50 to 160 μm in the most severely degraded portions of the blades, particularly in the leading-edge region. Bogard et al. [5] studied oxidized and deposited turbine vanes from two military aeroengines and reported that the surface character bore greater similarity to "scale" rather than the sand grit typically used to model roughness. Finally, Bons et al. [6] reported on a study of over 100 different used industrial turbine components, showing examples of deposition, corrosion, erosion, and coating spallation (see Fig. 6). Spatial nonuniformity of roughness character was the norm for all blades, and transitions between rough and smooth surface conditions were at times abrupt. Differences between the roughness signatures of various degradation mechanisms (e.g., deposition vs spallation) led the authors to conclude that no single characterization (such as sand, cones, or hemispheres) can accurately capture the range of diverse features exhibited by the various forms of surface roughness on serviced turbine blades. Given this assessment, it is

Fig. 6 Samples of erosion, deposition, and TBC coating spallation on turbine blading: a) erosion sample from suction surface leading-edge region (7 × 10 mm); b) fuel deposition sample from pressure surface trailing-edge region (3 × 4 mm); and c) spallation on turbine blade pressure surface.

perhaps inconceivable that any one roughness parameter, such as k_s or Re_k, could ever be effectively used to characterize roughness from such diverse samples. Bons [26] later reported, however, that both Λ_s and α_f provided reasonable correlations for c_f and St enhancement caused by roughness for a broad spectrum of surface features.

Observed spatial variations of roughness for inservice turbomachinery components are of particular significance because the sand-roughness experiments of Nikuradse [31] (and many since then) have studied the effects of uniform distributions of closely packed sand grains of a similar size. All three of these characterizations ("uniform," "closely packed," and "similar size") appear to be contradicted by actual surface measurements in gas turbines. Also, roughness modeling algorithms in commercially available CFD codes that apply a uniform k_s roughness level to all wetted surfaces can produce misleading results when compared to real turbine airfoils with degraded leading edges but smooth surfaces elsewhere. In addition, experimental work by Pinson and Wang [62] showed that abrupt changes in roughness size can have a significant effect on transition location. Specifically, abrupt streamwise changes from large roughness to a smooth wall condition exhibited earlier transition than changes from large to small roughness.

One final difficulty with defining a single, universal k_s correlation for all roughness characterizations is that while Schlichting defined k_s from the perspective of aerodynamic drag equivalence, it has since been adopted for convective heat-transfer equivalence as well. Fundamental issues with this ill-conceived "adoption" from aerodynamics to heat transfer are addressed in further detail later in this chapter.

VI. IS THERE A MINIMUM ACCEPTABLE LEVEL OF ROUGHNESS?

In many cases, the gas-turbine operator is solely concerned with insuring that the surface roughness is below the hydraulically smooth limit. Schlichting [35] plotted skin-friction data for turbulent boundary-layer flow over sand-roughened plates (again based on Nikuradse's pipe flow data) and noted the first evidence of roughness influence for $k_s \cong 100\nu/U_\infty$. For gas-turbine roughness on airfoils, it is most convenient to represent this admissible level of roughness ($k_{s,\text{adm}}$) relative to the blade (or vane) chord c:

$$\frac{k_{s,\text{adm}}}{c} \leq \frac{100}{Re_c} \quad \text{or} \quad Re_k \leq 100 \tag{12}$$

For a typical chord Reynolds number of 1×10^6 and blade chord of 5 cm, this represents a $k_{s,\text{adm}}$ of 5 μm. Given the range of correlations in Table 1 ($1 < k_s/Ra < 10$), this explains industry-wide specifications for surface finish $Ra \leq 1$ μm. The dimensionless roughness parameter $U_\infty k_s/\nu$ (sometimes called the "roughness Reynolds number" or Re_k) is often reported relative to the "100" threshold for admissible roughness. A review of gas turbine roughness literature shows a broad acceptance of Eq. (12) as a design guideline [45, 58, 63–66] although anecdotal evidence of temporary SFC reductions immediately following surface refurbishment on aeroengines well below the "100" threshold has been supported by recent research into "ultrapolishing" [67]. Nevertheless, the roughness Reynolds number Re_k is the most commonly reported roughness parameter in the open literature and is routinely used in the scaling of research facilities for roughness research [33, 61, 68, 69].

At issue, then, is how far above $k_{s,\text{adm}}$ the surface roughness can be tolerated before maintenance is necessary. Unfortunately, to answer this question requires an understanding of two of the most intractable problems in the fluid dynamics of wall-bounded flows, namely, boundary-layer separation and transition. The limited scope of this chapter cannot possibly do justice to these two subjects; it is hoped that the reader will forgive the meager discussion provided here.

Boundary-layer transition is influenced by a wide assortment of flow and surface parameters, in addition to surface roughness (including Reynolds number, freestream turbulence level and length scale, surface heating or cooling, pressure gradient, upstream history, blowing or suction, Mach number, shocks, and curvature). As an example of surface roughness effects on transition, Feindt [70] studied laminar boundary-layer transition on a sand-roughened wall and found the critical sandgrain dimension that induced early transition to be

$Re_k \approx 120$. The same threshold was noted for a modest range of adverse and favorable freestream pressure gradients. This criterion is very close to the $k_{s,\mathrm{adm}}$ relation for turbulent boundary layers [Eq. (12)]. Unfortunately, there are at least two complicating factors that must be taken into account. First, Feindt's data show a smooth wall transition Reynolds number ($Re_{x,\mathrm{tr}}$) of 6.6×10^5, which is considerably less than values that can be obtained experimentally using clean/quiet wind tunnels ($Re_{x,\mathrm{tr}} > 3 \times 10^6$). Thus, Feindt's transition criterion of $Re_k < 120$ must be adjusted for flows with other mitigating factors that influence transition (e.g., higher or lower freestream turbulence levels). The only component in a gas turbine engine that would be likely to have a more "disturbance-free" environment than Feindt's experimental facility is the engine inlet or nacelle. Thus, the roughness criterion for transition would be more stringent (less than 120) in this case. Embedded blade rows in compressors and turbines are rife with disturbances, like wakes and shocks, and thus can have significantly higher thresholds for roughness to effect transition. Braslow [40] reviewed boundary-layer development with distributed roughness and suggested a criterion of $k^+ > 19$ for roughness-induced transition. For the Re_c values typical of gas-turbine airfoils, this corresponds to a roughness Reynolds number criterion of $Re_k < 500$ for aerodynamically smooth surfaces. This is close to the threshold of 600 suggested by White [32] for an isolated roughness element. Indeed, transition is generally considered to be a significant factor in turbomachinery for chord Reynolds numbers above $2 - 4 \times 10^5$ [39, 45]. Roughness can also promote earlier transition in separated free shear layers, thus reducing the size of separation bubbles, as noted by Roberts and Yaras [71]. Prior to boundary-layer transition, roughness has no perceptible influence on local values of aerodynamic drag (Fig. 4) or heat transfer [19, 72].

Second, to prudently apply a minimum admissible roughness criterion, one must understand what happens once the threshold is exceeded. For a turbulent boundary layer, exceeding $k_{s,\mathrm{adm}}$ might mean a modest increase in blade profile losses (Fig. 4). For a laminar boundary layer, if exceeding $k_{s,\mathrm{adm}}$ causes transition, the profile loss could increase by up to a factor of 2 (and the local heat-transfer coefficient up to a factor of 8) [39, 19]. On the other hand, if the laminar boundary layer was already prone to separation, roughness-induced transition could minimize or eliminate separation and actually reduce profile losses (as in the golf ball flow) [73]. Thus, the effects of transition and separation are coupled in many turbomachinery flows.

VII. CAN ROUGHNESS EVER BE BENEFICIAL?

Because of the beneficial effect of surface roughness in overcoming laminar boundary-layer separation at low Reynolds numbers by inducing early transition, roughness has been explored as a form of passive flow control in many turbomachinery applications. Ishida et al. [74] applied 40-grit sandpaper to the hub-side only of a vaneless diffuser for a centrifugal blower and reported increased stall margin (up to 42%) because of the suppression of three-dimensional separation at low flow rates. The additional pressure drop due to the locally rough wall was less than 1% of the entire pressure rise for the blower. Boese and Fottner [75] studied the application of spanwise aligned v-groove type riblets to a highly loaded compressor cascade and found similar success controlling separation on the suction surface. Their optimal v-groove configuration reduced the total pressure loss coefficient up to a maximum of 5% within the range $1.5 \times 10^5 < Re_c < 11 \times 10^5$. Riblet heights in excess of 210 μm ($k/c = 1.2 \times 10^{-3}$ and $k^+ = 21$) caused turbulent boundary-layer separation, thus increasing profile losses. Optimal performance was for a rib with a height of $k^+ = 9$ and a spacing of $s^+ = 18$. Several passive flow control studies have been conducted on low-pressure turbine cascades using roughness elements including trip wires [76, 77], surface finish [73], spanwise ribs [73], scallops [78], v-grooves [79], and dimples [79]. All have shown reasonable success at low Reynolds numbers, as expected, though results are sometimes obscured with the addition of upstream wakes. For rotor tip flow control, Rao et al. [80] artificially roughened the outer rotor casing in their single-stage HP turbine facility and found a significant reduction in leakage mass flow rate, as well as a reduction in momentum deficit for the core of the tip vortex. They speculated that the observed change was caused by roughness-induced increases in the casing boundary-layer thickness, which would allow less mass and momentum flux through the tip gap. Although the same level of roughness could arise as a result of natural degradation processes, in many of these configurations it would be unlikely to occur in precisely the magnitude and location desired by the designer. Thus, strategically engineered roughness might become more commonplace in gas turbines of the future. Although unrelated to the topic of this study, engineered roughness elements in the form of cross-stream and angled ribs are routinely used to augment heat transfer in internal turbine cooling passages.

VIII. WHAT ARE THE SYSTEM-LEVEL IMPLICATIONS OF ROUGHNESS?

Unless frequent boroscopic evaluations are made to detect and repair surface damage, gas-turbine operators will generally see a gradual drop in performance (thrust or power output) as the first sign of degradation. Upon closer inspection, an increase in specific fuel consumption (SFC) will be evident as component efficiencies (for the compressor and/or turbine) decline. As degradation worsens, excessive temperatures can occur in the turbine as a result of coating loss or flow restriction. In the most severe case of excessive deposition buildup, loss of compressor stall margin can make safe operation of the gas turbine altogether impossible [12–17].

Because measures to prevent and repair surface degradation are costly and negatively impact availability, operators must be equipped with accurate cost and benefit estimates to maintain efficient gas-turbine operation. A number of recent studies have proposed system-level models aimed at determining optimum maintenance intervals and procedures to maximize gas-turbine availability and performance [16, 21, 23, 81–83]. These models account for system degradation through performance decrements associated with component efficiency loss and/or changes in mass flow (increase for erosion, decrease for deposition). Specific numerical values for these decrements are obtained from reported degradation losses available in the open literature [13, 14]. Zwebek and Pilidis, for example, account for turbine fouling (deposition) in their system-level analysis by assuming a 1% reduction in nondimensional mass flow and a 0.5% reduction in turbine efficiency [81, 82]. With this component-level input, their model predicts a drop in gas turbine power output and efficiency of 1.2 and 1%, respectively. Models have also been developed to predict specific component-level performance losses with degradation. For instance, Millsaps et al. [84] evaluated the effect of fouled airfoil surfaces due to deposits on a three-stage axial compressor. They assumed a doubling of blade profile losses and predicted a 1.5% drop in π_c and η_c with a 1% drop in mass flow. Component-level model predictions such as this can then be used as building blocks in a larger system-level model [81].

Although such models are useful to indicate trends, their ability to accurately predict changes in overall gas turbine system performance for a given installation is dependent on the accuracy of the numeric decrements employed. Determining these quantitative adjustments is complicated by the diversity of degradation sources and turbomachinery designs. There have been some attempts to test the accuracy of these system-level models using accelerated deposition testing with interim measurements of deposit thickness and surface roughness levels [1, 2, 15, 20, 85, 86]. Even with measurements of surface roughness in hand, however, there is still a significant leap from centerline averaged roughness *Ra* measurements to predicting compressor efficiency η_c. In practice, this chasm is often spanned using empirical correlations to convert measured roughness (e.g., *Ra* or *Rq*) to an "equivalent sandgrain" k_s value [Table 1 or Eqs. (10) and (11)]. This k_s value can then be used to predict local changes in boundary-layer parameters (e.g., θ and c_f) leading to profile loss estimates, again using empirical correlations, such as Eqs. (6–9) [17, 43, 46, 52, 63, 64]. Finally, summing profile losses through multiple stages (using a stage-stacking model, for example [86]) yields the desired estimate for an efficiency decrement $\Delta\eta_c$. The remainder of this chapter will review studies that have been conducted to bridge this gap between actual surface measurements and predicted performance decrements, specifically for the case of surface roughness in compressors and turbines.

IX. WHAT EFFECT DOES ROUGHNESS HAVE ON COMPRESSOR PERFORMANCE?

A review of the last 30 years of compressor roughness research shows a nearly equal interest in axial and centrifugal machines. Because the two devices experience different fluid mechanics issues, they will be treated separately here, though the primary focus is on axial machines. The role of the compressor is to increase the total pressure of the gas prior to combustion. This is done by imparting kinetic energy to the fluid and subsequently decelerating it to collect static pressure. Thus, compressor airfoils are designed with considerable care because of the inherent adverse streamwise pressure gradient. Overly aggressive blade camber can result in boundary-layer separation (particularly at low flow rates, i.e., high positive incidence angles), which reduces stage efficiency and can ultimately lead to compressor stall and/or surge. Circular arc camber lines with maximum thickness-to-chord ratios of less than 8% are therefore typical. Controlled diffusion airfoils (front-loaded airfoils that take advantage of laminar flow prior to the suction peak and control boundary-layer diffusion to avoid separation) are also widely used in multistage compressors. Because the pressure rise scales with rotor rotational frequency, substantial benefit can be derived from operating at higher speeds, creating the potential for transonic flow in the compressor passages. Supersonic compressor airfoils are designed with sharp leading edges to reduce shock-induced losses and are therefore sensitive to changes in leading edge shape. Because of the critical role of boundary-layer separation (both for sub- and supersonic airfoil designs) in determining profile losses, it is important to ensure that compressor roughness research is conducted with the appropriate range of dimensionless parameters: Reynolds and Mach number.

The critical role of Reynolds number is well-illustrated in the experimental work of Leipold et al. [61]. They studied the effects of typical blade forging roughness using a high-pressure cascade facility over a range of Re_c from

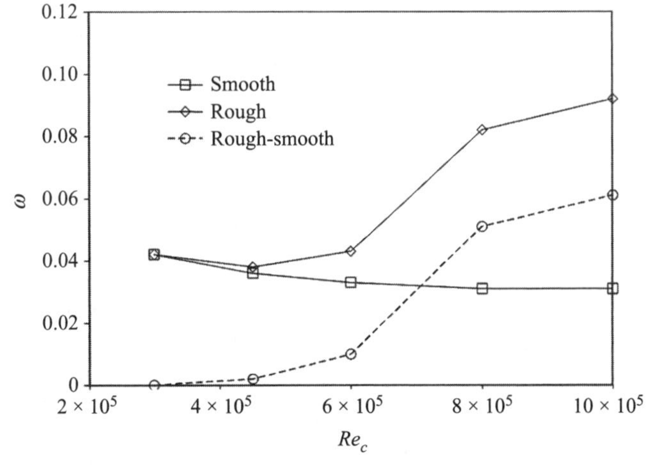

Fig. 7 Compressor cascade loss measurements for rough and smooth blades vs Re_c (Fig. 8 from Leipold et al. [61]).

3×10^5 to 1×10^6 (based on inlet conditions). The inlet Mach number to the cascade was set to 0.67, with a peak isentropic Mach number near 1 at the highest Re_c (i.e., the flow was shock free). Figure 7 shows their wake measurements of total pressure loss taken at midspan (span/chord = 1.67) for both the smooth and artificially roughened airfoils. The roughness-to-chord ratio was $Rz/c = 4 \times 10^{-4}$, which just exceeds the $k_{s,\text{adm}}$ threshold [Eq. (1)] at $Re_c = 2.5 \times 10^5$ (assuming $Rz \approx k_s$ for this case). The trend for the smooth airfoil in Fig. 7 follows the conventional wisdom that losses fall (efficiency rises) as Re_c increases up to a plateau around $Re_c = 5 \times 10^5$. The rough blade losses show the opposite trend of increasing losses above this Re_c threshold. Boundary-layer measurements confirmed that this rise in profile loss was caused by turbulent boundary-layer separation on the airfoil suction surface. As Re_c increases, the thinning boundary layer creates a more pronounced roughness effect (increasing k_s/δ), and roughness-induced momentum losses in the turbulent boundary layer at midchord cannot overcome the diffusion on the aft portion of the blade. Although roughness does eliminate a small laminar separation bubble at moderate Re_c, the added viscous drag more than makes up for this in the aggregate profile loss measurement. This finding is particularly relevant because a large aeroengine at cruise experiences core compressor Reynolds numbers from $1 - 2 \times 10^6$ (higher still at takeoff conditions). Schaffler [45] emphasized this same trend of increased loss with inlet Reynolds number using polytropic efficiency measurements from a five-stage high-pressure compressor rig. For blade surfaces with a roughness Reynolds number only 60% higher than the admissible threshold, Schaffler reported lower efficiencies (compared to smooth blades) beginning at $Re_c = 3.1 \times 10^5$ (first stage rotor Re_c). Because Re_c increased by a factor of two through Schaffler's five-stage high-pressure compressor, the last stage was clearly above the "knee" in the profile loss curve reported by Leipold et al. (Fig. 7). Thus, roughness effects that would not be evident from a low-Reynolds-number compressor cascade test can become critical during actual engine operation.

In 1972, Bammert and Milsch [63] published an oft-referenced study on a low-speed compressor cascade with five sand-grain roughness levels: $2.3 \times 10^{-4} < k_s/c < 5.6 \times 10^{-3}$. The inlet Reynolds number was fixed at $Re_c = 4.3 \times 10^5$, and the Mach number was 0.11. They reported increases in profile losses from 2 to 10% over this range of roughness, largely because of trailing-edge separation on the suction surface. They also noted reduced turning with increased roughness, which would certainly affect the efficiency of the subsequent blade row by altering the incidence angle. Thus, a follow-on study by Bammert and Woelk [87] using a three-stage axial compressor with comparable sand-grain roughness levels on all three stages of airfoils produced a 6–13% loss in efficiency and up to 30% loss in overall pressure ratio. Similar cascade profile loss increases were reported by Elrod and King [88] though their sand roughness ($2 \times 10^{-3} < k_s/c < 9 \times 10^{-3}$) was only applied to the leading 25% of the suction surface (to simulate regions typically affected by erosion). Elrod et al. also found that roughness effects were overshadowed at elevated freestream turbulence levels (up to 5%). This finding runs counter to the observations of Schreiber et al. [89] who studied roughness effects on transition in a compressor cascade using liquid crystals ($0.7 \times 10^6 < Re_c < 3 \times 10^6$). They noted that elevated freestream turbulence (up to 4%) actually increased the frequency of turbulent wedges forming in the wake of discrete roughness elements in an otherwise laminar boundary layer. Thus, the effects of roughness and freestream turbulence on transition were complementary or synergistic. Incidentally, Leipold et al. 's study was conducted with 2–3.5% turbulence, whereas Bammert and Milsch's facility had less than 1%. Schreiber et al. also visually corroborated previous findings that an increasing flow Re_c amplifies the effect of roughness on early transition, thus reducing laminar separation zones.

One of the few compressor roughness studies on a transonic rotor, by Suder et al. [18], concluded that roughness decreases performance by increasing shock losses. Roughness-laden paints were applied to a research rotor (relative tip Mach number = 1.48 at design speed) with roughness Reynolds numbers more than five times the threshold value of 100. At design speed, roughness and increased blade thickness (due to the paint thickness) both produced efficiency decrements of 3% (with a combined 9% reduction in pressure ratio). By systematically applying the paints to the rotor-blade leading edges and then trailing edges only, Suder et al. determined that the effect of roughness on the shock-boundary-layer interaction at the leading edge of the airfoil was responsible for the measured losses in performance. Roughness (and added paint thickness) over the first 2% of blade chord caused a thickening of the boundary layer. Subsequent interaction of this thickened boundary layer with the rotor passage shock created increased blockage and reduced diffusion in the blade passage. The effect of roughness was much less pronounced at lower operating speeds (only 1–2% effect on efficiency at 60% design), as would be expected due to the lower Re (lower k_s/δ) and the absence of shocks.

The critical role of compressor blade leading-edge modification as a result of degradation (erosion or deposition) has been well researched, usually without considering the complicating effects of surface roughness [2, 20, 65, 67, 90]. One exception to this is the recent work by Gbadebo et al. [33], who studied roughness effects on endwall separation in a low-speed, single-stage compressor facility ($Re_c = 2.7 \times 10^5$). The first 20% chord on the stator blades was modified by attaching emery paper with a grain size to match the roughness Reynolds number of a modern turbofan engine at takeoff with a typical surface condition of $Ra \approx 2\ \mu m$. Measurements showed significant increases in hub endwall losses (up to 15%) on the suction side of the airfoil as a result of three-dimensional endwall separation triggered by the leading-edge roughness. The affected region extended to 30% span (Fig. 8). At the same time, midspan profile losses and tip endwall losses were virtually unaffected. Roughness applied downstream of the suction peak had no perceptible effect on profile or endwall losses.

Centrifugal compressors have considerably higher single-stage pressure ratios (2–4) compared to axial compressors (1.2–1.5), and the flow is generally not

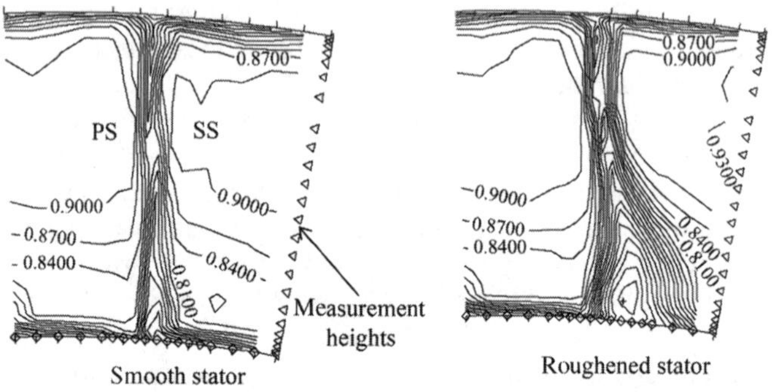

Fig. 8 Contours of total pressure rise coefficient at the exit of compressor stator for a) smooth and b) LE roughness only conditions; $Re_c = 2.7 \times 10^5$ (Fig. 2 from Gbadebo et al. [33]).

transonic. Efficient compressor performance is often dictated by the interaction of the impeller with the diffuser and collector. Useful one-dimensional models have been developed by Wiesner, Simon and Bulskamper, and Benra et al., among others [46, 64, 91]. Results show trends similar to those just cited, namely, an increased detrimental effect of roughness with increasing Reynolds number. Relative roughness size is referenced to the impeller discharge height b_2 or hydraulic diameter d_h, and roughness effects become significant for $k_s/d_h > 3 \times 10^{-4}$. Roughness influences centrifugal compressor loss through the same fluid mechanisms of earlier transition, increased boundary-layer losses, and flow separation.

X. WHAT EFFECT DOES ROUGHNESS HAVE ON TURBINE PERFORMANCE?

Axial-flow turbines are distinguished from compressors by several key design features. First, they are designed to extract work from an expanding flow. Because the static pressure is falling in the flow direction, boundary-layer separation is generally not as critical, and blade designs can be much more aggressive (i.e., higher turning) than in an axial-flow compressor stage. Consequently, there are far fewer turbine stages than compressor stages because the stagnation enthalpy drop through one turbine stage can be used to drive three to five compressor stages. Higher blade loading requires thicker airfoils, and turbine airfoils can have maximum thickness-to-chord ratios in excess of 25%. Because of this, airfoil trailing edges are thicker creating larger wake losses. Also, the flow in many first-stage (high-pressure) turbines is transonic, with choked nozzles during normal operation. Thus, maintaining minimum passage (throat) areas is critical for efficient operation at the design flow coefficient.

The other distinctive feature of turbines is their proximity to the combustor exit. Turbine inlet temperatures for modern gas turbines regularly exceed material limits of the most exotic single crystal Ni-based alloys. Thus, since the 1960s, hollow turbine vanes and blades have been fed with compressor discharge air to alleviate the effects of thermal fatigue. In the last two decades, this internal cooling has been augmented with elaborate film-cooling passages and external thermal barrier coatings (TBCs) to protect the structural metal. Designers now routinely depend on these "thermal management" systems to allow higher firing temperatures ($\sim$1800 K) than turbine metals can withstand. Because of these added complexities, turbines now account for a disproportionate amount of the overall gas-turbine cost both in terms of R&D and operating costs. It is perhaps no surprise then that most of the gas-turbine roughness research in recent years has been devoted to turbine applications. For example, over three times as many turbine roughness articles (compared to compressor roughness articles) were used in the preparation of this chapter, and roughly half of them are related to heat transfer as opposed to aerodynamic loss. The following is a summary of major findings from these articles. Surface roughness effects on turbine film cooling are addressed in the subsequent section.

Although compressors must deal with ingested foreign matter, including aerosols, airborne salts, pollutants, and foreign objects that can deposit or cause erosion, any large particles are generally broken up into micron-sized dust (mean diameter $<7\ \mu m$) by the time they reach the turbine [92]. This does not mean, however, that the turbine is a pristine environment without its own degradation mechanisms. Ash bearing fuels used in land-based turbines can rapidly clog cooling holes and reduce choked passage throats [93, 94]. Also, because of the high turbine inlet temperatures, many of the airborne contaminants that pass through the compressor can become molten in the turbine, agglomerating in significant quantities [92]. As deposits adhere to ceramic turbine coatings (TBCs), coating removal by delamination (or spallation) can occur, creating local roughness features comparable to the boundary-layer dimensions (0.2–0.5 mm) [6, 10, 95]. Thus, it is critical to understand both the aerodynamic and thermal penalties associated with surface roughness in turbines.

Perhaps because of the mitigating factors just outlined (elevated operating temperatures), there have been relatively few roughness studies on actual operating turbines, as compared to compressors. In 1972, Bammert and Sandstede [96] applied emery grain roughness to a 703-kW, four-stage research turbine and reported a 10% drop in overall efficiency for $k_s/c = 0.00765$. An additional 3% efficiency loss was attributed to the increased blade thickness (emery grains plus the adhesive layer). They recommended maintaining turbine surfaces to a finish below k_s/c of 0.001 ($Re_k < 500$). Recently, Yun et al. [97] conducted a similar study by adding sandpaper roughness to a single-stage 15-kW research turbine. By carefully matching the blade thickness for the smooth and rough test cases, they measured efficiency losses caused only by surface roughness that was slightly larger than the findings reported by Bammert and Sandstede (up to 19% drop in η for $k_s/c = 0.0067$). Another full turbine rig test with roughness was conducted by Boynton et al. [98] with the Rocketdyne Space Shuttle Main

Engine (SSME) two-stage fuel side turbine (axial chord ≈ 2 cm). Polishing the rotor and stator airfoil surfaces from $Rq = 10$ μm down to 0.76 μm yielded a 2.5% increase in turbine efficiency. A series of follow-on experiments were conducted by Blair [69] and Dunn et al. [99] to discern whether the same benefit could be expected for blade surface heat transfer. Blair applied large grain sand ($k_s/c \approx 0.004$) to a large scale ($c \approx 15$ cm) stage and a half research turbine and found dramatic increases in local heat transfer, compared to the smooth wall case. The most significant effect (100% increase in local St) was caused by early roughness-induced transition on the suction surface. Stanton-number increases in turbulent boundary-layer regimes (e.g., on the pressure surface) were a more modest 20–30%. Dunn et al. [99] subsequently measured surface heat transfer on a replica of the same SSME turbine used by Boynton et al. [98] but with a surface roughness of $Rq = 15$ μm and $Rz = 84$ μm ($Rz/c \approx 0.004$). Lacking a smooth-blade data set for direct comparison, Dunn et al. used a CFD calculation of Boyle and Civinskas [47] to conclude that the surface was not rough enough to produce a measurable increase in heat transfer. Though Taylor et al. [100] later argued that Dunn et al.'s use of smooth heat-flux gauges might be inappropriate for accurate rough surface heat-transfer measurement, the primary factors leading to this apparent contradiction between the results of Blair [69] and Dunn et al. [99] can be explained by dimensional scaling arguments. Though the sand-grain size employed by Blair ($Rz = k_s = 660$ μm) was selected to be an eight times geometric scaling of the measured turbine roughness used by Dunn et al. ($Rq = 15$ μm and $Rz = 84$ μm) to match the different blade chord ratios, Boyle and Civinskas [47] postulated that sand roughness and blade forging roughness do not have the same k_s/Rz conversion. In addition, the study by Blair operated at a higher Reynolds number (5.8×10^5) than the Dunn et al. experiment (3.8×10^5). Both Re_{cx} values were significantly lower than that used in the Boynton et al. aerodynamic study (8.9×10^5).

Thus, we again see the critical role that Reynolds number plays in determining the significance of surface roughness effects in turbines (as was just outlined for compressors). There are several excellent studies that document the synergies between Re_{cx} and roughness effects in turbines [49, 51, 54, 101]. Boyle and Senyitko [39] applied a ZrO-laden spray-on coating to a high-pressure turbine vane cascade. The average Rq was 1.6 and 18.2 μm for the smooth and rough vanes respectively (estimated $k_s/c = 0.00084$ for the rough vane). The test facility had the capability of varying Mach and Reynolds numbers independently. Figure 9 contains two of the wake loss coefficient plots from this study, for exit Mach numbers of 0.5 and 0.9. (Note: The Re_c based on true chord shown in the figure can be adjusted down to the Re_{cx} based on axial chord by multiplying by 0.5.) Symbols are shown for several levels of freestream turbulence: 1% (no grid), 4.5% (small grids no air), 8% (small grids with air), and 17% (large grids with air). All of the data show a trend of increasing roughness effect on vane losses with increasing Reynolds number (up to a 50–60% loss increase at the higher Re_c). The increased loss is attributed to both roughness-induced

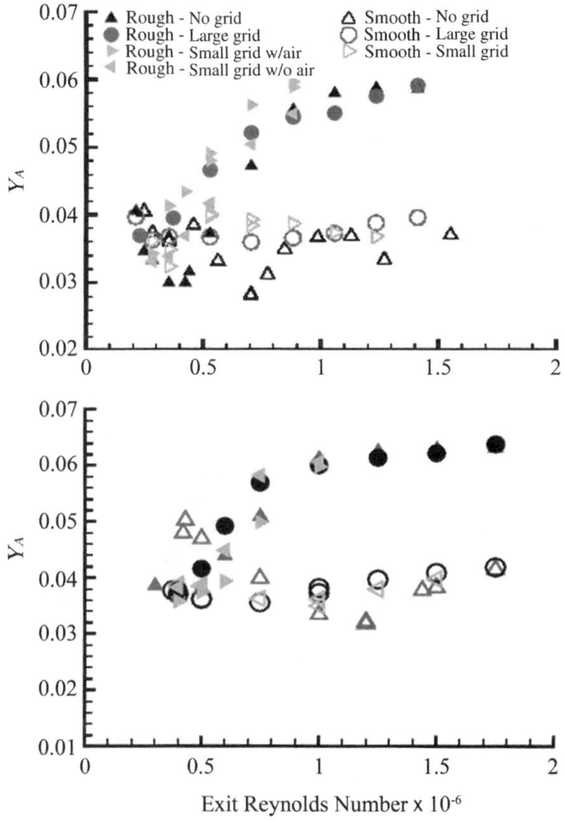

Fig. 9 Area-averaged loss coefficient vs Re_c for HP vane cascade at exit a) Mach number = 0.5 and b) 0.9 (Figs. 6a and 4a from Boyle and Senyitko [39]) ($Re_{cx} \approx 0.5\ Re_c$).

transition and increased turbulent boundary-layer momentum losses. For some low Re_c cases, roughness-induced transition actually prevented laminar boundary-layer separation on the suction surface, thus lowering the overall vane losses compared to the smooth-wall case. The threshold for a rough-wall loss benefit occurred near $Re_c = 5 \times 10^5$ for $M_{ex} = 0.9$ (slightly lower at $M_{ex} = 0.5$) (Fig. 9). Hummel et al. (58) explored a similar exit Re_{cx} range (5.6 to 11×10^5) in a linear rotor blade cascade with spanwise grooves applied to the blade surfaces. Their grooved surface roughness height was smaller ($0.8 \times 10^{-5} < Ra/c < 8 \times 10^{-5}$) than the range used by Boyle and Senyitko ($1.2 \times 10^{-5} < Ra/c < 14 \times 10^{-5}$), and they found a correspondingly smaller profile loss increase (20–50% vs 40–60%). At the lowest Reynolds number tested ($Re_{cx} = 5.6 \times 10^5$), Hummel et al. did not find a region of roughness benefit. The low Reynolds-number benefit only occurs if a smooth-wall laminar separation is prevented by roughness-induced transition, which depends strongly on the airfoil profile and loading. Low-pressure turbine airfoils are more prone to experience this benefit, as will be discussed later. The blade profiles used by Hummel et al. are significantly different from the nozzle guide vanes in Boyle and Senyitko's cascade, which might explain the different observations. It is possible that the low Reynolds-number roughness benefit might be muted when considering multiple blade row or multistage effects because neither Bammert and Sandstede [96] ($3 \times 10^5 < Re_c < 6 \times 10^5$) nor Yun et al. [97] ($Re_c \cong 2 \times 10^5$) found evidence of this phenomenon in their full rig tests just cited.

The Boyle and Senyitko [39] data shown in Fig. 9 also include the combined effect of roughness with freestream turbulence. For the conditions studied (1–17% turbulence), turbulence alone appears to have some of the same low and high Reynolds-number effects on losses as roughness although roughness effects dominate when the two are present simultaneously. This is not a universal finding as a number of other studies have shown. In flat-plate wind-tunnel studies, Bogard et al. [5] found the two effects (roughness and freestream turbulence) to be additive for heat-transfer augmentation, meaning that the increased heat transfer with both roughness and turbulence present was approximately equal to the sum of the heat-transfer increase measured with each one alone. Bons [9] describes a synergistic effect of combined roughness and turbulence on c_f due to the fuller turbulent boundary layer producing higher momentum loss on the protruding roughness peaks. Bunker [101] measured vane heat transfer in a linear cascade with both roughness and freestream turbulence (4–12%) at much higher Re_{cx} ($2–5 \times 10^6$) and found the effects to be at least additive. Zhang and Ligrani [56] ($Re_{cx} = 6 \times 10^5$) report similar results for cascade aerodynamic losses with combined roughness and freestream turbulence (up to 8%). The Boyle and Senyitko [39] data (Fig. 9) also show roughness-induced losses to be fairly insensitive to cascade exit Mach number. This finding is corroborated in the data sets of Yuan and Kind [59] and Zhang et al. [102] as well as the low-pressure turbine work of Vera et al. [73]. Thus, roughness effects deduced from low-speed experiments are relevant, as long as the Reynolds number, roughness scaling, and freestream turbulence levels are representative of engine conditions.

The aerodynamic loss studies just referenced universally identify the suction surface as the region most susceptible to roughness effects. In fact, several researchers applied roughness exclusively to the suction surface to demonstrate this fact conclusively [52, 97, 103]. As explained earlier, losses generated on the suction side are strongly dependent on the interplay between separation and transition, and roughness can have a significant effect on both of these phenomena. Although the efficiency of a turbine stage depends on the mass-averaged losses at the exit, heat transfer is a phenomenon of local concern. For example, if roughness-induced transition prevents separation on the suction surface at low Reynolds number, it might ultimately reduce the stage losses and improve efficiency. However, local levels of heat transfer at the transition location will see an increase of up to a factor of two to three compared to the smooth-wall condition where laminar separation is present.

The vane heat-transfer data from Stripf et al. [57] shown in Fig. 10 provides an excellent example of the local heat-transfer augmentation caused by roughness-induced transition. (Note: The Re_1 values shown in the figure are based on true chord and inlet velocity. It is estimated that the corresponding Re_{cx} values would be 60% larger.) The data were acquired in an HP vane cascade with truncated cones of various geometries and spacings etched onto the surface. The legend contains the identifier for each roughness case. The two numerical digits indicate the height of the cones in microns ($k = Rz$) while the

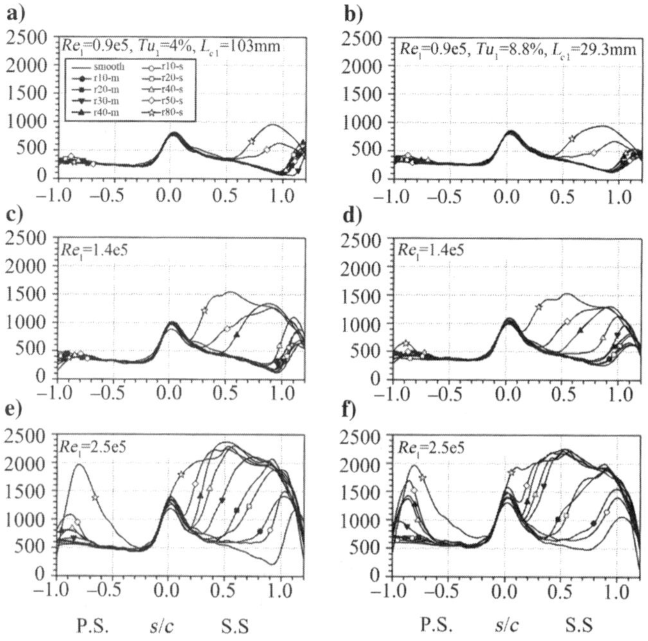

Fig. 10 Turbine vane *Nu* data for various *Re* and *Tu*: 1.6 $Re_1 \approx Re_\alpha$ (Fig. 6 from Stripf et al. [57]).

"-s" indicates a sparse separation and an "-m" is a closer spacing (thus, *r*40-s is roughness with 40-μm high truncated cones at sparse spacing). The vane chord was 94 mm, so the largest Rz/c ratio was 8.5×10^{-4}. Using a correlation derived from Waigh and Kind [104], Stripf et al. estimated $k_s/Rz \cong 2$ for the "-s" cases and $k_s/Rz \cong 5$ for the "-m" cases. The authors indicate that at low freestream turbulence, transition on the suction surface takes place via a short laminar separation bubble, while the flow is completely laminar on the pressure surface, because of the strong flow acceleration there. Roughness has the effect of eliminating the suction-side separation in favor of a bypass transition mode that moves upstream with roughness size. The effect is more pronounced at higher Reynolds numbers and freestream turbulence levels. Roughness contributed to early boundary-layer transition on the pressure surface only at the highest Re_1 tested (2.5×10^5). Stagnation point heat transfer also shows an increase with roughness, up to 25% for the largest roughness considered. Bunker [101] reported similar trends for an order of magnitude higher Reynolds number in his transonic linear vane cascade ($c = 12.75$ cm). Figure 11 shows heat-transfer coefficients measured at $Re_{cx} = 4.7 \times 10^6$ and $Tu = 9\%$. As with the Stripf et al. data set, transition moves forward on the suction surface as roughness level is increased by an order of magnitude from $Ra = 0.4$ μm

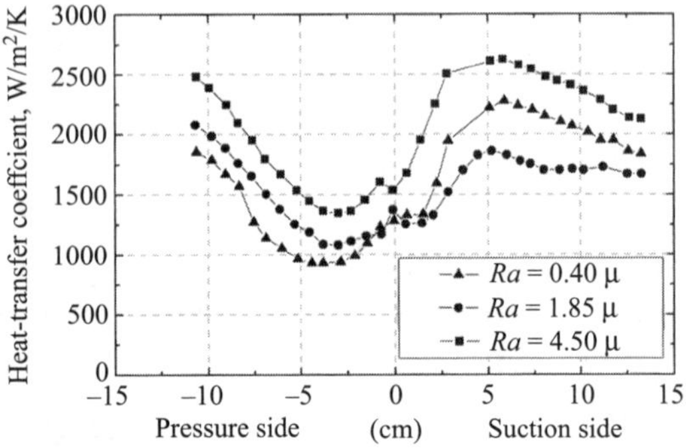

Fig. 11 Vane heat-transfer coefficient data for various *Ra* at $Re_{cx} = 4.7 \times 10^6$ and $Tu = 9\%$ (Fig. 12 from Bunker [101]).

($Rz/c = 1.8 \times 10^{-5}$) to 4.5 μm ($Rz/c = 2.2 \times 10^{-4}$). Stagnation heat transfer showed a slight increase, though this was within the experimental uncertainty. Outside of the leading edge and transition regions, Bunker found that heat transfer scaled roughly as $Rz^{0.25}$ for all of the turbulent boundary-layer results. The heat-transfer increases of 20–40% shown in Fig. 11 are consistent with the results of rough-wall turbulent boundary-layer heat-transfer studies conducted in numerous flat-plate wind-tunnel facilities [5, 9, 54, 100].

Two recent studies have explored the influence of roughness on end-wall secondary flows in turbines. Matsuda et al. [103] measured profile and endwall total pressure losses in a large-scale vane cascade with varying degrees of surface polish ($0.8 \times 10^{-5} < Rz/c < 8.4 \times 10^{-4}$). The freestream turbulence level in the tunnel was low (0.5%), and Re_c was varied from 0.3 to 1×10^6. Figure 12 contains a summary of profile loss measurements from their study. Roughness causes a significant rise in loss for $Rz/c > 2.5 \times 10^{-4}$. Incidentally, Matsuda et al. also found a roughness benefit at a low Reynolds number similar to that found by Boyle and Senyitko [39] for $Re_c < 4 \times 10^5$ (not shown). Contour maps of pressure loss (Fig. 13) also show a marked rise in endwall losses for the large roughness cases (a net increase of up to 50% for the largest roughness case). When roughness is added to the endwall (for the largest roughness case), an additional 30% net loss is measured. The authors

	C1	C2	C3	C4	C5	C6	C7
$Rz/c \times 10^5$	0.8	5.2	6	17	36	63	84
URZ/ν	10.4	67.7	78.1	224	468	817	1093

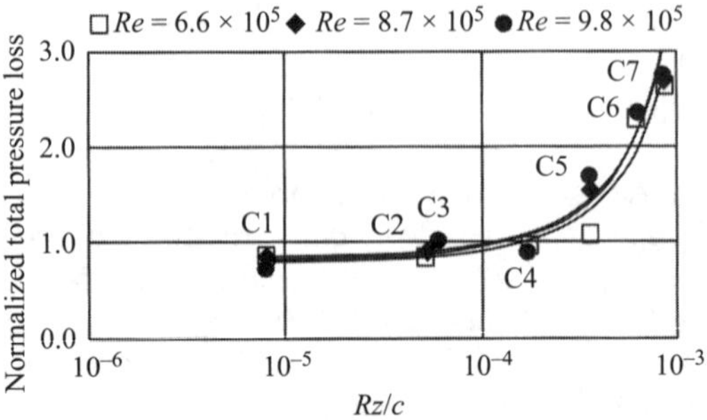

Fig. 12 Turbine vane total pressure loss data for various Rz/c for three values of Re_c and $Tu = 0.5\%$: $Re_{cx} \approx 0.7\, Re_c$ (Fig. 7 from Matsuda et al. [103]).

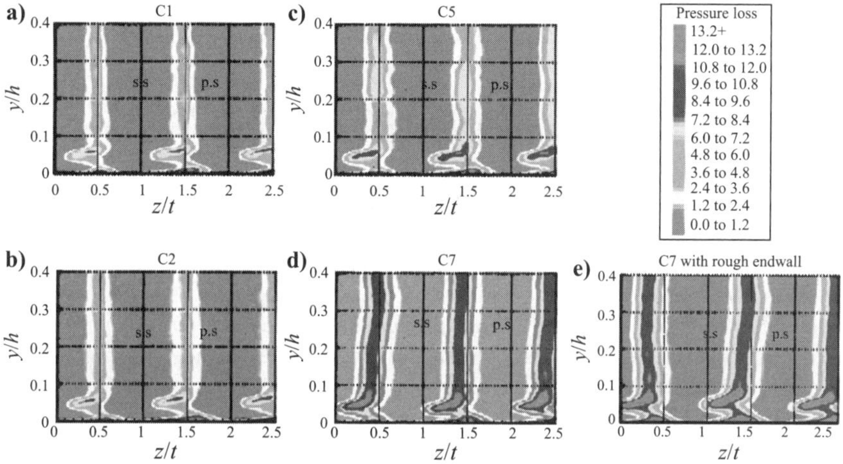

Fig. 13 Turbine vane total pressure loss data for $Re_c = 8.7 \times 10^5$ and $Tu = 0.5\%$: a)–d) rough vanes with smooth endwall, e) rough C7 vane with rough endwall: $Re_{cx} \approx 0.7\ Re_c$ (Figs. 9 and 15 from Matsuda et al. [103]).

attribute this increased loss to secondary flows (vortices) that roll up roughness-affected low momentum fluid from the boundary layer. The magnitude of endwall loss increase is larger than that reported by Gbadebo et al. [33] for compressors. It is unlikely that the effect is caused by separation losses on the endwall because Fig. 13 shows no appreciable movement of the endwall loss structure away from the hub (as was observed by Gbadebo et al., Fig. 8).

Using the linear vane cascade cited earlier [57], Stripf et al. [19] explored vane surface heat transfer in the endwall-affected region (with smooth endwalls and a spanwise boundary-layer trip near the vane leading edge). Like Matsuda et al., they found that vane roughness did not appreciably alter the size of the endwall-affected region [103]. Turbulent heat transfer in this region, however, increased up to 80%, with the largest vane roughness ($k/c > 3.2 \times 10^{-4}$), compared to 40% at midspan ($Re_{cx} \approx 4 \times 10^5$ and $Tu = 4$ & 8%). Without a leading-edge boundary-layer trip, the midspan heat-transfer increase was over 600% because of roughness-induced transition at this low Re_{cx} whereas the endwall heat-transfer increase was still only 80%. Roughness-induced endwall heat-transfer increases are of particular concern for turbine designers because the rotor endwall is a region of aggravated mechanical stress as well.

As with the compressor, many of degradation mechanisms that produce surface roughness in the turbine result in undesirable geometry modifications. Examples include rotor tip wear [105], leading-edge deposit buildup [106], and large surface steps created by TBC spallation [54]. Studies of the impact of these geometry modifications typically neglect the associated surface roughness effects because they are considered to be of secondary importance.

XI. WHAT EFFECT DOES ROUGHNESS HAVE ON TURBINE COOLING?

Film cooling is employed in high-pressure turbine stages to provide a layer of cooler air near the metal surface. Because the highest heat load occurs in the stagnation zone, multiple spanwise rows of compound angle holes are typically used in this region. Additional streamwise-aligned film-cooling rows are often found on the pressure surface and near the leading edge on the suction surface. Film hole exits can either be cylindrical or fan-shaped. The fan shape provides for additional diffusion of the coolant and thus better surface coverage over a wide range of coolant flowrates. Film cooling performance is evaluated using the film-cooling effectiveness parameter, $\eta = (T_s - T_\infty)/(T_c - T_\infty)$, which is generally spanwise or area-averaged in the region downstream of the film holes.

Figure 14 shows the effect of roughness on film-cooling effectiveness, as reported by Rutledge et al. [107]. For their study, Rutledge et al. used a distributed array of conical roughness elements ($k_s/c \approx 8.4 \times 10^{-4}$) on the surface of a film-cooled HP turbine vane in a low-speed linear cascade ($Re_{cx} \approx 5 \times 10^5$). The trends in the data of Fig. 14 are similar to the results obtained by others [48, 108–110]. At low blowing ratios ($M = \rho_c U_c / \rho_\infty U_\infty$), roughness reduces the film effectiveness whereas at high blowing ratios, roughness can actually improve film effectiveness by limiting jet liftoff from the surface.

Surface roughness degrades performance at low blowing ratios through two mechanisms. First, rough surfaces produce thicker boundary layers and thus lower near-wall velocities compared to smooth surfaces. This produces a higher "effective" blowing ratio for the roughness-thickened boundary layer, which can

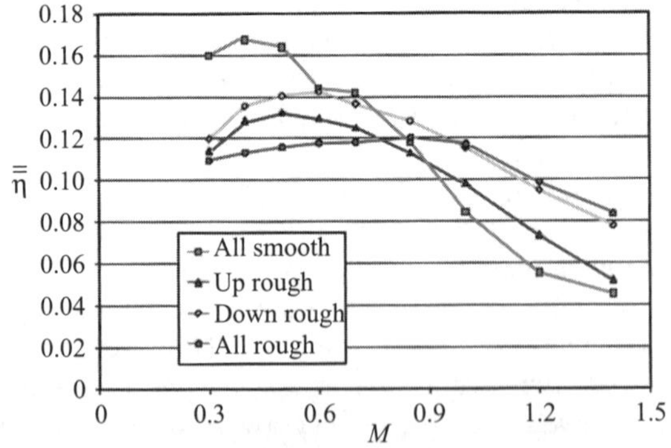

Fig. 14 Area-averaged film effectiveness vs blowing ratio for roughness elements upstream, downstream, or both upstream and downstream ("All") of film holes on vane cascade: $Re_c = 1 \times 10^6$, $Re_{cx} \approx 0.5\ Re_c$, and $Tu = 6\%$ (Fig. 5 from Rutledge et al. [107]).

lead to jet liftoff at lower values of M. Second, roughness generates significant near-wall turbulence that dissipates coolant more rapidly. Cardwell et al. [111] measured a 30% decrease in endwall film-cooling effectiveness due to a combination of these two roughness-induced effects using a vane cascade identical to Rutledge et al. [107]. Their cascade endwall was roughened with 32-grit sandpaper ($Ra = 0.23$ mm, $k_s \approx 0.64$ mm) while the vane surface was smooth. Film-cooling effectiveness alone does not capture the full effect of roughness on film cooling because roughness also enhances surface heat transfer. When the combined effects of film-cooling effectiveness and heat transfer are properly accounted for, the result can be a 30–70% increase in surface heat flux with roughness as compared to a smooth surface [107, 112]. Considerable attention has also been given to the damaging effect of flowpath obstructions on film effectiveness, either from deposits [112–114] or TBC residual [115].

XII. WHAT IS THE STATE OF THE ART IN ROUGHNESS MODELING FOR CFD?

Though the primary focus of this chapter is experimental measurements of roughness effects, a brief summary of roughness modeling is included here for the sake of completeness. For more detail, the reader can consult any of the references listed in this section. To accurately predict the effect of roughness on turbomachinery aerodynamics and heat transfer, models must account for roughness effects on boundary-layer development and transition. Three different modeling strategies have been employed to this end: 1) adding a roughness sensitivity to the turbulent eddy viscosity near the wall, 2) accounting for roughness blockage, heat transfer, and obstruction drag through a "discrete-element model," and 3) fully discretizing the roughness features.

The majority of turbulence models in use today incorporate the turbulent eddy viscosity μ_t in their formulation. Accordingly, the most common method of accounting for roughness is to make μ_t a function of roughness height. For example, Boyle and coworker [47, 116] incorporated Cebeci and Chang's [117] roughness model into a 2-D Navier–Stokes solver used for turbine heat-transfer predictions by modifying a version of the Baldwin–Lomax turbulent eddy viscosity model. For a rough surface, the turbulent lengthscale (and thus μ_t) in this zero-equation model increases with the equivalent roughness height k_s^+, effectively eroding the effect of viscous (vanDriest) damping near the wall. Others have used similar implementations [33, 69]. Alternative strategies incorporate roughness into the wall boundary condition for μ_t [59, 118, 119] or ω (in k-ω turbulence model [39, 65]) or through modified wall functions [50, 120]. Using the Cebeci and Chang model, Boyle and Senyitko reported reasonably accurate predictions for vane profile loss [39] but not for vane heat transfer [121]. This result was only achieved with fine-tuning of both a k_s/Rq correlation and a roughness-induced transition model. The fundamental weakness of these models is their

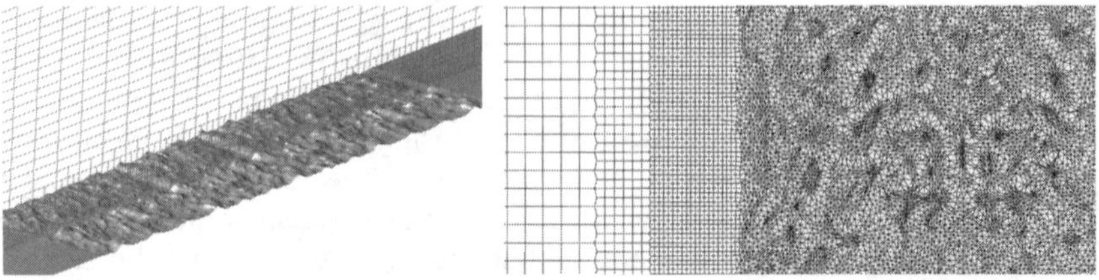

Fig. 15 a) Cutting plane showing the viscous adaptive Cartesian grid for a 240 × 60 mm patch of erosion roughness; b) surface grid on the erosion surface showing grid refinement near leading edge (from Bons et al. [128]).

reliance on k_s to characterize the roughness. The roughness "equivalence" (as proposed originally by Schlichting, and in itself still controversial) was only for skin friction c_f. Therefore, the use of k_s for predictions of heat transfer or boundary-layer transition in addition to c_f is inappropriate. This issue is reviewed by Aupoix and Spallart [118] and Stripf et al. [72]. In essence, extrapolating from roughness effects on c_f to St is problematic because Reynolds analogy ($2St/c_f \cong$ constant) is inappropriate for rough surfaces [26] and for gas-turbine flows in general [122].

A popular alternative to k_s-based roughness models is the "discrete element" approach. This method accounts for the roughness by extra terms in the governing equations that represent the flow blockage due to the roughnesses and the drag and heat flux on roughness elements [72, 100, 123–125]. As such, it is not dependent on the appropriateness of Reynolds analogy and does not require a sand-roughness equivalent. Though originally developed for ordered roughness elements (spheres, cones, etc.), this method has been successfully applied to

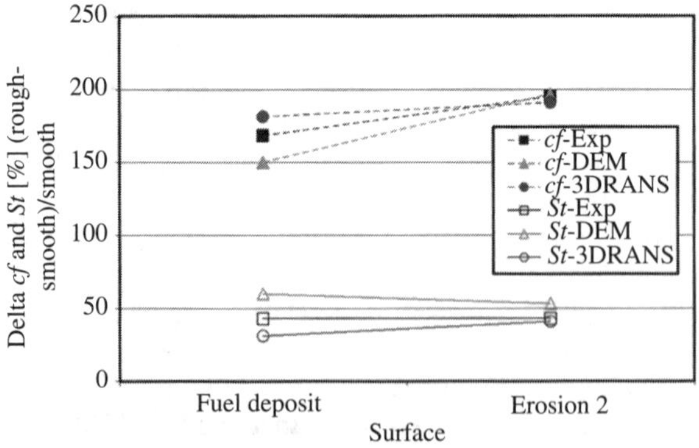

Fig. 16 Comparison of percentage change in c_f and St from experiment and computation (3-D RANS and DEM). Two different roughness surfaces: fuel deposit and erosion. Zero-pressure-gradient turbulent boundary layer with $Re = 1 \times 10^6$ (from Bons et al. [128]).

real, randomly rough surfaces as well [126, 127]. In its current form, the discrete element method is not formulated for three-dimensional, unsteady flowfields and thus has seen limited application to turbomachinery flowfields.

Theoretically, the reliance on roughness models could be eliminated if the roughness were fully resolved with the computational grid. Bons et al. [128] presented a comparison of c_f and St predictions for a 3-D RANS solver and a 2-D discrete element model. The 3-D RANS solver (with Spalart–Allmaras turbulence model) required over one million cells to adequately resolve the surface features on a 240×60 mm patch of scaled turbine roughness (Fig. 15). Figure 16 shows a comparison of the predicted percent increase in c_f and St for the 3-D RANS vs the discrete element model (DEM) vs an experimental measurement. The results show only marginal differences between the two models though the 3-D, high spatial fidelity RANS simulation requires two to three orders of magnitude more computational time for this flat-plate, turbulent boundary-layer test case. Clearly, fully resolving surface roughness features in complex turbomachinery flowfields is beyond the currently available computational resources. Even if the surface roughness is eventually accommodated by the computational grid, there will remain the issue of boundary-layer transition. Many of the roughness model implementations just cited fail to provide accurate predictions for c_f and St because they do not adequately model the effect of roughness on transition. Recently, several transition models have been proposed with roughness sensitivity [39, 71, 72, 129]. Stripf et al. [19] combined a discrete element roughness formulation with a two-layer model of turbulence and a

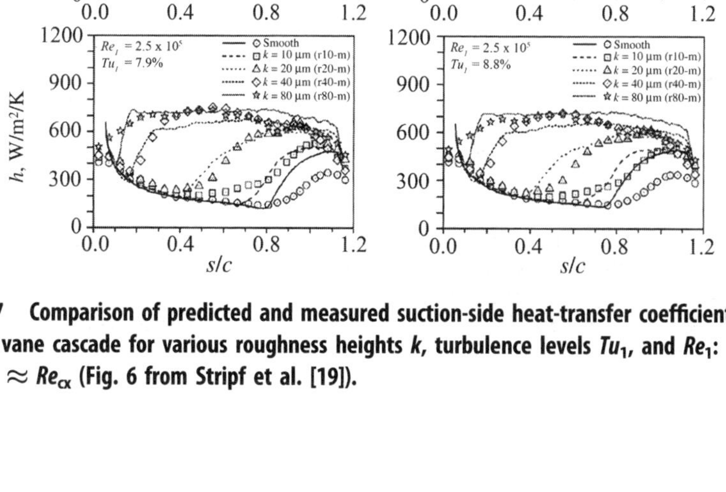

Fig. 17 Comparison of predicted and measured suction-side heat-transfer coefficient on a linear vane cascade for various roughness heights *k*, turbulence levels Tu_1, and Re_1: $1.6Re_1 \approx Re_{cx}$ (Fig. 6 from Stripf et al. [19]).

roughness-sensitive transition model to obtain very good agreement with suction surface heat-transfer coefficient distributions in a linear turbine vane cascade (Fig. 17).

XIII. CONCLUSIONS

Surface degradation in gas turbines is caused by a wide variety of operational and environmental factors. Some of the most common sources are ingested aerosols, namely, salt spray from marine applications [3–6], airborne dust, sand [2, 85], pollen, combustion products [130, 131], and even volcanic ash [7]. Occasionally, larger objects can also be ingested (e.g., inlet nacelle ice accumulation [132]) resulting in significant surface damage. From within the gas turbine itself, oil leaks, rust or scale, and even dirty fuels (for power turbines) [21, 24] can roughen the gas path surfaces with deposits. The adverse effects of such service-induced surface roughness have been studied for over half a century. Both compressors and turbines are affected by these phenomena. Although there is some consensus as to the general trends of performance with increasing roughness, accurate models for predicting roughness effects on profile losses and surface heat transfer are still beyond our present capability. This is largely because of the rich parameter space required to adequately characterize roughness (including size, shape, spacing, blade location). The broad use of the equivalent sand-grain roughness characterization k_s has hampered modeling because it does not account for different roughness effects on skin friction, heat transfer, and boundary-layer transition. Significant progress is being made to overcoming this shortcoming using discrete-element and high-resolution CFD models, but there is still significant work to be done in the area of roughness modeling. The proper accounting of roughness effects on boundary-layer transition is a critical aspect of this modeling conundrum. Finally, current roughness models are universally validated with experimental data from ordered roughness elements. Thus, the proper application of these models to the more random roughness characterizations typical of serviced gas turbine hardware deserves considerable attention.

REFERENCES

[1] Syverud, E., Brekke, O., and Bakken, L. E., "Axial Compressor Deterioration Caused by Saltwater Ingestion," *Journal of Turbomachinery,* Vol. 129, Jan. 2007, pp. 119–126.

[2] Ghenaiet, A., Tan, S. C., and Elder, R. L., "Particles Trajectories Through an Axial Fan and Performance Degradation due to Sand Ingestion," American Society of Mechanical Engineers, Paper 2001-GT-0497, June 2001.

[3] Tarada, F., and Suzuki, M., "External Heat Transfer Enhancement to Turbine Blading due to Surface Roughness," American Society of Mechanical Engineers, Paper No. 93-GT-74, 1993.

[4] Taylor, R. P., "Surface Roughness Measurements on Gas Turbine Blades," *Journal of Turbomachinery*, Vol. 112, Apr. 1990, pp. 175–180.

[5] Bogard, D. G., Schmidt, D., and Tabbita, M., "Characterization and Laboratory Simulation of Turbine Airfoil Surface Roughness and Associated Heat Transfer," *Journal of Turbomachinery*, Vol. 120, No. 2, April 1998, pp. 337–342.

[6] Bons, J. P., Taylor, R. P., McClain, S. T., and Rivir, R. B., "The Many Faces of Turbine Surface Roughness," *Journal of Turbomachinery*, Vol. 123, Oct. 2001, pp. 739–748.

[7] Dunn, M. G., Baran, A. J., and Miatech, J., "Operation of Gas Turbine Engines in Volcanic Ash Clouds," *Journal of Engineering for Gas Turbines and Power*, Vol. 118, Oct. 1996, pp. 724–731.

[8] Walsh, W. S., Thole, K. A., and Joe, C., "Effects of Sand Ingestion on the Blockage of Film-Cooling Holes," American Society of Mechanical Engineers, Paper GT2006-90067, 2006.

[9] Bons, J. P., "St and c_f Augmentation for Real Turbine Roughness with Elevated Freestream Turbulence," *Journal of Turbomachinery*, Vol. 124, Oct. 2002, pp. 632–644.

[10] Borom, M. P., Johnson, C. A., and Peluso, L. A., "Role of Environmental Deposits and Operating Surface Temperature in Spallation of Air Plasma Sprayed Thermal Barrier Coatings," *Surface and Coatings Technology*, Vol. 86–87, 1996, pp. 116–126.

[11] Ghenaiet, A., Tan, S. C., and Elder, R. L., "Numerical Simulation of the Axial Fan Performance Degradation due to Sand Ingestion," American Society of Mechanical Engineers, Paper GT-2002-30644, 2002.

[12] Ghenaiet, A., Tan, S. C., and Elder, R. L., "Study of Erosion Effects on an Axial Fan Global Range of Operation," American Society of Mechanical Engineers, Paper GT2004-54169, 2004.

[13] Diakunchak, I. S., "Performance Deterioration in Industrial Gas Turbines," *Journal of Engineering for Gas Turbines and Power*, Vol. 114, Apr. 1992, pp. 161–168.

[14] Kurz, R., and Brun, K., "Degradation in Gas Turbine Systems," *Journal of Engineering for Gas Turbines and Power*, Vol. 123, Jan. 2001, pp. 70–77.

[15] Caguiat, D. E., Zipkin, D. M., and Patterson, J. S., "Compressor Fouling Testing on Rolls Royce/Allison 501-K17 and General Electric LM2500 Gas Turbine Engines," American Society of Mechanical Engineers, Paper GT-2002-30262, 2002.

[16] Levine, P., and Angello, L., "Axial Compressor Performance Maintenance," American Society of Mechanical Engineers, Paper GT2005-68014, 2005.

[17] Syverud, E., and Bakken, L. E., "The Impact of Surface Roughness on Axial Compressor Performance Deterioration," American Society of Mechanical Engineers, Paper GT2006-90004, 2006.

[18] Suder, K. L., Chima, R. V., Strazisar, A. J., and Roberts, W. B., "The Effect of Adding Roughness and Thickness to a Transonic Axial Compressor Rotor," *Journal of Turbomachinery*, Vol. 117, Oct. 1995, pp. 491–505.

[19] Stripf, M., Schulz, A., and Bauer, H. J., "Surface Roughness and Secondary Flow Effects on External Heat Transfer of a HP Turbine Vane," International Symposium on Airbreathing Engines, No. 2005-1116, 2005.

[20] Kacprynski, G. J., Gumina, M., and Roemer, M. J., "A Prognostic Modeling Approach for Predicting Recurring Maintenance for Shipboard Propulsion

Systems," American Society of Mechanical Engineers, Paper 2001-GT-0218, June 2001.

[21] Basendwah, A. A., Pilidis, P., and Li, Y. G., "Turbine Off-Line Water Wash Optimization Approach for Power Generation," American Society of Mechanical Engineers, Paper GT2006-90244, 2006.

[22] Stalder, J.-P., "Gas Turbine Compressor Washing State of the Art: Field Experiences," *Journal of Engineering for Gas Turbines and Power*, Vol. 123, Apr. 2001, pp. 363–370.

[23] Veer, T., Haglerod, K. K., and Bolland, O., "Measured Data Correction for Improved Fouling and Degradation Analysis of Offshore Gas Turbines," American Society of Mechanical Engineers, Paper GT2004-53760, 2004.

[24] Basler, B., and Marx, D., "Heavy Fuel Operation at Limay Bataan Power Station," American Society of Mechanical Engineers, Paper 2001-GT-0213, 2001.

[25] Acharya, M., Bornstein, J., and Escudier, M., "Turbulent Boundary Layers on Rough Surfaces," *Experiments in Fluids*, Vol. 4, 1986, pp. 33–47.

[26] Bons, J., "A Critical Assessment of Reynolds Analogy for Turbine Flows," *Journal of Heat Transfer*, Vol. 127, 2005, pp. 427–485.

[27] Simpson, R. L., "A Generalized Correlation of Roughness Density Effects on the Turbulent Boundary Layer," *AIAA Journal*, Vol. 11, Feb. 1973, pp. 2424–2444.

[28] Dirling, R. B., Jr., "A Method for Computing Roughwall Heat Transfer Rates on Re-Entry Nose Tips," AIAA Paper 73–763, July 1973.

[29] Sigal, A., and Danberg, J., "New Correlation of Roughness Density Effect on the Turbulent Boundary Layer," *AIAA Journal*, Vol. 28, No. 3, 1990, pp. 554–556.

[30] van Rij, J. A., Belnap, B. J., and Ligrani, P. M., "Analysis and Experiments on Three-Dimensional, Irregular Surface Roughness," *ASME Journal of Fluids Engineering*, Vol. 124, Sept. 2002, pp. 1–7.

[31] Nikuradse, J., "Laws for Flows in Rough Pipes," *VDI-Forchungsheft* 361, Series B, Vol. 4, 1933 (English Translation NACA TM 1292, 1950).

[32] White, F. M., *Viscous Fluid Flow*, 2nd ed., McGraw–Hill, New York, 1991.

[33] Gbadebo, S. A., Hynes, T. P., and Cumpsty, N. A., "Influence of Surface on Three-Dimensional Separation in Axial Compressors," American Society of Mechanical Engineers, Paper GT2004-53619, June 2004.

[34] Schlichting, H., "Experimentelle Untersuchungen zum Rauhigkeitsproblem," *Ingenieur-Archiv*, Vol. 7, 1936, pp. 1–34.

[35] Schlichting, H., *Boundary Layer Theory*, 7th ed., McGraw–Hill, New York, 1979.

[36] Coleman, H. W., Hodge, B. K., and Taylor, R. P., "A Re-Evaluation of Schlichting's Surface Roughness Experiment," *ASME Journal of Fluids Engineering*, Vol. 106, 1984, pp. 60–65.

[37] Mills, A. F., *Heat Transfer*, 1st ed., Irwin, IL, 1992.

[38] Kays, W. M., and Crawford, M. E., *Convective Heat and Mass Transfer*, 3rd ed., McGraw–Hill, New York, 1993.

[39] Boyle, R. J., and Senyitko, R. G., "Measurements and Predictions of Surface Roughness Effects on Turbine Vane Aerodynamics," American Society of Mechanical Engineers, Paper GT2003-38580, 2003.

[40] Braslow, A. L., "Review of the Effect of Distributed Surface Roughness on Boundary-Layer Transition," NATO AGARD, Rept. 254, April 1960.

[41] Speidel, L., "Determination of the Necessary Surface Quality and Possible Losses due to Roughness in Steam Turbines," *Elektrizitätswirtschaft*, Vol. 61, No. 21, 1962, pp. 799–804.

[42] Forster, V. T., "Performance Loss of Modern Steam Turbine Plant due to Surface Roughness," *Proceedings of the Instrn. Mechanical Engineers*, 1966–67, Vol. 181, Pt. 1, 1967, pp. 391–405.

[43] Koch, C. C., and Smith, L. H., Jr., "Loss Sources and Magnitudes in Axial-Flow Compressors," *Journal of Engineering for Power*, Vol. 98, No. 3, July 1976, pp. 411–424.

[44] Bammert, K., and Sandstede, H., "Influences of Manufacturing Tolerances and Surface Roughness of Blades on the Performance of Turbines," *Journal of Engineering for Power*, Vol. 98, No. 1, Jan. 1976, pp. 29–36.

[45] Schäffler, A., "Experimental and Analytical Investigation of the Effects of Reynolds Number and Blade Surface Roughness on Multistage Axial Flow Compressors," *Journal of Engineering for Power,* Vol. 102, 1980, pp. 5–13.

[46] Simon, H., and Bülskämper, A., "On the Evaluation of Reynolds Number and Relative Surface Roughness Effects on Centrifugal Compressor Performance Based on Systematic Experimental Investigations," *Journal of Engineering for Gas Turbines and Power*, Vol. 106, April 1984, pp. 489–501.

[47] Boyle, R. J., and Civinskas, K. C., "Two-Dimensional Navier-Stokes Heat Transfer Analysis for Rough Turbine Blades," AIAA Paper 91-2129, 1991.

[48] Barlow, D. N., and Kim, Y. W., "Effect of Surface Roughness on Local Heat Transfer and Film Cooling Effectiveness," American Society of Mechanical Engineers, Paper 95-GT-14, 1995.

[49] Hoffs, A., Drost, U., and Boics, A., "Heat Transfer Measurements on a Turbine Airfoil at Various Reynolds Numbers and Turbulence Intensities Including Effects of Surface Roughness," American Society of Mechanical Engineers, Paper 96-GT-169, 1996.

[50] Guo, S. M., Jones, T. V., Lock, G. D., and Dancer, S. N., "Computational Prediction of Heat Transfer to Gas Turbine Nozzle Guide Vanes with Roughened Surfaces," *Journal of Turbomachinery*, Vol. 120, 1998, pp. 343–350.

[51] Abuaf, N., Bunker, R. S., and Lee, C. P., "Effects of Surface Roughness on Heat Transfer and Aerodynamic Performance of Turbine Airfoils," *Journal of Turbomachinery*, Vol. 120, July 1998, pp. 522–529.

[52] Kind, R. J., Serjak, P. J., and Abbott, M. W. P., "Measurements and Prediction of the Effects of Surface Roughness on Profile Losses and Deviation in a Turbine Cascade," *Journal of Turbomachinery*, Vol. 120, Jan. 1998, pp. 20–27.

[53] Boyle, R. J., Spuckler, C. M., Lucci, B. L., and Camperchioli, W. P., "Infrared Low-Temperature Turbine Vane Rough Surface Heat Transfer Measurements," *Journal of Turbomachinery*, Vol. 123, Jan. 2001, pp. 168–177.

[54] Bunker, R. S., "The Effects of Thermal Barrier Coating Roughness Magnitude on Heat Transfer with and Without Flowpath Surface Steps," American Society of Mechanical Engineers, Paper IMECE2003-41073, Nov. 2003.

[55] Shabbir, A., and Turner, M. G., "A Wall Function for Calculating the Skin Friction with Surface Roughness," American Society of Mechanical Engineers, Paper GT2004-53908, 2004.

[56] Zhang, Q., and Ligrani, P. M., "Aerodynamic Losses of a Cambered Turbine Vane: Influences of Surface Roughness and Freestream Turbulence Intensity," *Journal of Turbomachinery*, Vol. 128, July 2006, pp. 536–546.

[57] Stripf, M., Schulz, A., and Wittig, S., "Surface Roughness Effects on External Heat Transfer of a HP Turbine Vane," *Journal of Turbomachinery*, Vol. 127, 2005, pp. 200–208.

[58] Hummel, F., Lotzerich, M., Cardamone, P., and Fottner, L., "Surface Roughness Effects on Turbine Blade Aerodynamics," *Journal of Turbomachinery*, Vol. 127, July 2005, pp. 453–461.

[59] Yuan, L. Q., and Kind, R. J., "Measurements and Computations of Compressible Flow Through a Turbine Cascade with Surface Roughness," American Society of Mechanical Engineers, Paper GT2006-90018, May 2006.

[60] Traupel, W., *Thermische Turbomaschinen*, Vol. 1, 3rd ed., Springer-Verlag, New York, 1988.

[61] Leipold, R., Boese, M., and Fottner, L., "The Influence of Technical Surface Roughness Caused by Precision Forging on the Flow Around a Highly Loaded Compressor Cascade," *Journal of Turbomachinery*, Vol. 122, July 2000, pp. 416–425.

[62] Pinson, M. W., and Wang, T., "Effect of Two-Scale Roughness on Boundary Layer Transition over a Heated Flat Plate: Part 1 – Surface Heat Transfer," *Journal of Turbomachinery*, Vol. 122, 2000, pp. 301–307.

[63] Bammert, K., and Milsch, R., "Boundary Layers on Rough Compressor Blades," American Society of Mechanical Engineers, Paper 72-GT-48, 1972.

[64] Wiesner, F. J., "A New Appraisal of Reynolds Number Effects on Centrifugal Compressor Performance," *Journal of Engineering for Power,* Vol. 101, July 1979, pp. 384–396.

[65] Elmstrom, M. E., Millsaps, K. T., Hobson, G. V., and Patterson, J. S., "Impact of Non-Uniform Leading Edge Coatings on the Aerodynamic Performance of Compression Airfoils," American Society of Mechanical Engineers, Paper GT2005-68091, 2005.

[66] Harbeche, U. G., Riess, W., and Seume, J. R., "The Effect of Milling Process Induced Coarse Surface Texture on Aerodynamic Turbine Profile Losses," American Society of Mechanical Engineers, Paper GT2002-30333, 2002.

[67] Roberts, W. B., Prahst, P. S., Thorp, S., and Stazisar, A. J., "The Effect of Ultrapolish on a Transonic Axial Rotor," American Society of Mechanical Engineers, Paper GT2005-69132, 2005.

[68] McIlroy, H. M., Budwig, R. S., and McEligot, D. M., "Scaling of Turbine Blade Roughness for Model Studies," American Society of Mechanical Engineers, Paper IMECE2003-42167, 2003.

[69] Blair, M. F., "An Experimental Study of Heat Transfer in a Large-Scale Turbine Rotor Passage," *Journal of Turbomachinery*, Vol. 116, Jan. 1994, pp. 1–13.

[70] Feindt, E. G., "Untersuchungen uber die Abhangigkeit des Umschlages Laminar-Turbulent von der Oberflachenrauhigkeit und der Druckverteilung," Diss. Braunschweig 1956, Jb. 1956 Schiffbautechn. Gesellschaft 50, pp. 180–203.

[71] Roberts, S. K., and Yaras, M. I., "Boundary Layer Transition in Separation Bubbles over Rough Surfaces," American Society of Mechanical Engineers, Paper GT2004-53667, 2004.

[72] Stripf, M., Schulz, A., and Bauer, H. J., "Modeling of Rough Wall Boundary Layer Transition and Heat Transfer on Turbine Airfoils," *Journal of Turbomachinery*, Vol. 130, April 2008, 021003 (11 pp.).

[73] Vera, M., Zhang, X. F., Hodson, H., and Harvey, N., "Separation and Transition Control on an Aft-Loaded Ultra-High-Lift LP Turbine Blade at Low Reynolds Numbers: High-Speed Validation," *Journal of Turbomachinery*, Vol. 129, April 2007, pp. 340–347.

[74] Ishida, M., Sakaguchi, D., and Ueki, H., "Suppression of Rotating Stall by Wall Roughness Control in Vaneless Diffusers of Centrifugal Blowers," *Journal of Turbomachinery*, Vol. 123, Jan. 2001, pp. 64–72.

[75] Boese, M., and Fottner, L., "Effects of Riblets on the Loss Behavior of a Highly Loaded Compressor Cascade," American Society of Mechanical Engineers, Paper GT2002-30438, June 2002.

[76] Vera, M., Hodson, H. P., and Vazquez, R., "The Effects of a Trip Wire and Unsteadiness on a High Speed Highly Loaded Low-Pressure Turbine Blade," *Journal of Turbomachinery*, Vol. 127, Oct. 2005, pp. 747–754.

[77] Volino, R. J., "Passive Flow Control on Low-Pressure Turbine Airfoils," *Journal of Turbomachinery*, Vol. 125, Oct. 2003, pp. 754–764.

[78] Pacciani, R., and Spano, E., "Numerical Investigation of the Effect of Roughness and Passing Wakes on LP Turbine Blades Performance," American Society of Mechanical Engineers, Paper GT2006-90021, May 2006.

[79] Lake, J. P., King, P. I., and Rivir, R. B., "Low Reynolds Number Loss Reduction on Turbine Blades with Dimples and V-Grooves," AIAA Paper 2000-738, Jan. 2000.

[80] Rao, N. M., Gumusel, B., Kavurmacioglu, L., and Camcl, C., "Influence of Casing Roughness on the Aerodynamic Structure of Tip Vortices in an Axel Flow Turbine," American Society of Mechanical Engineers, Paper GT2006-91011, May 2006.

[81] Zwebek, A., and Plidis, P., "Degradation Effects on Combined Cycle Power Plant Performance, Part 1: Gas Turbine Cycle Component Degradation Effects," *Journal of Engineering for Gas Turbines and Power*, Vol. 125, July 2003, pp. 651–657.

[82] Zwebek, A., and Plidis, P., "Degradation Effects on Combined Cycle Power Plant Performance, Part 2: Steam Turbine Cycle Component Degradation Effects," *Journal of Engineering for Gas Turbines and Power*, Vol. 125, July 2003, pp. 658–663.

[83] Zwebek, A., and Plidis, P., "Degradation Effects on Combined Cycle Power Plant Performance, Part 3: Gas and Steam Turbine Degradation Effects," *Journal of Engineering for Gas Turbines and Power*, Vol. 126, Apr. 2004, pp. 306–315.

[84] Millsaps, K. T., Baker, J., and Patterson, J. S., "Detection and Localization of Fouling in a Gas Turbine Compressor from Aerothermodyanmic Measurements," American Society of Mechanical Engineers, Paper GT2004-54173, June 2004.

[85] Hamed, A. A., Tabakoff, W., Das, K., Rivir, R. B., and Arora, P., "Turbine Blade Surface Deterioration by Erosion," *Journal of Turbomachinery*, Vol. 127, July 2005, pp. 445–452.

[86] Tabakoff, W., Lakshminarasimha, A. N., and Pasin, M., "Simulation of Compressor Performance Deterioration due to Erosion," *Journal of Turbomachinery*, Vol. 112, Jan. 1990, pp. 78–83.

[87] Bammert, K., and Woelk, G. V., "The Influence of the Blading Surface Roughness on the Aerodynamic Behavior and Characteristics of an Axial Compressor," ASME *Journal of Engineering for Power*, Vol. 102, Apr. 1980, pp. 283–287.

[88] Elrod, R., and King, P. I., American Society of Mechanical Engineers, Paper 90-GT-208, 1990.

[89] Schreiber, H., Steinert, W., and Küsters, B., "Effects of Reynolds Number and Free-Stream Turbulence on Boundary Layer Transition in a Compressor Cascade," *Journal of Turbomachinery*, Vol. 124, Jan. 2002, pp. 1–9.

[90] Roberts, W. B., Armin, A., Kassaseya, G., Suder, K. L., Thorp, S. A., and Strazisar, A. J., "The Effect of Variable Chord Length on Transonic Axial Rotor Performance," *Journal of Turbomachinery*, Vol. 124, July 2002, pp. 351–357.

[91] Benra, F., Klapdor, V., and Schulten, M., "Sensitivity Study on the Impact of Surface Roughness due to Milling on the Efficiency of Shrouded Centrifugal Compressor Impellers," American Society of Mechanical Engineers, Paper GT2006-90499, May 2006.

[92] Kim, J., Dunn, M. G., Baran, A J., Wade, D. P., and Tremba, E. L., "Deposition of Volcanic Materials in the Hot Sections of Two Gas Turbine Engines," *Journal of Engineering for Gas Turbines and Power*, Vol. 115, July 1993, pp. 641–651.

[93] Wenglarz, R. A., "An Approach for Evaluation of Gas Turbine Deposition," *ASME Journal of Engineering for Gas Turbines and Power*, Vol. 114, April 1992, pp. 230–234.

[94] Wenglarz, R. A., and Fox, R. G., Jr., "Physical Aspects of Deposition from Coal-Water Fuels Under Gas Turbine Conditions," *Journal of Engineering for Gas Turbines and Power*, Vol. 112, Jan. 1990, pp. 9–14.

[95] Bunker, R. S., "Effect of Discrete Surface Disturbances on Vane External Heat Transfer," International Gas Turbine and Aeroengine Congress & Exhibition, Paper 97-GE-134, 1997.

[96] Bammert, K., and Sandstede, H., "Measurements Concerning the Influence of Surface Roughness and Profile Changes on the Performance of Gas Turbines," *Journal of Engineering for Power*, Vol. 94, No. 3, July 1972, pp. 207–213.

[97] Yun, Y. I., Park, I. Y., and Song, S. J., "Performance Degradation due to Blade Surface Roughness in a Single-Stage Axial Turbine," *Journal of Turbomachinery*, Vol. 127, 2005, pp. 137–143.

[98] Boynton, J. L., Tabibzadeh, R., and Hudson, S. T., "Investigation of Rotor Blade Roughness Effects on Turbine Performance," *Journal of Turbomachinery*, Vol. 115, July 1993, pp. 614–620.

[99] Dunn, M. G., Kim, J., Civinskas, K. C., and Boyle, R. J., "Time-Averaged Heat Transfer and Pressure Measurements and Comparison with Prediction for a Two-Stage Turbine," *Journal of Turbomachinery*, Vol. 116, Jan. 1994, pp. 14–22.

[100] Taylor, R. P., Taylor, J. K., Hosni, H. H., and Coleman, H. W., "Heat Transfer in the Turbulent Boundary Layer with a Step Change in Surface Roughness," *Journal of Turbomachinery*, Vol. 114, Oct. 1992, pp. 778–794.

[101] Bunker, R. S., "Separate and Combined Effects of Surface Roughness and Turbulence Intensity on Vane Heat Transfer," *International Gas Turbine and Aeroengine Congress & Exhibition*, Paper 97-GT-135, 1997.

[102] Zhang, Q., Goodro, M., Ligrani, P. M., Trindade, R., and Sreekanth, S., "Influence of Surface Roughness on the Aerodynamic Losses of a Turbine Vane," *Journal of Turbomachinery*, Vol. 128, May 2006, pp. 568–578.

[103] Matsuda, H., Otomo, F., Kawagishi, H., Inomata, A., Niizeki, Y., and Sasaki, T., "Influence of Surface Roughness on Turbine Nozzle Profile Loss and Secondary Loss," American Society of Mechanical Engineers, Paper GT2006-90828, May 2006.

[104] Waigh, D. R., and Kind, R. J., "Improved Aerodynamic Characterization of Regular Three-Dimensional Roughness," *AIAA Journal*, Vol. 36, No. 6, 1998, pp. 1117–1119.

[105] Bunker, R. S., and Bailey, J. C., "Effect of Squealer Cavity Depth and Oxidation on Turbine Blade Tip Heat Transfer," American Society of Mechanical Engineers, Paper 2000-GT-0155, June 2001.

[106] El-Batsh, H., and Haselbacher, H., "Numerical Investigation of the Effect of Ash Particle Deposition on the Flow Field Through Turbine Cascades," American Society of Mechanical Engineers, Paper GT2002-30600, June 2002.

[107] Rutledge, J. L., Robertson, D., and Bogard, D. G., "Degradation of Film Cooling Performance on a Turbine Vane Suction Side due to Surface Roughness," *Journal of Turbomachinery*, Vol. 128, July 2006, pp. 547–554.

[108] Goldstein, R. J., Eckert, E. R. G., Chiang, H. D., and Elovic, E., "Effect of Surface Roughness on Film Cooling Performance," *Journal of Engineering for Gas Turbines and Power*, Vol. 107, Jan.1985, pp. 111–116.

[109] Bogard, D., Snook, D., and Kohli, A., "Rough Surface Effects on Film Cooling of the Suction Side Surface of a Turbine Vane," American Society of Mechanical Engineers, Paper 2003-42061, 2003.

[110] Schmidt, D. L., Sen, B., and Bogard, D. G., "Effects of Surface Roughness on Film Cooling," American Society of Mechanical Engineers, Paper 96-GT-299, 1996.

[111] Cardwell, N. D., Sundaram, N., and Thole, K. A., "Effects of Mid-Passage Gap, Endwall Misalignment and Roughness on Endwall Film-Cooling," *Journal of Turbomachinery*, Vol. 128, Jan. 2006, pp. 62–70.

[112] Lewis, S., Barker, B., Bons, J. P., Ai, W., and Fletcher, T. H., "Film Cooling Effectiveness and Heat Transfer near Deposit-Laden Film Holes," American Society of Mechanical Engineers, Paper GT2009-59567, June 2009.

[113] Demling, P., and Bogard, D. G., "The Effects of Obstructions on Film Cooling Effectiveness on the Suction Side of a Gas Turbine Vane," American Society of Mechanical Engineers, Paper GT2006-90577, May 2006.

[114] Sundaram, N., and Thole, K. A., "Effects of Surface Deposition, Hole Blockage, and TBC Spallation on Vane Endwall Film-Cooling," American Society of Mechanical Engineers, Paper GT2006-90379, May 2006.

[115] Bunker, R. S., "Effect of Partial Coating Blockage on Film Cooling Effectiveness," American Society of Mechanical Engineers, Paper 2000-GT-0244, May 2000.

[116] Boyle, R. J., "Prediction of Surface Roughness and Incidence Effects on Turbine Performance," *Journal of Turbomachinery*, Vol. 116, Oct. 1994, pp. 745–751.

[117] Cebeci, T., and Chang, K. C., "Calculation of Incompressible Rough-Wall Boundary Layer Flows," *AIAA Journal*, Vol. 16, No. 7, 1978, pp. 730–735.

[118] Aupoix, B., and Spalart, P. R., "Extensions of the Spalart-Allmaras Turbulence Model to Account for Wall Roughness," *International Journal of Heat Transfer and Fluid Flow*, Vol. 24, No. 4, 2003, pp. 454–462.

[119] Lee, J., and Paynter, G. C., "Modified Spalart-Allmaras One Equation Turbulence Model for Rough Wall Boundary Layers," *Journal of Propulsion and Power*, Vol. 11, No. 4, 1996, pp. 809–812.

[120] Kang, Y. S., Yoo, J. C., and Kang, S. H., "Numerical Study of Roughness Effects on a Turbine Stage Preformance," American Society of Mechanical Engineers, Paper GT2004-53750, June 2004.

[121] Boyle, R. J., and Senyitko, R. G., "Effects of Surface Roughness on Turbine Vane Heat Transfer," American Society of Mechanical Engineers, Paper GT2005-68133, June 2005.

[122] Holley, B. M., and Langston, L. S., "Surface Shear Stress and Pressure Measurements in a Turbine Cascade," American Society of Mechanical Engineers, Paper GT2006-90580, May 2006.

[123] Tarada, F. H. A., "Heat Transfer to Rough Turbine Blading," Ph.D. Dissertation, Univ. of Sussex, Sussex, England, U.K., 1987.

[124] Tolpadi, A. K., and Crawford, M. E., "Predictions of the Effect of Roughness on Heat Transfer from Turbine Airfoils," American Society of Mechanical Engineers, Paper 98-GT-87, June 1998.

[125] Aupoix, B., "Modelling of Boundary Layers over Rough Surfaces," *Advances in Turbulence V*, edited by R. Benzi, Kluwer, Dordrecht, The Netherlands, 1994, pp. 16–20.

[126] Bons, J. P., and McClain, S. T., "The Effect of Real Turbine Roughness with Pressure Gradient on Heat Transfer," *Journal of Turbomachinery*, Vol. 126, July 2004, pp. 385–394.

[127] McClain, S. T., Hodge, B. K., and Bons, J. P., "Predicting Skin Friction and Heat Transfer for Turbulent Flow over Real Gas-Turbine Surface Roughness Using the Discrete-Element Method," *Journal of Turbomachinery*, Vol. 126, April 2004, pp. 259–267.

[128] Bons, J. P., McClain, S. T., Wang, Z. J., Chi, X., and Shih, T. I., "A Comparison of Approximate vs. Exact Geometrical Representations of Roughness for CFD Calculations of cf and St," *Journal of Turbomachinery*, Vol. 130, Apr. 2008, 021024 (10 pp.).

[129] Roberts, S. K., and Yaras, M. I., "Modeling of Boundary Layer Transition," American Society of Mechanical Engineers, Paper GT2004-53664, June 2003.

[130] Fielder, J., "Evaluation of Zero Compressor Wash Routine in RN Service," American Society of Mechanical Engineers, Paper GT2003-38887, June 2003.

[131] Haq, I. U., and Bu-Hazza, A. I., "Modeling and Computation of Fouling of a 36 MW Multistage Centrifugal Compressor Train Operating in a Cracker Gas Environment," American Society of Mechanical Engineers, Paper 2001-GT-0229, June 2001.

[132] Hefazi, H., Kaups, K., and Murry, R., "Ice Accretion on a Radial Inflow Turbine Blade," *Journal of Turbomachinery*, Vol. 118, July 1996, pp. 606–612.

CHAPTER 16

Effects of Alternative Fuels and Engine Cycles on Turbine Cooling

Douglas L. Straub* **and Geo A. Richards**†

U. S. Department of Energy, National Energy Technology Laboratory, Morgantown, West Virginia

ACRONYMS

C:H	carbon-to-hydrogen ratio in fuel
FT	Fischer–Tropsch
HRSG	heat recovery steam generator
IGCC	integrated gasification combined cycle
NG	natural gas
NOx	oxides of nitrogen
PDE	pulse detonation engine

This chapter focuses on possible developments in the field of turbomachinery heat transfer in the near future. One of the most important challenges facing the turbine industry is the desire to reduce carbon dioxide emissions. According to the 2009 Annual Energy Outlook [1], even without federal regulations to limit greenhouse gas emissions, local regulatory commissions and the investment community are applying pressure to pursue technologies that produce less greenhouse gas emissions. It appears that this issue will continue to impact gas-turbine designs and performance goals for years to come. This chapter considers innovative approaches that can be pursued to reduce CO_2 emissions and how these approaches could potentially impact aerothermal heat transfer and cooling designs. Approaches to this issue can be broadly categorized into the following areas.

First, alternative fuels are expected to attract growing attention as industries begin to emphasize domestic energy supplies and reduce CO_2 emissions. Stationary gas turbines can be developed to operate on a wide range of fuels [2] but typically require special combustor changes or adjustments [3]. More recently, alternative fuels for commercial airlines have become increasingly popular, as

*Mechanical Engineer, Energy Process Innovation Division.
†Focus Area Leader, Energy System Dynamics.

the entire industry moves toward decreasing carbon footprints by using renewable fuel sources. In February 2008, an Airbus 380 powered by Rolls-Royce Trent 900 engines flew from the United Kingdom to France on a synthetic fuel produced by a gas-to-liquid process [4]. Also, in February 2008, Virgin-Atlantic Airlines successfully completed a test flight from London to Amsterdam in a Boeing 747 that used a blend of biojet fuel in one of its General Electric engines. The fuel was derived from babassu and coconut oil, and no modifications were made to the engines [5]. Because these are "proof-of-principle" demonstrations, many different industry teams have formed to pursue the use of alternative fuels in aircraft engines. The heat transfer and cooling issues associated with these alternative fuels will be described in more detail in this chapter.

Second, in 2007, fossil fuels produced over 70% of U.S. domestic electricity. When categorized by business sector, the transportation sector and electric power sector are the two largest producers of CO_2 emissions in the United States [1]. At this time, it seems that the trend to reduce anthropogenic CO_2 emissions will continue, and this will have significant impacts on transportation and energy, both of which are important segments for the gas-turbine industry. In addition to increasing engine efficiency, which is limited by turbomachinery materials and cooling technology, new engine cycles are also being pursued. For example, in the stationary gas-turbine community, oxy-fuel engine cycles have been proposed, in which the air in the combustor is replaced by oxygen. Steam, or CO_2, can be injected into the combustion zone to control the gas temperatures, but the main advantage is the binary mixture of CO_2 and water in the exhaust. The water can be condensed, and the CO_2 can be sequestered in geological formations or used for enhanced-oil recovery. This chapter will discuss the implications of these "zero-emission" engine cycles on the heat transfer and cooling design for turbomachinery. Before describing these alternative engine cycles, and their associated heat-transfer issues, we outline the key heat-transfer processes that require consideration.

I. COOLING ISSUES

Working/coolant fluid properties: As will be seen, future engine cycles can use high-hydrogen gaseous fuels, resulting in exposure of the turbine gas path to much higher levels of water vapor than in the past, with attendant changes in gas-side thermal properties [6, 7]. Advanced "oxy-fuel" engine cycles will result in the expansion of gases that are almost entirely steam and CO_2, rather than mostly nitrogen, as in the case of combustion in air. Different cooling media might be required. For example, in order to maintain the binary (CO_2/H_2O) composition in the "oxy-fuel" cycle, it might be necessary to use CO_2 or H_2O as the cooling media.

Unsteady-bulk flow processes: Some engine cycles are being considered that incorporate unsteady bulk flow variations (such as pulse-detonation engines and pulse deflagration cycles). Although these novel cycles could result in significant improvements in thermodynamic efficiency, the issues surrounding the heat-transfer processes in these unsteady flowfields can become an increasingly important issue.

Reacting flows: Depending on the cycle design objective, future engine cycles might consider more compact combustor technologies and/or interstage turbine combustion. These types of designs would require a higher degree of integration between the coolant and reactant flows. This issue will be discussed in more detail in the section on reheat cycles.

Radiant heat transfer: Radiant heat transfer to combustor liners has typically been a small fraction of the heat load for gaseous fueled stationary turbines. According to Bunker [8], radiation is typically 5–10% of the heat flux, except in regions viewing the flame zone. Advanced oxy-fuel engine cycles, however, will experience quite different radiant heat-transfer phenomena because the gases are optically thick. Recent numeric predictions, however, suggest that the increase in radiative heat load for oxy-fuel vs conventional systems can be relatively small, less than a 5% increase [9].

These issues will be discussed in the remainder of this chapter.

II. ALTERNATIVE FUELS

A. STATIONARY POWER APPLICATIONS

Different fuels have been routinely used with stationary gas turbines for many decades. Prior to stringent emissions controls beginning around 1990, it was relatively simple to retrofit stationary engines to burn new fuel blends. Combustion modifications required to change from liquid to gaseous fuel were modest even with wide variations in fuel heating value and stoichiometry [10, 11].

Simple fuel flexibility ended with the advent of low-emission premixed combustors, which can be very sensitive to fuel type. Kurz [12] has noted that some engines that utilize lean premixed combustors can allow as much as a 10% variation in fuel heating value whereas others can accommodate less than 2–3%. Since 1997, a majority of the new stationary engines are low-emission (lean premixed) engines. Modern, high-efficiency turbines operate with turbine inlet temperatures near the limit of current nozzle cooling schemes. Because temperature variations on the order of 50°F can reduce the life of some hot gas path components by 50% [13], changes to the combustor that could potentially redistribute the temperature profile must be made with great caution. For example, switching from gaseous fuel to liquid fuel must achieve a similar temperature profile—both to minimize pollutant formation and to avoid hot streaks in the turbomachinery.

B. AIRCRAFT APPLICATIONS

For aircraft engines, fuel specifications are fairly strict. The fuel must not be prone to significant evaporation at sea-level conditions, and it must remain volatile for engine relight at very high altitudes. The minimum flashpoint for jet fuel is 38°C, and the maximum freeze point is −47°C. The fuel must also meet specifications for lubricity, static discharge, and seal compatibility. These requirements virtually ensure that any changes to fuels will be accompanied by stringent testing. As explained by Edwards [14], significant development has occurred over the last five decades to produce fuels that meet the current performance specifications. Nevertheless, interest in qualifying at least two alternative fuels is very high, and it is important to consider how these fuels may affect engine heat transfer.

A commonly discussed alternative fuel for aviation is the Fischer–Tropsch (FT)-derived jet fuel. Steynberg and Dry [15] review the history and current status of the FT process. The FT process has been practiced since the 1930s and converts a mixture of hydrogen and carbon monoxide (known as synthesis gas, or syngas) to a hydrocarbon blend that is further refined to various hydrocarbon products. The syngas can be supplied from many feedstocks, including gasified coal, biomass, natural gas, heavy oils, or even waste-streams. Thus, the appeal of the FT process is that it can convert many feedstocks to valuable jet fuel, reducing dependence on refined crude oil. In some scenarios, it is possible to combine biomass and coal feedstock with geological sequestration to create a fuel with a net CO_2 emission that is lower than conventional refined crude oil [16]. This explains the growing interest in FT liquids as an aviation fuel. In the United States, legislation passed in 2007 requires that government-purchased alternative fuels have a lower life-cycle CO_2 signature than conventional fuels [17].

Because the FT fuel is manufactured from a very pure mixture of CO and H_2, the fuels have virtually no sulfur or trace elements and are usually very low in aromatic hydrocarbons. The low aromatic content complicates the fuel use in engines because the aromatics are generally needed to ensure that fuel system "o-rings" seal. These elastomer rings are designed to swell in the presence of the aromatic hydrocarbons, so that simply exchanging an FT liquid for the existing fuel can lead to problems. This issue is being addressed with fuel additives [18].

From a heat-transfer standpoint, conversion to FT fuels can change several features of engine operation. Higher fuel volatility can lead to changes in the flame position, which, in turn, can change the mixing between combustion products and secondary airflow. Based on testing to date, however, this effect is minor, if not insignificant, in real engine applications [4]. The lower aromatic content also provides a benefit in reduced flame radiation, reducing the heat load on the combustion liner. As discussed by Lefebvre [19], the radiation heat transfer to the combustion liner can be correlated with the C:H ratio of the fuel, which is related to the aromatic content. As just noted and by Edwards [14], the aromatic content of FT-derived fuel is less than that of conventional refined petroleum, providing a potential for lower radiation heat transfer.

Another alternative fuel considered for flight engines is often termed "biojet fuel." These fuels can be derived from vegetable oils, nut oils, and even algae. The University of North Dakota and the U.S. Air Force have tested a blend of JP-8 and soybean-derived fuel in a T63 helicopter engine with no detrimental effect on the overall engine performance [20]. More recently, there have been reports that the University of North Dakota has developed a biojet fuel that meets many of the fuel specifications for JP-8, including aromatic content, freeze point, flash point, and energy content [21].

The effects of these new fuels on aerothermal heat transfer have not been thoroughly reported at this time, but it is reasonable to expect that variations in the C:H ratio of the fuel can impact other sections of the gas turbine. Although the differences between jet fuel and these alternative fuels are expected to be small, these factors can become more important when larger variations in the C:H ratio are pursued (as in high-hydrogen fuels). Changes in oxidant composition, as in oxy-fuel cycles or exhaust gas recirculation, might also have important effects [6, 7].

III. EMERGING ENGINE CYCLES

For aircraft engine applications, most near-term proposals to control CO_2 emissions involve increasing efficiency. Engine efficiency is linked to turbine pressure ratio and firing temperature, meaning that advances will require still higher operating pressure and temperature than in the past. The routes to improved cooling for these advanced operating conditions are the subject of other chapters in this book. This section will discuss cooling issues associated with emerging engine cycles for both stationary and aircraft engine applications. The potential impacts on aerothermal cooling will also be discussed.

A. COOLING WITH ALTERNATIVE HOT GAS PATH WORKING FLUIDS

1. HYDROGEN TURBINES

The hydrogen turbine is a primary option for power generation without CO_2 emissions. Hydrogen can be generated by gasifying coal to produce "syngas"—a mixture of CO and H_2. The CO can be further processed with the water-gas-shift reaction:

$$CO + H_2O \longrightarrow CO_2 + H_2$$

Because coal gasification is usually carried out at elevated pressure (>30 atm), the hydrogen can be separated from the CO_2 using physical solvents [22], which absorb the CO_2 from the high-pressure gas stream. The solvent can be regenerated by depressurization, where the CO_2 is desorbed and can be pumped to geological storage (sequestration) [23].

Various system studies have shown that this process of CO_2 separation and storage will reduce the plant efficiency by 6–10% percentage points (e.g., 40% base HHV efficiency without CO_2 removal drops to 34–30%) depending on details of the plant configuration [24]. The powerplant is typically configured with a combined gas-turbine/steam turbine power cycle, where the gas-turbine exhaust provides heat for the bottoming steam cycle. Coupled to the gasification process, this type of powerplant is known as integrated gasification combined cycle (IGCC) and is the subject of considerable interest and development [25]. In this type of process, the CO_2 is typically removed upstream of the gas turbine.

Of course, the engine must be configured to use hydrogen fuel. Hydrogen was the first fuel used to demonstrate a propulsion turbine engine by von Ohain [26] and is used today as a fuel in some stationary turbine engine applications [27]. Although these experiences show that hydrogen fuel can be used, it cannot currently be used in low-emission "premixed" combustors, where fuel and air are mixed upstream of the flame zone. Premixed combustors are routinely used in natural-gas-fueled powerplants and have achieved excellent control of NOx emissions compared to earlier "diffusion flame" designs.

In premixed systems, a more uniform fuel–air mixture is introduced into the combustion region. NOx is controlled by ensuring that no portion of the flame zone exceeds the temperatures at which NOx is formed. In contrast, the diffusion flame design simply injects fuel into the combustor primary zone, where fuel and air mix and burn at stoichiometric conditions. Given the high flame temperature of hydrogen (approximately 200 K higher than NG), undiluted diffusion flame designs for hydrogen would produce very high levels of NOx. At the present time, premix combustors have not been developed for hydrogen because the high flame speed of hydrogen makes it difficult to avoid flame flashback and attachment in the premixer. There have been some studies of premixed hydrogen combustors at the laboratory scale [28] and in a small microturbine [29], but the combined goal of low emissions and flame stability remains a research challenge.

To control emissions on engines that use hydrogen (or syngas with appreciable hydrogen), the existing strategy is to dilute the flame with either nitrogen or moisture. By reducing the fuel heating value with a diluent, the diffusion flame temperature is controlled so that much of the NOx can be avoided, and the problem of flame flashback is eliminated. Diluents do, however, present some disadvantages. To date, NOx levels from dilute diffusion flame combustors are not as low as in premixed designs. Also, the diluent addition requires that the engine must be configured to accommodate the added mass flow while avoiding compressor surge. This reduces the operational flexibility.

The use of diluent also has potential implications for turbomachinery cooling. Where moisture (steam) is used as a diluent with hydrogen, the moisture levels in the turbomachinery change the thermal properties of the working fluid, as compared to natural gas or distillate fueled engines. There is a considerable body of experience using steam injection in stationary gas turbines. As described by Larson and Williams [30], the steam generated by adding a heat recovery steam generator (HRSG) to some gas turbines provides a simple way to utilize a portion of the rejected turbine exhaust heat. Tuzson [31] describes experience with steam-injected gas turbines prior to 1992, defining water purity issues, and noting that steam injection levels are typically around 15% of the total mass flow through the turbine. Injecting steam in this manner is not the most efficient approach to utilizing the rejected exhaust heat because the latent heat of water vaporization is sent out the stack.

For larger gas turbines, it is usually more efficient to add a complete steam bottoming cycle [32] so that steam can be expanded across a larger pressure drop in the bottoming Rankine cycle. More power is extracted from the HRSG's steam because it is expanded to the vacuum of a steam condenser rather than atmospheric pressure as is the case when the steam is injected into the gas turbine flowpath.

Nevertheless, for IGCC applications, it is common to moisturize the fuel or inject moisture or steam into the compressor discharge. The choice to add moisture is the result of plant design considerations, including flame temperature reduction (for NOx emissions control), power augmentation, and integration of low-level heat from the gasifier [33]. In some instances, nitrogen from the air separation unit is used as a diluent or in combination with water.

Where moisture is added as a diluent, the thermal properties of water change the heat transfer in the turbomachinery. Chiesa et al. [34] consider how a model gas turbine could be configured to use hydrogen fuel, diluted with steam injection, to reduce NOx emissions. Compared to air (or nitrogen), steam has higher thermal conductivity and specific heat. The result is that the effective heat-transfer coefficient to cooled components is larger for steam-injected systems. Thus, attempts to raise firing temperature in steam-injected gas turbines face the double problem of higher temperature and somewhat greater heat transfer from the turbine working fluid.

2. OXY-FUEL CYCLES

To control CO_2 emissions, oxy-fuel power cycles have been proposed as an alternative to H_2 turbines. The basic idea is to replace combustion air with oxygen and recycled CO_2 or H_2O diluent. In this manner, any hydrocarbon (C_nH_m) can be directly burned to produce CO_2 and water as a product:

$$\mathrm{C_nH_m} + (\mathrm{n}+\mathrm{m}/4)\mathrm{O_2} + \{aCO_2 + bH_2O\} \longrightarrow \mathrm{nCO_2} + \mathrm{m}/2\mathrm{H_2O} + \{aCO_2 + bH_2O\}$$

$$\text{Fuel} + \text{oxygen} + \{diluent\} \longrightarrow \text{products} + \{diluent\}$$

In the preceding expression, the diluent coefficients a and b are set by the need to control temperature in the application. The advantage of oxy-fuel firing is that the exhaust stream can simply be cooled to produce CO_2 for sequestration. There is no need to generate hydrogen, or incorporate additional methods, to separate CO_2. The disadvantage is that oxygen must be supplied, which can be costly and can require significant energy. The oxy-fuel process has been proposed for boilers firing pulverized coal [35] and also for power cycles using turbomachinery, described next.

Oxy-fuel power cycles will use a combination of CO_2 and H_2O as the working fluid. The ratio of CO_2 to H_2O depends on the cycle, and various system studies have been conducted to compare cycles that use these working fluids [36–40]. A common feature of these studies is the need to develop turbomachinery operating at high temperatures on mixed steam/CO_2 working fluid. For example, Sans et al. [38] propose a novel cycle configuration that operates with a mixed working fluid, requiring a turbine expander capable of 1400°C (2550°F) operation. Anderson et al. [40] propose a steam-diluted oxy-fuel cycle, using reheat stages that include steam turbine expansion that is significantly higher than conventional steam turbines, which operate (uncooled) at temperatures around 600°C with advanced cycles proposed for 700°C [41].

B. COOLING DESIGNS REQUIRING DIFFERENT COOLING MEDIA

1. OXY-FUEL CYCLES

Development of steam or mixed steam/CO_2 cycles operating at these temperatures will therefore require a novel cooling scheme. An interesting possibility is the use of internal blade cooling flow with external film cooling, as is practiced on most modern gas turbines. To the authors' knowledge, no prior experiments have been completed on how the aerothermal cooling would change with the combination of different working fluid and cooling media. Nirmalan et al. [42] describe results using water-air cooling media, but the hot-gas path composition was not varied in these studies. The presence of atomized water in the cooling media had a dramatic effect on the cooling effectiveness, and this will be discussed in more detail later in this section.

If film cooling is used in an oxy-fuel turbine cycle, then the coolant must also be either steam or CO_2 because other coolants, such as air or nitrogen, would dilute the binary CO_2/H_2O exhaust stream. Separating CO_2 from dilute multicomponent gas mixtures requires more energy and capital cost, and so conventional coolants such as air are not preferred. If the coolant is steam, precautions must be taken to avoid conditions in the blade where the steam could condense and lead to irregular cooling, blade imbalance (for rotating parts), or blade erosion.

Figure 1 is a plot of the saturation pressure vs temperature for water and binary mixtures of water/CO_2 [43]. The pure steam curve is the standard vapor pressure curve for water. The remaining curves show the mixture pressure at which steam/CO_2 would condense based on experimental results from Patel [43]. These curves define a minimum coolant temperature needed to avoid condensation both on the inside of the blade and in the external flow. Assuming that advanced oxy-fuel cycles operate between 10 and 100 atm, Figure 1 shows the conditions that must be avoided to prevent condensation. The operating window appears to be very large.

Based on the conditions shown in Fig. 1, it does not appear likely that local conditions in the mixing layer would lead to condensation. Even at 100 atm with pure steam as the working fluid, temperatures in the mixing layer would

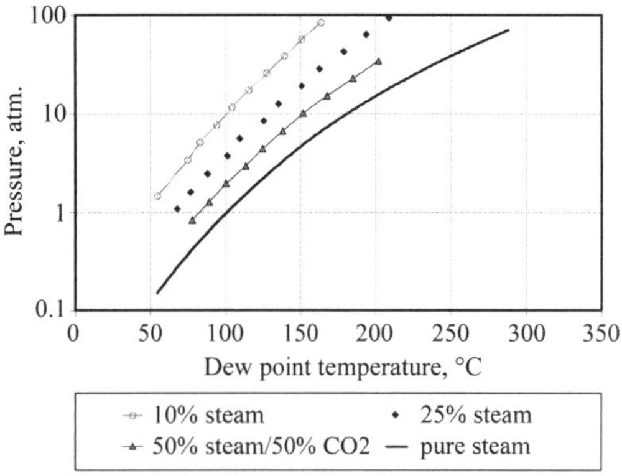

Fig. 1 Vapor pressure curves for binary mixtures of CO_2 and H_2O.

need to drop below 300°C in order for condensation to be an issue. Furthermore, if a steam/CO_2 coolant mixture is used, then the dewpoints are even lower. Although detailed studies of this type of cooling scheme have not been performed, it appears that film cooling in advanced oxy-fuel cycles would not be complicated by condensation of the working fluid around the blade. At the lower temperatures of the low-pressure stages, there would be no need for cooling, and the issue of condensation (if any) would be similar to that in existing low-pressure steam turbine designs.

2. TURBINES CYCLES WITH EXHAUST RECYCLE

Another engine cycle that is currently gaining interest involves recycling a portion of the gas-turbine exhaust through the turbomachinery to improve the CO_2 separation and recovery process. For a conventional gas-turbine powerplant without exhaust recycle, postcombustion removal of CO_2 must address significant increases in capital cost and thermodynamic penalties due to as a result of the low partial pressure of CO_2 in the exhaust. To understand the motivation for exhaust recycle in gas-turbine applications, it is important to realize that conventional amine-based CO_2 removal processes could reduce the turbine powerplant efficiency by as much as 10 percentage points (i.e., from 36 to 26%). Exhaust gas recycle could significantly reduce the cost of the CO_2 removal system and lower the efficiency penalty to only 5% [44]. Griffin et al. [44] have evaluated several different CO_2 mitigation approaches, and the exhaust gas recycle approach was preferred primarily because of the relatively lower risk.

If exhaust gas recycle options are realized, then the composition of the coolant could include significantly higher concentrations of CO_2 and H_2O, depending on the amount of recycle. In this scenario, it will be important to understand the implications of the different coolant properties for the hot-section components.

3. WATER/STEAM COOLING

As already mentioned, prior studies have considered both steam and liquid water cooling for turbine blades [42, 45–47]. In the early 1980s, water-cooled stators and rotating vanes were designed and tested in the Department of Energy (DOE)-sponsored High Temperature Turbine Technology Program investigating the development of turbomachinery for turbines fired with coal syngas. The program goals for firing temperature were aggressive at the time (2600–3000°F). By current standards, a 2600°F firing temperature is achievable in clean, natural-gas-fired turbines with air and steam cooling, but is high when trace coal impurities are present in the hot gas. Caruvana et al. [45] discuss the development and testing of a water-cooled turbine nozzle, as well as the rotating vane with open-circuit water cooling. Liquid water was supplied to the root of the vane from the turbine wheel, traveled along a blade cooling passage, and was ejected into the hot-gas path flow from the tip cavity. The cooling passage was designed to create a film of boiling water, which combines boiling heat transfer in the strongly accelerating (rotating) passage. Data on these heat-transfer processes are presented in Caruvana, with more details in Horner [46]. Some test results on water-cooled first-stage nozzles suggest that firing temperatures of 3000°F are possible with very modest metal temperatures. The drawback of this approach is the complexity of supplying liquid water to the hot section and insuring even cooling distribution over the whole engine load range.

Kikuza et al. [47] present a comparison of cooling schemes for hydrogen/oxygen-fueled gas turbines with a planned peak temperature of 1700°C. The authors suggest that a combination of closed-loop water cooling on the first-stage nozzle, and closed-loop steam cooling on the rotating vane would enable operation at these aggressive conditions. The combination of closed-loop water/steam cooling for the first-stage nozzle/vane was projected to have higher cycle efficiency than steam cooling of both parts because of the added demand for low-grade steam used in cooling both nozzle and vane. Open-loop steam cooling of both parts was projected to have much lower cycle efficiency, as indicated in Table 1. The relative effect of cooling strategy on efficiency is specific to the cycle arrangement studied, but it is included here to emphasize the significant

TABLE 1 COMPARISON OF COOLING STRATEGIES AND EFFECT ON CYCLE EFFICIENCY FROM KIKUZA ET AL. [47]

First-stage nozzle cooling strategy	First-stage vane cooling strategy	Predicted cycle efficiency (based on H_2 fuel HHV), %
Closed-loop water cooled	Closed-loop steam cooled	60.1
Closed-loop steam cooled	Closed-loop steam cooled	58.7
Open-loop steam cooled	Open-loop steam cooled	54.9

interaction between the choice of cooling strategy and overall cycle efficiency. Kikuza et al. [47] present the design approach for the water-cooled nozzle, using a copper alloy containing Cr and Zr, and possibly a thermal barrier coating. The rotating vane design was also considered, using a more conventional single crystal nickel alloy and thermal barrier coating (TBC) for thermal protection. A minimal amount of film cooling was needed, and so the steam cooling was not entirely closed loop. Predicted thermal flux through the TBC was 3.6 MW/m^2 reported as approximately three times the state-of-the art. Improved TBCs are suggested as a needed enabling technology. No experimental test data were available to confirm the performance of designs.

Okamura et al. [48] present an experimental evaluation of turbine blade cooling for a 1700°C hydrogen turbine system. As in Kikuza et al. [47], the target application is an oxygen-hydrogen turbine cycle. Closed-loop steam cooling was used in both the first-stage nozzle and rotating vane. Some steam coolant was used for film cooling at the leading edge, and so the concept is not entirely closed loop. An interesting comparison (based on numeric simulations) shows that the heat-transfer coefficients in the steam environment are about three times larger than in existing machines using air as a working media.

C. COOLING FOR UNSTEADY BULK FLOW CYCLES/PROCESSES

In recent years, unsteady heat-transfer studies have primarily focused on unsteadiness generated by wakes and shocks from upstream vanes and the subsequent interaction with downstream blade rows [49–53]. According to these studies, unsteady process can increase the heat-transfer coefficients on the hot-gas side by 15–25% [53], and the effectiveness of film cooling can be reduced by as much as 50%, depending on the Strouhal number [51].

Although these efforts illustrate the significant impact of unsteady flow effects on heat transfer, very little information is available on periodic bulk flow fluctuations like those that may occur as a result of thermoacoustic instabilities. In rig tests, combustion-driven oscillations can result in pressure amplitudes on the order of 1–10% of the operating pressure that occur at frequencies of 100–3000 cycles per second [54]. In engine applications, the allowable pressure amplitudes are much smaller because the turbomachinery components, particularly those that are cooled using low-pressure drop coolants (i.e., the first-stage components), are vulnerable. Pressure amplitudes that exceed a certain threshold could result in backflow of hot gases into the cooling channels, which could result in a catastrophic failure [55]. To reduce risk, the aerothermal cooling designer can choose an appropriate safety margin to account for the potential of encountering combustion-driven flow oscillations. The tendency to overcool components will have a detrimental impact on the overall efficiency of the cycle.

Unlike bulk flow variations that are produced as a result of unwanted combustion instabilities, other innovative cycles utilize these unsteady effects to improve cycle performance. For example, pulse-detonation engines are being pursued for aeropropulsion applications [56–59]. The pressure-pulse generated by the detonation can be converted directly to thrust, or it could (in principle) be used in a turbine. Some studies suggest that such a design could improve engine efficiency by approximately five percentage points relative to a Brayton cycle with the same temperature ratio [60]. Hutchins and Metghalchi [61] suggest that the improvement at lower pressure ratio conditions could be significantly larger. Because the time-averaged pressure rise appears as expansion work in the downstream turbomachinery, the loss associated with constant pressure combustion [62] is replaced by generation of pressure energy associated with constant volume combustion. Similar designs for oscillating deflagrations can also produce a modest pressure rise and boost engine efficiency [63]. Whether pulse detonation, or pulse deflagration, the challenge is to convert the unsteady flow to useful work either as direct jet propulsion or in turbomachinery.

From a heat-transfer standpoint, the difficulty of cooling components in a strongly oscillating flow has not been the subject of significant research. In the case of combustion-driven instabilities, the cooling issues associated with unsteady bulk flow processes have been avoided by careful combustor design. In the case of pulse-detonation engine cycles and pulse-deflagration cycles, the cooling issues have not been as significant as defining the actual performance advantage of these cycles. Some studies have shown that the performance advantage of PDEs is strongly dependent on various loss mechanisms, including heat transfer [64] originating in the unsteady flow [65–67]. When the performance benefits of these novel cycles have been demonstrated, then the cooling and heat-transfer processes in these unsteady processes will become an important barrier issue.

D. COOLING FOR COMPONENTS EXPOSED TO REACTING FLOW CONDITIONS

This issue has been described in other review articles on aerothermal cooling [8], but the motivation might be different. This section is divided into two subsections. First, an alternative cycle that can decrease specific fuel consumption is discussed. These in situ reheat cycles have been studied for both stationary and propulsion gas-turbine applications. In the second part, a different motivation for cooling components immersed in a reacting flow is discussed. Future combustor designs might require a better understanding of the interactions between cooling designs and reacting flows.

1. REHEAT CYCLES

Reheat engine cycles have been considered as a means to increase the performance of both stationary and propulsion gas turbines. Reheat engine cycles for stationary power are commercially available [68]. The basic idea is that heat is added by combustion between turbine stages or even during expansion. The temperature rise in this interturbine combustion region is small relative to a conventional combustor and is approximately the same as the temperature drop associated with the upstream turbine expansion.

In a conventional simple-cycle gas turbine, using reheat would *lower* the gas turbine efficiency because it adds heat to the cycle at a lower pressure than the main combustor. Because most turbines are limited in peak temperature, however, very efficient high-pressure turbine cycles cannot add as much heat as desired to the hot compressor discharge. This limits the specific power output (power per unit of airflow). The reheat approach allows more of the compressed oxidant (air) to be utilized, raising the power output per unit of air flow from a given machine. Improvements in cost or weight performance on a flight engine can offset any efficiency penalties.

The effect of reheat on overall efficiency is different for cycles that recover the heat rejected by the turbine. In a combined cycle, much of the heat exhausted from the turbine is recovered by the steam cycle. In a recuperated cycle, the heat from the turbine exhaust is used to heat the compressor discharge air. Sirignano and Liu [69, 70] considered reheat cycle performance of both stationary and propulsion turbines. For propulsion turbines, continuous heat addition during turbine expansion has significant advantages for increasing thrust with modest gains in specific fuel consumption (compared to conventional afterburning cycles). Chiu [71] has also studied a reheat, or isothermal turbine, concept for supersonic transport and concluded that twice as much cooling air could be utilized in a reheat approach for the same percentage of performance degradation as a conventional cycle without reheat.

Bachovchin et al. [72] have studied how interstage combustion might be implemented on an actual stationary power turbine, including analysis of the interstage combustion process, some experimental studies, and a cycle analysis. The cycle analysis shows that an improvement in cycle efficiency is possible using reheat for the combined-cycle application. Experimental and numerical combustion studies [73] were performed in a duct intended to simulate the flow conditions that would exist when fuel was injected downstream of a fixed turbine vane. The results suggest that a modest increase in NOx emissions would occur from diffusion flames established in the flowpath. Isvoranu and Cizmas [74] describe numeric simulations of the interstage combustion process. In summary, reheat cycles have been considered as options for both propulsion and stationary turbines in the past, but the implications for aerothermal cooling designs have not yet been adequately addressed in the open literature.

2. EFFECTS OF OTHER COMBUSTION STRATEGIES

Premixed combustors are routinely used in natural-gas-fueled powerplants and have achieved excellent control of NOx emissions as compared to earlier "diffusion flame" designs. On the other hand, aircraft engine applications use diffusion flame combustors. Intuitively, the length and timescales for these fundamentally different combustion approaches should differ, but to the authors' knowledge the significance of the combustion approach on downstream cooling designs is not discussed in great detail in the open literature. Dunn [75] discusses the fact that the flow profiles, turbulence, and length-scale properties of "real" gas turbine combustors are not well known.

More recently, there have been a few efforts to utilize combustor simulators upstream of the turbine stage to emulate the flow conditions at the combustor exit [49, 50, 76]. Although this recent work is a significant step forward in understanding exit flow conditions from the combustor, the combustor simulators in these studies are not reacting, and so the effects of heat release and flow acceleration across the flame region are not reproduced.

Other low-emission combustion designs, like the trapped vortex combustor described by Roquemore et al. [77] and others [78–81], might change the way combustors of the future are designed. These cavity-stabilized combustion approaches have many advantages that have been demonstrated in the literature. Cooling designs that protect surfaces in direct contact with the high-temperature reactions will, however, be required. In the case of film cooling, these cooling designs must also consider the impacts of the cooling media on the reactions. The coolant can impact the flame stability and emissions performance.

From a heat-transfer standpoint, interstage combustion and cavity-stabilized combustors would introduce new challenges in the cooling designs. There is relatively little prior experience with combined combustion and heat-transfer problems in accelerating flows typical of turbomachinery. These issues will need to be addressed, if future designs incorporate such approaches.

IV. CONCLUSIONS

Recognizing global concerns to reduce carbon dioxide emissions, this chapter describes some approaches that can be pursued in the gas-turbine industry to reduce greenhouse gas emissions. These approaches can impact the cooling design.

Alternative fuels for stationary and aircraft applications seem to be a very solid, near-term approach to reduce anthropogenic CO_2 emissions, but the impacts on the cooling design might not be fully optimized at this time.

Some of the alternative engine cycles described in this chapter might have a longer development cycle than the alternative fuels. Each concept has its inherent advantages and disadvantages, and the commercialization path will depend on both technical and commercial factors. Proactive research and development of turbomachinery cooling issues could significantly impact the perceived risk with these future engine cycles, particularly because the cooling design is such a critical component of the overall gas turbine performance.

In summary, it appears that cooling technologies will continue to play a significant role in future gas-turbine engine performance, but the specific challenges can change. It also seems reasonable to assume that global concerns over climate change will continue, and gas turbines will play an active role in the solution.

REFERENCES

[1] "2009 Annual Energy Outlook – Early Release," U.S. Dept. of Energy, Rept. #DOE/EIA-0383, Washington, D.C., March 2009; http://www.eia.doe.gov/oiaf/aeo/

[2] Meier, J. G., Hung, W. S. Y., and Sood, V. M., "Development and Application of Industrial Gas Turbines for Medium-BTU Gaseous Fuels," *Journal of Engineering Gas Turbines and Power*, Vol. 108, No. 1, 1986, pp. 182–190.

[3] Noble, D. R., Zhang, Q., Shareef, A., Tootle, J., Meyers, A., and Lieuwen, T., "Syngas Mixture Composition Effects upon Flashback and Blowout," American Society of Mechanical Engineers, Paper GT2006-90470, May 2006.

[4] Rizk, N., "Overview of the Relationship Between Fuel Properties and Engine Performance," International Civil Aviation Organization, ICAO Workshop, Aviation & Alternative Fuels, Paper 27, Feb. 2009.

[5] "Virgin Atlantic Uses Biojet Fuel Blend in 747 Test Flight; Babassu and Coconut Oil as Feedstocks," Green Car Congress, Seattle, WA, 24 Feb. 2008; http://www.greencarcongress.com/2008/02/virgin-atlantic.html.

[6] Mazzotta, D. W., Chyu, M. K., and Alvin, M. A., "Airfoil Heat Transfer Characteristics in Syngas and Hydrogen Turbines," American Society of Mechanical Engineers, Paper GT2007-28296, May 2007.

[7] Mazzotta, D. W., Chyu, M. K., Karaivanov, V. G., Slaughter, W. S., and Alvin, M. A., "Gas-Side Heat Transfer in Syngas, Hydrogen-Fired and Oxy-Fuel Turbines," American Society of Mechanical Engineers, Paper GT2008-51474, June 2008.

[8] Bunker, R. S., "Gas Turbine Heat Transfer: Ten Remaining Hot Gas Path Challenges," *Journal of Turbomachinery*, Vol. 129, 2007, pp. 193–201.

[9] Mazzotta, D. W., Chyu, M. K., and Alvin, M. A., "Airfoil Heat Transfer Characteristics in Syngas and Hydrogen Turbines," American Society of Mechanical Engineers, Paper GT2007-28296, May 2007.

[10] Hung, W. S. Y., "A Diffusion Limited Model that Accurately Predicts the NOx Emissions from Gas Turbine Combustors Including the Use of Nitrogen Containing Fuels," *Journal of Engineering for Power*, ASME Series A, Vol. 98, No. 3, 1976, pp. 320–326.

[11] Hung, W. S. Y., "The NOx Emission Level of Unconventional Fuels for Gas Turbines," *Journal of Engineering for Power*, ASME Series A, Vol. 99, No. 7, 1977, p. 94.

[12] Kurz, R., "Gas Turbine Fuel Considerations," Gas Machinery Conference, 2004.

[13] Han, J-C., Dutta, S., and Ekkad, S. V., *Gas Turbine Heat Transfer and Cooling Technlogy*, Taylor and Francis, New York, 2000.

[14] Edwards, T., "Advancements in Gas Turbine Fuels from 1943 to 2005," *Journal of Engineering for Gas Turbines and Power*, Vol. 129, No. 1, 2007, pp. 13–20.

[15] Steynberg, A., and Dry, M. (eds.), *Studies in Surface Science and Catalysis*, Vol. 152, Fischer-Tropsch Technology, Elsevier, Amsterdam, 2004.

[16] Tarka, T. J., "Affordable, Low-Carbon Diesel Fuel from Domestic Coal and Biomass," Dept. of Energy, National Energy Technology Lab., DOE/NETL-2009/1349, Morgantown, WV, Jan. 2009; www.netl.doe.gov.

[17] Energy Independence and Security Act of 2007, Sec. 526; http://frwebgate.access.gpo.gov/cgi-in/getdoc.cgi?dbname=110_cong_bills&docid=f:h6enr.txt.pdf.

[18] Link, D. D., Gormley, R. J., Baltrus, J. P., Anderson, R. R., and Zandhuis, P. H., "Potential Additives to Promote Seal Swell in Synthetic Fuels and Their Effect on Thermal Stability," *Energy and Fuels*, Vol. 22, No. 2, 2008, pp. 115–1120.

[19] Lefebvre, A. H., *Gas Turbine Combustion*, 2nd ed., Taylor and Francis, Philadelphia, PA, 1999.

[20] Corporan, E., Relch, R., Monroig, O., DeWitt, M. J., Larson, V., Aulich, T., Mann, M., and Seames, W., "Impacts of Biodiesel on Pollutant Emissions of a JP-8 Fueled Turbine Engine," *Journal of the Air and Waste Management Association*, 2005, Vol. 55, No. 7, pp. 940–949.

[21] "EERC 100% Renewable Biojet Fuel Meets Key JP-8 Standards [corrected]," Green Car Congress, 29 Sept. 2008; http://www.greencarcongress.com/2008/09/eerc-100-renewa.html.

[22] Kohl, A., and Nielsen, R., *Gas Purification*, 5th ed., Gulf Publishing, Houston TX, 1997.

[23] "Fossil Energy Cost and Performance Baseline Studies, Volume 1," National Energy Technology Lab., NETL Report #2007-1281, Morgantown, WV; http://www.netl.doe.gov/energy-analyses/baseline_studies.html.

[24] Gasification Technologies; http://www.gasification.org/library/overview.aspx

[25] "Fossil Energy IGCC," U.S. Dept. of Energy; htpp://www.fe.doe.gov.

[26] Conner, M., *Hans Von Ohain: Elegance in Flight*, AIAA, Reston, VA, 2001.

[27] Moliere, M., "Hydrogen-Fueled Gas Turbines: Status and Prospects," 2nd CAME-GT Conference, April 2004.

[28] Dobbeling, K., Eroglu, A., Winkler, D., Sattelmayer, T., and Keppel, W., "Low NOx Premix Combustion Mbtu Fuels in a Research Burner," *ASME Journal of Engineering for Gas Turbines and Power*, Vol. 119, No. 3, 1997, pp. 553–558.

[29] McDonell, V., Therkelsen, P., Werts, T., and Samuelsen, S., "Analysis of Nox Formation in a Hydrogen Fueled Microturbine Generator," American Society of Mechanical Engineers, Paper GT2008-50841, June 2008.

[30] Larson, E. D., and Williams, R. H., "Steam-Injected Gas Turbines," *ASME Journal of Engineering Gas Turbines and Power*, Vol. 109, No. 1, 1987, pp. 55–63

[31] Tuzson, J., "Status of Steam Injected Gas Turbines," *ASME Journal of Engineering Gas Turbines and Power*, Vol. 114, No. 4, 1992, pp. 682–686.

[32] Bolland, O., and Stadaas, J. F., "Comparative Evaluation of Combined Cycles and Gas Turbine Systems with Water Injection, Steam Injection, and Recuperation," *ASME Journal of Engineering Gas Turbines and Power*, Vol. 117, No. 1, 1995, pp. 138–145.

[33] Todd, D. M., "Gas Turbine Improvements Enhance IGCC Viability," 2000 Gasification Technologies Conference, GE Power Systems, Oct. 2000; www.gasification.org.

[34] Chiesa, P., Lozza, G., and Mazzocchi, L., "Using Hydrogen as Gas Turbine Fuel," *ASME Journal of Engineering Gas Turbines and Power*, Vol. 127, No. 1, 2005, pp. 73–80.

[35] Darde, A., Prabhakar, R., Tranier, J-P., and Perrin, N., "Air Separation and Flue Gas Compression and Purification Units for Oxy-Coal Combustion Systems," *Energy Procedia*, Vol. 1, No. 1, pp. 527-534.

[36] Staicovici, M. D., "Further Research Zero CO2 Emission Power Production: The COOLENERG Process," *Energy*, Vol. 27, No. 9, 2002, pp. 831–844.

[37] Chiesa, P., and Lozza, G., "CO2 Emission Abatement in IGCC Power Plants by Semiclosed Cycles: Part A – with Oxygen Blown Combustion," *ASME Journal of Engineering for Gas Turbines and Power,* Vol. 121, No. 4, 1999, pp. 635–641.

[38] Sans, W., Jericha, H., Moser, M., and Heitmeir, F., "Thermodynamic and Economic Investigation of an Improved Graz Cycle Power Plant for CO_2 Catpure," American Society of Mechanical Engineers, Paper GT2004-53722, June 2004.

[39] Jackson, A. J., Neto, A. C., Whellens, M. W., and Audus, H., "Gas Turbine Performance Using Carbon Dioxide as Working Fluid in Closed Cycle Operation," American Society of Mechanical Engineers, Paper 2000-GT-153, May 2000.

[40] Anderson, R., Brandt, H., Doyle, S., Mueggenburg, H., Taylor, J., and Viteri, F., "A Unique Process for Production of Environmentally Clean Electric Power Using Fossil Fuels," 8th International Symposium on Transport Phenomena and Dynamics of Rotating Machinery, March 2000.

[41] Henderson, C., *Understanding Coal-Fired Power Plants, IEA Clean Coal Center, London*, 2004; www.iea-coal.org.uk.

[42] Nirmalan, N. V., Weaver, J. A., and Hylton, L. D., "An Experimental Study of Turbine Vane Heat Transfer with Water-Air Cooling," *Journal of Turbomachinery*, Vol. 120, 1998, pp. 50–60.

[43] Patel, M. R., Holste, J. C., Hall, K. R., and Eubank, P. T., "Thermophysical Properties of Gaseous Carbon Dioxide-Water Mixtures," *Fluid Phase Equilibria*, Vol. 36, 1987, pp. 279–299.

[44] Griffin, T., Bucker, D., and Pfeffer, A., "Technology Options for Gas Turbine Power Generation with Reduced CO2 Emissions," *Journal of Engineering for Gas Turbines and Power*, Vol. 130, No. 4, 2008, pp. 041801.

[45] Caruvana, A., Rose, R. S., Alderson, E. D., and Cincotta, G. A., "Design and Test of a 73-MW Water-Cooled Gas Turbine," American Society of Mechanical Engineers, Paper 80-GT-112, 1980.

[46] Horner, M. W., "High-Temperature Technology Program Turbine Subsystem Design Report – Low BTU Gas," General Electric Co., Contract AC01-76ET10340, DOE/ET/10340-T3, Schenectady, NY, Dec. 1980.

[47] Kizuka, N., Sagae, K, Anzai, S., Marushima, S., Ikeguchi, T., and Kawaike, K., "Conceptual Design of the Cooling System for 1700C-Class Hydrogen Fueled Combustion Gas Turbines," American Society of Mechanical Engineers, Paper 98-GT-345, June 1998.

[48] Okamura, T., Koga, A., Itoh, S., and Kawagishi, H., "Evaluation of 1700C Class Turbine Blades in Hydrogen Fueled Combsution Turbine System," American Society of Mechanical Engineers, Paper 2000-GT-615, May 2000.

[49] Haldeman, C. W., Mathison, R. M., Dunn, M. G., Southworth, S. A., Harral, J. W., and Heitland, G., "Aerodynamic and Heat Flux Measurements in a Single-Stage Fully Cooled Turbine – Part 1: Experimental Approach," *Journal of Turbomachinery*, Vol. 130, No. 2, 2008, pp. 021015-1–021015-10.

[50] Haldeman, C. W., Mathison, R. M., Dunn, M. G., Southworth, S. A., Harral, J. W., and Heitland, G., "Aerodynamic and Heat Flux Measurements in a Single-Stage Fully Cooled Turbine – Part 2: Experimental Results," *Journal of Turbomachinery*, Vol. 130, No. 2, 2008, pp. 021016-1–021016-11.

[51] Womack, K. M., Volino, R. J., and Schultz, M. P., "Measurements in Film Cooling Flows with Periodic Wakes," *Journal of Turbomachinery*, Vol. 130, No. 4, 2008, pp. 041008-1–041008-13.

[52] Allan, W. D., Ainsworth, R., and Thorpe, S., "Unsteady Heat Transfer Measurements from Transonic Turbine Blades at Engine Representative Conditions in a Transient Facility," *Journal of Engineering for Gas Turbines and Power*, Vol. 130, No. 4, 2008, pp. 041901-1–041901-12.

[53] Du, H., Ekkad, S., and Han, J.-C., "Effect of Unsteady Wake with Trailing Edge Coolant Ejection on Detailed Heat Transfer Coefficient Distributions for a Gas Turbine Blade," *Journal of Heat Transfer*, Vol. 119, No. 2, 1997, pp. 242–248.

[54] Straub, D. L., and Richards, G. A., "Effect of Fuel Nozzle Configuration on Premix Combustion Dynamics," American Society of Mechanical Engineers, Paper 98-GT-492, June 1998.

[55] Lieuwen, T., and McManus, K., "Introduction: Combustion Dynamics in Lean Premixed Prevaporized (LPP) Gas Turbines," *Journal of Propulsion and Power*, Vol. 19, No. 5, p. 721.

[56] Kailasanath, K., "Recent Developments in the Research on Pulse Detonation Engines," *AIAA Journal*, Vol. 144, No. 2, 2003, pp. 145–159

[57] Ma, F., Choi, J.-Y., and Yang, V., "Propulsive Performance of Airbreathing Pulse Detonation Engines," *Journal of Propulsion and Power*, Vol. 22, No. 6, 2006, pp. 1188–1203.

[58] Wintenberger, E., and Shepherd, J. E., "Model for the Performance of Airbreathing Pulse Detonation Engines," *Journal of Propulsion and Power*, Vol. 22, No. 3, 2006, pp. 593–603.

[59] Roy, G. D., Frolov, S. M., Borisov, A. A., and Netzer, D. W., "Pulse Detonation Engines: Challenges, Current Status, and Future Perspective," *Progress in Energy and Combustion Science*, Vol. 30, No. 6, 2004, pp. 545–672.

[60] Goldmeer, J., Tangirala, V., and Dean, A., "System-Level Performance Estimation of a Pulse Detonation Based Hybrid Engine," *Journal of Engineering for Gas Turbines and Power*, Vol. 130, 2008, pp. 011201-1–011201-8.

[61] Hutchins, T. E., and Metghalchi, M., "Energy and Exergy Analysis of the Pulse Detonation Engine," *Journal of Engineering for Gas Turbines and Power*, Vol. 125, pp. 1075–1080.

[62] Nishida, K., Takagi, T., and Kinoshita, S., "Analysis of Entropy Generation and Exergy Loss During Combustion," *Proceedings of the Combustion Institute*, Vol. 29, 2002, pp. 869–874.

[63] Richards, G. A., and Gemmen, R. S., "Pressure-Gain Combustion: Part II-Experimental and Model Results," *Journal of Engineering for Gas Turbines and Power*, Vol. 118, No. 3, 1996, pp. 469–473.

[64] Radulescu, M. I., and Hanson, R. K., "Efffect of Heat Loss on Pulse-Detonation-Engine Flow Fields and Performance," *Journal of Propulsion and Power*, Vol. 21, No. 2, 2005, pp. 274–285.

[65] Heiser, W. H., and Pratt, D. T., "Thermodynamic Cycle Analysis of Pulse Detonation Engines," *Journal of Propulsion and Power*, Vol. 18, No. 1, 2002, pp. 68–76.

[66] Allgood, D., Gutmark, E., Rasheed, A., and Dean, A. J., "Experimental Investigation of a Pulse Detonation Engine with a Two-Dimensional Ejector," *AIAA Journal*, Vol. 43, No. 2, 2005, pp. 390–398.

[67] Allgood, D., Gutmark, E., Hoke, J., Bradley, R., and Shauer, F., "Performance Measurments of Multi-Cycle Pulse-Detonation-Engine Exhaust Nozzles," *Journal of Propulsion and Power*, Vol. 22, No. 1, 2006, pp. 70–77.

[68] Alstom; http://www.power.alstom.com

[69] Sirignano, W. A., and Liu, F., "Performance Increases for Gas-Turbine Engines Through Combustion Inside the Turbine," *Journal of Propulsion and Power*, Vol. 15, No. 1, 1999, pp. 111–118.

[70] Liu, F., and Sirignano, W.A., "Turbojet and Turbofan Engine Performance Increases Through Turbine Burners," *Journal of Propulsion and Power*, Vol. 17, No. 3, 2001, pp. 695–705.

[71] Chiu, Y., "A Performance Study of a Super-Cruise Engine with Isothermal Combustion Inside the Turbine," Ph.D. Dissertation, Virginia Polytechnic Inst. and State Univ., Blacksburg, VA, 2004.

[72] Bachovchin, D. M., Lippert, T. M., Newby, R. A., and Cizmas, P. G., "Gas Turbine Reheat Using in situ Combustion," Dept. of Energy, Final Rept., DOE Contract DE-FC26-00NT40913, Orlando, FL, May 2004.

[73] Chambers, S., Flitan, H., Cizmas, P., Bachovchin, D., Lippert, T., and Little, D., "The Influence of in Situ Reheat on Turbine-Combustor Performance," *Journal of Engineering for Gas Turbines and Power*, Vol. 128, No. 3, 2006, pp. 560–572.

[74] Isvoranu, D., and Cizmas, P., "Numerical Simulation of Combustion and Rotor-Stator Interaction in a Turbine Combustor," *International Journal of Rotating Machinery*, Vol. 9, No. 5, 2003, pp. 363–374.

[75] Dunn, M. G., "Convective Heat Transfer and Aerodynamics in Axial Flow Turbines," *Journal of Turbomachinery*, Vol. 123, No. 4, 2001, pp. 637–686.

[76] Barringer, M. D., Thole, K. A., and Polanka, M. D., "Experimental Evaluation of an Inlet Profile Generator for High-Pressure Turbine Tests," *Journal of Turbomachinery*, Vol. 129, No. 2, 2007, pp. 382–393.

[77] Roquemore, W. M., Shouse, D. T., and Hsu, K. Y., "Combustor Flame Stabilizing Structure," U.S. Patent 5,857,339, 1999.

[78] Roquemore, W. M., Shouse, D., Burrus, D., Johnson, A., Cooper, C., Duncan, B., Hsu, K. Y., Katta, V. R., Sturgess, G. J., and Vihinen, I., "Trapped Vortex Combustor Concept for Gas Turbine Engines," AIAA Paper 2001-0483, Jan. 2001.

[79] Burrus, D. L., Johnson, A. W., Roquemore, W. M., and Shouse, D. T., "Performance Assessment of a Prototype Trapped Vortex Combustor Concept for Gas Turbine Application," American Society of Mechanical Engineers, Paper 2001-GT-0087, June 2001.

[80] Straub, D. L., Casleton, K. H., Lewis, R. E., Sidwell, T. G., Maloney, D. J., and Richards, G. A., "Assessment of RQL Trapped Vortex Combustor for Stationary Gas Turbines," *Journal of Engineering for Gas Turbines and Power*, Vol. 127, No. 1, 2005, pp. 36–41.

[81] Edmonds, R. G., Steele, R. C., Williams, J. T., Straub, D. L., Casleton, K. H., and Bining, A., "Low NOx Advanced Vortex Combustor," *Journal of Engineering for Gas Turbines and Power*, Vol. 130, No. 3, pp. 034502-1–034502-4.

INDEX

Note: Page numbers are followed by *f* or t (indicating figures or tables).

SUPPLEMENTAL MATERIALS

A complete listing of titles in the Progress in Astronautics and Aeronautics series is available from AIAA's electronic library, Aerospace Research Central (ARC) at arc.aiaa.org. Visit ARC frequently to stay abreast of product changes, corrections, special offers, and new publications.

AIAA is committed to devoting resources to the education of both practicing and future aerospace professionals. In 1996, the AIAA Foundation was founded. Its programs enhance scientific literacy and advance the arts and sciences of aerospace. For more information, please visit www.aiaafoundation.org.